Vision in Fishes

New Approaches in Research

NATO ADVANCED STUDY INSTITUTES SERIES

A series of edited volumes comprising multifaceted studies of contemporary scientific issues by some of the best scientific minds in the world, assembled in cooperation with NATO Scientific Affairs Division.

Series A: Life Sciences

Volume 1 – Vision in Fishes: New Approaches in Research
edited by M. A. Ali

Volume 2 – Nematode Vectors of Plant Viruses
edited by F. Lamberti, C. E. Taylor, and J. W. Seinhorst

Volume 3 – Genetic Manipulations with Plant Material
edited by Lucien Ledoux

Volume 4 – Phloem Transport
edited by S. Aronoff

The series is published by an international board of publishers in conjunction with NATO Scientific Affairs Division

A	**Life Sciences**	Plenum Publishing Corporation
B	**Physics**	New York and London
C	**Mathematical and Physical Sciences**	D. Reidel Publishing Company Dordrecht and Boston
D	**Behavioral and Social Sciences**	Sijthoff International Publishing Company Leiden
E	**Applied Sciences**	Noordhoff International Publishing Leiden

Vision in Fishes

New Approaches in Research

Edited by

M. A. Ali
Départment de Biologie
Université de Montréal
Montréal, Canada

PLENUM PRESS • NEW YORK AND LONDON
Published in cooperation with NATO Scientific Affairs Division

Library of Congress Cataloging in Publication Data

NATO Advanced Study Institute on New Approaches to the Study of Vision in Fishes, Bishop's University, 1974.
Vision in fishes.

(NATO advanced study institutes series: Series A, life sciences; v. 1)
"Co-sponsored by the Université de Montreal and the National Research Council of Canada."
Includes bibliographies and index.
1. Sense-organs–Fishes. 2. Vision. 3. Fishes–Physiology. I. Ali, Mohamed Ather, 1932- II. Montreal. Université. III. National Research Council, Canada. IV. Title. V. Series.
QL639.1.N37 1975 597'.01 823 75-8570
ISBN 0-306-35601-5

Lectures presented at the 1974 NATO Advanced Study Institute on New Approaches to the Study of Vision in Fishes held at Bishop's University, Lennoxville, Quebec, Canada, August 11-24, 1974

Co-sponsored by the Université de Montréal and the National Research Council of Canada

Organizing Committee: M. A. Ali, Director, Montreal
W. R. A. Muntz, Sussex
J. P. Raynauld, Montreal

A Division of Plenum Publishing Corporation
227 West 17th Street, New York, N.Y. 10011

United Kingdom edition published by Plenum Press, London
A Division of Plenum Publishing Company, Ltd.
Davis House (4th Floor), 8 Scrubs Lane, Harlesden, London, NW10 6SE, England

Printed in the United States of America

Preface

No more than a fish loves water. - Is not this a strange fellow, my lord, that so confidently seems to undertake this business, which he knows is not to be done; damns himself to do, and dares better be damn'd than to do't?

All's Well That Ends Well Act III, Sc. 6

This volume is the direct result of a NATO-Advanced Study Institute of the same title, held at Bishop's University, Lennoxville, Québec, Canada, August 1974, under the joint sponsorship of the NATO-Scientific Advisory Committee, National Research Council of Canada and the Université de Montréal. It is not, however, strictly restricted to the lectures and seminars presented at the ASI. Contributions have been included from two workers who found at a fairly late stage that they could not attend and also included are a table giving the visual pigments and another dealing with the distribution and development of retinomotor responses. I encouraged the authors to prepare speculative reviews highlighting their own research or that of their immediate colleagues and a number of them have done so. Several contributors, notably those who were asked to give lectures of a general nature at the ASI have written reviews of somewhat greater scope.

The result is a collection of papers representing a great variety of approaches to the study of vision in fishes. Fishes form the largest class of vertebrates and occupy a wide variety of habitats ranging from the shallowest of streams to the deepest of oceans. The physical and chemical characteristics of their environment vary greatly. Adaptation to these different habitats is reflected in the variations of their retinal structure and function. Light, of course is the primary stimulus for vision and the aquatic milieu presents considerably greater variations in the quality and quantity of light than do the aerial and terrestrial habitats.

The last two decades have witnessed a surge in the study of piscine vision. While many of these investigations have merely employed the fish as a convenient material with which to work in order to solve problems of a general visual nature, there nevertheless have been a good number of studies dealing with questions concerning fish vision and the adaptive radiation of the fish eye. However, much emphasis has been placed on electrophysiology and visual pigments. Other aspects seem to have been neglected except by a handful of workers. One of the main purposes of this volume is to stimulate the interest of investigators, especially the younger ones, in the lesser explored fields. Another aim is to bring to the attention of workers in this field, present and future, the significance of the electrophysiological and visual pigment studies to the life and habits of fishes. This is done by presenting an overall picture of the structure of the eye and the optic tectum, electrophysiological responses and visual pigments in fishes with reference to their habitats and behaviour. The purpose is to give a new and broader perspective to these popular fields of study. In addition, an attempt has been made to point out the possibilities in the investigation of lesser known aspects such as neurotransmitter substances, contractility of the photoreceptor myoids, enzymology, relationship with biolaminescence, histochemistry, and histophysiology. Detailed descriptions of methods and techniques involved in the study of some of these lesser known aspects are also given.

In addition to attending each and every one of the lectures and seminars given at the ASI, I have had to read all the papers in this volume carefully in my editorial capacity. This has considerably broadened my understanding of vision in general and that of fishes in particular. I appreciate having had this opportunity and hope that the readers of this volume will benefit in the same way.

The idea of approaching the NATO-Scientific Affairs Division came up when Bill Muntz was spending the summer of 1973 with me in Montreal. We immediately wrote to Dr. T. Kester of the NATO-Scientific Affairs Division and received an encouraging response after which preparations were started. I am grateful to Dr. Kester for his encouragement and help and to Dr. Maurice L'Abbé, Vice-Rector of Research of my University, for moral and material support. Financial help provided by the National Research Council of Canada is gratefully acknowledged.

The authorities of Bishop's University coöperated in every possible way making our meeting enjoyable and I am thankful to Dr. Dennis Healey, the Principal and Vice-Chancellor, Monsieur J.L. Grégoire, the Comptroller, Mr. Ivan Saunders, the Director of Buildings and Grounds, and his assistant Mr. Ralph Reid for their courtesy. My colleague Pierre Couillard, though not officially a member of the organising committee, contributed as much as if he had been one and I am thankful to him. He deserves particular mention for arranging the excursion to Québec city so well.

I have put in a quotation which I considered either apt or humourous at the beginning of each article. Some of these were provided by Professor H.J.A. Dartnall. I am grateful to Professor Dartnall also for the useful advice he gave with regards to the editing, and Professor P.K. Menon for his help. I appreciate greatly the help provided by my librarian Miss Margaret Pertwee, who edited the bibliographies, prepared the indices, and provided some of the quotations.

I thank Mesdames M. Lafrance and J.D. Larocque as well as Mademoiselle N. Valiquette for typing most of the camera-ready manuscript. Since the volume was prepared by non-English speaking persons some words have been incorrectly divided at the end of lines. I have tried to correct as many of these as possible but some still remain, I therefore ask the authors' and readers' indulgence.

On behalf of the authors and myself I wish to acknowledge the permission given by various authors and publishers for the reproduction of figures. These have been mentioned also individually in the legends. Finally, Miss Phyllis Straw and Mr. Stephen Dyer of the Plenum Press are to be thanked for the help they provided in the preparation of this book.

Montreal
November, 1974

Contents

1. PHOTOCHEMISTRY

PHOTOCHEMICAL EFFICIENCY OF CHROMOPROTEINS. ROLE OF THE MOLECULAR ENVIRONMENT

Jacques AGHION

Département de Botanique, Université de Liège

Sart Tilman, B-4000 Liège, Belgium

I cannot synthesise a bun,
By simply sitting in the sun.
J.B.S. Haldane

The biological role of light can be that of a trigger for enzymatic mechanisms deriving their energy from, say, respiratory ATP, as in vision, phototaxis, biological rhythms: it is not conceivable that visible light gives unicellular algae all the energy they need to rush towards it or away from it. On the other hand, in photosynthesis light seems to be an actual source of energy and the phenomena initiated by it are such that they actually promote the synthesis of some ATP.

Rhodopsin, the chromoprotein of vision, is the association of a 40,000 molar weight protein (opsin) with a 275 molar weight lipid molecule (retinal). Chlorophyll, the chromophoric group responsible for photosynthesis, is a 900 molar weight hydrophobic molecule, attached to unknown proteins. Rhodopsin is probably oriented in vivo, in such a way that its photon capture section is multiplied by 1.4 relative to the randomly oriented rhodopsin in solution (Daemen, 1973). It is now classical knowledge that by orientation and aggregation in vivo, the light-processing properties of chlorophyll are very much enhanced compared to those it has in solution (Weier et al., 1967).

When rhodopsin is hit by efficient photons, it is believed that retinal is isomerised (but remains in its reduced form) and detached from the opsin molecule which is relieved from steric constraints and therefore, can slightly change its higher-order structure (Abrahamson and Ostroy, 1967). What happens when chlorophyll is hit by a photon of the right energy? It immediately creates an electron-hole pair, sending an electron to an acceptor and apt to receive one from a donor (see Witt, 1971). In sum, in both photophysiological processes, the redox environment of the chromoprotein has to be strictly controlled if the whole machine is to work. In addition, the steric relationships between protein and chromophore on one hand, between the whole chromoprotein and the environment on the other are of utmost importance. In effect, the result of the excitation by light has bo be perceived: the acceptor molecules, besides being in the right redox conditions have to be present within a reasonably short distance of the pigment, in a concentration which is a function of the number of photons which have to be processed. In photosynthesis, the system must accomodate 4, 8 or 12 photons depending on the theories (see Gregory, 1971, p. 103) in order that one mole of O_2 be evolved or one mole of CO_2 absorbed. Similarly, depending on the accuracy of the measurements, the quantum efficiency of the retinal detachment is 50 to 60 percent: two quanta are needed for such an event to occur. But if one wants to compare the quantum yield of vision with that of the evolution of O_2 in photosynthesis, one has to know how many retinal molecules have to be detached in order to produce a "visual event" (conversion into synaptic activity?).

Because the primary events seem to be similar in vision and in photosynthesis and because it is easier to quantify photosynthesis where light produces chemical energy than vision where the variations of chemical energy are much more elusive, it was decided to study the physical-chemical conditions which permit the functioning of the primary photosynthetic acts.

A reminder of the primary electron transfers in photosynthesis can be found in Figure 1; Figure 2 shows isolated chloroplasts, the organelles in which photosynthesis occurs. Details about the reactions going on in the chloroplasts and the models which try to mimick them can be found in Boardman (1970), in Katz (1973) and in the references mentioned by those two authors.

The results I wish to describe have been obtained with pure materials. They are ambitious attempts at describing ways of studying the relatively gross photochemistry of photosynthetic membrane-models which could, hopefully, be used for other photon-receiving membranes (with corrections).

Technical information can be found in Strain and Svec (1966), Dijkmans and Aghion (1974), and Dinant and Aghion (1973).

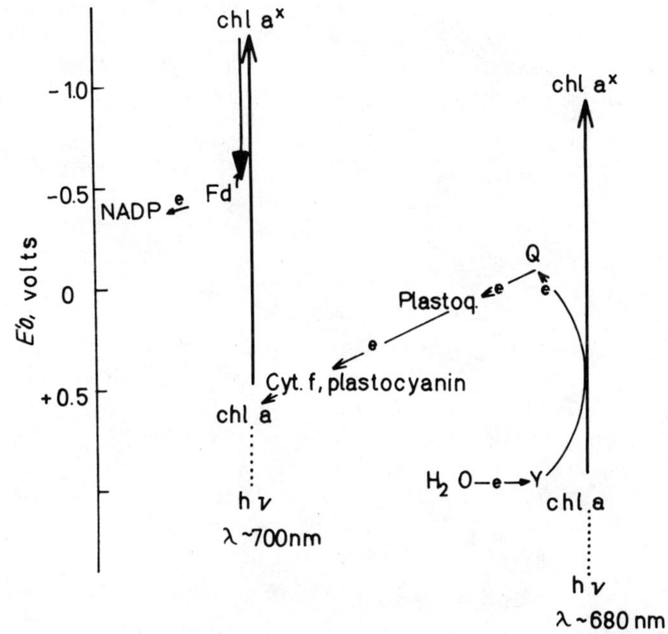

[1] ferredoxin

Fig. 1. Path of electrons from water to NADP in photosynthesis, mediated by the excitation of chlorophyll a (chl a → chl a^x).

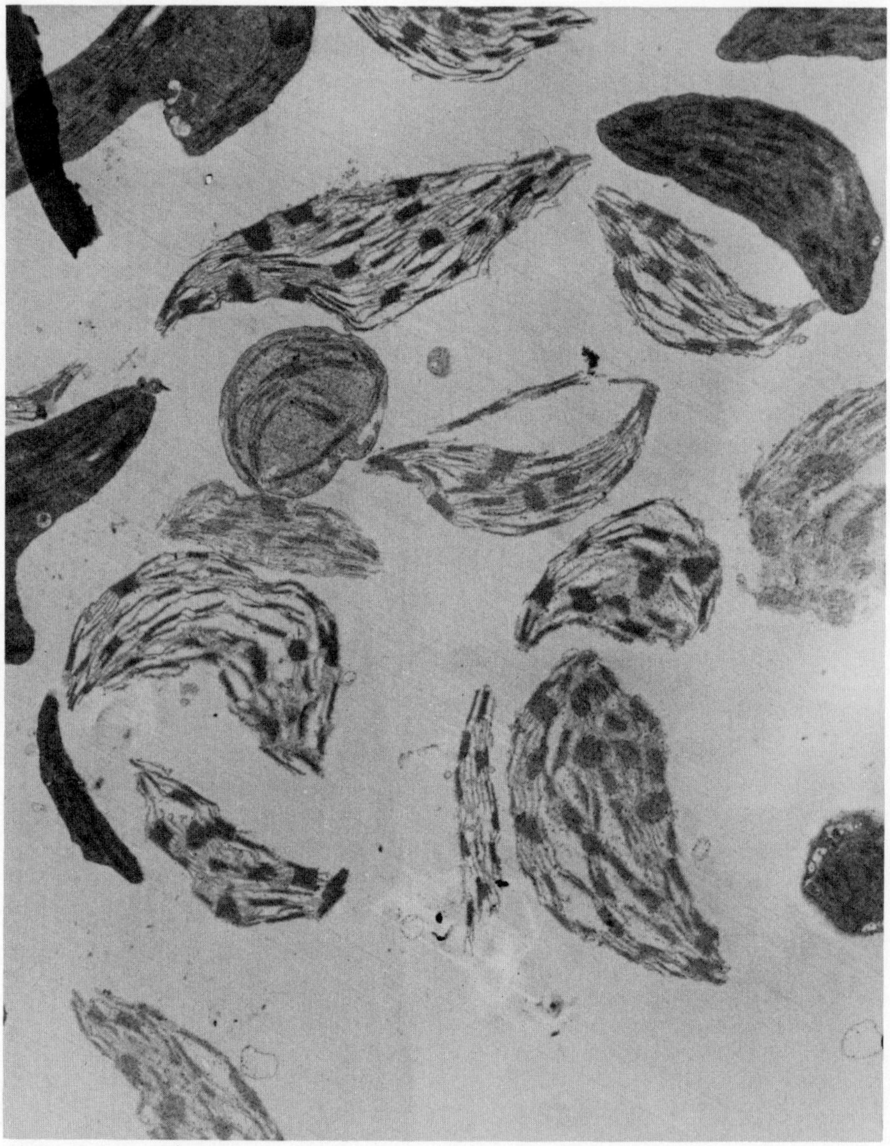

Fig. 2. Chloroplasts extracted from Spinach leaves (Stroma lamellae) and their stacking into grana). Courtesy of Dr. K. Kaminski. Photo by Professor R. Bronchart. x 7200.

EXPERIMENTAL RESULTS

Pigments dispersed in aqueous solutions of polar solvents

a) Chlorophylls a and b. In all the tested solvents (methanol, ethanol, 1-propanol, tert.-butanol and acetone) both chlorophylls dissolve readily (concentration of the order of 10^{-5} mole/litre) and remain dissolved when water is added, up to a value f_i of its mole fraction in the solvent. When the mole fraction of water is higher than another value f_s, both chlorophylls form colloids. The f values between f_i and f_s form the "twilight zone" (Bannister, 1963) where the chlorophylls aggregate into microcrystals and/or small oligomers. It is seen, in absorbance spectroscopy, by a decrease of the red maximum and the appearance and increase of bands at higher wavelengths: 700 nm and 750 nm for chlorophyll a (depending on f), 680 nm for chlorophyll b (Figure 3).

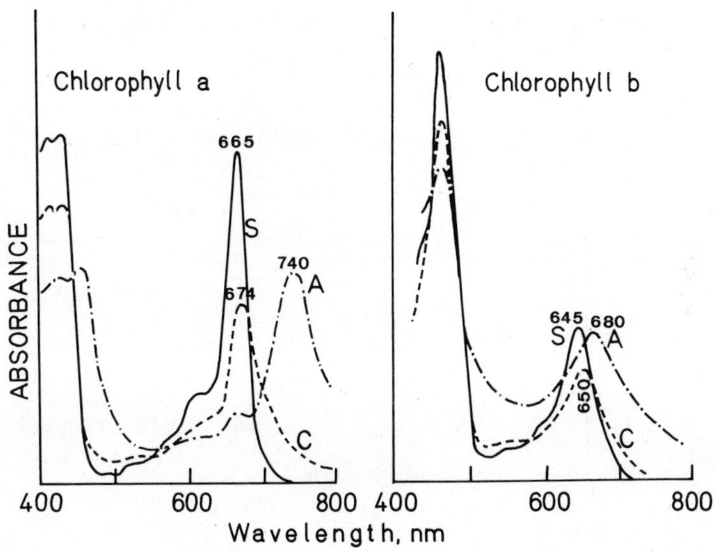

Fig. 3. Absorption spectra of chlorophyll a and of chlorophyll b when they are solvated (S), colloidal (C) and aggregated in the twilight zone (A).

In methanol, the twilight zone of chlorophylls a and b are respectively limited by f_i = 0.4 and 0.45, and f_s = 0.7 and 0.65: they are approximately the same. The values of f_i and f_s are different in the different solvents (Dijkmans and Aghion, 1974). Moreover, the molar refraction α^o (see Smyth, 1955) decreases linearly when f increases from 0 to 1. Whatever the solvent is, α^o reaches the same value between f_i and f_s.

The spectral variations observed here are reversible. It appears that, to a given value of f in a given solvent, corresponds a determined absorption spectrum.

b) β-Carotene (Aghion, in preparation). The experiments were done with methanol and ethanol containing different mole fractions of water. (Carotene concentration 5 x 10^{-6} mole/litre). Absorption spectra were taken between 400 nm and 580 nm: they did not change appreciably over a period of six days in the dark at room temperature. When there is more than 10 percent (v/v) water in methanol or 20 percent in ethanol, the β- carotene spectra exhibit no sharp bands. Hence, instead of measuring the variation of intensity of any individual band (leading to huge errors) the area S limited by the wavelength axis and the spectrum was measured between 400 nm and 580 nm. When f is between 0.3 and 0.6 in methanol, there is an accident in the (otherwise) monotonous variation of S (Figure 4). The same happens in ethanol in its own twilight zone.

As is the case with the chlorophylls, the changes are reversible i.e. a given value of f in a determined solvent determines a value of S whatever the direction of variation of f is.

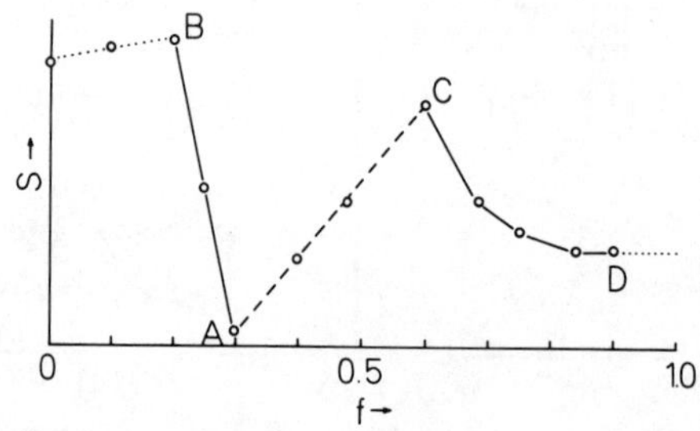

Fig. 4. Probability of meeting between β-carotene molecules and photons (400 nm $\leqslant \lambda \leqslant$ 580 nm) as a function of f, the mole fraction of water in methanol.

Since S is proportional to the probability that a photon of wavelength between 400 nm and 580 nm will meet a molecule of β-carotene, the curve of figure 4 can be accounted for by a sieve effect (see Rabinowitch, 1956) due to an increasing aggregation of the pigment molecules as f increases. The accident occurring when f is between 0.3 and 0.6 can be described as an inversion of the sieve effect. If such an inversion were due to the solvation of β-carotene, the well-characterised spectrum of the solvated species would reappear; a more probable reason is: due to the molar refraction of the solutions, the individual β-carotene molecules are re-oriented respective to each other, increasing the individual extinction coefficient of each aggregate. Thus the accident zone is comparable with the twilight zone defined with chlorophylls.

c) Significance of the twilight zone. It is, here, considered as a property of the solvent systems because the solutes are much too dilute, compared with the solvent molecules, to have any influence on the physical properties of the solvents (checked with refractive indices, surface tension and specific gravity). It is known (see Adamson, 1967) that when f increases, the surface tension γ of a water-alcohol solution remains more or less the same up to a certain value of f, then it rises sharply. This rise begins in the twilight zone for methanol and for ethanol which leads one to think that the twilight zone is linked with the critical alcohol micelle concentration in water. The same is true with the capillary constant $a^2 = 2\gamma/g.d.$ where g is the acceleration of gravity. If a^2 is the capillary rise in a tube having a unit radius, it can be considered as a volume. In it, the total number of water and of alcohol molecules, N/a^2, reaches a maximum identical value in methanol as well as in ethanol and in 1-propanol between f_i and f_s (Figure 5).

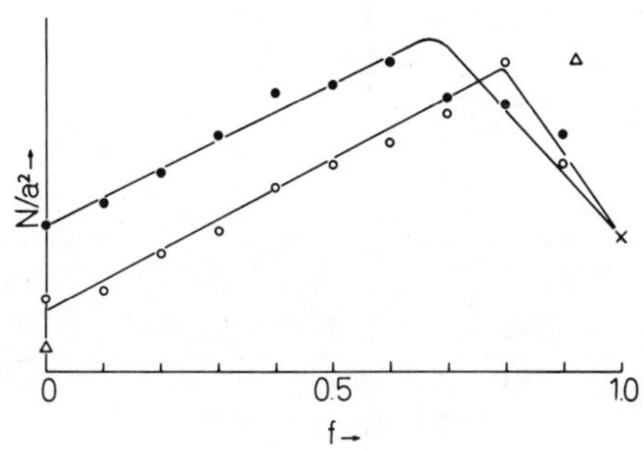

Fig. 5. Number of water + alcohol molecules in the volume defined by the capillary constant a^2, with methanol (•——•), ethanol (○——○) and 1-propanol (Δ) as a function of f, the mole fraction of water in alcohol.

Chlorophylls fixed upon lipoproteinic substrates

a) The fixation (Dinant, unpublished results). When the acetone used to precipitate the milk lipoproteins contains chlorophyll or carotene, the lipoproteins precipitate (as globules having a 1.5μ diameter and a weight of 1 picogram) with some of the solute fixed upon them. Figure 6 cannot help to decide whether the pigments are truly adsorbed on the lipoproteins or separated between the acetone-water phase and the lipoproteinic phase. When r_g is smaller than 7 x 10^{-3}, the absorption spectra have red maxima at 671 nm with chlorophyll a, at 653 nm with chlorophyll b. When r_g is higher than 0.05, there appears a new band at 750 nm with the chlorophyll a globules. When r_g is higher than 7 x 10^{-3}, the chlorophyll b globules have a new band at 690 nm and the chlorophyll a globules a very slight band at 700 nm which can be demonstrated only by difference spectroscopy. The presence of β-carotene in the acetone-milk initial solution, together with chlorophylls, in a constant carotene/protein weight ratio diminishes the fixation of chlorophyll a but not that of chlorophyll b (Figure 6). Since chlorophyll a fixation follows a saturation curve and chlorophyll b fixation does not, it seems that chlorophyll a and β-carotene compete for a finite number of fixation sites and chlorophyll b does not. This is further confirmed by the fact that the presence of either chlorophyll in addition to the other in the initial acetone solution does not hinder its fixation on the globules. In the absence of precise knowledge of the mode of β-carotene- and of chlorophyll-fixation, it can be assumed that β-carotene is tied to the globules through hydrophobic bonds; therefore chlorophyll a would be more hydrophobic than chlorophyll b.

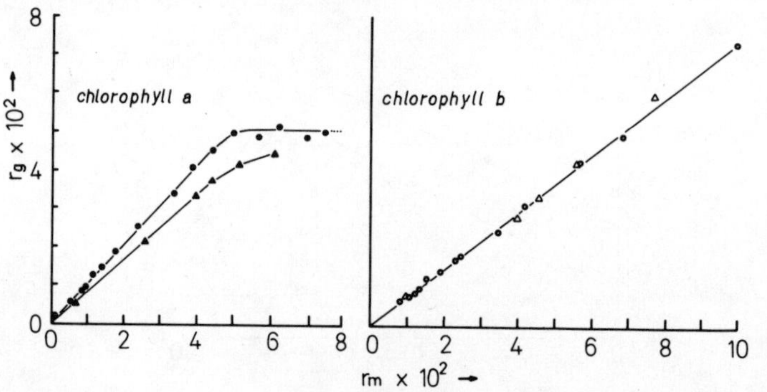

Fig. 6. Fixation of chlorophyll a (●——●) and chlorophyll b (○——○) out of an acetone solution onto lipoprotein globules. Role of β-carotene on their fixation (▲——▲ chlorophyll a; △——△ chlorophyll b). r_g is the chlorophyll/protein ratio (w/w) on the globules and r_m is the same ratio in the mixture of acetone and milk. The ratio of β-carotene to protein is 0.033 by weight.

The 750 nm band of chlorophyll a appears very slowly (24 hours) but the 690 nm band of chlorophyll b appears immediately upon precipitation of additional chlorophyll. Since those bands are probably witnesses to the presence of chlorophyll microcrystals upon or inside the globules, the lag in the appearance of the 750 nm band might be assigned to movements of the chlorophyll a molecules, relative to each other and to the lipoproteins, leading to the building of a crystalline lattice.

The fluorescence of the chlorophyll-lipoprotein system can be excited around 445 nm. The chlorophyll a-systems have emission bands at 680 nm and at 740 nm and the chlorophyll b ones at 660 nm and 715 nm. When the two chlorophylls are combined upon the same globules, their fluorescence is always a chlorophyll a fluorescence (Figure 7) whether the system is excited via its chlorophyll a component, its chlorophyll b component or both. This shows that there

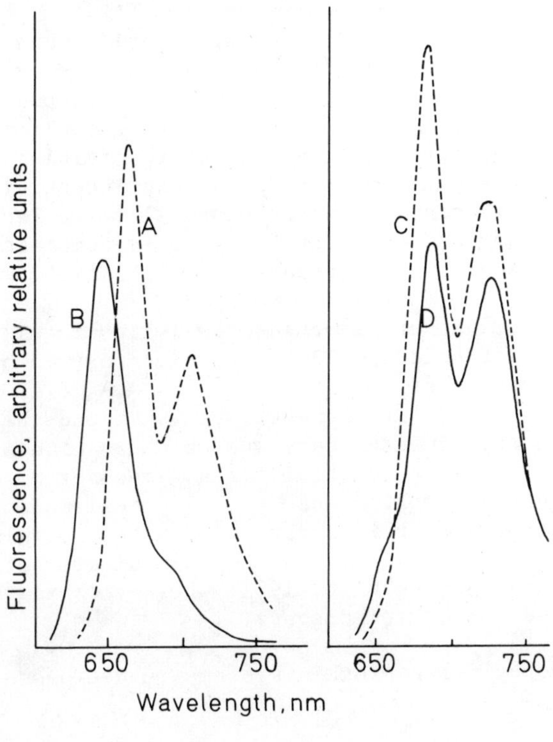

Fig. 7. Emission spectra (at 77 K) of a chlorophyll a-lipoprotein system (A), a chlorophyll b-lipoprotein system (B) and of a mixed system of chlorophyll a and b with lipoprotein, excited via chlorophyll a (C) or via chlorophyll b (D).

is a transfer of excitation energy from chlorophyll b to chlorophyll a. The efficiency of this transfer has not been measured with any accuracy but, considering that a given mixed preparation of chlorophylls a and b with lipoproteins yields 40 fluorescence units when excited via chlorophyll a at 430 nm, and ca 30 units when excited via chlorophyll b at 460 nm, it can be said that the efficiency of the transfer is of the order of 1.0 because the absorbance at 460 nm is between 0.5 and 0.75 times that at 430 nm. However, figure 7 shows that the yield is surely less than 1 since the fluorescence of the mixed preparation excited at 460 nm shows a slight shoulder characteristic of chlorophyll b.

b) The photobleaching (Dinant unpublished results). The chlorophyll a systems as well as the chlorophyll b ones, when they are illuminated undergo bleaching (Dinant and Aghion, 1973). This is monitored by the decrease of absorbance at the red maximum in the 600 nm to 700 nm region. The bleaching needs air and light simultaneously, which leads to the conclusion that it is a photo-autoxidation by atmospheric O_2. When the chlorophyll a-system is illuminated in a Clarke oxygen electrode, it absorbs oxygen during illumination either by absorption of O_2 upon the globules in the light (which would imply a modification of the globule surface by light) or by an actual chemical reaction (irreversible) between chlorophyll a and oxygen. Since the O_2 absorption persists in the light even after 90 percent of the chlorophyll molecules are bleached, there must be an oxidation of the globule components, perhaps sensitised by the remaining chlorophyll a molecules.

In acetone solutions and in chlorophyll-lipoprotein systems, the photo-oxidation of chlorophyll by dissolved oxygen has been described as a second order reaction ($A_o - A_t = A_o \times A_t \times k \times t$, see Dinant and Aghion, 1973). However, it was found that a better fit of curves with experimental data was obtained with $A_o - A_t = A_o \times t/(t + u)$ where u is a constant for a given set of conditions. Such a type of kinetics implies that two identical molecules (of chlorophyll?) react together in order to form the (oxidised) product. Chlorophyll, in this case, would have to form dimers before being able to react to oxygen (see Witt, 1971, where the existence of a dimer is postulated for other reasons).

In the chlorophyll-lipoprotein systems, u decreases when A_o increases.

In acetone solutions of chlorophyll a, u is independent of A_o but it decreases when f increases at least as long as f is lower than f_i, as long as chlorophyll a is solvated. In acetone and water mixtures, with chlorophyll a, and when f is lower than 0.7, u = -1.3 f

+ 1.05. It was shown above, that the molar refraction α^0 of the solvent mixture is a linear function of f: $\alpha^0 = -9f + 13$. Until further experimentation is done, it is therefore considered here that, for chlorophyll a dissolved in acetone, u is a measure of α^0 by $u = 0.14\alpha^0 + 0.85$, and in the case of the chlorophyll a-lipoproteins, the variation of u could represent that of the local molar refraction of the sites of chlorophyll a fixation.

Finally, using interference filters, approximate quantum yields of bleaching were measured. With both chlorophylls, this is about 10 times higher when the chlorophylls are fixed upon globules (of the order of 0.05) than when they are dissolved. A similar finding had been reported with chlorophyll a monolayers where the absorbing molecules are also more oriented than in a solution (Aghion et al., 1969).

DISCUSSION

A system is described, in which chlorophylls are associated with lipoproteins in such a way as to have absorbance spectra very similar to those of photosynthetic systems in vivo and to display excitation energy transfers between chlorophyll molecules. The possibility exists, also, that some redox reactions could take place between the chlorophyll molecules and their substrates. Since, in addition to those features, the systems show a definite improvement of photochemical efficiency over the solvated pigments it seems they are encouraging ones to study further.

The physical state of the chlorophylls, represented by their visible absorption spectra, is linked to the local molar refraction of their environment. Their aggregation could be determined by some sort of squeezing of the chlorophyll molecules into an ordered lattice: such a squeezing, in solution, could be caused either by a definite proportion of water in solvent (Gurinovitch et al., 1968) such that N/a^2 is maximum, or by very high solute concentrations (Brody, 1958). The aggregation of chlorophyll upon globules appears, in effect, to be a result of concentration. In both cases, it is probable that water molecules are trapped in between the chlorophyll molecules, permitting their crystallisation.

It can now be proposed, as a working hypothesis, that photoacts I and II of photosynthesis are performed by the same group of molecules of chlorophylls a and/or b linked to the adequate redox enzymes. The traps (chl a on Figure 1) could be formed locally and temporarily according to local conditions favoring a movement of chromophore molecules towards greater proximity and the insertion of water between them, i.e. wherever and whenever the local conditions are the same as those created by the alcohol and water solutions in the twilight zone. Those conditions can be described in terms of molar refraction or in terms of N/a^2 and it can be imagined that the active sites of the actually enzymatic lipoprotein in vivo play a role in the gathering of those conditions. Indeed the excitation of photoreceptors of vision diminishes when T decreases (Arden, 1969). The dependence of α^0 upon T is of the shape $\alpha^0 = (a/T) + b$; when T decreases, α^0 increases and it would be interesting to relate the excitability of the photoreceptors to α^0 rather than to T.

The redox reactions, by modifying the local properties of the photosynthetic membranes could modify the absorbance properties of the chlorophylls linked to them and make them act as water-splitting chlorophylls (system II) or as CO_2-reducing ones (system I) alternately. Such an idea accounts for the fact that the two photosystems could never be separated physically and it might also explain that substituted hydroquinones can be electron donors to photosystem I if they are made hydrophobic by their substituent and to photosystem II if they are hydrophilic (Kitajima et al., 1973). Their substituents make them apt to react with the photosystem closest to them on the membranes (on an hydrophobicity scale). This idea could also help understand the changes occurring in greening plants (ageing plants) where photosystem II appears after photosystem I (Strasser, 1973) and where the physical-chemical properties of the chlorophyll-lipoprotein complexes of young plants are different from those of older plants (De Koning and Aghion, 1967; Rorive and Aghion, 1972).

In conclusion, the physical state of the photoreceptor molecules and the packing of surrounding (non) photoreceptor molecules seem unequivocally tied in the case of photosynthesis. In this same symposium, Delmelle (this volume) notices a change of fluidity at the surface of the rod outer segment membranes in the light. Harosi (this volume), studying the photobleaching of the visual pigments, notices that it is always accompanied by a change of their absorbance spectra. Alone, those two sets of observations would justify an investigation of the physical-chemical properties of the visual photon-capting membranes, along the lines described here.

REFERENCES

Abrahamson, E.W. and Ostroy, S.E. (1967). The photochemical and macromolecular aspects of vision. Progress in Biophysics and Mole. Biol. 17: 179-215.

Adamson, A.W. (1967). Physical Chemistry of Surfaces, p. 11. Interscience publishers, New York.

Aghion, J., Broyde, S.B. and Brody, S.S. (1969). Surface reactions of chlorophyll a monolayers at a water-air interface. Biochem. 8: 3120-3126.

Arden, G.B. (1969). The excitation of photoreceptors. Progress in Biophy. and Mol. Biol. 19: 372-421.

Bannister, T.T. (1963). Les propriétés physiques et photochimiques de la chlorophylle colloidale. Physiol. Vég. 1: 115-127.

Boardman, N.K. (1970). Physical separation of the photosynthetic photochemical systems. Ann. Rev. Plant Physiol. 21: 115-140.

Brody, S.S. (1958). New excited state of chlorophyll. Science 128: 838-839.

Coumans, M., Strasser, R.J. and Bronchart, R. (1972). Bochum 6th International Congress on Photobiology, abstract 244.

Daemen, F.J.M. (1973). Vertebrate rod outer segment membranes. Biochim. Biophys. Acta 300: 255-288.

De Koning, H.W. and Aghion, J. (1967). Etude des complexes de pigments chlorophylliens d'*Euglena gracilis* au cours de la croissance et à l'état adulte. Plant Cell Physiol. 8: 129-139.

Dijkmans, H. et Aghion, J. (1974). Conditions de formation d'agrégats de chlorophylle a dans des solvants polaires aqueux. Polarité du milieu. Plant Cell Physiol. 15: (in press).

Dinant, M. et Aghion, J. (1973). Agrégation des chlorophylles in vitro. Photodécoloration de la chlorophylle a adsorbée sur des particules de lipoprotéines extraites du lait. Photochem. Photobiol. 17: 25-30.

Gregory, R.P.F. (1971). Biochemistry of photosynthesis. Wiley Interscience, London.

Gurinovitch, G.P., Sevchenko, A.W. and Solov'ev, K.N. (1968). Spectroscopy of chlorophyll and related compounds, Minsk. English Translation by U.S. Joint Publications Research Service, Washington 1971.

Katz, J.J. (1973). Chlorophyll. In: Inorganic Biochemistry, vol. 2, p. 1022-1066. Elsevier, New York.

Kitajima, M., Ogawa, T., Inoue, Y. and Shibata, K. (1973). Sites of electron donation by alkyl-hydroquinones in the electron transport chain of spinach chloroplasts. Plant Cell Physiol. 14: 787-790.

Rabinowitch, E.I. (1956). Photosynthesis and Related Processes, II. Interscience, New York.

Rorive, B. et Aghion, J. (1972). Variation de quelques propriétés des chlorophylles de *Marchantia polymorpha* L. au cours du passage d'un stade jeune à l'état adulte. Physiol. Vég., 10: 665-670.

Smyth, C.P. (1955). Dielectric Behavior and Structure. McGraw Hill, New York.

Strain, H.H. and Svec, W.A. (1966). Extraction, Separation, Estimation and Isolation of the Chlorophylls. In: The Chlorophylls, edited by L.P. Vernon and G.R. Seely, Academic Press, New York, p. 21-66.

Strasser, R.J. (1973). Induction phenomena in green plants when the photosynthetic apparatus starts to work. Arch. Intern. Physiol. Biochim. 81: 935-955.
Weier, T.E., Stocking, C.R. and Shumway, L.K. (1967). The photosynthetic apparatus in chloroplasts of higher plants. In: Energy Conversion by the photosynthetic apparatus, Brookhaven National Laboratories, Upton, New York, p. 353-373.
Witt, H.T. (1971). Coupling of quanta, electrons, fields, ions and phosphorylation in the functional membrane of photosynthesis. Quart. Rev. Biophysics 4: 365-477.

ACKNOWLEDGMENTS

Work done in fulfilment of the conditions for a grant from the Belgian F.R.F.C. (number 10 118).

PHOTOPHYSICAL PROCESSES IN VISUAL PIGMENTS

Roger M. LEBLANC

Groupe de Recherche en Biophysique

Université du Québec, Trois-Rivières, Québec, Canada

Canst thou catch any fishes, then?
Pericles Act II, Sc. 1

INTRODUCTION

The photoreceptor cells use the energy of light quanta as transducers, converting light energy into neural signals. The first process in these cells is absorption which produces an electronically excited state. Their great light sensitivity involves an efficient absorption of visible light.

A photon absorbed by a cell activates only one molecule in the primary excitation step of a photochemical sequence. Following this absorption process, one can ask the question: What is the detailed fate of the excitation pumped into this molecule by electromagnetic radiation? To answer this question, we need an understanding of the primary processes, which is the aim of photophysical research.

A great part of the research directed to the photophysical problems tends to take a synthetic approach. The idea is to study the pure visual pigments and their chromophores in simple laboratory systems, such as solution, micelle, film, black lipid membrane (BLM), monolayer and solid; to compare these systems with the photoreceptor cell properties; and to increase the complexity of the system by the incorporation of other photoreceptor components until the properties of the laboratory and the *in vivo* systems begin to coincide. One believes that this approach will give some insight into the primary processes in vision.

PHOTOPHYSICAL PROCESSES

We define a photophysical process as a physical process, resulting from the electronic excitation of a molecule by photon but involving no chemical change. It is most convenient to represent the photophysical processes in terms of the energy level of molecules. Fig. 1 shows a typical energy level diagram (The horizontal scale has no physical meaning). We use simple notation for the electronic states,

S_o : ground singlet state,
S_1 : first excited singlet state,
S_m : higher excited singlet states, $(m > 1)$,
T_1 : first excited triplet state,
T_n : higher excited triplet states, $(n > 1)$.

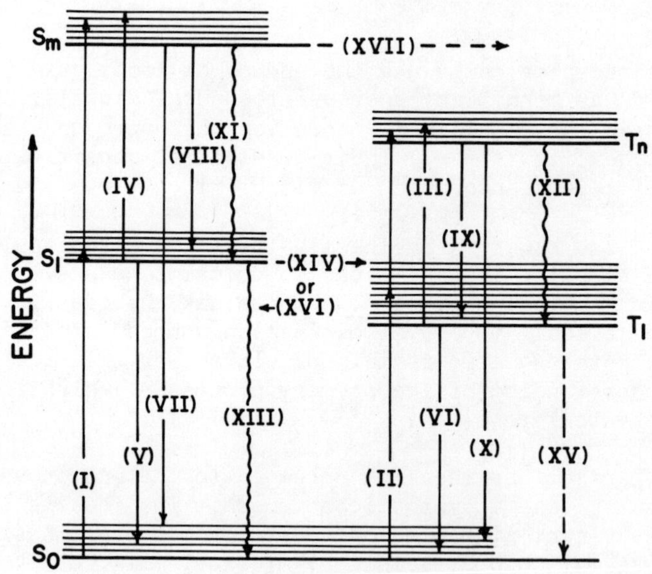

Fig. 1. Schematic state energy level diagram. The fine horizontal lines represent vibrational levels. Arrows represent absorption of radiation, luminescence transitions, or nonradiative pathways.

In the nomenclature of photophysics, the terms singlet and triplet excited states appear very often. According to the exclusion principle, no two electrons having the same spin can occupy the same orbital. The multiplicity (M) of a state is given by the formula

$$M = 2 \, |S| + 1,$$

$|S|$ is the absolute net spin. Thus in the ground state of molecules there are precisely two electrons per orbital and the absolute total spin is zero ($|S| = |+ \frac{1}{2} - \frac{1}{2}|$), so that the multiplicity will be one, hence the designation singlet. If two electrons have parallel spins, which can occur in an excited state without violating the exclusion principle, the absolute net spin will be $|S| = 1$, so that $2\,|S| + 1 = 3$, and the state is designated a triplet.

The singlet and triplet states of the same molecules do not possess in themselves very different chemical properties. However, a triplet state has a lifetime much longer than the corresponding singlet state; the triplet state thus has a better chance of undergoing chemical reactions.

Following Birks's (1970) classification of photophysical processes, they can be enumerated as

A. Unimolecular photophysical processes,
B. Biphotonic and multiphotonic photophysical processes,
C. Bimolecular photophysical processes.

A. Unimolecular photophysical processes

Three categories of unimolecular processes are involved in an isolated molecule (very dilute solution in a transparent medium), namely, absorption transitions, luminescence transitions and radiationless transitions.

a) Absorption transitions: The molecule is raised to an excited state electronically, and usually vibrationally as well, from which any one of a number of photophysical processes may take place.

The different types of radiative excitation transitions can be summarised in the following way,

(i) $S_0 - S_1$ or $S_0 - S_m$ absorption process is represented in Fig. 1 by an arrow which corresponds to the frequency of the photon absorbed by the molecule. Experimentally, these transitions correspond to the main electronic absorption spectrum.

Most of what is known or inferred about the function of visual pigments in the photoreceptor cell is based on absorption spectroscopy in the visible. For rhodopsin, the absorption maxima are observed between 433 nm and 562 nm, whereas the α-peaks of porphyropsins are distributed from 510 nm to 543 nm (Bridges, 1972). The absorption spectra of the chromophores of visual pigments and their Schiff bases were studied recently by Becker and co-workers (Schaffer et al, 1974).

(ii) $S_o - T_1$ or $S_o - T_n$ is a spin-forbidden transition. However, the high photon flux available from a laser (Moore, 1971) and the perturbation method (Raubach and Guzzo, 1971) can be advantageously used to study this type of process. The latter method has used the oxygen perturbation studies to determine the singlet-triplet absorption spectrum of all-trans retinal. The lowest energy band was assigned as the 0-0 singlet-triplet transition ($S_o - T_1$) at 803 nm.

(iii) $T_1 - T_n$ is spin-allowed transition observed by flash photolysis methods. The principal feature of many conventional flash photolysis apparatus consists in exciting the reactants by a short intense pulse of light, duration of about 10^{-5} sec (microsecond time scale), and then examining the mixture for any transient species formed (Boag, 1968). Flash photolysis of retinal shows a very strong band, maximum at approximately 450 nm, with an extinction coefficient of $\sim$80,000. The spectroscopic properties favoured a triplet-triplet absorption for this intermediate (Abrahamson et al, 1959; Grellman et al, 1962; Dawson and Abrahamson, 1962; Truscott et al, 1973).

It should be mentioned that transient species have been observed and their rate of thermal decay have been measured using a conventional flash system (Porter, 1950). This is the case with prelumirhodopsin, lumirhodopsin, metarhodopsin I and metarhodopsin II, the transients from the flash excitation of rhodopsin, (Grellman et al, 1962, 1964; Sengbusch and Stieve, 1971 a, b).

(iv) $S_1 - S_m$ absorption has been observed for aromatic compounds by nanosecond (ns) flash photolysis (Porter and Topp, 1968). The flash duration of conventional flash photolysis limited the observation of transients in the μs time scale. The appearance of Q switched laser gave the opportunity to reduce flash pulses between 10 ns and 300 ns and thus enable us to study the absorption of excited singlet states whose lifetimes are in the nanosecond region (Hodgkinson and Munro, 1973).

Studies of the initial stages of rhodopsin photochemistry in solution at room temperature have been made possible with the nanosecond laser photolysis system (Rosenfeld et al, 1972). A transient around 580 nm was observed with a 50 ns delay which was attributed to the prelumirhodopsin.

The advent of picosecond (psec) spectroscopy, a mode locked Nd^{+3} glass laser generating psec pulses, facilitated many investigations in photophysics (Rentzepis, 1968). The first application to biophysics was oriented to the detection of transients absorbing at longer wavelengths than rhodopsin solution at room temperature, on time scales from 6 psec to 0.1 μsec. The psec measurements indicate that a single transient intermediate is being observed in a time that is at least as short as the experimental time resolution (about 6 psec). The transient species absorbs strongly at 560 nm and decays with a lifetime of about 30 ns. Busch et al (1972) suggest that the extremely fast risetime of prelumirhodopsin strongly supports the hypothesis that this species is the product of the primary photoprocesses in rhodopsin.

b) Luminescence transitions: Once the molecule is in an excited state, the molecule is de-excited to a lower electronic energy level by the emission of a photon.

There are two types of photon emission: fluorescence and phosphorescence. Fluorescence is a radiative transition between states of the same multiplicity, while phosphorescence is the transition between states of different multiplicity.

(v) $\underline{S_1 - S_0}$ fluorescence transition is a frequently observed emission (Lifetime for fluorescence is about 10^{-6} - 10^{-8} sec). All-trans, 9-cis, 13-cis retinals exhibit fluorescence in 3-methylpentane at 77°K. No emission was observed for the 11-cis retinal (up to 820 nm) in the temperature range from 77 to 208°K (Becker et al, 1971). Fluorescence excitation spectra of some retinyl polyenes were examined recently by an excitation spectrometer which permits to measure a true excitation spectrum (Christensen and Kohler, 1974).

Guzzo and Pool (1968, 1969) have reported a very weak

fluorescence for rhodopsin at about 570 nm, either at 77°K or 276°K. They have made further fluorescence studies on prelumirhodopsin, lumirhodopsin, metarhodopsin I, and metarhodopsin II in both the rod outer segment and in the digitonin extract. However, these results have not been duplicated to date (See Busch et al, 1972).

In opsin, tryptophan is the principal fluorescing amino acid. Excitation of the rhodopsin at 280 nm shows a fluorescence emission maximum at 331 nm. Examining the emission maximum of rhodopsin, one can suggest that most of the tryptophans are in a non-polar environment (Ebrey, 1972). Fluorescence properties of opsin were used as a probe of molecular events following the electronic excitation (Guzzo, 1973).

Since very weak fluorescence is observed for visual pigments and their chromophores, we think that the use of a laser spectrofluorimeter will give more sensitivity and reproducibility in the fluorescence measurements (Leblanc et al, 1974).

(vi) $\underline{T_1 - S_o}$ phosphorescence has a longer duration (10^{-3} - 10 sec) than the fluorescence since the transition is spin-forbidden. Phosphorescence has been observed for many aromatic compounds, however, this emission does not occur for visual pigments and their chromophores.

(vii) $\underline{S_m - S_o}$, (viii) $\underline{S_m - S_1}$, (ix) $\underline{T_n - T_1}$ transitions have been observed only for a few compounds. No emission was detected for retinyl polyenes.

(x) $\underline{T_n - S_o}$ phosphorescence is a very improbable process.

c) Radiationless transitions: These transitions occur between isoenergetic vibrational levels of different electronic states. They are the result of conversion of electronic energy into vibrational energy.

When a radiationless process proceeds between states of the same multiplicity, it is known as internal conversion. A radiationless transition between states of different multiplicity is described as intersystem crossing.

(xi) $\underline{S_2 - S_1}$ and $\underline{S_m - S_{m-1}}$, (xii) $\underline{T_2 - T_1}$ and $\underline{T_n - T_{n-1}}$ internal conversion transitions occur rapidly (10^{-11} - 10^{-14} sec). The psec spectroscopy seems to indicate that the transformation from electronically excited rhodopsin to prelumirhodopsin may involve such electronic relaxation (Busch et al, 1972).

(xiii) $S_1 - S_o$ internal conversion to the ground state competes with the fluorescence process (v) and the intersystem crossing transition (xiv). However, its rate constant (or lifetime $10^{-5} - 10^{-7}$ sec) does not favour primarily this process.

(xiv) $S_1 - T_1$ intersystem crossing ($\sim 10^{-8}$ sec) has been studied for all-trans retinal. The overall mean value for the quantum yield of this process was 0.60 which means that 60% of the molecules populate triplet state (Bensasson et al, 1973). This value is consistent with the value of 0.67 for the quantum efficiency of photobleaching of rhodopsin (Dartnall, 1968).

(xv) $T_1 - S_o$ intersystem crossing competes with the phosphorescence process. This $T_1 - S_o$ transition may be important for all the retinyl polyenes.

(xvi) $T_1 - S_1$ and (xvii) $S_m - T_n$ intersystem crossing transitions may occur by thermal activation of T_1 for the $T_1 - S_1$ process, whereas the process $S_m - T_n$ is probable between isoenergetic vibration levels of the different upper electronic excited states.

B. Biphotonic and multiphotonic photophysical processes

Irradiation of a molecular system by an intense laser beam gives numerous sorts of biphotonic and multiphotonic processes. For example, these processes are involved in laser damage to the retina (Marshall and Mellerio, 1970).

Some biphotonic photophysical processes have already been discussed: $S_o - S_1$ absorption followed by S_1 T_1 intersystem crossing, and $T_1 - T_n$ absorptions; $S_o - T_1$ absorption followed by $T_1 - T_n$ absorption; $S_o - S_1$ absorption followed by $S_1 - S_n$ absorption.

The resulting electronically excited molecule may undergo biphotonic ionisation via S_1 or T_1. This process can be relevant to the photoconductivity processes in visual pigments and the retinyl polyenes chromophores (Rosenberg, B., 1966).

C. Bimolecular photophysical processes

We must expect different interactions in concentrated or aggregated systems compared to a very dilute solution. Not much work has been done actually in concentrated systems since the techniques to study bimolecular photophysical processes have only just come into the field of photophysics.

The various bimolecular processes for visual pigments and their chromophores may be divided into broad categories such as perturbation processes (Raubach and Guzzo, 1971); energy transfer processes (Kropf, 1967; Guzzo and Pool, 1969, 1973); complex formation by two unexcited molecules (Moore and Song, 1973); and complex formation by an excited molecule and an unexcited molecule. The complex formed between identicals molecules is called an excimer and that between different molecules an exciplex. The photophysical processes of an excimer or an exciplex exhibit their own characteristics. No evidence, however, for this kind of complexes has been observed for visual pigments. The last category is the interaction between two excited molecules. An example of this interaction is the triplet-triplet annihilation.

ACKNOWLEDGEMENT

This work was supported by the Defence Research Board of Canada (Grant CRD 9530-117-UG).

REFERENCES

Abrahamson, E.W., Adams, R.G. and Wulff, V.J. (1959). Reversible spectral changes in retinene solutions following flash illumination. J. Phys. Chem. 63: 441-443.

Becker, R.S., Inuzuka, K., King, J. and Balke, D.E. (1971). Comprehensive investigation of the spectroscopy and photochemistry of retinals. II. Theoretical and experimental consideration of emission and photochemistry. J. Am. Chem. Soc. 93: 43-50.

Bensasson, R., Land, E.J. and Truscott, T.G. (1973). The triplet state of retinal: is it involved in vision? Photochem. Photobiol. 17: 53-55.

Birks, J.L. (1970). Photophysics of Aromatic Molecules, Wiley, London.

Boag, J.W. (1968). Technique of flash photolysis. Photochem. Photobiol. 8: 565-577.

Bridges, C.D.B. (1972). The rhodopsin - porphyropsin visual system. In: Handbook of Sensory Physiology, Vol. VII/1, chapter 11, edited by H.J.A. Dartnall, Springer-Verlag, New York.

Busch, G.E., Applebury, M.L., Lamola, A.A. and Rentzepis, P.M. (1972). Formation and decay of prelumirhodopsin at room temperatures. Proc. Nat. Acad. Sci. USA 69: 2802-2806.

Christensen, R.L. and Kohler, B.E. (1974). Excitation spectroscopy of retinal and related polyenes. Photochem. Photobiol. 19: 401-410.

Dartnall, H.J.A. (1968). The photosensitivities of visual pigments in the presence of hydroxylamine. Vision Res. 8: 339-358.

Dawson, W.R. and Abrahamson, E.W. (1962). Population and decay of the lowest triplet state of polyenes with conjugated heteroatoms: retinene. J. Phys. Chem. 66: 2542-2547.

Ebrey, T.G. (1972). The fluorescence from the tryptophans of rhodopsin. Photochem. Photobiol. 15: 585-588.

Grellman, K.H., Memming, R. and Livingston, R. (1962). Some flash-photolytic and photochemical studies of retinene and related compounds. J. Am. Chem. Soc. 84: 546-548.

Guzzo, A.V. and Pool, G.L. (1968). Visual pigment fluorescence. Science 159: 312-314.

Guzzo, A.V. and Pool, G.L. (1969). Fluorescence spectra of the intermediates of rhodopsin bleaching. Photochem. Photobiol. 9: 565-570.

Guzzo, A.V. and Pool, G.L. (1969). Energy transfer to the triplet level of all-trans retinal. J. Phys. Chem. 73: 2512-2515.

Guzzo, A.V. (1973). Light induced changes in the protein fluorescence of rhodopsin. Vision Res. 13: 2581-2586.

Hodgkinson, K.A. and Munro, I.H. (1973). Excited singlet and triplet states of aromatic hydrocarbons determined by a laser photolysis technique. J. Mol. Spectrosc. 48: 57-71.

Kropf, A. (1967). Intramolecular energy transfer in rhodopsin. Vision Res. 7: 811-818.

Leblanc, R.M., Galinier, G., Tessier, A. and Lemieux, L. (1974). Laser spectrofluorimetry of chlorophylls. Can. J. Chem. (in press).

Marshall, J. and Mellerio, J. (1970). Laser irradiation of retinal tissue. Br. Med. Bull. 26: 156-160.

Moore, C.B. (1971). Lasers in chemistry. Ann. Rev. Phys. Chem. 22: 387-428.

Moore, T.A. and Song, P.S. (1973). Molecular interactions in the ground and excited states of retinal. Nature New Biol. 243: 30-32.

Porter, G. (1950). Flash photolysis and spectroscopy. A new method for the study of free-radical reactions. Proc. R. Soc. Lond. Ser. A, 200: 284-300.

Porter, G. and Topp, M.R. (1968). Nanosecond flash photolysis and the absorption spectra of excited singlet states. Nature 220: 1228-1229.

Raubach, R.A. and Guzzo, A.V. (1971). Singlet-triplet absorption spectrum of all-trans retinal. J. Phys. Chem. 75: 983-984.

Raubach, R.A. and Guzzo, A.V. (1973). Photoisomerization pathways in the visually important polyenes. I. The retinals. J. Phys. Chem. 77: 889-892.

Rentzepis, P.M. (1968). Lasers in chemistry. Photochem. Photobiol. 8: 579-588.

Rosenberg, R. (1966). A physical approach to the visual receptor process. In: Advances in Radiation Biology, edited by L.G. Augenstein, R. Mason and M.R. Zelle, Vol. II, p. 193-241. Academic Press, New York.

Rosenfeld, T., Alchalal, A. and Ottolenghi, M. (1972). Nanosecond laser photolysis of rhodopsin in solution. Nature 240: 482-483.

Schaffer, A.M., Waddell, W.H. and Becker, R.S. (1974). Visual pigments. IV. Experimental and theoretical investigations of the absorption spectra of retinal Schiff bases and retinals. J. Am. Chem. Soc. 96: 2063-2068.

Sengbusch, G.V. and Stieve, H. (1971 a). Flash photolysis of rhodopsin. I. Measurements on bovine rod outer segments. Z. Naturforsch. 26b: 488-489.

Sengbusch, G.V. and Stieve, H. (1971 b). Flash photolysis of rhodopsin. II. Measurements on rhodopsin digitonin solutions and fragments of rod outer segments. Z. Naturforsch. 26b: 861-862.

Truscott, T.G., Land, E.J. and Sykes, A. (1973). The in vivo photochemistry of biological molecules. III. Absorption spectra, lifetimes and rates of oxygen quenching of the triplet states of β-carotene, retinal and related polyenes. Photochem. Photobiol. 17: 43-51.

MONOLAYER STUDIES OF VISUAL PIGMENTS

Surat HOTCHANDANI, Christophe KOSSI & Roger M. LEBLANC

Groupe de Recherche en Biophysique

Université du Québec, Trois-Rivières, Québec, Canada

Give us a light there, ho!
Macbeth, Act III, Sc. 3

ORDERED STRUCTURES

One of the most interesting and fascinating facts about biology is the existence of ordered structures in the organelles. The electron microscopic and X-ray diffraction studies have shown that in living systems the molecules are, in fact, arranged in strict order and are present in aggregates.

In many organelles these structures are two dimensional producing a membrane. The stacking of membrane sheets provides the third dimension and this imparts to the gross organelle a lamellar form. Lamellar form has been found in rod and cone outer segments (Sjostrand, 1949, 1953).

In order to study the properties of these organelles the synthesis of aggregates of definite order is of prime importance. The monolayer techniques help us achieve the necessary orientation and aggregation among the molecules. With these techniques:

- One may try to imitate simple biological systems which is the aim of monolayer studies at air/water interface;
- One may build any planned arrangement of molecules of interest, which is the aim of monolayer studies at air/solid interface.

MONOLAYERS AT THE AIR/WATER INTERFACE

a) Introduction: The requirements that enable a chemical compound to exists in the form of an insoluble monolayer at air/water interface are (1). A paraffin chain or some other group virtually insoluble in water and (2). A polar group, such as the carboxyl, hydroxyl or aldehyde, having considerable affinity for water. The monolayer forming molecules have their polar "heads" immersed, while the nonpolar hydrocarbon "tails" stick out of water.

b) Surface pressure: The extreme thinness of monolayer, often 10 - 30 Å, makes direct methods of measurement very difficult. Various physical properties like surface pressure (π) have, therefore, been used to study the monolayers. The Langmuir float method and the Whilmey plate continue to be the standard methods for surface pressure measurement (Gaines, p. 44, 1966).

c) Surface pressure-Area (π-A) curve: The characteristic description of an insoluble monolayer is usually in terms of pressure-area (π-A) curve, i.e. the relationship between the surface pressure observed and the area occupied on the liquid surface by the molecules of the film. The technique consists of simultaneous measurements of the film area and the corresponding surface pressure as the barrier compresses the film.

d) Types of monolayers: There are three main types of monolayers. 1. Condensed monolayers 2. Gaseous monolayers 3. Liquid or expanded monolayers.

Condensed monolayers: In these, the molecules are essentially close-packed and well oriented at the air/water interface. The molecules stand nearly upright. The pressure-area (π-A) curves are nearly straight and very steep (Fig. 1). The steepness indicates the low compressibility of the condensed monolayers. The examples are those of stearic acid and other lipids. The area per molecule is about 20 $Å^2$ (Gaines, 1966, p. 187). In condensed films of homologous series the π-A curves are very similar and the area per molecule is independent of chain length.

Gaseous monolayers: In these monolayers the molecules are floating about in the surface layer far apart so that they exert relatively little force on one another. The surface pressures for many gaseous monolayers are extremely low (less than 0.1 dyne/cm) at room temperature and study of the gaseous monolayers is thus experimentally very difficult. The examples are the protein films in low pressure region.

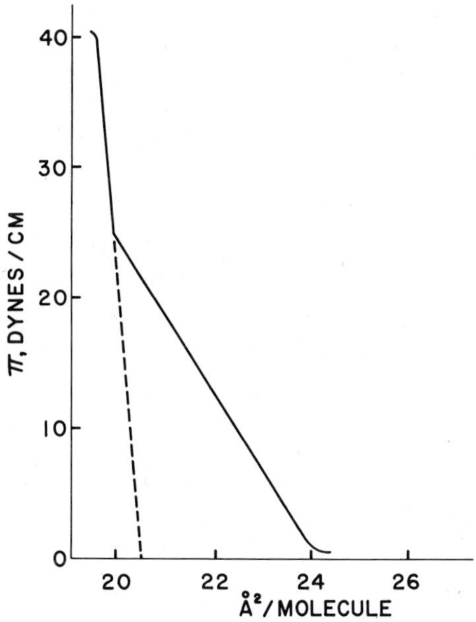

Fig. 1. Pressure-area isotherm of stearic acid (Gaines, 1966, p. 187)

Liquid or expanded monolayers: These occupy an intermediate position between the condensed and gaseous monolayers. The π-A curves of these films show considerable curvature. The molecular areas are roughly two or three times as large as the molecular area of the condensed film. Fig. 2 shows π-A curves of a few retinyl polyenes. Their shape is typical of expanded film. The limiting area per molecule ($A_{\pi \rightarrow 0}$) of the pigments reported by Hotchandani, Leblanc and Boucher (in press).

all-trans retinal = 59 Å²

9-cis retinal = 54 Å²

all-trans retinol = 44 Å².

The large values of the molecular area as compared to the condensed film of fatty acids, are believed to be due to the presence of ionone ring in the hydrophobic part of the retinyl polyenes and the fact that the unsaturation interferes with the close packing of molecules.

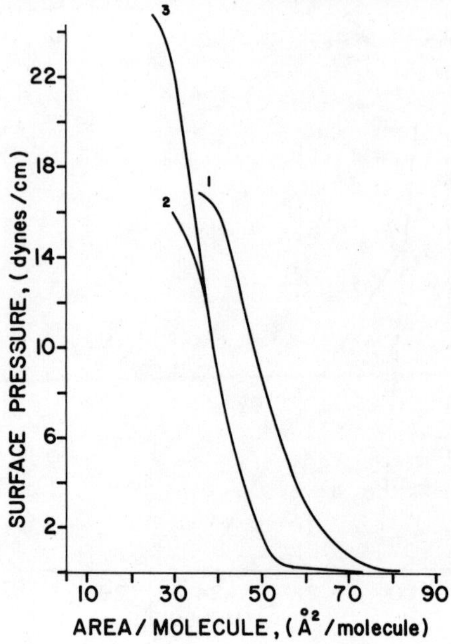

Fig. 2. Pressure-area isotherms of all-trans retinal (1), 9-cis retinal (2) and all-trans retinol (3) at air/water interface; pH = 8.0 - 8.1 and temperature = 22°C.

e) Retinal monolayers at air/water interface: Brody et al (1971, 1973) have studied the complexation of retinal isomers with β-mercaptoethylamine (MEA), lysine and L-cysteine. Since the molecules in monolayers are oriented in well defined order with respect to the subphase, the monolayer is expected to react with materials dissolved in the subphase only in a very specific manner. For example Brody et al (1971) reported that 9-cis retinal at air/water interface complexes with MEA but does not complex with lysine.

f) Rhodopsin monolayer at air/water interface: The π-A curve is of expanded type (Hyono, Kuriyama, Tsuji and Hosoya, 1962). Azuma and Takagi (1966) studied the rhodopsin film at air/water interface and found that their results did not agree with Hyono et al. Their results of the spectrophotometric studies indicate that every rhodopsin molecule spread at air/water interface is not bleached.

Bonting and Bangham (1967) and Depont et al (1968) in the

course of their studies of rhodopsin monolayer observed an expansion of the film on illumination. This was interpreted as a result of penetration into the rhodopsin monolayer of retinaldehyde liberated during the photolysis of rhodopsin. The different results on the rhodopsin film at air/water interface reported by various research groups are probably due to the varying degrees of purity of rhodopsin.

MONOLAYER AT AIR/SOLID INTERFACE

The absorption spectroscopy of monolayers of retinyl polyenes deposited on solid support has been studied by S. Hotchandani, C. Kossi and Roger M. Leblanc (in press). The monolayers were deposited on hydrophilic quartz slide and on hydrophobic cadmium arachidate base. The absorption spectra of monolayers of all-trans retinal and all-trans retinol together with their solution spectra are shown in Fig. 3. The spectrum of all-trans retinal monolayer is red shifted (~ 17 nm) as compared to its solution spectrum whereas that of all-trans retinol monolayer is blue shifted (~ 35 nm) relative to its solution spectrum. This shift has been explained on the basis of interaction among the chromophores. The deposition of multilayers of all-trans retinol on both hydrophilic and hydrophobic bases has been achieved and the optical density has been found to vary linearly with the number of layers.

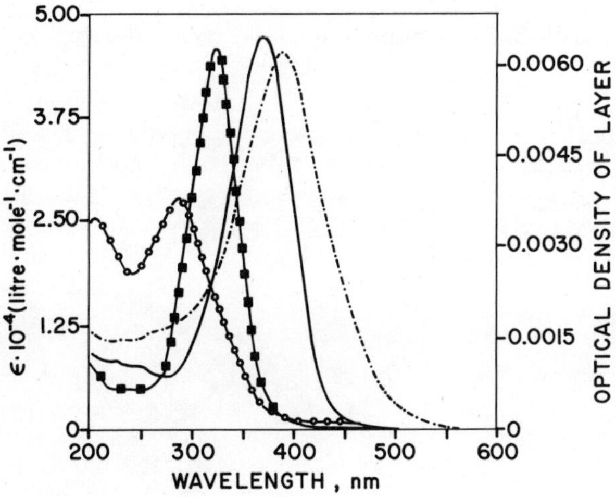

Fig. 3. Absorption spectra of all-trans retinal in ether (-), all-trans retinol in ether (-■-■-■-), all-trans retinal in monolayer at air/lipid interface (----) and all-trans retinol in monolayer at air/lipid interface (-ooo-).

REFERENCES

Azuma, K. and Takagi, M. (1966). Rhodopsin film at air/water interface. Ann. Rep. Biol. Works. Fac. Sci. Osaka Univ. 14: 73-81.

Bonting, S.L. and Bangham, A.D. (1967). On the biochemical mechanism of the visual process. Exp. Eye Res. 6: 400-413.

Brockmann, R.E. and Brody, S.S. (1971). Photoreactions and complex formation of retinal monolayers at water-air interface. Z. Naturforsch. 26b: 119-125.

Brody, S.S. (1973). Reactions of retinals in a model membrane system. Z. Naturforsch. 28c: 157-164.

DePont, J.J.H.M., Daemen, F.J.M. and Bonting, S.L. (1968). Biochemical aspects of the visual Process II. Schiff Base formation in Phosphatidylethanolamine monolayers upon penetration by retinaldehyde. Biochim. Biophys. Acta 163: 204-211.

Gaines, G.L. Jr. (1966). Insoluble monolayers at liquid-gas interfaces. Interscience Publishers, New York, p. 44, 187.

Hotchandani, S., Leblanc, R.M. and Boucher, F. (Vision Res., in press)

Hyono, A., Kuriyama, S., Tsuji, K. and Hosoya, Y. (1962). Monolayer film of rhodopsin at the air/water interface. Nature 193: 679-680.

Sjöstrand, F.S. (1949). An electron microscope study of the retinal rods of the guinea pig eye. J. Cell. Comp. Physiol. 33: 383-403.

Sjöstrand, F.S. (1953). The ultrastructure of the outer segments of rods and cones of the eye as revealed by the electron microscope. J. Cell. Comp. Physiol. 42: 15-44.

MARQUAGE PAR SPIN DES MEMBRANES DE BATONNETS

M. DELMELLE and M. PONTUS

Institut de Physique, Université de Liège

Sart Tilman, 4000 Liège, Belgique

Ce début n'est pas mal; et contre le prochain
La conversation prend un assez bon train.
Le Misanthrope, Acte II, Sc. 4

INTRODUCTION

Au cours des dernières années, il est apparu clairement que pour comprendre le mécanisme de la transduction, il était indispensable d'acquérir des connaissances supplémentaires sur l'organisation moléculaire des membranes qui forment les segments externes des bâtonnets. Aujourd'hui, les recherches sur les photorécepteurs font intervenir, notamment, les rayons X et la microscopie électronique (Blasie et al., 1969, Chabre et al., 1972, Chen et Hubbell, 1973) ainsi que les techniques les plus raffinées de la spectroscopie (Shichi, 1971).

Nous avons entrepris l'étude de cette question en faisant usage de la méthode du marquage par spin et nous voudrions présenter, ici, quelques-uns de nos résultats.

LA RESONANCE DE SPIN ELECTRONIQUE

Le marquage par spin s'insère comme une application de la technique de résonance paramagnétique électronique (RPE). Peut-être est-il utile d'en rappeler brièvement les principes fondamentaux (voir, par exemple, Wertz et Bolton, 1972).

Cette méthode de résonance a, notamment, pour objet l'étude des radicaux libres. Si un électron non apparié est placé dans un champ magnétique, il s'oriente soit parallèlement, soit antiparallèlement au champ, ce qui correspond à deux niveaux d'énergie distincts, l'orientation parallèle correspondant à l'état le plus stable. Dans la technique RPE, on envisage l'observation des transitions entre ces sous-niveaux en fournissant à l'échantillon, l'énergie nécessaire pour induire ces transitions. La relation qui lie la fréquence de cette radiation électromagnétique à la valeur du champ magnétique externe, est donnée par la formule: $h\nu = g\beta H$, où h est la constante de Planck, ν la fréquence, g le rapport gyromagnétique, β le magnéton de Bohr et H le champ magnétique. Dans l'hypothèse où l'électron libre subit l'interaction d'un noyau doué de propriétés magnétiques, le spectre RPE devient plus complexe, mais, de ce fait, plus utile. Limitons-nous au cas d'un électron libre qui interagit avec un noyau comme l'azote qui possède un spin nucléaire I = 1. Cet exemple présente une importance capitale dans le cas des molécules qui servent au marquage par spin. En présence de ce moment nucléaire, chaque niveau se décompose en trois sous-niveaux qui correspondent respectivement à une orientation parallèle, perpendiculaire et antiparallèle du moment magnétique nucléaire par rapport à celui de l'électron libre. En tenant compte des règles de sélection, le spectre se compose de trois raies également séparées et de même intensité (fig. 1).

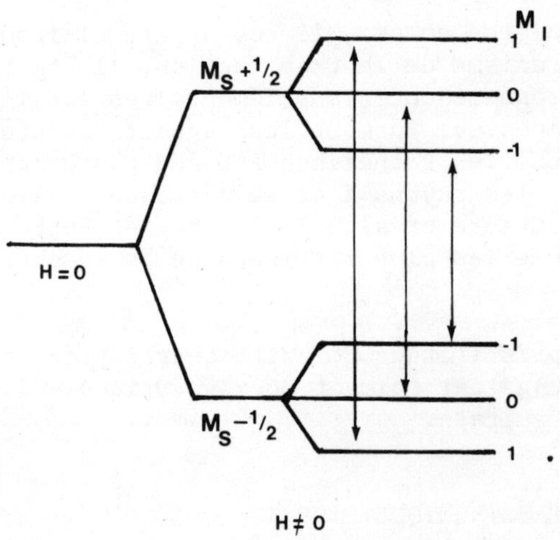

FIG. 1. Schéma des niveaux d'énergie pour un électron libre. La présence d'un champ magnétique H produit un dédoublement ($M_S = -1/2 + 1/2$). Par suite de l'interaction avec l'atome d'azote, chaque sous-niveau se décompose en trois. Les flèches indiquent les transitions permises.

LE MARQUAGE PAR SPIN

Cette technique a été introduite en 1965 par le Professeur McConnell et ses collaborateurs (Stone et al., 1965). On introduit, au sein d'un système biologique, un marqueur, en l'occurence une molécule possédant un radical libre stable, constitué, le plus souvent, par un groupement nitroxide. Ce radical est greffé sur une molécule dont la structure chimique présente un intérêt particulier pour l'étude en cause. La molécule marquée peut, soit jouir d'une réactivité chimique préférentielle vis à vis de groupements spécifiques de certaines biomolécules, soit avoir une structure suffisamment semblable à une biomolécule particulière pour remplacer cette dernière au sein d'un système biologique complexe.

De nombreux marqueurs ont été synthétisés et les applications multiples de cette technique sont apparues dans une abondante littérature consacrée à cette question (revue par McConnell et McFarland, 1970).

Notre travail se limite à l'étude des membranes des segments externes des bâtonnets et il s'oriente dans deux directions: la première envisage l'analyse des protéines marquées, en particulier, la rhodopsine, la seconde a pour objet l'étude de la phase lipidique de la membrane.

Nous avons utilisé un matériel extrait de rétines de bovidés. La technique d'extraction a été décrite par de Grip et al., (1972). Quant aux marqueurs, il s'agissait, dans la plupart des cas, de produits commerciaux (Syva, Palo Alto Californie), (fig. 2).

MARQUAGE DE LA RHODOPSINE

Nous avons fait usage, pour la rhodopsine, d'un marqueur dont la formule chimique s'apparente à celle du maleimide (fig. 2A). Cette substance a la propriété de réagir préférentiellement avec les groupements sulfhydryles pour former des liaisons covalentes (Gregory, 1955). Comme chaque molécule de rhodopsine présente, au sein de la membrane, au moins deux fonctions SH (de Grip et al., 1973), nous pouvons espérer, en marquant la protéine, tester les propriétés de la rhodopsine elle-même. Le spectre correspondant est repris dans la figure 3A. Sa forme diffère fondamentalement de celle que l'on observe avec une solution aqueuse de ce marqueur, à température ordinaire (fig. 3B). Cette différence résulte du fait que, dans l'eau, les molécules de nitroxide subissent une réorientation rotationnelle rapide ($\approx 10^{-10}$ sec) alors que, dans le cas des membranes, une proportion importante des molécules de maléimide se trouvent fixées rigidement sur la rhodopsine, ce qui limite leur mouvement. Plusieurs informations peuvent être obtenues à partir de ce spectre.

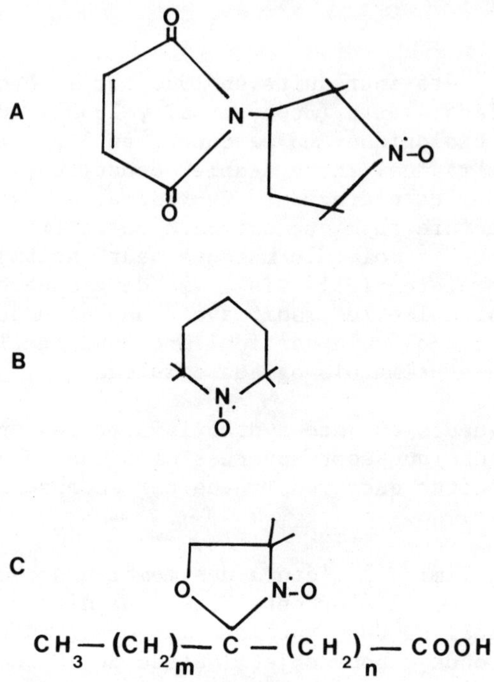

I(m,n)

FIG. 2. Formules chimiques des marqueurs utilisés:
A- 3 maléimido-2,2,5,5 tétraméthyl-1-pyrrolidinyloxyle
B- 2,2,6,6 tétraméthyl-pipéridine-1-oxyle
C- dérivés oxazolidinyloxyles de l'acide stéarique. Les marqueurs employés avaient pour formule I (12-3), I(5-10) et I (1-14).

Ainsi, Lassman et al., (1973) ont montré comment une mesure quantitative de l'hydrophobicité du milieu entourant le marqueur, pouvait être réalisée. En adoptant le formalisme proposé par ces auteurs, nous représenterons par h le paramètre qui mesure cette propriété. Il associe les écarts entre les extrema des spectres (2 A_{zz}), obtenus respectivement, dans le cas d'un solvant polaire (H_2O), dans le cas d'un solvant hydrophobe pris comme référence (décaline) et l'écart caractéristique observé dans le cas de la protéine marquée.

$$h = \frac{A_{zz}\ (\text{eau}) - A_{zz}\ (\text{rhodopsine})}{A_{zz}\ (\text{eau}) - A_{zz}\ (\text{décaline})}$$

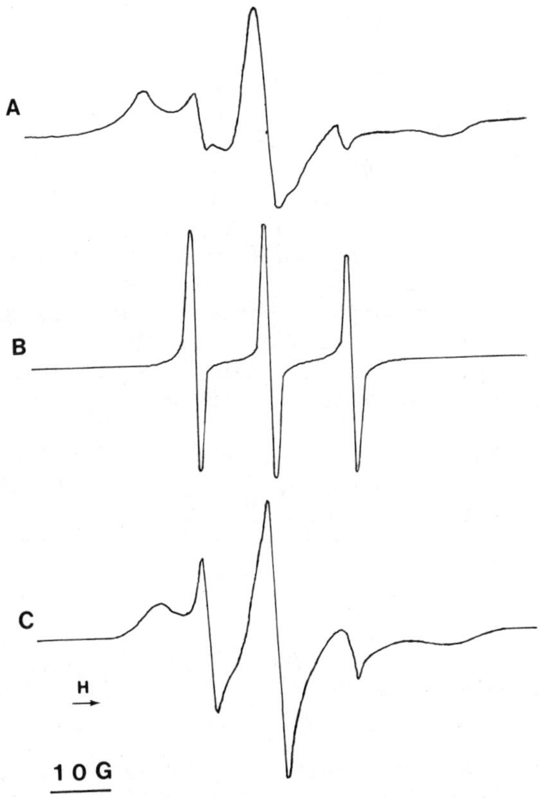

FIG. 3. A- Spectre d'une suspension de membranes de bâtonnets marquéespar le 3 maléimido-2,2,5,5 tétraméthyl-1-pyrrolidinyloxyle.
B- Spectre de ce marqueur en solution aqueuse.
C- Spectre d'une solution de rhodopsine dans le Triton X-100, traitée par le même marqueur.

Si le milieu est fortement polaire, h ≃ 0. Par contre, si les forces hydrophobes sont importantes, h augmente et est voisin de l'unité lorsque l'hydrophobicité se compare à celle de la décaline. Pour la rhodopsine, nous avons mesuré une valeur de h égale à 1,3. Les groupements SH de la rhodopsine sont donc localisés dans un environnement hydrophobe. Outre cette information, nous avons tenté d'estimer le temps de corrélation de cette protéine, c'est-à-dire le temps nécessaire pour qu'elle effectue une rotation autour d'un axe. La technique du marquage par spin permet, en principe, la mesure des temps de corrélation dans un domaine s'étalant entre 5×10^{-11} et 2.5×10^{-7} sec (Shimshick et McConnell, 1972). Le détail de notre analyse a été publié récemment (Delmelle et Pontus, 1974). En fait, la molécule de rhodopsine est caractérisée par un temps de corrélation supérieur à 0.25 μsec, au sein de la membrane, alors que, si la protéine est solubilisée par un détergent (nous avons fait usage du triton X-100) le temps de corrélation devient aussi petit que 70 nsec.

Nous avons également étudié les modifications que la lumière était susceptible d'apporter dans l'allure du spectre (Pontus et Delmelle, 1974). En fait, lorsqu'une molécule de rhodopsine marquée est soumise à l'illumination (λ > 320 nm), le spectre n'est pas modifié de manière appréciable. Il ressort de cette observation que la transformation rhodopsine-opsine ne peut induire qu'une modification partielle de la conformation de la protéine au sein de la membrane. Une conclusion semblable a été atteinte à partir de mesures de dichroisme circulaire (Shichi, 1971).

Par contre, la solubilisation de cette protéine membranaire par des détergents, s'accompagne, le plus souvent, d'importantes modifications structurales. Ainsi, dans la figure 3C, on montre que le Triton X-100 accroît, de manière importante, l'amplitude des raies qui correspondent à des molécules de marqueur faiblement immobilisées.

MARQUAGE DE LA PHASE LIPIDIQUE

Le "Tempo"

Le tempo (2,2,6,6 tétraméthyl-pipéridine-1-oxyle), de composition chimique relativement simple (fig. 2B), possède la propriété d'être soluble, à la fois, en phase hydrophile et en phase hydrophobe. Pour une suspension de membranes, le spectre en cause résulte de la superposition de ces deux composantes (fig. 4). Dans le cas des bâtonnets, on constate que l'amplitude du spectre décroît en fonction du temps. Nous avons pu établir (Pontus et Delmelle, 1973) que cette disparition de radicaux avait pour origine, d'une part, une réaction chimique entre les molécules de nitroxide et les groupements sulfhydryles des protéines, et, d'autre part, une réaction entre les molécules de marqueur et les produits d'oxydation d'acides gras non saturés. Il est important de noter que ce phénomène de réduction s'amplifie lors de l'illumination (Pontus et Delmelle, 1974). Ceci pourrait bien être lié, d'une part, aux observations de Kagan et al., (1973) qui ont mis en évidence la formation d'hydroperoxides induite par la lumière, et,d'autre part, à celles de Noell et al., (1966) qui ont établi que des altérations pouvaient être induites, in vivo, au niveau des photorécepteurs, probablement par action photodynamique.

Outre ce phénomène, nous avons également observé un accroissement de la fluidité de la membrane sous l'action de la lumière visible. Cet effet se traduit par une augmentation de l'importance relative de la composante hydrophobe du spectre, d'environ 10% (Pontus et Delmelle, 1974).

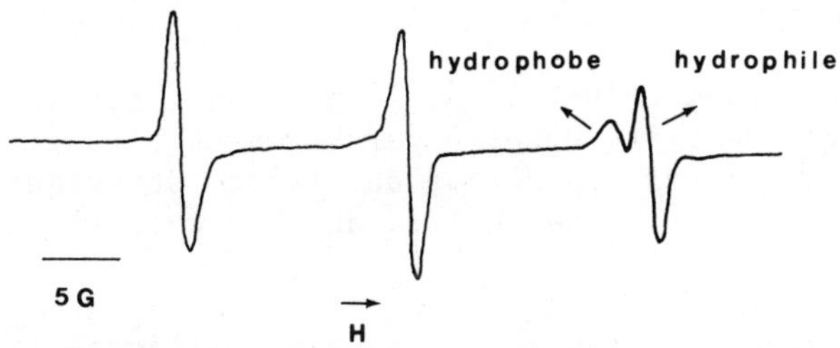

FIG. 4. Spectre d'une suspension de membranes de bâtonnets marquées par le "tempo".

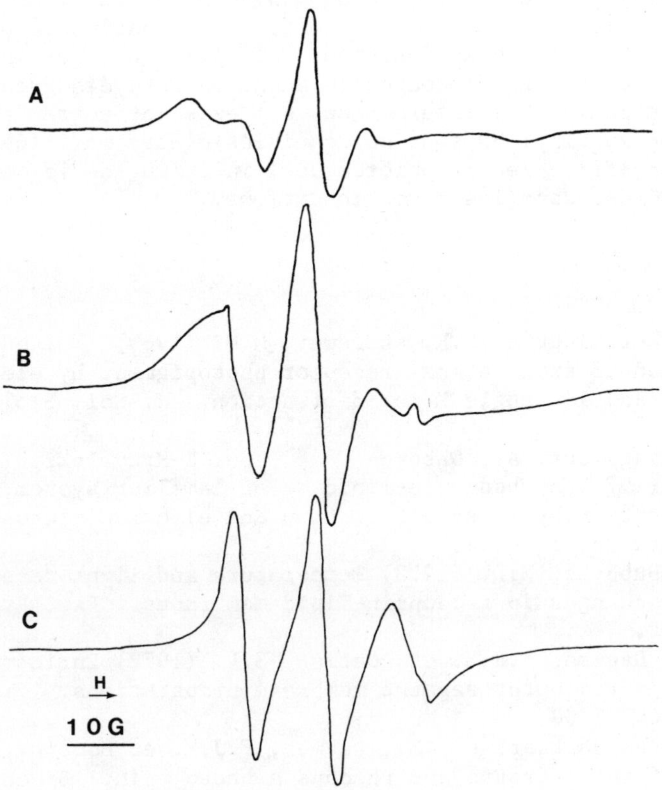

FIG. 5. Spectre d'une suspension de membranes de bâtonnets marquées par: A) I(12-3), B) I(5-10), C) I(1-14).

LES DERIVES D'ACIDES GRAS

L'accroissement de la fluidité de la membrane, produit par illumination, a également été observé par Verma et al., (1973) qui ont utilisé, comme marqueur, un dérivé de l'acide stéarique contenant le radical libre en bout de chaîne, du côté polaire de celle-ci.

Par contre, lorsque le radical libre est positionné le long de la chaîne, aucune modification significative des spectres n'est observable sous l'action de lumière. Cependant, selon la position du marqueur, le spectre correspondant est fort caractéristique. La figure 5 le démontre pour les marqueurs I(12-3), I(5-10) et I (1-14). On constate que la fluidité de la membrane est la plus faible aux abords de la surface polaire. Dans ce cas (I(12-3)), les mouvements de la chaîne sont fortement anisotropes (Seelig,1970). Cette anisotropie se réduit fortement avec le marqueur I(5-10), et, lorsque le radical libre se situe en bout de chaîne, du côté non polaire, (I(1-14)), il est soumis à un mouvement virtuellement isotrope. Nous concluons donc que les membranes des bâtonnets présentent un profil de fluidité en fonction de l'épaisseur, que la surface la plus rigide se situe incontestablement du côté des groupes polaires et que seules les interactions qui existent entre les têtes polaires des phospholipides sont susceptibles d'être modifiées par la lumière. En effet, les propriétés de l'intérieur de la membrane semblent inaltérées dans les mêmes conditions.

REFERENCES

Blasie, J.K., Worthington, C.R. et Dewey, M.M. (1969) Molecular localization of frog retinal receptor photopigment by electron microscopy and low angle X-ray diffraction. J. Mol. Biol. 39: 407-416.

Chabre, M., Cavaggioni, A., Osborne, H.B., Gulik-Krzywicki, T. et Olive, J. (1972) A rhodopsin-lipid-water lamellar system: its characterisation by X-ray diffraction and electron microscopy. FEBS Letters 26: 197-202.

Chen, Y.S. et Hubbell, W.L. (1973) Temperature and light dependent structural changes in rhodopsin-lipid membranes. Exp. Eye Res. 17: 517-532.

de Grip, W.J., Daemen, F.J.M. et Bonting, S.L. (1972) Enrichment of rhodopsin in rod outer segment membrane preparations. Vision Res. 12: 1697-1707.

de Grip, W.J.,van de Laar, G.L.M., Daemen, F.J.M. et Bonting, S.L. (1973) Sulfhydryl groups and rhodopsin photolysis. Biochim.

Biophys, Acta 325: 315-322.
Delmelle, M. et Pontus, M. (1974) Magnetic resonance study of spin labelled rhodopsin. Biochim. Biophys. Acta 365: 47-56.
Gregory, J.D. (1955) The stability of N-ethylmaleimide and its reaction with sulfhydryl groups. J. Am. Chem. Soc. 77: 3922-3923.
Kagan, V.E., Shvedova, A.A., Novikov, K.N. et Kozlov, Y.P. (1973) Light induced free radical oxidation of membrane lipids in photoreceptors of frog retina. Biochim. Biophys. Acta 330: 76-79.
Lassman, G. Ebert, B. Kuznetsov, A.N. et Damerau, W. (1973) Characterisation of hydrophobic regions in protein by spin labeling technique. Biochim. Biophys. Acta 310: 298-304.
McConnell, H.M. et McFarland, B.G. (1970) Physics and chemistry of spin labels. Q. Rev. Biophys. 3: 91-136.
Noell, W.K., Walker, V.S., Kang, B.S. et Berman, S. (1966) Retinal damage by light in rats. Invest. Ophthal. 5: 450-473.
Pontus, M. et Delmelle, M. (1973) Réduction de radicaux nitroxides par des membranes de cellules photoréceptrices. C.R. Acad. Sc. Paris 277: 1069-1072.
Pontus, M. et Delmelle, M. (1974) Light induced changes in rod outer segment membrane fluidity. Vision Res. 14 (sous presse).
Seelig, J. (1970) Spin label studies of oriented smectic liquid crystals (a model system for bilayer membranes). J. Am. Chem. Soc. 92: 3881-3887.
Shichi, H. (1971) Circular dichroism of bovine rhodopsin. Photochem. Photobiol. 13: 499-502.
Shimshick, E.J. et McConnell, H.M. (1972) Rotational correlation time of spin labeled α-chymotrypsin. Biochem. Biophys. Res. Commun. 46: 321-327.
Stone, T.J., Buckman, T., Nordio, P.L. et McConnell, H.M. (1965) Spin labeled biomolecules. Proc. Nat . Acad. Sci. 54: 1010-1017.
Verma, S.P., Berliner, L.J. et Smith, I.C.P. (1973) Cation dependent light induced structural changes in visual receptor membranes. Biochem. Biophys. Res. Commun. 55: 704-709.
Wertz, J.E. et Bolton, J.R. (1972) Electron spin resonance: elementary theory and practical applications. McGraw Hill, New York.

MICROSPECTROPHOTOMETRY: THE TECHNIQUE AND SOME OF ITS PITFALLS

Ferenc I. HÁROSI

NINDS, National Institutes of Health

Bethesda, Md. 20014, U.S.A.

Be plain, good son, and homely in thy drift.
Romeo and Juliet, Act II, Sc. 2

HISTORICAL OUTLINE

Caspersson, by combining the procedures of microscopy and spectrophotometry, founded the technique of microspectro-photometry (MSP) nearly forty years ago. Whereas the light microscope affords the visual identification of subcellular elements, spectroscopy, by virtue of the fact that many substances possess characteristic absorption spectra, allows the determination of the presence and concentration of known substances by the measure of light fluxes transmitted at various wavelengths through structures of known thickness. Thus, he reasoned, the absorption of light by cellular components may be used as a tool for their qualitative and quantitative chemical analyses. Caspersson (1940, 1950) succeeded not only in designing the first microspectrophotometer and setting down the principles of the method but also in identifying nucleic acids ($\lambda_{max} \simeq 260$ nm) in the nucleus and proteins with cyclic amino acids ($\lambda_{max} \simeq 280$ nm) in the cytoplasm of isolated cells.

The technological development of MSP and spectrophotometry went hand-in-hand thereafter. Automatically recording instruments with great sensitivity and rapid operation appeared (Chance, 1951; Yang and Legallais, 1954) which permitted the study of reaction kinetics of enzymes in solution. To measure enzymes in living cells, Chance et al. (1959) developed a recording microspectrophotometer in which light is chopped into two beams. The historical significance of this instrument is that it stimulated work in the field of MSP and that it served as an advanced example for subsequent designers of several dual-beam ratio-recording microspectrophotometers which proved to be important in the study of visual cells.

The pioneering attempt to apply the MSP technique to the study of isolated vertebrate photoreceptors is credited to Hanaoka and Fujimoto (1957). Although the spectral absorption curves they obtained from the outer segments of frog and carp rods and cones could not be reproduced, they nevertheless demonstrated the feasibility of detecting the bleachable visual pigments directly in the cell structures that contain them. Several groups of investigators pursued this idea with newly constructed instruments, including Brown (1961), Liebman and Entine (1964), and Marks (1963, 1965) in collaboration with MacNichol (1964). For a particularly lucid description of a dual-beam ratio-recording instrument, the reader is referred to MacNichol (1964), and for further discussions on the same subject, to MacNichol et al. (1973).

A CLASSIFICATION OF PHOTORECEPTOR MICROSPECTROPHOTOMETRY

Photoreceptor MSP may be classified according to several criteria . By cell type: (A) vertebrate or (B) invertebrate; by method of sample preparation: (A1) single cell, (A2) retinal patch, or (A3) sonicated cell suspension; and by method of measurement: (A11) transverse or (A12) axial. While such a classification is somewhat arbitrary and is likely to be incomplete, it provides a convenient framework for the discussion to follow. The MSP of invertebrate photoreceptors will not be dealt with here; the interested reader may consult the recent review by Goldsmith (1972).

Retinal patch MSP is defined for the present as a method whereby spectral information is simultaneously gathered from many retinal cells by photometric means. The initiator of this method was Denton. Although the aforesaid classification excludes the photomicrographic method of Denton and Wyllie (1955), the photomicrographic film was replaced with a sensitive photomultiplier in the apparatus of Denton and Walker (1958) and Denton (1959), who studied the spectral absorption of retinas of conger eel, deep-sea fish, rudd, and pike as well as several amphibia. The method has since been frequently used. For example, Brown (1961) adopted it to measure frog retinal patches; Brown and Wald (1963) applied it to human and monkey retinas; and, more recently, Muntz (1973) revived the method for the study of retinal light absorption of Amazonian fishes.

The combination of the MSP technique with suspension of sonicated photoreceptors was apparently used only by Murray (1968). He designed a special microspectrophotometer with which he was able to measure the spectral absorption of small volumes (10^{-7}l) of sonicates of cone outer segments removed from the central foveal portions of primate retinas.

Axial MSP of single cells may be thought of as an extension of retinal patch MSP in which the spot size of the light is reduced to single-cell dimensions. Although this method of measurement has infrequently been used (presumably because of the attendant technical difficulties), it yielded valuable information in the hands of Marks et al (1964) and Brown and Wald (1964) about the light absorbing properties of parafoveal cones and rods of human and monkey retinas.

In the opinion of this writer, transverse single-cell MSP appears more desirable than the axial method for measuring most, if not all, vertebrate visual cells. The advantages of the transverse method are: (1) the path of light in the sample is fairly well defined; (2) the cellular structure of interest is accessible to the measurement without the interference of other cells or structures; (3) the method affords the relative ease of making thin preparations in which the long rod-like or cone-shaped outer segments are sideways oriented. The fact that most of the reproducible and spectrally accurate data have been obtained with transverse MSP lends support to the above contention. The method has also found wide application; from frog rods (Brown, 1961; Liebman, 1962) through goldfish cones (Marks, 1963, 1965; Liebman and Entine, 1964) to primate rods and foveal cones (Dobelle et al. 1969). In the recent review by Liebman (1972), the reader may find many more examples for the application of transverse single-cell MSP.

BASIC RELATIONS OF ABSORPTION SPECTROSCOPY

The radiant flux absorbed (Φa) by a sample is not directly available for detection, so that its value must be inferred from the magnitudes of incident flux (Φi), transmitted flux (Φt), reflected or refracted flux (Φr), and scattered flux (Φs). Thus,

$$\Phi_a = \Phi_i - \Phi_t - \Phi_r - \Phi_s.$$

The fundamental condition that must be met in absorption spectroscopy is that the flux lost from detection due to reflection, refraction and scattering be negligibly small compared to the incident and transmitted fluxes, or must be corrected for; for example, by obtaining a bleaching difference spectrum (cf., Caspersson, 1940; MacNichol et al. 1973). Hence the expression for absorptance as a function of wavelength, A(λ), reduces to the simple forms of

$$A(\lambda) = \frac{\Phi_a(\lambda)}{\Phi_i} = \frac{\Phi_i - \Phi_t(\lambda)}{\Phi_i} = 1 - T(\lambda) \quad . \; . \; . \; 1,$$

where $T(\lambda) = \dfrac{\Phi_t(\lambda)}{\Phi_i}$ is the transmittance. The decadic logarithm of the reciprocal of transmittance, called optical density (or absorbance or extinction), is expressible as

$$D(\lambda) = \log_{10}[\frac{\Phi_i}{\Phi_t(\lambda)}] = \varepsilon(\lambda)\ c\ l \quad . \quad . \quad . \quad . \quad .2,$$

where $\varepsilon(\lambda)$ is the molar extinction, c is the molar concentration of the absorbing substance, and l is the path length of light through the sample. The two expressions (eqs. 1 and 2) are completely equivalent; absorptance and density are readily interconvertible quantities. Thus,

$$A(\lambda) = 1 - 10^{-D(\lambda)} \quad \text{and} \quad D(\lambda) = \log_{10}[\frac{1}{1-A(\lambda)}].$$

Whereas transmittance (absorptance) is more directly measurable, densities have the useful property of being additive and their spectral curves invariant in shape for any concentration and path length. Absorptances, on the other hand, are not directly additive and their spectral curves have variable shape, broadening as the concentration or path length increases. These important distinctions must be understood before any interpretation of spectral recordings is attempted. For more detailed development of these relationships, derivations, discussions, and illustrations, see Dartnall, (1957, 1962) or MacNichol et al. (1973).

NECESSARY CONDITIONS AND PITFALLS

Although some general rules are available for us to follow in setting up the necessary optical conditions of MSP (cf., Caspersson, 1940), the photosensitive nature of visual pigments requires special attention.

It has been recognised for some time (e.g., Marks, 1963; 1965) that the duration and intensity of light exposure of visual cells is limited by visual pigment photolysis. However, photometric measurements at low light fluxes are limited by the statistical uncertainty of detection. Thus, a critical balance must be reached between the extent of destruction of pigment by the measurement and the detectability of light. In other words, the signal-to-noise ratio must be made adequate at the expense of pigment bleaching. Aside from the fact that each visual cell has only a limited number of pigment molecules present, there are two problems of concern. First, excessive bleaching during a spectral scan tends to distort the absorption spectrum by shifting the λ_{max} toward that part of the spectrum that was measured earlier. Second, photolised molecules do not disappear from the measuring light beam but undergo stages of photochemical transformations that are commonly referred to as photoproducts. Because the photoproducts themselves absorb visible and near-ultraviolet light, their various spectra may affect the ongoing measurement by distorting the true spectral character of the parent pigment. These problems are further discussed by Harosi

(1971), Liebman (1972), MacNichol et al. (1973), and Harosi and MacNichol (1974b).

The relatively low light fluxes used in measuring photoreceptors, necessitated by the photo-labile nature of the vertebrate visual pigments, require high quantum-efficiency and broadly-sensitive light detectors (Marks, 1963,1965). It is also important to use highly transparent, good quality objectives (Liebman and Entine, 1964); stable light sources; low-noise signal amplification and recording; and sturdy construction of the optical sections of the instrument to prevent vibrations, deformations, or creep.

The numerical aperture (NA)of the condenser and objective lenses should be carefully considered in their selection for MSP. First, the condenser must have a high enough NA to ensure optical resolution to dimensions below that ot the smallest cellular structures of interest in the presence of light diffraction (cf., Harosi, 1971). Second, the objective should have a substantially higher NA than that of the condenser for good light collection and to decrease the effect of wavelength dependent light scattering (Harosi, 1971; MacNichol et al., 1973). Third, we should keep in mind that as the NA of the condenser is increased, so does the angle of convergence of the rays impinging upon the specimen. Although this effect may be corrected for on theoritical grounds (Harosi and Malerba, 1974), if uncorrected, the use of high NA condensers may lead to erroneous interpretations of measured results. The most satisfactory microscope objectives we found for our instrument are two oil-immersion type from the Zeiss Ultrafluar series used in the following combination: a 32x, NA = 0.4, as the condenser, and a 100x, NA = 0.85, as the objective (Harosi, 1971; Harosi and MacNichol, 1974a, b).

It appears that the use of oil-immersion type condenser and objective is important in reducing reflection and refraction losses at optical interfaces (due to large changes in index of refraction), thus providing better quality images for light stop and cell detail. The thinness of preparation is important in reducing light scattering within the specimen; this was listed above as one of the advantages of the transverse MSP technique as compared with the axial method (where the whole retinal thickness must be accommodated by the holder). The making of a thin preparation in which relatively intact rod and cone cells lie sideways, free of debris or obstructing cellular parts, immobile, and with surrounding clear areas uncluttered by tiny particles driven by Brownian motion is probably the most difficult single step in the process.

The most important pitfall to avoid is the generation of spurious spectral curves. After mastering the technical problems in electronics instrumentation and optics, we must be alert for signs of creep in both microscope and preparation. The cell under test, for example, may move laterally, rotate slightly, or jitter almost imperceptibly. Also, the suspending medium may contain small fragments that move incessantly and scatter more or less light depending on their size, their relation to the reference or measuring beam, and the wavelength of light. Spectral distortions resulting from these effects are often difficult to detect, partly because no objective criteria exist by which they can be recognised. If, however, the principle of spectroscopy is followed that chemically pure substances have characteristic absorption spectra and if it is assumed that the spectral properties (such as width or shape) of visual pigments are known by the spectra of purified extracts, then one may expect correspondence between the spectral recordings of MSP and those of spectrophotometry of pigments in solution. Since we have reasons to believe that specialised structures of vertebrate photoreceptors, the outer segments of rods and cones, contain spectroscopically pure visual pigments, singly or in a limited number of combinations (when dark-adapted), and nothing else that absorbs visible and near-ultraviolet light (except their own photoproducts), the expected correspondence of the two techniques also appears reasonable. Because good correspondence has indeed been obtained in some cases, it is possible to establish some guidelines.

We have been using two criteria for the detection of distortions in our spectral recordings: one, concerning the shape of the absorption curve and the other, the zero-level absorption. The first is based on the principle established by Dartnall (1953), who found that the shape of visual pigment spectra, when relative extinction is plotted on a reciprocal wavelength (wave number) scale, is nearly invariant within a pigment family regardless of λ_{max}. This permits us to construct standard curves for the shape. The second criterion is based on the observation that the longwave limbs of absorption spectra should fall to zero; for example, frog rhodopsin, with λ_{max} = 502 nm, has a relative absorbance spectrum that dips below the 1% line above about 620 nm. Thus, we demand of our single-cell recordings of absorptance to have a region at longer wavelengths where the spectrum falls to about ±1% of the baseline (cf. Harosi and MacNichol, 1974a).

THE DICHROIC MICROSPECTROPHOTOMETER

A further advantage of the transverse MSP of photoreceptors is that it permits studying the linear dichroism of their outer segment.. This is the main subject of another paper (Harosi, this volume). It is mentioned here merely to point out that our instrument, called dichroic microspectrophotometer (DMSP), was partly designed to explore this phenomenon. A description of it is given below insofar as it pertains to the general problems of MSP and our method of overcoming them. A fuller description of the instrument is provided elsewhere (Harosi and MacNichol, 1974b).

The DMSP is outlined in Fig. 1. Briefly described, it is a single-beam, wavelength-scanning and recording photometer for microscopic samples. The source of the measuring light (S_1) is a quartz-iodine lamp. Its filament is focussed by a concave mirror (M_1) at the entrance slit of the monochromator (MO). An iris diaphragm (D_1) limits the cone of light that fills the collimating mirror (M_2) of the monochromator. The flat diffraction grating (G) is affixed to a rotatable shaft. Another concave mirror (M_3) focusses the light dispersed by the grating in the plane of the exit slit of the monochromator, which is normally blocked by an electromagnetic shutter (SH_1). When this is energised, a narrow beam of neraly monochromatic light can pass through the tilted quartz plates (QP) to illuminate an adjustable rectangular diaphragm (D_2). This stops down the light to accommodate the acceptance angle of the rotating polariser (P).

A stationary quartz beam splitter (QB) is used to provide background illumination from an auxiliary source (S_3) or bleaching light from a flash bulb (S_4). The final field stop is provided by another adjustable rectangular diaphragm (D_3). Its opening is imaged by the condenser (CO) by way of a flat mirror (M_4) in the plane of the specimen (SP). The transmitted light is collected by the objective (OB) either to a photomultiplier (PM) via the opening of an electromagnetic shutter (SH_2) or, when the sliding prism (PR) of the microscope is in the way, to the viewer (V).

The auxiliary mirror (AM) and a spectral lamp (S_2) are inserted only for wavelength calibration. F_1 and F_3 are heat filtres. The colour of the background illumination (can be infrared) is set by another filter (F_2). The two achromatic lenses (L_1 and L_2) serve to image the filaments of S_3 and S_4 near the back focal plane of the condenser. The swinging mirror (SM), having an electromagnetic release mechanism, is manually cocked in position for photolysis. When searching for cells, SM is uncocked, S_3 energised, and D_3 slid out of the way to illuminate larger areas of the specimen.

The DMSP differs from the dual-beam ratio-recording instruments in several respects. Two such features are visible in Fig. 1, the use of a single beam and linear polarisation by a (rapidly) rotating polariser. Whereas the latter feature permits the assessment of dichroism, the former is important in avoiding some of the most nagging problems of both the optics and the preparation. Since the main optical path has only two lenses, the condenser and the objective, which can be centred with respect to the single beam, all measurements are carried out at the optical axis of the instrument. Consequently, all the optical aberrations are reduced to their minimum value, practically eliminating them (with the exception of

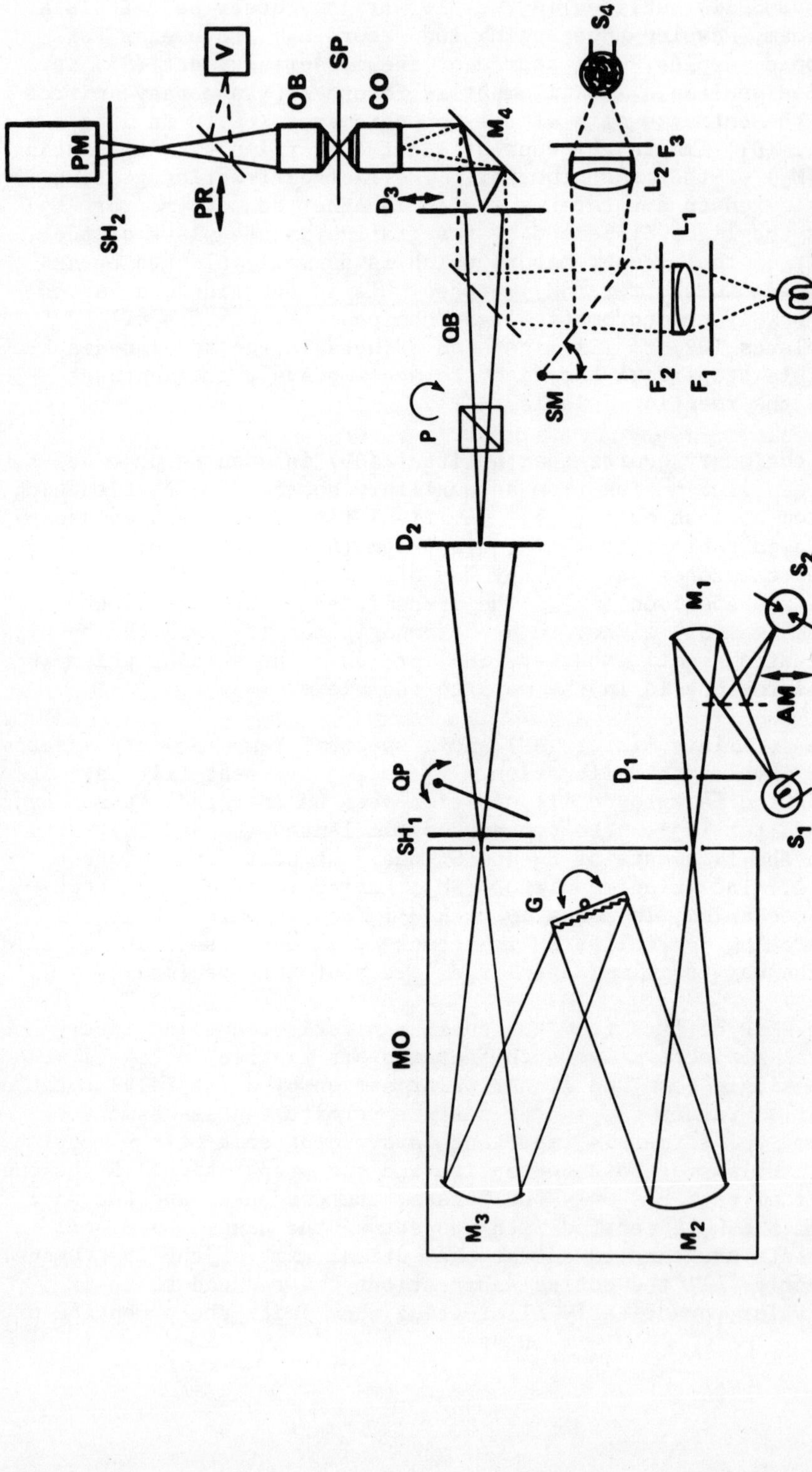

FIG. 1. Scheme for the optical arrangement of the dichroic microspectrophotometer (modified from Harosi and MacNichol, 1974b). See text for details.

the longitudinal chromatic aberrations).

The demand on the preparation is also reduced by the use of one beam, because the reference measurement can be made through clear areas of the preparation anywhere and any time. Moreover, the reference and sample measurements are carried out under identical optical conditions with respect to both instrument and preparation. As a result, baselines are obtained flat and they are probably close to being the true and natural reference levels.

The method of sequentially recording the transmitted fluxes of reference and sample, of course, requires good stability and a long-term memory, two properties so eminently possessed by digital computers. The DMSP is equipped with a small computer, providing the basis for additional novel features. For example, the method of rapid, repeated, and bidirectional spectral scanning and computer averaging of transmitted fluxes at corresponding wavelengths reduces the spectral distortions discusses above that the measurement may introduce into visual pigment absorption spectra. For further information, the reader is referred to Harosi (1971) and Harosi and MacNichol (1974b). Some results obtained using this instrument are discussed in the following paper (Harosi, this volume).

MICROSPECTROPHOTOMETRY OF FISH VISUAL CELLS

In view of the abundance of living fish species and the great varieties of their visual system, it is surprising how infrequently this vertebrate class has been studied by the method of single-cell lateral MSP. While good records are difficult to obtain because of the relatively small size of most fish receptors, the explanation may, in part, be the lack of awareness by the general public of the wealth of information locked in the eyes of fish.

The rather short list of teleosts, whose photoreceptors have been measured this way, is headed by the carp (Hanaoka and Fujimoto, 1957). Its relative cyprinid, the goldfish, was next studied by Marks (1963,1965), by Liebman and Entine (1964) and by Svaetichin et al (1965), who obtained similar and reproducible absorption spectra for the three cone pigments of its retina. In addition to goldfish, Svaetichin et al (1965) investigated eight other species of fish photoreceptors, thereby carrying out the most extensive study on fishes to date. The rainbow trout was used in a study by Liebman (1969); he presented two spectral records of its retinal rods. A shallow-water tropical marine fish (*Eugerres plumieri*) was investigated by Laufer and Millan (1970), who found three spectrally distinct cone types in its retina. Recently, Liebman (1973) presented the absorption spectra of a rod and an accessory cone of the bullhead catfish (*Ictalurus melas)*, and reported to have measured the principal cones of sunfish (*Lepomis microlophus)*, of chenel catfish (*Ictalurus punctatus)* as well as the cones of the tench (*Tinca tinca)*. His records, however, are as yet unpublished.

Although differences in optical and spectroscopic properties of goldfish cones have been found (Harosi and MacNichol, 1974a) with respect to frog rods (Harosi and MacNichol, 1974b), the differences may be ascribed to the difference in the chromophores they employ (see Harosi, this volume). Thus, the observed differences appear not to be peculiar to goldfish cones, for similar variations can also be found among the rods of amphibia (Harosi, 1974). Therefore, fish photoreceptors could be expected to converge with other vertebrate visual cells in their major characteristics, and yet, more experimental data are needed before we can generalise such a principle of convergence. The similarities found so far between fish and other retinal receptors must not, however, be used as an excuse for inactivity in this field. The reader, after scanning the literature for absorption spectra of isolated fish photoreceptors, will probably arrive at the conclusion that published information is upsettingly scarce. If so, my hope is that some of you become sufficiently upset about the situation to decide to give us a helping hand.

REFERENCES

Brown, P.K. (1961). A system for microspectrophotometry employing a commercial recording spectrophotometer. J. Opt. Soc. Am. 51: 1000-1008.

Brown, P.K. and G. Wald (1963). Visual pigments in human and monkey retinas. Nature 200: 37-43.

Brown, P.K. and G. Wald (1964). Visual pigments in single rods and cones of the human retina. Science 144: 45-52.

Caspersson, T.O. (1940). Methods for the determination of the absorption spectra of cell structures. J.R. Microsc. Soc. 60: 8-25.

Caspersson, T.O. (1950). Cell growth and cell function. A cytochemical study. Norton, New York.

Chance, B. (1951). Rapid and sensitive spectrophotometry. I. The accelerated and stopped-flow methods for the measurement of the Reaction Kinetics and spectra of unstable compounds in the visible region of the spectrum. Rev. Sci. Instr. 22: 619-627.

Chance, B., R. Perry, L. Akerman and B. Thorell (1959). Highly sensitive recording microspectrophotometer. Rev. Sci. Instr. 30: 735-741.

Dartnall, H.J.A. (1953). The interpretation of spectral sensitivity curves. Br. Med. Bull. 9: 24-30.

Dartnall, H.J.A. (1957). The visual pigments. Methuen & Co., Ltd., London; John Wiley & Sons, Inc., New York.

Dartnall, H.J.A. (1962). The photobiology of visual processes.

In: The Eye, Vol. 2, p. 323-533, edited by H. Davson. Academic Press, New York.
Denton, E.J. (1959). The contributions of the orientated photosensitive and other molecules to the absorption of whole retina. Proc. R. Soc. Lond. Ser. B. 150: 78-94.
Denton, E.J. and M.A. Walker (1958). The visual pigment of the conger eel. Proc. R. Lond. Sér. Soc. B. 148: 257-269.
Denton, E.J. and J.H. Wyllie (1955). Study of the photosensitive pigments in the pink and green rods of the frog. J. Physiol. 127: 81-89.
Dobelle, W.H., W.B. Marks, and E.F. MacNichol, Jr. (1969). Visual pigment density in single primate foveal cones. Science 166: 1508-1510.
Goldsmith, T.H. (1972). The natural history of invertebrate visual pigments. In: Handbook of Sensory Physiology, Vol. VII/1. Photochemistry of Vision. Chapter 17, p. 685-719, edited by H.J.A. Dartnall, Springer-Verlag, New York.
Hanaoka, T. and K. Fujimoto (1957). Absorption spectrum of a single cone in carp retina. Jap. J. Physiol. 7: 276-285.
Harosi, F.I. (1971). Frog rhodopsin in situ: orientational and spectral changes in the chromophores of isolated retinal rod cells. Ph.D. Thesis. The Johns Hopkins University, Baltimore, Md.
Harosi, F.I. (1974). Absorption spectra and linear dichroism of some amphibian photoreceptors. (In preparation).
Harosi, F.I. and E.F. MacNichol, Jr. (1974a). Visual pigments of goldfish cones. Spectral properties and dichroism. J. Gen. Physiol. 63: 279-304.
Harosi, F.I. and E.F. MacNichol, Jr. (1974b). Dichroic microspectrophotometer: A computer-assisted, rapid, wavelength-scanning photometer for measuring linear dichroism in single cells. J. Opt. Soc. Am. 64: 903-918.
Harosi, F.I. and F.E. Malerba (1974). Plane-polarised light in microspectrophotometry. Vision Res. 14:(in press).
Laufer, M. and E. Millan (1970). Spectral analysis of L-type S-potentials and their relation to photopigment absorption in a fish (*Eugerres plumieri)* Retina. Vision Res. 10: 237-251.
Liebman, P.A. (1962). In situ microspectrophotometric studies on the pigments of single retinal rods. Biophys. J. 2: 161-178.
Liebman, P.A. (1969). Microspectrophotometry of retinal cells. Ann. N.Y. Acad. Sci. 157: 250-264.
Liebman, P.A. (1972). Microspectrophotometry of photoreceptors. In: Handbook os Sensory Physiology, Vol. VII/1. Photochemistry of Vision, Chapter 12, p. 481-528, edited by H.J.A. Dartnall, Springer-Verlag, New York.
Liebman, P.A. (1973). Microspectrophotometry of visual receptors. In: Biochemistry and Physiology of Visual Pigments, p. 299-305, edited by H. Langer, Springer-Verlag, New York.
Liebman, P.A. and G. Entine (1964). Sensitive low-light-level

microspectrophotometer: detection of photosensitive pigments of retinal cones. J. Opt. Soc. Am. 54: 1451-1459.

MacNichol, E.F., Jr. (1964). Three-pigment colour vision. Scientific American 211: 48-56.

MacNichol, E.F., Jr., R. Feinberg and F.I. Harosi (1973). Colour discrimination processes in the retina. In: Colour 73 Proceedings for the Second Congress of the International Colour Association, p. 191-251. Adam Hilger, Rank Precision Industries Ltd., London.

Marks, W.B. (1963). Difference spectra of the visual pigments in single goldfish cones. Ph.D. Thesis. The Johns Hopkins University, Baltimore, Maryland.

Marks, W.B. (1965). Visual pigments of single goldfish cones. J. Physiol. 178: 14-32.

Marks, W.B., W.H. Dobelle and E.F. MacNichol, Jr. (1964). Visual pigments of single primate cones. Science 143: 1181-1183.

Muntz, W.R.A. (1973). Yellow filters and the absorption of light by the visual pigments of some Amazonian fishes. Vision Res. 13: 2235-2254.

Murray, G.C. (1968). Visual pigment multiplicity in cones of the primate fovea. Ph.D. Thesis. The Johns Hopkins University, Baltimore, Maryland.

Svaetichin, G., K. Negishi, and R. Fatehchand (1965). Cellular mechanisms of a Young-Hering visual system. In: Ciba Foundation Symposium on Colour Vision: Physiology and Experimental Psychology, p. 178-207, edited by A.V.S. DeReuck and J. Knight. Little, Brown, Boston.

Yang, C.C. and V. Legallais (1954). A rapid and sensitive recording spectrophotometer for the visible and ultraviolet region. I. Description and performance. Rev. Sci. Instr. 25: 801-807.

LINEAR DICHROISM OF RODS AND CONES

Ferenc I. HÁROSI

NINDS, National Institutes of Health

Bethesda, Md. 20014, U.S.A.

> O, be some other name! What's in a name!
> That which we call a rose
> By any other name would smell as sweet.
> Romeo and Juliet, Act II, Sc. 1

INTRODUCTION

Though Boll (1876) was the first to note the bleaching of the reddish colour of the freshly-removed frog retina and Kühne (1878) was the first to extract the light-sensitive coloured substance that he named rhodopsin, it was Schmidt (1938) who introduced a microscopic technique for studying the optical properties of the pigment-bearing cells. Using the polarising microscope, he not only discovered the birefringent and dichroic properties of the frog retinal rod outer segment but accounted for them as well, qualitatively, at least, in terms of the molecular models of cell structure that he proposed. From his observation of dichroism of the rhodopsin-containing rods, he concluded that the pigment molecules must be oriented in the outer segment. Moreover, since transversely polarised light is strongly absorbed by the side-illuminated isolated cells, in the spectral region where rhodopsin absorption is strongest, but not when the light is polarised along the length of the rod, he assigned a transverse orientation to the rhodopsin molecules within the rod cells.

Schmidt's discovery of frog rod dichroism was subsequently confirmed and extended to several species of vertebrate retinal rods by Denton (1954 a,b; 1959), who used preparations of the folded retina in his demonstration. Microspectrophotometry (MSP) of isolated photoreceptors was also used to determine dichroism in the outer segments of frog rods (Liebman, 1962; Wald et al., 1963, Harosi, 1971; Harosi and MacNichol, 1974b), of goldfish cones (Marks, 1963, 1965; Liebman and Entine, 1964; Harosi and MacNichol, 1974a), of frog cones (Liebman and Entine, 1968), and of cones in nine species of fish by Svaetichin et al (1965). With the demonstration (Dobelle et al., 1969) that primate rod and foveal cone outer segments (when dark-adapted) also preferentially absorb the light when the plane of polarisation is across the axis, there should be little doubt about the widespread occurrence of dichroism in vertebrate photoreceptors. It is just as certain that the dichroism is imparted to the cells by the visual pigment they contain and that the similarities in the extent of dichroism among the various cells imply similarities in molecular order of their pigment.

DEFINITION OF LINEAR DICHROISM

The passage of plane-polarised light through optically anisotropic media may encounter two refractive indices (birefringence) or two coefficients of light absorption (dichroism). Since the absorption coefficient is proportional to the imaginary part of the refractive index in electro-magnetic theory, these two phenomena are related. Linear (as distinguished from circular) dichroism refers to preferential absorption of plane-polarised light in a certain direction (plane) by a specimen. The two planes of polarisation, in which absorption is either at a maximum or at a minimum, define the principal directions of the sample. To characterise this property, it is customary to determine the dichroic ratio as the ratio of the optical densities measured along the two principal directions in the sample (e.g., Oster, 1955).

Photoreceptor dichroism (more precisely, transverse linear dichroism) is observable when an unbleached rod or cone outer segment is illuminated with plane-polarised light (of proper wavelength) sideways; that is, transversely with respect to its physiological direction of light propagation. Because vertebrate visual cells have their principal directions aligned perpendicular to and parallel with the long (z) axis of symmetry, their dichroism can be associated with cellular features. Accordingly, the dichroic ratio of a photoreceptor absorption band means a ratio of the optical density for light polarised perpendicular to its length ($D_{\perp}$) to the optical density measured with light polarised parallel with this (z) direction ($D_{||}$). Although dichroism is always associated with an absorption band, the dichroic ratio may vary from band to band, and in a multi-component spectrum (such as those of visual pigments) it may be wavelength dependent. Thus, the wavelength at which the density ratio is determined ought to be specified. A natural choice for this is the λ_{max} (wavelength of peak absorption of the main band).

RECENT MICROSPECTROPHOTOMETRIC RESULTS

Single-cell lateral MSP appears well suited for studying not only the spectroscopic properties of the visual pigment stored within the photoreceptor but also the optical properties of the whole subcellular structure. Two such optical properties are (1) the specific optical density of the cell and (2) the linear dichroism at the peak of the main absorption band. The first is a measure of the absorptivity per unit path length (thickness) which is related to molar concentration and extinction coefficient of the pigment as well as to the prevailing molecular order. The second is only related to the anisotropy of the cell caused by the orderly arrangement of the absorbing molecules. Thus, if the specific density and dichroic ratio are known and if prior information exists about the molar extinction, the _in situ_ pigment concentration can be calculated. Conversely, the molar extinction of the pigment may be estimated to the same extent as its molar concentration in the cell is known.

It has been commonly believed until recently that all vertebrate rod and cone outer segments are nearly identical as far as their specific density and peak dichroic ratio are concerned (e.g. Marks, 1965; Liebman, 1972). Using the DMSP (see Harosi, this volume), however, systematic variations have been found in these quantities (Harosi and MacNichol, 1964 a,b). It now appears that there are at least two large groups of cells with regard to specific density and dichroic ratio: those with higher specific density and higher dichroic ratio, probably containing vitamin A_1-based pigments, such as in frog, toad, and adult tiger salamander; and those with lower specific density and lower dichroic ratio, probably containing mostly vitamin A_2-based pigments, such as in goldfish, mudpuppy, and larval tiger salamander (Harosi, 1974). Figs. 1 and 2 show one example of each of these groups.

The two components of the absorptance spectrum of adult salamander rod outer segments (with average diameter, d = 9.8 μm) reveal (in Fig. 1) peak values (at λ_{max} = 502 ± 1 nm) of $A_{\perp}$ = 0.3291 and $A_{\parallel}$ = 0.0840, corresponding to $D_{\perp}$ = 0.1733 and $D_{\parallel}$ = 0.0381. The dichroic ratio, therefore, is 4.55, and the specific density for transversely polarized light ($D_{\perp}/d$) is 0.0177/μm. The two components of the absorptance spectrum obtained from mudpuppy rod outer segments (with average diameter, d=9.7μm), as depicted in Fig. 2, show peak values (at λ_{max} = 523 ± 1nm) of $A_{\perp}$ = 0.2485 and $A_{\parallel}$ = 0.0874, corresponding to $D_{\perp}$ = 0.1241 and $D_{\parallel}$ = 0.0397. Thus the dichroic ratio is 3.12, and the transverse specific density is

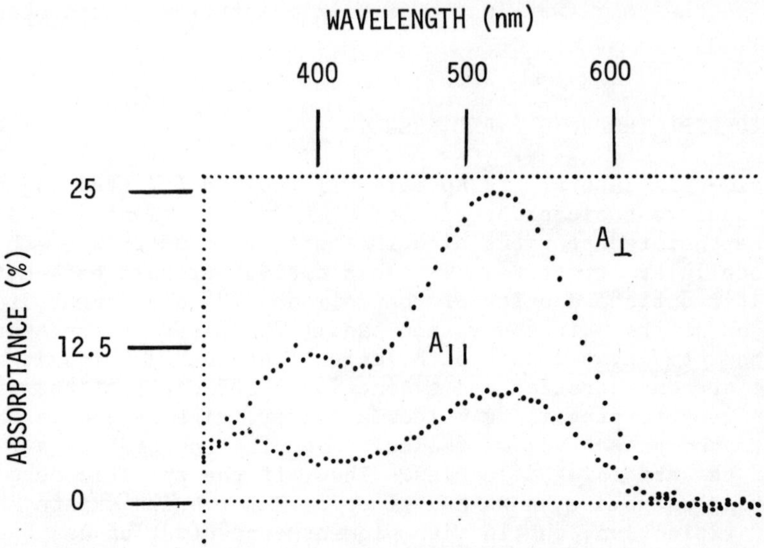

FIG. 1. Lateral absorption of polarized light by (red) rod outer segments of the adult tiger salamander (*Ambystoma tigrinum*). The spectrum is the average of 6 single-cell recordings, each consisting of 32 bidirectional spectral scans between 325 and 695 nm (from Harosi, 1974).

0.0128/μm. It is to be noted that the specific densities just obtained are uncorrected for the loss of pigment due to bleaching by the measuring light. When we add the estimated losses of 10% in case of the salamander rods and 7% in case of the mudpuppy rods (based on earlier determinations on frog and goldfish receptors by Harosi and MacNichol, 1974a, b), the two values become 0.0195/μm and 0.137/μm.

INTERPRETATION OF LINEAR DICHROISM

Probably the simplest interpretation of linear dichroism of vertebrate photoreceptors obtains if we assume an electronic transition dipole moment vector to be associated with the chromophore of each pigment molecule while collimated, linearly polarised light is used for measuring the cellular extinctions. The presence of identical molecules is further assumed, residing in the transverse (x-y) planes in such a manner that each contributes equally to extinction in the axial (z) direction of the cell and takes up no preferential position with respect to the x or y coordinate axes (Model I of Liebman, 1962; Harosi and Malerba, 1974). In analytical terms, these conditions mean that the transverse extinction com-

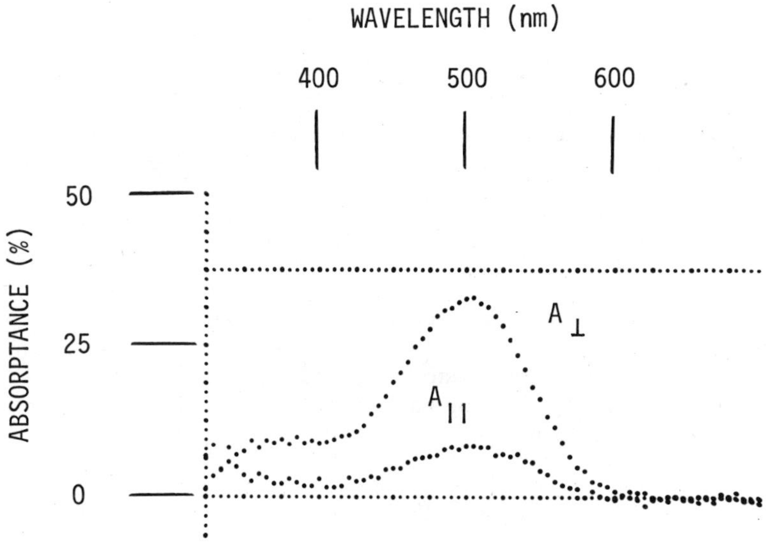

FIG. 2. Lateral absorption of polarised light by rod outer segments of the mudpuppy (*Necturus maculosus*). The spectrum is the average of 10 single-cell recordings, each consisting of 32 bidirectional spectral scans (from Harosi, 1974).

ponents $E_x = E_y$ when the total extinction is $E = E_x + E_y + E_z$. If the z component of the moment is defined as $M_z = E_z/E$ and the dichroic ratio as $R = E_x/E_z$, then

$$M_z = \frac{E_z}{E} = \frac{E_z}{2E_x + E_z} = \frac{1}{1 + 2R} \quad . \; . \; . \; . \; . \qquad 1.$$

Thus, eq. 1 relates the measured cellular dichroic ratio (R) to a fraction (M_z) that expresses the relative proportion of each transition moment being ineffective in absorbing light that propagates along the physiological (z) direction of the cell. The evaluation of eq. 1 for the two dichroic ratios obtained above for salamander and mudpuppy rods are $M_z = 0.099$ (for $R = 4.55$) and $M_z = 0.138$ (for $R = 3.12$). When corrected for the numerical aperture of the condenser actually used in those measurements, the two numbers are 0.093 and 0.133, respectively (cf., Harosi and Malerba, 1974). Based on these results we may conclude that the visual pigment (rhodopsin) in the adult salamander rods has an average molecular orientation such as to render about 9.3% of each molecular extinction ineffective in catching light that travels along the cell axially. The mudpuppy rod pigment (porphyropsin) appears somewhat worse, for the ineffective portion of its molecular extinction is about 13.3%.

INTERPRETATION OF SPECIFIC DENSITY

Based on the assumptions presented in the foregoing section, it is easy to show that the transverse cellular density per unit path length is proportional to the molecular extinction (E), the number of molecules per unit volume (N = Ac), and the dichroic ratio (implicit in M_z) as

$$\frac{D_\perp}{l} = cAE\left(\frac{1-M_z}{2}\right) \quad . . . . 2,$$

where c is the molar concentration and A is Avogadro's Number (cf., Harosi and Malerba, 1974). If the cell were filled with a random collection of the same light-absorbing molecules, the measured transverse density would (since each moment had 1/3 chance of being effective in one direction) be

$$\frac{D_\perp}{l} = c\,\varepsilon_{max} = \frac{cAE}{3} \quad . . . . 3.$$

Following the substitution of eqs. 1 and 3 in eq. 2, an expression is obtained relating equivalent molar concentration (c) and extinction (ε_{max}) to transverse specific density ($D_\perp/l$) and dichroic ratio (R) as

$$(c\,\varepsilon_{max}) = \left(\frac{1+2R}{3R}\right)\left(\frac{D_\perp}{l}\right) \quad . . . . 4.$$

Eq. 4, therefore, permits the calculation of the molar concentration of the oriented pigment, knowing the molar extinction coefficient, or vice versa, by making use of the two measured optical properties of the cell.

For example, if we take ε_{max} = 40,000 litre/mol cm and ε_{max} = 30,000 litre/mol cm as the molar extinction coefficients of rhodopsin and porphyropsin (cf., Wald et al., 1963) to be valid for the pigments of adult salamander and mudpuppy rods, respectively, we can estimate their *in situ* pigment concentration. Thus, we obtain (with the corrected specific densities) c = 3.61 mM as the actual concentration of rhodopsin in salamander rod outer segments and c = 3.53 mM as the actual porphyropsin concentration in mudpuppy rod outer segments. When we also correct for the effect of the condenser's numerical aperture used in the measurement, the concentrations are 3.58 mM and 3.51 mM, respectively. These values are somewhat higher than those previously reported (cf., Liebman and Entine, 1968). Nevertheless, they are consistent with this supposition: although vertebrate visual cells may belong to different groups as far as their specific density and dichroic ratio are concerned, their dark-adapted pigment concentration are nearly the same. For relevant results and discussions, see Harosi and MacNichol (1974a, b) and Harosi (1974).

MODEL FOR IN SITU ORIENTATION OF CHROMOPHORES

The ultimate information to be derived from the study of linear dichroism concerns the orientation of the visual pigment chromophore in the cell before and after it undergoes an electric dipole moment transition in the course of absorbing a photon. Indeed, the description of this event in detail might enable new progress concerning the excitation process. A challenging first step in this regard is to account for the dichroic ratio variations found among the vertebrate photoreceptors.

As far as we know, the chromophore of a visual pigment is either the aldehyde of vitamin A_1 (retinal$_1$) or the aldehyde of vitamin A_2 (retinal$_2$). With them the two large pigment families are formed: rhodopsins and porphyropsins. The essential difference between these two families appears to be the chemical difference between the chromophores they employ, manifested mainly in retinal$_2$ which lacks 2 of the 28 hydrogen atoms of retinal$_1$ but possesses instead an extra double bond in the six-membered ring (e.g., Wald, 1968). A stereo-model of the chromophore of the rhodopsin-type pigments, 11-cis, 12-s-cis retinal$_1$, as found in the crystalline state by Gilardi et al. (1971), is shown in Fig. 3. The unmarked free ends of rods and tubes in the model represent centers of hydrogen atoms, whereas centres of black junctions represent the atomic nuclei of carbon (C_1 - C_{20}). Double bonds are designated by two white bands separated by a black band. (The clamps appearing on five rods representing single bonds merely served to stabilise the model).

The proposed orientation of this molecule in vertebrate photoreceptors is shown in Fig. 4 (Harosi, 1974). The hypothetical arrangement is based on two fundamental assumptions. First, the electronic structure and conformation of 11-cis,12-s-cis retinal$_1$, as found in the crystalline state by x-ray diffraction, closely resemble those of the chromophores of all visual pigments. Second, the electronic transition dipole moment (corresponding to the main absorption band of the pigment) has a spatial relation with the atoms of the chromophore that reflects the average direction of travel of the π electrons ($C = C$) along the conjugated carbon atoms. Although we have no experimental evidence to prove their validity, they appear to be quite reasonable assumptions. Moreover, the hypothesis itself seems to enable us to explain several seemingly unrelated experimental results in a coherent manner. For example, the main transition moment of the molecule, shown in Fig. 4, would not be expected to lie in the x-y plane because of the ring double bond ($C_5 = C_6$), but rather to incline at a small angle which, in turn, could account for the 9-10% axial extinction found in rhodopsin-containing cells. If, on the other hand, an extra double bond were added to lengthen the conjugation in the ring ($C_3 = C_4$),

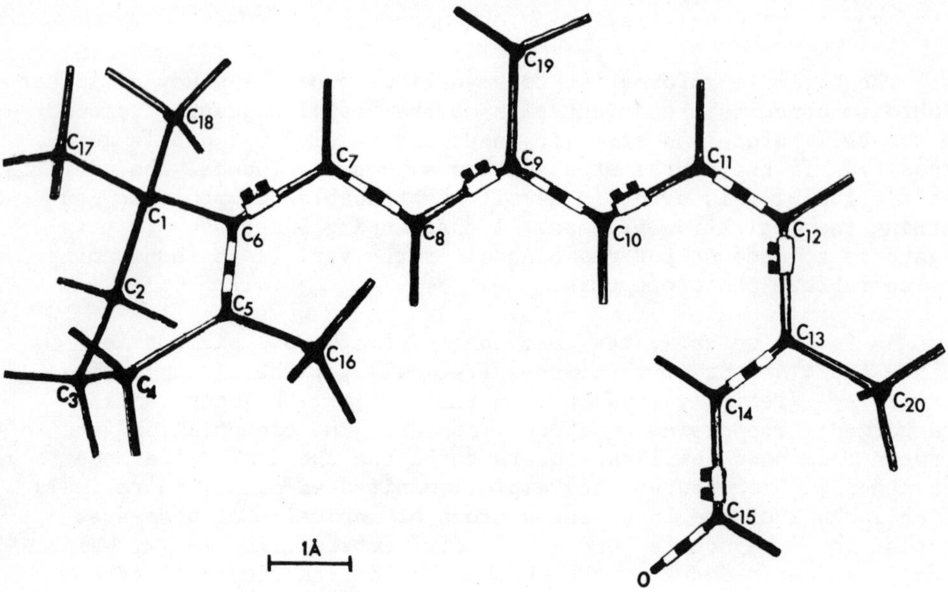

FIG. 3. Perspective view of a stereo-model of 11-cis,12-s-cis retinal$_1$, assembled according to the x-ray data of Gilardi et al. (1971). For details, see Harosi (1974).

the transition moment would be expected to incline at a steeper angle with respect to the x-y plane and hence cause a larger z contribution. Such an increase may explain the 13-14% axial extinction found in porphyropsin-containing cells. Other uses of the model and further discussions are provided elsewhere (Harosi, 1974).

In summary, transverse MSP with polarised light enabled the identification of two groups of rod and cone cells with respect to specific density and dichroic ratio. These groups appear to correlate with the presence of retinal$_1$ and retinal$_2$. The group of cells measuring lower in both specific density and dichroic ratio bear the porphyropsins. The suggested explanation for the lower density is the lower molar extinction of porphyropsin as compared to rhodopsin. The dichroism is accounted for by a model in which mainly the ring portion of conjugation contributes to the axial extinction. Thus, the reduced dichroic ratio in retinal$_2$-

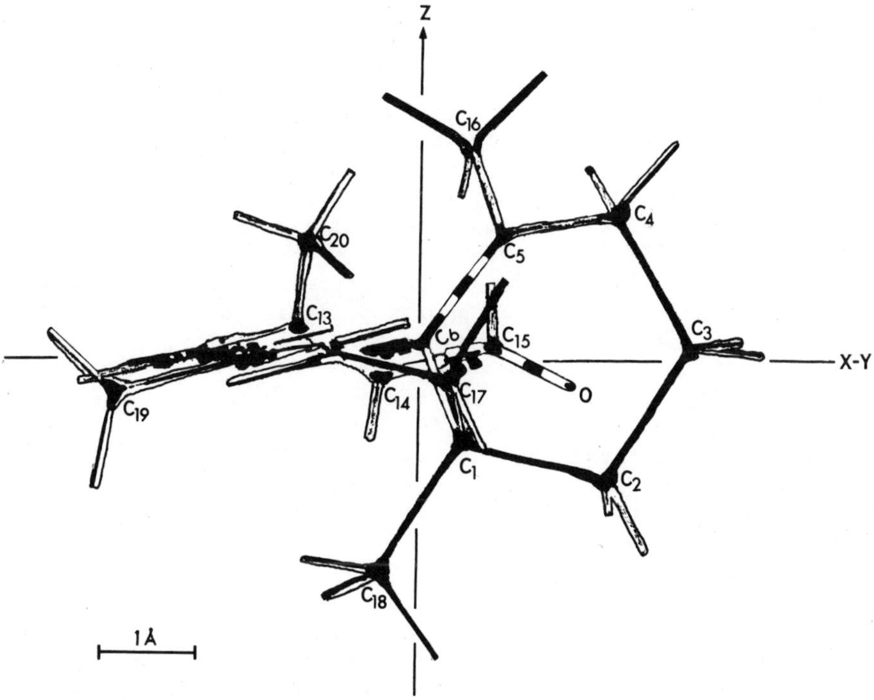

FIG. 4. Proposed orientation of 11-cis,12-s-cis $retinal_1$ in photoreceptors. The model is viewed from the six-membered ring side in the x-y plane signifying the plane of a transverse membrane (from Harosi, 1974).

bearing cells is a consequence of the increased ring conjugation. In spite of these differences, all rod and cone outer segments tested contain about the same concentration of visual pigment, equivalent to a solution of ca. 3.5 mM strength.

REFERENCES

Boll, F. (1876). Zur Anatomie und Physiologie der Retina. Monatsber. Kön. Preus. Akad. Wissensch. Berlin. Nov. 783-788.

Denton, E.J. (1954a). A method of easily observing the dichroism of the visual rods. J. Physiol. 124: 16-17P.

Denton, E.J. (1954b). On the orientation of molecules in the visual rods of *Salamandra maculosa*. J. Physiol. 124: 17-18P.

Denton, E.J. (1959). The contributions of the orientated photosensitive and other molecules to the absorption of whole retina. Proc. R. Soc. Ser. B. 150: 78-94.

Dobelle, W.H., W.B. Marks and E.F. MacNichol, Jr. (1969). Visual pigment density in single primate foveal cones. Science 166: 1508-1510.

Gilardi, R., I.L. Karle, J. Karle, and W. Sperling (1971). Crystal structure of the visual chromophores, 11-cis and all-trans retinal. Nature 232: 187-189.

Harosi, F.I. (1971). Frog rhodopsin in situ: orientational and spectral changes in the chromophores of isolated retinal rod cells. Ph.D. Thesis. The Johns Hopkins University, Baltimore, Md.

Harosi, F.I. (1974). Absorption Spectra and linear dichroism of some amphibian photoreceptors. (In preparation).

Harosi, F.I. and E.F. MacNichol, Jr. (1974a). Visual pigments of goldfish cones. Spectral properties and dichroism. J. Gen. Physiol. 63: 279-304.

Harosi, F.I. and E.F. MacNichol, Jr. (1974b). Dichroic microspectrophotometer: A computer-assisted, rapid, wavelength-scanning photometer for measuring linear dichroism in single cells. J. Opt. Soc. Am. 64: 903-918.

Harosi, F.I. and F.E. Malerba (1974). Plane-polarised light in microspectrophotometry. Vision Res. 14 (in press).

Kühne, W. (1878). On the photochemistry of the retina and on visual purple, edited by M. Foster. Macmillan, London.

Liebman, P.A. (1962). In situ microspectrophotometric studies on the pigments of single retinal rods. Biophys. J. 2: 161-178.

Liebman, P.A. (1972). Microspectrophotometry of photoreceptors. In: Handbook of Sensory Physiology, Vol. VII/1. Photochemistry of Vision, ed. H.J.A. Dartnall, Chap. 12, p. 481-528. Springer, N.Y.

Liebman, P.A. and G. Entine (1964). Sensitive low-light-level microspectrophotometer: detection of photosensitive pigments of retinal cones. J. Opt. Soc. Am. 54: 1451-1459.

Liebman, P.A. and G. Entine (1968). Visual pigments of frog and tadpole (*Rana pipiens*). Vision Res. 8: 761-775.

Marks, W.B. (1963). Difference spectra of the visual pigments in single goldfish cones. Ph.D. Thesis. The Johns Hopkins University, Baltimore, Md.

Marks, W.B. (1965). Visual pigments of single goldfish cones. J. Physiol. 178: 14-32.

Oster, G. (1955). Birefringence and Dichroism. In: Physical Techniques in Biological Research. Vol. I. Optical Techniques. Chapter 8, pp. 439-460, edited by G. Oster and A.W. Pollister, Academic Press, New York.

Schmidt, W.J. (1938). Polarisations-optische Analyse eines Eiweiss-Lipoid-Systems, erläutert am Aussenglied der Sehzellen. Kolloid-Z. 85: 137-148.

Svaetichin, G., K. Negishi, and R. Fatehchand (1965). Cellular mechanisms of a Young-Hering visual system. In: Ciba Foundation Symposium on Colour Vision: Physiology and Experimental Psychology, p. 178-207, edited by A.V.S. DeReuck and J. Knight, Little, Brown, Boston.

Wald, G. (1968). The molecular basis of visual excitation. Nature 219: 800-807.

Wald, G., P.K. Brown, and I.R. Gibbons (1963). The problem of visual excitation. J. Opt. Soc. Am. 53: 20-35.

2. ELECTROPHYSIOLOGY

MICROELECTRODE STUDY OF THE PHYSIOLOGY OF NEURONS IN THE FISH RETINA

Tsuneo TOMITA

Dept. of Ophthalmol. & Visual Science, Yale University

New Haven, Conn. 06510, U.S.A.

Acu rem tetigisti* (proverb)

The microelectrode technique was first applied to the retina by Granit and Svaetichin (1939) for the analysis of the nervous activity of single ganglion cells. Later this technique was utilised for recording electrical activities from within the retinal tissue (Tomita, 1950). Intraretinal recording made possible the more direct localisation of ERG components in retinal layers and the detection of potentials arising in limited regions within the retina. In the following two decades, microelectrode techniques have evolved to the level when intracellular recording from every type of retinal cell is possible, as well as marking after recording for later histological identification of cell type (Kaneko, 1970).

Because of the small size of most of the retinal cells, it is not always easy to obtain their response intracellularly. Retinas having large cells have to be chosen, and the microelectrodes should be extremely fine. In addition, auxiliary techniques are used to help penetration, such as the jolter (Tomita et al., 1967). The table on which the retina is mounted is vibrated periodically to propel the retina into a vertically held, slowly advancing microelectrode. In another auxiliary technique, the microelectrode itself is propelled into the retina by short trains of oscillatory potentials applied to the electrode, which probably behaves as an electromechanical transducer. In yet another, the microelectrode is sharpened at the tip by beveling (Kripke and Ogden, 1974).

---

* "Thou hast touched the matter with a needle".

In this paper I would like to give a survey of the electrical activity of the receptor, the horizontal cell and the bipolar cell, and discuss the role of these neurons in forming the receptive field in the ganglion cell. Observations on retinas other than those of fishes will be freely referred to, since there is ample evidence of similarities in function.

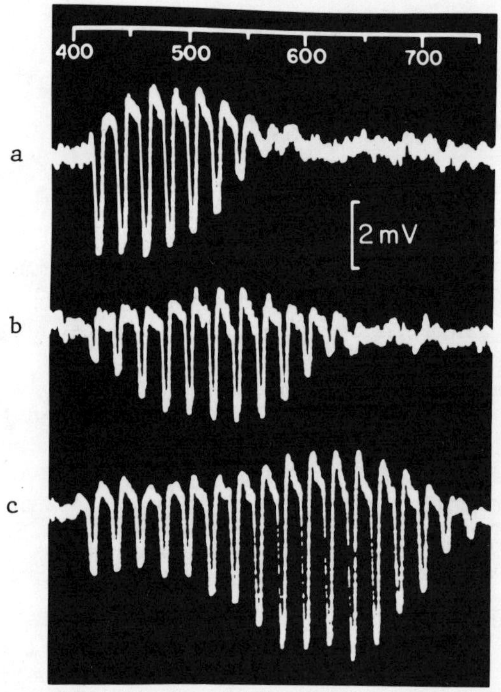

FIG. 1. Sample records from single carp cones, demonstrating three of spectral responses. Scanning of the spectrum was made in steps of 20 nm with monochromatic light adjusted to equal quantum flux (2×10^5 photons/$\mu m^2 \cdot$ sec). A downward deflection indicates negativity. Recording was made with an RC-coupled amplifier having a time constant of 0.5 sec. Spectral scale is given in nm at the top of the figure. A dominant peak occurs at blue in (a), green in (b), and red in (c). (Tomita et al., 1967).

RECEPTORS

The response to light of single vertebrate photoreceptors has been the subject of extensive study during the past 10 years. Fig. 1 shows three sample records of the responses to spectral light of single carp cones, obtained by scanning the spectrum from 400 nm through 750 nm in 20 nm steps (Tomita et al., 1967). The records represent those that were maximally sensitive at different wavelengths. Statistical analysis of a number of such recordings showed three definite groupings of the carp cones in terms of their maximally sensitive wavelengths.

Besides the carp, intracellular recording from single photoreceptors has been made for several other animals. The most important and consistent observation in these studies is that the vertebrate photoreceptors, both rods and cones, are hyper-polarised by photic stimulation. This is in strong contrast with most of the invertebrate photoreceptors which are known to be de-polarised by light. The membrane mechanisms for the hyper-polarising response were studied by Toyoda, Nosaki and Tomita (1969) and Sillman, Ito and Tomita (1969a,b) with the following conclusion: The primary effect of light in causing hyper-polarisation is one of decreasing the membrane permeability to sodium which is kept high in the dark. On the other hand de-polarisation of most invertebrate photoreceptors by light is caused by an increase in membrane permeability, mainly to sodium, as has been well established. The membrane mechanisms in these two types of photoreceptors are schematically illustrated in Fig. 2 by two electrically equivalent models.

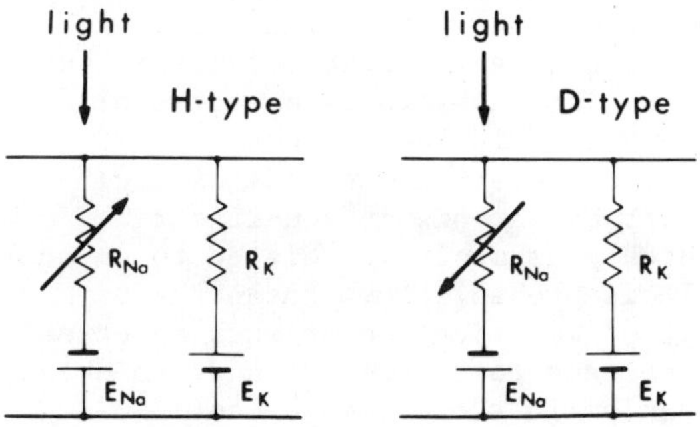

FIG. 2. Electrical models of photoreceptor cells of vertebrate type (left, H-type) in which R_{Na} is increased by light, and of invertebrate type (right, D-type) in which a decrease in R_{Na} is caused by light.

The result was quite unexpected but was basically supported by the later work of Baylor and Fuortes (1970) in single turtle cones, of Hagins, Penn and Yoshikami (1970) on rods in slices of the albino rat retina, of Korenbrot and Cone (1972) who measured the sodium influx by osmotic behaviour of isolated frog rod outer segments, of Zuckerman (1972) in frog rods, and of Brown and Pinto (1974) in single rods of *Bufo marinus*.

One would expect that further work would reveal more details of the membrane mechanism of the photoresponse. Indeed, recent investigations including some of the above are disclosing additional information, concerning rod function in particular. Lasansky and Marchiafava (1974), working on single cones and rods of the tiger salamander, with microelectrodes placed in their inner segments, observed that the cone photoresponse was accompanied by an increase in the input resistance as reported by previous authors, but that the rod photoresponse was accompanied by a decrease in the input resistance. Their observation on cones, therefore, confirmed the current view that the photoresponse is the consequence of a reduction in the permeability to sodium. Their observation on rods, however, led them to an alternative view that, according to my interpretation, in rods K^+-channels or Cl^--channels or both are activated by light, while Na^+-channels remain fixed, resulting in a hyper-polarisation associated with a reduction in the input resistance. The alternative view on rods is quite contradictory with the current view, but the contradiction appears to me to be only superficial upon consideration of Werblin's latest observations (1974) on single mudpuppy rods. Applying a voltage-clamp technique to this material, Werblin observed on illumination an increase in the input resistance in the rod outer segment just as in the cone. He also found that the rod inner segment has a highly non-linear I-V (current-voltage) characteristic; the input resistance would either increase or decrease with hyper-polarisation by extrinsic current in the dark. A regenerative, negative resistance region existed at potentials just hyper-polarised from the dark level. These observations of Werblin suggest that the photoresponse is generated in the rod outer segment in a manner much the same as in the cone outer segment but is modified when entering the rod inner segment by the highly non-linear I-V characteristic of this region, which presumably is related to an activation of K^+ or Cl- or both. It is probable that Lasansky and Marchiafava observed the result of modification to such an extent that the activation of these ions came to the foreground, to obscure the original photoresponse mechanism in the outer segment.

HORIZONTAL CELLS

The horizontal cells and bipolar cells are two cell types that are directly postsynaptic to the receptors (Stell, 1967). It should therefore be of great interest to know how the transmission of information occurs from the receptors to these postsynaptic neurons. Trifonov (1968) and Toyoda, Nosaki and Tomita (1969) assumed that there is a continuous release of transmitter from receptor terminals in the dark where the receptors remain de-polarised and that the release is suppressed when the receptors are hyper-polarised by light. One way of testing this assumption was to examine the effect of media of low calcium or high magnesium content, known to suppress the transmitter release. The assumption seems to have been substantiated by our preliminary experiment with carp horizontal cells in low calcium and high magnesium (Tomita et al., 1973) and by the more elaborate work of Dowling and Ripps (1973) with the skate in high magnesium, and of Cervetto and Piccolino (1974) with the turtle in low calcium, high magnesium and cobalt. The horizontal cell was hyper-polarised in these media and the response to light was suppressed.

Horizontal cells, the response of which is called the S-potential, are classified into two gross types; the luminosity type (L-type) which responds to light with hyper-polarisation irrespective of wavelength, and the chromaticity type (C-type) the response polarity of which is wavelength-dependent (MacNichol and Svaetichin, 1958). Fig. 3 shows an L-type (a) and two C-types (b,c) of S-potential in the carp. These three response curves resemble the three fundamental response curves of Hering's opponent colour type, obtained in humans by Hurvich and Jameson (1960) from their psychophysical study. A transformation from Young's trichromatic type to Hering's opponent colour type at a certain level in the visual pathway has been suggested by several pioneer investigators. It is now obvious that the site of the transformation is in the outer plexiform layer.

Little is known, however, concerning the underlying mechanism for such a transformation. Attempts by many investigators to elucidate the mechanism by means of bridge records of the input resistance and polarisation experiments for a reversal potential have led to quite inconsistent results. Nelson (1973), working on the mudpuppy retina, observed that the reversal potential for the L-type S-potential is well outside the physiological range. He suspects that the horizontal cell might respond in a glia-like manner to light-modulated changes in extracellular potassium concentration brought about by photoreceptor activity. Trifonov, Byzov and Chailahian (1974) argue against Nelson, based on their observation in the pike retina that the L-type S-potential does reverse its polarity when adequately de-polarised to about zero potential levels. They also found that the horizontal cell exhibits a highly non-linear I-V characteristic particularly at potentials hyper-polarised from the dark level. They consider that the non-linearity is due mainly to activation of K^+ or Cl^- or both at the "non-synaptic" membrane of the horizontal cell. It is noted that the horizontal cell body resembles in its membrane properties the rod inner segment.

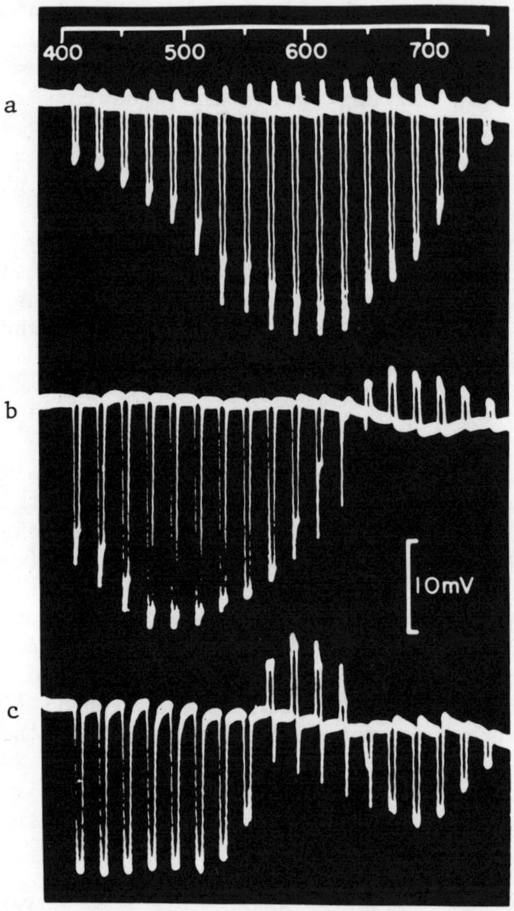

FIG. 3. Three spectral response types of S-potentials in the carp retina. (a), luminosity type; (b), biphasic chromaticity type; (c), triphasic chromaticity type. A downward deflection indicates negativity. Spectral scale is given in nm at the top of the figure. (Tomita, 1965).

Despite the non-linear I-V characteristic of the horizontal cell membrane, the following rule appears to apply to S-potentials induced by a moderate intensity of light: Whenever the response is a hyper-polarisation this is accompanied by a resistance increase, whereas if the response is a de-polarisation this is accompanied by a resistance decrease (Tomita et al., 1973). Furthermore, our polarisation experiments suggest that both de- and hyper-polarising responses of the C-type horizontal cell have an equilibrium potential toward that of sodium. It is probable that both inputs from receptors for de- and hyper-polarisation of horizontal cells regulate channels for the same kind of ions such as sodium but in opposite directions.

BIPOLAR CELLS

There are two types of bipolar cells: the on-centre type which responds to a spot of light on the recording site with a de-polarisation, and the off-centre type which responds to the same stimulus with a hyper-polarisation. These bipolar cells respond to an annulus of light with the reversed polarity, so that they make an on-centre/off-surround type and an off-centre/on-surround type. The result of bridge records by Toyoda (1973) of the on-centre and off-centre responses in the absence of the surround effect showed a simple relation that the de-polarising and hyper-polarising responses, elicited by a spot of light, are accompanied by a decrease and increase in the input resistance respectively and that both responses are enhanced by hyper-polarising extrinsic current and suppressed by de-polarising extrinsic current. The relation is similar to that for the de- and hyper-polarising C-type S-potentials. Here, again, is an indication that the hyper-polarising and de-polarising responses are the result of regulation of channels for the same kind of ions, most likely sodium, but in opposite directions. The concept is summarised in Fig. 4 by a pair of hypothetical models.

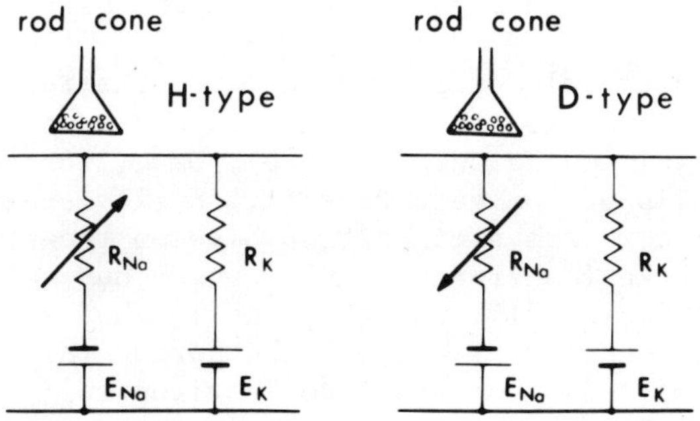

FIG. 4. Electrical models of proposed two types of sub-synaptic membrane in the on and off-centre bipolar cells. Explanation in text.

A synaptic terminal of rod or cone containing vesicles is in contact with the subsynaptic membrane of a bipolar cell where the Na^+-channel either remains open when there is no transmitter release and is closed by release of transmitter (left, H-type), or remains closed when there is no transmitter release and is opened by release of transmitter (right, D-type). On the assumption that there is a continuous release of transmitter in the dark and that this is suppressed by light, the H-type should represent the mechanism of the on-center response and the D-type that of the off-center response. This same pair of models might also apply to the de- and hyperpolarising responses in C-type S-potentials.

RECEPTIVE FIELD ORGANISATION

Each receptor is influenced predominantly by light falling directly upon it, but a study in the turtle (Baylor, Fuortes and O'Bryan, 1971) has revealed that the receptors interact through electrical connexions between them (the synergistic near-surround effect) and also by a negative feedback from horizontal cells (the antagonistic far-surround effect). The field for the negative feedback should then correspond to that of horizontal cells which is known to have a span of millimetres.

The horizontal cells not only feed information back to receptors but also feed forward to proximal neurons, evidently through horizontal-bipolar synapses. Naka (1971) studied the receptive field mechanism in the catfish by a method of injecting current in L-type horizontal cells and observing the effect upon the ganglion cell activity. The result was consistent in that a hyper-polarisation of horizontal cells by current gives rise to a response in ganglion cells similar to that elicited by an annulus of light. Apparently, the horizontal cell plays an important role in forming the surround field in both bipolar and ganglion cells.

In retinas capable of sensing colour, the receptive field organisation should be far more complex, as has been demonstrated by Wagner, MacNichol and Wolbarsht (1960), Daw (1968) and Kaneko (1973). The details of how such receptive fields are generated are unknown, but there are similarities between receptive field formation in the pike (Maksimova, 1969), which has colour vision and the catfish (Naka, 1971), which has not. Thus Maksimova has shown that de-polarisation of L-type pike horizontal cells activates the red response of red-blue sensitive ganglion cells while hyperpolarisation activates the blue response in these center-surround organised cells. However, the problem of how such responses are initiated and organised from the receptors on through to the ganglion cells remains to be solved.

ACKNOWLEDGMENT

This work was supported in part by Research Grants EY00764 and EY00017 from the National Eye Institute, U.S. Public Health Service.

REFERENCES

Baylor, D.A. and Fuortes, M.G.F. (1970). Electrical responses of single cones in the retina of the turtle. J. Physiol. 207: 77-92.

Baylor, D.A., Fuortes, M.G.F. and O'Bryan, P.M. (1971). Receptive fields of cones in the retina of the turtle. J. Physiol. 214: 265-294.

Brown, J.E. and Pinto, L.H. (1974). Ionic mechanism for the photoreceptor potential of the retina of *Bufo marinus*. J. Physiol. 236: 575-591.

Cervetto, L. and Piccolino, M. (1974). Synaptic transmission between photoreceptors and horizontal cells in the turtle retina. Science 183: 417-418.

Daw, N.W. (1968). Colour-coded ganglion cells in the goldfish retina: extension of their receptive fields by means of new stimuli. J. Physiol. 197: 567-592.

Dowling, J.E. and Ripps, H. (1973). Effect of magnesium on horizontal cell activity in the skate retina. Nature 242: 101-103.

Granit, R. and Svaetichin, G. (1939). Principles and technique of the electrophysiological analysis of colour reception with the aid of micro-electrodes. Upsala Lakareforenings Forh. 65: 161-177.

Hagins, W.A., Penn, R.D. and Yoshikami, S. (1970). Dark current and photocurrent in retinal rods. Biophys. J. 10: 380-412.

Hurvich, L.M. and Jameson, D. (1960). Perceived color, induction effects, and opponent-response mechanisms. J. Gen. Physiol. 43: Pt. 2, 63-80.

Kaneko, A. (1970). Physiological and morphological identification of horizontal, bipolar and amacrine cells in goldfish retina. J. Physiol. 207: 623-633.

Kaneko, A. (1973). Receptive field organization of bipolar and amacrine cells in the goldfish retina. J. Physiol. 235:133-153.

Korenbrot, J.I. and Cone, R.A. (1972). Dark ionic flux and the effects of light in isolated rod outer segments. J. Gen. Physiol. 60: 20-45.

Krike, B.R. and Ogden, T.E. (1974). A technique for beveling fine micropipettes. EEG and Clin. Neurophysiol. 36: 323-326.

Lasansky, A. and Marchiafava, P.L. (1974). Light-induced resistance changes in retinal rods and cones of the tiger salamander. J. Physiol. 236: 171-191.

MacNichol, E.F. and Svaetichin, G. (1958). Electric responses from isolated retinas of fishes. Am. J. Ophthal. 46: Pt. 2, 26-40.

Maksimova, E.M. (1969). The effect of intracellular polarization of horizontal cells on the activity of ganglion cells of fish retina. (in Russian) Biophysica 14: 537-544.

Naka, K.-I. (1971). Receptive field mechanism in the vertebrate retina. Science 171: 691-693.

Nelson, R. (1973). A comparison of electrical properties of neurons in *Necturus* retina. J. Neurophysiol. 36: 519-535.

Sillman, A.J., Ito, H. and Tomita, T. (1969a). Studies on the mass receptor potential of the isolated frog retina. I. General properties of the response. Vision Res. 9: 1435-1442.

Sillman, A.J., Ito, H. and Tomita, T. (1969b). Studies on the mass receptor potential of the isolated frog retina. II. On the basis of the ionic mechanism. Vision Res. 9: 1443-1451.

Stell, W.K. (1967). The structure and relationships of horizontal cells and photoreceptor-bipolar synaptic complexes in goldfish retina. Am. J. Anat. 121: 401-424.

Tomita, T. (1950). Studies on the intraretinal action potential. Part I. Relation between the localization of micropipette in the retina and the shape of the intraretinal action potential. Japan. J. Physiol. 1: 110-117.

Tomita, T. (1965). Electrophysiological study of the mechanisms subserving color coding in the fish retina. Cold Spring Harbor Symp. Quant. Biol: 30, 559-566.

Tomita, T., Kaneko, A., Murakami, M. and Pautler, E.L. (1967). Spectral response curves of single cones in the carp. Vision Res. 7: 519-531.

Tomita, T., Miller, W.H., Hashimoto, Y. and Saito, T. (1973). Electrical response of retinal cells as a sign of transport. Exp. Eye Res. 16: 327-241.

Toyoda, J. (1973). Membrane resistance changes underlying the bipolar cell response in the carp retina. Vision Res. 13: 283-294.

Toyoda, J., Nosaki, H. and Tomita, T. (1969). Light-induced resistance changes in single photoreceptors of *Necturus* and *Gekko*. Vision Res. 9: 453-463.

Trifonov, Yu. A. (1968). Study of synaptic transmission between photoreceptors and horizontal cells by means of electric stimulation of the retina (in Russian). Biophysica 13: 809-817.

Trifonov, Yu. A., Byzov, A.L. and Chailahian, L.M. (1974). Electrical properties of subsynaptic and nonsynaptic membranes of hori zontal cells in fish retina. Vision Res. 14: 229-241.

Wagner, H.G., MacNichol, E.F. and Wolbarsht, M.L. (1960). The response properties of single ganglion cells in the goldfish retina. J. Gen. Physiol. 43: Pt. 2, 45-62.

Werblin, F.S. (1974). Electrical properties of the rod membranes. Abstr. ARVO Spring Meeting, HI HII-10, 14.
Zuckerman, R. (1973). Ionic analysis of photoreceptor membrane currents. J. Physiol. 235: 333-354.

STRUCTURAL STUDIES OF FUNCTIONAL PATHWAYS IN GOLDFISH RETINA

William K. STELL*

Jules Stein Eye Institute, UCLA Medical Center

Los Angeles, California 90024, U.S.A.

But to speak that in words which his eye hath disclos'd ---
Love's Labour Lost, Act II, Sc. 1

INTRODUCTION

Fish retinas have long been favoured for electrophysiological investigations of single cells. Their popularity is due largely to their long survival in vitro, the size of their cells, and the ability of many of them to respond selectively to light of different wave-lengths. Various investigators have recorded from ganglion cells (Wagner et al, 1960; Daw, 1968; Beauchamp and Daw, 1972; Raynauld, 1972; Spekreijse et al, 1972), amacrine and bipolar cells (Kaneko, 1970, 1973; Toyoda, 1973), horizontal cells (Tomita, 1963, 1965; Kaneko, 1970; Kaneko and Yamada, 1972), and cone cells (Tomita et al, 1967) in carp and goldfish with some facility. Yet their recordings to date have revealed only the barest outlines of retinal pathways for specific rod and cone information.

The limitations of microelectrode techniques have led me to investigate functional pathways in goldfish retina by morphological methods. The task is twofold: first, to recognise and characterise specific morphological types within general classes of cells; and second, to analyse the patterns and fine structure of inter-connexions between cells of known types. The technique which is most useful for this -- the silver chromate impregnation method of Golgi -- is one of the oldest. It is at the same time new and

* Supported in part by Research Grant # 01190 and Vision Research Centers Grant # 00331 from the National Institutes of Health.

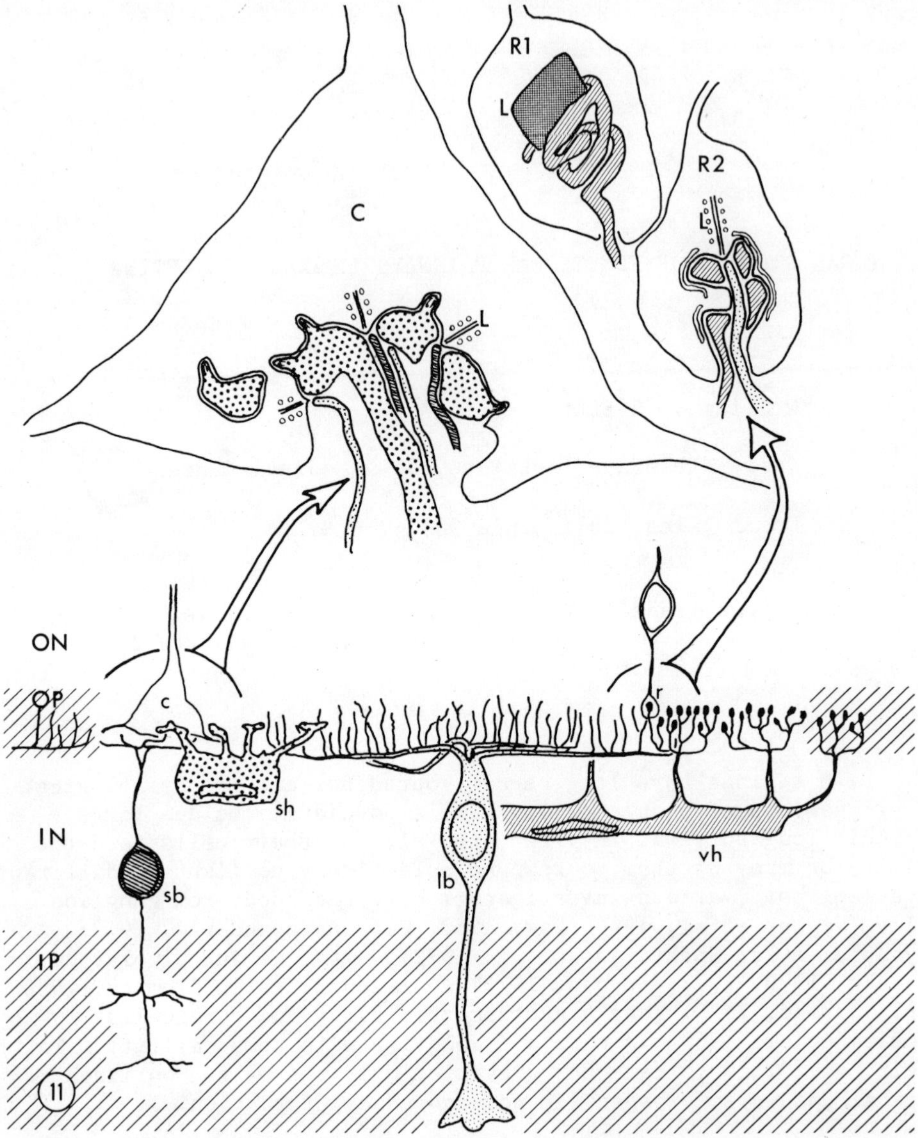

FIG. 1. Diagram summarising rod- and cone-horizontal cell contacts in goldfish retina. Below: cell structure as seen in Golgi preparations; above: fine structure of cell contacts. Left: cone pedicles (C,c) and cone horizontal cell (sh); right: rod spherules (R,r) and rod horizontal cell (vh). Reproduced from Stell (1967) by permission of Wistar Institute Press.

powerful, however, when modified and supplemented by modern techniques. In this paper I shall summarise our progress in understanding the outer synaptic layer of goldfish retina by means of this venerable, yet "new", method.

MATERIALS AND METHODS

All of the studies to be discussed here were performed on common goldfish, *Carassius auratus*, about 10-12 cm long from tip of nose to base of tail. Eyes were removed and hemisected; then either whole eye cups were fixed for sectioning, or retinas were isolated and spread flat for whole-mount Golgi impregnation. Individual cells impregnated with silver chromate were studied in whole mounts, vertical sections 90 μm thick, and serial sections 1 μm thick, by light microscopy; or in ultrathin sections, by electron microscopy. Details are given by Stell (1965, 1974) and Stell and Lightfoot (1974). Cone pigments and structures were investigated in a microspectrophotometer (Harosi and MacNichol, 1974; Stell and Hárosi, 1974).

OBSERVATIONS AND DISCUSSION

Since at least the time of Cajal (1892), it has been known that cyprinid retinas contain both rods and cones. Cajal himself observed several different kinds of horizontal cells (HC's) in vertically thick-sectioned, Golgi-impregnated carp retinas. He described these as "external", "intermediate", and "internal". His external HC's appeared to send dendrites to the level of the cone pedicles, while his intermediate HC's appeared to send dendrites to the level of the rod spherules. Stell (1964, 1965, 1967) confirmed, by Golgi staining and electron microscopy, that Cajal's "externals" are indeed cone HC's and that his "intermediates" are rod HC's (figure 1). Cajal's "internal HC's", which appeared to send no dendrites at all to the outer synaptic layer (and, according to earlier accounts, had no nucleus), remained mysterious.

The identity of Cajal's "internal HC's" was clarified recently by observations of completely impregnated cells in whole mounts (Stell, 1973, 1974). Favourable preparations (figure 2) revealed that probably all cone HC's have axons. A thin fibre, about 200 μm long and 0.5 μm in diameter, descends obliquely from the cell body or a large dendrite into the middle of the inner nuclear layer. There it expands gradually to a maximum diameter of 5-7 μm, tapers again, and terminates in a fine filament. This fusiform terminal expansion, about 200 μm long, is identical to Cajal's "internal HC". It is of interest that unlike the cone HC's of mammals, whose axons terminate in contact with rod spherules (Kolb, 1970), the cone HC axons of

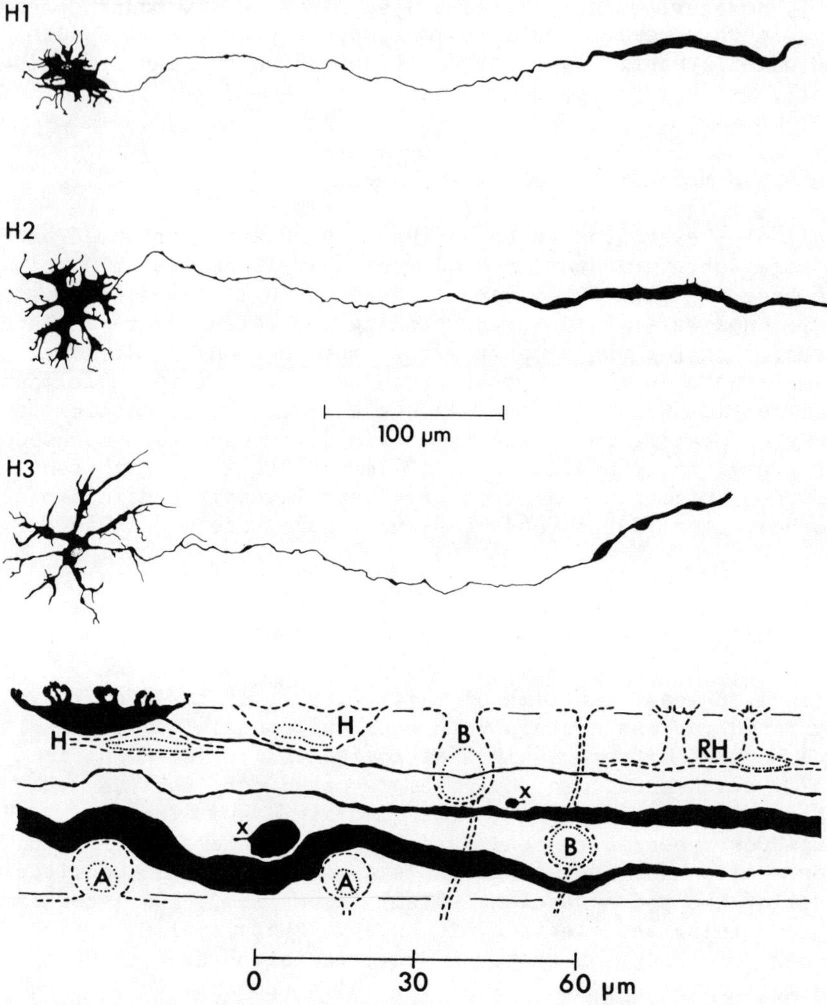

FIG. 2. Drawings of cone horizontal cells and their axons in Golgi preparations of goldfish retina. Above, three types as seen in flat whole mounts. Below, representative cell (H1) as seen in vertical sections (entire cell represented in 3 separate segments). From Stell (1974).

goldfish do not go near the outer synaptic layer. They must make their connexions instead with cells or processes within the inner nuclear layer. They are provided with short, slender horizontal projections which appear to be likely sites of synaptic contact.

The cone HC's seen in flat mounts appeared to comprise three distinct categories. This intrigued me, because microelectrode recordings had revealed three functional types of cone-dependent horizontal cells (e.g., Tomita, 1963, 1965). David Lightfoot and I were able to establish that there are indeed three types of cone HC's, which differ in form, dimensions, and patterns of cone contacts. A functional interpretation of these structural characteristics was made possible by correlating cone structure and visual pigment content, in collaboration with Dr. Ferenc Hárosi. I should like now to present our findings, which have been reported in a preliminary form by Stell, Hárosi and Lightfoot (1974).

By ordinary light microscopy we can discern at least 5 distinct types of cones in goldfish. These are the unequal double cones, which comprise a long and a short member, and the long, short, and miniature single cones. By microspectrophotometry we have shown that

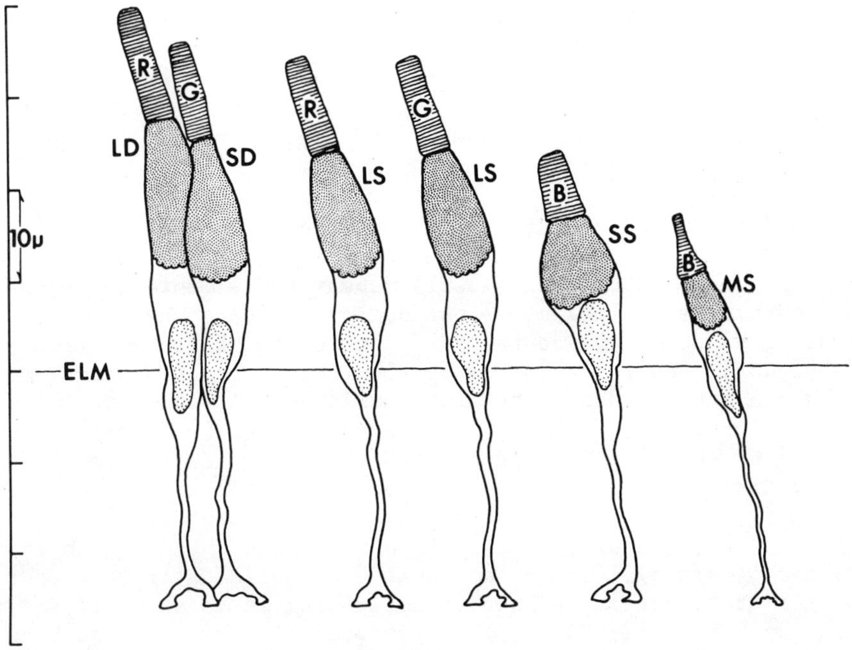

FIG. 3. Cone types in goldfish retina, reconstructed from serial vertical sections at 1 μm: long member of double (LD), short member of double (SD), long single (LS), short single (SS), and miniature single (MS). Region of peak spectral absorption indicated by letter in outer segment: R (red, 625 nm), G (green, 530 nm), B (blue, 455 nm). From Stell and Lightfoot (1974).

the long members of double cones and probably some long single cones contain a red-sensitive visual pigment (λ max ≅ 625 nm), that the short members of double cones and probably other long single cones contain a green-sensitive pigment (λ max ≅ 530 nm), and that the short and probably miniature single cones contain a blue-sensitive pigment (λ max ≅ 455 nm). A full account of these observations, summarised in figure 3, is given by Stell and Hárosi (1974). Similar but more fragmentary observations have been made previously by Marks (1965), Svaetichin et al (1965), and Scholes and Morris (1973).

Since the axons of goldfish cones are large enough to be seen in 1 μm vertical sections, cone pedicles may be identified as to cone type by following along the axon to the inner and outer segments. We have selected typical cone horizontal cells in vertical sections 90 μm thick, re-sectioned them serially at 1 μm, and examined them in the light microscope (Stell and Lightfoot, 1974). The approach is similar to Kolb's (1970), except that we have found it advantageous not to use the electron microscope when analysing such large cells, and we are able to distinguish, not just rods and cones, but cones of different chromatic type by their structure. We find that cone HC's of one type ("H1") contact cones of all types, that cone HC's of a second type ("H2") contact all except the long members of double cones and some long single cones, and that cone HC's of the third type ("H3") contact only short and miniature single cones (figure 4).

It can be argued that the H1's should give L-type spectral responses, the H2's biphasic C-type responses, and the H3's triphasic C-type responses. Thus the H1 cells appear to be contacted by red-, green- and blue-sensitive cones but dominated functionally by the red system; H2 cells are contacted by green- and blue-sensitive cones but dominated functionally by the red and green systems; and H3 cells are contacted only by blue-sensitive cones but are influenced functionally by all 3 cone systems. The functional properties of the cone HC's cannot, therefore, be determined entirely by direct synapses of cones upon horizontal cells. It seems likely that cone HC's themselves act as interneurons (e.g., red cone → H1 → H2, causing depolarisation of biphasic C-cells to red light; see figure 5). We are now analysing serial ultrathin sections in an attempt to determine whether horizontal cells act directly upon one another (H1 → H2 → H3), or indirectly by way of feedback onto cone pedicles (H1 → G cone → H2 → B cone → H3).

The function of the cone HC axons is obscure. Responses seem to be conducted from cell body to axon terminal without significant decrement (Marmarelis and Naka, 1973). This suggests that the axons either have unusually resistive membranes or propagate responses by some active, although non-spiking, mechanism. In any case, the axons can be assumed to conduct. The axon terminals apparently are con-

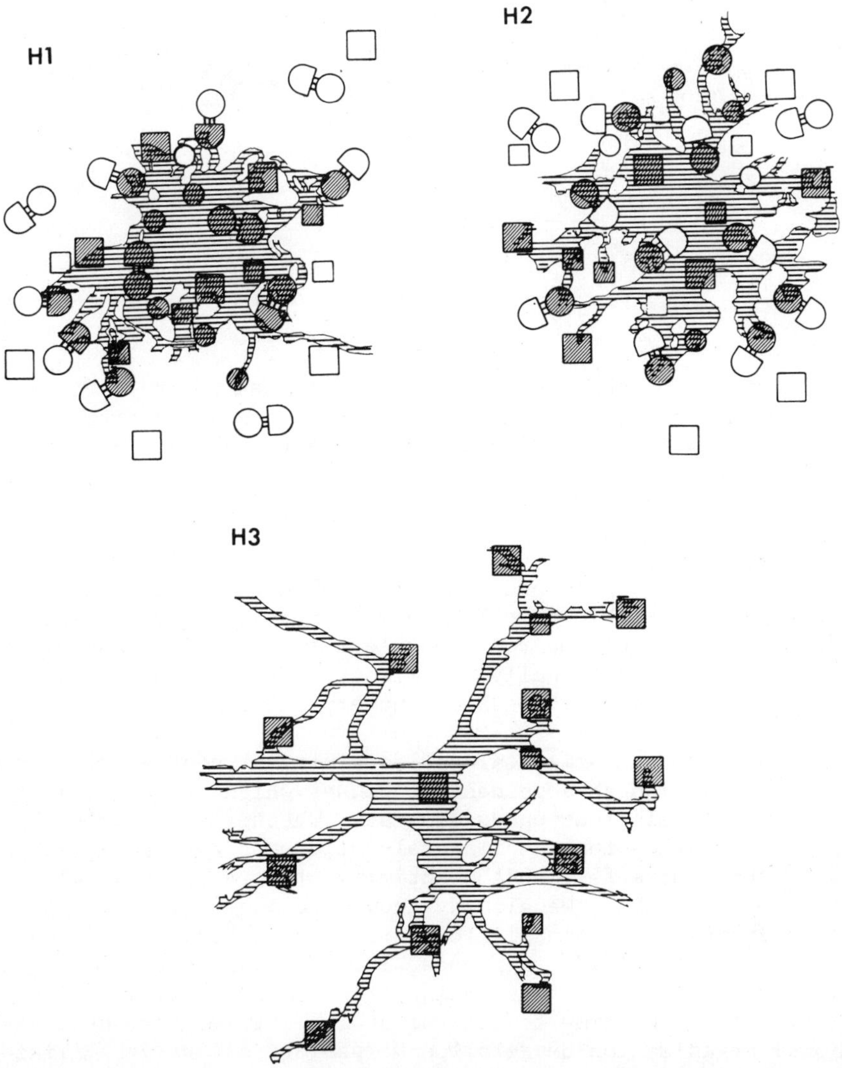

FIG. 4. Reconstructions and contact patterns of cone horizontal cells, viewed in horizontal projection. H1 contacts all cones (cross-hatched symbols) within its field; H2 contacts all except long members of doubles (half-circles) and some long singles (small circles); H3 contacts only short singles (large squares) and miniature singles (small squares). From Stell and Lightfoot (1974).

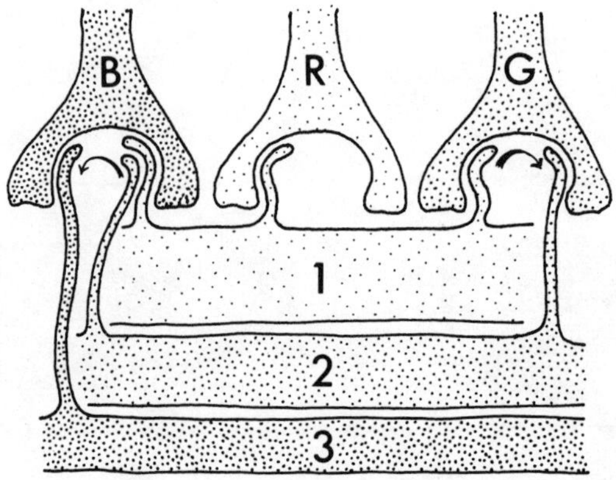

FIG. 5. Summary of cone-HC contacts in goldfish retina. Probable region and direction of information transfers from H1 → H2 and H2 → H3 are shown by arrows. From Stell and Lightfoot (1974).

cerned with functional interconnexions within the inner nuclear layer, and although this property sets them apart from mammalian HC axons, which act at the rod spherules, the relative ease of studying HC axons functionally in teleosts may make them useful models of their less approachable counterparts in mammals.

The role of the horizontal cells is not, of course, merely to generate the wavelength-dependent responses which are so diverting to the experimental electrophysiologist. Whether directly or by way of the photoreceptors, horizontal cells must eventually express themselves by some effect on the activity of more proximal elements, such as bipolar cells. We already know more about bipolar cells structurally than functionally (compare, e.g., Parthe, 1972, and Scholes and Morris, 1973, vs Toyoda, 1973, and Kaneko, 1973). Considering the difficulty of recording from bipolar cells, it is likely that for some time to come we shall learn most about their functional organisation by detailed morphological analyses, like those of Scholes and Morris (1973). Such analyses are in progress in my laboratory.

A most curious feature of the goldfish retina is the multiplicity of cone types. Why should there be two structurally distinctive cones for each visual pigment, each with the same pattern of HC connexions? It seems likely that the two types for each pigment differ in some functional property (e.g., threshold, gain, kinetics) or neural interconnexions (e.g., cone → cone or cone → bipolar). It is clear that much remains to be learned about the interconnexions and physiology of cells in the complex and fascinating retina of the lowly goldfish.

REFERENCES

Beauchamp, R.D. and Daw, N.W. (1972). Rod and cone input to single goldfish nerve fibers. Vision Res. 12: 1201-1212.
Cajal, S.R.y (1892). La rẽtine des vertẽbrẽs. La Cellule 9: 119-255.
Daw, N.W. (1968). Colour-coded ganglion cells in the goldfish retina: Extension of their receptive fields by means of new stimuli. J. Physiol. 197: 567-592.
Hárosi, F.I. and MacNichol, E.F., Jr. (1974). Visual pigments of goldfish cones. Spectral properties and dichroism. J. Gen. Physiol. 63: 279-304.
Kaneko, A. (1970). Physiological and morphological identification of horizontal, bipolar, and amacrine cells in goldfish retina. J. Physiol. 207: 623-633.
Kaneko, A. (1973). Receptive field organization of bipolar and amacrine cells in the goldfish retina. J. Physiol. 235: 133-153.
Kaneko, A. and Yamada, M. (1972). S-potentials in the dark-adapted retina of the carp. J. Physiol. 227, 261-273.
Kolb, H. (1970). Organization of the outer plexiform layer of the primate retina: Electron microscopy of Golgi-impregnated cells. Philos. Trans. R. Soc. Lond. Ser. B. 258: 261-283.
Marks, W.B. (1965). Visual pigments of single cones. Ciba Foundn. Sympos. on Colour Vision. Little, Brown, Boston. p. 208-216.
Marmarelis, P.Z. and Naka, K.-I. (1973). Spatial distribution of potential in a flat cell. Application to the catfish horizontal cell layers. Biophys. J. 12: 1515-1532.
Raynauld, J.-P. (1972). Goldfish retina: Sign of the rod input in opponent color ganglion cells. Science 177: 84-85.
Scholes, J. and Morris, J. (1973). Receptor-bipolar connectivity patterns in fish retina. Nature 241: 52-54.
Spekreijse, H., Wagner, H.G. and Wolbarsht, M.L. (1972). Spectral and spatial coding of ganglion cell responses in goldfish retina. J. Neurophysiol. 35: 73-86.
Stell, W.K. (1964). Correlated light and electron microscope observations on Golgi preparations of goldfish retina. J. Cell Biol. 23: 89A.
Stell, W.K. (1965). Correlation of retinal cytoarchitecture and ultrastructure in Golgi preparations. Anat. Rec. 153: 389-398.
Stell, W.K. (1967). The structure and relationships of horizontal cells and photoreceptor-bipolar synaptic complexes in goldfish retina. Am. J. Anat. 121: 401-424.
Stell, W.K. (1973). Horizontal cells in a cyprinid fish retina. Abstrs. Ann. Mtng. A.R.V.O. p. 75, #9.

Stell, W.K. Horizontal cell axons and axon terminals in goldfish retina. J. Comp. Neurol. (in press, 1974).

Stell, W.K. and Hárosi, F.I. Cone structure and visual pigment content in the retina of the goldfish. Manuscript in preparation.

Stell, W.K., Hárosi, F.I. and Lightfoot, D.O. (1974). Cones and the color-coding of cone-horizontal cell contacts in the goldfish retina. Abstrs. Ann. Mtng. A.R.V.O.

Stell, W.K. and Lightfoot, D.O. (1974). Color-specific interconnexions of cones and horizontal cells in the retina of the goldfish. J. Comp. Neurol. (in press, 1974).

Svaetichin, G., Negishi, K. and Fatehchand, R. (1965). Cellular mechanisms of a Young-Hering visual system. Ciba Foundation Symposium, on Colour Vision. Little, Brown, Boston. pp. 178-203.

Tomita, T. (1963). Electrical activity in the vertebrate retina. J. Opt. Soc. Am. 53: 49-57.

Tomita, T. (1965). Electrophysioligical study of the mechanisms subserving color coding in the fish retina. Cold Spr. Harb. Symp. Quant. Biol. 30: 559-566.

Tomita, T., Kaneko, A., Murakami, M. and Pautler, E.L. (1967). Spectral response curves of single cones in the carp. Vision Res. 7: 519-531.

Toyoda, J.-I. (1973). Membrane resistance changes underlying the bipolar cell response in the carp retina. Vision Res. 13: 283-294.

Wagner, H.G., MacNichol, E.F., Jr. and Wolbarsht, M.L. (1960). The response properties of single ganglion cells in the goldfish retina. J. Gen. Physiol. 43: (6, Part 2), 45-62.

A MODEL FOR THE GANGLIONIC RECEPTIVE FIELD ORGANISATION

Jean-Pierre RAYNAULD

Département de physiologie, Université de Montréal

Montréal 101, Qué. Canada

How from the finny subject of the sea
These fishes tell the infirmities of men;
And from their watery empire recollect
All that may men approve or men detect !
Pericles Act II, Sc. 1

In 1960, Wagner, MacNichol and Wolbarsht trying to find at the ganglionic level the correlate of the opponent colour behaviour observed by Svaetichin (1956) in the slow potential cell system, recorded in the goldfish retina from a class of ganglion cells which also showed an opponent colour response. The study of this class of ganglion cells has yielded, over the years, an impressive set of results which has furthered our knowledge of the functioning of the retina.

I would like to review in this paper those findings relevant to the organisation of the receptive field and with the help from the recent results of Kaneko present a model which accounts for most findings and points to further research.

Testing with small spots, Wagner, MacNichol and Wolbarsht (1963) were able to show that this class of ganglion cells had a centre-surround organisation. However, because of the different spectral sensitivity of the two systems, it was possible to see an important characteristic: the superposition of the two receptive fields. A long wavelength sensitive field was superposed on a short wavelength sensitive field of larger dimensions. At the time this was believed to be the complete receptive field organisation.

Using annuli as stimuli, Daw (1968) showed that the field was indeed more complex and that an another antagonistic surround was present. This surround also had colour opponent characteristics making this cell a very efficient detector for simultaneous colour contrast.

The receptive field dimensions which are important for the model to be presented later are as follows:

WAVELENGTH	CENTRE	SURROUND
650 nm	1.35 ± 0.36 mm	5.4 ± 1.55 mm
500 nm	1.6 ± 0.84 mm	5.8 ± 1.45 mm

Tested in the dark-adapted state, the same class of ganglion cell was found to have a rod input which had an interesting characteristic: the rod input whether excitatory or inhibitory had always the same sign as the red cone input both in the centre and periphery of the receptive field. This finding (Raynauld 1969, 1972) has been confirmed by Beauchamp and Daw (1972) and by Adams and Afanador (1971). Beauchamp and Lovasik (1973) have furthered our knowledge of the receptive field organisation by presenting data which indicates that the blue cone input is of the same sign as that of the green cone input to these cells.

Recently Kaneko (1973) presented two important results on the field organisation of another type of cell found in the goldfish retina, the bipolar cells:

1- There were only two types of bipolar cells from which he was able to record: colour-coded and non-colour-coded cells.

2- The receptive field was of the centre-surround type and the centre was always red sensitive. The surround was red sensitive in non-colour-coded cells and green sensitive in colour-coded cells. In colour-coded cells it was possible to show with chromatic adaptation that the red and the green systems overlapped in the centre of the field. The receptive field dimensions for the bipolar cells were as follows: the centre was 200 microns large and the overall diameter about 1 mm.

Another important finding of Kaneko (1973) is the recording from a class of amacrine cells showing a graded potential, an opponent colour response, and summating properties over a large area. Given the data presented above, I will now attempt to show how colour opponent bipolar cells can converge onto a ganglionic cell to generate a field similar to the one described by Wagner et al. (1963). Furthermore, by adding the action of the graded potential amacrine of Kaneko (1973), we can account for the double opponent cell described by Daw (1968).

Let us assume that the red sensitive centre of the ganglion cell's receptive field is constructed of closely packed but non-overlapping colour-coded bipolar fields. Along the horizontal axis seven red sensitive bipolar field centres measuring 200 μ in diameter are required to make up the red ganglion cell's receptive field centre of 1.35 mm. This layout is represented to scale in Fig. 1.

It is seen that this organisation generates a green sensitive centre field of larger dimension. This is due to the action of the surround of those bipolar cells whose centres are situated on the border of the red ganglionic field centre. In this model the overall diameter for the green mechanism is 2.2 mm which is within the limits set by Daw of 1.6 ± 0.84 mm. Up to now, the model agrees very well for the ganglionic receptive field organisation described by Wagner, MacNichol and Wolbarsht (1963).

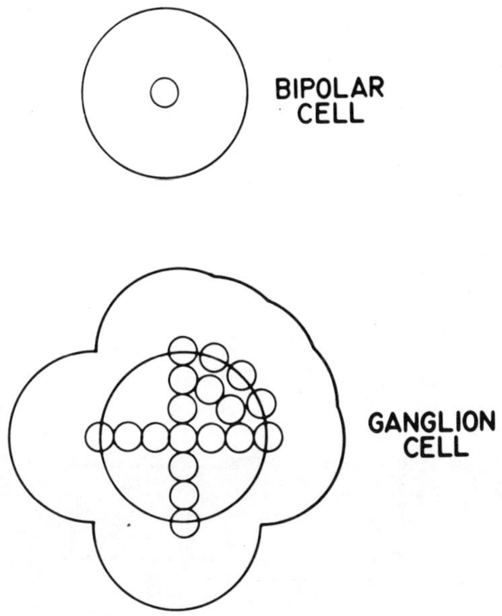

FIG. 1. Scale model showing the type of ganglionic receptive field generated by the simple convergence of bipolar cells. For the ganglionic cell, the drawing is complete only in the first quadrant in order to illustrate the role of the bipolar field surround.

An important aspect of this receptive field is the annular part which is green sensitive. From an operational point of view, the annular part could be called the surround, since it has an antagonistic action to the more central (red sensitive) part of the field. This antagonistic effect of the surround is expected to be greater in a non-colour-coded field because of the identical spectral sensitivitees of the centre and surround. This is shown in Fig. 2 which presents a theoretical area-sensitivity curve for a field of similar organisation. What I would like to stress here, is that this portion of the surround of the ganglionic field is one which is carried over from the bipolar centre-surround organisation and does not require additional neural processing other than the one required for the generation of the central part of the field. It is suggested that this part of the surround be called the "carried" surround since it is carried from the bipolar cell organisation.

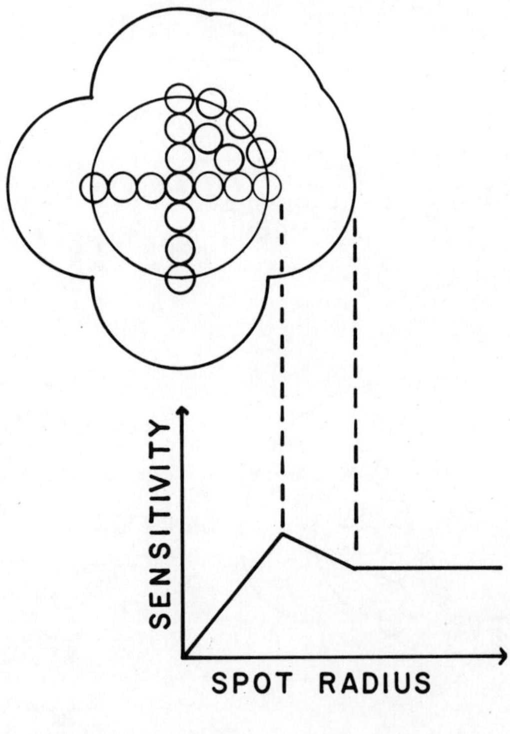

FIG. 2. Predicted area-sensitivity curve for a ganglionic cell having for an organisation the simple convergence presented in Fig. 1. Such a curve would hold for white light only if the bipolar field is colour-coded and for all wavelengths for non-colour-coded bipolar fields.

On the other hand, the surround found by Daw (1968) using annuli which were colour-coded with each colour system antagonistic to the central system of the same colour requires further neural processing. The cell responsible for this processing is most likely the colour-coded amacrine cell described by Kaneko (1973) which has all the required properties. This would be the "intrinsic" surround. In other words the ganglion cell has two surrounds: on carried over from the centre-surround organisation of the bipolar cell and one which truly belongs to the ganglion cell.

Though this model of the receptive field organisation accounts fairly well for the known properties of double opponent ganglion cell in the goldfish one can question its validity in the case of non-colour-coded ganglionic field. The model depends only on the existence of a centre-surround organisation at the level of the bipolar and on the existence of amacrines having summating properties over large area and an inhibitory action on the ganglion cell. The presence of a centre-surround organisation in non-colour-coded bipolar cells is well documented (Kaneko 1973, Dowling and Werblin 1969), but the required type of amacrine is still hypothetical.

If the model holds, one can make some predictions regarding the area-sensitivity of the ganglion cell. Such a theoretical curve is show in Fig. 3. The decrease in sensitivity which has classically been taken as the surround action can be seen to be made of two parts: one part where the sensitivity is decreasing rapidly due to the synergistic action of the bipolar surround and of the amacrine cells. The other part would be due to the sole action of the amacrines. A break in the sensitivity curve in the region of decrease sensitivity would indicate the presence of the two mechanisms.

The presence or the absence of a surround in the dark adapted state when the rods are the active receptors may also be considered in the light of the model presented above. In the fish, because of the convergence of rod and cone onto the same bipolars (Stell 1967, Scholes and Morris 1973) it is most likely that the colour-coded amacrine will also have a rod input of the same sign as that of the red cone input. Given this organisation it is not surprizing to find that, in the dark-adapted state goldfish ganglion cells were found to have a surround (Raynauld 1969, Beauchamp and Daw 1972). In the cat, Barlow, Fitzhugh and Kuffler (1957) have reported the disappearance of the surround in the dark-adapted cat retina. The model presented above allows us to ask whether both the "carried" and the "intrinsic" surrounds have disappeared or only one of them.

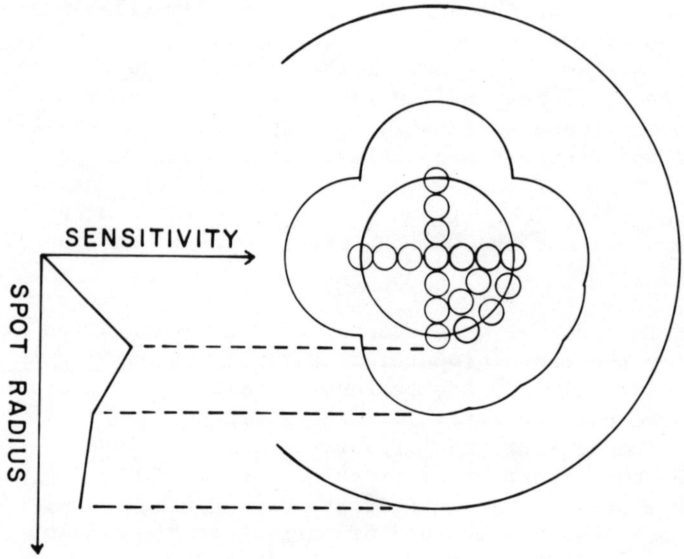

FIG. 3. Predicted area-sensitivity curve for a ganglionic cell having the organisation presented in Fig. 1 plus the inhibitory action of an amacrine cell whose receptive field is limited by the outer circle. Such a curve would hold for white light only in the case of a double opponent ganglion cell and for all wavelengths for non-colour-coded ganglion cell.

There is however an element which is not accounted for in the model: it is the colour-coded horizontal cell which was at the origin of these investigations. Indeed there is good evidence that the horizontal cells are responsible for the surround of the bipolar cells (Byzov 1968, Werblin and Daw 1969, Toyoda 1973, Naka 1972, Naka and Witkovsky 1972). If so, one should find bipolar cells whose surround is colour-coded but to date no such cells have been found. To explain this discrepancy there are a number of possibilities:

1- These bipolars exist but have not yet been recorded.

2- These colour opponent responses, first believed to belong to the cones, and then to the horizontal cells belong in reality to the amacrine cells since the amacrine described by Kaneko have a very similar behaviour. However histological controls made by many investigators conflict with this hypothesis.

3- The colour-coded responses really belong to the horizontal cell system but their opponent colour behaviour is an artefact due to feed back from the centers of the colour-coded bipolar cells.

In conclusion, the double opponent colour cell found in the goldfish retina has to date been considered the result of an organisation much more complex that the simple non-colour-coded centre-surround type. The model presented above and based on the recent findings of Kaneko (1973) proposes that this may not be true. It shows that by simple substitution of a colour-coded bipolar with a non-colour-coded one it is possible to switch from a double opponent ganglion cell to a cell having spatial characteristics very similar to a classical centre-surround type. The only requirement is the existence of a non-colour-coded amacrine having a large summation area and an inhibitory action on the ganglion cell.

Another aspect of the receptive field organisation that this model brings into focus is the distinction between the features which are "carried" from the previous level and those which are "intrinsic" to the level under study.

REFERENCES

Adams, A.J. and Afanador, A.J. (1971). Ganglion cell receptive field organization at different levels of light adaptation. Am. J. Optom. and Arch. Am. Acad. Optom. 48: 889-896.

Barlow, H.B., Fitzhugh, R. and Kuffler, S.W. (1957). Change of organization in the receptive fields of the cat's retina du during dark adaptation. J. Physiol. 137: 338 - 354.

Beauchamp, R.D. and Daw, N.W. (1972). Rod and cone input to single goldfish optic nerve fibers. Vision Res. 12: 1201-1212.

Byzov, A.L. (1968). The component analysis of electroretinogram in the retina of cold blooded vertebrates and the regulative function of horizontal cells. In: Advances in Electrophysiology and Pathology of the Visual System (6 Iscerg Symposium), p. 217-229, Veb Georg Thieme, Leipzig.

Daw, N.W. (1968). Color-coded ganglion cells in the goldfish retina: extension of their receptive fields by means of new stimuli. J. Physiol. 223: 449-460.

Dowling, J.E. and Werblin, F.S. (1969). Organization of the retina of the mudpuppy, Necturus Maculosis. II. Intracellular recording. J. Neurophysiol. 32: 339-355.

Kaneko, A. (1973). Receptive field organisation of bipolar and amacrine cells in the goldfish retina. J. Physiol. 235: 133-153.

Naka, K.I. (1972). The horizontal cells. Vision Res. 12: 573-588.

Naka, K.I. and Witrovsky, P. (1972). Dogfish ganglion cell discharge resulting from extrinsic polarization of the horizontal cells. J. Physiol. 223: 449-460.

Raynauld, J.P. (1969). Rod and cone responses of ganglion cells in goldfish retina: a microelectrode study. Ph.D. Thesis, Johns Hopkins University, Baltimore, Md.

Raynauld, J.P. (1972). Goldfish retina: sign of the rod input in opponent color ganglion cells. Science 177: 84-85.

Scholes, J. and Morris, J. (1973). Receptor-bipolar connectivity patterns in fish retina. Nature 241: 52-54.

Svaetichin, G. (1956). Spectral response curves from single cones. Acta Physiol. Scand. 39: supp. 134, 16-46.

Toyoda, J.U. (1973). Membrane resistance changes underlying the bipolar cell response in the carp retina. Vision Res. 13: 283-294.

Wagner, H.G., MacNichol, E.F. Jr. and Wolbarsht, M.L. (1963). Functional basis for "ON"-center and "OFF"-center receptive fields in the retina. J. Opt. Soc. Am. 53: 66-70.

GIANT RETINAL GANGLION CELLS IN DOGFISH (*MUSTELUS*): ELECTROPHYSIOLOGY OF SINGLE ON-CENTRE UNITS

W.K. STELL*, P.B. DETWILER, H.G. WAGNER and M.L. WOLBARSHT

UCLA Medical Center; National Institutes of Health; Duke University Medical Center; Marine Biological Laboratory

Pucelle or puzzle, dolphin or dogfish,
Your hearts I'll stamp out with my horse's heels,
And make a quagmire of your mingled brains.
King Henry the Sixth, Pt. I, Act I, Sc. 4

INTRODUCTION

The ganglion cells are the most typical neurons in the vertebrate retina. With their axons they form the only pathway from the retina to the brain. Because their axons are so accessible to electrophysiological study, they are still the best known single units in the retina. Even so, rather little is understood about the cellular structure and synaptic organisation which must account for the variety of functions these cells can perform. Ganglion cells have been shown to generate responses which depend in part on the spatial distribution of the stimulating light on the retina. The region within which light influences the response, called the "receptive field" (RF) by Hartline (1940), typically comprises concentric regions of opposite sign: a centre within which light evokes responses of one type (e.g. firing at ON), and a surround within which light evokes responses of complementary type (firing at OFF). Such cells, described in the cat by Kuffler (1952), were found to undergo changes in surround organisation going from the light- to the dark-adapted state (Barlow et al, 1957).

*Jules Stein Eye Institute, UCLA Medical Center, Los Angeles, California 90024, U.S.A.

In many cases, the diameters of the largest RF centres and the broadest ganglion cell dendritic trees are nearly equal, whereas the diameters of the RF surrounds exceed the extent of even the largest ganglion cell dendritic trees. For this reason it has been proposed that the extent of the RF centre is determined by the extent of the ganglion cell's dendrites, i.e., by the direct receptor-bipolar-ganglion cell pathway, while the RF surround is determined by lateral pathways through horizontal or amacrine cells (Gallego, 1954; Dowling and Boycott, 1969; Naka and Nye, 1971). In an attempt to further clarify the functional organisation of ganglion cells, we undertook a series of studies on the structure and function of cells in the retina of the smooth dogfish, *Mustelus canis*. This benign shark appears to use mainly olfaction to locate the bottom-dwelling

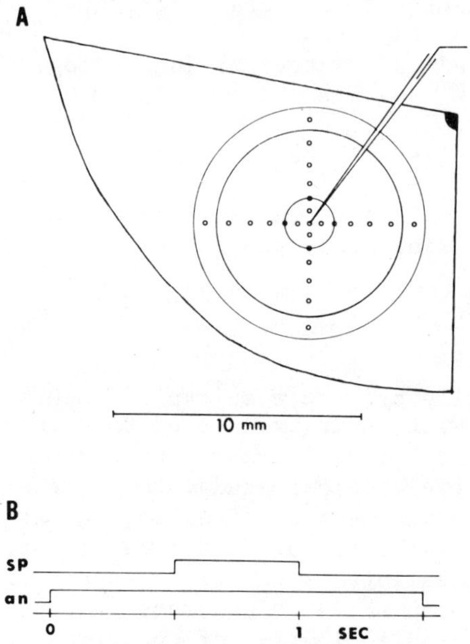

FIG. 1. Diagram of arrangement for stimulating and recording from ganglion cells in dogfish retina.
A shows piece of eye cup, right temporoventral quadrant, with optic nerve head at upper right and microelectrode placed against vitreal surface in centre. Spots 0.2 mm in diameter (small circles) were projected on points along vertical and horizontal axes of the eye. Centred spots of various diameters (larger circles) and annuli (a typical annulus, I.D. = 7.6 mm and O.D. = 9.5 mm, is defined by two largest circles) also could be delivered. B shows temporal relations of centred spot (sp) and surrounding annulus (an) used to test for peripheral inhibition.

crustaceans upon which it feeds. Its retina is dominated by rods; probably fewer than 1% of its receptors are cones (Stell and Witkovsky, 1973b). It has been shown electrophysiologically to contain ganglion cells with large receptive fields, whose inputs are mostly if not entirely rod-dependent (Stell et al, 1970,1971), and structurally to contain ganglion cells with large dendritic fields (Stell and Witkovsky, 1973a). This report concerns the spectral sensitivity, RF shape and summation properties at threshold for ON-type ganglion cells with large RF's in *Mustelus*.

MATERIALS AND METHODS

Pieces of eye cup (temporoventral quadrant) from visually responsive animals were placed in a chamber, suffused with moist 100% oxygen at 20°-22° C, and stimulated with steps of light from a monochromator or tungsten lamp (100 W, operated at 2850° K). The dual-beam optical system (Wagner et al, 1960; Spekreijse et al, 1972) generated circular spots and annuli ranging in diameter from about 0.2-10.0 mm, which could be positioned anywhere in a circular stimulus field 10 mm in diameter (Fig. 1). Ganglion cell responses were recorded extracellularly with blunt-tipped glass-coated Pt-Ir microelectrodes centred in the stimulus field and advanced against the vitreal surface of the retina without penetrating it. Signals were conventionally

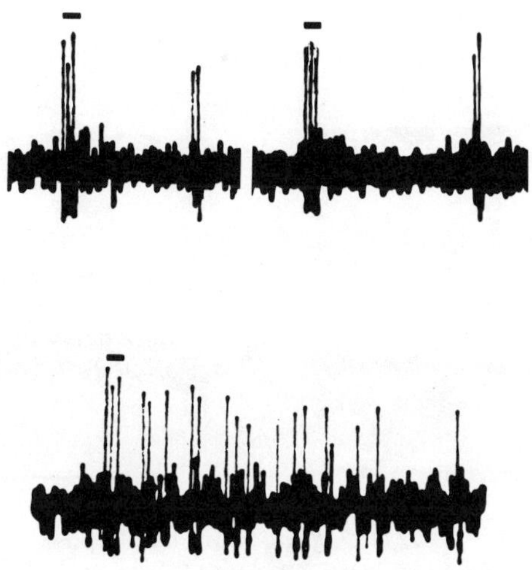

FIG. 2. Discharges of nerve impulses by *Mustelus* ON-centre ganglion cell in response to a small spot of white light centred in the receptive field. Above, stimulus intensity near threshold. Below, stimulus intensity 2 logarithmic units greater than threshold. Horizontal bars = 10 msec.

amplified and filtered (typical band-pass 140-1600 Hz), and thresholds detected with the aid of CRO and audio monitors. A few units were recorded on magnetic tape.

OBSERVATIONS

In healthy preparations, units were quickly isolated and could be held in stable condition for as long as eight hours. Since most of the units studied did not fire spontaneously in the steady light- or dark-adapted state, thresholds were easily determined as the lowest intensity at which at least one spike was evoked by 2 out of 3 successive stimuli. A total of 175 units were isolated; 90 of these were studied in detail. This report concerns only the 80% of those units which had ON-centres.

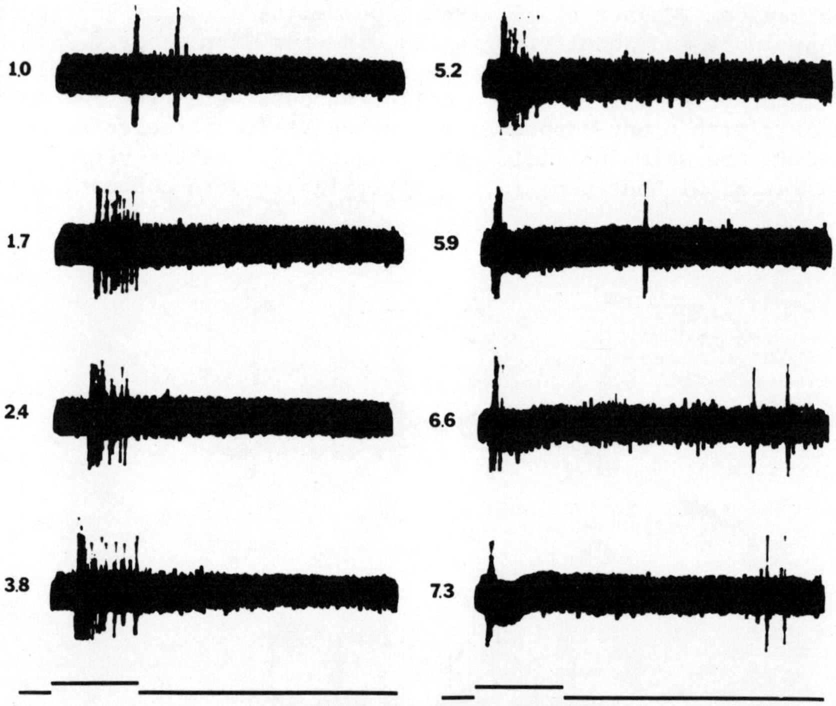

FIG. 3. Responses of *Mustelus* ganglion cell to white light of different intensities. Recording trace begins at onset of 0.5 sec. light pulse, indicated by upward deflection of signal trace. Numbers at left indicate logarithmic units of intensity above threshold. The stimulus was a circular spot, 0.8 mm in diameter, in the centre of the receptive field.

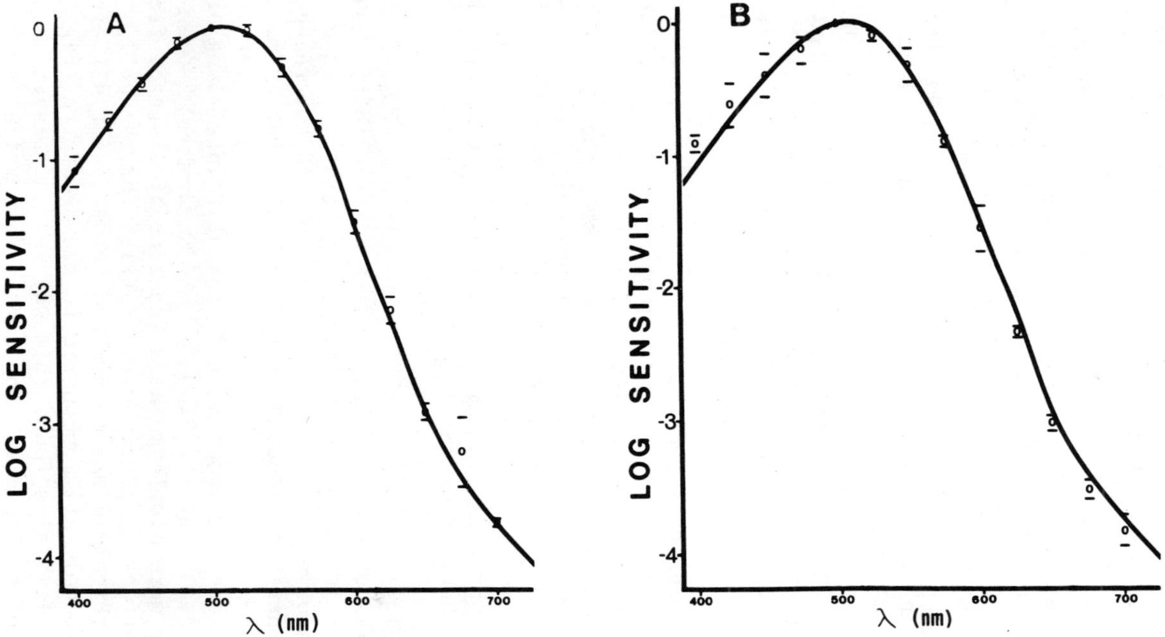

FIG. 4. Relative spectral sensitivities of *Mustelus* ON-centre ganglion cells, normalised to equality at 500 nm. (A) Sensitivity function for excitation by centred spots, diameters $\leq$ 2 mm. Continuous curve is drawn to fit the mean $\pm$ standard error. (B) Sensitivity function for inhibition of central ON-response by concentric annuli, I.D. $\geq$ 2mm. Same curve as in (A) fits the means $\pm$ S.E.

If a light spot no larger than 2 mm in diameter was centred in the receptive field, these units responded at threshold with one to several bursts of 2-4 nerve impulses each. (Fig. 2). The minimum interspike interval in these bursts was 3-5 msec. We observed a non-random coincidence of some bursts in two units recorded simultaneously on tape, suggesting either that the spike clusters are generated by bursts of presynaptic (bipolar or amacrine cell) activity or that some ganglion cells are functionally coupled. At threshold the latent period to the first burst of impulses was several hundred msec. (Fig. 3). As stimulus intensity increased, at first the latency decreased, and the discharges became less bursty and more prolonged. At intensities 4-6 logarithmic units above threshold, the response became transient at ON and then resumed at OFF, as if an inhibitory mechanism also had been activated. In most of these units, very large spots (6-10 mm diameter) evoked weak OFF-responses near threshold, and no activity at all at higher intensities. The sensitivity of the RF centre to a small spot of light was reduced in the presence of a maintained annulus of light (1.D $\geq$ 2 mm). An annulus alone, however, evoked either no response or a feeble OFF-response.

The threshold spectral sensitivity function for the ON-response to small, centred spots (Fig. 4A) was monotonic, and resembled the absorption curve for a $retinal_1$ visual pigment with a peak around 505 nm. The shape and λ_{max} were not altered by either blue (438 nm) or red ($\geq$ 680 nm) background lights which reduced sensitivity by about 1 log unit; stronger adapting lights were not used. The spectral sensitivity function for inhibition of response to a standard central spot by an annulus of variable wave length (Fig. 4B) was not distinguishable from the function for excitation of the centres.

The properties of the ON-centre and the OFF/inhibitory surround were further investigated by determining thresholds of response to white light spots of various sizes and locations in the receptive field. The RF centre was located by finding the point of maximum sensitivity to a 1-2 mm spot. Then the threshold was determined for spots 0.2 mm in diameter, located at different points along orthogonal axes through the centre of the RF (Fig.1). The threshold sensitivities found in this way were displayed as a function of distance from the RF centre to construct a "sensitivity profile" for the cell. Alternatively, the threshold was determined for spots of different diameters (0.2-9.5 mm), all concentric to the RF centre (Fig. 1). The threshold sensitivities found this way were displayed as a function of area of the spot to construct a "summation (area-intensity) function". In most cases the sensitivity profile (Fig. 5, 7) showed that the RF centre was a plateau 1-2 mm in diameter, everywhere equally sensitive, with a sharp drop in sensitivity at the edges. Although ON-responses sometimes could be evoked by spots

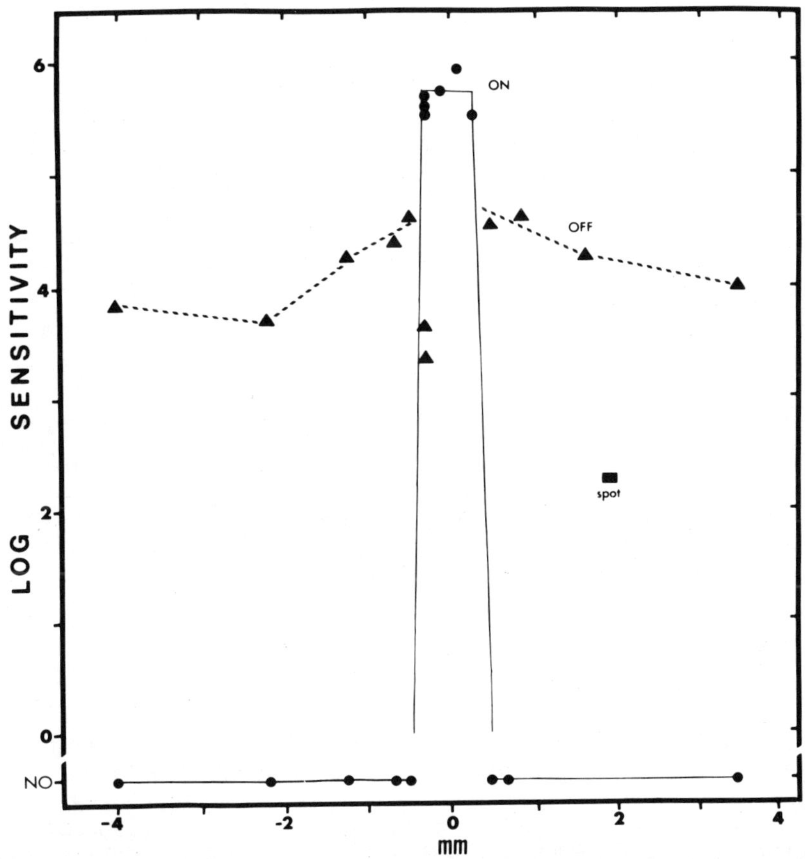

FIG. 5. Threshold sensitivity profile for dark-adapted ON-centre ganglion cell in *Mustelus*, stimulated with a spot of white light 0.2 mm in diameter (bar indicates spot size) located at various points along one axis through receptive field centre. The size of the ON centre was sharply defined; even a light nearly 6 log units brighter gave no ON response in the surround. This cell has a prominent OFF-surround.

located as far as 4 mm from the centre, the difference in threshold (3-4 log units) allowed the possibility that the centre itself was activated by scattered light. With concentric spots ("summation function" method) the log threshold sensitivity increased, usually linearly, up to a diameter of 1-3 mm (Fig. 6); with a further increase in diameter the sensitivity sometimes showed no further change, but more often it decreased. This decrease in threshold at large diameters tended to be most prominent in units which showed the strongest OFF/inhibitory-surrounds. As noted previously, the spots sometimes gave OFF- (or ON-OFF) responses near threshold.

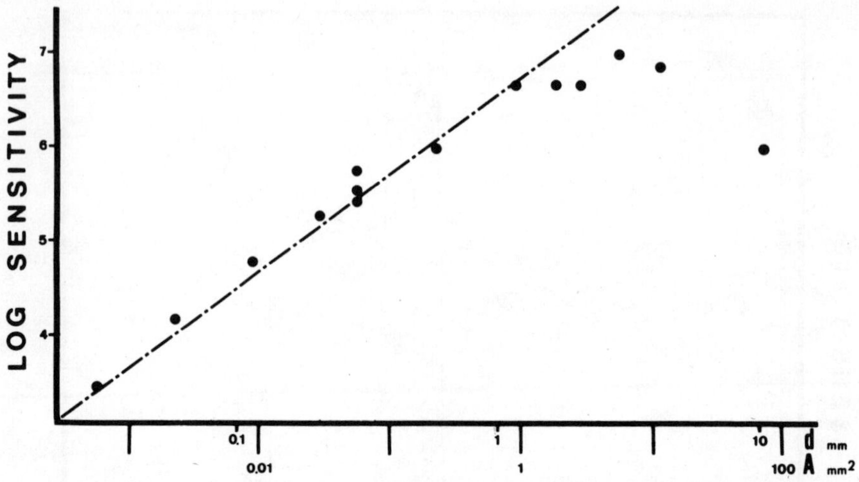

FIG. 6. Summation function for the ON-response of an ON-centre ganglion cell in dark-adapted *Mustelus* retina. Increase in threshold is linear with area (A) up to a spot diameter (d) of about 1 mm. There is some decline in sensitivity for the largest spot.

The dimensions of the RF centres varied considerably from one unit to another. In every case in which both dark-adapted sensitivity profiles and summation functions were obtained, however, the diameter of the centre as judged by the limit of linear summation was significantly larger than the diameter as judged by the size of the field of equal sensitivity to a small spot.

The effect of adaptation on RF organisation was investigated by determining sensitivity profiles and summation functions first in dark-adapted eye cups, as just described, and then in the presence of a full-field steady white background light. When this light was turned on, most cells responded at first with a maintained spike discharge and showed a marked increase in threshold. Within 10-15 minutes the spontaneous discharge ceased and the threshold stabilised at a lower level than when the light was first turned on. Background light intensities were chosen which gave stable light-adapted thresholds 1-2 log units higher than the dark-adapted thresholds to the same stimuli. The RF centre still responded at ON to a small spot at this new, elevated threshold. But now the diameter of the equal-sensitivity centre was usually reduced (Fig. 7, 9), as was the diameter within which summation is linear (Fig. 8, 9). The change in summation diameter was more marked, so that it approached the profile diameter in the light-adapted retina (Fig. 9). A flashed annulus (7.6 mm I.D.) had the same effect on summation diameter as a steady

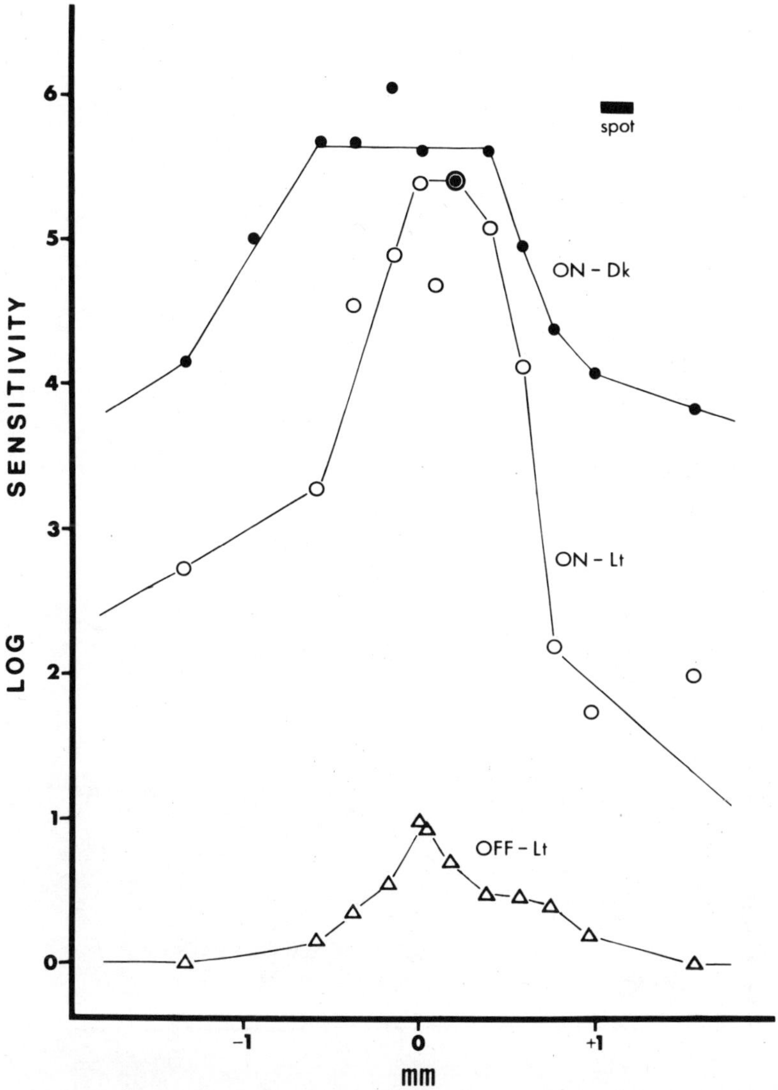

FIG. 7. Threshold sensitivity profiles for an ON-centre ganglion cell in *Mustelus*, stimulated with a spot of white light 0.2 mm in diameter (bar indicates spot size) located at various points along one axis through receptive field centre. The ON-centre is narrower and less sensitive against a steady diffuse white background ("ON-Lt") than in the dark ("ON-dark"). An OFF-response is evident only in the presence of the background ("OFF-Lt").

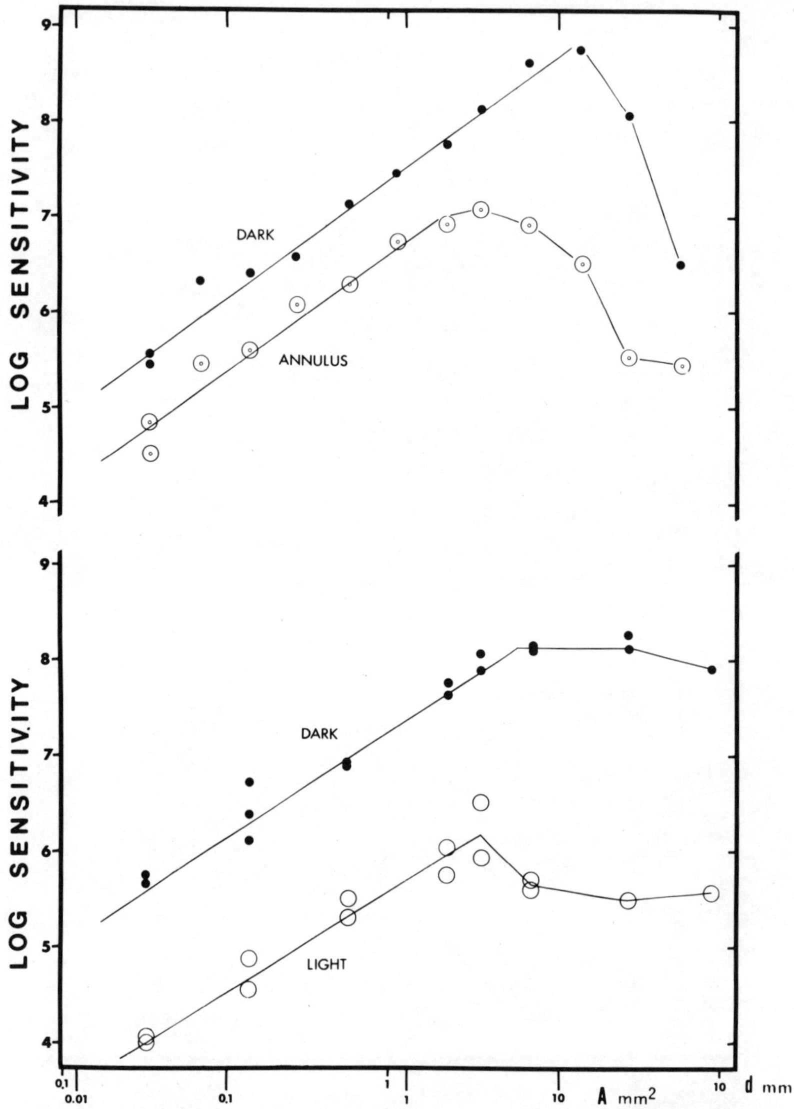

FIG. 8. Summation functions of ON-centre ganglion cells in *Mustelus* retina. Both a flashed annulus (above) and a steady background light (below) reduce threshold sensitivity at all spot diameters (d). Both also reduce the diameter within which summation of sensitivity is linear with spot area (A).

Fig. 9. Curves summarising receptive field properties of an ON-centre ganglion cell in *Mustelus* retina, tested with spots of white light. Upper two curves are conventional plots of threshold sensitivity versus spot diameter in mm (upper abscissa). Lower two curves are plots of threshold sensi-

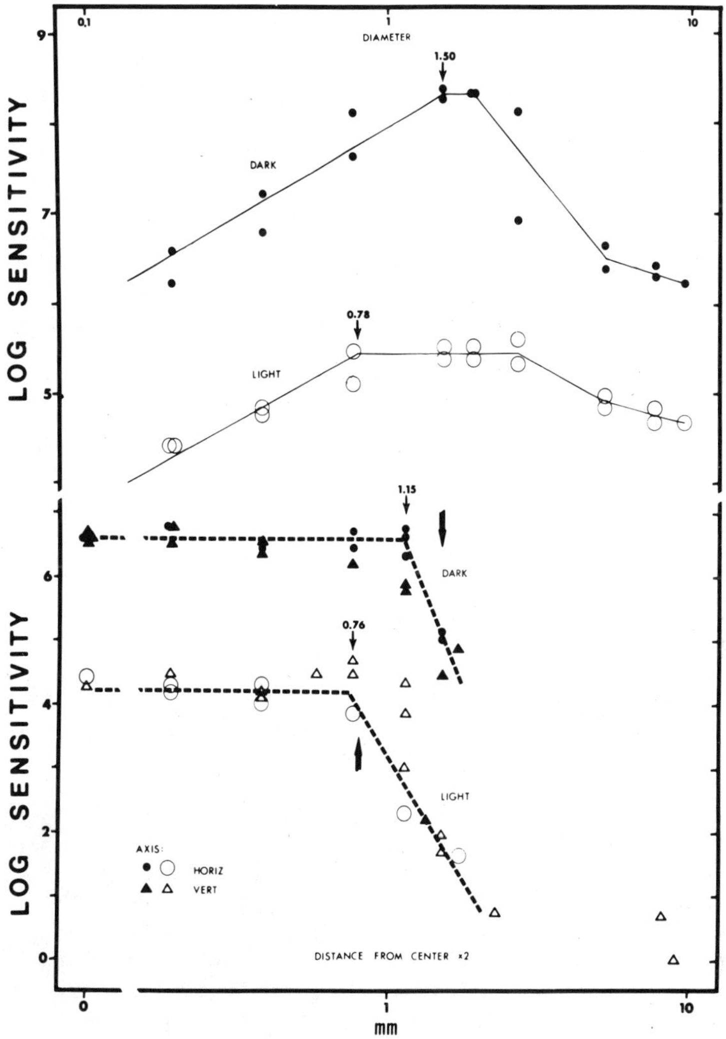

tivity to a spot 0.2 mm in diameter moved along both horizontal axis (circles) and vertical axis (triangles); for comparison to upper curves, the distance of the spots in mm (x2) from the RF centre is expressed on a logarithmic scale (lower abscissa). Filled symbols identify measurements in dark-adapted retina; open symbols identify measurements against a steady, diffuse white background. Small arrows with numbers indicate the edge and diameter of the centre in each curve; large arrows near the lower curves indicate these parameters of the corresponding upper curves, for easier comparison. These data illustrate (1) the difference in apparent size of the RF centre judged by the two methods in the dark-adapted retina, and (2) the reduction in diameter of the RF centre in the light-adapted retina.

full-field light (Fig. 8). Finally, a small spot flashed against the steady background usually evoked a threshold OFF-response in the RF surround, as often did a centred spot of large diameter which extended far into the surround.

DISCUSSION

Large-tipped metal microelectrodes appear to be biased toward recording from large neurons (Wiesel, 1960; Stone, 1973). It is probable, therefore, that we have recorded frequently from giant ganglion cells like those described by Stell and Witkovsky (1973a), which have dendritic field diameters of at least 1.5-2.0 mm. The diameters of RF centres, as determined by sensitivity profiles, coincide rather well with these dimensions. This might suggest that the RF centres are determined by the GC dendrites. On the other hand, we sought but were unable to find a physiological correlate to the elliptical form of the dendritic fiels, since there was no consistent difference in sensitivity profiles along horizontal and vertical axes. Furthermore, as noted previously there is no unique physiological parameter for RF centre, since the diameter differs according to the method employed (sensitivity profile vs. summation).

Quantitative data on ganglion cell dendritic trees (Stell and Witkovsky, 1973a) also fail to clarify the role of ganglion cell dendrites in determining RF properties. Since the density of giant ganglion cell dendrites appears to decline exponentially with distance from the cell body, while sensitivity is everywhere equal, the dendritic tree itself cannot account for the constancy of sensitivity and the linearity of summation which we have observed in the RF centre. It follows most probably either that the distribution of bipolar cell inputs to ganglion cell dendrites is much less homogeneous than generally assumed, or that many of the spatial properties of the ganglion cell's receptive field are determined by cells or networks distal to the ganglion cell.

What might such cells or networks be? Horizontal cells, shown to be electrically coupled in dogfish (Kaneko, 1971), are certainly capable of wide spatial summation and are capable of influencing ganglion cell firing (Naka and Witkovsky, 1972). Amacrine cells, likewise believed by many to be involved in horizontal transmission within the retina, also must be considered; some amacrines in dogfish are known to extend at least several millimetres (Stell and Witkovsky, 1973a). A third major horizontal pathway is comprised by bipolar cells; in dogfish the axons of some of them are not only unusually long but probably electrically coupled to one another (Witkovsky and Stell, 1973a,b). Finally, recent studies have revealed that rod cells themselves are functionally coupled in a turtle (Schwartz, 1973). Coupling of rods might constitute a fourth horizontal pathway in dogfish. Thus there is no dearth of possible routes for horizontal interactions, although the specific mechanisms in dogfish remain to be elucidated.

We know that cones are present in *Mustelus* (Stell and Witkovsky, 1973b), but we know nothing about their function or role in vision. The threshold properties of the giant ganglion cells, at least, are controlled by rods in the dark-adapted state, since their spectrum matches reasonably well that of dogfish rhodopsin (H. Ripps, personal communication). Whether these cells receive any signals from the cone system cannot be determined from our data.

The ganglion cells of dogfish resemble those of the domestic cat in the effect of light-adaptation on both summation diameter and surround sensitivity (Barlow et al, 1957) as well as the sensitivity profile of the RF centre (Yoon, 1972). We do not know how a background light reduces the size of the RF centre. But such a reduction, even in units with so large a receptive field, may serve as a model for a means of increasing acuity at the level of ganglion cells concurrent with light adaptation.

REFERENCES

Barlow, H.B., Fitzhugh, R., and Kuffler, S.W. (1957). Change of organization in the receptive fields of the cat's retina during dark adaptation. J. Physiol. 137: 338-354.

Dowling, J.E. and Boycott, B.B. (1969). Retinal ganglion cells: A correlation of anatomical and physiological approaches. In: The Retina. Morphology, Function, and Clinical Characteristics. UCLA Forum in Medical Sciences #8. Edited by B.R. Straatsma, M.O. Hall, R.A. Allen and F. Crescitelli. University of California Press, Berkeley and Los Angeles. pp. 145-161.

Gallego, A. (1954). Conexiones transversales retinianas. An. Inst. Farmacol. Esp. 3: 31-39.

Hartline, H.K. (1940). The receptive fields of optic nerve fibers. Am. J. Physiol. 130: 690-699.

Kaneko, A. (1971). Electrical connexions between horizontal cells in dogfish retina. J. Physiol. 213: 95-105.

Kuffler, S.W. (1952). Neurons in the retina: Organization, inhibition and excitation problems. Cold Spring Harbor Symp. Quant. Biol. 17: 281-292.

Naka, K.-I. and Nye, P.W. (1971). Role of horizontal cells in organization of the catfish retinal receptive field. J. Neurophysiol. 34: 785-801.

Naka, K.-I. and Witkovsky, P. (1972). Dogfish ganglion cell discharge resulting from extrinsic polarization of the horizontal cells. J. Physiol. 223: 449-460.

Schwartz, E.A. (1973). Responses of single rods in the retina of the turtle. J. Physiol. 232: 503-514.

Spekreijse, H., Wagner, H.G. and Wolbarsht, M.L. (1972). Spectral and spatial coding of ganglion cell responses in goldfish retina. J. Neurophysiol. 35: 73-86.

Stell, W.K., Detwiler, P.B., Wagner, H.G. and Wolbarsht, M.L. (1971). Spatial organization and adaptional changes of ON-OFF ganglion cells in *Mustelus* retina. Biol. Bull. (Woods Hole) 141: 403-404.

Stell, W.K., Wagner, H.G. and Wolbarsht, M.L. (1970). Receptive field organization of ganglion cells in the retina of the smooth dogfish, *Mustelus canis*. Biol. Bull. (Woods Hole) 139: 437-438.

Stell, W.K. and Witkovsky, P. (1973a). Retinal structure in the smooth dogfish, *Mustelus canis:* General description and light microscopy of giant ganglion cells. J. Comp. Neurol. 148: 1-32.

Stell, W.K. and Witkovsky, P. (1973b). Retinal structure in the smooth dogfish, *Mustelus canis:* Light microscopy of photoreceptor and horizontal cells. J. Comp. Neurol. 148: 33-46.

Wagner, H.G., MacNichol, E.F., Jr. and Wolbarsht, M.L. (1960). The response properties of single ganglion cells in the goldfish retina. J. Gen. Physiol. 43: (6, part 2), 45-62.

Wiesel, T.N. (1960). Receptive fields of ganglion cells in the cat's retina. J. Physiol. 153: 583-594.

Witkovsky, P. and Stell, W.K. (1973a). Retinal structure in the smooth dogfish, *Mustelus canis*: Light microscopy of bipolar cells. J. Comp. Neurol. 148: 47-60.

Witkovsky, P. and Stell, W.K. (1973b). Retinal structure in the smooth dogfish, *Mustelus canis:* Electron microscopy of serially sectioned bipolar cell synaptic terminals. J. Comp. Neurol. 150: 147-168.

Yoon, M. (1972). Influence of adaptation level on response pattern and sensitivity of ganglion cells in the cat's retina. J. Physiol. 221: 93-104.

CENTRAL PROJECTIONS OF THE RETINA AND VISION

H.O. SCHWASSMANN

Department of Zoology, University of Florida

Gainesville, Fla. 32611, U.S.A.

That makes me see, and cannot feed mine eye?
All's Well That Ends Well, Act I, Sc. 1

INTRODUCTION

Vision is a dominant sense in most teleost fish, as is indicated by the presence of large eyes with a well developed retina and by the size of the dorsal mid-brain which is very large relative to the forebrain and the cerebellum. The majority of optic nerve fibres terminates in the mid-brain roof, the optic tectum. Additional minor projections have been demonstrated by silver impregnation techniques. In *Eugerres plumieri* these are the lateral geniculate nucleus, the cortical nucleus, the pretectal area, and the optic nucleus in the hypothalamus (Vanegas and Ebbesson, 1973).

Since most electrophysiological and behavioural work has been concerned with the role of the mid-brain tectum, it is difficult to assign any specific visual function to the minor optic terminations. There is considerable variation in the external appearance of the optic tectum. In their simplest and most commonly encountered form, the tecta appear like regularly curved hemispheric lateral bulges, with their dorsomedial edges being adjacent throughout most of their longitudinal extent. When viewed from above, about two thirds of the tectal surface is visible while the remaining third continues ventro-lateral and is hidden underneath the lateral bulge. In carp and a few other cyprinids a very large valvula cerebelli inside the optic ventricle is the cause for the considerable lateral and ventral displacement of the posterior tectal portions. Even more striking morphological specialisation is found in the optic tecta of several tuna-related families where the dorso-medial and ventro-lateral fascicles of the optic tract spread over separate anterior tectal bulges. The resulting deep horizontal indentation of the tectum appears to be a consequence of excessive surface enlargement (Schwassmann and Kruger, 1968).

With respect to the particular role played by the optic tectum in vision, there have been studies investigating response properties of single units, and studies concerned with the spatial relations of the retinal projection upon the mid-brain roof. The latter approach has yielded detailed retino-tectal projection maps in several species of teleost fish.

RETINO-TECTAL TOPOGRAPHY IN TELEOSTS

Spatial orderliness in the projection of sensory surfaces upon relay nuclei and central receiving areas is a well documented fact the far-reaching implications of which have not always been clearly recognised. The geometric correspondences between visual field, or retina, and a central projection area are most easily demonstrated when the latter is easily accessible to electrode mapping, and also provides an unfolded surface as it is the case in the mid-brain roof of most teleost fish.

Earlier experimental degeneration studies, utilising the Marchi method, resulted in evidence favouring a systematic projection of retinal quadrants onto the mid-brain of fish (Lubsen, 1921; Ströer, 1939; Akert, 1949). Electrophysiological methods brought one great advantage over the histological lesion-requiring technique; a single experiment could yield a great number of point-to-point relationships between the periphery and the centre. Julia Apter's study of the cat's superior colliculus by "evoked potential" method was one of the first successful attempts (Apter, 1945). Further improvement resulted from the use of sturdy metal-microelectrodes, especially developed for extra-cellular recordings from inside nervous tissue, and produced retino-tectal maps for a variety of vertebrates, but the most detailed of such maps exist for the mid-brain tectum of teleost fish (Jacobson and Gaze, 1964; Schwassmann, 1968; Schwassmann and Krag, 1970; Schwassmann and Kruger, 1965a, b).

Electrophysiological mapping of retino-tectal dependencies is generally used in several laboratories, especially for studies of the mechanisms underlying neuronal specificities during axonal regeneration. Since the desirable detail of results from this perimetric technique depends chiefly on most discrete and distorsion-free photic stimulation of the retina, a brief account of the presently used methods seems appropriate.

Methods of Visual Perimetry in Teleosts

Our perimeter has been designed to ensure discrete photic stimulation at constant distance of the retina of a submersed eye from any point of a 180 degree wide visual field. Distortions due to refraction of light by interfaces of media of differing optical densities are practically eliminated. The perimeter (Fig. 1) is made of transparent acrylic plastic (Plexiglass) of 1/4 to 1/2 inch thickness and consists of a vertically placed hemisphere of 60 cm diameter, the rim of which has been milled down by about one cm and then glued on to a large plane sheet of plastic. This closed hemisphere is filled with distilled water. The centre of curvature lies about 0.5 cm outside of the plane sheet. A faint scratch mark on the plastic, indicating the centre, facilitates alignment of the pupil for mapping. The plane vertical sheet forms one wall of the rectangular tank that is filled with water and in which the fish is held on an adjustable lucite tray. The fish is supported laterally by plastic holders and is firmly attached by wires with its protractile jaws to a rubber-covered mouthpiece of plastic tubing.

A system of visual field coordinates, consisting of a horizontal and a vertical meridian and their ten degree parallels, is marked on the hemispheric surface by thin black tape. Photic stimuli are displayed and moved over this surface; delimited responsive areas are marked on it with a wax pencil.

The fish is anesthetised in a solution of tricaine methane sulfonate (Finquel, Ayerst, New York) in continuously aerated water. After several minutes in the proper concentration the fish loses equilibrium and opercular movements become faint. Before the latter cease completely, the animal is mounted on the tray and its jaws are fitted over the mouth-piece through which a continuous flow of anesthetic solution is maintained. Dorsal craniectomy by small circular saw and rongeurs exposes the mid-brain; any perimeningeal tissue, often mixed with fat globules, inside the cranial cavity is removed by gentle suction, or by wiping with moist cotton. No foreign substance is needed on the tectal surface to prevent drying since the normally slight seepage of cerebro-spinal fluid serves this purpose adequately. The tray with the fish is then placed in the tank where the left eye is aligned with the hemisphere's centre. The water level control is so adjusted that the eye remains submerged but the cranial opening is above water. A re-circulating system returns the water from a lower collecting vessel back to the upper reservoir. For simple mapping,the tricaine solution can be left at full strength. Since tricaine depresses spontaneous activity of units, it is diluted, or replaced by new water, for single-unit studies after the fish has been given an adequate dose of a neuromuscular blocking agent. Under either of these drugs the eye will always relax in the same position, and the eye movements are absent, which is easily ascertained by re-mapping several of the earliest points.

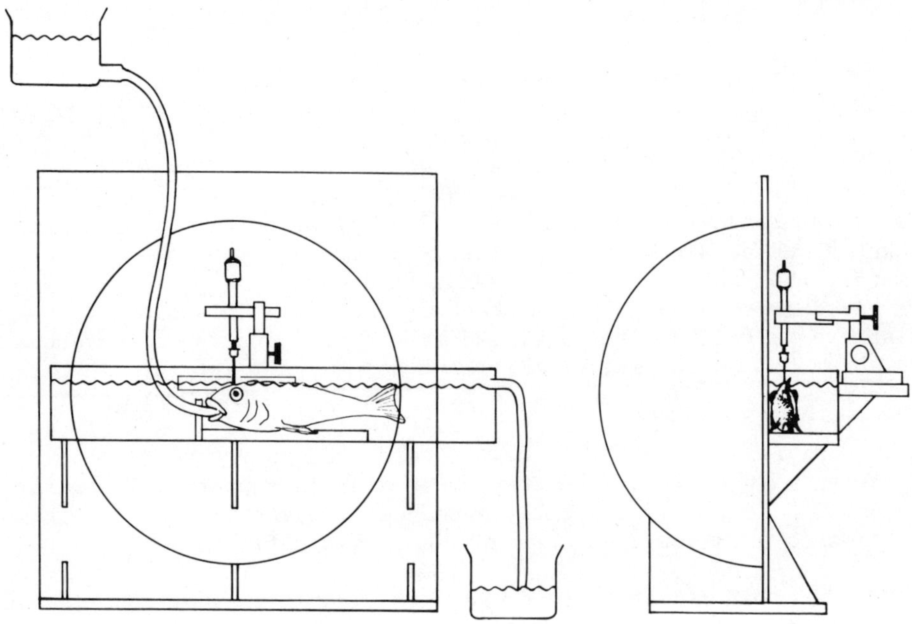

FIG. 1. Perimetric apparatus for mapping retino-tectal correspondences in fish, in side view on the left, and shown from the rear on the right. The left eye of the fish is positioned in the centre of the water-filled hemisphere on the curved surface of which photic stimuli are displayed and receptive fields of visual units are marked.

Sharpened and insulated steel electrodes (Green, 1958) or large-tipped metal-filled glass capillaries (Dowben and Rose, 1953) are used. They are mounted above the mid-brain on the vertical shaft of a micrometer that is attached to a micromanipulator. This system permits calibrated movements of the electrode in three planes. Conventional amplifying and recording methods are employed; audio-monitoring is especially useful for efficient work.

The electrode tip is placed on the dorsal surface of the right tectum and is slowly advanced in depth until maximum unit activity is found in response to gross photic stimulation of the left eye. A flash-light beam, or hand-movement over the hemispheric surface, is a convenient stimulus for depth adjustment. A small discrete stimulus is then moved over the hemispheric surface until an area is found from which unit activity can be influenced. This multi-unit "receptive field" is outlined on the hemisphere and numbered with the particular electrode location on the tectum. The dorsal tectal surface is mapped by moving the electrode in 0.5 mm steps, or in smaller increments if greater detail is required. A four by five mm large tectum will yield from 50 to 60 active recording sites when 0.5 mm steps are used. These recording sites will correspond to receptive fields in the dorsal half of the hemisphere, and only a few will extend below the horizontal meridian. The ventral visual field projects to the ventro-lateral portion of the tectum, which is explored by penetrating with the electrode through the overlying tectum, utilising the same lateral and antero-posterior coordinates. A complete map of the visual field results only if the entire surface of the optic tectum is explored in this fashion.

After corrections for distortions due to non-perpendicular orientation of the electrode to the responsive surface at the outer tectal margins, the visual field coordinates can be transposed onto the surface of the contralateral tectum by utilising the receptive field-recording site correspondences.

Linear and Non-Linear Projection Patterns

Representatives of several vertebrate classes have been studied and it is evident from the data that the general orientation of the retino-tectal projection is essentially alike in vertebrates. With the exception of most mammals, the direct projection is strictly contralateral. This was also found to be the case in the rat, which apparently represents a rather primitive stage in the evolution of the mammalian visual pathway (Siminoff, Schwassmann and Kruger, 1966). In the cat a considerable degree of binocular overlap is then correlated with the complex binocular tectal projection (Feldon, Feldon and Kruger, 1970).

In the retino-tectal projection of a generalised vertebrate with laterally oriented eyes, as a bluegill, the dorsal visual field (ventral retina) projects to medial tectum; the ventral visual area to the latero-ventral tectum. Rostral visual field positions (temporal retina) are represented anteriorly on the mid-brain roof, and the temporal visual field corresponds to posterior locations on the tectum.

An early investigation of several fresh-water fish, large-mouth bass, bluegill, carp, and goldfish, indicated a linear projection pattern on the mid-brain roof where the mapped visual field coordinates were approximately evenly spaced (Fig. 2A) (Schwassmann and Kruger, 1965a). Regional specialisation of the retina results in non-linear projection patterns. The first clear

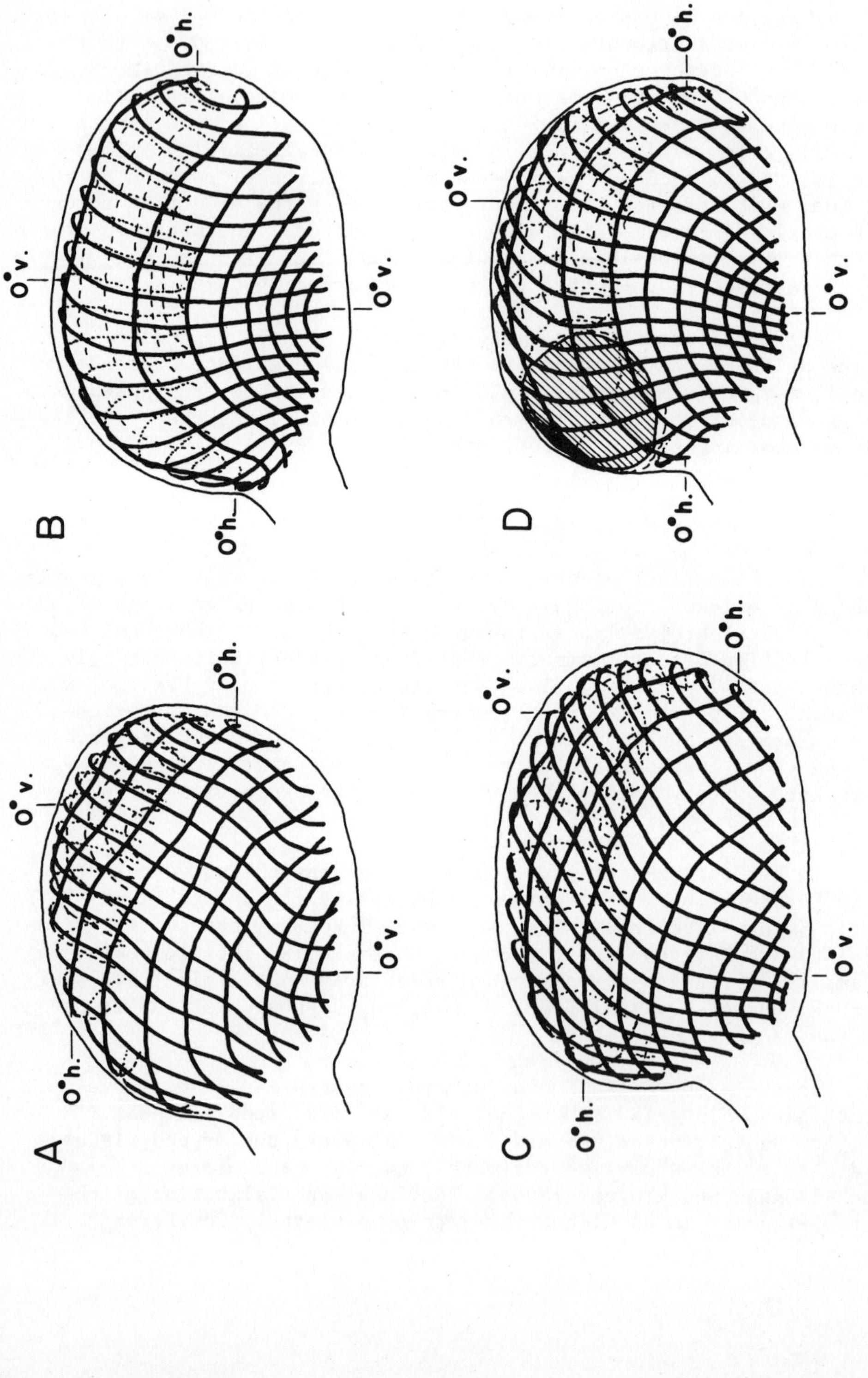
A
B
C
D
0°v.
0°h.

FIG. 2. Projection of the left visual field on the right optic tectum in four species of teleosts. Horizontal meridian (0°h), vertical meridian (0°v.) and their 10° parallels are shown as solid lines on the dorsal tectum surface, and as broken and dotted lines on the latero-ventral tectum. The orientation of all four optic tecta is the same, anterior is to the left, medial is down. A: approximately linear projection in the large-mouth bass (*Micropterus salmoides)*; B: magnification of the area above the horizontal meridian in *Anableps microlepis*; C: magnified area temporalis projecting to the antero-ventral tectum in the Pacific opal-eye (*Girella nigricans)*; D: greatly magnified projection of a temporal fovea (hatched area) in a serranid bass (*Centropristes striatus)*.

example of a teleost with retino-tectal magnification of a horizontally oriented and about 20 degrees wide area was the "four-eyed fish", *Anableps microlepis* (Fig. 2b) (Schwassmann and Kruger, 1965b). The retinal specialisation consists of a greater density of cones, interneurons, and ganglion cells. The relatively greater number of optic axons emanating from this horizontal streak apparently requires a correspondingly larger central area for synaptic contact with tectal elements. The probable adaptive significance lies in superior visual capability within the region of the visual field just above the water surface. These fish normally swim at the surface; the water line bisects the eye at the level of the iris flaps between the larger aerial and smaller aquatic pupil (Schwassmann and Kruger, 1965b).

All marine teleosts that were studied exhibited differing degrees of non-linearities in their visual projection, mostly consisting of temporal retinal areas that had magnified tectal representations. One such example of a moderately specialised area temporalis is shown in Fig. 2c. These temporal areas of higher cone and ganglion cell density also coincide with the direction of accommodative lens movement, and they specify the "visual axis", (Kahmann, 1934; Tamura, 1957). In several species of the predominantly marine serranid family, a deep depression is found in the centre of the area temporalis. This fovea temporalis of *Paralabrax* spp., and of *Centropristes* spp. is not exactly concentric but oblong in its horizontal extent. When seen in vertical cross section, its shape is distinctly convexiclivate (Fig. 3) as in the central foveae of some reptiles and some birds. There are no rods in the foveal centre; but, unlike conditions in the sauropsidian and primate fovea, the ganglion cells of the serranid fovea are not laterally displaced. This pure cone fovea in a typical duplex retina resembles very much the fovea described in the pure rod retina of the deep-sea fish *Platytroctes* (Locket, 1971). The similarities concern the morphology of the

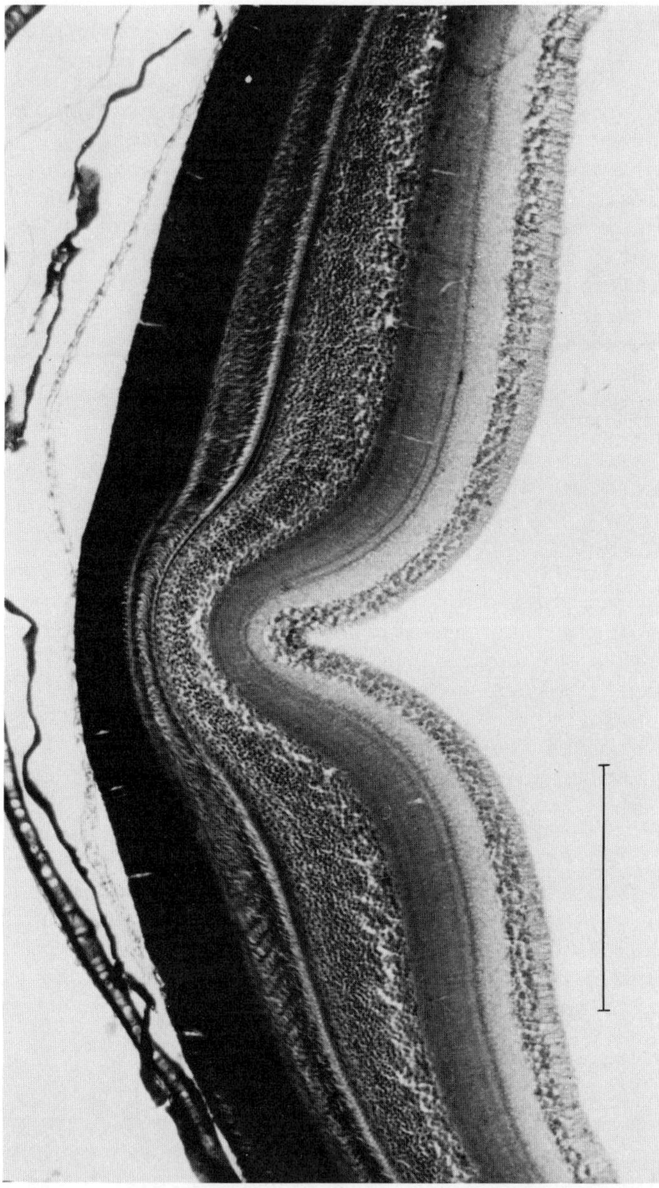

FIG. 3. Photomicrograph of a parasagittal section through the temporal part of the retina of a kelp bass, *Paralabrax clathratus*, showing the convexiclivate foveal depression. There is some reduction in thickness of several neural layers in the foveal centre and a thickening of the same layers at the foveal shoulders. The scale bar represents 200 μm.

foveal depression, its temporal location, and a noticeable central reduction with corresponding thickening at the shoulders of several retinal layers. The retino-tectal map in the foveate serranids (Fig. 2d) exhibits a more than five-fold magnification of that ten degree-wide area that surrounds the visual axis pointing into the anterior visual field (Schwassmann, 1968).

FUNCTIONAL SIGNIFICANCE OF PRECISE RETINO-TECTAL TOPOGRAPHY

Concerning the precise nature of the electrical response recorded from the optic tectum of fish, a recent study could successfully distinguish between pre- and postsynaptic unitary activity (Vanegas, Essayag-Millan and Laufer, 1971). However, since precise quantification of the stimulus necessitates electrical stimulation of the optic nerve, and since only one unit at a time can be investigated, meaningful parameters in terms of visual function that might distinguish between pre- and post-synaptic neuron activity are not easily derived from these studies. In view of the mixed composition of the multi-unit response that is utilised for visual field perimetry, it is surprising to find a high degree of uniformity in response characteristics as well as in size of single or multi-unit receptive fields. Occasional easily identifiable post-synaptic units within the massive multi-unit response seem to be less frequent than should be assumed from the proportions indicated by Vanegas et al. (1971). It could be that the use of the tricaine anesthetic and also the method of discrete photic stimulation of small retinal areas, necessary for precise mapping, may be the cause for activating only a negligible percentage of post-synaptic elements. We feel justified in assuming that the unit activity encountered in mapping experiments reflects predominantly the activity of retinal ganglion cells, the axons of which are making synaptic contact with dendrites of tectal neurons near the electrode tip.

A most striking observation is the high degree of precision and reproducibility in the point-to-point relations between visual field and contralateral tectum. Under optimal experimental conditions where the eye is emmetropic in water, and in the absence of distorting interfaces of non-perpendicular light incidence, lateral displacements of the electrode of 50 μ, or even less, result in detectable shifts of the responsive areas in the visual field. These shifts, in spite of some overlapping of the small fields, are always in the direction predicted by retino-tectal topography.

With regard to the functional importance of this precise retinotopic organisation, the question arose whether the anatomical orderliness is a prerequisite for normal vision, especially spatial information, or whether it might be merely incidental to normal development, and information about spatial relations might be in terpreted, coded, and transmitted to higher centres by different mechanisms. Such possibility cannot be discounted since there is substantial evidence from our own work and from published accounts (Jacobson and Gaze, 1964; Cronly-Dillon, 1964) for the rather common occurrence of directional sensitivity in tectal units. Furthermore, because of our spatially restricted method of stimulation in this kind of perimetry and our limitation to monitoring one unit at one time,the presence of electrical activity at a particular place on the mid-brain tectum must not necessarily be regarded as a unique equivalent of vision correlated to only one area of the visual field. There could also exist a certain less restrictive mechanism in which a greater or lesser proportion of optic axons with rather widespread synaptic extent in the tectum might convey information about stimulus location, and even achieve a hight degree of accuracy by differential interaction with the fibre spread from adjacent retinal regions.

An attempt to solve some of these uncertainties about the functional meaning of retino-tectal topography was made by Schwassmann and Krag (1970). Employing classical conditioning of heart rate, and using a small spot of light of one second duration as conditional stimulus, they mapped the extent of scotomata which resulted from lesions set in the optic tectum of marine teleosts. These scotomata, areas of the visual field from where no conditioned response could be obtained, were always much larger than those regions of the visual field noted by an absence of tectal response found with the electrode mapping method. The results of the experiments are in favour of a strict parallel between the anatomical orderliness of retino-tectal topography, accurately revealed by electrophysiological mapping, and functional vision as it can be assayed by the certainly higher-level conditioned response technique. Whenever the lesion was small and restricted to the superficial fibre layer, the scotoma was small, and regeneration with restitution of vision was noted within five to seven weeks. Deeper and more extensive lesions caused relatively large scotomata when perimetry by cardiac conditioning was done immediately; and they resulted in apparent blindness in half the visual field, hemianopia, when from one to five weeks had passed after setting the lesion, even though the electrically unresponsive region remained relatively small and apparently confined to the actual extent of tectal damage. As shown by recently published work, extensive tectal damage can result in reorganisation and final restitution of a slightly distorted and compressed retino-tectal projection (Gaze and Sharma, 1970; Yoon, 1971, 1972; Sharma, 1972a, b). It seems possible that during this reorganisation process electrical activity on the tectum in response to photic stimulation of the eye is present, but that higher-level visual function is absent during this time, as is indicated by our results (Schwassmann and Krag, 1970).

CONCERNING A POSSIBLE MECHANISM UNDERLYING RETINO-TECTAL SPECIFICITY

Several theories have been proposed to account for the local specificity of tectal termination of retinal ganglion cells (Stone, 1948; Sperry, 1951; Székely, 1954) and a molecular basis for neuronal selective growth has been suggested (Weiss, 1962; Sperry, 1965). At least in amphibians it appears that specification of the ultimate tectal connexions of ganglion cells occurs in two distinct steps; determination along the horizontal axis precedes that in the vertical axis (Jacobson, 1967, 1968).

If the optic tract is cut in adult teleosts, regeneration results in re-establishment of a "normal" retino-tectal topography and restitution of vision suggesting that neuronal local specificity is a characteristic of retinal ganglion cells that is not only present during early development but also persists throughout the life of the animal. It was uncertain whether local specificity of optic axons was expressing itself by interaction with the central substrate only, or whether the final retinotopic pattern represents the result of continually acting factors during growth, or regrowth, along the optic pathway. The second alternative should mean strict segregation of axons within optic nerve and tract. There is no question about this in the case of the partially decussating optic nerves of mammals where Marchi studies demonstrated such precise segregation (Polyak, 1941, 1957). Concerning fish, Ströer (1939) claimed a retinotopic arrangement of optic nerve fibres in the Atlantic salmon, whereas Herrick (1941) reported random intermingling of optic axons in a fresh water catfish.

An organised segregation of fibres in the optic nerve could be demonstrated by electro-physiological mapping of squirrel fish (Holocentridae). These fish possess an interdigitating optic chiasm where 7-9 fascicles of one nerve interdigitate with a similar number from the other eye. Removal of one eye made possible the selective interruption of any number of fascicles coming from the other eye. As shown by the results from subsequent electrode-mapping, coherent unresponsive regions on the tectal surface corresponded to precisely delimited blind areas in the visual field (Fig. 4). Optic nerve fibres at the optic chiasm of squirrel fish are segregated into fascicles according to their origin in the retina as well as to the region of tectal synaptic contact. Taking into account the inter-individual variation of the extent of each fascicle's mapped projection area, corresponding to noticeable size differences among the different bundles, this tendency of optic axons from adjacent retinal areas to remain grouped together may be insufficient to account for the high degree of spatial precision attained by their tectal synaptic terminations. On the other hand, the results of recent work by Yoon (1971, 1972) and Sharma (1972 a, b) would seem to de-emphasise the importance of at least an intact

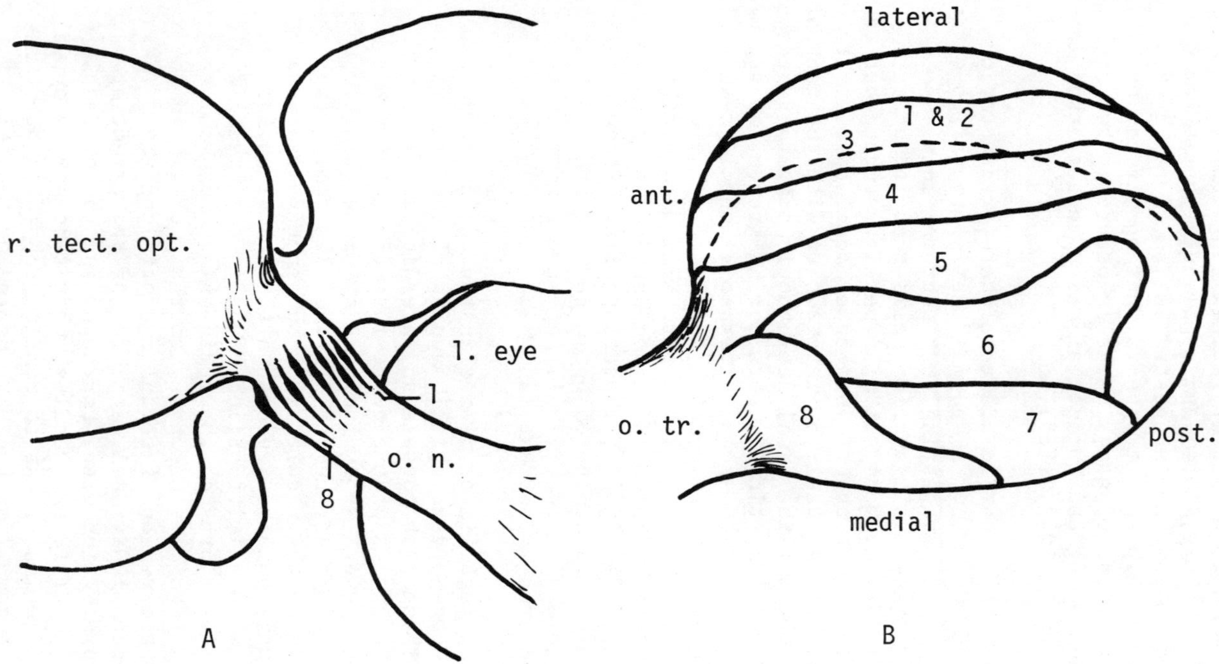

FIG. 4. The interdigitating optic chiasm and the projection of the individual fascicles on the optic tectum in the holocentrid *Myripristis leiognathos*. A: after removing the right eye and optic nerve, the left optic nerve (o.n.) from the left eye is seen separated into eight bundles (1-8) by corresponding bundles of the other nerve. Proximal to the chiasm the bundles unite and innervate the right optic tectum (r. tect. opt.). B: the areas of termination of optic nerve fibres within each of the eight separate fascicles are outlined on the right optic tectum (o. tr.: optic tract).

entirety of the tectal substrate for determining the retinotopic pattern.

REFERENCES

Akert, K. (1949) Der visuelle Greifreflex. Helv. Physiol. Acta. 7: 112-134.
Apter, J.T.(1945) Projection of the retina on superior colliculus of cats. J. Neurophysiol. 8: 123-134.
Cronly-Dillon, J.R. (1964) Units sensitive to direction of movement in goldfish optic tectum. Nature (London) 203: 214-215.
Dowben, M. and Rose, J.E. (1953) A metal-filled microelectrode. Science (N.Y.) 118: 22-24.
Feldon, S., Feldon, P. and Kruger, L. (1970) Topography of the retinal projection upon the superior colliculus of the cat. Vision Res. 10: 135-143.
Gaze, R.M. and Sharma, S.C. (1970) Axial differences in the reinnervation of the goldfish optic tectum by regenerating optic nerve fibres. Exp. Brain Res. 10: 171-181.
Green, J.D. (1958) A simple microelectrode for recording from the central nervous system. Nature (London) 182: 962.
Herrick, C.J. (1941) Development of the optic nerves of *Amblystoma*. J. Comp. Neur. 74: 473-543.
Jacobson, M. (1967) Retinal ganglion cells: Specification of central connections in larval *Xenopus laevis*. Science (N.Y.) 155: 1106-1108.
Jacobson, M. (1968) Development of neuronal specificity in retinal ganglion cells of *Xenopus*. Develop. Biol. 17: 202-218.
Jacobson, M. and Gaze, R.M. (1964) Types of visual response from single units in the optic tectum and optic nerve of the goldfish. J. Exp. Physiol. 49: 199-209.
Kahmann, H. (1934) Über das Vorkommen einer Fovea centralis im Knochenfischauge. Zool. Anz. 106: 50-55.
Locket, N.A. (1971) Retinal structure in *Platytroctes apus*, a deep-sea fish with a pure rod fovea. J. Mar. Biol. Ass. U.K. 51: 79-91.
Lubsen, J. (1921) Over de projectie van het netvlies op het tectum opticum bij een beenvisch. Ned. Tydschr. Geneeskd. 2: 1258-1261
Polyak, S. (1941) The Retina. Univ. Chicago Press, Chicago, III.
Polyak, S. (1957) The Vertebrate Visual System. Univ. Chicago Press, Chicago, III.
Schwassmann, H.O. (1968) Visual projection upon the optic tectum in foveate marine teleosts. Vision Res. 8: 1337-1348.
Schwassmann, H.O. and Krag, M.H. (1970) The relation of visual field defects to retinotectal topography in teleost fish. Vision Res. 10: 29-42.

Schwassmann, H.O. and Kruger, L. (1965a) Organization of the visual projection upon the optic tectum of some freshwater fish. J. Comp. Neurol. 124: 113-126.

Schwassmann, H.O. and Kruger, L. (1965b) Experimental analysis of the visual system of the four-eyed fish *Anableps microlepis*. Vision Res. 5: 269-281.

Schwassmann, H.O. and Kruger, L. (1968) Anatomy of visual centers in teleosts. In: The Central Nervous System and Fish Behavior edited by D. Ingle, p. 3-16, Univ. Chicago Press, Chicago III.

Sharma, S.C. (1972a) Reformation of retinotectal projections after various tectal ablations in adult goldfish. Exp. Neurol. 34: 171-182.

Sharma, S.C. (1972b) Redistribution of visual projections in altered optic tecta of adult goldfish. Proc. Nat. Acad. Sci. USA 69: 2637-2639.

Siminoff, R., Schwassmann, H.O. and Kruger, L. (1966) An electrophysiological study of the visual projection to the superior colliculus of the rat. J. Comp. Neurol. 130: 329-342.

Sperry, R.W. (1951) Regulative factors in the orderly growth of neural circuits. Symp. Soc. Study Develop. Growth 10: 63-87.

Sperry, R.W. (1965) Embryogenesis of behavioral nerve nets. In: Organogenesis, edited by R.L. DeHaan and H. Ursprung, p. 161-186, Holt, New York.

Stone, L.S. (1948) Functional polarization in developing and regenerating retinae of transplanted eyes. Ann. N.Y. Acad. Sci. 49: 856-865.

Ströer, W.F.W. (1939) Zur vergleichenden Anatomie des primären optischen Systems bei Wirbeltieren. Z. Anat. Entwickl. 110: 301-321.

Székely, G. (1954) Zur Ausbildung der lokalen funktionellen Spezifizität der Retina. Acta Biol. Acad Sci. Hung. 5: 157-167.

Tamura, T. (1957) A study of visual perception in fish, especially on resolving power and accommodation. Bull. Jap. Soc. Sci. Fish. 22: 536-557.

Vanegas, H. and Ebbesson, S.O.E. (1973) Retinal projections in the perch-like teleost *Eugerres plumieri*. J. Comp. Neurol. 151: 331-358.

Vanegas, H., Essayag-Millan, E. and Laufer, M. (1971) Response of the optic tectum to stimulation of the optic nerve in the teleost *Eugerres plumieri*. Brain Res. 31: 107-118.

Weiss, P. (1962) From Cell to Molecule. In: The Molecular Control of Cellular Activity, edited by J.M. Allen, pp. 1-72. McGraw-Hill, New York.

Yoon, M. (1971) Reorganization of retinotectal projection following surgical operations on the optic tectum in the goldfish. Exp. Neurol. 33: 359-411.

Yoon, M. (1972) Reversibility of the reorganization of retinotectal projection in adult goldfish. Exp. Neurol. 35: 565-577.

REPONSES VISUELLES UNITAIRES CHEZ LA TRUITE

G. GALAND & B. LIEGE

Laboratoire de Neurophysiologie et Psychophysiologie

40, avenue du Recteur Pineau, 86022-POITIERS, France

C'est une chose admirable, que
tous les grands hommes ont toujours
du caprice, quelque petit grain de folie
mêlé à leur science.
Le Médecin malgré lui. Acte I, Sc. 4

INTRODUCTION

De très nombreuses investigations neurophysiologiques ont été effectuées à différents niveaux des voies visuelles chez les mammifères ou chez la grenouille. Chez les poissons, par contre, de telles investigations sont encore assez peu nombreuses. Ce sont les raisons pour lesquelles il nous a semblé intéressant d'entreprendre une étude électrophysiologique unitaire des voies visuelles chez un poisson téléostéen: la truite. De courts résumés partiels ont déjà été publiés antérieurement (Galand et al., 1969; Liège et Galand, 1970, 1971a et b).

MATERIEL ET METHODES

Les méthodes générales employées dans ce travail diffèrent peu de celles précédemment décrites dans des études effectuées chez la la grenouille (Liège et Galand, 1972; Morin et al., 1974). Nous allons les résumer en précisant les points qui diffèrent.

L'animal est anesthésié par immersion dans une solution d'uréthane ou de MS 222. Ensuite il est curarisé et placé dans un appareil de contention comportant un système d'alimentation en eau convenablement oxygénée. Le corps de l'animal est entièrement dans l'air. Par conséquent, les rayons lumineux qui pénètrent dans l'oeil traversent le dioptre air-cornée. Aucune correction dioptrique n'est effectuée. La cornée est protégée de la dessication par un film d'huile de parafine.

Les électrodes d'enregistrement sont de deux types et correspondent soit à celui décrit par Hubel (1957), soit à celui décrit par Gesteland et al. (1959).

Les stimulations visuelles sont effectuées de différentes manières: a) stimulations on-off localisées effectuées à l'obscurité avec un petit ($0{,}5^0$ à 2^0 environ) spot lumineux immobile; b) stimulations off par passage transitoire à l'obscurité générale; c) stimulations avec un jeu de mires de carton noir ou blanc que l'expérimentateur déplace à la main sur un écran placé devant l'oeil de l'animal. C'est au niveau de cet écran que sont déterminés les contours du RRF (responsive receptive field) défini par Maturana et al. (1960).

Les stimulations non visuelles sont diverses: a) stimulation mécanique des téguments au niveau des flancs et du dos de l'animal avec le jet d'eau d'une seringue ou l'extrémité d'un petit fil de fer déplacé à la main par l'expérimentateur; b) stimulation électrique des téguments au niveau de la ligne latérale; c) stimulations acoustico-vibratoires en frappant légèrement l'armature de l'appareil stéréotaxique avec une petite règlette métallique.

REPONSES RETINIENNES

Les réponses rétiniennes unitaires sont enregistrées dans les couches superficielles du tectum, c'est-à-dire au niveau de l'arborisation terminale des axones des cellules ganglionnaires. Des enregistrements de contrôle ont été effectués au niveau du nerf optique. A côté d'unités on soutenues et d'unités off soutenues (certaines sont directionnelles), on rencontre des unités on-off. Elles sont très nombreuses et représentent la majeure partie des unités rétiniennes. Elles répondent au mouvement.

Les unités on-off non directionnelles, très nombreuses, ont des caractéristiques assez banales. Leurs champs récepteurs sont en général de petite taille (1 à 6^0 environ), pouvant parfois être plus grand (jusqu'à 17^0).

Les unités on-off directionnelles présentent des composantes on et off assez antagonistes pour inhiber plus ou moins partiellement toute réponse de la part d'une unité stimulée par une brusque variation de l'illumination générale. Quant à la directionnalité, elle peut se manifester de plusieurs manières. On ne rencontre que très peu d'unités directionnelles pures, c'est-à-dire d'unités qui ne répondent que pour un déplacement dans une direction donnée et selon un sens donné, et cela quelque soit le contraste. Ce que l'on rencontre beaucoup plus fréquemment, ce sont des unités dont la directionnalité est moins stricte. Par exemple, on rencontre des unités répondant selon les deux sens d'une direction donnée: ce sont des unités unidirectionnelles, et bi orientées. C'est le cas des unités répondant uniquement à un mouvement vertical. On rencontre aussi des unités répondant selon deux directions généralement perpendiculaires entre elles, l'une des directions pouvant être bi orientée: on a affaire alors à des unités bi directionnelles et tri orientées. Dans cette dernière catégorie on a les unités répondant par exemple à tout mouvement de direction verticale, quel que soit le sens, et, dans la direction horizontale, au seul mouvement temporo-nasal.

Il existe d'autres unités dont le comportement s'écarte notablement des précédentes. Ces unités sont relativement peu nombreuses. C'est ainsi qu'on rencontre, à côté d'unités spontanées, des unités qui présentent une activité permanente. La fréquence de décharge varie en sens inverse du niveau de la luminosité ambiante. Le comportement de telles unités évoque celui des unités de la classe V décrit par Maturana et al. (1960) chez la grenouille. Enfin, il existe des unités dont le niveau d'activité semble refléter le niveau de luminosité ambiante. Ces unités sont à rapprocher de certaines unités de la classe III décrites par Maturana et al. (1960) chez la grenouille; nous les désignerons par: unités de la classe III'.

REPONSES MESENCEPHALIQUES

Le tectum représente un aire de projection de la rétine contralatérale. Chez la truite, la partie caudale d'un tectum correspond au champ visuel temporal de l'oeil contralatéral, tandis que la partie rostrale correspond au champ nasal. De la sorte, nous confirmons les localisations obtenues initialement par Buser et Dussardier (1953), ainsi que les cartographies effectuées par Jacobson et Gaze (1964) et par Schwassmann et Kruger (1965).

Les réponses visuelles unitaires enregistrées dans le mésencéphale peuvent être classées en trois catégories: a) unités d'origine rétinienne enregistrées au niveau des couches superficielles du tectum (voir précédemment); b) unités enregistrées au niveau des couches profondes du tectum. Il s'agit de la réponse des cellules tectales; c) unités enregistrées dans le mésencéphale ventral, c'est-à-dire après que la pointe de l'électrode ait traversé la cavité ventriculaire. Il n'existe pas de différences marquées entre les unités tectales proprement dites et les unités mésencéphaliques ventrales. Nous allons les décrire rapidement.

Unités à grand champ et répondant au mouvement. - Ces unités sont très nombreuses. Elles répondent aux déplacements de mires de tailles et de formes variables. Les réponses sont non directionnelles. La caractéristique principale, c'est la grandeur du champ récepteur: de 30^0 à 60^0, et même plus. Parmi ces unités, on en rencontre qui présentent une certaine activité permanente, généralement modifiée lorsqu'on allume ou que l'on éteint la lumière de la pièce. C'est parmi ces unités à grand champ et répondant au mouvement qu'on rencontre des unités, dont le comportement correspond aux unités de similitude (sameness neurons) décrites chez la grenouille par Lettvin et al. (1961).

Unités de nouveauté. - Ces unités sont nombreuses. Elles répondent au mouvement avec un comportement d'habituation décrit par Lettvin et al. (1961) chez la grenouille (newness neurons).

Unités à activité permanente et ne répondant pas au mouvement. - Il existe des unités ne répondant pas (ou très mal) au mouvement et présentant une activité permanente aussi bien à la lumière qu'à l'obscurité. Cette activité est généralement irrégulière. Les variations de la luminosité générale produisent parfois une variation transitoire de la fréquence.

Unités binoculaires. - Ces unités sont activables par l'intermédiaire des deux yeux. Chaque champ monoculaire est de grande taille (45^0 à 120^0). La partie binoculaire commune est large et située devant l'animal. Les propriétés de chaque champ monoculaire sont semblables aux unités précédentes: réponses aux mouvements avec possibilité d'habituation, absence de directionnalité. Les réponses contralatérales sont plus importantes que celles obtenues avec l'oeil ipsilatéral.

Unités visuelles à convergences hétérosensorielles. - Il s'agit d'unités activables non seulement avec des stimulations visuelles, mais aussi avec des stimulations non visuelles. La composante visuelle présente les propriétés suivantes: champ récepteur de grande taille dont les limites sont difficiles à déterminer avec précision, réponses aux mouvements avec parfois phénomènes d'habituation, réponses plus ou moins nettes aux stimulations de type on-off. La composante non visuelle correspond à l'un des différents types de stimulations non visuelles employées au cours de cette étude.

Autres unités. - En dehors de ces unités qui représentent la majeure partie des réponses visuelles mésencéphaliques n'ayant pas une origine rétinienne directe, on peut rencontrer d'autres unités visuelles. Il existe par exemple des unités off de type soutenu. Certaines de ces unités peuvent avoir des champs récepteurs de taille assez grande. On peut rencontrer aussi des unités de type on soutenu, des unités on-off directionnelles à grand champ et présentant des phénomènes d'habituation. Il y a aussi des unités appartenant aux classes III' et V. Enfin, il existe des unités non visuelles: spontanées, vibro sensibles et unités correspondant aux téguments de la région du museau.

AUTRES REGIONS DU SNC

Les unités visuelles enregistrées en dehors du mésencéphale sont beaucoup moins nombreuses. Cependant, il est possible d'obtenir des réponses visuelles unitaires au niveau de toutes les parties de l'encéphale: télencéphale, diencéphale, cervelet, partie métencéphalique du tronc cérébral. Il ne semble pas y avoir de type particulier pour une région donnée de l'encéphale. Ces unités correspondent à celles déjà enregistrées au niveau du mésencéphale: unités à grand champ et répondant au mouvement, unités de nouveauté, unité à activité permanente et ne répondant pas au mouvement, unités visuelles à convergence hétérosensorielle. Pour ces dernières, on a une nette prépondérance des unités répondant à la stimulation de la ligne latérale.

DISCUSSION

Les différents types d'unités rétiniennes enregistrées chez la truite au cours de ce travail ne diffèrent pas beaucoup de ceux décrits chez le poisson rouge (Cronly-Dillon, 1964; Jacobson et Gaze, 1964; Daw et Beauchamp, 1972; Wartzok et Marks, 1973), chez le brochet (Zenkin et Pigarev, 1969), et même chez la plie (Hammond, 1968). Si l'on compare les réponses des cellules ganglionnaires de la truite avec celles de la grenouille, on peut remarquer qu'à côté de l'absence des détecteurs des classes I et II (Maturana et al., 1960). il existe une grande richesse en unités directionnelles. A ce sujet, il faut signaler la présence d'unités off directionnelles. De telles unités sont très rares. On en rencontre cependant quelques-unes dans le nerf optique du triton (Cronly-Dillon et Galand, 1966). En conclusion, la rétine de la truite présente un certain équipement en détecteurs. Cette particularité est peut-être en rapport avec l'organisation synaptique originale rencontrée au niveau de cette rétine (Chèze et Gas, 1973).

Au niveau du tectum, les fibres du nerf optique ne se distribuent pas selon une stratification fonctionnelle comme chez la grenouille (Maturana et al., 1960). Les réponses cellulaires enregistrées dans le tectum et le mésencéphale ventral de la truite présentent des caractéristiques qu'on rencontre parfois chez le poisson rouge: Page et Sutterlin (1970), Callens et al. (1967), Guthrie et Banks (1974). Certaines de ces réponses ont été également enregistrées chez la grenouille: Lettvin et al. (1961), Fite (1969), Grüsser et Grüsser-Cornehls (1970), Liège et Galand (1972). Quant aux unités hétérosensorielles qu'on rencontre le plus fréquemment chez la truite, elles concernent surtout la convergence avec la stimulation des téguments au niveau de la ligne latérale.

En dehors du mésencéphale, les réponses visuelles sont difficiles à obtenir chez les poissons. Par stimulation électrique du nerf optique, on peut enregistrer des potentiels évoqués, par exemple au niveau du télencéphale ou du cervelet: Veselkin et Kovachevich (1969), Bertharion et Peres (1971), Peterson et Prosser (1972). Quelques unités visuelles ont été enregistrées dans le cervelet du poisson rouge (Ingle, 1971); certaines présentaient des phénomènes de convergence (Callens et al., 1967).

Du point de vue comportement, les poissons présentent un réflexe d'origine visuelle leur permettant de conserver leur position lorsqu'ils sont soumis à un courant d'eau qui tend à les entraîner. On peut penser que les unités directionnelles participent à ce réflexe. En effet, dans la rétine de la truite, 80% des unités directionnelles horizontales sont orientées dans le sens temporo-nasal.

Par ailleurs, dans la détection du mouvement chez le poisson rouge, Ingle (1966) a été amené à distinguer deux types de mouvements conduisant à des comportements différents. La détection d'une dérive de l'environnement visuel déclenche une réaction motrice de natation comme on vient de le voir précédemment. Par contre, le mouvement rapide d'un objet peut induire un comportement de poursuite et d'attaque. Les données électrophysiologiques obtenues chez la truite semblent confirmer une telle distinction: parmi les unités répondant au mouvement enregistrées aussi bien au niveau de la rétine que du mésencéphale, il en existe qui sont spécifiquement activées par des mouvements relativement lents, tandis que d'autres le sont par des mouvements rapides.

REMERCIEMENTS

Ce travail a été réalisé grâce à des subventions financières en provenance du CNRS et de l'INSERM.

SUMMARY

Different types of retinal ganglion cells have been recorded in the trout: on units; off units, some of which are directionally sensitive; directionally and non directionally sensitive on-off units; spontaneous units; continuously activated units, the activity of some of which being function of the light intensity, as in frog. In the tectum and the deep mesencephalon, there are: newness units; sameness units; large field units responding to movement and on-off stimulations; continuously active units non responding to movement; binocular units; units responding to visual and non visual (tegumental, lateral-line, vibro-acoustic) stimulations; on or off units; class V units; spontaneous units. In the other parts of the central nervous system (telencephalon, diencephalon, metencephalon), visual units have been recorded. The main types of these units are: newness units; large field units responding to movement; continuously active units; heterosensorial (lateral line) visual units. Such results show that heterosensorial visual convergence phenomenons are largely encountered in the trout CNS.

REFERENCES

Bertharion, G. et Peres, G. (1971). Approche électrophysiologique du cervelet de la petite roussette (*Scyliorhinus canicula*). Ann. Instit. Michel Pacha, n^{0} 4: 35-38.

Buser, P. et Dussardier, M. (1953). Organisation des projections de la rétine sur le lobe optique, étudiée chez quelques Téléostéens. J. Physiol., Paris, 45: 57-60.

Callens, M., Van Denbussche, E. and Greenway, Ph. (1967). Convergence of retinal and lateral line stimulation on tectum opticum and cerebellar neurons. Arch. Int. Physiol. Biochem. 75: 148-150.

Chèze, G. et Gas, N. (1973). Organisation synaptique originale des cellules visuelles de la truite (*Salmo irideus R.*). C.R. Acad. Sci. 276D: 2821-2822.

Cronly-Dillon, J.R. (1964). Units sensitive to direction of movement in goldfish optic tectum. Nature, 203: 214-215.

Cronly-Dillon, J.R. et Galand, G. (1966). Analyse des réponses visuelles unitaires dans le nerf optique et le tectum du triton, *Triturus vulgaris*. J. Physiol., Paris, 58: 502-503.

Daw, N.W. and Beauchamp, R.D. (1972) Unusual units in the goldfish optic nerve. Vision Res. 12: 1849-1856.

Fite, K.V. (1969). Single-unit analysis of binocular neurons in the frog optic tectum. Exp. Neurol. 24: 475-486.

Galand, G., Liège, B. et Marillaud, A. (1969). Réponses visuelles unitaires chez la Truite; résultats préléminaires. J. Physiol. Paris, 61: Suppl. 1: 132-133.

Gesteland, R.C., Howland, B., Lettvin, J.Y. and Pitts, W.H. (1959). Comments on microelectrodes. Proc. Inst. Radio Engrs. N.Y. 47: 1856-1862.

Grüsser, O.J. and Grüsser-Cornehls, U. (1970). Die Neurophysiologie visuell gesteuerter Verhaltensweisen bei Anuren. Verh. Zool. Ges. 64: 201-218.

Guthrie, D.M. and Banks, J.R. (1974). Input characteristics of the intrinsic cells of the optic tectum of teleost fish. Comp. Biochem. Physiol. 47A: 83-92.
Hammond, P. (1968). Spectral properties of dark-adapted retinal ganglion cells in the plaice (*Pleuronectes platessa, L.*). J. Physiol., London, 195: 535-556.
Hubel, D.H. (1957). Tungsten microelectrode for recording from single units. Science, 125: 549-550.
Ingle, D.J. (1966). Effect of direction on movement detection by goldfish. Proc. Am. Psychol. Assoc. 2: 151-152.
Ingle, D. (1971). Comments on the article by Vesselkin, Agayan and Nomokonova. Brain, Behav. Evol. 4: 307-309.
Jacobson, M. and Gaze, R.M. (1964). Types of visual response from single units in the optic tectum and optic nerve of the goldfish. Q. J. Exp. Physiol. 49: 199-209.
Lettvin, J.Y., Maturana, H.R., Pitts, W.H. and McCulloch, W.S. (1961). Two remarks on the visual system of the frog. In Rosenblith, W.A. ed. Sensory communication, p. 757-776. Wiley, New York.
Liège, B. et Galand, G. (1970). Phénomènes de convergence au niveau de certaines unités visuelles du SNC de la truite. J. Physiol., Paris 62: Suppl. 3, 405-406.
Liège, B. and Galand, G. (1971a). Types of single-unit visual responses in the trout's optic tectum. In: Visual information processing and control of motor activity, p. 63-65. Publishing House of the Bulgarian Academy of Sciences, Sofia.
Liège, B. et Galand, G. (1971b). Réponses visuelles unitaires chez la Truite. Proc. Internat. Union Physiol. Sci. 9: 346.
Liège, B. and Galand, G. (1972). Single-unit visual responses in the frog's brain. Vision Res. 12: 609-622.
Maturana, H.R., Lettvin, J.Y., McCulloch, W.S. and Pitts, W.H. (1960). Anatomy and physiology of vision in the frog (*Rana pipiens*). J. Gen. Physiol. 43: Suppl. 129-176.
Morin, G., Gaillard, F., Liège, B. et Galand, G. (1974). Réponses proprioceptives unitaires à l'abaissement de la mâchoire enregistrées dans le tronc cérébral de la grenouille. Interactions hétéro-sensorielles visuelles. J. Physiol., Paris 68: 121-144.
Page, C.H. and Sutterlin, A.M. (1970). Visual-auditory unit responses in the goldfish tegmentum. J. Neurophysiol. 33: 129-136.
Peterson, R.H., and Prosser, C.L. (1972). The effects of cooling on electrical responses of goldfish (*Carassius auratus*) central nervous system. Comp. Biochem. Physiol. 42A: 1019-1037.
Schwassmann, H.O. and Kruger, L. (1965). Organization of the visual projection upon the optic tectum of some freshwater fish. J. Comp. Neurol. 124: 113-126.
Veselkin, N.P. et Kovachevich, N. (1969). Projections visuelles dans le télencéphale et le cervelet chez les Poissons téléostéens. Zh. Evol. Biokhim. Fiziol. 5: 416-418 (en russe).

Wartzok, D. and Marks, W.B. (1973). Directionally selective visual units recorded in optic tectum of the goldfish. J. Neurophysiol. 36: 588-604.

Zenkin, G.M. and Pigarev, I.N. (1969). Detector properties of the ganglion cells of the pike retina. Biophysics. 14: 763-772.

ELECTROPHYSIOLOGICAL ANALYSIS OF RETINO-TECTAL SYNAPSES IN TELEOSTS

Horacio VANEGAS

Departamento de Fisiologia, IVIC

Apartado 1827, Caracas 101, Venezuela

The eye obeys exactly the action of the mind.
Emerson

INTRODUCTION

In general terms, there are two visual systems in vertebrates: the retino-tectal system and the retino-geniculo-striate system. More than half of the vertebrate species are fishes, and lack a retino-geniculo-striate system. In these animals, as well as in amphibians, reptiles and birds, the optic tectum is by far the main recipient of the retinofugal axons. Hence, in an overwhelming majority of vertebrates the manner in which the tectal neurons respond to retinofugal impulses constitutes a fundamental mechanism in visual function. Some of our findings regarding this mechanism will be reviewed in the present article.

Percidae is one of the biggest family of fishes and our experiments have been done on the percomorph, *Eugerres plumieri*. In this fish, the waveform of the field potential evoked by optic nerve stimulation is similar to that found in siluroids (*Bagrus, Ameiurus)* and cyprinoids (carp, goldfish, tench) (Buser, 1955; Konishi, 1960; Sutterlin and Prosser, 1970). The tectal cytoarchitecture and ultrastructure of the berycomorph, *Holocentrus* (Schroeder et al., 1974) are similar to those of *Eugerres* (Laufer and Vanegas, 1974; Vanegas et al., 1974b). It is assumed, therefore, that the results reported herein are applicable to teleosts in general.

EXPERIMENTAL SET-UP

The fish is rigidly held outside the water while a recirculation system conveys aerated sea-water into its buccopharynx. The optic nerve is electrically stimulated just behind the eyeball, and the field and unitary potentials so elicited are recorded from the optic nerve, the optic tract and the optic tectum by means of silver macroelectrodes, stainless-steel microelectrodes, and high-impedance (20-60 megohm) micropipettes. In all recordings, positive is upwards. (Vanegas et al., 1971a).

THE INPUT TO THE TECTUM

The compound action potential of the optic nerve and tract indicates that they are made of three groups of fibres (I, II and III, Fig. 1A) with mean approximate conduction velocities of 20, 10 and 5m/sec respectively (Vanegas et al., 1971a). There is evidence (Vanegas et al., 1973) that the 20 m/sec fibre group is made of axons which originate in the optic tectum and normally conduct towards the eye (see below). It is thus considered that retino-tectal afferents account for the 10 and 5 m/sec fibre groups.

EXCITATION OF VERTICAL NEURONS

The arrival of optic nerve impulses at the tectum is signaled (Fig. 1B) by three negative peaks (waves 1, 2 and 3) which correspond to the three fibre groups just mentioned. The field potential then shows a large deflection with two peaks (waves 4 and 5), which is negative within -and above- the layers where retino-tectal afferents largely terminate (cf. Vanegas, this volume), that is stratum opticum (SO) and stratum fibrosum et griseum superficiale (SFGS) (Fig. 1C). This indicates (Vanegas et al., 1971a) that the arrival of the retino-tectal impulses sets up an active inward current at -and a depolarisation of- the apical segments of neurons which extend vertically within the tectum (Fig. 1C). This inward current at the level of the SFGS is "seen" as a negativity, and causes an outward current to flow from the deeply lying segments of these neurons, which is in turn "seen" as a positivity within the stratum griseum centrale (SGC) and deeper layers (Fig. 1C). Waves

Fig. 1. A) Compound action potential of optic tract. B) Field potential evoked in optic tectum by electrical stimulation of optic nerve. Figures indicate the various deflections. C) Depth profile of field potential. Matching letters indicate depths at which recordings were obtained. The diagramme represents current flow elicited in vertical neurons by the arrival of retino-tectal impulses. D) The

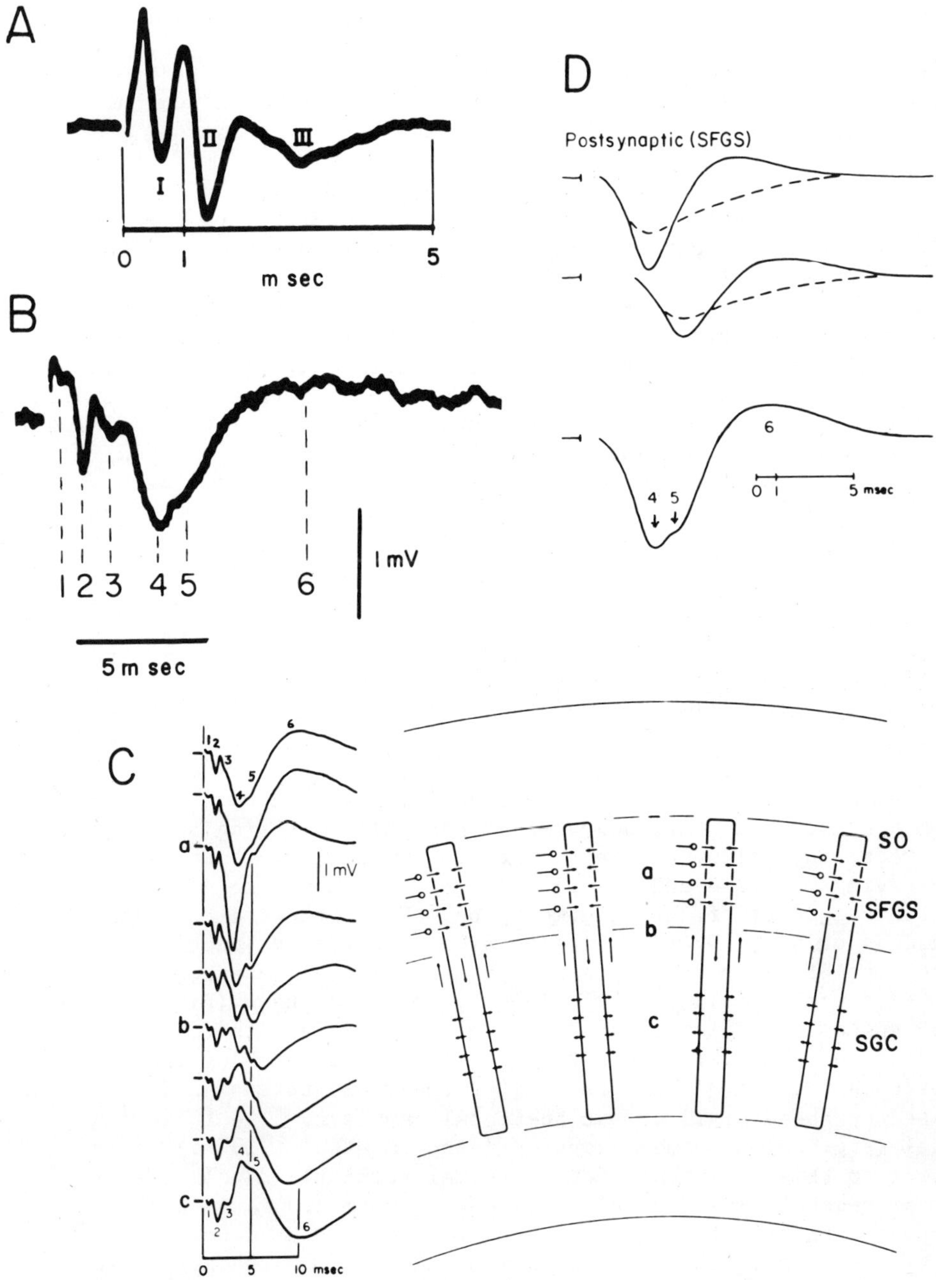

Fig. 1. (contd.) two postulated excitatory events elicited by group II and III axons. Dashed lines: graded potentials. Whole lines: regenerative potentials. Bottom trace is the sum of the two whole lines, and mimicks the field potential. (From Vanegas, 1974, and Vanegas et al., 1971a, 1971b, 1974a.)

4 and 5 thus represent two successive excitatory events (Fig. 1D) elicited by the arrival of two successive afferent volleys conducted by the axons of groups II and III (Vanegas, 1974). These two excitatory surges are also evidenced as two separate peaks of unitary neuronal discharges (Vanegas et al., 1974a).

Wave 4 can be shown (Vanegas et al., 1974a) to be made of two different components (Fig. 1D): a) a graded, EPSP-type phenomenon, which constitutes most of the initial limb of wave 4 and gives rise to b) a regenerative potential, which constitutes the peak and final limb of wave 4. The graded potential appears at low stimulus strengths, grows linearly with the stimulus and summates with a similar phenomenon elicited by a second stimulus. The regenerative potential, on the other hand,appears at a higher threshold stimulus strength, grows very steeply to a saturation level as the stimulus strength is increased, and is followed by a refractory period. The regenerative spike falls off, while the graded potential persists, at high stimulus repetition rates. There are reasons to belive that wave 5 is also made of a graded plus a regenerative potential. The phenomena which underlie waves 4 and 5, as they overlap in time, would account for the final shape of the field potential, as shown in Fig. 1D (Vanegas, 1974). It is important to emphasize that, in either case, both the graded and the regenerative potentials take place within, or very close to, the SFGS. At deeper layers only mirror images are found (Fig. 1C).

There are indications in many authors'data that, when a fine micropipette tip is recording extracellularly at a short distance from a neuron (without, however, mechanically altering the membrane), the recording faithfully represents the excitatory state of the underlying membrane and its relationship to the state of other, distant segments of the same membrane. Under these circumstances (Vanegas et al., 1974a), unitary spikes recorded at tectal layers below the SFGS generally are 3-7 mV in amplitude and show a brief, positive phase, followed by a longer, negative phase (Fig. 2A). Such spikes are elicited at latencies compatible with monosynaptic excitation. This brings us to the question of which of the tectal neurons are the ones responsible for the field potential deflections caused by the arrival of retino-tectal impulses. The following model (Fig. 2B) has been proposed (Vanegas et al., 1974a) as an answer to this question: Retino-tectal terminals depolarize the apical segments (within the SFGS) of fusiform and large pyriform neurons (Vanegas, this volume), and this fires the nearby axon 's initial segment. These two phenomena account for the graded and the regenerative components of the field potential, respectively. The regenerative potential at apical segments is a fast inward current which causes an outward current to flow from neuronal regions located at deeper tectal layers, and this is "seen" as the positive phase of the unitary action potential. Subsequently, the depolarization invades (regeneratively or electrotonically) the deeply located neuronal region, thus giving rise to the slower, negative phase of the unitary action potential.

Also the pyramidal neurons are synapsed upon by retino-tectal afferents (Laufer and Vanegas, 1974) and they most likely contribute to the field potential, since they are vertically elongated across the tectum and their axon takes off from within the SFGS.

TECTO-OCULAR NEURONS

Within the SFGS, neurons are found (Vanegas et al., 1973) which are fired antidromically by the optic nerve stimulus, thus showing that their axons normally conduct towards the eye (Fig. 2C). The conduction velocity of such axons is 20 m/sec approximately, and they thus account for group I of optic nerve fibres. Subsequent to the antidromic action potential, these neurons show an EPSP and action potential which correspond in latency to wave 4, i.e., to the earliest postsynaptic events of the field potential evoked by optic nerve stimulation. Tecto-ocular neurons thus constitute a monosynaptic feedback loop which reaches the eye at twice the velocity of the fastest retinofugal impulses.

CONCLUDING REMARKS.

In summary, then, the arrival of retino-tectal impulses along fast (10 m/sec) and slow (5 m/sec) axons excites: 1) The apical segments of deep-lying vertical neurons. The evidence suggests that these neurons are the fusiform and large pyriform, whose axons probably terminate in extra-tectal structures (Vanegas, this volume). This axon originates from the apical shaft within the territory of retino-tectal synapses, and this fact probably provides a large safety factor for the relay of visual information. 2) The pyramidal neurons. The axons of these neurons distribute intratectally in broad horizontal fields at mid-SGC (Vanegas, this volume). And 3) the tecto-ocular neurons. These are few in number, and their anatomical identity is still unknown.

Monosynaptic excitation of tectal neurons sets up inhibitory and excitatory loops, the action of which upon tectal neurons is apparent from 10 to 100 msec after the optic nerve stimulus (Vanegas, 1974).

ACKNOWLEDGEMENTS

The studies summarised herein were done in association with J.

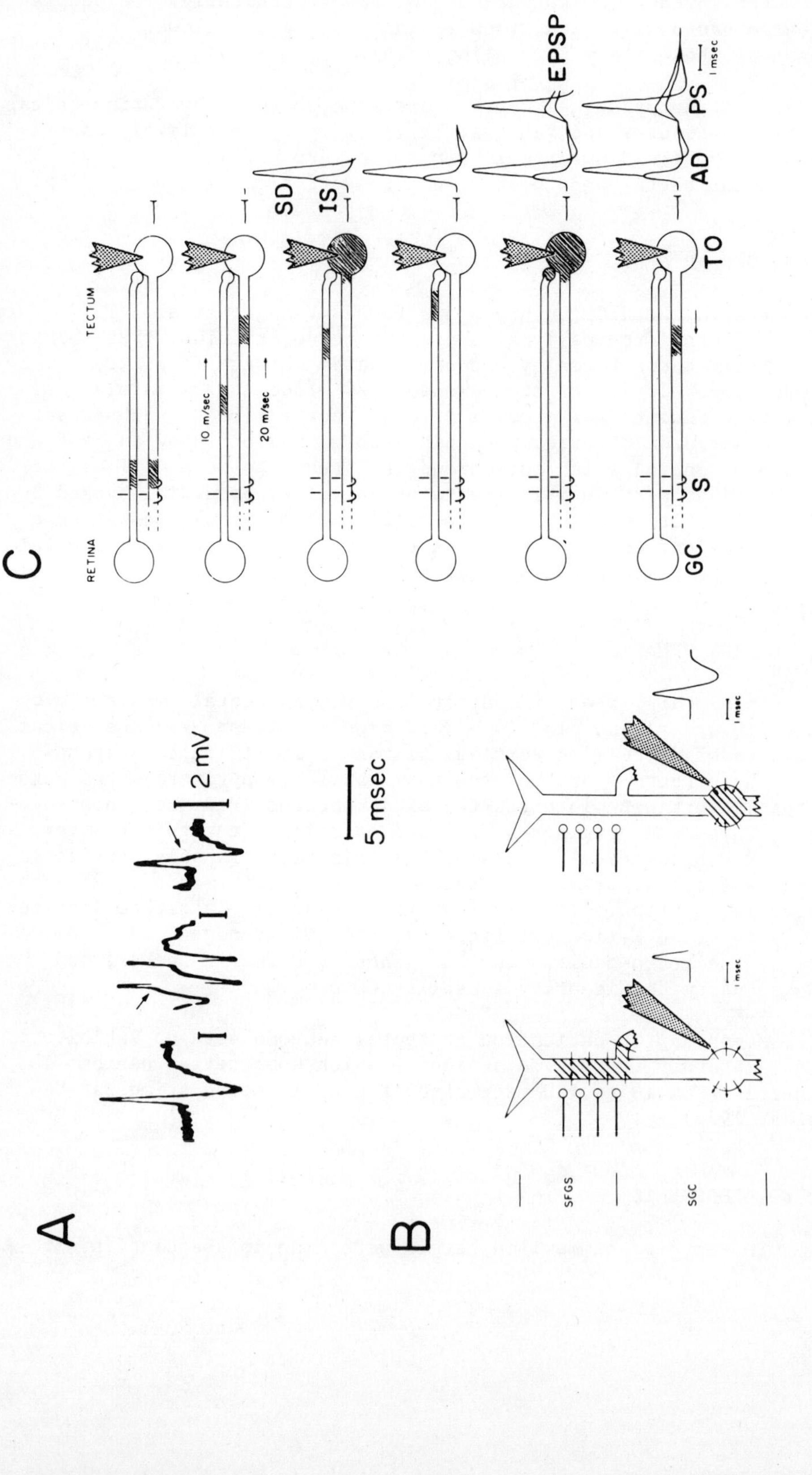
A
2 mV
5 msec
B
SFGS
SGC
1 msec
1 msec
C
RETINA
TECTUM
10 m/sec
20 m/sec
SD
IS
EPSP
AD
PS
1 msec
GC
S
TO

Fig. 2. A. Unitary extracellular action potentials recorded at deep tectal layers by means of fine micropipettes. B. The diagram represents a fusiform or a large pyriform neuron (Vanegas, this volume). By means of a fine micropipette (shaded) a recording (adjacent trace) is obtained from the neuron's perisomatic region. Retinotectal terminals depolarise the apical dendritic shaft and fire the axon's initial segment (cross-hatched). Subsequently, the deep neuronal segments are depolarised (cross-hatched). C.Diagram to show elicitation of antidromic (AD) and postsynaptic (PS) spikes in a tecto-ocular (TO) neuron. GC: retinal ganglion cell. S: electrodes stimulating the optic nerve. Action potentials (cross-hatched) are triggered on group I and II axons. A micropipette (shaded) records from TO neuron and the sequence of events (from top downwards) is recorded in the adjacent traces. IS: spike of axon's initial segment. SD: spike of somatodendritic membrane (From Vanegas, 1974, and Vanegas et al., 1974a).

Amat, E. Essayag-Millan and M. Laufer. To them all I wish to express my gratitude. Part of this work was supported by grand DF-S1-0313 from CONICIT, Venezuela.

REFERENCES

Buser, P. (1955) Analyse des réponses électriques du lobe optique à la stimulation de la voie visuelle chez quelques vertébrés inférieurs. (These) Mason, Paris.

Konishi, J. (1960) Electric responses of visual center to optic nerve stimulation in fish. Jap. J. Physiol. 10: 28-41.

Laufer, M. and Vanegas, H. (1974) The optic tectum of a perciform teleost. II. Fine structure. J. Comp. Neurol. 154: 61-96.

Schroeder, D.M., Butler, A.B. and Ebbesson, S.O.E. (1974) Cytoarchitecture and ultrastructure of the optic tectum in the teleost *Holocentrus* (In manuscript).

Sutterlin, A.M. and Prosser, C.L. (1970) Electrical properties of goldfish tectum. J. Neurophysiol. 33: 36-45.

Vanegas, H. (1974) Basic relationships for visual information processing in optic tectum of fish. Acta Cient. Venez. (in press).

Vanegas, H., Amat, J. and Essayag-Millan, E. (1973) Electrophysiological evidence of tectal efferents to the fish eye. Brain Res. 54: 309-313.

Vanegas, H., Amat, J. and Essayag-Millan, E. (1974a) Postsynaptic

phenomena in optic tectum neurons following optic nerve stimulation in fish. Brain Res. 77: 25-38.
Vanegas, H., Essayag-Millan, E. and Laufer, M. (1971a). Response of the optic tectum to electrical stimulation of the optic nerve in the teleost *Eugerres plumieri*. Brain Res. 31: 107-118.
Vanegas, H., Essayag-Millan, E. and Laufer, M. (1971b). Laminar profile analysis of the tectal evoked response in the teleost *Eugerres plumieri*. Acta Cient. Venez. 22:82-85.
Vanegas, H., Laufer, M. and Amat, J. (1974b). The optic tectum of a perciform teleost. I. General configuration and cytoarchitecture. J. Comp. Neurol. 154: 43-60.

RETINOTECTAL SPECIFICITY IN ADULT GOLDFISH

S.C. Sharma

Department of Ophthalmology, New York Medical College

New York, New York 10029, U.S.A.

O, there has been much throwing about of brains.
Hamlet, Act II, Sc. 2

The concept of neuronal specificity has evolved historically around the possible developmental mechanisms of the nervous system, in particular the formation of nerve circuits (Sperry, 1951, 1965; Gaze, 1970; Jacobson, 1970). The most widely held view of the mechanism of selective connexions between nerve fibres is Sperry's hypothesis of neuronal chemoaffinity. This states that during development the cells of the retina and optic tectum acquire individual cytochemical properties, that those neurons with like or matching cytochemical specificities become linked together, and that synaptic connexions are thus formed in a topographical order. Sperry (1945, 1951) further suggested that the cytochemical labelling of individual neurons is achieved through a polarized field type of differentiation of the retina and the optic tectum separately along the two main embryonic axes. Evidence supporting the polarized field type of specification of retinal neurons came from behavioural study of animals whose eye cup had been rotated 180° at various stages of development (Stone, 1944, 1960; Székély, 1954). These experiments on amphibians showed that the animal acquires the nasotemporal axis of its eye before it is specified along the dorsoventral axis and the results have been confirmed electrophysiologically (Jacobson, 1967, 1968).

Sperry's hypothesis requires that the tectum as well as the retina be specified and in a complementary fashion. Regarding the question of the time of tectal specification, no experimental confirmation is available to date, to my knowledge.

Anatomical and electrophysiological studies have revealed that there is a continuous projection of the retinal surface on to the tectal surface by optic nerve fibres. Thus, one locus on the retinal surface projects fibres to one corresponding place on the tectal surface. The techniques employed to study the retinal projections over the optic tectum have been elegantly described by Dr. Schwassman earlier in these proceedings. The simplicity of this visuotectal mapping enables us to investigate the mode of formation and reformation of the retinotectal connexions. After sectioning of the optic nerve in goldfish, the retinal axons regenerate and an orderly re-establishment of the normal topographic retinotectal projection occurs (Sperry, 1942 and Weiller, 1966, behaviourally; Attardi and Sperry, 1963, histologically; Gaze, 1970, electrophysiologically).

RETINOTECTAL SIZE DISPARITY IN GOLDFISH

Attardi and Sperry (1963) introduced a controlled abnormality into the relationship between the eye and the optic tectum in adult goldfish by removing part of the retina and cutting the optic nerve. Using histological techniques, the authors showed that after regeneration, the remaining retinal fibres innervated the appropriate region of the tectum, in certain situations, bypassing the inappropriate tectal areas.

The electrophysiological evidence supporting terminal specific regeneration came from the studies of Jacobson and Gaze (1965), in which they created size disparity between the optic tectum and the retina in goldfish by removing the lateral or medial optic tectum and allowing all optic axons to regenerate into the remaining tectum. The results indicated that fibres returned only to the appropriate positions on the tectum.

However, with longer survival periods following surgery, the experiments on the goldfish visual system provided strong suggestions of plasticity in retinotectal connexion (Sharma, 1967; Gaze and Sharma, 1970). After removal of the caudal half of the optic tectum, it was found that a complete and retino-topically organised visual map could be formed over the remaining rostral half tectum. The visual maps were compressed by a factor of two into the rostrocaudal axis of the tectum (Sharma, 1972a; Yoon, 1971). The evidence that plasticity of retinotectal connexion exists in both the rostrocaudal and medio-lateral axes of the goldfish tectum was shown by Yoon (1971) and Sharma (unpublished observations).

Yoon (1972a) further extended these experiments on goldfish and demonstrated that compression of the visual projection map is a reversible phenomenon. By inserting a barrier to separate the two halves of the tectum, the compressed map was found in the rostral tectum. After removal of the barrier, the optic fibers gradually extended and expanded the projection to their normal limits.

The differences in the results of Jacobson and Gaze (1965) and Sharma (1972a) and Yoon (1972a) could be attributed to the time allowed to elapse between the surgery and the final electrophysiological recordings. Sharma (1972c) examined the projection maps of adult goldfish at different intervals after cutting the optic nerve and the removal of a longitudinal strip of rostrocaudal tectum. Visual projections mapped one month after the surgery revealed scotoma corresponding to the ablated tectum. Subsequent maps recorded four to five months after operations showed complete representation of the visual fields on to the residual tectum. Yoon (1972b) further showed that the compression and expansion phenomenon is a gradual process. These results suggested that optic nerve fibres had initially returned to their proper places in the tectum (as shown by Attardi and Sperry, 1963, and by Jacobson and Gaze, 1965), but later the optic axons redistributed in such a manner as to accommodate all optic axons into the residual tectum. Complementary studies by Horder (1971) and Yoon (1972b), in which half of the retina was ablated, showed that the remaining half-retina redistributed over the whole tectum.

Jacobson and Hunt (1973) succeeded in projecting a transplanted embryonic eye and a normal eye into the same tectum (also see Sharma, 1972d), thus mingling the fibres of the two optic nerves, which shared similar tectal positions. It is possible with this technique to explore the retinal specificities of the experimental eye and compare them to the specificities of a normal eye. The evidence that regenerating optic axons can share a common tectal locus terminal in adult goldfish came from the recent work of Sharma (1973). After complete removal of one optic tectum in adult goldfish, the regenerating optic axons entered the remaining tectum and innervated it retinotopically; normal and ipsilateral retinal projections shared the common tectal locus terminals (Sharma, 1973).

The conclusion which can be drawn from size disparity experiments on retinotectal connexion in adult goldfish is that regenerating axons, resulting in either compression or expansion of the visuotectal map, retain a correct polarity.

The mechanism underlying polarity is completely unclear, although the acquisition of new specificities by tectal or retinal cells following experimental ablation is proposed by Yoon (1972b).

The experiments discussed above raise numerous questions. A few of them are:

1. Are optic axonal connexions, initially formed after regeneration, exchanged for new sites to allow the compression or expansion to occur?
2. Are new cells formed, following removal of the tectal pieces, to allow replacements?
3. Are new connexions synaptically functional?
4. What are the functional consequences of visual compression in the half tectum?

It is important to mention that the methods used so far to explore retinotectal connexion, especially electrophysiological methods, show only the relative orderly distribution of the retinal axons. Present techniques cannot determine whether or not new properties are acquired by the tectal cells following experimental lesions or ablations. Similarly, experimental work to date has not explored the changes (either electrophysiological or anatomical) which might occur during the process of regeneration. Present data are solely based on the presynaptic recordings of the optic terminals in the tectum. Future experiments should be designed to record single units in the optic terminals for the details of types of units in the regenerated axons and also post-synaptic units from the tectal cells. To further understand the rules of regrowth of optic axons into the tectum, implanted electrode techniques to record repeatedly from one or more positions would be necessary.

EVIDENCE OF TECTAL SPECIFICITY

Regarding the development of topographic polarization of the optic tectum, no evidence to date supports the notion of independent tectal specification. However, indications that the adult optic tectum does have topographic polarity stem from the data reported below.

Gaze, Jacobson and Sharma (1966) and Sharma (1967) showed that when a part of the optic tectum was excised and reimplanted in situ in adult goldfish, the reimplanted tectal tissue became innervated by optic axons, thereby restoring the normal retinotopic projection over the tectal graft. We further reported (Sharma and Gaze, 1971) that when tectal pieces were reimplanted by 90° clockwise rotation, the reinnvervated visual projection in the graft exhibited 90° rotated visuotectal maps. These experiments have been confirmed recently in adult goldfish by Yoon (1973). Furthermore, reimplantation of the tectal pieces with 180° rotation resulted in the inversion of both axes in the regenerated map over the graft as compared to the normal map outside the graft (Yoon, 1973; Sharma, 1967 and unpublished observations). These experiments confirm the notion that tectal polarization existed before the tectal reimplantations were performed and that the tectal reimplants retained their original polarity after rotation.

In both retinotectal size disparity experiments and tectal reimplant experiments, the topographic order of original polarity is retained by the regenerating axons. The mechanisms responsible for the retention of topographic polarity in retinotectal connexion, leading to the reformation of visuotectal maps by the regenerating optic axons, are only speculative at present.

Author is supported by N.S.F. Grant No. G. B. 43506.

REFERENCES

Attardi, D.G. and R.W. Sperry. (1963) Preferential selection of central pathways by regenerated optic fibers. Exp. Neurol. 7: 46-64.

Gaze, R.M. (1970). Formation of nerve connections. Academic Press, New York.

Gaze, R.M., M. Jacobson, S.C. Sharma. (1966). Visual responses from the goldfish brain following excision and reimplantation of the optic tectum. J. Physiol. (Lond.) 183: 38-39.

Gaze, R.M. and S.C. Sharma. (1970). Axial differences in the reinnervation of the goldfish tectum by regenerating optic nerve fibers. Exp. Brain Res. 10: 171-181.

Horder, T. (1971). The course of recovery of the retinotectal projection during regeneration of the fish optic nerve. J. Physiol. (Lond.) 217: 53-54.

Jacobson, M. (1967). Retinal ganglion cells: specification of central connections in larval Xenopus laevis. Science. 155: 1106-1108.

Jacobson, M. (1968). Development of neuronal specificity in retinal ganglion cells of Xenopus. Dev. Biol. 17: 202-218.

Jacobson, M. (1970). Developmental Neurobiology. Holt, Rinehart and Winston, New York.

Jacobson, M. and R.K. Hunt. (1973). The origin of nerve cell specificity. Sci. Am. 228: 26-35.

Jacobson, M. and R.M. Gaze. (1965). Selection of appropriate tectal connections by regenerating optic nerve fibers in adult goldfish. Exp. Neurol. 13: 418-430.

Sharma, S.C. (1967). Neural development and regeneration in the visual system of teleosts. Ph.D. Thesis, University of Edinburgh, Scotland.

Sharma, S.C. (1972 a). The reformation of retinotectal projections after various tectal ablations in adult goldfish. Exp. Neurol. 34: 171-182.

Sharma, S.C. (1972 b). The restoration of visual projections following tectal lesions in goldfish. Exp. Neurol. 35: 358-365.

Sharma, S.C. (1972 c). Redistribution of visual projection in altered optic tecta of adult goldfish. Proc. Nat. Acad. Sci. 69: 2637-2639.
Sharma, S.C. (1972 d). Retinotectal connections of a heterotopic eye. Nature (New Biology) 238: 286-287.
Sharma, S.C. (1973). Anomalous retinal projections after removal of contralateral optic tectum in adult goldfish. Exp. Neurol. 41: 661-669.
Sharma, S.C. and R.M. Gaze. (1971). The retinotopic organization of visual responses from tectal reimplants in adult goldfish. Arch. Ital. Biol. 109: 357-366.
Sperry, R.W. (1942). Reestablishment of visuomotor coordination by optic nerve regeneration. Anat. Rec. 84: 470.
Sperry, R.W. (1945). Restoration of vision after uncrossing of optic nerves and after contralateral transposition of the eye. J. Neurophysiol. 8: 15-28.
Sperry, R.W. (1951). Mechanism of neural maturation. PP. 236-280 in Handbook of Experimental Psychology. S.S. Stevens (ed.) Wiley; New York.
Sperry, R.W. 1965. Embryogenesis of behavioral nerve nets. In: Organogenesis. Edited by R.L. DeHaan and H. Ursprung. Holt, Rinehart and Winston, New York.
Stone, L.S. (1944). Functional polarisation in retinal development and its reestablishment in regenerating retina of rotated grafted eyes. Proc. Soc. Exp. Biol. Med. 57: 13-14.
Stone, L.S. (1960). Polarization of the retina and development of vision. J. Exp. Zool. 145: 85-96.
Székély, G. (1954). Zür ausbildung der lokalen functionellen specifitat der retina. Acta. Biol. Hung. 5: 157-167.
Weiler, I.J. (1966). Restoration of visual acuity after optic nerve section and regeneration in Astronotus ocellatus. Exp. Neurol. 15: 377-386.
Yoon, M. (1971). Reorganization of retinotectal projection following surgical operations on the optic tectum in goldfish. Exp. Neurol. 33: 395-411.
Yoon, M. (1972 a). Reversability of the reorganization of retino-tectal projection in goldfish. Exp. Neurol. 35: 565-577.
Yoon, M. (1972 b). Transposition of the visual projection from the nasal hemiretina into the foreign rostral zone of the optic tectum in goldfish. Exp. Neurol. 37: 451-462.
Yoon, M. (1973). Retention of the original topographic polarity by 180° rotated tectal reimplants in young adult goldfish. J. Physiol. (Lond.) 233: 575-588.

CYTOARCHITECTURE AND CONNEXIONS OF THE TELEOSTEAN OPTIC TECTUM

Horacio VANEGAS

Departamento de Fisiologia, IVIC

Apartado 1827, Caracas 101, Venezuela

In God's Name, What art thou?
King Richard The Third, Act I, Sc. 1V

INTRODUCTION

More than a plain "roof" the teleostean optic tectum is an ample double-dome whose sides are supported by the midbrain tegmentum and whose front edges lean upon the diencephalon. It is made of several thin concentric strata (Fig. 1) which are known as: stratum marginale (SM), stratum opticum (SO), stratum fibrosum et griseum superficiale (SFGS), stratum griseum centrale (SGC), stratum album centrale (SAC) and stratum periventriculare (SPV). In general terms, three types of neurons are found within such strata (Fig. 1): vertical neurons, which have a long axis oriented radially with respect to the dome's curvature; horizontal neurons, which extend in the plane of the layers; and multipolar neurons. Needless to say, the handling of information which the tectum receives from the retina, the confrontation of such information with data which arrive on - line from other neural areas plus data which are retrieved from past experience, and the addressing of messages which will ultimately result in the animal's behaviour, are all based upon the manner in which tectal neurons extend and interconnect, upon the distribution of afferent terminals among the neurons's processes, and upon the targets which finally receive the tectal outflow. Such aspects will be reviewed in the present article.

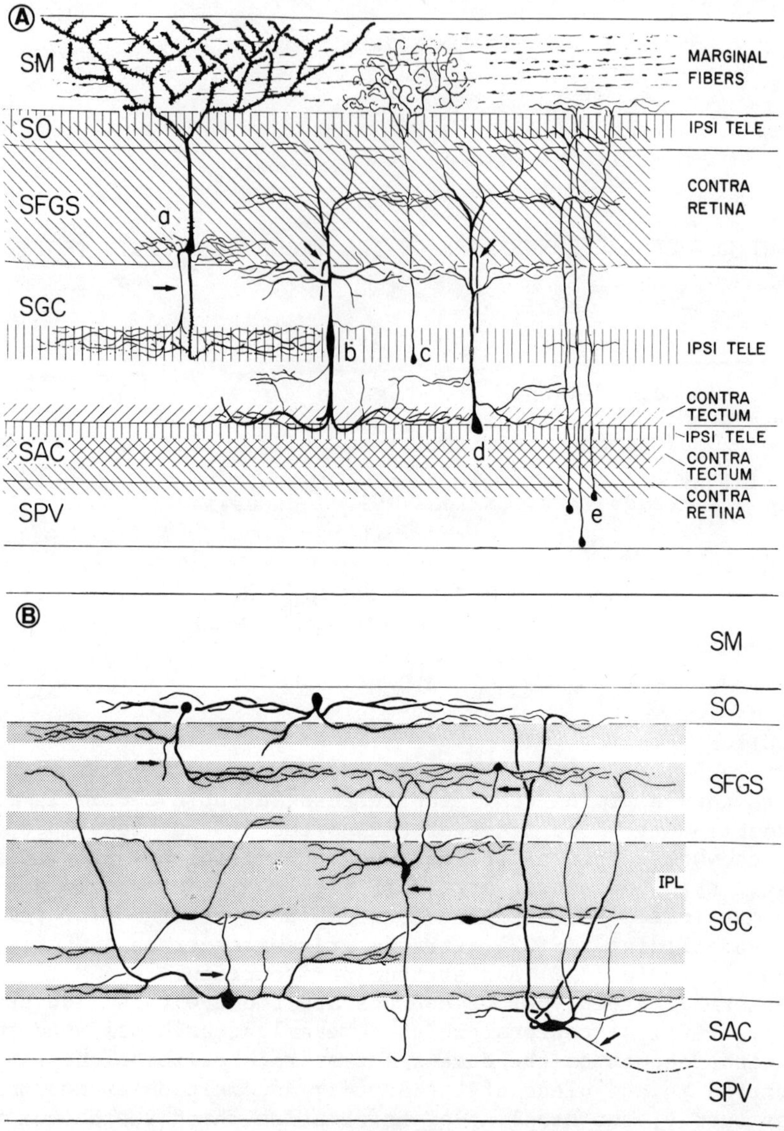

Fig. 1. A. The main types of vertical neurons in the optic tectum of *Eugerres*, and the distribution of some afferent axons and terminals in *Eugerres* and *Holocentrus*. a: pyramidal neuron. b: fusiform neuron. c: small pyriform neuron. d: large pyriform neuron. e: periventricular neuron. The sources of afferents are indicated on the right. B. The main types of horizontal and multipolar neurons, and the location of plexiform layers (shaded) in the optic tectum of *Eugerres*. Diagrammes A and B overlap completely. IPL: inner plexiform layer. SAC: stratum album centrale. SFGS: str. fibrosum et griseum superficiale. SGC: str.

CYTOARCHITECTURE

The following account is based mostly upon studies made on the perciform *Eugerres* by means of Golgi methods (Vanegas et al., 1974) b). The description is applicable, in general, to *Barbus* (Cajal, 1955), *Carassius* (Leghissa, 1955), *Holocentrus* (Schroeder et al., 1974) and *Bagrus* (Vanegas, unpublished).

The main types of vertical neurons are the pyramidal, the fusiform, the small and the large pyriform, and the periventricular neurons (Fig. 1A).

The pyramidal neuron's soma (Fig. 1A) is located just above the deep boundary of the SFGS. From this soma stems a long and thick apical dendritic shaft which reaches the SO, divides into two branches, and then gives rise, by dicotomy, to an impressive dendritic tree which spreads through the SM towards the pia mater. The apical dendritic shaft is naked except near the soma, where spine-like processes can be seen which probably correspond to the retinotectal terminals observed with the electron-microscope (Laufer and Vanegas, 1974a, 1974b). The first two dendritic branches are also naked, but the rest of the dendritic branches contain abundant spines. Such spines are synapsed upon by the passing marginal fibres (see below). From the soma's deep pole stems the basal dendritic shaft. This is a spiny shaft, thinner than its apical counterpart, which branches out horizontally at the mid-level of the SGC, just at the deep boundary of Leghissa's (1955) inner plexiform layer. Also from the soma's deep pole, thin basal dendrites stretch out horizontally. The axon originates from the soma's basal pole, from a basal dendrite, or from the basal dendritic shaft. This axon swings downwards, becomes very thin, then regains its previous thickness, and branches out horizontally within the same territory where the branches of the basal dendritic shaft are distributed.

The fusiform neuron (Fig. 1b) has an elongated soma located within the SGC. Its apical dendritic shaft extends into the SFGS as far as the SO, and gives rise to vertical, as well as horizontal, branches. The basal dendritic shaft divides into thick horizontal branches at the boundary between the SGC and the SAC. The axon stems from the apical dendritic shaft within, or very close to, the SFGS, and has downward arciform course.

The soma of the large pyriform neuron (Fig. 1d) is located in the deep region of the SGC or in the SAC. The long and thick apical

←

griseum centrale. SM: str. marginale. SO: str. opticum. SPV: str. periventriculare. Arrows indicate axons. (Combined data from Vanegas and Ebbesson, 1973; Ebbesson and Vanegas, 1974; Vanegas and Ebbesson, 1974; and Vanegas et al., 1974b).

dendritic shaft reaches into the SFGS, sometimes as far as the SO. Several branches are issued from it into the plexiform layers of the tectum. The axon originates form the dendritic shaft within, or very close to, the SFGS, and arches downwards, sometimes giving off an ascending collateral.

The axons of fusiform and large pyriform neurons have been seen to leave the optic tectum (Cajal, 1955; Leghissa, 1955).

The small pyriform neuron's soma (Fig. 1c) is located within the SGC. It sends a slender shaft which reaches the SM, where it terminates in profuse slender and tortuous branches. No axon has been seen to arise from this neuron.

Finally, the periventricular neurons' somata (Fig. 1e) are small and tightly packed together. Each soma issues a thin process which terminates in horizontal branches at levels that vary from mid-SFGS to the superficial boundary of the SO. In *Eugerres* no axon has been seen to arise from typical periventricular neurons.

Horizontal neurons (Fig. 1B) are found particularly in the boundaries SO/SFGS and SGC/SAC, as well as in the mid-levels of SFGS and of SGC. These neurons branch out extensively and contribute significantly to the plexiform layers' meshwork. The destination of their axons is unknown.

The multipolar neurons (Fig. 1B) are found at the SGC and the SAC, and their dendrites may reach into various tectal strata, usually contributing to several plexiform layers. An axon can generally be recognised, but the destination of such axons is not known.

AFFERENT CONNECTIONS

The following account is based only upon results obtained by means of Nauta and Fink-Heimer techniques as applied to *Holocentrus* (Campbell and Ebbesson, 1969; Vanegas and Ebbesson, 1974), *Carassius* (Sharma, 1972) and *Eugerres* (Vanegas and Ebbesson, 1973; Ebbesson and Vanegas, 1974). The description is, in general, also applicable to *Gymnothorax* and *Opsanus* (Ebbesson, 1968).

The retino-tectal axons (Fig. 1A) are totally crossed and enter the tectum via the dorsomedial and ventrolateral optic tracts. These axons directly invade, and terminate at, the SO and the SFGS, but do not reach the SGC. The retino-tectal terminals make Gray type-I synapses upon the pyramidal neurons of the SFGS (Laufer and Vanegas, 1974a) as well as on several dendritic profiles which probably belong to deeply lying vertical neurons. Electro-physiological data suggest (Vanegas et al., 1974a) that the fusiform and large pyriform neurons are, in fact, synapsed upon by retino-tectal afferents within the SFGS. A few retino-tectal afferents course along the SAC and terminate within or near this layer. Retino-tectal afferents are not found in the SM or the SPV except in the immediate vicinity of the terminal areas mentioned above.

Another source of afferents to the optic tectum is, of course, the contralateral tectal hemisphere (Fig. 1A). Such commissural fibres course along the SAC exclusively, and terminate in the neighbouring, deep region of the SGC in a zone which is homotopic to the fibres' origin (Ebbesson and Vanegas, 1974).

The telencephalon projects onto the ipsilateral optic tectum (Fig. 1A). A small contigent of axons runs along, and terminates within, the SO. The majority of axons, however, course along the SAC; some axons terminate within this layer but most of them ascend and terminate in the mid-level of the SGC, just on the deep boundary of Leghissa's inner plexiform layer (Vanegas and Ebbesson, 1974).

One of the most characteristic fibre systems of the teleostean optic tectum is that of the marginal fibres (Laufer and Vanegas, 1974a; Vanegas et al., 1974b). These are unmyelinated axons which course the SM (Fig. 1A) and make Gray type-I synapses of the spines of the pyramidal neurons' apical dendritic branches. The origin of the marginal fibres is still uncertain, although Leghissa (1955) states that they arise from small neurons located in the SFGS. Such neurons have not been found in *Barbus* (Cajal, 1955), *Eugerres* (Vanegas et al., 1974b) or *Holocentrus* (Schroeder et al., 1974).

In summary, then, different afferent systems terminate mainly at different tectal levels. Therefore, vertical neurons, which extend like a column across many tectal layers, potentially receive specific synaptic inputs upon specific "bands" of their somatodendritic membrane. The synaptic actions brought about by each of these input systems would thus be integrated lenghtwise along the neuron in a way that will finally determine the neuron's output. Also, if different input systems have different synaptic transmitters, such substances -as well as their related metabolic systems - should be detectable at specific tectal layers.

EFFERENT CONNEXIONS

The efferent connexions of the teleostean optic tectum (Fig. 2) have been studied by means of Nauta and Fink-Heimer techniques in *Holocentrus* and *Eugerres* (Ebbesson and Vanegas, 1974).

The majority of tectofugal axons funnel down into the ipsilate-

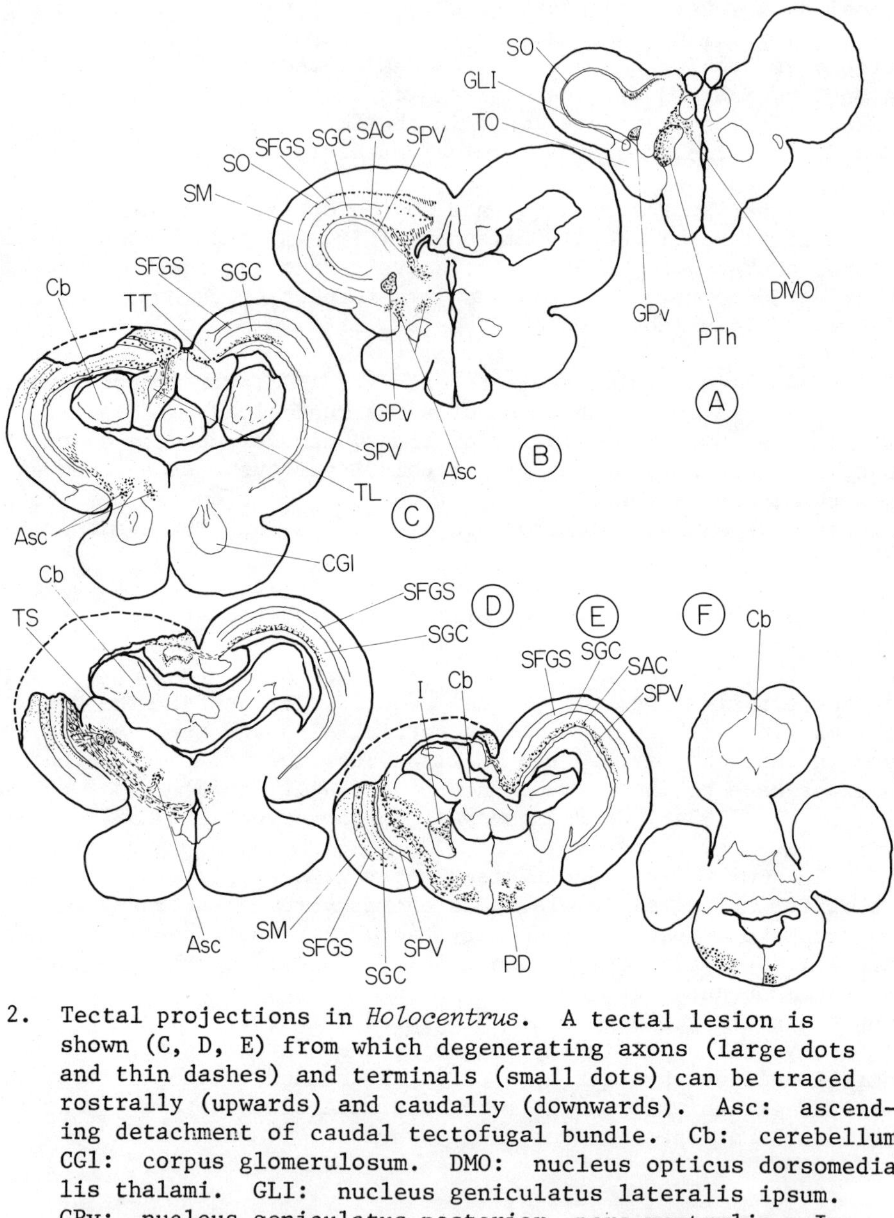

Fig. 2. Tectal projections in *Holocentrus*. A tectal lesion is shown (C, D, E) from which degenerating axons (large dots and thin dashes) and terminals (small dots) can be traced rostrally (upwards) and caudally (downwards). Asc: ascending detachment of caudal tectofugal bundle. Cb: cerebellum. CGl: corpus glomerulosum. DMO: nucleus opticus dorsomedialis thalami. GLI: nucleus geniculatus lateralis ipsum. GPv: nucleus geniculatus posterior, pars ventralis. I: nucleus isthmi. PD: predorsal bundle. PTh: nucleus prethalamicus. SAC: str. album centrale. SFGS: str. fibrosum et griseum superficiale. SGC: str. griseum centrale. SM: str. marginale. SO: str. opticum. SPV: str. periventriculare. TL: torus longitudinalis. TO: tractus opticus. TS: torus semi-circularis. TT: tecto-tectal fibres. (Modified from Ebbesson and Vanegas, 1974).

ral midbrain tegmentum (Fig. 2D) and many of them terminate here upon dorsolateral cell groups, upon a large nucleus which has been provisionally labelled "nucleus isthmi" (E) and upon the midbrain reticular formation (D, E). One contingent of axons leaves the mainstream (D), ascends towards the diencephalon and terminates in the posterior geniculate nucleus, pars ventralis (B) and the prethalamic nucleus (A). The bulk of descending fibres, however, proceeds medial - and caudalwards (E), and divides into an ipsilateral and a contralateral component. The ipsilateral component descends as far as the medulla and gives terminals to the lateral reticular formation all along its course (E, F). The contralateral descending component (the predorsal bundle) terminates upon the medial reticular formation of the pons and the medulla (E, F).

Some efferents ascend along the SO and the SAC, enter the pretectum and the dorsomedial thalamus, and terminate upon the dorsomedial optic nucleus of the thalamus and other, smaller cell groups (A, B). Many such ascending axons extend rostrally and terminate upon the prethalamic nucleus(A).

Finally, a distinct fibre group extends along the SAC towards the midline and gives abundant terminals to the torus longitudinalis (C). Then it crosses the midline, runs along the contralateral SAC, and terminates on the deep boundary of SGC (C, D, E), as mentioned above.

ACKNOWLEDGEMENTS

Most of the studies summarised herein were done in association with J. Amat, S.O.E. Ebbesson and M. Laufer. To them all I wish to express my gratitude. Part of this work was supported by grant DF-S1-0313 from CONICIT, Venezuela.

REFERENCES

Cajal, S. Ramon y (1955). Histologie du Système Nerveux de l'Homme et des Vertébrés, Vol. II, Consejo Superior de Investigaciones Cientificas, Madrid.

Campbell, C.B.G. and Ebbesson, S.O.E. (1969). The optic system of a teleost: *Holocentrus* re-examined. Brain Behav. Evol. 2: 415-430.

Ebbesson, S.O.E. (1968). Retinal projections in two teleost fishes (*Opsanus tau* and *Gymnothorax funebris*). An experimental study with silver impregnation methods. Brain Behav. Evol. 1: 134-154.

Ebbesson, S.O.E. and Vanegas, H. (1974). Projections of the optic tectum in two teleost species. (In manuscript).

Laufer, M. and Vanegas, H. (1974a). The optic tectum of a perciform teleost. II. Fine structure. J. Comp. Neurol. 154: 61-96.
Laufer, M. and Vanegas, H. (1974b). The optic tectum of a perciform teleost. III. Electronmicroscopy of degenerating retino-tectal afferents. J. Comp. Neurol. 154: 97-116.
Leghissa, S. (1955). La structtura microscopica e la cito-architettonica del tetto ottico dei pesci teleostei. Z. Anat. Entwicklungsgesch. 118: 427-463.
Schroeder, D.M., Butler, A.B. and Ebbesson, S.O.E. (1974). Cytoarchitecture and ultrastructure of the optic tectum in the teleost *Holocentrus*. (In manuscript).
Sharma, S. (1972). The retinal projections in the goldfish. An experimental study. Brain Res. 39: 213-223.
Vanegas, H., Amat, J. and Essayag-Millan, E. (1974a). Postsynaptic phenomena in optic tectum neurons following optic nerve stimulation in fish. Brain Res. 77: 25-38.
Vanegas, H. and Ebbesson, S.O.E. (1973). Retinal projections in the perch-like teleost *Eugerres plumieri*. J. Comp. Neurol. 151: 331-358.
Vanegas, H. and Ebbesson, S.O.E. (1974). Telencephalic projections in two teleost species. (In manuscript).
Vanegas, H., Laufer, M. and Amat, J. (1974b). The optic tectum of a perciform teleost. I. General configuration and cytoarchitecture. J. Comp. Neurol. 154: 43-60.

ELECTRORETINOGRAPHY AS A TOOL FOR STUDYING FISH VISION

M.A. ALI and W.R.A. MUNTZ

Université de Montréal, Canada and

University of Sussex, Brighton, U.K.

> Sirs, take you to your tools. You, cousins, shall
> Go sound the ocean, and cast your nets:
> Happily you may catch her in the sea.
> Titus Andronicus, Act IV, Sc. 3

INTRODUCTION

It has been known since the middle of the 19th century that there is a large and relatively constant potential of several millivolts across the vertebrate eye, with the cornea normally positive with respect to the fundus (Du Bois Raymond, 1849). With the apparatus available at that time it was not possible to determine whether this potential was affected by light, but later Holmgren (1865) detected, in frog eyes, transient changes at both the onset and cessation of light stimuli, showing that this was indeed the case. Since that time a very large number of experiments of increasing accuracy and sophistication have been undertaken by many workers on a wide variety of species, and the details of the response, which has come to be known as the electroretinogram or ERG, are now well known (see for example Granit 1947, Brindley 1970, for reviews). Many of these experiments have been concerned with the physiological mechanisms underlying visual function, and where fishes have been used this has been only incidental and because of their convenience as experimental subjects. Experiments of this type fall outside the scope of this paper, and will therefore not be considered further here.

Ever since its discovery, however, the ERG has also been used in comparative studies, and for understanding the influence of environmental factors. Dewar and McKendrick (1873), for example, re-discovered the ERG without knowing about Holmgren's work, and went on to obtain data from a variety of fishes from different habitats. They commented as follows: "On comparing these results, it will be seen that in the active alert stickleback the variations occurred rapidly; in the more sedate goldfish they occurred more slowely, while in the case of the sluggish rockling, which hides beneath stones near low-water mark, the variations occurred very slowely" (quoted in Day 1915; the variations referred to are variations in potential, i.e. the ERG). This comparative approach was continued by early workers, perhaps most notably by Piper (1905, 1910), whose experiments on a variety of animals were one of the first experimental tests of Duplicity Theory, and has continued to this day (eg Kobayashi 1962, Easter and Hamasaki 1973).

The ERG was the first electrical response to light to be recorded from vertebrate eyes. In 1927 however, Adrian and Mattews (1927a, b, 1928) succeeded in recording from small groups of fibres in the optic nerve of the eel. This was followed by Hartline's (1938) recordings from single optic nerve fibres in the bullfrog retina, and by the development and application of microelectrodes by Granit (1947), Barlow (1953), Svaetichin (1956), Tomita (1963), and many others (see paper by Tomita in this volume for a review). Since the ERG is a complex phenomenon which is even now not fully understood, whereas comparatively simple unit responses from single ganglion cells could be obtained with microelectrodes, the ERG declined in popularity and became largely superseded by the single-unit recording techniques. Nevertheless it still offers considerable advantages for certain applications, particularly in comparative work, and should not be abandoned.

Many of these advantages are a consequence of the fact that the ERG is a large response, so that it can be readily recorded from unopened eyes in intact undamaged animals. This characteristic is of course the reason why it is so frequently used with human subjects, but it is also of value in animal work since it allows the same animals to be used over and over again, over long periods of time, thus greatly facilitating studies of the effects of age, diet, seasons, and other environmental factors. Another significant advantage is that the experiments can be done on a limited number of animals without killing them, which is important if the species is rare or valuable (Ali and Kobayashi, 1968a). Finally the technique is simple, and easy to set up and carry out, which can be important in comparative work where subjects are often scare and laboratory facilities primitive (Kobayashi and Ali, 1971).

SOME EXAMPLES OF ERG WORK ON FISHES

(i) Effects of temperature. It has been known for some time that temperature can affect the visual functions of cold-blooded animals (eg Denton and Pirenne 1954). Such effects could well be important in fishes, resulting, for example, in seasonal changes in the animals' visual capabilities, and a number of ERG studies on this topic have been carried out. Thus Hanyu and Ali (1963), using goldfish, demonstrated that the ERG flicker fusion frequency increases linearly with the logerithm of stimulus intensity (the Ferry-Porter law) up to a maximum, and that it also increases with temperature. These data are in line with previous work on carp by Tamura and Hanyu (1959) using the ERG, and by Motokawa et al (1958) using intraretinal slow potentials measured with microelectrodes. The general relationship between flicker fusion frequency and temperature was subsequently confirmed for salmon (Hanyu and Ali 1964) and sunfish (Ali and Kobyashi 1967); there were, however, marked differences between the absolute values obtained, which are summarised in Table I. Recently Schellart et al. (1974) have obtained Bode plots using the ERG in goldfish, and shown that these are shifted along the frequency axes by temperature.

TABLE I. Maximum fusion frequency values for different fishes at different temperatures. From Hanyu & Ali, 1963, 1964; Ali & Kobayashi, 1967.

Temperature	25°	20°	15°	10°	5°
Goldfish (*Carassius auratus*)	67.2 ± 4.2		43.4 ± 1.3		24.4 ± 1.9
Salmon (*Salmo salar*)	95.8 ± 6.3		73.5 ± 4.3		47.7 ± 5.6
Sunfish (*Lepomis gibbosus*)		79.9 ± 2.0		51.1 ± 4.5	

The ERG flicker fusion data for the sunfish (*Lepomis gibbosus*) do not agree with behavioural data obtained by Crozier et al (1936) on the closely related species, *Enneacanthus gloriosus*. These authors used the optomotor response, and obtained flicker fusion values at temperatures of 12.4°, 21.5°, and 27.3° C of about 50 cps in each case. Apart from being independent of temperature, such values are lower than those obtained with the ERG, suggesting the possibility that although the eye can detect frequencies up to about 80 cps, by the time the peripheral, central, and motor nervous systems, and the muscles of the body have mediated the responses of the eye, the maximum value becomes lower.

Temperature also affects the form of the ERG to single flashes, which in general becomes larger and more rapid at higher temperatures, especially under scotopic conditions (Hanyu and Ali 1964, Thorpe 1973). Temperature did not affect the spectral sensitivity of the ERG goldfish as determined with the ERG (Thorpe 1973, Schellart et al 1974), although it has been reported to affect behaviourally determined spectral sensitivity data (Thorpe 1971).

(ii) Spectral sensitivity. Another field in which a considerable amount of ERG work has been done is that of spectral sensitivity. The general procedure involves selecting some criterion ERG response, and then finding the intensities of different monochromatic lights necessary to evoke this response. Under dark-adapted conditions we should expect the resulting action spectrum to agree with the absorption spectrum of the visual pigment of the rods, and in many cases this expectation is fulfilled. Crouzy and Ali (1966), for example, using the goldfish, obtained good agreement between the ERG spectral sensitivity and the absorption of the extractable visual pigment, and Kobayashi and Ali (1971) obtained similarly good agreement in the case of the brook trout (*Salvelinus fontinalis*). Easter and Hamasaki (1973) have shown that the shape of the ERG scotopic spectral sensitivity curves of three marine teleosts agree with the standard shape of the absorption curves of A_1-based pigments (Dartnall's (1953) nomogram), so that although the visual pigments of the species used in this study are unknown, it seems probable that here also the ERG spectral sensitivity is directly related to the absorptive properties of the visual pigment.

In other cases, however, the agreement is less good. Sometimes the ERG scotopic sensitivity curve is considerably broader than Dartnall's nomogram (Burkhardt 1966 and Thorpe 1973 for goldfish; Witkovsky 1968 for *Cyprinus carpio*, a carp). Some of this broadening may well be caused by the high density at which visual pigments often occur in fish retinas (eg Muntz and Northmore 1973), but it seems clear that a residual contribution from the cones is also often involved, either because the animals are insufficiently dark-adapted, or because the comparatively high stimulus intensities usually necessary in ERG work exceed the cone threshold. Many fish eyes possess coloured intraocular filters and reflectors which must affect sensitivity (see Muntz 1973 for a review), and these may also cause deviations from expectations based on our knowledge of visual pigments. The ERG spectral sensitivity curves of albino brook trout, for example, differ markedly from those of pigmented brook trout, mainly under photopic conditions, presumably because of reflectance from retinal blood vessels (Kobayashi and Ali 1971).

ERG spectral sensitivity curves obtained under photopic conditions tend to be more complex than those obtained under scotopic conditions. In a variety of fishes more or less smooth bell-shaped curves have been obtained, displaced towards long wavelengths compared to the scotopic ERG sensitivity curve of the same species, indicating the presence of cones as well as rods (Kobayashi 1962, Easter and Hamasaki 1973). It is not, however, necessarily the case that such smooth bell-shaped functions can be directly related to the animals' photopic spectral sensitivities, for where detailed ERG and behavioural data are both available for the same species the agreement is poor. Thus Burkhardt's (1966, 1968) photopic ERG spectral sensitivity curves for goldfish, which are fairly smooth and maximal at about 575nm, bear little obvious resemblence to the behavioural sensitivity curves of either Cronly-Dillon and Muntz (1965) or Yager (1967) on the same species, and similarly the photopic ERG curves of Northmore and Muntz (1970) for *Scardinius erythrophthalmus* (the rudd) do not resemble any of the four photopic behavioural sensitivity curves that have been obtained with this fish (Northmore and Muntz 1974, Muntz this volume). In general, light-adapted spectral sensitivity curves obtained behaviourally show several distinct maxima that are lacking in ERG data. One exception, however, may be found in Kobayashi and Ali's (1971) ERG photopic curve for the brook trout (*Salvelinus fontinalis*), which shows three maxima at about 425, 545, and 595 nm, and resemble quite closely Cronly-Dillon and Muntz's (1965) behavioural curve for goldfish. These maxima also correspond roughly to the λ_{max} of the goldfish cone pigments (Marks, 1965).

The ERG has also been used on several occasions in attempts to determine the spectral sensitivities of the different receptor mechanisms underlying colour vision in fishes. One approach has been to measure the spectral sensitivities of the different component waves of the ERG separatly (eg Witkovsky 1968, Thorpe 1973), another to obtain spectral sensitivity curves in the presence of various chromatic adapting backgrounds (Burkhardt 1968, Witkovsky 1968, Northmore and Muntz 1970, Thorpe 1973). Finally, spectral sensitivity curves have been obtained by holding the spectral composition and intensity of the stimulus constant, and then finding the intensity of different monochromatic backgrouds necessary to reduce the size of the response by a given amount (Burkhardt 1966, 1968; Northmore and Muntz 1970). The spectral characteristics of the stimulus are chosen

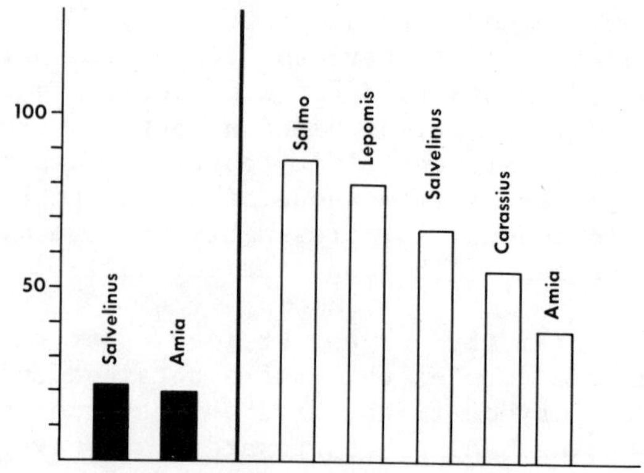

FIG. 1. Maximum ERG-flicker fusion frequencies (c.p.s.) in dark- and light-adapted fishes of different species. The data were obtained in the same laboratory using the same method. Black bars indicate dark-adapted state. (After Gramoni and Ali, 1970).

so as to stimulate a single class of receptor much more strongly than any other class: if this can be achieved the relative effectiveness of the different background stimuli in suppressing the response should give an accurate measure of the spectral sensitivity of this receptor type. The results of all these experiments are relatively complex and difficult to interpret, though some separation of the different mechanisms is achieved in all cases. The red mechanism in particular has been isolated and characterised with considerable accuracy by the background sensitivity method.

(iii) Environmental variations. A number of ERG studies have been specifically devoted to studying differences in visual function between fishes occupying different habitats, among which that of Kobayashi (1962) is notable for its comprehensiveness. This author studied a wide variety of teleost and cartilagenous fishes from both freshwater and marine environments. Among other findings, he obtained results suggesting that deep-water and nocturnal species have ERG responses that saturate at relatively low intensities and low ERG flicker fusion rates, as compared to active, diurnal and shallow-water species; the absolute sensitivity, on the other hand, was higher in the former group. He also reported that the ERG scotopic spectral sensitivity curves were shifted towards shorter wavelengths in the deep-water forms, a finding that is in line with results from visual pigment studies (see papers by Dartnall and Muntz in this volume). Ali and his collaborators have also compared the visual function of species from different habitats, by measuring the ERG flicker fusion rate (Hanyu and Ali 1963, Ali and Kobayashi 1967,

1968b, Gramoni and Ali 1970). The results of these studies, which are summarised in Fig. 1, are strikingly reminiscent of the early observations by Dewar and McKendrick quoted at the beginning of this chapter, with high fusion rates in active alert fish such as the salmon, lower rates in the more sedate goldfish, and yet lower rates in the sluggish "living fossil", *Amia*. Interesting and useful correlations can also be established between electroretinographic responses and retinal structure, the latter of course being adapted to the milieu (see papers by Locket, Nicol and Wagner in this volume).

SUMMARY

As with other techniques for studying vision, the ERG has both advantages and disadvantages. The disadvantages include the fact that ERG results do not always reflect behavioural data, and that even now the physiological mechanisms underlying the response are not fully understood. The advantages are largely based on its convenience: the response is large, the technique easy to apply, and it is not necessary to operate on or kill the animals used. These advantages are particularly important in comparative work, where specimens are often in short supply and laboratory facilities may not be good. However, as with data obtained by other techniques, ERG results should not be considered in isolation, but in the light of other available information.

REFERENCES

Adrian, E.D., and Matthews, R. (1927a). The action of light on the eye. I. The discharge of impulses in the optic nerve and its relation to the electric changes in the retina. J. Physiol. 63: 378-414.

Adrian, E.D. and Matthews, R. (1927b). The action of light on the eye. II. The processes involved in retinal excitation. J. Physiol. 64: 279-301.

Adrian, E.D. and Matthews, R. (1928). The action of light on the eye. III. Nervous conduction in retina. J. Physiol. 65: 273-298.

Ali, M.A. and Kobayashi, H. (1967). Temperature: influence on the electroretinogram-flicker fusion frequency of the sunfish (*Lepomis gibbosus* L.). Rev. Can. Biol., 26: 341-345.

Ali, M.A. and Kobayashi, H. (1968a). Electroretinogram of albino and pigmented trout *Salvelinus fontinalis* (Mitchill). Rev. Can. Biol., 27: 145-161.

Ali, M.A. and Kobayashi, H. (1968b). Electroretinogram-flicker fusion frequency in albino trout. Experientia 24: 454-455.

Barlow, H.B. (1953). Summation and inhibition in the frog's retina J. Physiol. 119: 69-88.

Brindley, G.S. (1970). Physiology of the retina and visual pathway. Arnold: London.

Burkhard, D.A. (1966). The goldfish electroretinogram: relation between photopic spectral sensitivity functions and cone absorption spectra. Vision Res. 6: 517-532.

Burkhardt, D.A. (1968). Cone action spectra: Evidence from the goldfish electroretinogram. Vision Res. 8: 839-853.

Cronly-Dillon, J.R. and Muntz, W.R.A. (1965). The spectral sensitivity of the goldfish and the clawed toad tadpole under photopic conditions. J. Exp. Bio. 42: 481-493.

Crouzy, R. et Ali, M.A. (1966). Relation entre la sensibilité électrorétinographique et le spectre d'absorption du pigment visuel scotopique chez le poisson rouge. Nombre minimum de quanta absorbées. Bull. Mus. Nat. Hist. Nat. Mars., 2e série, 38: 730-743.

Crozier, W.H., Wolf, E. and Zerrahn-Wolf., G. (1936). Temperature and critical illumination for reaction to flickering light. II. Sunfish. J. Gen. Physiol. 20: 411-431.

Dartnall, H.J.A. The interpretation of spectral sensitivity curves. (1953). Br. Med. Bull. 9: 24-30.

Day, E.C. (1915). Photoelectric currents in the eye of the fish. Am. J. Physiol. 38: 369-398.

Denton, E.J., and Pirenne, M.H. (1954). The visual sensitivity of the toad *Xenopus laevis*. J. Physiol. 125: 181-207.

Dewar, J. and McKendrick, J.G. (1873). Recent researches on the physiological action of light. Nature 8: 204-205.

DuBois Reymond, E. (1849). Untersuchungen über thierische Elektricität. Vol. II., G. Reimer, Berlin. p. 256-257.

Easter, S.S. and Hamasaki, D.I. (1973). Electroretinographically Determined scotopic spectral sensitivity of some marine fish. Vision Res. 13: 1175-1181.

Gramoni, R. and Ali, M.A. (1970). L'électrorétinogramme et sa fréquence de fusion chez *Amia calva* (Linné). Rev. Can. Biol. 29: 353-363.

Granit, R. (1947). Sensory mechanisms of the retina. Oxford University Press, London. p. 412.

Hanyu, I., and Ali, M.A. (1963). Flicker fusion frequency electroretinogram in light-adapted goldfish at various temperatures. Science 140: 662-663.

Hanyu, I. and Ali, M.A. (1964). Electroretinogram and its flicker fusion frequency at different temperatures in light-adapted salmon (*Salmo salar*). J. Cell. Comp. Physiol. 63: 309-322.

Hartline, H.K. (1938). The response of single optic nerve fibres of the vertebrate eye to illumination of the retina. Amer. J. Physiol. 121: 400-415.

Holmgren, F. (1865-1866). Method att objectivera effecten av ljusintryck på retina. Upsala Läkaref. Förh. 1: 177-191.

Kobayashi, H (1962). A comparative study on electroretinogram in fish, with special reference to ecological aspects. J. Shimonoseki Coll. Fish., 11: 407-538.

Kobayashi, H. and Ali, M.A. (1971). Electroretinographic determination of spectral sensitivity in albino and pigmented brook trout (*Salvelinus fontinalis*, Mitchill). Can. J. Physiol. Pharmacol. 49: 1030-1037.

Marks, W.B. (1965). Visual pigments of single goldfish cones. J. Physiol. (London) 178: 14-32.

Motokawa, K. Oikawa, T., Tasaki, T. and E. Yamashita. (1958). Flicker and fusion of potentials recorded from fish retina with a microelectrode. Tohoku J. Exptl. Med. 68: 249-256.

Muntz, W.R.A. (1973). Inert absorbing and reflecting pigments (1973). In "Handbook of Sensory Physiology", Vol. VII/I, "Photochemistry of Vision", edited by H.J.A. Dartnall. Springer-Verlag: Heidelberg, New York. p. 529-565.

Muntz, W.R.A., and Northmore, D.P.M., (1973). Scotopic spectral sensitivity in a teleost fish (*Scardinius erythrophthalmus*) adapted to different daylengths. Vis. Res. 13: 245-251.

Northmore, D.P.M. and Muntz, W.R.A. (1970). Electroretinogram determinations of spectral sensitivity in a teleost fish adapted to different daylengths. Vis. Res. 10: 799-816.

Northmore, D.P.M. and W.R.A., Muntz, W.R.A. (1974). Effects of stimulus size on spectral sensitivity in a fish (*Scardiniis erythrophthalmus*), measured with a classical conditioning paradigm. Vision Res. 14: 503-514.

Piper, H. (1905). Untersuchungen über das elektromotorische Verhalten der Netzhaut bei Warmblütern. Arch. Anat. Physiol., Leipzig, Suppl. p. 133-192.

Piper, H. (1910). Die Aktionsströme der Vogel- und Säugernetzhaut bei Reizung durch kurzdauernde Belichtung und Verdunkelung. Arch. Anat. Physiol., Leipzig, Suppl. p. 461-466.

Schellart, N.A.M., Spekreijse, and van den Borg, T.J.T.P. (1974). Influence of temperature on retinal ganglion cell response and ERG of goldfish. J. Physiol. 238: 251-267.

Svaetichin, G. (1956). Spectral response curves from single cones. (1956). Acta physiol. scand. 39: Suppl. 134: 17-46.

Tamura, T. and Hanyu, I. (1959). The flicker Electroretinogram of the carp eye. Bull. Jap. Soc. Scient. Fish. 25: 624-631.

Thorpe, S.A. (1971). Behavioural measures of spectral sensitivity in goldfish at different temperatures. Vision Res. 11: 419-433.

Thorpe, S.A. (1973). The effects of temperature on the psychophysical and electroretinographic spectral sensitivity of the chromatically-adapted goldfish. Vision Res. 13: 59-72.

Tomita, T. (1963). Electrical activity in the vertebrate retina. J. Opt. Soc. Am. 53: 49-57.

Witkovsky, P. (1968). The effect of chromatic adaptation on color sensitivity in the carp electroretinogram. Vision Res. 8: 823-837.

Yager, D. (1967). Behavioural measures and theoretical analysis of spectral sensitivity and spectral saturation in the goldfish. Vision Res. 7: 707-727.

3. NEUROTRANSMITTERS

METHODS OF RESEARCH FOR NEUROTRANSMITTER SUBSTANCES: GENERAL INTRODUCTION

Herbert H. JASPER

Groupe de Recherche en Sciences Neurologiques du
Conseil des Recherches Médicales du Canada
Université de Montréal, Canada

Sir, it is
A charge too heavy for my strength; but yet
We'll strive to bear it, for your worthy sake,
To th' extreme edge of hazard.
All's Well That Ends Well Act III, Sc. 3

I would like first to thank Babu Ali for kindly inviting us to participate in this NATO Advanced Study Institute on vision in fishes. We consider this a special privilege since none of the speakers in this morning's panel qualify for membership in this very distinguished and prestigious club of international experts on vision in fish. We are particularly pleased to be able to help welcome so many foreign visitors to Canada, and especially to this beautiful setting at Bishop's University. We expect to learn much from the splendid programme that has been organised. In order to earn this privilege, we have offered to speak of some of the work we are doing on the mammalian nervous system hoping that the methods we shall describe may be of value also in studies of the nervous system in fish.

We have chosen the topic of "Methods of Research for Neurotransmitter Substances" first since it has been a subject of major interest to several members of the Centre de Recherche en Sciences Neurologiques for many years so that presumably we should know something about it from personal experience. Secondly, it is an area of the neurosciences which has witnessed most remarkable advances during recent years partly due to refinements in our methods of study. These advances have covered a variety of special disciplines, often used in combination, in wide ranging multidisciplinary attacks on common problems. The fundamental basis lies in neurochemistry and neuropharmacology. These are two subjects which often have me completely mystified and I presume there may be a few others in this group who may be able to sympathise with me in this regard. I have noticed that neurochemists and pharmacologists are also frequently mystified as well. As a neurophysiologist, however, I have long been convinced that electrophysiological and behavioural methods can only lead us so far in our understanding. Ultimately we must understand the chemical and biophysical generators of the electrical properties of nerve cells and synapses. The relationships and modes of communication between cells is of particular importance. We now know that such communication may be either electrical or chemical, or both in some synapses. New forms of chemical communication are also being brought to light, some of which may not be associated with electrical signs familiar to classical electrophysiology. In the retina we have examples of visual information transmitted from receptor cells to ganglion cells without action potentials.

The economy of nature has simplified our problem by using the same chemical substances as neurotransmitters throughout the animal kingdom both at interneuronal synapses and at effector junctions in muscle and glands. However their distribution and action may be different in different parts of the nervous system and in different species. Complications arise in the central nervous system, and in some peripheral ganglia, by the fact that a given transmitter may have more than one kind of action upon the same or different nerve cells. Acetylcholine, for example, may have two forms of excitatory action, one of the rapid depolarising form, the so-called nicotinic action as at the striate neuromuscular junction and in Renshaw cells of the spinal cord, and a much slower and long lasting excitatory action, called muscarinic, which is characteristic of its action upon cortical neurones. The slower long lasting action is apparently not associated with a decrease in membrane conductance, depolarisation being possibly related to increased flux of K according to the work of Krnjević. Depending upon the nature of the cholinergic receptor substance in the postsynaptic membrane, ACh may also cause inhibition and hyperpolarisation of postsynaptic membranes in the central nervous system of some species as it does in the heart.

The principal transmitter substances with which we shall be concerned this morning are Acetylcholine, the monoamines (Noradrenaline, Dopamine, and Serotonin), and the physiologically active amino acids, particularly glutamic acid, gamma-aminobutyric acid or GABA, and Glycine. The emphasis is to be upon methods of study as applied to these substances, some of which have a more general application to the study of other possible neurotransmitter substances which may prove to be of equal importance in the future.

Nico van Gelder will first speak about methods for the neurochemical study of amino acids. This will be followed by a critical presentation of the technique of microiontophoresis by Ernie Puil, who has recently joined our neurosciences research group following post-doctoral studies in Europe. Jacques de Champlain, who spent several years in post-doctoral research with Axelrod in Bethesda, will describe histofluorescent and recently developed ultra microtechniques for the study of biogenic amines.

The use of high resolution electronmicroscopic autoradiography for the identification and study of neurones containing biogenic amines in the central nervous system will be described by Laurent Descarries, who did some of his post-doctoral work with Droz in Paris. The beautiful techniques he will describe are very much his own, however, some now in the process of perfection and development at the Université de Montréal.

It will not be necessary, I feel sure, in this sophisticated audience for me to discuss the various criteria presumed necessary to be proven before a chemical substance can be accepted as a physiological transmitter. These have been repeatedly emphasised by many authors. In the central nervous system it is seldom possible to satisfy all of these criteria so that certain compromises have to be made in many instances and the precise mode of action upon the postsynaptic membrane in order to selectively increase permeability to certain ions, and/or to modify the action of certain metabolic pumps, or to control the action of certain metabolic enzymes, such as adenyl cyclase and cyclic adenosine monophosphate, cyclic AMP, are still largely unknown for any transmitter substance. The most important essential criteria are the following, in my opinion:

1. Presence of the substance in higher concentration in synaptosomes, together with its specific synthesising enzymes.
2. Demonstration that the substance is actually liberated at synaptic sites during normal physiological activity.
3. Demonstration by histochemical methods, including electronmicroscopic radioautography, of the location of a given transmitter substance with precautions against nonspecific artefacts.
4. Demonstration that the physiological action of small quantities of a given substance, administered by microiontophoreses, produces a characteristic excitatory or inhibitory action upon individual nerve cells or their dendrites identical in all respects to that produced by normal physiological activation of the same synapses. Ideally this should include intracellular microelectrode studies in order to observe directly the changes in level of membrane polarisation together with

measures of changes in membrane conductance. Final proof requires demonstration that the ionic flux changes are the same, and the equilibrium potential for reversal of the synaptic potential is also the same, but only in rare instances has it been possible to achieve these final proofs in the central nervous system.

5. Demonstration of specific processes of inactivation of liberated transmitter substance, either through binding, re-uptake, diffusion, or specific enzyme transformation.
6. The discovery of specific pharmacological blocking agents such as atropine or curare for acetylcholine, and the more recently studied specific cholinergic blocking agent alpha bungarotoxin obtained from the venom of the Formosan krait *Bungarus multicinctus* have proven most valuable, particularly now for the chemical isolation of specific receptor proteins. Relatively specific blocking agents for the biogenic amines and for the amino acids, such as picrotoxin or bicuculine for GABA may not be sufficiently specific. Also blocking agents all have secondary actions which may make interpretation difficult.

Another complication of studies of transmitter substances in the central nervous system is the existence of presynaptic as well as postsynaptic actions which cannot be analysed by the directly intracellular microelectrode recording techniques mentioned above. In the spinal cord, for example, there is now considerable evidence to suggest that GABA acts largely as a presynaptic inhibitory substance. There are also many other complications to be mentioned by the other speakers on this panel, including the problem of the reversing action of interneurones which will be discussed by Ernie Puil with illustrations from some recent experiments we have done together.

My own work has been largely concerned with the measurement of the rate of liberation of acetylcholine and the amino acids during physiological activation of central synapses.

OUR WORK IS SUPPORTED BY A GROUP GRANT OF THE MEDICAL RESEARCH COUNCIL OF CANADA.

METHODS TO INVESTIGATE THE FUNCTION OF CERTAIN AMINO ACIDS IN THE CNS

N.M. van GELDER

Groupe de Recherche en Sciences Neurologiques
du Conseil Médical du Canada
Département de Physiologie, Université de Montréal, Can.

Master, I marvel how the fishes live in the sea.
Pericles Act I, Sc. I

INTRODUCTION

Aside from participating in metabolic processes, certain amino acids in the CNS may serve a more specialised physiological purpose in one of the three ways proposed below which, it should be remembered, are however not mutually exclusive. Thus an amino acid serving a certain role in the CNS, may well be used in a somewhat altered function by a specialised tissue such as the retina.

I - AMINO ACIDS AS SYNAPTIC TRANSMITTERS

Trans-synaptic release of a substance provides a direct link by which the activity of one neurone releases a chemical at its terminal which alters the behaviour of the target cells (see Krnjevic, 1974; Gerschenfeld, 1973; Werman, 1966). A number of biochemical parameters must be fulfilled before an amino acid can be classified as a synaptic transmitter:

(i) Pre-synaptic synthesis and/or rapid transport into the terminals from perikarya or from the immediate environment:

If an amino acid serves uniquely to chemically link the physiological activity in nerve terminals to that of the innervated cells, it is clear that such terminals must maintain a steady supply of the amino acid. The presence of a synthesising system locally in the terminals would be one direct means by which a steady supply may be assured. When a similar synthesising system is present in the perikaryon or in surrounding glia, its sole function should be to replenish the terminals by means of axonal transport or by release from glial elements in close proximity to the terminals.

(ii) Storage:

Since the amino acid must be continuously available to be released on demand, the presence of a depot or reservoir at the nerve terminals seems essential. Thus, a fraction of the total amount of amino acid present in the terminals must exist in a form from which it is released only according to precise physiological conditions, i.e., the nerve impulse (depolarisation) plus calcium (Douglas and Rubin, 1961; Katz and Miledi, 1967). Moreover, only small and equal quantities should be released from the storage site, proportional to the number of impulses arriving at the terminals.

(iii) Inactivation:

If an amino acid functions as a transmitter, each quantity released on the arrival of the impulse will, within certain limits, exert an influence on the target cell. At least a certain fraction of the amount liberated must therefore be destroyed or inactived between impulses in order to prevent continuous maximal activation of the responsive mechanisms post-synaptically.

To date uptake of an amino acid in its role as a transmitter by either post- or pre-synaptic structures appears to represent the most likely route of physiological inactivation. Pre-synaptically the amino acid may re-enter the stores to be released again or it may be subsequently metabolised. Post-synaptically, the amino acid may be taken up either by the neuronal perikarya or by surrounding glia.

II - AMINO ACIDS AS HORMONES

In contrast to a chemical released at nerve terminals, hormones may be loosely classified as substances synthetised by specialised cells and secreted into the circulatory system (see Pitot and Yatvin, 1973). Besides the vascular system, the cerebrospinal fluid, or the aqueous humour, the neuroglia may also perhaps be considered as serving in part a similar circulatory function. Close apposition of glial processes, as well as their extensive insertion among neurones, would favour wide distribution of a substance once taken up into glia (Schon and Kelly, 1974; Henn and Hamberger, 1971).

Other probable criteria may include:

(i) its occasional presence in appreciable quantities in one of the circulatory systems, (ii) a slow, tonic action and (iii) no specialised endplate regions on target cells.

III - AMINO ACIDS AS MODULATORS

A modulator has been considered to exhibit physiological and biochemical properties compatible with a role as one or the other class of substances. One may propose that the following criteria apply to an amino acid functioning as a modulator:

Stimulation of a neurone causes release of the amino acid not only from the terminals but also from perikarya. Biological action is therefore not confined to specific cells innervated by the terminals, but is also apparent in adjacent neuronal and glial structures, including other nerve terminals. An excitatory substance thus released would tend to synchronise activity of neurones adjacent to the stimulated cell or, alternatively, may activate inhibitory terminals, thereby limiting the duration of excitation. On the other hand, release of an inhibitory substance may produce surround inhibition. In addition, the resulting differences in potential between the affected region of the CNS and other areas, e.g. between a perikaryal and dendritic layer, may further rapidly synchronise the activity of large groups of neurones. The advantages of this type of action are obvious, especially in the retina and in regions of the CNS with a relatively uniform cell population e.g. the cerebellum.

Unlike a transmitter, affinity sites for a modulator on target cells may be uniformly distributed, but similar to a transmitter its physiological action is rapid. Storage sites should be present equally in perikarya and at the terminals. Furthermore, release from such sites at both locations will occur under the same physiological conditions.

It is clear that classification of a physiologically active amino acid as belonging to one or another of the three groups outlined above will depend on how its biochemical and physiological properties fulfil certain criteria. In the following discussion of methods no details of techniques have been provided since these may be obtained by consulting specific references.

I - RECEPTORS

Much of the work on the specificity of the amino acid receptors, their localisation on the neuronal membrane and the physiological response elicited on activation, have been accomplished by sophisticated electrophysiological techniques (e.g. Zieglgänsberger and Puil, 1973; Takeuchi and Takeuchi, 1966). However, an essential corollary to these types of studies is the need for structural analogues which may respectively block the receptors or mimic the action of the amino acids. Successful design of such analogues requires knowledge of receptor-amino acid interaction, emphasising the role played by reactive groups of the amino acids, the structural conformation of the amino acid to the receptor as well as the tertiary structure of the receptor itself, especially as it pertains to changes in ionic permeabilities of the membrane (Curtis and Watkins, 1963; van Gelder, 1971). Elucidation of receptor structure itself is far more difficult and may not be possible until such receptors are isolated as pure fractions (Changeux et al., 1970; Miledi et al., 1971).

II - BINDING, TRANSPORT AND STORAGE

These three events are often studied as a single phenomenon but in fact represent distinct biochemical processes, not always linked to one another. Numerous studies suggest three principle mechanisms responsible for the final distribution of an amino acid between two compartments (physicochemical parameters aside) which may be briefly summarised:

1 - Passive diffusion: This process is highly dependent on initial concentrations in the two compartments and is independent of energy. Following equilibration of concentrations between compartments, the rate of exchange of a molecule in one compartment with an identical molecule in the other compartment assumes a characteristic temperature dependency and is not influenced by the presence of structurally related compounds. Equilibration between compartments is determined by the concentrations of free (unbound) amino acid.

2 - Carrier-mediated exchange (diffusion): The distribution of an amino acid depends on the presence of specific transport proteins in the membrane, is energy-dependent and is influenced by the presence of structurally-related compounds as well as certain ions (Collins, 1974; Hammerstead and Cutler, 1972, Martin and Smith, 1972). Flux across the membrane may be partially independent of the initial concentrations in the compartments but remains dependent on an unbound form of the amino acid.

3 - Active transport: This mechanism represents movement of an amino acid against a concentration gradient, requires energy and specific transport proteins, and is influenced by competition with structurally related compounds (Pardee, 1968). Most importantly, the flux is unidirectional either because the membrane

transports only in one direction, or very often, because the amino acid is metabolised or sequestered in one of the compartments. The latter process appears to involve calcium (Kennedy and Voaden, 1974; Martin and Smith, 1972). Rate of uptake is determined by the amount of transport protein available and becomes constant when all is in use. The actual quantity transported and stored, assuming unlimited availability, depends on saturation of the stores. Following saturation, carrier-mediated exchange (diffusion) may replace active transport, possibly using the same membrane proteins.

The movement of all amino acids has been demonstrated to occur by an energy-dependent, structurally-competitive process and a temperature constant compatible with an active process. Binding sites on the membrane may be the transport protein itself if an analogy is permitted between this mechanism in the CNS and the sodium, energy-dependent movement of amino acids across the membranes of certain other cells (Riggs et al., 1972).

Rapid and short-term time course studies using varying exogenous concentrations of the isotopically labelled amino acid, will allow a fairly accurate estimation of transport rates. In addition, certain preparations of nervous tissue such as the sympathetic ganglia or dorsal root ganglia have also been shown to transport a number of these amino acids across the membrane of some cells (Schon and Kelly, 1974; Bowery and Brown, 1972). Since in a number of instances these preparations normally do not contain an appreciable concentration of certain amino acids, they may be eminently suited to study carrier-mediated diffusion of an amino acid exogenously applied, possibly in its pure form or in association with exchange of structurally related substances present in cells of such preparations.

The determination of active uptake, in contrast to carrier-mediated diffusion, is often simplified when the amino acid being studied is not isotopically labelled, despite a certain loss of additional information which can be obtained if the amino acid is labelled (Starr, 1973; Neal and Starr, 1973; Ehinger, 1972). It is evident that the total amount of labelled amino acid in the tissue following incubation depends on a number of factors, difficult to control. For example, a decrease in specific activity may occur by dilution of radioactive amino acid with its unlabelled homologue, derived from endogenous stores or continuous synthesis. Such dilution would be enhanced in the presence of compounds inhibiting its enzymic removal. These processes would tend to underestimate the amount of amino acid taken up. On the other hand if the exogenous amino acid fails to enter endogenous stores because of inaccessibility or saturation of such sites, active uptake may be easily confused with carrier-mediated diffusion: a small, unsequestered (free) endogenous fraction of the amino acid may rapidly exchange with the exogenous label without bringing about an actual alteration in concentration. This free portion may also exhibit a different rate of turnover than the bound fraction. Especially when the free fraction initially is small, inhibition of destruction results in an increase of the free pool which would enhance carrier-mediated exchange (e.g. Kennedy and Voaden, 1974). Thus, active transport i.e. a net accumulation of an amino acid within membrane-enclosed organelles, requires that the actual tissue concentration of the amino acid be measured (Elliott and van Gelder, 1958).

III - SELECTIVE DISTRIBUTION

Beside the demonstration that certain amino acids exert a powerful physiological action on neurones and that considerable stores are maintained in certain components of nervous tissue, inequal distribution in structural layers or regions of the nervous system provide additional evidence that these substances may perform a specific physiological function in such regions:

1 - DISSECTION

This procedure allows isolation of various homogeneous regions in nervous tissue, or of cell types and their processes, according to specific function, e.g. inhibition or excitation. Because dissection often yields only small quantities of tissue, it is essential that methods for determining the presence of a particular amino acid be sensitive and precise (Cohen et al., 1973; Otsuka and Miyata, 1972).

Circumscribed lesions in anatomical pathways with a known function followed by determination of changes in amino acid concentration along the length of the pathway, can be considered another form of dissection. Accumulation or no change in amino acid concentration above the lesions, in combination with diminishing concentrations in regions where the axons terminate, may represent evidence for a specialised function of amino acid in such a pathway (Fonnum and Walberg, 1973; Davidoff et al., 1967).

2 - ENZYME DISTRIBUTION

Instead of measuring actual amino acid concentrations it is occasionaly more practical to study the distribution of the enzymes associated with the metabolism of an amino acid, especially those associated with glutamic acid and GABA (Saito et al., 1974; Molinoff and Kravitz, 1967; van Gelder, 1965).

3 - RADIOAUTOGRAPHY

Among the methods available for visual localisation of amino acids as well as associated enzymes, radioautography appears to be the most universally applicable. With the aid of this procedure it has been possible to demonstrate that exogenous labelled GABA, glutamic acid and glycine are selectively distributed among cells of the retina and the CNS (e.g. Voaden et al., 1974; Ehinger, 1972; Neal and Iversen, 1972; Henn and Hamberger, 1971). Since the tissue must be submitted to some form of fixation media, it is puzzling however that these procedures do not remove the amino acids from the tissue.

IV - INACTIVATION

Under normal physiological conditions, only a small percentage, if any, of the physiologically active amino acids are present in the extracellular spaces. Many methods are available to study possible mechanisms of elimination ranging from following the rates of disappearance of amino acids in cerebrospinal fluid (Snodgrass and Lorenzo, 1973; Levin et al., 1966), glial uptake by means of radioautography (Schon and Kelly, 1974; Ehinger, 1973, 1972), or by determining whether characteristic parameters of transport and storage (Bennett et al., 1973; Levi and Raiteri, 1973) as well as mechanisms of enzymic inactivation (Buu and van Gelder, 1974) differ at suspected sites of release from those found at sites of inactivation.

V - RELEASE

Determination of the exact site of release of an amino acid, together with the physiological conditions when this occurs and the resulting response, have been studied using preparations of both invertebrate and the mammalian nervous system. In invertebrates, where cells of known function can be isolated together with the axons and the target cells, such investigations have proven particularly useful in examining the exact physiological action of an amino acid (see Iversen, 1970).

Whereas the above method provides information regarding the physiological effect of an amino acid on neurones, an opposite purpose is served when defined anatomical pathways (Jasper and Koyama, 1969), tissue slices (Pasantes-Morales et al., 1974; McIlwain et al., 1969) or isolated nerve terminals (Bradford et al., 1973) are stimulated, and the perfusates or effluents from such preparations are collected. In this case the action of the amino acid in the CNS is presumed known. This type of technique can yield suggestive evidence in assigning a physiological role to certain anatomical pathways.

CONCLUDING REMARKS

Although many methods are available to study biological properties of the amino acids, a clear interpretation of results is difficult because of the complex composition of the nervous tissue (including the retina). Methods designed to study selective distribution of an amino acid or the biological processes governing its function, in many cases require separation of the tissue into homogeneous components such as perikarya, nerve terminals, dendrites, specific glial cell types, etc. However, these procedures obviate a number of cooperative events which surely take place between various nervous structures or between regions of the CNS. In this respect the use of non-mammalian nervous systems and tissue slices, which are structurally more simple or more accessible to experimental manipulations, often provide a valuable adjunct to *in vivo* studies of the mammalian CNS.

REFERENCES

Bennett, J.P., Logan, W.J. and Snyder, S.H. (1973) Amino acids as central nervous transmitters: The influence of ions, amino acid analogues and ontogeny on transport systems for L-glutamic and L-aspartic acids and glycine into central nervous synaptosomes of the rat. J. Neurochem. 21: 1533-1550.

Bowery, N.G. and Brown, D.A. (1972) γ-Aminobutyric acid uptake by sympathetic ganglia. Nature New Biol. 238: 89-91.

Bradford, H.F., Bennett, G.W. and Thomas, A.J. (1973) Depolarizing stimuli and the release of physiologically active amino acids from suspensions of mammalian synaptosomes. J. Neurochem. 21: 495-505.

Buu, N.T. and van Gelder, N.M. (1974) Differences in biochemical properties of γ-aminobutyric acid aminotransferase from synaptosome-enriched and cytoplasmic mitochondria-enriched subcellular fractions of mouse brain. Can. J. Physiol. Pharmacol. 52: 674-680.

Changeux, J.P., Kasai, M. and Lee, C.-Y. (1970) Use of snake venom toxin to characterize the cholinergic receptor protein. Proc. Natl. Acad. Sci. U.S., 67: 1241-1247.

Cohen, A.I., McDaniel, M. and Orr, H. (1973) Absolute levels of some free amino acids in normal and biologically fractionated retinas. Invest. Opthalm. 12: 686-693.

Collins, G.G.S. (1974) The spontaneous and electrically-evoked release of ^{3}H -GABA from the isolated hemisected frog spinal cord. Brain Res. 66: 121-137.

Curtis, D.R. and Watkins, J.C. (1963) Acidic amino acids with strong excitatory actions on mammalian neurones. J. Physiol. (Lond.) 166: 1-14.
Davidoff, R.A., Graham, L.T., Shank, R.P., Werman, R. and Aprison, M.H. (1967) Changes in amino acid concentrations associated with loss of spinal interneurons. J. Neurochem. 13: 1025-1031.
Douglas, W.W. and Rubin, R.P. (1961) The role of calcium in the secretory response of the adrenal medulla to acetylcholine. J. Physiol. 159: 40-57.
Ehinger, B. (1972) Cellular location of the uptake of some amino acids in the rabbit retina. Brain Res. 46: 297-311.
Ehinger, B. (1973) Glial uptake of taurine. Brain Res. 60: 512-516.
Elliott, K.A.C. and van Gelder, N.M. (1958). Occlusion and metabolism of γ-aminobutyric acid by brain tissue. J. Neurochem. 3: 28-40.
Fonnum, F. and Walberg, F. (1973) An estimation of the concentration of γ-aminobutyric acid and glutamate decarboxylase in the inhibitory Purkinje axon terminals in the cat. Brain Res. 54: 115-127.
Gerschenfeld, H.M. (1973) Chemical transmission in invertebrate central nervous systems and neuromuscular junctions. Physiol. Review 53-1-119.
Hammerstead, J.P. and Cutler, R.W.P. (1972) Sodium ion movements and the spontaneous and electrically-stimulated release of ^{3}H-GABA and ^{14}C-glutamic acid from rat cortical slices. Brain Res. 47: 401-413.
Henn, F.A. and Hamberger, A. (1971) Glial cell function. Uptake of transmitter substances. Proc. Nat. Acad. Sci. 68: 2686-2690.
Iversen, L.L. (1970) Neurotransmitters, neurohormones and other small molecules in neurons. In: The Neurosciences, Second Study Program, edited by Francis O. Schmidt, Rockefeller Univ. Press, N.Y., pp. 768-782.
Jasper, H.H. and Koyama, I. (1969) Rate of release of amino acids from the cerebral cortex in the cat as affected by brainstem and thalamic stimulation. Can. J. Physiol. Pharmacol. 47: 889-905.
Katz, B. and Miledi, R. (1967) The timing of calcium action during neuromuscular transmission. J. Physiol. 189: 535-544.
Kennedy, A.J. and Voaden, M.J. (1974) Factors affecting the spontaneous release of ^{3}H-γaminobutyric acid from the frog retina in vitro. J. Neurochem. 22: 63-71.
Krnjevic, K. (1974) Chemical nature of synaptic transmission in vertebrates. Physiol. Review 54: 418-540.
Levi, G. and Raiteri, M. (1973) GABA and glutamate uptake by subcellular fractions enriched in synaptosomes: critical evaluation of some methodological aspects. Brain Res. 57: 165-185.
Levin, E., Garcia, Argiz, C.A. and Nogueira, G.T. (1966) Ventriculocisternal perfusion of amino acids in cat brain. J. Neurochem. 13: 979-988.
Martin, D.L. and Smith, A.A. (1972) Ions and the transport of gamma-aminobutyric acid by synaptosomes. J. Neurochem. 19: 841-855.

McIlwain, H., Harvey, J.A. and Rodriguez, G. (1969) Tetrodotoxin on the sodium and toehr ions of cerebral tissues, excited electrically or with glutamate. J. Neurochem. 16: 363-370.

Miledi, R., Molinoff, P. and Potter, L.T. (1971) Isolation of the cholinergic receptor protein of Torpedo electric tissue. Nature (Lond.), 229: 554-557.

Molinoff, P.B. and Kravitz, E.A. (1967) The metabolism of γ-aminobutyric acid (GABA) in the lobster nervous system - glutamate decarboxylase. J. Neurochem. 15: 391-409.

Neal, M.J. and Iversen, L.L. (1972) Autoradiographic localisation of ^{3}H-GABA in rat retina. Nature New Biol. 235: 217-218.

Neal, M.J. and Starr, M.S. (1973) Effect of inhibitors of γ-aminobutyrate aminotransferase on the accumulation of ^{3}H-γ-aminobutyric acid by the retina. Br. J. Pharmac. 47: 543-555.

Otsuka, M. and Miyata, Y. (1972) Application of enzymatic cycling to measurement of gamma-aminobutyric acid in single neurons of the mammalian nervous system. Advances in Biochem. Psychopharmacol. 6: 61-74.

Pardee, A.B. (1968) Membrane transport proteins. Science, 162: 632-637.

Pasantes-Morales, H., Klethi, J., Urban, P.F. and Mandel, P. (1974). The effect of electrical stimulation, light and amino acids on the efflux of ^{35}S-taurine from retina of domestic fowl. Exp. Brain Res. 19: 131-141.

Pitot, H.C. and Yatvin, M.B. (1973) Interrelationship of mammalian hormones and enzyme levels in vivo. Physiol. Review 53: 228-325.

Riggs, T.R., Pan, M.W. and Feng, H.W. (1972) Transport of amino acids into estrogen-primed uterus: III: Effect of the level of extracellular sodium ion. J. Biol. Chem. 247: 7128-7134.

Saito, K., Barber, R., Wu, J.-Y., Vaughn, J.E. and Roberts, E. (1974) Immunohistochemical localisation of glutamic acid decarboxylase in rat central nervous system at light microscopic level. Proc. Am. Soc. Neurochem. 5: 113-114.

Schon, F. and Kelly, J.S. (1974a) Autoradiographic localisation of [^{3}H] GABA and [^{3}H] glutamate over satellite glial cells. Brain Res. 66: 275-288.

Snodgrass, S.R. and Lorenzo, A.V. (1973) Transport of GABA from the perfused ventricular system of the cat. J. Neurochem. 20: 761-769.

Starr, M.S. (1973) Effects of changes in the ionic composition of the incubation medium on the accumulation and metabolism of ^{3}H-γ-amino-butyric acid and ^{14}C-taurine in isolated rat retina. J. Neurochem. 22: 1693-1700.

Takeuchi, A. and Takeuchi, N. (1966) A study of the inhibitory action of γ-aminobutyric acid on neuromuscular transmission in the crayfish. J. Physiol. 183: 418-432.

Van Gelder, N.M. (1965) A comparison of γ-aminobutyric acid metabolism in rabbit and mouse nervous tissue. J. Neurochem. 12: 239-244.

Van Gelder, N.M. (1971) Molecular arrangement for physiological action of glutamic acid and γ-aminobutyric acid. Can. J. Physiol. and Pharm. 49: 513-519.

Voaden, M.J., Marshall, J. and Murani, N. (1974) The uptake of [^{3}H] γ-aminobutyric acid and [^{3}H] glycine by the isolated retina of the frog. Brain. Res. 67: 115-132.

Werman, R. (1966) Criteria for identification of a central nervous system transmitter. Comp. Biochem. Physiol. 18: 745-766.

Zieglgänsberger, W. and Puil, E.A. (1973) Action of glutamic acid on spinal neurons. Exp. Brain. Res. 17: 35-49.

THE MICROIONTOPHORETIC APPROACH

E. PUIL

Centre de Recherche en Sciences Neurologiques
Université de Montréal
Montréal 101, Canada

O day and night, but this is wondrous strange!
Hamlet, Act 1, Sc. V

1. BASIC PRINCIPLES

The term "microiontophoresis" refers to the passage of an electric current through an ionised solution within a glass micropipette resulting in the ejection of a substance from the tip of the micropipette. Comprehensive reviews of this method of drug application have been written by Curtis (1965), Salmoiraghi and Weight (1967) and Krnjević (1964; 1971). For solutions of high conductivity, the amount of ionised substance ejected (Q) is a simple function of the total ionic current flowing through the solution:

$$Q = \frac{It\tau_n}{zF}$$

where Q is the number of ion equivalents, I is current in amperes (A), t is time in seconds, z is the valence, F is Faraday's constant and τ_n is the transport number. The transport number of a particular ion through the orifice is dependent on the concentration of ions within and just outside the tip of the micropipette. An ideal transport number is 1.0, but experimentally, values ranging from 0.01 to 0.6 have been obtained for pharmacologically active ions released *in vitro* (see Table 1, Krnjević 1971). Since these values serve only as rough estimates, the apparent rate of iontophoretic release *in vivo* can be only approximated by measuring the current applied to the solution. Electroosmosis becomes more important than iontophoresis when a potential is applied to solutions of low ionic strength. Such bulk flow depends on the presence of the boundary potential at the glass-solution interface, i.e., on the zeta potential. This phenomenon can be used to advantage when ejection from solutions of weak ionic strength is desirable. If strong ionization can be assured by adjusting the drug solution to an appropriate pH (see list of pK's compiled by Perrin, 1965), a current of 10^{-8} - 10^{-9} A may eject as little as 10^{-13} equivalents/sec.

Some estimates of the amount of a substance required, microiontophoretically, to excite central neurones have been made by Herz et al., (1969). Local concentrations around 10^{-6} M of glutamic acid were achieved by iontophoresis with very high currents and diminished over distances more than 200 m from the tip of the application electrode. These data depend on the heterogeneity of the medium, the amount of nonpermeable tissue present and the uptake of the substance by the tissue. In fact, there is no simple way of knowing the exact concentration achieved in the tissue, the extent of diffusion, or even where the application electrode is situated with relation to the cell under study. Nevertheless, the microiontophoretic technique has the major advantage that it permits the study of the predominantly postsynaptic effects and actions of drugs ejected in a controlled manner onto single neurones or onto a relatively small number of cells. Another feature of the method is that it circumvents the problem encountered with pharmacological agents or suspected neurotransmitters which do not easily reach central neurones by other routes of administration. For example, iontophoretic application of dopamine depresses spontaneous or illumination-evoked firing of retinal ganglion cells (Straschill and Perwein 1969). The action of this putative retinal neurotransmitter can therefore be studied directly without resort to the administration of its precursor, L-DOPA, by another route.

2. EXTRACELLULAR STUDIES

The simplest use of microiontophoresis is in combination with conventional extracellular recording. This technique can be employed for the study of the effects of possible synaptic transmitters or drugs (for recent reviews, see McLennan (1970), Phillis (1970), Curtis and Johnston (1974) and Krnjević (1974)). The typical effects of three neurotransmitter candidates, ejected from a five-barrelled micropipette during concomitant extracellular recording from the central barrel in a microiontophoretic study of cortical neurones, are shown in Fig. 1. The excitatory effect of glutamate (Fig. 1a), which is quite brief both in onset and in termination, is in contrast with that of acetylcholine (Fig. 1c), which is comparatively slow to start and exhibits a characteristically prolonged "after-discharge". Larger amounts of an excitant amino acid such as glutamate very often produce spike inactivation

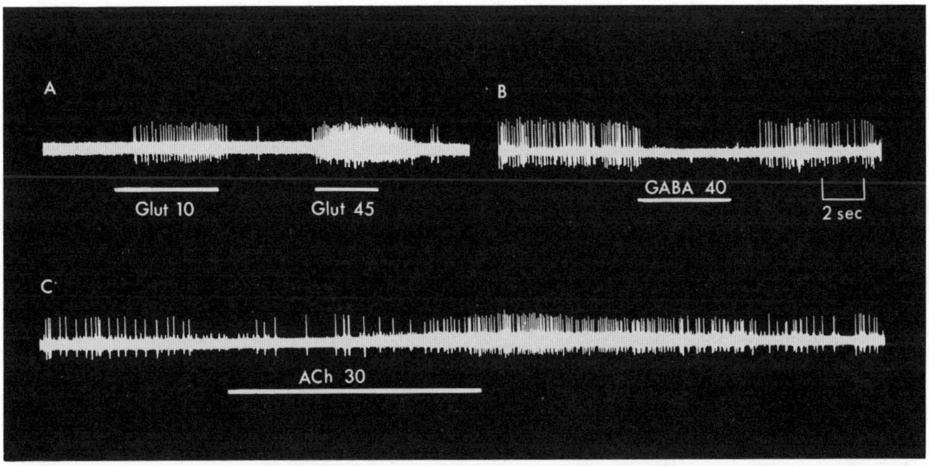

Fig. 1 Microiontophoretic studies recorded on oscilloscope and moving film. Cortical neurones showing excitation by L-glutamate (a) or ACh (c), and inhibition of ongoing discharge by GABA (b). The substances were applied during times indicated by white lines; the doses in this and other figures are in nA (10^{-9} A). Reprinted from "Microiontophoresis" by K. Krnjević in Methods of Neurochemistry, Vol. 1, edited by R. Fried, p. 149, by courtesy of Marcel Dekker, Inc.

and a reversible depression of unit firing. The depressant effect of GABA on spontaneous firing, or on discharge evoked by a concomitant application of glutamate or acetylcholine, is as brief in action as its excitant analogue. Both glutamate and GABA have also been found to have these effects on most other central neurones including those in the retina (Noell 1959; Straschill and Perwein 1969). However, the excitation of retinal ganglion cells by acetylcholine is almost as brief in onset and termination as that of glutamate (Noell and Lasansky 1959). Unlike the muscarinic effect illustrated in Fig. 1c, it would appear that the effect of acetylcholine in the retina is more akin to nicotinic excitation which acetylcholine produces in Renshaw cells (Curtis and Eccles 1958).

A. Possible artefacts in extracellular recording

Before a "true effect" can be ascertained, it is usually necessary to perform adequate controls by passing current through a barrel containing Na^+ or Cl^- at a comparable pH. For example, although the predominant effect of norepinephrine on central neurones, including retinal ganglion cells (Straschill and Perwein 1969), is a depression of their firing, excitation is sometimes observed (see review by Krnjević (1974). Frederickson et al., (1971) have demonstrated that an artefactual excitation of cortical cells occurs when norepinephrine is released from acidic solutions. This has been attributed to an excess of H^+ released simultaneously with norepinephrine from solutions adjusted to a pH of less than 4.0. In addition, some cells e.g., retinal ganglion cells (Noell and Lasansky 1959), are very sensitive to iontophoretic currents. Current artefacts (Fig. 2) are usually characterised by a practically instantaneous onset of a maximal effect, which often wanes,

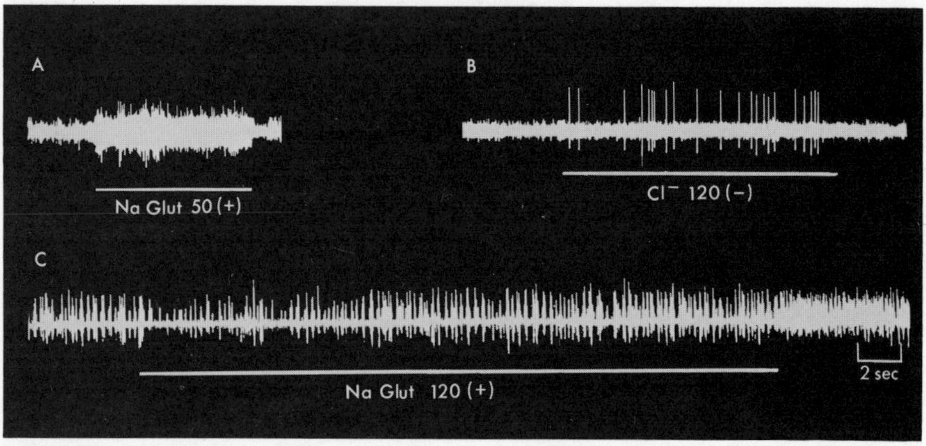

Fig. 2 Some current artifacts recorded with extracellular microelectrodes in cerebral cortex. (a) Excitation of cortical unit by an outward current, through Na glutamate (pipet positive). (b) Excitation by an inward current, through NaCl (pipet negative). (c) Depression of spontaneous firing by an outward current through Na glutamate. Note rebounds of excitability at end of current flow in (a) and (c). Reprinted from "Microiontophoresis" by K. Krnjević in Methods of Neurochemistry, Vol. 1, edited by R. Fried, p. 156, by courtesy of Marcel Dekker, Inc.

and terminates abruptly when the iontophoretic current is turned off. Neurones tend to be excited when the electrode tip is made positive and depressed when it is negative if the electrode is relatively close to the cell. In most cases, opposite effects are observed when the application current is reversed. Also, a rebound of excitability occurs when the current is switched off

(Fig. 2a and 2c). Current effects may be reduced by returning the iontophoretic current flow through an adjacent NaCl-containing barrel (see Salmoiraghi and Stefanis 1967).

B. Multicellular responses to drug application

There exists no compelling evidence that an effect observed upon microiontophoresis of a drug is solely a postsynaptic response of one neurone. For example, the depression of retinal ganglion cells by D-amphetamine administered microiontophoretically (Straschill and Perwein 1969) might be at least partly due to a presynaptic effect, such as a release of norepinephrine from presynaptic stores (cf. Goodman and Gilman 1970). The effect of a drug on many cells is, of course, often observed when the substance is added to a tissue bath, but it may sometimes be observed after iontophoretic application. As an example, harmaline, which is a potent hallucinogen (Naranjo 1967), causes a rapid postural tremor when administered intravenously in high doses. This has been attributed to a production of rhythmic activity within the inferior olive, which is the source of climbing fibre input to the cerebellum (de Montigny and Lamarre 1973). Harmaline also evokes the appearance of rhythmic discharge when administered microiontophoretically in the inferior olive (Biscoe et al., 1973; Lamarre and Puil 1974; Fig. 3a). One of the problems with this sort of study (see also

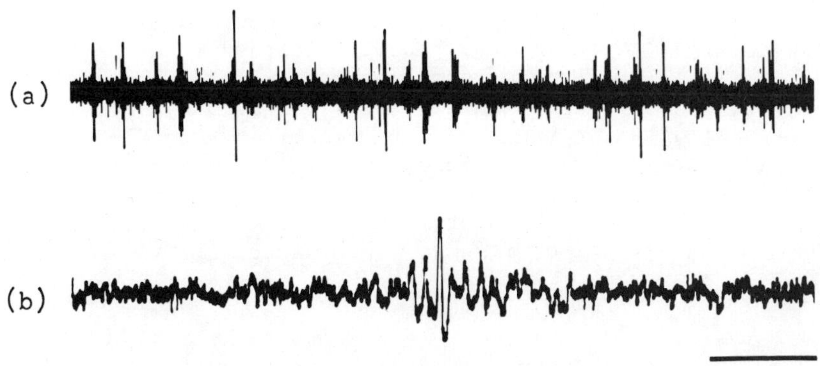

Fig. 3 Extracellular burst activity evoked by harmaline in the inferior olive. (a) Bursts appeared a few seconds after termination of a 75 nA application of harmaline. (b) Composition of bursts shown on faster time scale indicates participation of more than one cell. Time scales: 1s in (a), 20 ms in (b).

Puil et al., 1974) is the difficulty in discriminating the activity coming from one cell from that coming from a group of cells. However, it is probable that this effect can be attributed to the excitation of a group of olivary cells discharging in synchrony particularly in view of the recent demonstration (Llinás et al., 1974) of strong electrical coupling between olivary cells. Clearly, more definitive evidence about the mechanism of action of harmaline might be obtained from intracellular studies which are difficult because of poor recording conditions in the inferior olive.

It is sometimes possible to obtain reasonable proof, extracellularly, that more than one cell is involved in a response to a drug applied iontophoretically. In the cerebral cortex, for example, acetylcholine may sometimes cause a depression of unit discharge (Fig. 4). The depressant effect might be due to a reduction of

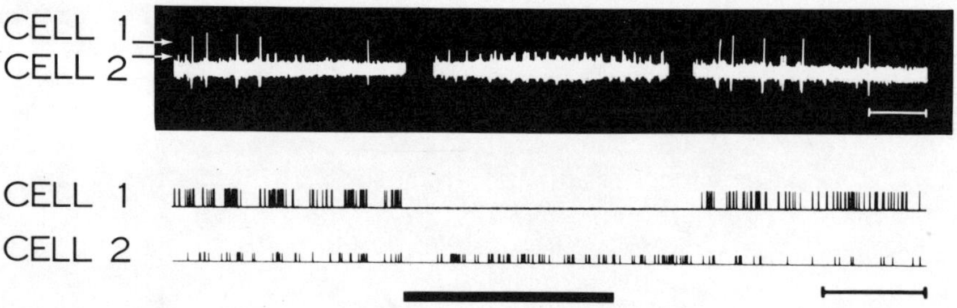

Fig. 4 Effect of acetylcholine on extracellular firing of cortical neurone (depth 270 μm) excited by glutamate application (20 nA). Upper records are comprised of four superimposed oscilloscope sweeps taken before (left) during (center) and after (right) acetylcholine application (40 nA). (b) Penwriter recording of standard pulses triggered by spikes discriminated at two levels as above. Time scales: 50 ms in upper traces, 5 s in lower record. From experiments in collaboration with Dr. H.H. Jasper.

anionic glutamate release by a cationic current applied to the adjacent acetylcholine barrel (e.g., see onset of application in Fig. 4 below) or a result of an inhibitory transmitter-like action, particularly in the superficial layers (see review by Krnjević 1974). Neither explanation seems to be adequate for the results presented in Fig. 4, where two different populations of spikes are present. If the two amplitudes are discriminated and trigger standard pulses which are printed out on separate channels (Fig. 4

below), it becomes evident that the discharge of the larger spike is blocked during smaller spike discharge. Such data suggest that the action of acetylcholine in this case is excitatory on a smaller, or more distant neurone, which has an inhibitory influence on the cell which discharged the larger spike (see Fig. 5a; cf., Krnjević 1964). A similar type of mechanism has been considered for some effects of glutamate in the spinal cord (Zieglgänsberger and Herz 1971).

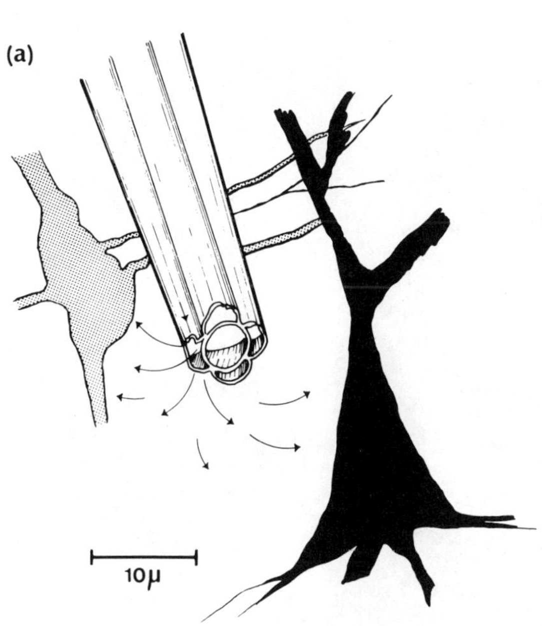

Fig. 5 Schematic representation of recording arrangement involving microiontophoresis in the cerebral cortex. (a) Illustrates the possibility that the release of a substance from one barrel (arrows) may affect a smaller, or more distant cell, which has inhibitory connections with larger cell from which the largest spike is recorded extracellularly via the central barrel. (b) Parallel electrode assembly for extracellular drug application from three different barrels during intracellular recording from single pipette is shown in relation to impaled cell.

3. INTRACELLULAR STUDIES

One micropharmacological method which can be used to determine the mechanism of action of a drug on a single neurone is the combination of microiontophoresis with conventional intracellular recording (Krnjević and Schwartz 1967; Curtis 1968; Zieglgänsberger and Puil 1973a; Fig. 5b). Using this technique, changes in neuronal membrane properties produced by drug application can be measured (Fig. 6) via the intracellular pipette, which can also be

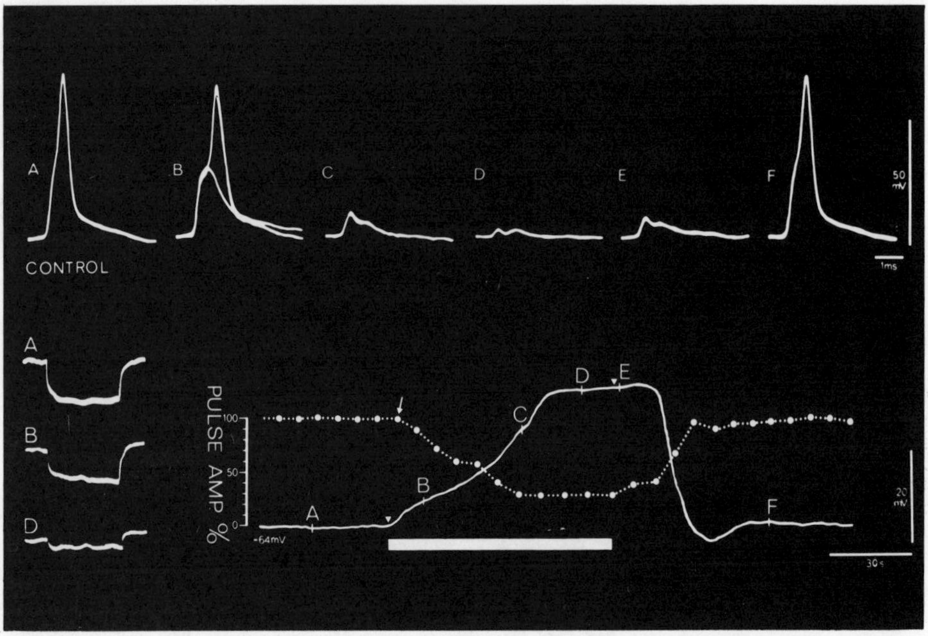

Fig. 6 Intracellular recording from a biceps-semitendinosus motoneurone. 200 nA of GLUT was applied. Intertip distance, 85 µm. An antidromic spike is evoked (two traces superimposed). A, control; B-D, recordings taken at different levels of depolarisation caused by GLUT; E-F, during recovery. Note that a depolarisation has already occurred before the point (arrow) where the conductance begins to change. Right bottom: penwriter recording of the membrane potential; letters refer to the recordings above. Bar indicates period of application. Dotted line shows simultaneous measurements of conductance changes (%). The amplitude of 4 consecutive current pulses were averaged. Left bottom: hyperpolarising test pulses (duration 20 msec). A, B, D taken before the corresponding antidromic spikes. Resting potential, -64 mV; initial membrane resistance, 1.2 MΩ. Stimulus frequency, 1/s. From Zieglgänsberger and Puil (1973a), by permission of Springer Verlag, Inc.

used for drug, ion, or dye injection. The possibility that one is recording from an impaled nerve fibre must at first be eliminated. Nerve fibres can be identified by the fast rise time of their action potentials, the lack of membrane potential swing, and very often their anatomical location. Also, fibres are not affected by glutamate even when very large amounts are iontophoresed for long periods of time (Zieglgänsberger and Puil 1973a; see also Straschill and Perwein 1969). During glutamate iontophoresis, both the membrane potential and resistance of a cell (as measured by injecting hyperpolarising pulses through the recording barrel - see left of Fig. 6), are observed to decrease. If the conductance increase produced by the application is large, failure of antidromic spike invasion can occur, resulting in a depolarisation block (Fig. 6 above). Both glutamic and aspartic acids produce a similar type of inactivation of horizontal cells (Cervetto and MacNichol 1972) but have no effect on photoreceptors in the retina (Murakami et al., 1972).

Substances that depolarise central neurones can decrease membrane conductance. Acetylcholine, for example, causes a depolarisation of cortical (Krnjević et al., 1971) and spinal (Zieglgänsberger and Reiter 1974) neurones, in association with a rise in membrane resistance. The mechanism underlying the depolarisation in the cortex is believed to be a consequence, although not a simple one, of a reduction in membrane permeability to K^+ (Godfraind et al., 1971; Krnjević et al., 1971a; 1971b).

In some cases a hyperpolarisation may be observed upon application of a putative excitatory transmitter. Iontophoretic applications of glutamate can cause an initial hyperpolarisation (Bernardi et al., 1972; Zieglgänsberger and Puil 1973a). This effect is not infrequently observed when the intertip distances of the micropipette assembly are long. It may possibly be due to excitation of a neurone which has an inhibitory influence on the neurone under study (see above and Fig. 5b). A second interesting possibility is that glutamate may activate on the same neurone, other types of receptors, such as the nonsynaptic glutamate receptors that increase chloride conductance in muscle membrane (Cull-Candy and Usherwood 1973). Acetylcholine is known to activate more than one type of receptor in some systems (Kehoe 1972).

A. Possible artefacts

The depolarisation produced by an iontophoresed substance may be partially masked by phoretic current (Fig. 7). When glutamate is applied as an anion, the membrane potential jumps in the hyperpolarising direction (a in Fig. 7) and exhibits a greater jump upon termination of the iontophoretic current (b in Fig. 7). The reason for this difference in coupling is that the membrane resistance was

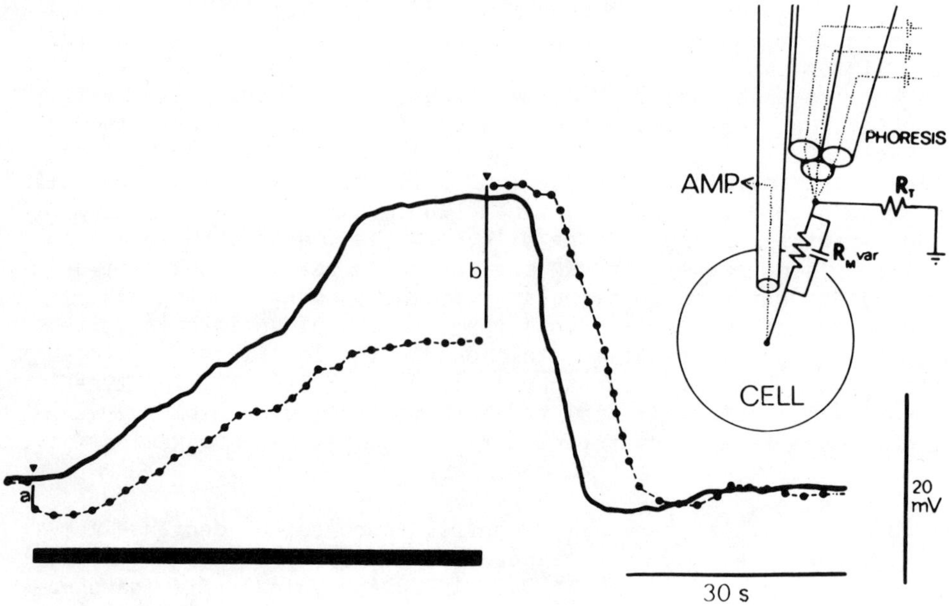

Fig. 7 Recordings of the changes in the membrane potential of a motoneurone after GLUT application by different modes. 300 nA of GLUT was applied. Intertip distance, 80 µm. Solid line: the phoretic current was compensated for by a current of same magnitude but of opposite polarity passed through an adjacent NaCl-filled barrel. Dotted line: uncompensated phoretic current. (a) and (b) show the coupling of the phoretic current at "on" and "off" respectively. Insert: schematic drawing illustrating the recording and application arrangement. R_m represents the membrane resistance which is altered by GLUT application. R_t is the resistance of the nervous tissue which is assumed to remain fairly constant during application. Note that the cell is slightly more depolarised and recovery takes longer after application by the uncompensated mode. From Zieglgänsberger and Puil (1973a), by permission of Springer Verlag, Inc.

reduced by the action of glutamate and hence the "true" membrane potential was less attenuated when the current was switched off. The current effect which prevented the real time course of depolarisation from being seen can be overcome, at least partially, by passing a current of the same magnitude, but of opposite polarity, through an adjacent NaCl-containing barrel.

B. Reversal potential determination

Although there seems to be plenty of evidence to suggest that GABA and dopamine, among other substances (see Descarries, de Champlain and van Gelder in this volume), may be inhibitory transmitters in the retina, crucial evidence that their mechanisms of action are identical to the physiological transmitters has not yet been presented. The most sophisticated and most difficult method of determining the mechanism of action of a drug in the central nervous system is the measurement of reversal potential for its action. Since the effect of a drug is abolished at this potential, knowledge about what ions are involved in a conductance change may be obtained from a comparison of the reversal potential with the ionic equilibrium potentials (see also Werman et al., 1968). Also, strong proof for the identity of a neurotransmitter may be obtained if the reversal potentials of a membrane transient and the suspected transmitter substance are the same (see Werman 1966). The usual procedure is to polarise the cell by current injection through the recording barrel, and during drug application, note the potential at which drug action is abolished. The second method is to cause a state of isopotential by iontophoresing large amounts of a drug onto the cell, until further drug application produces no more change in membrane potential. Unfortunately, there are difficulties associated with this technique, such as poor recording stability or the failure of the iontophoresed drug to reach all ramifications of a neurone. The experimental requirement of having to inject more current into the soma in order to reverse the effect of a drug or a neurotransmitter originating elsewhere on the neurone tends to limit the usefulness of reversal potential determinations.

C. Concomitant use of other agents

The use of other drugs with known actions may help to uncover the mechanism of action of a substance in question. Conclusive evidence for a specific antagonism of an agonist must include the demonstration that the antagonist itself causes no changes in membrane properties which could account for its blocking effect. Thus, evidence exists that the diethylester of glutamate changes neither membrane potential nor conductance but blocks the glutamate response at a neuromuscular junction (Lowagie and Gerschenfeld 1974). However, it has been shown that the diethylester of glutamate (Curtis et al., 1972) and α-methylglutamate block the extracellular responses of central neurones evoked by glutamate and not those evoked by acetylcholine (Haldeman et al., 1972) but these substances change the membrane resistance and potential of spinal neurones (Zieglgänsberger and Puil 1973b). Hence, the "antagonism" could be attributed to the fact that a cell depolarised by glutamate does not reach the firing level during hyperpolarisation often caused by the "blocker". More importantly, however, the lowered membrane resistance would provide an effective shunt of the current required for spike generation. This would further decrease the firing probability of neurones depolarised by substances with glutamate-like action, i.e., by agents or transmitters which act by increasing membrane conductance. Clearly, the action of α-methylglutamate or the diethylester of glutamate cannot be considered "true antagonism" in the mammalian central nervous system.

An example of a drug which has become a powerful pharmacological tool in the study of excitable tissue is tetrodotoxin (TTX). This substance blocks the specific increase of Na^+ conductance during spike genesis leaving the delayed K^+ conductance change relatively unaltered (Kao 1966). TTX, however, does not very effec-

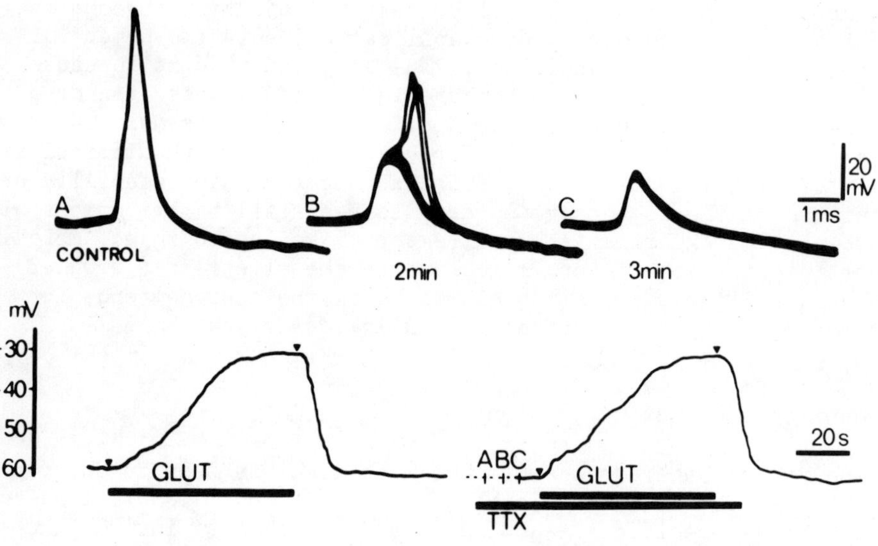

Fig. 8 Effects of TTX on an antidromic spike evoked in a motoneurone and on the depolarising action of glutamate. Upper part: antidromic spike (A-C, five superimposed oscilloscope sweeps each). 80 nA of TTX was applied from an electrode 100 µm distant from the tip of the recording electrode (applications indicated by bars). Lower part: glutamate application (GLUT, 200 nA) before and during TTX electrophoresis. A-C indicate where recordings given above were taken. Second glutamate application was started when the spike was blocked (C). Note that the cell is depolarised to about the same level in both cases. From Zieglgänsberger and Puil (1972) by permission of Nature.

tively block the influx of Na^+ into brain incubates stimulated with glutamate (McIlwain et al., 1969; Okamoto and Quastel 1970). When TTX is applied microiontophoretically to central neurones, there is a pronounced rise in the threshold amount of injected current required to evoke a spike (Zieglgänsberger and Puil 1972). Although the electrical excitability of the neurone can be blocked with TTX (see Fig. 8 above), the depolarising action of glutamate persists (Fig. 8 below).

These data provide presumptive evidence that two types of Na^+ channels may exist and may be affected by different agents.

4. CONCLUDING REMARKS

The technique of microiontophoresis, which is relatively simple in principle, has been used in studies involving the effects and actions of drugs on single neurones in the central nervous system. The method possesses the advantage of permitting controlled drug ejection onto a limited area. Also, substances of neurophysiological interest which do not gain access to the tissues of the central nervous system by other routes of administration can be applied directly by microiontophoresis. The postsynaptic responsiveness or other neuronal membrane properties may be tested, with a high degree of reproducibility in a large population of neurones. These features may be advantageous in investigations of the retina where the technique has only been rarely employed.

REFERENCES

Bernardi, G., Zieglgänsberger, W., Herz, A. and Puil, E.A. (1972). Intracellular studies on the action of L-glutamic acid on spinal neurones of the cat. Brain Res. 39: 523-525.

Biscoe, T.J., Duggan, A.W., Headley, P.M. and Lodge, D. (1973) Rhythmical field potentials induced in the inferior olive complex by iontophoretically applied harmaline and other unrelated alkaloids. Br. J. Pharmacol. 49: 174P-175P.

Cervetto, L. and Mac Nichol, E.F. (1972). Inactivation of horizontal cells in turtle retina by glutamate and aspartate. Science 178: 767-768.

Cull-Candy, S.G. and Usherwood, P.N.R. (1973). Two populations of glutamate receptors on locust muscle fibres. Nature (New Biol.) 246: 62-64.

Curtis, D.R. (1964). Microelectrophoresis. Physical Techniques in Biological Research, Vol. V, Electrophysiological Methods, Part A, edited by W.L. Nastuk, Academic Press, New York, pp. 144-190.

Curtis, D.R. (1968). A method for assembly of "parallel" micropipettes. Electroenceph. clin. Neurophysiol. 24: 587-589.

Curtis, D.R., Duggan, A.W., Felix, D., Johnston, G.A.R., Tebēcis, A.K. and Watkins, J.C. (1972). Excitation of mammalian central neurones by acidic amino acids. Brain Res. 41: 283-301.

Curtis, D.R. and Eccles, R.M. (1958). The excitation of Renshaw cells by pharmacological agents applied electrophoretically. J. Physiol. 141: 435-445.

Curtis, D.R. and Johnston, G.A.R. (1974). Amino acid transmitters in the mammalian central nervous system. Ergeb. Physiol. 69: 97-188.

de Montigny, C. and Lamarre, Y. (1973). Rhythmic activity induced by harmaline in the olivo-cerebello-bulbar system. Brain Res. 53: 81-95.

Frederickson, R.C.A., Jordan, L.M. and Phillis, J.W. (1971). The action of noradrenaline on cortical neurones: effects of pH. Brain Res. 35: 556-560.

Godfraind, J.M., Kawamura, H. and Krnjević, K. and Pumain, R. (1971). Actions of dinitrophenol and some other metabolic inhibitors on cortical neurones. J. Physiol. 215: 199-222.

Goodman, L.S. and Gilman, A. (1970). The Pharmacological Basis of Therapeutics. 4th ed. Macmillan, New York, p. 504-505.

Haldeman, S., Huffman, R.D., Marshall, K.C. and McLennan, H. (1972). The antagonism of the glutamate-induced and synaptic excitations of thalamic neurones. Brain Res. 39: 419-425.

Herz, A., Zieglgänsberger, W. and Färber, G. (1969). Microelectrophoretic studies concerning the spread of glutamic acid and GABA in brain tissue. Exp. Brain Res. 9: 221-235.

Kao, C.Y. (1966). Tetrodotoxin, saxitoxin, and their significance in the study of excitation phenomena. Pharmacol. Rev. 18: 997-1049.

Kehoe, J. (1972). The physiological role of three acetylcholine receptors in synaptic transmission in Aplysia. J. Physiol. 225: 85-172.

Krnjević, K. (1964). Micro-iontophoretic studies on cortical neurones. Int. Rev. Neurobiol. 7: 41-98.

Krnjević, K. (1971). Microiontophoresis. Methods of Neurochemistry, Vol. 1, edited by R. Fried, Marcel Dekker, New York, p. 129-172.

Krnjević, K., Pumain, R. and Renaud, L. (1971a). Effects of Ba^{2+} and tetraethylammonium on cortical neurones. J. Physiol. 215: 223-245.

Krnjević, K., Pumain, R. and Renaud, L. (1971b). The mechanism of excitation by acetylcholine in the cerebral cortex. J. Physiol. 215: 247-268.

Krnjević, K. and Schwartz, S. (1966). The action of γ-aminobutyric acid on cortical neurones. Exp. Brain Res. 3: 320-336.

Lamarre, Y. and Puil, E. (1974). Induction of rhythmic activity by harmaline. Can. J. Physiol. Pharmacol. 52: 905-908.

Llinás, R., Baker, R. and Sotelo, C. (1974). Electrotonic coupling between neurons in the cat inferior olive. J. Neurophysiol. 37: 560-571.

Lowagie, C. and Gerschenfeld, H.M. (1974). Glutamate antagonists at a crayfish neuromuscular junction. Nature 48: 533-535.

Noell, W.K. (1959). The visual cell: Electric and metabolic manifestations of its life processes. Am. J. Ophthalmol. 48: 347-370.

Noell, W.K. and Lasansky, A. (1959). Effects of electrophoretically applied drugs and electrical currents on the ganglion cell of the retina. Fed. Proc. 18: 115.

Mc Ilwain, H., Harvey, J.A. and Rodriguez, G. (1969). Tetrodotoxin on the sodium and other ions of cerebral tissues, excited electrically and with glutamate. J. Neurochem. 16: 363-370.

McLennan, H. (1970). Synaptic Transmission, 2nd ed. Saunders, Philadelphia.

Murakami, M., Ohtsu, K. and Ohtsuka, T. (1972). Effects of chemicals on receptors and horizontal cells in the retina. J. Physiol. 227: 899-913.

Naranjo, C. Psychotropic properties of harmala alkaloids. Ethnopharmacologic search for psychoactive drugs. U.S. Public Health Serv. Pub. No. 1645, edited by D.H. Efron, B. Holmstedt and N.S. Kline. Gov. Pr. Of., Washington, D.C., p. 385-391.

Okamoto, K. and Quastel, J.H. (1970). Tetrodotoxin-sensitive uptake of ions and water by slices of rat brain _in vitro_. Biochem. J. 120: 37-47.

Perrin, D.D. (1965). Dissociation constants of organic bases in aqueous solution, Butterworths, London.

Phillis, J.W. (1970). The pharmacology of synapses, Pergamon Press, New York.

Puil, E., Reiffenstein, R.J. and Triggle, C. (1974). Epileptiform after discharges and chemical responsiveness of cortical neurones. Electroencephalogr. Clin. Neurophysiol. 36: 265-273.

Salmoiraghi, G.C. and Stefanis, C.N. (1967). A critique of iontophoretic studies of central nervous system neurones. Int. Rev. Neurobiol. 10: 1-30.

Salmoiraghi, G.C. and Weight, F. (1967). Micromethods in neuropharmacology: an approach to the study of anesthetics. Anesthesiology 28: 54-64.

Straschill, M., and Perwein, J. (1969). The inhibition of retinal ganglion cells by catecholamines and γ-aminobutyric acid. Pflügers Arch. 312: 45-54.

Werman, R. (1966). Criteria for identification of a central nervous system transmitter. Comp. Biochem. Physiol. 18: 745-766.

Werman, R., Davidoff, R.A. and Aprison, M.H. (1968). The inhibitory action of glycine on spinal neurons in the cat. J. Neurophysiol. 31: 81-95.

Zieglgänsberger, W. and Herz, A. (1971). Changes of cutaneous receptive fields of spino-cervical-tract neurones and other dorsal horn neurones by microelectrophoretically administered amino acids. Exp. Brain Res. 131: 111-126.

Zieglgänsberger, W. and Puil, E.A. (1972). Tetrodotoxin interference of CNS excitation by glutamic acid. Nature (New Biol.) 239: 204-205.

Zieglgänsberger, W. and Puil, E.A. (1973a). Actions of glutamic acid on spinal neurones. Exp. Brain Res. 17: 35-49.

Zieglgänsberger, W. and Puil, E.A. (1973b). Intracellular investigations on the effect of microelectrophoretically applied glutamate antagonists upon spinal neurones of the cat. Naunyn - Schmiedebergs Arch. Pharmakol. 275: Suppl. R89. 99

Zieglgänsberger, W. and Reiter, Ch. (1974). A cholinergic mechanism in the spiral cord of cats. Neuropharmacology 13: 519-527.

METHODS OF STUDY FOR BIOGENIC AMINES

Jacques de Champlain

Centre de Recherche en Sciences Neurologiques

Université de Montréal, Canada

> Thou hast no speculation in those eyes
> which thou dost glare with!
> Macbeth, Act III, Sc. 4

Due to the complexity of the organisation of the central nervous system, the role of numerous putative neurotransmitters is still poorly understood and the mechanisms of neurotransmission have often been extrapolated from knowledge obtained in peripheral organs. Due to its regular cytoarchitecture and to its accessibility to electro-physiological recording, the retina constitutes probably the best and simplest preparation for the study of synaptic transmission in nervous tissues. With the use of recent developments and refinements in the techniques of study of neurotransmitters, this model could be used more efficiently for the purpose.

In recent years, histochemical, pharmacological and biochemical studies have supported the existence of a network of dopaminergic neurons and fibres at the junction of the inner nuclear and inner plexiform layers in the retina of several species. Although considerable data have been obtained on the anatomical distribution of those fibres, relatively little is known on the physiologic and functional role of this neurotransmitter. Since the retina is a scanty tissue, the sensitivity limit of the techniques currently used for the measurement of catecholamines has prevented precise studies on the role of dopamine in the mechanisms of vision.

Although catecholamines have been known to serve as neurotransmitters in the peripheral sympathetic systems for over 50 years, most of our knowledge of the function of those fibres has been obtained only in the last decade mainly due to the development of highly sensitive and specific biochemical and histochemical techniques, to the availability of radioactive amine with a high specific activity and also to the development of highly specific adrenergic drugs. In recent years, the fundamental studies of von Euler, Axelrod and many others have contributed to the refinement of our concepts on adrenergic mechanisms (for most recent advances, see Usdin and Snyder, 1973).

A variety of highly specific technical approaches have been designed over the years for the study of the various nerve functions. It is now possible to investigate with precision the uptake, release, storage, metabolism, turnover and synthesis rate of catecholamines. However, since most of these technical advances are quite recent, relatively few organs have been studied using them and considerable amount of information will be forthcoming when these are applied to certain preparations.

The introduction of the histofluorescence technique of Falck and Hillarp (1962), has constituted by far the greatest advance in catecholamine research. The technique made it possible for the first time, to visualise neurotransmitters in their natural location, and map the exact localisation of aminergic fibers in the central nervous system and various peripheral organs.

Briefly, this method involves the treatment of freezedried tissues with formaldehyde gas, followed by their imbedding in paraffin, sectioning and examination in a fluorescence microscope. Certain thin membranes, such as the iris, atria and mesenterium can be prepared as a stretch preparation on glass slides, dryed for a few hours, treated with formaldehyde gas and studied immediately with the fluorescence microscope. The complete dehydration of the tissue is a critical step for the formaldehyde reaction and for the preservation in situ of the fluorescent products resulting from the reaction with the neurotransmitters. A freeze dryer has been specifically designed for insuring the best results with this histochemical technique (Olson and Ungerstedt, 1970). When enclosed in a dry protein layer, certain primary catecholamines (such as noradrenaline and dopamine) and certain tryptamines (such as serotonin) are easily converted to tetradydroisoquinolines and beta-carbolines respectively. These compounds are then dehydrogenated in a protein and/or amino-acid catalysed reaction to the corresponding intensely fluorescent 3-4, dihydro compound. Secondary catecholamines (such as adrenaline) can also be condensed with formaldehyde in an analogous reaction. But the dehydrogenation step, which gives a quartenary 3, 4, -dihydro-isoquinoline, requires more energy and necessitates longer incubation time in the presence of paraformaldehyde vapors. This technique is highly specific for these amines and their metabolites do not interfere with the reaction. The specificity of the technique has also been demonstrated by the total disappearance of the fluorescence after treatment with catecholamine depleting drugs or after denervation. (For more details on the use of the technique, see Norberg, 1967).

On direct microscopic examination, it is impossible to distinguish between the fluorescent compounds derived from noradrenaline and those derived from dopamine since both give a yellow-green fluorescence. However, with the help of microspectrofluometers, it has been possible to differentiate between these two compounds.

The sensitivity of the technique allows the detection of concentration of catecholamine higher than 2 μg/gram of tissue. Since the amine concentration is generally higher than this level in the cell body and in the terminal fibres, these two parts of the neuron can generally be visualised with this technique. The axon is generally not visible however, because the concentrations are lower than the limit of sensitivity. The axon becomes visible if the concentration of amine is increased in that structure by various procedures (Norberg, 1967).

The histofluorescence technique has been extensively used for the anatomical study of the retina of various animal species (Ehinger et al., 1966, 1967, 1969a, 1969b; Haggendal and Malmfors, 1965; Laties and Jacobowitz, 1966; Malmfors, 1963; Nichols et al. 1967; Sano et al. 1968). All these studies have uncovered a system of dopamine containing amacrine cells (about 10% of the amacrine cell population) localised at the junction of the inner plexiform and inner nuclear layer. Moreover, occasionnal alloganglion fluorescent cells have also been demonstrated. The dopamine containing cells give several processes which form one to three layers of dopaminergic fibers terminating within the inner plexiform layer depending on the species. Interestingly, the position of these layers was found to be the same in mammals, birds and fishes. The importance of these various sublayers varied according to the animal. The innermost layers are best developed in fishes whereas the outermost layer is more important in mammals. Moreover, a plexus of fluorescent fibers forming baskets of terminals of synaptic character was demonstrated around the middle horizontal cells (Ehinger et al. 1969a).

In addition to being useful for the anatomical localisation of aminergic fibers, the histofluorescence method has been used more dynamically in combination with drugs to study problems related to the synthesis, storage, release and metabolism of the amines in neurons and their terminals (Norberg, 1967). Although this dynamic approach has been used extensively in some preparation (such as the iris), it has not been used often for the studies on the retina. The only study in which this type of approach was used is that of

Nichols et al. (1967) in which they observed greater fluorescence of cells and fibres in the retina of rabbit and guinea pig after one hour in light than after 24 hours in total darkness, this suggests an accumulation of dopamine in these neurons during light exposure. The use of the histofluorescence technique is far from being exhausted for the study of the retina and much more information could be obtained on the dynamics of this dopaminergic network of neurons.

Recently, Axelsson, Bjorklund and co-workers (1972, 1973) have devised a new fluorescent histochemical technique using glyoxylic acid vapours instead of paraformaldehyde vapours for the formation of fluorophores. They have shown that the yield of fluorophores is considerably higher in the glyoxylic acid vapour reaction than in the formaldehyde vapour reaction of the standard Falck-Hillarp method. The major advantage with this technique is that it can be processed on 20 micron sections of fresh tissue made with a vibrotome. After drying in a dessicator for a few hours, the tissue is exposed to glyoxylic vapours thus eliminating the freeze drying procedure of the Falck and Hillarp technique, which usually takes 3 to 4 days. Moreover, this glyoxylic acid technique was reported to be more sensitive and to yield more details on the fibers distribution. This technique can certainly be used favourably to study, in greater details, the distribution of the dopaminergic system of fiber in the retina as well as their synaptic relation with the various neurons of the retina.

A variety of spectrofluorometric biochemical methods have been devised for the measurement of catecholamines. Although most of these are highly specific and rather sensitive, their inferior limit of sensitivity is in the order of 5 to 10 nanograms which does not allow for the accurate measurement of catecholamine in individual retina. Although a few studies have reported the presence of dopamine in the retina, the amount of information has been restricted by the sensitivity limits of these methods. The recent development of enzymatic radioisotopic techniques has permitted to push the limit of sensitivity to less than 50 picograms of catecholamine. The method reported by Coyle and Henry (1973) for the measurements of catecholamine seems to be the most sensitive so far available as well as the shortest. The principle of this technique is based on the conversion of noradrenaline, adrenaline and dopamine to their tritiated methylated metabolites in the presence of purified liver catechol-O-methyl-transferase and tritiated S-adenosyl-methionine as a labelled methyl donor. Through the extraction procedures, the metabolites derived from noradrenaline and adrenaline can be completely separated from those derived form dopamine so that either type of catecholamines can be measured accurately without interference in the same tissue sample. To achieve this separation, the purified metabolites are incubated in the presence of periodate which splits the side-chain with a hydroxyl group in position beta (such as with the methylated metabolites of adrenaline and noradrenaline) to yield tritiated vanillin which can be extracted from the mixture with toluene. The remaining methoxytyramine which is the methylated metabolite of dopamine is extracted in isoamyl and counted in a separate vial. The amount of tritiated metabolites formed is linearly proportional to the amount of substrate for concentrations varying between 50 picograms and 20 nanograms thus allowing for the determination of a wide range of concentrations. This technique is highly specific for catecholamine, and the presence of catecholamine precursors or metabolites does not interfere with the sensitivity or the specificity of the technique. The only drawback with this technique, is that it is not possible, at present, to estimate separately the concentration of noradrenaline and adrenaline.

With such sensitive technique, it has become possible to measure the catecholamine content in individual retina, or even in small portions of the retina. Using the technique, we have been able to detect easily the presence of catecholamine and dopamine in the retina of Goldfish (Table 1). The concentration of dopamine was found to be greater than catecholamine in this tissue. The presence of catecholamine in that tissue could be due to the presence of sympathetic fibres localized in blood vessels which might have been present in the dissected retina, or else it could correspond to the existence of noradrenergic or adrenergic neurons in the retina from that species. Interestingly the dopamine concentration was found to be significantly greater in the retina of fishes exposed to light for 24 hours than in fishes kept in darkness for the same period, thus suggesting a role for that transmitter in the mechanism of vision. On the other hand, catecholamine levels did not differ significantly in the dark or light adapted retina. These results are in agreement with those of Nichols et al (1967) who reported an increase in dopamine levels in the retina of light exposed rabbits, guinea-pigs and rats. It is difficult however, to understand the exact significance of those findings by simply measuring endogenous levels. For instances, it could mean that the synthesis of dopamine is increased in keeping with an increased release rate, or else that the endogenous stores increased due to a decrease in release. In order to understand the mechanisms and significance of these changes, studies on the synthesis and turnover rates should be made. With the use of the more sensitive microtechniques, it is possible to follow the decline rate of endogenous dopamine after synthesis blockade with alpha methyl-p-tyrosine in the retina exposed to various physiological situations. Theoritically, it would also be possible to estimate the release of dopamine or catecholamine from retina exposed to various stimuli during in vitro incubations.

Labelled dopamine has been used only by one group of investigators (Kramer, 1971; Kramer et al., 1971) in a physiological preretinal perfusion set-up on the cat retina. These studies have

TABLE 1

Catecholamines and dopamine concentration in Goldfish retina.

	Catecholamines (ng/mg)	Dopamine (ng/mg)
DARK ADAPTED	0.138 ± 0.012	0.194 ± 0.034
LIGHT ADAPTED	0.114 ± 0.003	0.296 ± 0.012*

*$p < 0.05$ vs dark adapted retina.

The numbers represent the mean ± S.E. of groups of four animals.

yielded important informations on the uptake and release of dopamine. It was found that flases released larger quantities of dopamine in the light adapted preparation, than in the dark adapted retina. This type of experimental set-up could favourably be used for the study of endogenous dopamine release using the enzymatic microtechnique.

The number of pharmacological studies using adrenergic drugs has been rather restricted on that preparation probably due to the existence of a blood-retinal barrier. In recent years, 6-hydroxydopamine has been found to have the unique capacity to destroy selectively and specifically adrenergic and dopaminergic fibres in the central and peripheral system (See De Champlain and Nadeau, 1971). The specificity of that compound resides in the fact that it is concentrated specifically into the nerve ending through the amine uptake mechanism so that it reaches toxic concentrations only at those sites. The use of that compound to destroy specifically the dopaminergic fibres network in the retina should prove to be a very valuable tool for the study of the role of those fibres.

In conclusion, it appears that recent advances in catecholamine research could better be applied to the study of the role of transmitters in the mechanisms of vision. Several powerful tools are now available and only await the enthousiasm of some investigators. There is certainly a need and also a possibility to establish sound correlation between physiological parameters and the biochemical changes in these transmitters. The gathering of anatomists, pharmacologists, biochimists and neurophysiologists around the same retinal preparation could yield exciting new informations for a better understanding of neurotransmitter functions.

Acknowledgement: The author is grateful for the skillful and efficient assistance provided by Ms. Solange Imbeault, Lise Farley, Johanne Le Querrier and Marie-Colette Renaud in the preparation of this manuscript.

REFERENCES

Axelsson, S., Bjorklund, A., Falck, B., Lindvall, O. and Svensson, L.A. (1973) Glyoxylic acid condensation: a new fluorescence method for the histochemical demonstration of biogenic monoamines. Acta. Physiol. Scand. 87: 57-62.

Bjorklund, A., Lindvall, O. and Svensson, L.A. (1972) Mechanism of fluorophore formation in the histochemical Glyoxylic Acid method for monoamines. Histochimie 32: 113-131.

Coyle, J.T. and Henry, D. (1973) Catecholamines in fetal and newborn rat brain. J. Neurochem. 21: 61-67.

De Champlain, J. and Nadeau, R.A. (1971) 6-Hydroxydopamine, 6-Hydroxydopa and degeneration of adrenergic nerves. Fed. Proc. 30: 877-885.

Ehinger, B. (1966) Distribution of adrenergic nerves in the eye and some related structures in the cat. Acta. Physiol. Scand. 66: 123-128.

Ehinger, B. (1967) Adrenergic nerves in the avian eye and ganglion. Z. Zellforsch. 82: 577-588.

Ehinger, B., Falck, B., and Laties, P.M. (1969a) Adernergic neurons in teletost retina. Z. Zellforsch 97: 285-297.

Ehinger, B. and Falck, B., (1969b) Adrenergic retinal neurons of some new world monkeys. Z. Zellforsch 100: 364-375.

Falck, B., Hillarp, N-A., Thieme, G. and Torp, A. (1962) Fluorescence of catecholamines and related compounds condensed with formaldehyde. J. Histochem. Cytochem., 10: 348-354.

Haggendal, J., and Malmfors, T. (1965) Identification and cellular localisation of the catecholamines in the retina and the choroid plexus of the rabbit. Acta. Physiol. Scand. 64: 58-66.

Kramer, S.G. (1971) Dopamine: a retinal neurotransmitter I. Retinal uptake, storage and light stimulated release of H^3-dopamine in vivo. Invest. Ophtalmol. 10: 438-452.

Kramer, S.G., Potts, A.M., and Mangnall, Y. (1971) Dopamine: a retinal neurotransmitter II. Autoradiographic localisation of H^3-dopamine in the retina. Invest. Ophtolmol. 10: 617-624.

Laties, A.M. and Jacobowitz, D. (1966) A comparative study of the autonomic innervation of the eye in monkey, cat and rabbit. Anat. Rec. 156: 383-396.

Malmfors, T. (1963) Evidence of adrenergic neurons with synaptic terminals in the retina of rats demonstrated with fluorescence and electron microscopy. Acta Physiol. Scand. 58: 99-100.

Nichols, C.W., Jacobowitz, D. and Hottenstein, M. (1967) The influence of light and dark on the catecholamine content of the retina and choroid. Invest. Ophtalmol. 6: 642-646.

Norberg, K.-A. (1967) Transmitter. Histochemistry of the sympathetic adrenergic nervous system. (1967). Brain Res. 5: 125-170.

Olson, L. and Ungerstedt, V. (1970) A simple high capacity freeze-dryer for histochemical use. Histochemie 22: 8-19.

Sano, Y., Yoshikawa, H. and Konishi, M. (1968) Fluorescence microscopic observations on the dog retina. Arch. Histol. Jap. 30: 75-81.

Usdin, E. and Snyder, S. (1973) Frontiers in Catecholamine Research. Pergamon Press, Oxford.

HIGH RESOLUTION RADIOAUTOGRAPHY OF NORADRENERGIC AXON TERMINALS IN THE NEOCORTEX

Laurent DESCARRIES

Centre de recherche en sciences neurologiques

Université de Montréal, Canada

> I know not what the success will be, my lord; but the attempt I vow.
>
> All's Well That Ends Well, Act. III, Sc. 6

INTRODUCTION

The existence of high affinity transport systems in the retina has been described for many substances being considered as neurotransmitters (for discussion, see van Gelder, this volume). These and other studies have demonstrated the usefulness of the "retinal model" for investigating biochemical properties of nervous tissue. They have also prompted several attemps to localise uptake sites by means of light microscope radioautography, in its highly ordered, though relatively simple cytoarchitecture (see Appendix I). However, no comparable investigations have yet been carried out at the electron microscopic level, despite the fact that tritiated dopamine, GABA and glycine, for example, which have been shown to be preferentially accumulated in certain cells of the retina, are also known to be retained in situ by the chemical fixation procedures required for high resolution radioautography. Thus, neither the fine structural features of the transmitter-containing elements, nor the intracellular distribution of the labelled neurotransmitters have been analysed in the retina, as in other parts of the central nervous system. The present report will therefore attempt to illustrate some of the principles and applications of high resolution radioautography in the identification and study of neuronal elements specified by their neurotransmitter content, using rat cerebral cortex as a model for the retina.

Noradrenergic (NA) axons have been demonstrated throughout mammalian cerebral iso- and allo-cortex by means of the fluorescence technique of Falck and Hillarp (Falck et al., 1965; Fuxe, 1965; Blackstad et al., 1967; Fuxe et al., 1968; Nyström et al., 1972). These presumably originate from "central norepinephrine neurons" situated in the lower brain stem (Dahlström et al., 1964; Anden et al., 1965, 1966, 1967). In the rat, Ungerstedt (1971b) has shown that most cortical NA fibres reach their destination via an ascending "dorsal NA bundle" arising mainly from the *locus coeruleus*, that is form a homogeneous group of some 1400 norepinephrine-containing nerve cell bodies closely packed on each side of the fourth ventricle (Descarries and Saucier, 1972). To seek characterisation of this NA system at the synaptic level, high resolution radioautography has been used after topical application of tritiated DL-norepinephrine-^{3}H (NA-^{3}H) on rat cerebral cortex (Descarries and Lapierre, 1973; Lapierre et al., 1973)*.

I. RADIOAUTOGRAPHIC VISUALISATION OF CORTICAL NA TERMINALS

A) Detection of Reactive Axon Terminals

Three hours after administration of 330-450 μCi of NA-^{3}H (specific activity: 4.18 - 12.5 Ci/mM) in adult rats pretreated with monoamine oxidase inhibitor, the tracer is highly concentrated within minute portions of the cortical neuropil beneath areas of application. In light microscope radioautographs, these reactive sites appear as small and dense aggregates of silver grains, prominent against an ubiquitous diffuse reaction (Fig. 1). Electron microscope radioautography demonstrates that all the identifiable structures where radioactivity is accumulated correspond to profiles of unmyelinated axons exhibiting enlargements containing synaptic vesicles (Figs. 2-5). When viewed in consecutive thin sections, labelling of this type invariably designates the same axonal processes.

B) Specificity of the Axonal Reaction

The intense axonal reactions can be attributed to specific uptake and retention of NA-^{3}H within presynaptic terminals normally containing endogenous norepinephrine. As previously established for NA nerve endings labelled with NA-^{3}H in paraventricular regions of the brain (Descarries and Droz, 1970), the reaction of cortical NA terminals, imputable to the amine itself rather than tritiated metabolites, is markedly enhanced by monoamine oxidase inhibition and reaches its maximum 3 h after the administration of the tracer. Moreover, under these conditions, the configuration and repartition of the reactive sites bears a close resemblance to the distribution

*For a detailed description of the radioautographic technique, see Appendix II.

pattern of fluorescent NA fibres in the frontal cortex, as described by Fuxe et al. (1968).

It is unlikely that other types of monoaminergic fibres are responsible for significant radioautographic reactions. Indeed, presumptive dopamine-containing axons would be restricted to the deepest layers of the neocortex (Hökfelt et al., 1974). As for serotonine-containing fibres, which might take-up NA-^{3}H in the particular conditions required for radioautography, their nonspecific labelling is readily excluded by repeating the experiments in animals treated, more than one month earlier, with intra-centricular 6-hydroxydopamine. Following administration of this drug in a dosage known to result in selective destruction of all NA terminals in cortex (Ungerstedt, 1971a), and eventual disappearance of most NA nerve cell bodies from the *locus coeruleus* (Descarries and Saucier, 1972), axonal reactions are no longer detectable after applications of NA-^{3}H.

C) Ultrastructural features of labelled NA terminals

In material fixed by perfusion with glutaraldehyde and post-fixed in osmium tetroxide, the mean diameter of cortical reactive terminals, calculated from surface measurements obtained by planimetry, may be estimated at 1.15 μ. Following conventional development of electron microscope radioautographs with Microdol-X (Fig. 2), the ultrastructure of the labelled profiles is sometimes difficult to discern beneath dense accumulations of silver grains. The silver grains, often distributed in clusters, overlap any of the recognisable components: synaptic vesicles of various sizes, shape and content, smooth endoplasmic reticulum, axoplasm and the axonal membrane.

Processing of radioautographs with the physical developer para-phenylenediamine (Caro et al., 1962) allows better examination of the fine structural characteristics of reactive terminals (Figs. 3-5). Typical junctional zones of synaptic contact are frequently visible between labelled cortical nerve endings and small dendritic profiles (Fig. 5). No reactive axon terminals are observed in close apposition to neuronal perikarya or making synaptic contact with trunks of apical dendrites. Most labelled presynaptic enlargements contain clear synaptic vesicles, often in association with several large dense-dore vesicles (Figs. 2 and 4). These are also seen, however, in unlabelled synaptic terminals.

It is noteworthy that recent radioautographic studies of rat brain slices or homogenates incubated in vitro with tritiated GABA (Iversen and Bloom, 1972) and serotonin (Kuhar and Aghajanian, 1973) have also failed to reveal distinct morphological features, which might exclusively belong to reactive axons. A similar conclusion had already been reached for NA nerve endings labelled in vivo within paraventricular regions of the brain (Descarries and Droz, 1970).

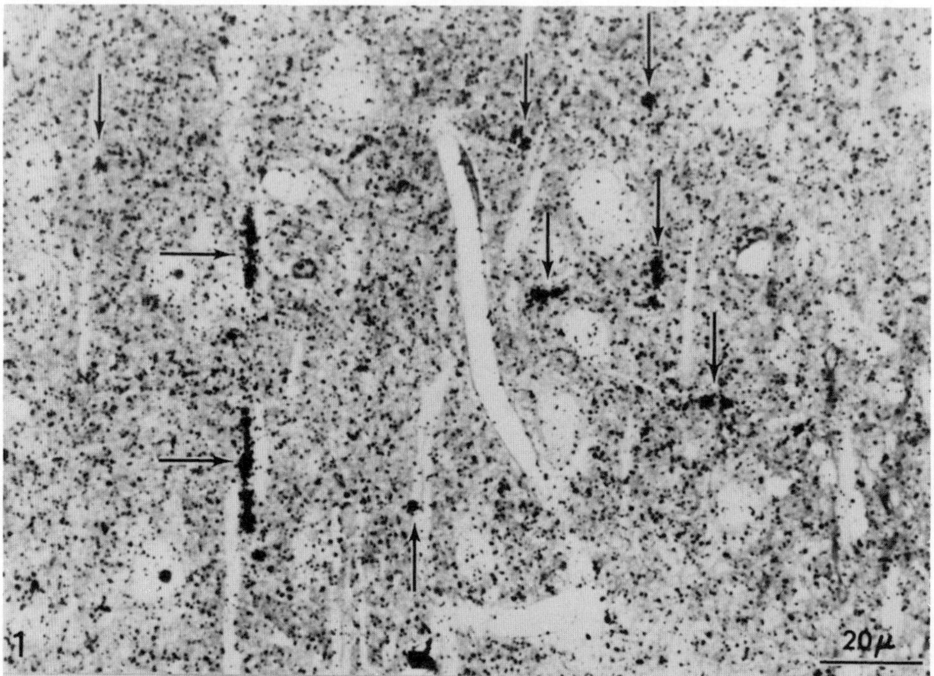
1
20μ

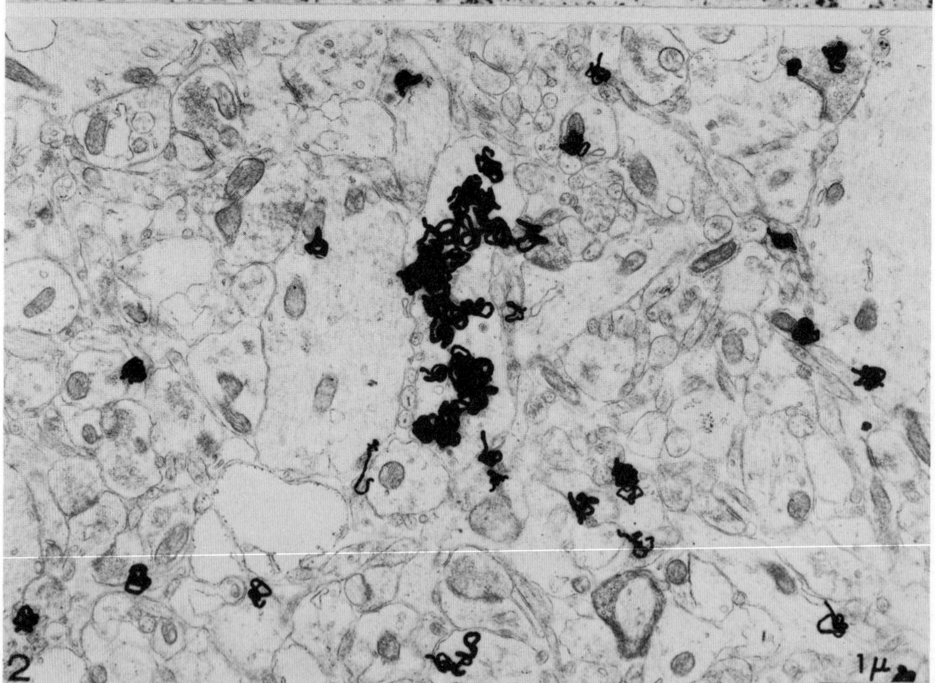
2
1μ

FIGS. 1-5 - Light and electron microscope radioautographs of the frontal cortex (Figs. 1 and 2-5, respectively), 3 hours after topical application of DL-norepinephrine-^{3}H in rats pretreated with monoamine oxidase inhibitor.

FIG. 1. Small and dense aggregates of silver grains (arrows) designate axonal enlargements where the tracer has been specifically taken up and retained. In this sample from layer II, some accumulations of radioactivity are aligned in rows, perpendicular to the cortical surface and adjacent to portions of apical dendrites (horizontal arrows). Such trajectories correspond to longitudinal segments of labelled noradrenergic axons. Exposure time, 2 weeks. X700. (From Descarries and Lapierre, 1973, with permission of the publisher.

FIG. 2. The radioactivity is highly concentrated within a presynaptic axonal enlargement exhibiting a few large dense-core vesicles. In the surrounding neuropil, isolated silver grains participate in a ubiquitous diffuse reaction. Microdol-X development. Exposure time, 6 weeks. X14,700.

FIG. 3. Paraphenylenediamine development: reduced silver grains are smaller. A labelled presynaptic axon contains several agranular vesicles of various sizes and shapes. Exposure time, 3 months. X22,400.

FIG. 4. This reactive axon terminal is in close apposition to a dendritic branch also in synaptic contact with two unlabelled nerve endings. Exposure time, 3 months. X17,000.

FIG. 5. In this favourable place of section, grouping of synaptic vesicles near an area of membrane differentiation is typical of a synaptic contact between a labelled axonal enlargement and a small dendritic process. Within the reactive terminal, most synaptic vesicles are small, round and clear. Exposure time, 3 months. X17,000.

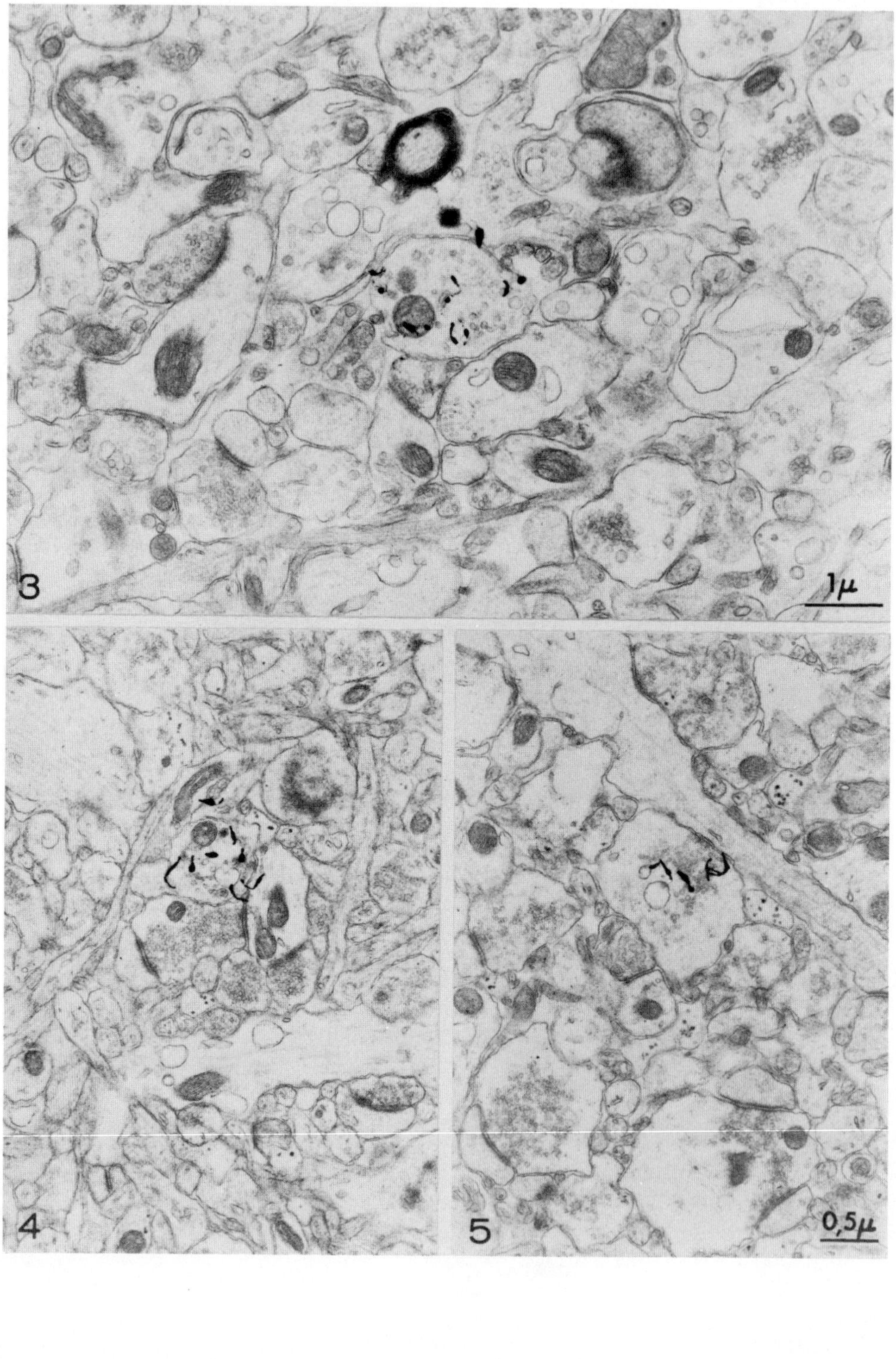
3
1μ
4
5
0,5μ

II. QUANTITATIVE DATA REVEALED BY RADIOAUTOGRAPHY

A) Intra-Cortical Distribution of Labelled NA Terminals

Following optimal labelling, all silver grain aggregates designating reactive NA nerve endings in light microscope radioautographs of the frontal cortex (0.5 µ-thick sections) were mapped with the aid of a camera lucida. The histogram of distribution (Fig. 6) confirms that the greatest proportion of NA terminals occupies molecular layer I. In layers II to IV, labelled nerve endings are less numerous and appear somewhat equally distributed. Some are visible in the upper portion of layer V, but few are found in the lower part of this layer or within layer VI. For most practical purposes, reactive terminals are confined within 1 mm of the cortical surface. In contrast with the number of labelled presynaptic axons, the intensity of diffuse reaction in the surrounding neuropil shows little diminution with increasing distance from the cortical surface (Fig. 6).

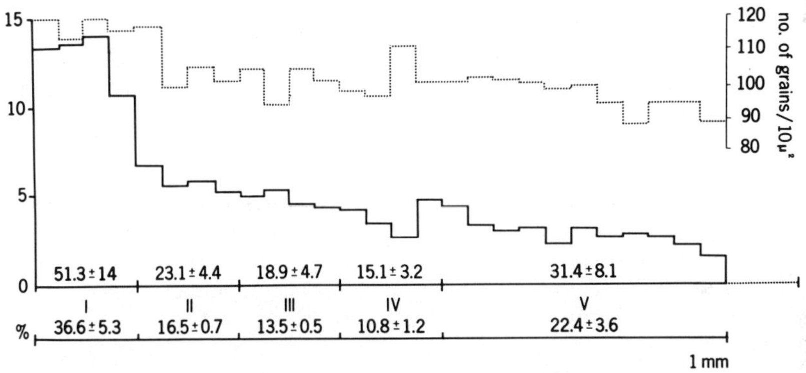

FIG. 6. Distribution of labelled noradrenergic axon terminals in rat frontal cortex. More than 35% of the reactive nerve endings belong to molecular layer I. Within deeper layers, no predominant localisation is apparent. Data collected from 21 0.5 µ-thick section processed by light microscope radioautography (means numbers ± s.e. for 3 animals corrected for 1 mm^2 of radioautograph). The dotted line represents the intensity of diffuse reaction, which decreases less than 25% within 1 mm of the cortical surface. (From Descarries, L. and Lapierre, Y., Characterisation of cortical noradrenergic axon terminals with high resolution radioautography, in "Frontiers in Catecholamine Research", E. Usdin and S. Snyder edits., Pergamon Press, New York, 463-465, 1973, with permission of the publisher.

The relative predominance of cortical NA nerve endings in molecular layer I was previously described using fluorescence histochemistry and was interpreted as an indication that NA fibres could be mainly in contact with arborisations of apical dendrites issuing from pyramidal nerve cell bodies (Blackstad et al., 1967; Fuxe et al., 1968). The present radioautographic results appear to confirm both facts and interpretation. They also suggest that scattered NA afferents whose synapses are situated far from pyramidal nerve cell bodies or axon hillocks might exert a modulatory rather than direct influence on cortical activity.

B) Density and Incidence of Cortical NA Terminals

Assuming that most NA nerve endings are detected, combined data obtained by light and electron microscope radioautography allow an approximation of the overall density and incidence of NA terminals in rat frontal cortex prerequisite to a better understanding of NA mechanisms operative at the synaptic level. Based on the mean number of reactive sites per mm^2 of 0.5 μ-thick sections, the total number of NA terminals is estimated to be 96,600 per mm^3. According to current evaluations of the total number of synapses in the neocortex of rodents (8.5×10^8 per mm^3, Cragg, 1967; 14×10^8 per mm^3, Aghajanian and Bloom, 1967), this figure corresponds to a mean incidence of 1 NA terminal per $8.8 - 14.5 \times 10^3$ cortical synapses. In molecular layer I, the proportion increases to 1 NA terminal per $3.6 - 5.9 \times 10^3$ synapses.

Although such approximations must be envisaged with caution, they do provide an order of magnitude which escapes the qualitative assessment of tissue sections examined by fluorescence microscopy and light or electron microscope radioautography. The scarcity of NA terminals in cortex does not preclude an important function for these widely dispersed afferents; it could in fact imply that they have strategic synaptic relationships and/or a particular mode of action.

Generalised to the entire cerebral cortex, the present values remain compatible with the proposition that most cortical NA terminals have a common origin in the nerve bodies of the *locus coeruleus* (Ungerstedt, 1971b): such being the case, each of the 2,800 noradrenergic perikarya of this nucleus would supply an average of 19,300 NA terminals to rat cerebral cortex (5.6 cm^2).

C) Estimates of Endogenous Norepinephrine Content per NA Terminal

Since cortical norepinephrine appears exclusively confined to axonal processes, and mainly concentrated within presynaptic nerve endings, the average amount of endogenous norepinephrine per cortical

NA terminal may be roughly estimated by dividing the amine concentration in the neocortex by the number of reactive nerve endings in the frontal region. Based on biochemical determinations of 225 ng/gm for the neocortex, the mean content of endogenous norepinephrine per terminal approximates 2.33×10^{-3} pg. Using a spherical volume of 0.8 μ^3 and a specific weight of 1 for NA nerve endings, the average concentration of norepinephrine per g wet weight of NA terminals corresponds to 2,900 µg or 0.3%.

This content of endogenous norepinephrine per cortical NA nerve ending appears significantly lower than that of peripheral sympathetic terminals: 4.2×10^{-3} pg in iris and 5.9×10^{-3} pg in vas deferens (Dahlström et al., 1966). However, the concentration of the amine is similar, presumably due to the smaller volume of NA terminals in cortex. The present figures are also in keeping with semi-quantitative results reported by Lindbrink and Jonsson (1971) who found a linear correlation between endogenous norepinephrine concentration and the intensity of formaldehyde-induced fluorescence in NA axon terminals of rat cerebral cortex.

CONCLUSION

Noradrenergic axon terminals of rat cerebral cortex may be comprehensively defined by their origin, morphological constituents, fine structural relationships and neurotransmitter content. Similar characterisation of retinal elements could not only provide crucial information on their functional identity in terms of transmitter specificity but also help to understand basic aspects of the intracellular distribution of certain neurotransmitters.

REFERENCES

Aghajanian, G.K. and Bloom, F.E. (1967) The formation of synaptic junctions in developing brain: a quantitative electron-microscopic study. Brain Research 6, 716-727.

Anden, N.-E., Dahlström, A., Fuxe, K. and Larsson, K. (1965) Mapping out of catecholamine and 5-hydroxytryptamine neurons innervating the telencephalon and diencephalon. Life Sci. 4, 1275-1279.

Anden, N.-E., Dahlström, A., Fuxe, K., Larsson, K., Olson, L, and Ungerstedt, U. (1966) Ascending monoamine neurons to the telencephalon and diencephalon. Acta physiol. scand. 67: 313-326.

Anden, N.-E., Dahlström, A., Fuxe, K., Olson, L. and Ungerstedt, U. (1966) Ascending noradrenaline neurons from the pons and the medulla oblongata. Experientia (Basel) 22: 44-45.

Anden, N.-E., Fuxe, K. and Larsson, K. (1966) Effect of large mesencephalic-diencephalic lesions on the noradrenaline, dopamine and 5-hydroxytryptamine neurons of the central nervous system. Experientia (Basel) 22: 842-843.

Anden, N.-E., Fuxe, K. and Ungerstedt, U. (1967) Monoamine pathways to the cerebellum and cerebral cortex. Experientia (Basel) 23: 838-839.

Blackstad, T.W., Fuxe, K. and Hökfelt, T. (1967) Noradrenaline nerve terminals in the hippocampal region of the rat and the guinea pig. Z. Zellforsh. 78: 463-473.

Caro, L.G., can Tubergen, R.P. and Kolb, J.A. (1962) High-resolution radioautography. I. Methods. J. Cell. Biol. 15: 173-188.

Cragg, B.G. (1967) The density of synapses and neurones in the motor and visual areas of the cerebral cortex. J. Anat. 101: 639-654.

Dahlström, A., Fuxe, K., Olson, L. and Ungerstedt, U. (1964) Ascending systems of catecholamine neurons from the lower brain stem. Acta physiol. scand. 62: 485-486.

Dahlström, A. Häggendal, J. and Hökfelt, T. (1966) The noradrenaline content of the varicosities of sympathetic adrenergic nerve terminals in the rat. Acta physiol. scand. 67: 289-294.

Descarries, L. and Droz, B. (1970) Intraneuronal distribution of exogenous norepinephrine in the central nervous system of the rat. J. Cell Biol. 49: 385-399.

Descarries, L. and Saucier, G. (1972) Disappearance of the *locus coeruleus* in the rat after intraventricular 6-hydroxydopamine. Brain Research 37: 310-316.

Descarries, L. and Lapierre, Y. (1973) Noradrenergic axon terminals in the cerebral cortex of rat. I. Radioautographic visualization after topical application of DL-norepinephrine-^{3}H. Brain Research 51: 141-160.

Falck, B., Mchedlishvili, G.I. and Owman, C. (1965) Histochemical demonstration of adrenergic nerves in cortex-pia of rabbit. Acta pharmacol. et Toxicol. 23: 133-142.

Fuxe, K. (1965) Evidence for the existence of monoamine neurons in the central nervous system. IV. Distribution of monoamine nerve terminals in the central nervous system. Acta physiol. scand. 64, Suppl. 247: 37-85.

Fuxe, K., Hamberger, B. and Hökfelt, T. (1968) Distribution of noradrenaline nerve terminals in cortical areas of the rat. Brain Research 8: 125-131.

Hökfelt, T., Fuxe, K., Johansson, O. and Ljungdahl, A. (1974) Pharmaco-histochemical evidence of the existence of dopamine nerve terminals in the limbic cortex. Europ. J. Pharmacol. 25: 108-112.

Iversen, L.L. and Bloom, F.E. (1972) Studies of the uptake of ^{3}H-GABA and ^{3}H-glycine in slices and homogenates of rat brain and spinal cord by electron microscope radioautography. Brain Research 41: 131-143.

Kuhar, M.J. and Aghajanian, G.K. (1973). Selective accumulation of ^{3}H-serotonin by nerve terminals of raphe neurones: an autoradiographic study. Nature New Biol. 241: 187-189.

Lapierre, Y., Beaudet, A., Demianczuk, N. and Descarries, L. (1973). Noradrenergic axon terminals in the cerebral cortex of rat. II. Quantitative data revealed by light and electron microscope radioautography of the frontal cortex. Brain Research 63:175-182.

Lidbrink, P. and Jonsson, G. (1971) Semiquantitative estimation of formaldehyde-induced fluorescence of noradrenaline in central noradrenaline nerve terminals. J. Histochem. Cytochem. 19: 747-757.

Nyström, B., Olson, L. and Ungerstedt, U. (1972) Noradrenaline nerve terminals in human cerebral cortices: first histochemical evidence. Science 176: 924-926.

Ungerstedt, U. (1971a) Histochemical studies on the effect of intracerebral and intraventricular injections of 6-hydroxydopamine on monoamine neurons in the rat brain. In 6-Hydroxydopamine and Catecholamine Neurons. Edited by Malmfors, T. and Thoenen, H., North-Holland Publ., Amsterdam, pp. 101-127.

Ungerstedt, U. (1971b). Stereotaxic mapping of the monoamine pathways in the rat brain, Acta physiol. scand. Suppl. 367: 1-48.

APPENDIX I

RADIOAUTOGRAPHIC LOCALISATION OF PUTATIVE NEUROTRANSMITTERS IN THE RETINA (Analytical bibliography updated in August 74)

LABEL	*SPECIES*		*SITES OF PREFERENTIAL ACCUMULATION*	*REF. NO.*
DOPAMINE	Rabbit	*(in vivo)*	"Junctional cells" and IPL (3 sublayers)	4
	Cat	*(in vivo)*	"Junctional cells", "alloganglion cells", IPL outer border	6
GABA	Rabbit	*(in vivo)*	Amacrine and other cells in INL, IPL, ganglion cells, nerve fibre layer	2, 3, 4
	Rabbit	*(in vitro)*	Müller cells	5
		(in vivo)	Amacrine cells	5
	Rat	*(in vitro)*	Müller cells	8, 9
	Goldfish	*(in vitro)*	Horizontal cells, amacrine cells	7
		(in vivo)	Horizontal cells, amacrine cells, ganglion cells, nerve fibre layer	7
	Frog	*(in vitro)*	Horizontal cells, amacrine and other cells in INL, IPL, nerve fibre layer	10
GLYCINE	Rabbit	*(in vivo)*	Amacrine and other cells in INL, IPL, ganglion cells	3, 4
	Rabbit	*(in vitro)*	Amacrine cells, IPL, ganglion cells	1
	Rat	*(in vitro)*	Amacrine cells, IPL	8
	Frog	*(in vitro)*	Amacrine cells, IPL, ganglion cells, nerve fibre layer, retinal blood vessels	10
β-ALANINE	Rabbit	*(in vivo)*	Amacrine cells, IPL, ganglion cells, nerve fibre layer	3
ASPARTATE GLUTAMATE	Rabbit	*(in vivo)*	Müller cells	3, 4

REFERENCES (APPENDIX I)

1 Bruun, A. and Ehinger, B. (1972). Uptake of the putative neurotransmitter, glycine, into the rabbit retina. Invest. Ophtalmol. 11, 191-198.

2 Ehinger, B. (1970). Autoradioautographic Identification of Rabbit Retinal Neurons that take up GABA. Experientia 26, 1063-1064.

3 Ehinger, B. (1972). Cellular location of the uptake of some amino acids into the rabbit retina. Brain Research 46, 297-311.

4 Ehinger, B. and Falck, B. (1971). Autoradiography of some suspected neurotransmitter substances: GABA, glycine, glutamic acid, histamine, dopamine, and L-Dopa. Brain Research 33, 157-172.

5 Iversen, L.L. and Schon, F.E. (1973). The use of autoradiographic techniques for the identification and mapping of transmitter specific neurons in CNS. In New Concepts in Neurotransmitter Regulation, A.J. Mandell edit., Plenum Press, New York - London, 153-193.

6 Kramer, S.G., Potts, A.M. and Mangnall, Y. (1971). Dopamine: a retinal neurotransmitter. II. Autoradiographic localization of H^3-dopamine in the retina. Invest. Ophtalmol. 10, 617-624.

7 Lam, D.M.K. and Steinman, L. (1971). The Uptake of $\{\gamma-^3H\}$ Aminobutyric Acid in the Goldfish Retina. Proc. Nat. Acad. Sci. USA 68, 2777-2781.

8 Marshall, J. and Voaden, M. (1974). An Investigation of the Cells Incorporating $\{^3H\}$ GABA and $\{^3H\}$ Glycine in the Isolated Retina of the Rat. Exp. Eye Res. 18, 367-370.

9 Neal, M.J. and Iversen, L.L. (1972). Autoradiographic Localization of ^3H GABA in Rat Retina. Nature (New Biol.) 235, 217-218.

10 Voaden, M.J., Marshall, J. and Murani, N. (1974). The uptake of $\{^3H\}$ γ-amino butyric acid and $\{^3H\}$ glycine by the isolated retina of the frog. Brain Research 67, 115-132.

APPENDIX II: A TECHNIQUE FOR HIGH RESOLUTION RADIOAUTOGRAPHY

> "Une recette est un thème sur lequel une cuisinière intelligente exécute chaque fois une variation différente"
>
> Jehane Benoit

The following technique, used in our laboratory to prepare resin embedded material for high-resolution radioautography, is mainly based on the original discoveries of Lacassagne et al. (1925), Bélanger and Leblond (1946), Kopriwa and Leblond (1962), Caro and Van Tubergen (1962), Granboulan (1965) and Salpeter and Bachmann (1965). It has been previously described in detail by Larra and Droz (1970).

The procedure is carried out in two steps: 1) *light microscopic radioautography* on "thick" sections allows extensive sampling of the material, screening and comparison of results under various experimental conditions, in terms of detection and localisation of tracers in tissue, visualisation of preferentially labelled constituents, regional distribution and intensity of reactions; 2) *electron microscopic radioautography* carried out on thin sections of specimens carefully selected from the same blocks for purposes of representativity, exact anatomical location and proper orientation, permits precise identification of labelled cellular elements, characterization of their fine structural features, and examination of the intracellular disposition of tracers.

LIGHT MICROSCOPE RADIOAUTOGRAPHY

Cleaning Glass Slides:

- Soak 10 min in chromic acid.
- Wash 10 min in running tap water.
- Rinse in 3 changes of distilled water.
- Dry and keep in dust free atmosphere.

Preparing Sections for Light Microscope Radioautography:

- One micron-thick sections of Epon embedded material are routinely used, since they provide maximal detection of tritiated compounds, excellent resolution and good photographic documentation of results.
- For each block, several sections are deposited on a slide, at approximately one third the distance from its end. The specimen number is inscribed at the opposite end, on the same side as the tissue sections, using a diamond pencil.
- Prior to coating with emulsion, the slides are placed in a 60°C oven for a minimum of 24 h, to ensure adherence of sections to

glass and avoid slippage and/or wrinkling during subsequent development and staining.

- Once removed from the oven, slides are conveniently stored in clean plastic slide boxes without liners. By convention, slides are always orientated so that the specimen number is on the right, facing a Rotex label affixed to the lower portion of the box. The raised lettering of the label ensures correct orientation of the box, and therefore slides, during darkroom manipulations.

- As a general rule, a minimum of 2 or 3 slides are prepared from each block. An extra 6 to 10 slides serve as tests for exposure-times, staining etc.
- Two series of slides plus a few test slides are usually coated simultaneously. The third series is saved for use as a substitute, in case of technical problems or future needs.

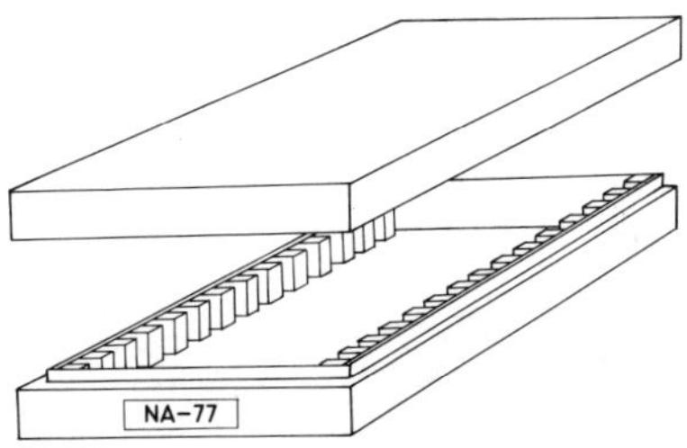

Coating with Emulsion:

- The following material is required in a photographic darkroom.
  Safelight with a 15 watt lightbulb and a Wratten #2 red filter.
  Water bath providing a constant 40°C temperature.
  Emulsion flask (Ilford K-5) in its lightproof container.
  Kleenex tissue or filter paper.
  4 porcelain spoons or glass stirring-rods.
  10 unused clean slides.
  Section-bearing slides in their respective boxes.
  Drying racks with slots allowing 80° angulation of slides (see drawing).
  Black plastic electrical tape and scissors.

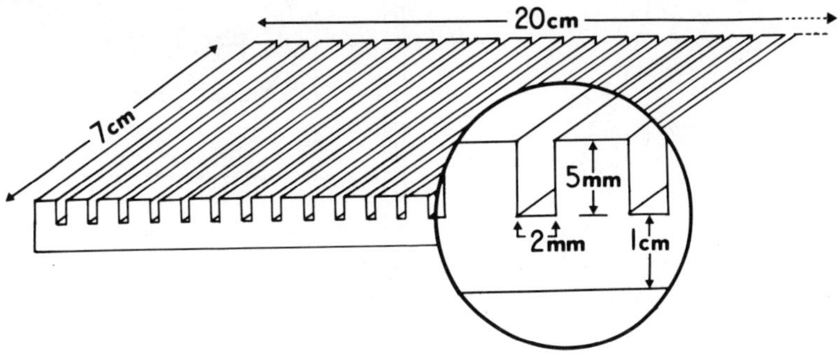

Dipping apparatus (see drawing).

- Place into the water bath a 250 ml beaker containing a Borel tube, a Jolly tube and a thermometer; stabilise the whole assembly at 40°C. Either tube can serve as an emulsion container (Jolly for up to 20 slides, Borel for up to 40 slides). The emulsion container should bear two black marks (one 3 cm, the other 6 cm from its base) and be filled with distilled water to the 3 cm mark, from the beginning of stabilisation. The other tube is filled with water and contains the thermometer. The beaker should contain water up to the 6 cm mark.
- Once the temperature of the system is stabilised at 40°C, all non-safelight illumination is turned off. Under safelight illumination only, the gelatinous emulsion is ladled into the emulsion container with one of the porcelain spoons until the water level reaches the 6 cm mark (providing a 1:1 dilution). The safelight is then turned off and the emulsion left to melt in the dark for 60-75 min. The emulsion flask is immediately returned to the refrigerator.
- At 20 min intervals, the melting emulsion is stirred thoroughly with a clean porcelain spoon: the stirring should be gentle enough to avoid introducing air bubbles.
- After 60 min, the coating may begin under safelight illumination. Just prior to dipping section-bearing slides, 8-10 clean unused slides are dipped in the emulsion to remove any air bubbles and/or dust particles. To verify the homogeneity of the emulsion, these coated slides may be held against the safelight.
- Section-bearing slides are manipulated by holding their numbered end only.

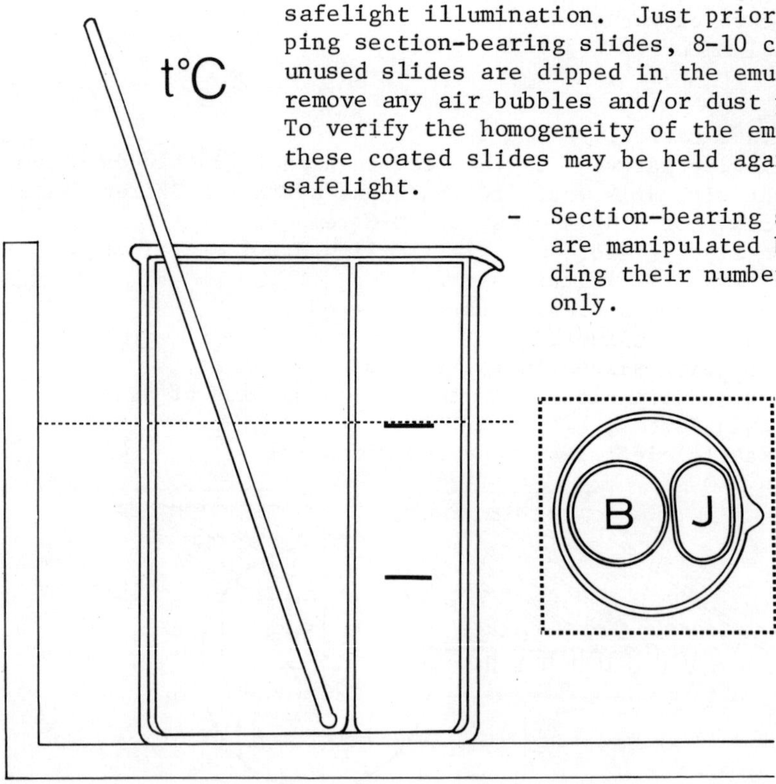

- Each slide is dipped to the bottom of the emulsion container and withdrawn vertically with a slow and steady movement of 1-2 sec duration to ensure uniform coating.
- The bottom of each slide is touched upon a Kleenex tissue or piece of filter paper to remove excess emulsion. The slides are then placed on the drying rack to stand at an angle of 80°, with the specimen number, and therefore sections, facing upward. Each series is placed on a different drying rack for subsequent identification.
- After all slides are dipped, they are left to dry in the darkroom for 3-4 h, with the safelight off. The position of each slide series should be noted and the slide boxes removed from the darkroom upon leaving.
- Just before re-entering the darkroom a small (20 g) packet of P_2O_5 wrapped in Kleenex tissue and partially enveloped in aluminum foil is placed in each box. This package is introduced between the labelled end of the box and a clean unused slide which serves as a guard.
- In the darkroom, each series of dry coated slides is stored in its respective box, in the same orientation as before coating. Each box is sealed light-tight with black plastic electrical tape.
- During exposure, the boxes, standing on their labelled end, are kept in a dry enclosure at room temperature.
- With tritiated compounds, slides may be stacked in every slot of slide box; with stronger isotopes, such as S^{35}, it is necessary to leave an interval of at least 1 cm between slides, to avoid radiation rebound.

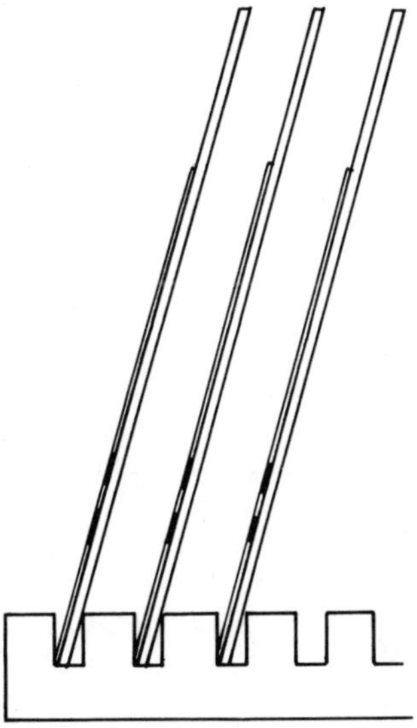

- K-5 emulsion, kept in its original packaging and stored in a refrigerator, should be used within 3 months of its manufacturing date.
- It may be useful to test each new batch of K-5 for background, prior to coating section-bearing slides; if necessary, background may be eradicated immediately after coating, by the method described by Caro and Van Tubergen (1962).

Development of Light Microscope Radioautographs:

- In the darkroom, development is carried out in a water-filled plastic tray containing 6 glass staining jars. The following solutions are freshly made:

  D-19 : Mix in order in 250 ml of distilled water: 0.5 g Metol, 2.25 g Quinol (hydroquinone), 18 g sodium sulphite, 12 g sodium carbonate and 1 g potassium bromide

Fix : Dissolve 75 g of sodium thiosulphate in 250 ml of distilled water

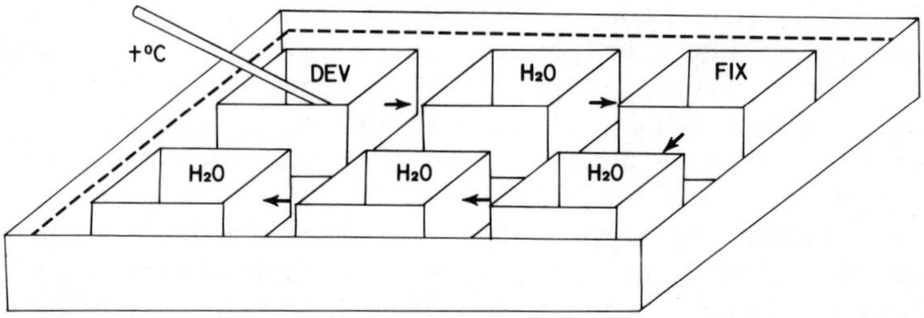

- Development is carried out at 18°C; this is done simply by lowering the temperature of the whole tray with ice and removing the developer when this temperature has been reached.
- Arrange the radioautographs in a staining rack and dip 4 min in D-19, 10 sec in water, 5-10 min in Fix and 5 min in each of 3 water baths. A maximum of 2 series of slides may be processed before replenishment of all solutions.
- All slides are dried on racks in a dust-free atmosphere.

Staining Radioautographs with Toluidine Blue:
- Stock solution: 1% toluidine blue in 1% sodium tetraborate.
- Filter stock solution and dilute 1:10 with 70% ethanol.
- Place slides on a hot plate (50-60°C) and stain for 45-60 sec with a sufficient amount of solution to cover the whole slide.
- Rinse and differentiate with 70% ethanol, wipe off excess alcohol, and evaporate any residue by heat.
- Clear the sections with a drop of xylene and mount with Permount or Eukit.

ELECTRON MICROSCOPE RADIOAUTOGRAPHY

Preparation of Celloidin Coated Slides:
- Prepare a 2% solution of celloidin or Parlodion in iso-amyl acetate. Both take 24 h to dissolve, may be re-used, and will keep indefinitely in a closely stoppered container.
- Cross-mark slides as shown, using a diamond scribe: these marks will later indicate the position of sections placed on the opposite surface.
- Clean slides as described earlier.
- Mark the outside of a Borel tube 5 cm from its internal base and filter the celloidin solution into it, until the mark is reached.

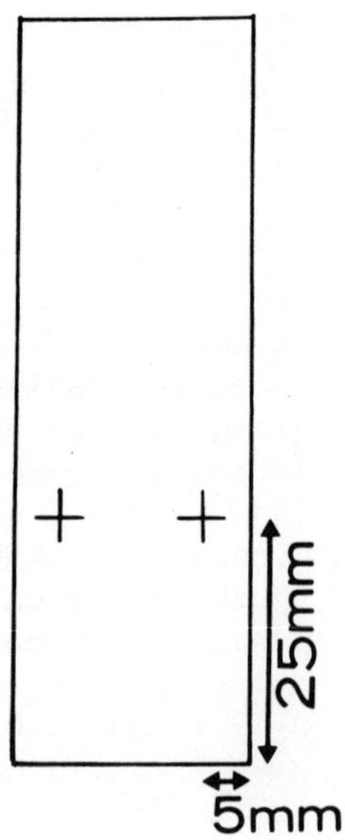

- Dip the scribed end of each slide, drain briefly, and allow to dry in a dust free atmosphere at an angle of 30° with the scribed surface facing downward.
- After thorough drying (2-3 h), re-dip the lower end of each slide to a depth of ½-1 cm and dry as before. This will reinforce the film at the re-dipped end and help prevent peeling.
- Celloidin coated slides can be stored indefinitely in dust-free boxes.

Transferring Sections to Slides:

- Ribbons of silvery-gold electron microscope sections are removed from the cutting through with a platinum loop (we use the "special thin loop" no 290 from Ladd Research Industries, Bultington, Vermont).
- Over the opposite surface of each inscribed cross, the loop containing the sections is gently lowered onto the celloidin film.
- The edge of the ring is touched with a small piece of filter paper to draw off the cutting solution, thereby depositing the floating sections onto the celloidin. Deposition is sometimes facilitated by putting a small drop of cutting solution on the slide just prior to placement of the sections. During the whole operation, care must be taken not to tear the celloidin film.
- The block number is inscribed on the side bearing the sections, at the uncoated end of the slide.
- The slides are stored in a labelled dust free box, orientated in the same manner as those for light microscope radioautography.

Staining the Sections:

- Prepare a stock solution of modified Reynold's lead citrate solution by dissolving separately 2.66 g lead nitrate in 30 ml distilled water and 2.52 g sodium citrate in 30 ml distilled water; mix the two solutions together and stir for 30 min, shaking intermittently; add 16 ml fresh 4% NaOH and bring the volume to 100 ml with distilled water.
- Just before use, dilute the required amount of stock solution 1:1 with 0.01 N NaOH.
- Test the stain on grid mounted sections from the same material to be exposed.
- Stain (usually 3-4 min) in Jolly jars containing not more than a 3 cm depth of solution and do not use the jar and contents for more than 2 slides. Wash the area in contact with the stain thoroughly, using a stream of distilled water meticulously directed to avoid the upper end of the celloidin film.
- Allow the slides to dry upright in racks, for a minimum of 1 h, in a dust free atmosphere.

Carbon Coating:

- It is essential to coat the slides with a thin layer of carbon. This will favour uniform layering of the emulsion, protect against negative chemography, avoid destaining during radioautographic processing, and provide greater stability of sections under the electron beam.
- The carbon layer should be as thin as possible. This may be controlled during coating by placing with the slides a small piece of paper, half of which is covered with high-vacuum grease. The thickness of carbon desired should produce the faintest shade of grey detectable when comparing the two parts of the paper.

Coating with Emulsion:

- Technique identical to that described above, except that emulsion Ilford L-4 is used at a 1:4 dilution (Jolly jar bearing marks at 12 mm and 60 mm from the bottom).
- The Ilford emulsion L-4 is more stable than K-5 and may be kept for longer periods of time in the refrigerator. It may be tested by light microscope radioautography in the same manner as the K-5, using a 1:1 dilution and a development of 5 min in D-19.

Processing of Electron Microscope Radioautographs:

- Expose in conditions similar to those previously described. The duration is approximately 10 times that required for light microscope radioautography to obtain reactions of similar intensity.
- Change P_2O_5 monthly in the slide boxes.
- Note that for quantitative comparisons between different experiments, all samples should be processed simultaneously.
- The developer is at 18°C, as before, using Microdol X (Kodak); this developer can be mixed in advance, will keep for 1 month, and should be filtered before use.
- A development of 4 min provides optimal size of silver grains. Under-developed grains may be preferred in certain circumstances and may be obtained by shortening the development time to $3\frac{1}{2}$ min.
- The technique of physical development described by Caro and Van Tubergen produces much smaller grains; however, it is not as sensitive and Microdol X should be preferred for routine purposes.
- When convenient, developed slides may be kept in the last rinsing bath and stored in the refrigerator for several hours.
- When ready to remove the preparations from the slides, stand all slides vertically on a drying rack, covered by an inverted jar, with a soaked Kleenex under the jar to maintain a high humidity.

- Assemble the following material:
  Fine forceps.
  Mounted needle.
  Scissors.
  Copper grids in Petri dish.
  Razor blade.
  12.5 cm Whatman #1 filter papers cut in half.
  Pencil.
  Flask of distilled water.
  Glass rod.
  Peeling jar (see drawing).

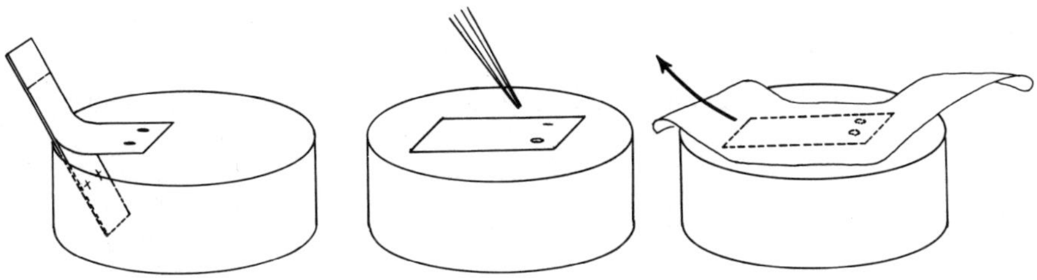

- The peeling assembly consists of a large container filled to the brim with water, resting in a tray lined with black paper. The water surface is constantly kept free of debris by adding fresh water and passing a glass rod over it.
- Detach the thick end of the celloidin film with a razor blade, and scrape both sides of the slide while holding the slide with the sections facing downward.
- Hold the slide against the edge of the glass container in an almost vertical position, sections facing up. Lower the slide into the water. The celloidin film detaches from the slide and floats onto the surface of the water. A mounted needle may aid in stripping.
- With fine forceps, place a 150 mesh grid, shiny side down, over each group of sections on the floating celloidin film.
- Write the block number on a half piece of Whatman #1 filter paper.
- Breathe on the floating film to hydrate the surface. From above, bring the filter paper in full contact with the whole celloidin film. As soon as water reaches the center of the filter paper, pull out the attached film in the direction of the thinner end of the celloidin.
- Cut the excess paper around the celloidin, and place to dry, paper down, in a Petri dish.
- Put the Petri dish in a 40°C oven for 1-2 h, partially covered, until the preparations are completely dry. These may be kept indefinitely at this stage, if stored in a dust free atmosphere.
- Thinning the celloidin film will promote contrast in the electron

microscope. The procedure is carried out in small glass Petri dishes, lined with filter paper at the bottom and half-filled with iso-amyl acetate. Grids are freed from the celloidin film by puncturing with fine-pointed forceps, immersed in the iso-amyl acetate for 2-2½ min, and dried in a 40°C oven for at least 1 h before electron microscopic examination.

- Electron microscope radioautographs are conveniently stored and treated in the same manner as regular thin sections.

MAJOR REFERENCES ON TECHNIQUE*

Bélanger, L.F. and Leblond, C.P. (1946). A method for locating radioactive elements in tissue by covering histological sections with photographic emulsion. Endocr. 39: 8-13.

Caro, L.G. and Van Tubergen, R.P. (1962). High resolution radioautography. I. Methods. J. Cell Biol. 15: 173-188.

Granboulan, P. (1965). Comparison of emulsions and techniques in electron microscope radioautography. In: The use of radioautography in investigating protein synthesis. Leblond, C.P. and Warren, K.B. edits., Academic Press, New York, 43-63.

Kopriwa, B.M. and Leblond, C.P. (1962). Improvements in the coating technique of radioautography. J. Histochem. Cytochem. 10: 269-284.

Lacassagne, A., Lattes, J. and Lavedan, J. (1925). Etude expérimentale des effets biologiques du polonium introduit dans l'organisme. J. Radiol. Electrol. 9: 1-14.

Larra, F. and Droz, B. (1970). Techniques radioautographiques et leur application à l'étude du renouvellement des constituants cellulaires. J. Microscopie 9: 845-880.

Salpeter, M.M. and Bachmann, L. (1965). Assessment of technical steps in electron microscope autoradiography. In: The use of radioautography in investigating protein synthesis. Leblond, C.P. and Warren, K.B. edits., Academic Press, New York, 23-39.

*For an extensive bibliography, see Larra and Droz (1970).

CHEMICAL NEUROTRANSMISSION IN THE RETINA OF THE TELEOST *EUGERRES PLUMIERI*

Boris D. DRUJAN and M. JODLOWSKY

Laboratory of Neurochemistry, I.V.I.C.

Apartado 1827, Caracas 101, Venezuela

> Without his roe, like a dried herring: - O flesh
> flesh, how art thou fishified!
> Romeo and Juliet, Act II, Sc. 3

INTRODUCTION

The transfer of information from one cell to another is mediated by the liberation of a small quantity of a chemical transmitter substance. This compound diffuses to the post-synaptic membrane where it combines with specific receptor sites, and initiates changes in the ionic permeability. The involved ions depend on the nature of the transmitter compound, and of the receptor sites for a particular synapsis. Finally, the transmitter substance is removed by specific mechanisms, and the whole procedure may repeat itself many times. This process we call chemical neurotransmission.

Neurotransmitter qualities are being attributed to an ever growing number of substances present in the nervous system. Werman (1966) suggests that a substance, in order to be considered a neurotransmitter, must fulfill the following conditions:

a. It must be present in the nerve endings.
b. The synthesizing enzymes of this substance must be present in the nerve tissue.
c. There must exist a mechanism to enable the action of this compound to reach the postsynaptic membrane.
d. The stimulation of the neurons must produce its liberation into the extracellular fluid.

e. The action of the exogenous transmitter must be identical, in any case, to that of the endogenous transmitter.

Very few, and rigourously none of the substances which are generally considered as neurotransmitters or chemical modulators do fulfill all of the above mentioned requirements. This is why, the consideration of a substance as a neurotransmitter depends on how rigidly we take the existing definitions. In this work, we shall discuss substances the behaviour of which allows us to accept them as putative neurotransmitters or modulator substances.

The photochemical adaptation mechanism which is based entirely on the bleaching and recovery of rod pigments, cannot possibly account for the enormous range of light intensities (10 Log units) in which the retina is able to adapt its sensitivity in order to be able to cope with its task to process the information obtained from the receptors and to codify signals of brightness, hue, colour, saturation and contrast with such an astonishing accuracy.

Results of a number of studies suggest the presence of an automatic control system in the retina, possibly of a metabolic nature. According to Svaetichin et al. (1965a and 1965b) the feedback control system in the retina consists of a liner receptor input, neuronal forward conductor line and horizontal cells as the feedback elements. These findings permit to assume that the cone pedicle with its synaptic triads, Stell (1967) contains an excitatory synaptic transmitter and all types of bipolar cells receive from the receptors excitatory signals only.

Several investigators have observed changes in the levels of certain neurotransmitter substances with excitatory as well as inhibitory characteristics in the retina, under the influence of light and darkness. This suggests their possible participation on the non-photochemical adaptation control mechanism of the visual system.

The purpose of this article is not to make an exhaustive review of this subject. The authors will rather limit themselves to discuss some selected works and original studies to give an overall idea of this field and evidence of the presence and changes in the activity of the cholinergic and adrenergic systems in the retina, as well as in the levels of certain amino acids with suspected neurotransmitter characteristics under the influence of light and darkness.

CHOLINERGIC TRANSMISSION

Ames and Pollen (1969) have studies the effects of numerous putative neurotransmitter substances on the electric activity of ganglion cells in the retina. These authors have found that the cell responses may be potentiated by Eserine at a concentration of 6×10^{-7}M. Also, the fact that this tissue reacts to other cholinergic agents can be considered as an evidence of the existence of cholinergic synapses in the retina.

The observation that nicotine increases activity (although not always in "off" cells) at a concentration of 2×10^{-7}M, but on the contrary, blocks the responses at a 2×10^{-5}M concentration indicates the existence of nicotinic receptors. The same authors have found that relatively high dosages of atropine are required to produce any appreciable effect, discovery which implies that the muscarinic synapses are of little importance in the retina. Due to this fact, some researchers, Nichols and Koelle (1967, 1968) have shown the activity of cholinesterase in amacrine and ganglion cells, but could not demonstrate its presence in the outer plexiform layer.

Ames and Pollen concluded that the cholinergic synapses are found in the inner plexiform layers, where the amacrine and ganglion cells are located. This conclusion also accords with our findings, obtained in measuring the activity of cholinesterase in isolated retinal cells from the fish *Eugerres plumieri*, Drujan and Svaetichin (1972).

TABLE I. Cholinesterase activity in the isolated single retinal cells.

CELL TYPE	ChE ACTIVITY (μl x 10^{-3} CO_2 / hr / cell)
Receptors	0.005
Horizontal	0.07 - 0.80
Amacrine	1.40 - 4.30
Ganglion	2.50 - 5.10

By the use of light microscopic histochemistry, we have localised cholinesterase in the same layer as was reported by Nichols and Koelle (1967, 1968).

The type of ChE present in the visual system of our experimental animal was established by the use of selective inhibitors for the two different ChE's. Eserine was used as general inhibitor, whereas Iso-OMPA and BW as selective inhibitors for the non-specific and specific cholinesterase respectively. The optimal concentration for their inhibitory action is approximately 10^{-5} to 10^{-4} M equal for all three compounds. Our observation that in the teleost retina we are dealing apparently with a ChE of the specific type, e.g. AChE, coincides with the studies by Eränkö et al., (1967), and Francis (1953).

Light induced changes in the activity of ChE seem to reflect the adaptational state of the visual system.

TABLE II. Activity of Cholinesterase in the dark-and light-adapted retina of the teleost *Eugerres plumieri* expressed in units per gram-tissue wet/weight. 1U = Hydrolysis of 10 μm AThCh/20 min at pH: 8.0 and 37°C.

EXP N°	CONDITION	N	AChE ACTIVITY * $\bar{X} \pm SD$	AChE ACTIVITY * $\bar{X}$ LIGHT - $\bar{X}$ DARK
1	DARKNESS	6	46.3 ± 10.3	-3.4 P ≤ 0.025
	LIGHT	6	42.4 ± 3.6	
2	DARKNESS	8	48.6 ± 5.8	-7.3 P ≤ 0.02
	LIGHT	8	41.4 ± 4.6	
3	DARKNESS	8	46.0 ± 6.1	-6.1 P ≤ 0.05
	LIGHT	8	39.9 ± 7.0	
4	DARKNESS	7	47.1 ± 7.8	-7.2 P ≤ 0.05
	LIGHT	8	39.9 ± 7.0	
5	DARKNESS	8	50.0 ± 9.1	-10.4 P ≤ 0.02
	LIGHT	8	39.6 ± 4.8	

* EXPRESSED IN U/G TISSUE WET WEIGHT. 1U= HYDROLYSIS OF 10 μG AThCh/20MIN. AT pH 8.0 AND 37°

Figure 1 shows that AChE activity is slightly (15%) but constantly higher in the dark-adapted retina. Trofimov (1968 and later also Negishi (1973) reported that the receptor activity is higher in the dark than in the light. This observation, as well as the work by Lam (1972) who demonstrated that from a number of substances studied, the isolated receptors possess the capability to synthesize ACh exclusively permits us to assume that there must be an increase in the liberation of an excitatory transmitter in darkness, apparently ACh. The so obtained raise in the levels of free ACh could account for the observed increase in AChE activity in the visual system in a dark-adapted state.

Yet, when we stimulate one eye of a dark-adapted animal with a very dim light (0.1-10.0 Lux) we can observe a large increase in ChE activity in the retina of the illuminated eye as compared with that in the scotopic retina of the other eye of the same animal. When one eye is illuminated with a light intensity to which only cones, and not rods, are sensitive (approx. 600 Lux) the effect upon AChE activity is the same but less pronounced. At light in-

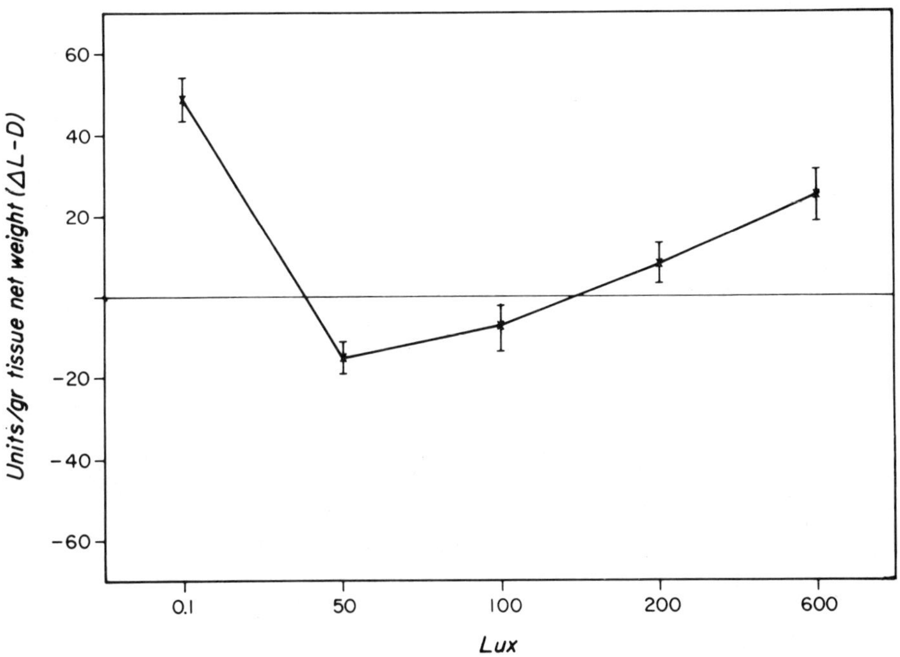

Fig. 1. AChE activity in the retina of the teleost *Eugerres plumieri* as a function of light intensity expressed in units per gram-tissue, wet/weight. The value of the dark-adapted condition is considered as 100%.

tensities to which the cones as well as the rods are sensitive, the effect upon AChE is cancelled.

Most of the vertebrate retinas contain two photoreceptor systems. One is highly sensitive, consisting of rods, and operates in dim illumination. The other, by far less sensitive, consisting of cones and active only in bright light. Parallel with the increase or decrease in illumination rod-vision is switched to cone vision and vice versa. This phenomenon is called the Duplex Theory of Vision, Hecht (1937) and is of basic importance in the understanding of the relationship between structure and function in the visual system of vertebrates.

Our observation that the activity of AChE depends upon the intensities of light stimulation shows a remarkable correlation between the activity of this enzyme in the retina and the theory of duplex vision.

ADRENERGIC TRANSMISSION IN THE RETINA

It has been found that the predominant retinal catecholamine varies from species to species. We had shown that the retina of the rabbit contains exclusively dopamine, Drujan et al., (1965), Häggendal and Malmfors (1965), while frog, toad and fish retinas mainly contain adrenaline, Drujan et al. (1965), Drujan and Diaz Borges (1972). Our studies show that adrenaline is the only catecholamine in the retina of *Eugerres plumieri*.

The localisation of retinal catecholamines at a sub-cellular level was performed by applying histochemical fluorescence techniques by Falk, Laties and Jacobowitz (1966, a,b) which showed that catecholamines are mainly localised in the amacrine cells and their processes.

Drujan et al. (1965 a,b), Nichols et al. (1967), and Drujan and Ciarletta (1973) have shown that the levels of catecholamines in the retina may be modified by the effects of light. An immediate depletion of the retinal adrenaline occurs when toads or fish adapted to darkness are exposed to a strong light (5,000 Lux) reaching a maximum depletion in approximately five minutes. The repletion of adrenaline occurs faster when animals, after having been exposed to light for five minutes are returned to the dark. In a period of 24 hours, in darkness, adrenaline reached approximately 85% of its original value. On the other hand, when animals are kept exposed to the light, repletion is slower, and the retinal adrenaline reaches a maximum of 65% of its level in the dark-adapted state. On the contrary, when an animal in mesopic conditions is exposed to a less intense illumination (300 Lux), an initial increase followed by depletion in the retinal adrenaline is observed, Ciarletta and Drujan (1973).

These findings suggest a possible participation of catecholamines in the non-photochemical adaptation mechanisms of the retina.

AMINO ACIDS AS POSSIBLE NEUROTRANSMITTERS IN THE RETINA

In recent years, an increasing interest has been developing regarding the physiologically active amino acids and their role as synaptic transmitters or modulators in the central nervous system, Krnjevic (1965), Curtis and Watkins (1965), Roberts and Kuriyama (1968). Iontophoretic administration permits to show the existence of nerve cells in all of the central nervous system, which are very sensitive to small quantities of certain amino acids injected locally. Their action is fast and ceases immediately after finishing the application of the current. Krnjevic (1964), Crawford and Curtis (1964).

After reviewing the studies of a number of workers, Curtis and Watkins came to the conclusion that there exists a Structure-Activity Relationship of the physiologically active amino acids.

$$Y-(CH_2)_n-\overset{\displaystyle X}{\overset{|}{CH}}-NH_2$$

Inhibitory Effect: When X=H and Y=$COOH^-$, SO_2H^- or SO_3H^-.

Excitatory Effect: When both X and Y=$COOH^-$.

The excitatory as well as the inhibitory effects diminish with the increase of <u>n</u>.

GLUTAMIC ACID

The excitatory effects of glutamate were first shown by Hayashi (1956) on the brain cortex. The effect is so intense and so rapid that it may not be recognised as an excitation. Negishi et al. (1973) showed that glutamate, when applied to the isolated retina, depolarises the resting potential of the horizontal cell, eliminates the S-potential and the intra and transretinal responses, increases the discharges from the ganglion cells, and excerts no effect on the transreceptoral response.

The main difficulty in accepting glutamate as a neurotransmitter is that it is uniformly distributed in all of the nervous system. Besides, it is an amino acid which does not clearly display the role of a neurotransmitter because it is present in very high concentrations in the nerve tissue, it has a very active metabolic turnover. Also, there exist several different compensation mechanisms to control the levels of glutamate in the nerve tissue, like protein synthesis, Roberts (1963), transamination, decarboxylation forming GABA, and others. The above mentioned reasons may explain, partly, the difficulty to detect changes in the concentration of glutamate as a result of a nervous stimulation. We believe that in any case there are no sufficient reasons to refute the hypothesis that glutamate may potentially be a neurotransmitter substance. Its influence on the electrical behaviour of the nervous system, and the fact that there are specific receptors for glutamate in the excitatory nerve endings, Takeuchi & Takeuchi (1964) constitute evidence in favour of the hypothesis being questioned. But the experimental results presented here are not convincing enough as to clarify the existing controversy on the possible neurotransmitter function of glutamate.

ASPARTIC ACID

Aspartate, together with glutamate, was postulated as an excitatory synaptic transmitter in the central nervous system of mammals by Curtis & Johnson (1970). Negishi et al. (1973) found that the activity of aspartate is similar to that of glutamate.

γ- AMINOBUTYRIC ACID (GABA)

The first evidence of the presence of GABA in the retina was presented by Kojima et al. (1958). Its uneven distribution in the eye of vertebrates has been reported in the frog, Graham et al. (1968), and in the rabbit, Kuriyama et al. (1968). GABA applied iontophoretically in the retina, or added to the medium of incubation in vitro, inhibits the responses of the ganglion retinal cells, Ames and Pollen (1969), Kishida and Naka (1967, Noell (1959). In a one day old chicken, GABA inhibits the retinal responses, blocking the b wave of the electroretinogram, (ERG), Kramer et al. (1967), Sholes and Roberts (1964). It has been suggested that the layer of bipolar cells of the retina from which the b wave seems to originate is the main location of visual adaptation. Dowling (1967), Graham et al. (1970) showed that GABA concentrates mainly between the cellular layers which are rich in amacrine cells, and that the levels of GABA decrease more than 40% in the layers of bipolar and ganglion cells in the retina of the dark-adapted frog.

Starr and Voaden (1972) found that the incorporation of GABA occurs in similar quantities in both states of adaptation, and that it occurred in three stages, possibly representing the selective incorporation of GABA in three separated cellular compartments. The same authors, Voaden and Starr (1972) found that rat retinas charged with ^{14}C-GABA in vitro discharged more GABA and less of their metabolites in the case of the retinas adapted to darkness, in comparison to those adapted to light. Lam and Steiman (1971) showed, by means of autoradiography that after incubating gold fish retinas for a period of one hour with γ-^{3}H aminobutyric acid, the product was localised in the outer and inner horizontal cells. Retinas stimulated by light were more intensively labelled than the corresponding cells from retinas kept in the dark. Lam (1972) in studying the biosynthesis and the content of GABA in the retina found that the retina of the goldfish incubated with L-glutamate -^{14}C (UL) synthesized γ-aminobutyric acid (GABA -^{14}C). The accumulation of synthesized GABA was increased by the physiological stimulation of the retina with flicker, and this increase was found to be directly proportional to the logarithm of the luminic intensity. The total content of GABA was also greater in the stimulated retinas than in those kept in the dark. We reported that the levels of GABA in the illuminated retina, and in the optic tectum were 2.58 and 3.04 μM/g, respectively, while the non illuminated retina contains 1.98 μM/g and the corresponding tectum 1.76 μM/g, that represents an induced increase by illumination of 24% and 42% in the retina and optic tectum respectively. This effect depends also on the intensity of the stimulating light, and may be reversed with a light intensity of 5,000 Lux.

CONCLUSION

We have considered several chemical entities as possible neurotransmitter agents, and have briefly discussed their possible participation in the retinal mechanisms.

The reviewed literature provides conclusive evidence of the presence of cholinergic as well as adrenergic system in the retina. There is also evidence of the presence of some amino acids like GABA, Glutamate, and Aspartate, some of them with inhibitory and others with excitatory properties.

The capacity of these compounds to modify the electrophysiological behaviour of the retina, and also the changes in their levels under different states of adaptation suggest that they may participate in the processes of neurotransmission, and maybe also in the mechanisms of the non-photochemical adaptation control of the visual system.

REFERENCES

Ames III, A. and Pollen, D.A. (1969). Neurotransmission in central nervous tissue: a study of isolated rabbit retina. J. Neurophysiol. 32: 424-442.

Ciarletta, E. and Drujan, B.D. (1973). Results non published yet.

Crawford, J.M. and Curtis, D.R. (1964). The excitation and depression of mammalian cortical neurones by amino acids. Br. J. Pharmacol. 23: 313-329.

Curtis, D.R. and Johnston, G.A.R. (1970). Amino Acid Transmitters. Handbook of Neurochemistry, edited by A. Lajtha, Vol. 4, p. 115-134, Plenum Press, New York.

Curtis, D.R. and Watkins, J.C. (1965). The pharmacology of amino acids related to GABA. Pharmacol. Rev. 17: 347-391.

Dowling, J.E. (1967). The site of visual adaptation. Science, 155: 273-279.

Drujan, B.D. and Ciarletta, E. (1973). The effect of light and darkness on the metabolism of adrenaline, γ-aminobutyrate, glutamate, and aspartate in the visual system of the teleost *Eugerres plumieri*. Proc. 4th. Int. Congr. Neurochem. p. 318.

Drujan, B.D. and Diaz Borges, J.M. (1972). Adrenaline induced changes in the metabolism of glycogen in the teleost retina. J. Fish. Biol. 4: 79-85.

Drujan, B.D., Diaz Borges, J.M. and Alvarez N. (1965a). Relationship between the contents of adrenaline, noradrenaline and dopamine in the retina and its adaptional state. Life Sciences, 4: 473-477.

Drujan, B.D., Diaz Borges, J.M. and Alvarez N. (1965b). The possible participation of catecholamines in the retinal inhibitory mechanism. Proc. 23rd. Int. Congr. Physiol. Sciences, 839.

Drujan, B.D. and Svaetichin, G. (1972). Characterization of different different classes of isolated retinal cells. Vision Res. 12: 1777-1784.

Eränkö, O. (1967). Histochemistry of nervous tissues: Catecholamines and Cholinesterases. Annu. Rev. Pharmacol. 7: 203-222.

Francis, C.M. (1953). Cholinesterase in the retina. J. Physiol. 120: 435-439.

Falk, B.,Laties A.M. and Jacobowitz, D. (1966a). Histochemical studies of monoamine containing cells in the monkey retina. J. Histochem. Cytochem. 14: 823-824.

Falk, B., Laties, A.M. and Jacobowitz, D. (1966b). A comparative study of the autonomic innervation of the eye in monkey, cat and rabbit. Anat. Record, 156: 383-395.

Graham, L.T.Jr., Lolley, R.N. and Baxter, C.F. (1968). Effect of illumination upon levels of GABA and glutamic acid in frog retina in vivo. Fed. Proc., 27: 463.

Graham, L.T.Jr., Baxter, C.F. and Lolley, R.N. (1970). In vivo influence of light and darkness on the GABA system in the retina of the frog (*Rana pipiens*). Brain. Res. 20: 379-388.

Häggendal, J. and Malmfors, T. (1965). Identification and cellular localization of the catecholamines in the retina and the choroid of the rabbit. Acta Physiol. Scand. 64: 58-66.

Hecht, S. (1937). Rods, cones and the chemical basis of vision. Physiol. Rev. 17: 239-290.

Kishida, K. and Naka, K.I. (1967). Amino acids and the spikes from the retinal ganglion cells. Science, 156: 648-650.

Kojima, K., Mizumo, K. and Miyazaki, M. (1958). Biochemical studies of the ocular melanine. Nature, (London) 181: 1200.

Kramer, S.Z., Sherman, P.A. and Seifter, J. (1967). Effects of gamma aminobutyric acid (GABA) and sodium L-glutamate on the visual system and egg of chicks. Int. J. Neuropharmacol., 6: 463-472.

Krnjevic, K. (1964). Micro-iontophoretic studies on cortical neurones. Int. Rev. Neurobiol., 7: 41-98.

Kuriyama, K., Sisken, B., Haber, B. and Roberts, E. (1968). The γ-aminobutyric acid system in rabbit retina. Brain Research 9: 165-168.

Lam, D.M.K. (1972). The byosynthesis and content of GABA in the gold fish retina. J. Cell.Biol. 54: 225-231.

Lam, D.M.K. (1972). Biosynthesis of acetylcholine in turtle photoreceptors. Proc. Nat. Acad. Sci. U.S.A., 69: 1987-1991.

Lam, D.M.K. and Steinman, L. (1971). The uptake of γ-3H aminobutyric acid in the goldfish retina. Proc. Nat. Acad. Sci. U.S.A. 68: No. 11, 2777-2781.

Negishi, K., Sugawara, K. and Kato, S. (1973). Supl. Anais da Academia Brasileira de Ciencias. (In press)

Nichols, C.W., Jacobowitz, D. and Hottenstein, M. (1967). The influence of light and dark on the catecholamine content of the retina and choroid. Invest. Ophthalmol., 6: 642-646.

Nichols, C.W. and Koelle, G.B. (1967). Acetylcholinesterase: Method for demonstration in amacrine cells of rabbit retina. Science, 155: 477-478.

Nichols, C.W. and Koelle, G.B. (1968). Comparison of the localization of acetylcholinesterase and non-specific cholinesterase activities in mammalian and avian retinas. J. Comp. Neurol. 133: 1-8.

Noell, W.K. (1959). The visual cell, electric and metabolic manifestations of its life processes. J. Ophthalmol. 48: 347-370.

Roberts, E. (1963). Some thoughts about GABA in the nervous system. Nutr. Rev. 21: 161,165.

Roberts, E. and Kuriyama, K. (1968). Biochemical-Physiological Correlations in studies of the γ-aminobutyric acid system. Brain Research 8: 1-35.

Sholes, N.W. and Roberts, E. (1964). Pharmacological studies of the optic system of the chick: Effect of GABA and pentobarbital. Biochem. Pharmacol. 13: 1319-1329.

Starr, M.S. and Voaden, M.J. (1972). The uptake of 14C-GABA by the isolated retina of the rat. Vision Res. 12: 549-557.

Stell, W.K. (1967). The structure and relationships of horizontal cells and photoreceptor - bipolar synaptic complexes in goldfish retina. Am. J. Anat. 121: 401.

Svaetichin, G., Negishi, K., Fatenchand, R., Drujan, B.D. and Selvin de Testa, A. (1965a). Nervous function based on interactions between neuronal and non-neuronal elements. In: Progress in Brain Research, 15. Biology of Neuroglia,p. 243. Edited by E.D.P. de Robertis and R. Carrera, Elsevier, Amsterdam.

Svaetichin, G., Negishi, K. and Fatenchand, R. (1965b). Cellular mechanisms of a young-hering visual system. Ciba Symp. on Color Vision: Physiology and Experimental Psychology, p. 178. Edited by A.V.S. de Reuck and J. Knight, Churchill, London.

Takeuchi and Takeuchi, N. (1964). The Effect on Crayfish Muscle of Iontophoretically Applied Glutamate. J. Physiol. (London), 170: 296-317.

Voaden, M.J. and Starr, M.S. (1972). The efflux of radioactive GABA from rat retina in vitro. Vision Res. 12: 559-566.

Werman, R. (1966). Criteria for the identification of a central nervous system transmitter. Comp. Biochem. Physiol. 18: 745-766.

THE RETINA AS A MODEL OF THE BRAIN

H.H. JASPER & J.-P. RAYNAULD

Centre de Recherche en Sciences Neurologiques de

l'Université de Montréal, Canada

Will it serve for any model to build mischief on?
Much Ado About Nothing, Act I, Sc.3

Because of its distance from the brain, the retina can be considered a sense organ at the same level as that of other sense organs like the ear, the muscle stretch receptor, or the pressure sensitive receptor of the skin. This view, however, would be a gross oversimplification, since the retina because of its layered structure closely resembles in its organisation the cerebral cortex itself. In a normal penetration of the retina you will record from up to five types of cells all of which will have their receptive field located in the same region in space. From this fact, which we all considered obvious in the retina, could we have predicted finding similar results in a normal penetration of the visual cortex and thus predicted the "columnar organisation" found by Hubel and Wiesel? Certainly not, if you consider the retina only as a sense organ; perhaps, if you think of it as a model of the brain.

In historical perspective let us first consider the pioneer work of Hartline, Granit and Kuffler, who first developed the techniques for study of the retina at the level of single ganglion cells or single optic nerve fibres. Stephen Kuffler has recently provided us with a most readable review of this subject in the Proctor Award Lecture published in Investigative Opthalmology, Vol. 12, No. 11, 1973. Kuffler recounts how excited and puzzled he was to find, in his work with Talbot in Baltimore, that uniform light was a very poor way to excite single ganglion cells, a small spot of light being far more effective. He was further surprised to find that the same spot of light could have opposite effects depending upon its exact position on the retina. An excitation or "on response" could be converted into an inhibitory or "off response" simply by shifting the spot by a millimetre or less across the retinal surface. By plotting the receptive fields of hundreds of ganglion cells the basic principle of retinal organisation was discovered: the presence of "on centre" cells which were excited only from the centre of the receptive field, inhibited from the surrounding area, and "off centre" cells which were inhibited from the centre of their receptive fields and excited from the surround. In each the centre spot and the surround were always antagonistic, either excitatory or inhibitory. Kuffler and Granit both emphasised at this time (about 25 years ago) the importance also of spontaneous activity in ganglion cells, especially in the dark-adapted retina and even in complete absence of light. Inhibitory phenomena which were found to be of such importance as a result of light stimulation thus gained a new significance when viewed on a background of spontaneous activity.

The foundation for the original work of Kuffler had already been laid by Adrian's *Physical Background of Perception* (1946) and by Granit in his classical work *The Sensory Mechanisms of the Retina* published in 1947. Hartline had also begun his lifelong and well known study of the compound eye of the *Limulus*. He and his co-workers were able to make much more precise studies by the illumination of single photoreceptors, the ommatidium, and to quantify by both intracellular and extracellular microelectrode studies the principle of surround inhibition and its importance in increasing contrast at the borders of light and dark areas of illumination, explaining in physiological terms Mach's bands long known to physiological optics.

Intracellular microelectrode studies of vertebrate retinal cells had to await the pioneering work of Svaetichin and Tomita who, together with later colleagues such as Kaneko have made possible truly revolutionary advances in the cellular physiology of the retina (see papers by Raynauld, Stell,and Tomita in this volume).

Following the work of these early pioneers the cellular physiology of the retina, coupled with microelectrode studies of the entire visual system has been developing by leaps and bounds. It has already had a remarkable impact upon our understanding of general principles of organisation of the brain as a whole, principles which apply not only to other sensory systems, but also to the cerebellum, the hippocampus, and motor systems of cortex, brain stem, and spinal cord. We would like now to draw attention to some of the most important of these principles, and to some derived from the retina which may be important in the future.

1. Receptor Potentials

Photoreceptor generator potentials correspond in many respects in relation to discharge in sensory nerve fibres as do other specialised sensory receptors, but with important differences which suggest principles of wider application. The discovery that they may be hyperpolarising in response to light, instead of depolarising as in most receptors (and as in the ommatidium of the *Limulus* eye) provides us with a new principle of transduction even though the precise ionic mechanisms involved at the membrane level may still be obscure. Intracellular studies of nerve cells in the brain have also shown an arrest of firing with depolarisation and a return of repetitive discharge during polarisation under certain conditions, although the mechanism here may be quite different. Other examples, however, should be sought in the central nervous system.

2. Information transfer without action potentials

The discovery that bipolar and horizontal cells of the retina transfer visual information to the ganglion cells without the benefit of action potentials, only by graded changes in polarisation provides another model which should be sought for in the CNS. We know very little indeed about the function of interneurones in the brain, especially those which may be without action potentials. We do record frequently from "silent" cells in the brain which show only variations in resting potential without spike discharge, but until now these have been considered to be glial cells. With the benefit of intracellular marking techniques, such as procion yellow, we should now be able to explore more precisely the possibility that there may be interneurones in the brain which function without action potentials.

The possibility that interneurones may be silent from the point of view of action potentials, i.e. not regenerative, and yet perform a very important function for the transfer of information and for the integration of assemblies of neurones is very important to consider in relation to cortical function. It seems highly likely that although some interneurones have been recorded from and shown to have rapid spike discharge, that many others may actually function in an electrotonic way without spikes and this principle should be kept in mind in exploring integrative functions of the brain itself in many regions.

3. Electrical and chemical modes of transmission

Gap junctions and electrical interconnexions have been shown between horizontal cells in the upper plexiform layer of the retina and possibly between other cells and even between cones at least in the turtle. Typical chemical synapses with transmitter substances common to the brain have also been established, though a great deal has yet to be done before we understand their precise synaptic distribution and significance for retinal function. Electrical and chemical transmission have also now been shown to exist side by side in parts of the brain of fishes and are now being found in parts of the brain in higher mammals.

4. Efferent control systems

The eye shares with all other sensory systems, but with a higher degree of organisation, several forms of efferent control, particularly with respect to pupillary diameter and eye movements, but also a direct sensory control via efferent fibres from the midbrain ending upon amacrine cells. This efferent control from the brain stem makes possible some control on receptor properties of the retina either inhibitory or facilitatory or possibly reorganising receptor field properties themselves. This is another general principle which is shown for the retina and gives insight to possible functional significance of efferent control systems which have been shown in the somatic, auditory, and the olfactory sensory systems.

5. Feature detection of biological significance

Retinal ganglion cells have been shown to respond to biologically important features of the visual stimulus in a highly specific but complex manner. Features involve really three aspects of the visual stimulus,

A. Temporal features, that is, on and off responses and with some evidence that the on and off responses are specific to certain cell types while other cells are more sensitive to continuous illumination when studying ganglion cells that is.

B. Spacial features. The principle of centre surround organisation for increasing contrast is a very important development beginning in the retina and now shown to apply throughout all sensory systems of the nervous system. The principle of collateral inhibition for increasing contrast in vision as well as in somatic sensory and even in motor functions has been a very great advance in our understanding of the organisation of the brain.

C. Movement direction sensitive organisation. This is a third feature selector which appears to have its foundation in retinal organisation which again is most important for an understanding of how the brain processes sensory information.

One is reminded also of the fly detector cells discovered by Maturana and his colleagues in the frog retina. Cells with colour coded receptive fields have also been found, with antagonist surrounds to increase colour contrast.

Following the principle of synaptic organisation for feature detection through various synaptic networks in the brain has been most rewarding, as in the well known work of Hubel and Wiesel, Phillips, Barlow and Blakemore, and now many others. Cells in the inferior temporal region of the monkey have been found by Gross to respond only to the image of the monkey's hand.

In conclusion, Jean-Pierre Raynauld and I would like to extend your thoughts and discussion somewhat beyond the subject of vision in fishes. We do not mean to imply that fishes are not in and of themselves of sufficient importance for such an Advanced Study Institute, representing as they do a large and important part of the biological family of which we are a part. Many of us, however, are not professional icthyologists but are studying vision in fishes because it provides a good model for the study of visual processes in general, the basic principles of which may have more general applications, even to an understanding of visual mechanisms in higher mammals, including man.

Finally we would like to suggest that present research on the retina, much of which has been reviewed in this volume, may have far reaching significance in our future search for a better understanding of some of the many as yet unsolved problems of brain function.

4. OPTICAL AND MECHANICAL ASPECTS

THE STRUCTURE AND PHYLOGENY OF IRIDESCENT CORNEAS IN FISHES

J.N. LYTHGOE

MRC Vision Unit, University of Sussex

Falmer, Brighton, BN1 9QG. U.K.

> The jewell that we find, we stoop and take't
> Because we see it; but what we do not see
> We tread upon, and never think of it.
> Measure for Measure, Act II, Sc. 1

INTRODUCTION

The iridescence so frequently seen in benthic and demersal marine teleosts shows little superficial difference between species. The structures involved are too small to be resolved by the light microscope, but the electron microscope shows that iridescence is produced by several anatomically distinct structures in different taxonomic groups of fishes.

The function of these structures in the visual life of fishes has been discussed in the previous paper in this volume. The most important function seems to be that the iridescent layer acts as a sophisticated sunshade that protects the eye from flare originating from the very bright downwelling light under water, but does not reduce the intensity of the visually important image forming light that passes through the cornea near to the optic axis.

For some reason that is not easy to explain, the iridescence of fish corneas has escaped scientific study in the past. The phenomenon is often clearly visible in aquarium fishes and is also visible in recently dead specimens if the main source of light reaches the eye from a dorsal direction. Walls (1967) has described layers of uncertain phylogeny in the corneal substantia propria of gobies and plectognaths (trunkfishes, puffers, ocean sunfish, etc) with histologically peculiar intercalations. These almost certainly correspond to the iridescent layers described below. Unfortunately, the light microscope is not very revealing in these studies since the individual lamellae that make up the layers are less than a wavelength of light in thickness and cannot be resolved with the light microscope.

The appearance and reflexion characteristics of corneal iridescence was described by Lythgoe (1971) for four Indian Ocean fishes and the fine structure of one of these, Nemanthias carberryi, was investigated by Locket (1972). He found that the iridescence was produced in the corneal endothelium where each of the single layer of cells is stacked with parallel lamellae of rough endoplasmic reticulum.

It is now known that regularly arranged endoplasmic reticulum is only one of several anatomically distinct structures that can produce iridescence in the cornea. The colour of the iridescence differs with the angle of incidence and reflexion of the light, but when seen under natural conditions may be pinkish-bronze, yellow-green, green, blue-green or blue (Lythgoe, In press, a). There is no obvious correlation with the type of multilayer responsible nor with the colour of the water in which the fish lives.

THE TELEOST CORNEA

The teleost cornea is a much more complex and variable structure than in the mammals or in the frog, four basic layers can be recognised (Fig. 1). In contact with the outside environment is an epithelium several cells thick. The substantia propria or stroma, which is composed of an orderly arrangement of collagen fibrils, occupies the bulk of the cornea. The innermost layer in contact with the aqueous humour is the single cell layer of the endothelium. Separating the endothelium and the stroma is a thin layer without obvious fine structure known as Descemet's membrane. In mammals another layer is recognised which lies between the epithelium and the stroma. This is Bowman's layer and is recognised because the collagen fibrils are not arranged in regular bundles but are essentially random in orientation. The only structures that are always present in all the vertebrate corneas so far examined are the stroma, the epithelium and the endothelium.

The structures found in the teleost cornea that do not seem to be present in mammals are the following:

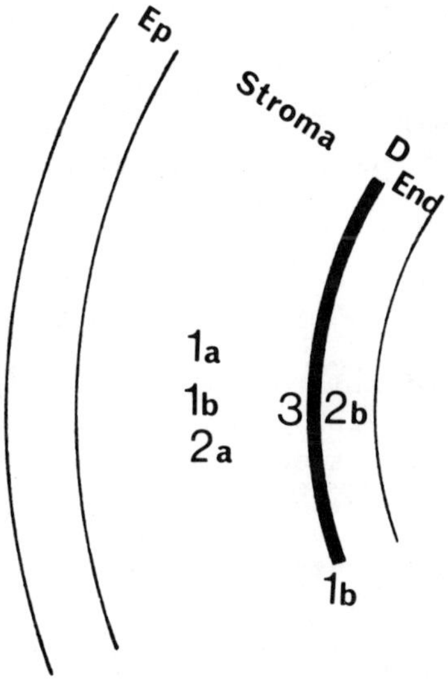

FIG. 1. Diagrammatic cross-section of the teleost cornea showing the locations of the various iridescent layer types mentioned in the text.

1. The secondary spectacle

According to Walls (1967), the cornea can be divided embryologically into two parts, the scleral part that is continuous with the cartilageneous sclera and the dermal part continuous with the dermis. The scleral and dermal parts are jointly homologeous to the entire cornea in chondrosteans and elasmobranchs. In teleosts, however, the two layers are often separated to form a secondary spectacle and the cornea proper. The secondary spectacle is particularly common in bottom-dwelling marine teleosts whose eyes frequently come into contact with abrasive material on the bottom.

2. The pigment layer

Many teleosts possess a layer containing yellow pigment in their cornea (Walls, 1967; Moreland & Lythgoe, 1968; Muntz, 1973). All or nearly all fishes that possess yellow corneas are diurnal but their exact function is problematical. Many species that possess the yellow layer also possess an iridescent layer (Lythgoe, in press, a).

3. Iridescent layers

These have been found in the scleral part of the cornea, but not so far in the spectacle. There are at least six different types of iridescent layer and a provisional classification is attempted below.

CLASSIFICATION OF THE CORNEAL LAYER

The main divisions of this classification is according to the types of tissue involved, although an alternative classification might be erected on the iridescent layer's position in the cornea (Fig. 1). It should be noted, however, that this classification, based on tissue types, is not always consistent. For instance, the iridescent layers in Siganidae, Plectognaths and Balistidae are obviously related, but in Siganus vulpinus the arrangement of the cytoplasm into plates separated by connective tissue more resembles that found in the gobies and their relatives.

A. Connective tissue only

The major part of the corneal stroma is made up of parallel collagen fibrils running through an amorphous mucopolysaccharide matrix. It remains an unresolved question whether the collagen fibrils have a different refractive index from the matrix (Maurice, 1957, 1970) or whether the two refractive indices are nearly the same (Smith, 1969, 1970). In this context, it is interesting to note that additional electron-dense staining material distributed in a regular pattern is present in the matrix, suggesting that the refractive indeces are not sufficiently dissimilar to produce iridescence without the addition of another compound, perhaps of higher refractive index. A possible exception to this is in the Zeus faber cornea, although here the sections do not allow a definite decision to be made.

Type 1a - Found only in the John Dory (Zeus faber) (Plate 1); plates of dense-staining material which are sometimes continuous with the irregularly arranged collagen fibrils form small "books" of lamellae in clear areas of the matrix.

Type 1b - In both the Lizard fish (Synodus intermedius) (Plate 2) and the Sargassum fish (Histrio histrio) (Plate 3) there is a layer that corresponds in position with Descemet's layer but has clearly marked densely staining lamellae running through it. These densely staining lamellae have no visible structure. The clear areas in the Sargassum fish have dark-staining fibrils running irregularly through it. In the Lizard fish the light staining areas have a mottled ap-

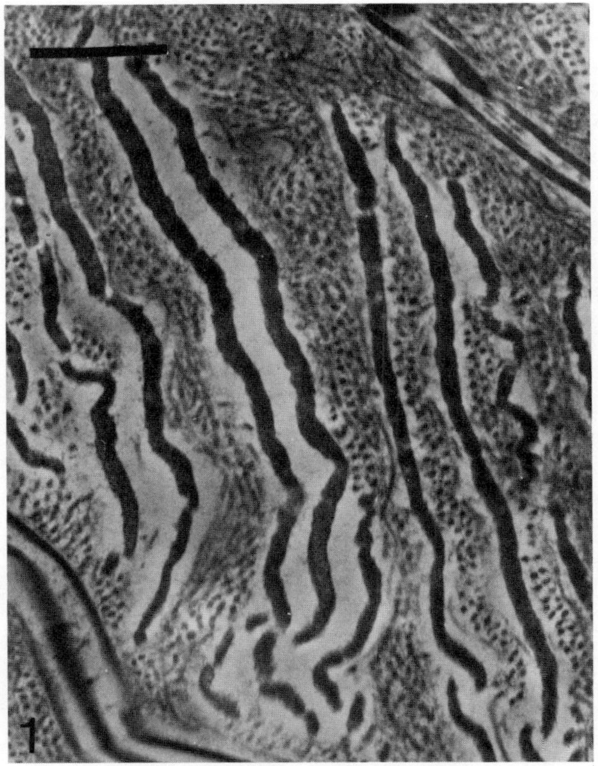

Plate 1. *Zeus faber*, type 1a

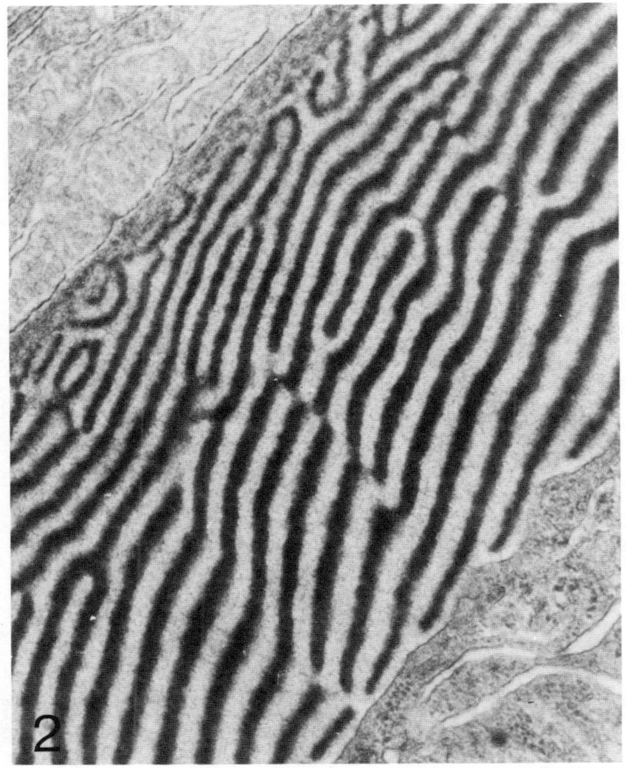

Plate 2. *Synodus intermedius*, type 1b

Plates 1 - 12. Fine structure of the various types of iridescent layer at present known in the corneas of teleost fishes. A classification of the various types is outlined in the text. Magnification: all plates are reproduced to the same scale; the black bar in Plates 1, 4, and 8 represents 500 nm.

Plate 3. _Histrio histrio_, type 1b

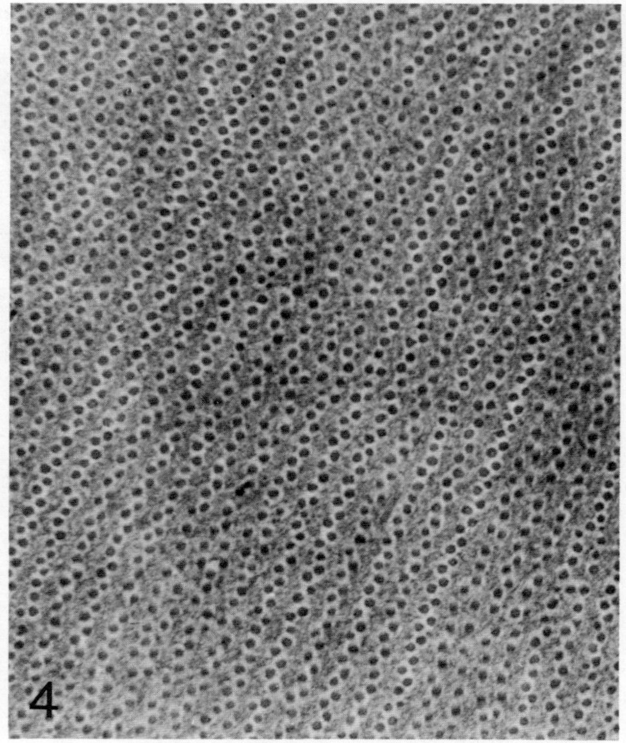

Plate 4. _Trigla lucerna_, type 1c

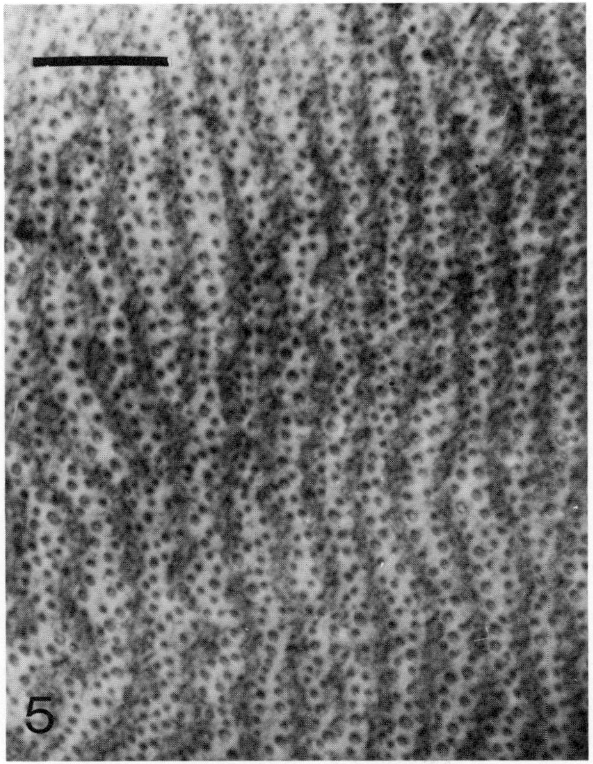

Plate 5. Trigla cuculus, type 1c

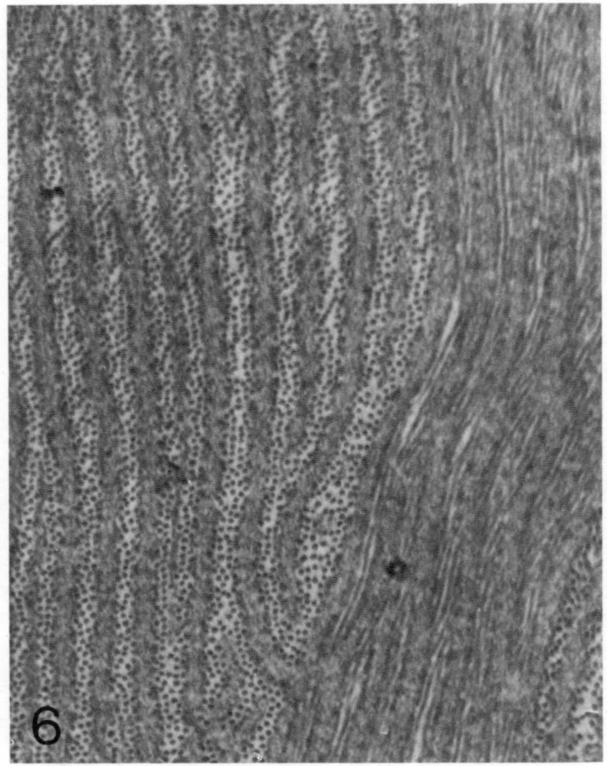

Plate 6. Cottus bubalis, type 1c

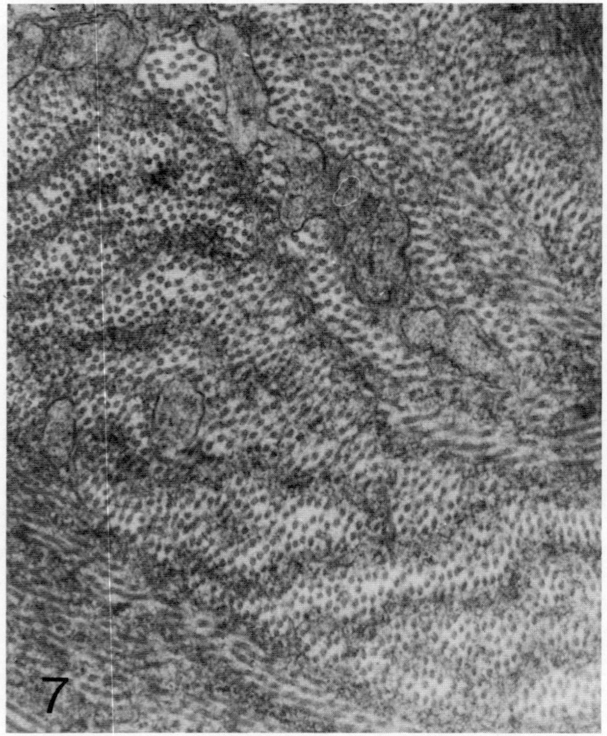

Plate 7. *Pholis gunellus*, type 1c

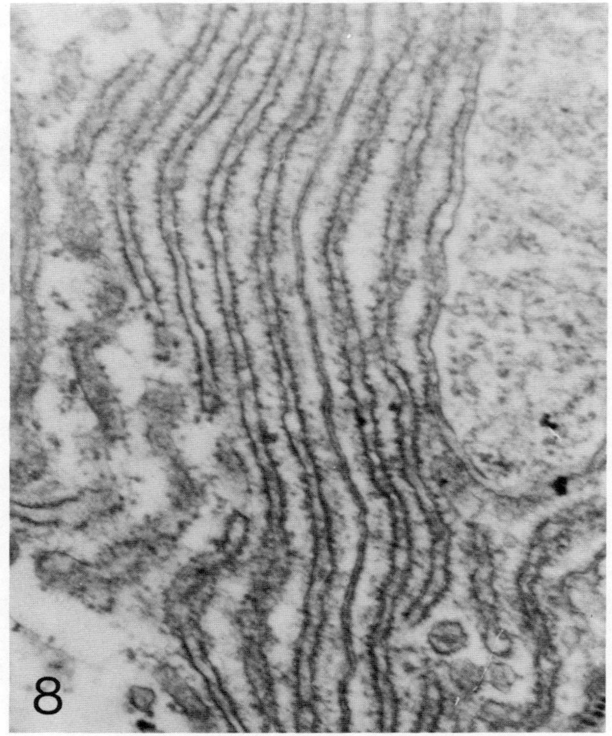

Plate 8. *Tetraodon samphongsi*, type 2a

Plate 9. Siganus (Lo) vulpinus, type 2a

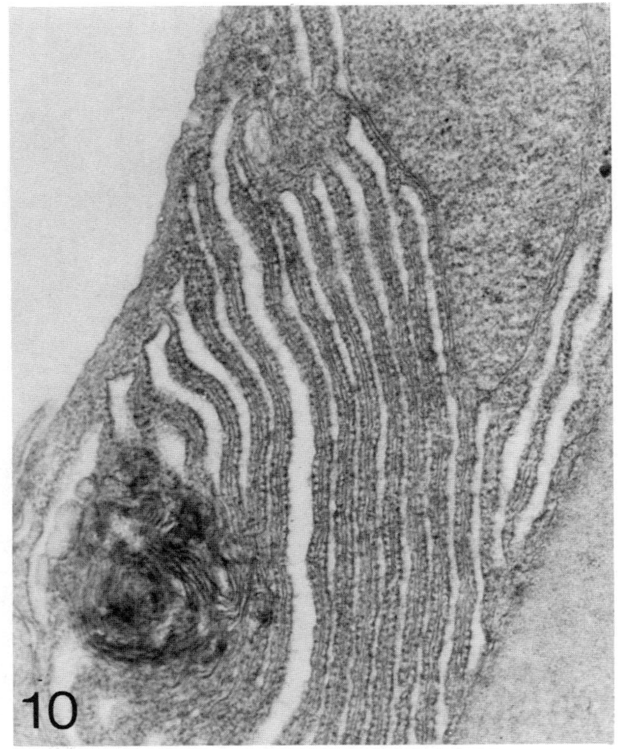

Plate 10. Anthias sp., type 2b

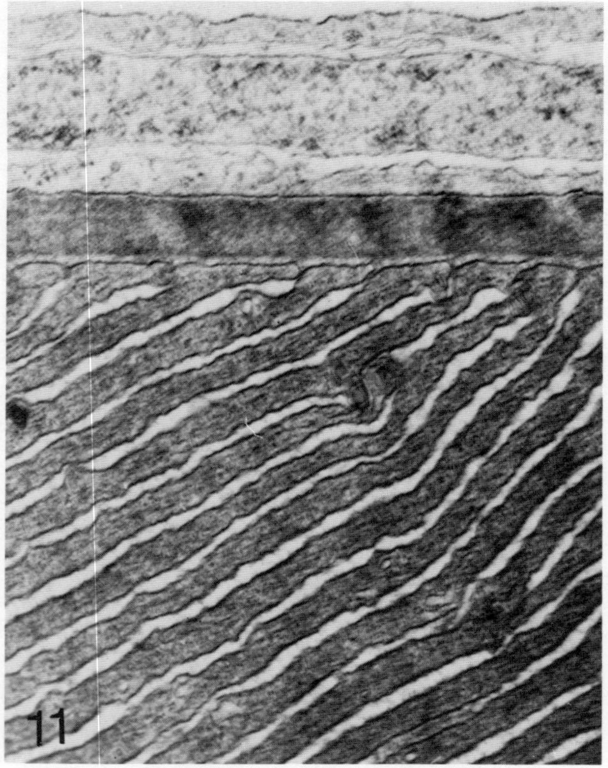

Plate 11. <u>Pomatoschistus</u> <u>minutus</u>, type 3

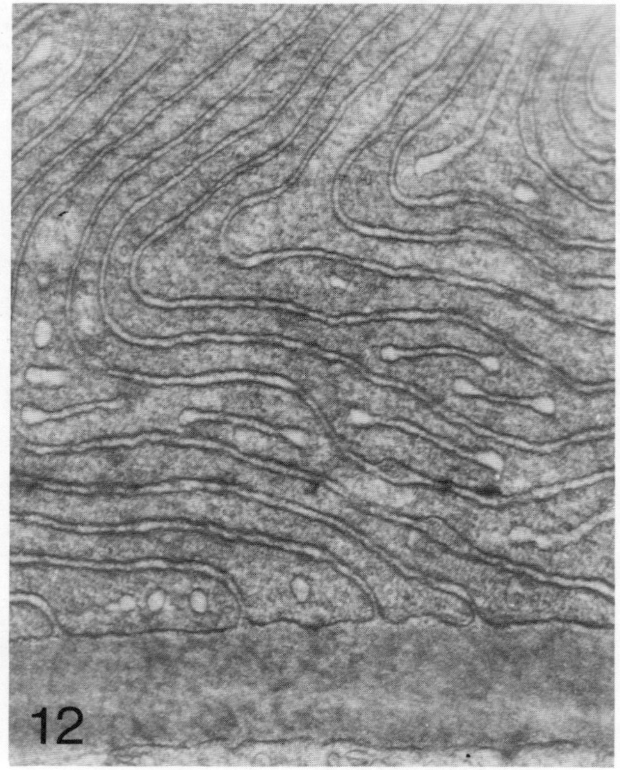

Plate 12. <u>Trachinus</u> <u>vipera</u>, type 3

pearance but in these preparations at least no discrete fibrils are visible.

Type 1c - Basically, part of the corneal stroma. Two or three layers of parallel collagen fibrils are modified to make an ordered array. Three types, all found in the order Scorpaeniformes, are illustrated. In the Red gurnard, Trigla cuculus, Plate 4, the collagen fibrils look more or less normal, but running through the matrix that separates them are dense staining bands of another material. In the Sapphirine gurnard, Trigla lucerna, Plate 5, the fibrils are approximately twice the diameter of the collagen fibrils in the normal part of the stroma. These large fibrils are arranged in regular ranks and are separated by bands of more darkly staining material. The Sea scorpion, Cottus bubalis, Plate 6, has bands of collagen stroma of normal appearance separated by bands of more darkly staining material without apparent structure.

It is believed that in two species belonging to the sub-order Blennioidei, namely, the Shanny, Blennius pholis, and the Gunnel, Pholis gunellus, Plate 7, there is an iridescent layer composed of modified collagen stroma.

B. Endoplasmic reticulum

Rough endoplasmic reticulum can be responsible for iridescent reflexions. There appear to be two main types distinguished by their position in the cornea and on the degree of specialisation of the reticulum.

Type 2a - The iridescent layer is situated within the stroma. This category is perhaps the least well defined and some modification of the iridescent layer classification will be required in the future. Three species have been found that possess this type of iridescence. These are the Green spotted puffer, Tetraodon samphongsi, Plate 8, Yellow trigger fish, Balistapus undulatus, and the Rabbit fish, Siganus vulpinus, Plate 9. Of these the ordered lamellae of the endoplasmic reticulum in Tetraodon is most clearly defined. Here each cell is packed with about 20 repeating units of endoplasmic reticulum forming a continuous layer.

Less well defined is the iridescent layer in Balistapus. Here the cells are much narrower and small islands of apparently normal fibril-containing stroma appear between them. The endoplasmic reticulum is much less well defined than in Tetraodon and it is possible that it does not contribute significantly to the iridescence.

Siganus vulpinus contains very thin cells in the stroma and each cell is completely separated from its fellows by fibril-containing stroma. The individual components of the iridescent stack are almost certainly collagen stroma and the cytoplasm of the cell. In this respect they more nearly resemble type 3 described below, and were it not for the clear affinity with Tetraodon would be classified as type 3.

Type 2b - This is an extremely well defined type, Plate 10. First described by Locket (1972) for Nemanthias carberryi, it has now also been seen in the Barred hamlet, Hypoplectrus puella. This type of layer has been described in detail by Locket (1972). The epithelium contains highly ordered lamellae of modified rough endoplasmic reticulum.

C. Whole cell

Type 3 - This is the most commonly encountered type of iridescence. The relevant layer is situated between Descemet's membrane and the stroma and thus corresponds in position to Walls' (1967) autochthonous layer. The simplest type is found in the gobies, of which the Sand goby, Pomatoschistus microps, Plate 11, has been the most studied. Here plates of cytoplasm, sometimes containing mitochondria and nuclei, are spaced by layers of electron-rare material without obvious structure that is continuous with the matrix surrounding the collagen fibrils of the stroma.

In the Sand goby, the plates are flat and unfolded. However, in many other species, such as the Weaver, Trachinus vipera, Plate 12, they are refolded on themselves in a complex manner that has not been studied in detail.

THE PHYLOGENY OF THE IRIDESCENT LAYER

There are indications that closely related species possess anatomically similar iridescent layers and that the layer might help to trace the phyletic relationships of different groups of teleost fishes.

Greenwood, Rosen, Weitzman & Myers (1966) have developed a provisional classification of Teleostean fishes that is based on phyletic relationships. It is within this framework that the various types of iridescent layer are discussed below. A tentative phyletic scheme based on the structure of the iridescent layer is set out in Fig. 2.

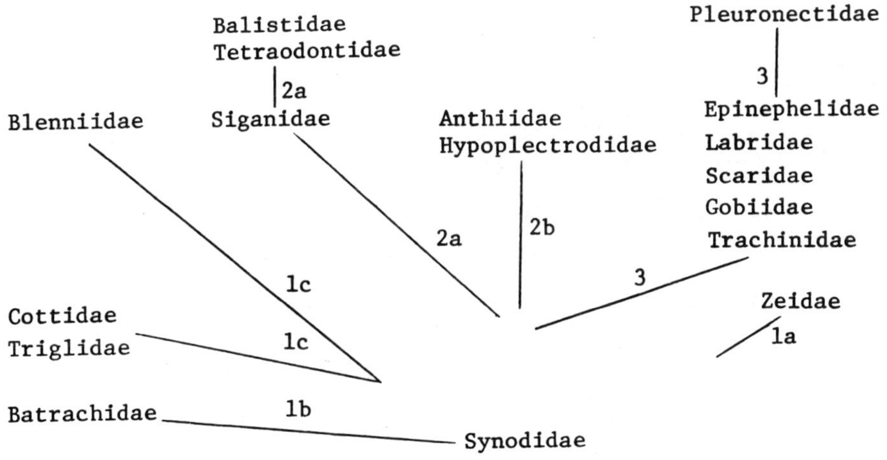

FIG. 2. Hypothetical scheme of the phyletic relationships of the teleosts based upon the morphology of the iridescent layers in their corneas and partly upon Greenwood *et al.*'s 1966 provisional evolutionary relationships amongst the teleosts.

Phyletic Summary of Iridescent Cornea Types

Type 1a - The phyletic relationships of the Zeidae are unclear and no other species with this corneal type has yet been found.

Type 1b - The Myctophoids in general do not have iridescent corneas, the only known exceptions are the Lizard or Snake fishes, (Synodidae). The only member of the Batiachoidiformes yet examined is the Sargassum fish, *Histrio histrio*. According to Greenwood *et al.* (1966) the Batiachoidiformes might be derived via the Neoscopelid-like fishes from the Myctophoids (Fig. 2).

Type 1c - Many species of the Scorpaeniformes possess iridescence, but to date only three species have been examined (*Trigla lucerna*, *T. cuculus* and *Cottus bubalis*). When there is a small overhead light source, the corneal iridescence is not evenly distributed over the pupillary area, but instead forms files of bright spots or lines. This pattern is visible in photographs of the three species mentioned and is also seen in Agonos cataphractus, Cyclopterus lumpus and Myxocephalus scorpius.

This type of iridescent pattern has not been noted outside the Scorpaeniformes.

However, a similar anatomical type has been seen in the Blenniiformes (Blennius pholis and Pholis gunellus). The Blennioidei are classified with the Perciformes. It is interesting to note in this context how different is the iridescent layer in the superficially similar blennies and gobies.

Type 2a - Romer (1955) classifies the Tetraodontidae, Siganidae and Balistidae in the same sub-order; the Plectognathi. Greenwood et al. (1966) consider the Siganidae to belong to the Perciformes and the Balistidae and Tetraodontidae, which they classify in a separate order (Tetraodontiformes), to be derived from the Perciformes.

Type 2b - Anthias, Nemanthias and Hypoplectrus all possess this type of cornea and are classified by Greenwood et al. in the same family, the Serranidae of the sub-order Percoidei.

However, two other members of this family, Cephalopholis fulva and Epinephelus guttatus, possess type 3 iridescence. It is interesting to note in this connexion that in a personal communication Lovett Smith (1974) points out that Anthias and Hypoplectrus are synchronous hermaphrodites whilst Epinephelus (including Cephalopholis) are protogynous.

Type 3 - This type of cornea is known to exist in the Perciformes and the Pleuronectiformes (Pleuronectes platessa) which is derived from them. The occurrence of the layer in Cephalopholis and Epinephelus has been discussed in the previous section. Type 3 iridescence is also known from several other members of the order Perciformes. Classified into families these are:

Labridae: Crenilabrus melops, C. rupestris, Halichoeres bivittatus.
Scaridae: Scarus taeniopterus
Trachinidae: Trachinus vipera
Gobiidae: Gobius paganellus, G. niger, Pomatoschistus minutus, P. microps

DISCUSSION

Amongst shallow water marine teleosts that are either benthic or demersal, corneal iridescence is a common phenomenon. Superficially the iridescence looks very much the same in most species and it is likely that its function is usually restricted to camouflage and display, and perhaps more important, as a type of sunshade that protects the eye from flare derived from very bright downwelling light, as discussed in the previous paper (Lythgoe, 1975) and Lythgoe (in press, b).

Nevertheless, examination under the electron microscope shows that the structures that produce the iridescent reflexions are very different in different groups of teleosts. It is interesting to note that it is amongst the more advanced teleosts that iridescence is most common and that it is unknown in the Osteoglossomorpha, Clupeomorpha and Elopomorpha. These facts, taken together, lead to the conclusion that corneal iridescence has been "invented" several times in the course of teleost evolution and presents a remarkably good example of convergent evolution.

Greenwood et al. (1966) have pointed out that the phyletic classification of the teleosts is still in its infancy. At this time it is not possible to "fit" the corneal structure into an accepted evolutionary scheme although there are indications that this may become possible. It is not yet possible to know whether it is the position of the layer within the cornea or the organelles involved that would be the best taxonomic indicator. Probably both must be considered together in any general evolutionary scheme based on corneal structure.

REFERENCES

Greenwood, H.P., Rosen, D.E., Weitzman, S.H. & Meyers, G.S. (1966) Phyletic studies of teleostean fishes, with a provisional classification of living forms. Bull. Ann. Mus. Nat. Hist. 131: 341-455.

Locket, N.A. (1972) The reflectiong structure in the iridescent cornea of the serranid teleost Nemanthias carberryi. Proc. R. Soc. B. 182; 249-254.

Lythgoe, J.N. (1971) Iridescent corneas in fishes. Nature (Lond.), 233: 205-207.

Lythgoe, J.N. The ecology, function and phylogeny of iridescent multilayers in fish corneas. In: Light as a ecological factor II. Edited by R. Bainbridge, and G.C. Evans (in press, a).

Lythgoe, J.N. The structure and function of iridescent corneas in Teleost fishes. Proc. R. Soc. B. (in press, b).

Maurice, D.M. (1957) The structure and transparency of the cornea. J. Physiol. 136: 263-286.

Maurice, D.M. (1970) The transparency of the corneal stroma. Vis. Res. 10: 107-108.

Moreland, J.D. & Lythgoe, J.N. (1968) Yellow corneas in fishes. Vis. Res. 8: 1377-1380.

Muntz, W.R.A. (1973) Yellow filters and the absorption of light by the visual pigments of some Amazonian fishes. Vis. Res. 13: 2235-2254.

Romer, A.S. (1955) Vertebrate Palaeontology. Univ. of Chicago Press, Chicago.

Smith, R.W. (1969) The transparency of the corneal stroma. Vis. Res. 9: 393-396.

Smith, R.W. (1970) The transparency of the corneal stroma. Vis. Res. 10: 109-110.

Walls, G.L. (1967) The Vertebrate Eye, Hafner, New York.

THE IRIDESCENT CORNEA OF THE SAND GOBY *POMATOSCHISTUS MINUTUS* (Pallas)

J.N. LYTHGOE

MRC Vision Unit, University of Sussex

Falmer, Brighton, BN1 9QG. U.K.

He's blind with too much light
Philip Massinger, Great Duke
of Florence, Act II, Sc. 1

INTRODUCTION

Many species of shallow-living marine teleost fishes possess regularly arranged lamellae in their cornea. Each of these lamellae are less than a wavelength of light in thickness and when the eye is illuminated from above but viewed from the side, the cornea shows coloured iridescent reflexions.

Iridescence occurs when a light beam travels from one medium through a thin layer of another which has a different refractive index (Fig. 1). If the layer is of the order of a wavelength of light in thickness, light rays reflected from the top and the bottom of the layer may mutually interfere. If the two waveforms are in phase they will interfere constructively, i.e. they will reflect light strongly at the relevant wavelength and the transmitted fraction will be correspondingly reduced. If, on the other hand, the two waveforms are out of phase, they will interfere destructively, there will be no reflexion and transmission losses will be minimal. Land (1972) has reviewed the various iridescent structures known in animals and the underlying physical theory, and Huxley (1968) has provided a mathematical treatment for calculating the reflectance from a regular multilayer stack provided the refractive index and thickness of the layers are known.

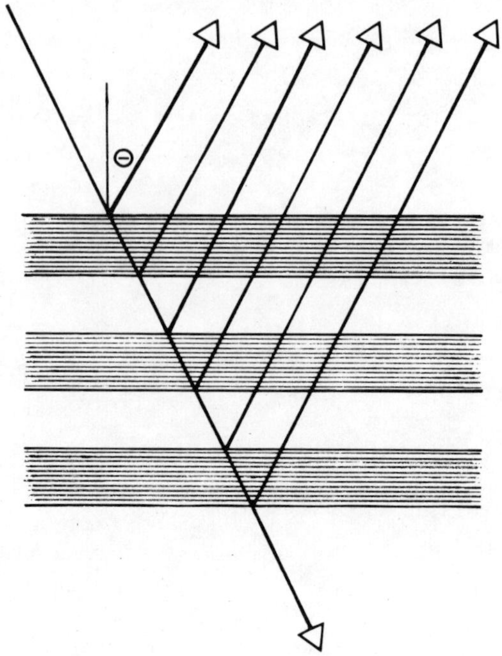

FIG. 1. Diagram of the reflexion of light in an iridescent multilayer. Light incident on the multilayer stack at an angle θ to the normal is reflected at each interface separating the layers. The light reflected from each successive interface suffers a phase delay. When the reflected rays are in phase there will be constructive interference and reflexion will be enhanced; when the reflected rays are out of phase reflexion is suppressed. The wavelengths that are reflected depend upon θ and the refractive index and thickness of each layer. See Land (1972) for a review.

Corneal iridescence is frequently visible in the corneas of the compound eyes of insects. Through the pioneering work of Bernard & Miller (1968) the natural history and optics of these structures have been well studied, although their function, as in the fishes, is not always easy to understand.

The appearance of the iridescence is much the same in all the species that possess it, but the multilayers that are responsible are constructed from different morphological structures and occupy various locations in the cornea depending upon the taxonomic position of the fish. The obvious conclusion to be drawn from this is that the iridescent layer has some useful function and that it has been independently 'discovered' several times in the evolution of the teleosts.

The fact that it is the bright downwelling light that is responsible for the iridescence led Lythgoe (1971) and Locket (1972) to postulate that the iridescent layer acts in some way as a 'sunshade' to protect the eye from the very bright downwelling light under water.

The Sand Goby is a bottom-living species that is particularly abundant on shallow sandy bottoms. Direct underwater observation shows that in life the eye is strongly iridescent but since the fish rarely exceeds 100 mm in length, detailed study in the field is difficult.

Under the electron microscope the layer responsible for the iridescent reflexions can be identified with reasonable certainty between the stroma and Descemet's membrane. It consists of 15 - 20 cytoplasmic plates that sometimes contain nuclei and mitochondria separated by an amorphous layer that is continuous with the matrix material separating the collagen fibrils of the stroma. The plates are oriented in a similar manner to the slats of a venetian blind (Plate 1) in the top two-thirds of the cornea, but their regular organisation becomes lost in the ventral one-third of the cornea where no iridescence can be discerned.

Reflected Light

The technique used for measuring the spectral distribution of the light reflected from the cornea is described elsewhere (Lythgoe, in press[a]) and will not be detailed here. Briefly, a collimated beam of monochromatic light is shone into the cornea at a known angle, and the reflected light is collected by the objective of a microscope. A light guide is inserted into the image plane of the eyepiece of the microscope and conducted to a photomultiplier.

The spectral distribution curve of the reflected light is rather broad at all angles of incidence, but as is to be expected (Land, 1972), the most strongly reflected light moves to shorter wavelengths as the angle of incidence increases (Fig. 2). The reflected light is strongly plane-polarised but no specific measurements have been made on this.

These reflection measurements give little information about the light that is transmitted by the cornea since the photocell does not receive all the light that is reflected and the orientation of the plates and the curvature of cornea makes it difficult to estimate the fraction of the incident light that is transmitted.

Plate 1. Section of the Sand Goby cornea. This is a young fish and the layer is not fully developed. Arrow indicates the approximate dorsal direction. end - endothelium; D - Descemet's membrane; ir - iridescent layer; str - stroma. The cornea shown here is separated from the spectacle which is not shown. aq - aqueous humor. The bar represents 1000 nm.

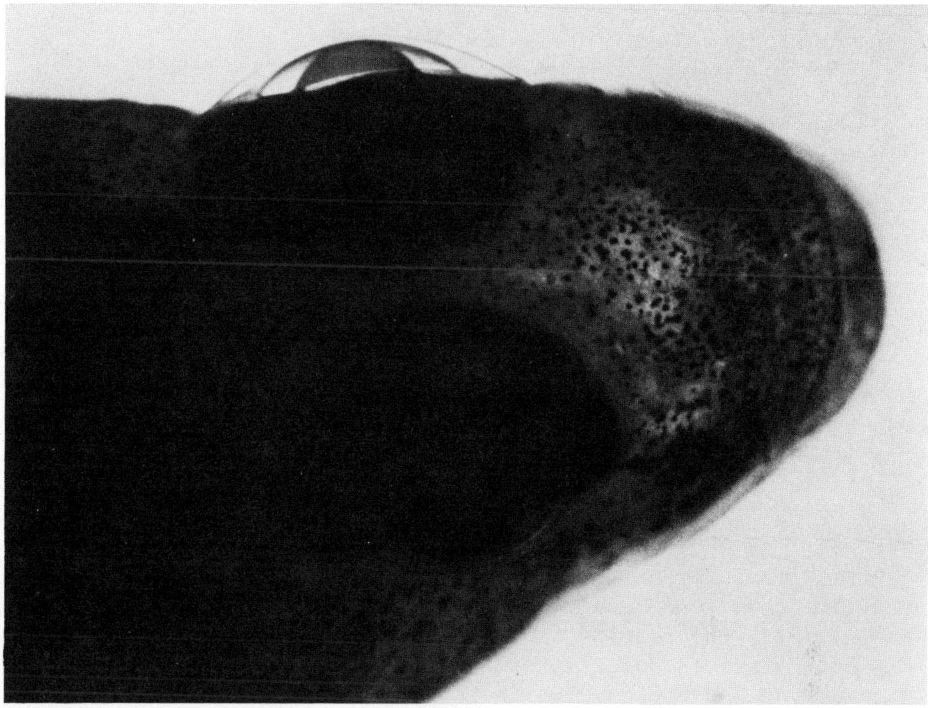

Plate 2. The cornea and spectacle of the Goby cornea photographed from above and seen in silhouette against a bright background. The spectacle, which is thicker than the cornea, is scarcely visible, whereas the cornea, which contains the iridescent layer, is very conspicuous.

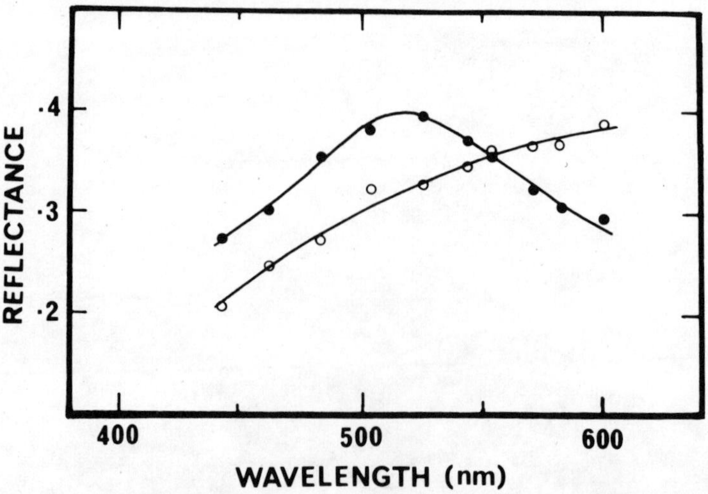

FIG. 2. The spectral distribution of reflected light from the Sand Goby cornea at two angles of collection. The light beam was directed onto the eye from a dorsal direction at an angle of 69° to the optic axis of the eye. The light was collected at angles of 40° (open circles) and 26° (filled circles) to the optic axis.

Direct measurements of optical density of the cornea at normal incidence were made using Muntz' 'Mini-spectrophotometer' (Muntz, 1973). The results of one such measurement are shown in Fig. 3. There is possibly some increase in optical density between 480 nm and 600 nm, but the effect is very small and it does not seem likely that the reduction will have a significant visual effect either through its colour filtering properties, its birefringent properties or as a neutral density filter.

However, when the eye is illuminated from above there are very strong reflexions in a ventral direction (Fig. 4), which in the Sand Goby are blue in colour.

The curvature of the cornea and the very large angles of incidence involved make the measurement of such reflections and consequent transmission losses difficult to measure. Photographs in other species confirm that it is the iridescent layer that

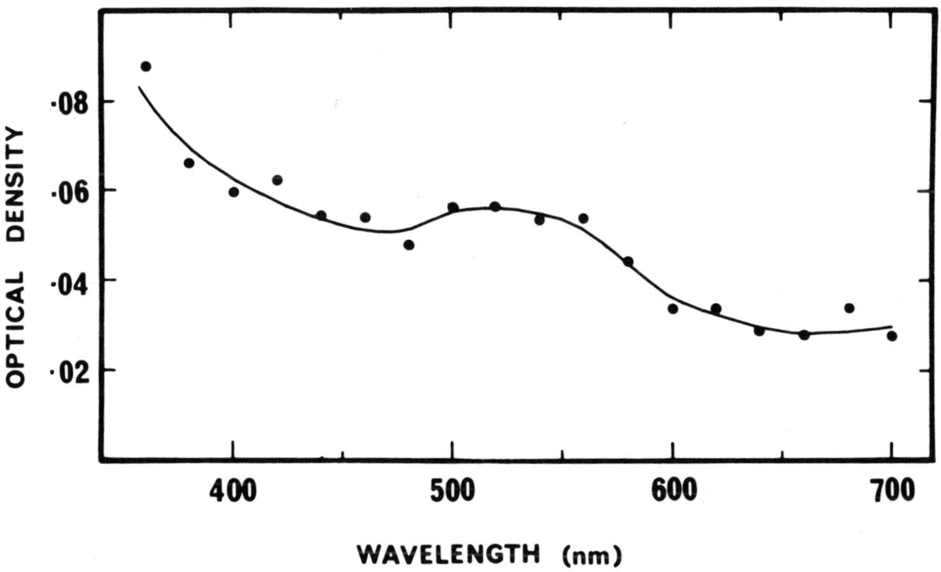

FIG. 3. The spectral transmission of Sand Goby cornea at normal incidence.

becomes highly reflecting and when viewed in silhouette the cornea shows up as optically dense whereas the spectacle, which is similar in construction to the cornea is very difficult to see (Plate 2).

The strong reflectivity at oblique angles of incidence can be predicted from Fresnell's equations and are not an interference effect at all. Fresnell's equations should be taken as a guide rather than an exact description of reflexions at large angles of incidence since they apply strictly only to layers that are considerably more than a wavelength of light in thickness; neither do they take into account interference phonemena (Born & Wolfe, 1970).

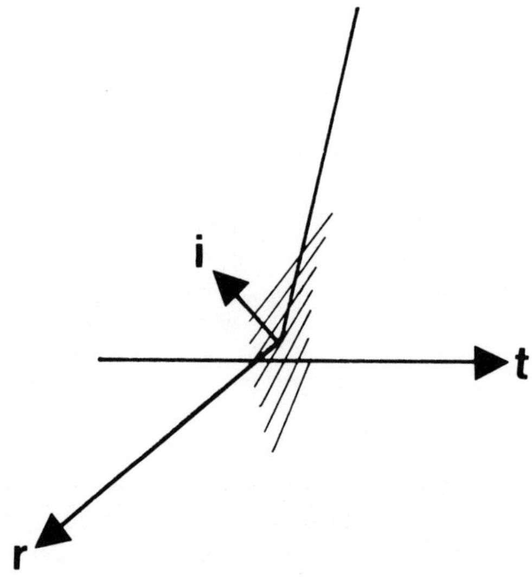

FIG. 4. Diagram of the reflexion properties of the oblique lamellae in the Sand Goby cornea. Light incident from above is reflected in the direction i. This light is highly coloured but the proportion of the incident light reflected out in this direction is relatively small. Light, r, is also reflected out of the eye at an oblique angle, and although not measured, is comparable to the incident light in brightness. In the Sand Goby this light is an unsaturated blue in colour. Light, t, passing through the eye near to the optic axis suffers little loss due to reflexion since at this angle of incidence there is destructive interference between the reflected rays.

The expected reflexions from regular stacks of thin plates can be calculated using Huxley's (1968) method, but these depend critically on exact knowledge of the refractive index and thickness of each layer type. Direct measurements of the araldite-embedded Goby cornea indicate that the protoplasmic layer has a thickness near to 143 nm and the matrix layer a thickness near to 49 nm. These measurements are probably accurate for the two layers as they appear embedded in araldite, but whether their dimensions are the same in life is very much an open question (Lythgoe, in press[a]). The principal uncertainty comes not in the fixation since the iridescence of unchanged colour can be seen in properly fixed material, but in the subsequent dehydration necessary to impregnate the material with resin.

Preliminary work using Pease & Peterson's (1972) technique of embedding in a urea/glutaraldehyde polymer, which does not involve dehydration, indicates that so far as frog collagen stroma is concerned the dimensions and packing of collagen fibrils are different in urea/glutaraldehyde fixed material than it is in araldite fixed material. Thus in araldite the cross-sectional area of 500 stroma fibrils was 100 - 900 nm^2 (mean 516 nm^2), and in 500 fibrils embedded in urea/gluteraldehyde it was 300 - 1200 nm^2 (mean 804 nm^2). The volume fraction occupied by the fibrils was 0.19 in araldite and 0.41 in urea/glutaraldehyde. As yet we have not got satisfactory urea/glutaraldehyde embedded material from iridescent corneas, but we would not expect it to have exactly the same dimensions as the araldite embedment that we have used in this study.

It is worth noting that if the urea/glutaraldehyde measurements are correct they tend to support Smith's (1969, 1970) contention that the fibrils and the matrix have essentially the same refractive index and it might be unnecessary to postulate that the quasi-regular arrangement of the fibrils results in destructive interference of the light in all directions except the forward direction (Maurice, 1957, 1970; Cox, Farrel, Hart & Langham, 1970).

In the Goby cornea the material which represents the cytoplasmic plates is continuous with the collagen matrix material in the stroma and perhaps has a similar refractive index of 1.374 (Cox, Farrel, Hart & Langham, 1970). The wavelength of maximum reflexion at near normal incidence to the lamellae is about 600 nm and from this it can be calculated, using Land's formula (Land, 1972) that the refractive index of the cytoplasmic layers is near 1.62. If the thickness of two such layers were such as to give the maximum constructive interference, a stack of the 18 layer pairs present in the Sand Goby would mean that nearly all the incident light was reflected and essentially none was transmitted. Indeed, Land (1972) has shown that an 'ideal' multilayer stack having layers of refractive index 1.33 and 1.56 would be nearly opaque for 20 layer pairs.

Obviously interference phenomena are present in the iridescent Goby cornea, yet the cornea is nearly transparent. It follows that the thickness and refractive index of the plates must be such as to result in destructive interference of the reflected ray and it is possible that at angles of incidence near the optic axis light losses due to reflexion are considerably reduced. We can, therefore, imagine a system (Fig. 4) analagous to a venetian blind whereby light reaching the layer from above is reflected out of the eye, but unlike a venetian blind, the light coming from more horizontal directions is not lost since destructive interference suppresses the losses due to reflexion.

Flare

A constant feature of the underwater light climate is that the downwelling radiance is very much brighter than light coming from other directions. In shallow water, when the sun is out, light coming from the direction of the sun may be at least 3 log units brighter than light arriving from a horizontal direction (Fig. 5).

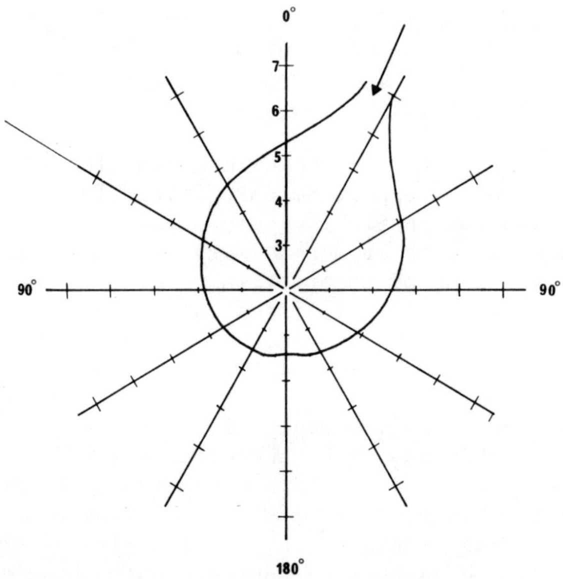

FIG. 5. Polar distribution of the underwater spacelight at 4.24 m depth in a freshwater lake on a sunny day. Data plotted on a log scale from Tyler & Preisendorfer, 1962.

The underwater photographer knows that with the very wide angle lenses now in common use, flare in the negative can only be avoided by carefully protecting the lens from direct overhead light. This flare probably originates from multiple reflexions between the various elements of the camera's optical system. It is suggested here that a similar effect (albeit of lesser magnitude) occurs in the eye of shallow water teleosts, and that the iridescent cornea is a device for reducing it.

Measurement of Flare

For the measurement of intra-ocular flare it is necessary to remove the eye and to cut away the back in such a way that the lens can be directly observed from behind. The eye is then mounted in the apparatus shown in Fig. 6 and light is shone onto the cornea at nearly 90° to the optic axis. The vessel containing the eye can then be rotated so that the eye can be illuminated from the dorsal, ventral, rostral or caudal direction and, indeed, from any direction at 90° to the optic axis.

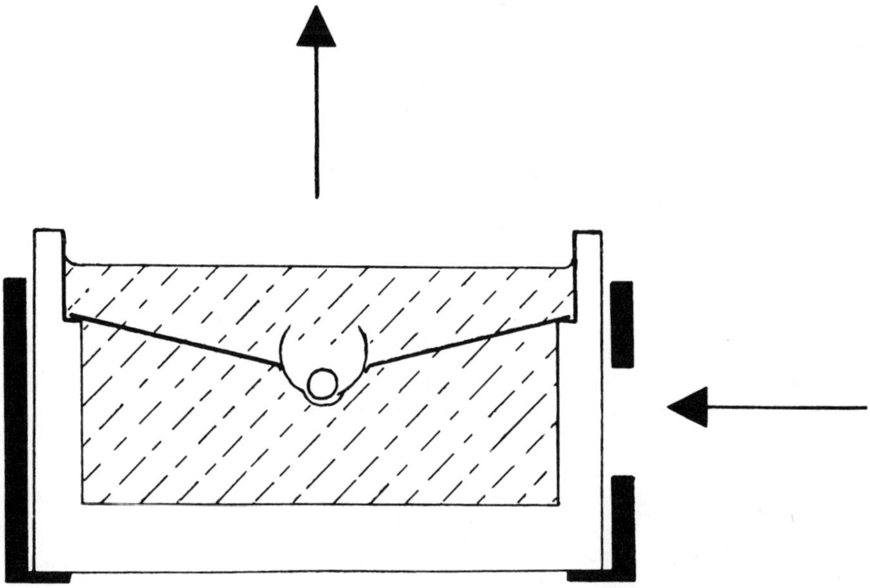

FIG. 6. The arrangement used to mount the eye during measurements of intra-optical flare. The cylindrical vessel is made of transparent perspex and contains a pH 7.4 phosphate buffer with osmolarity adjusted with sucrose to preserve corneal iridescence. The vessel is free to rotate within a fixed brass cylinder.

When this is done with a fish such as the Rudd, which has no iridescent layer, flare measured by focussing the collecting objective at the centre of the lens is nearly independent of direction. However, when the same procedure is carried out using a fish with an oblique multilayer within the cornea, such as the Sand Goby, the Weaver, Trachinus vipera, and an Anthias sp., it is found that light from the dorsal direction is responsible for much less flare than light from the ventral direction (Fig. 7).

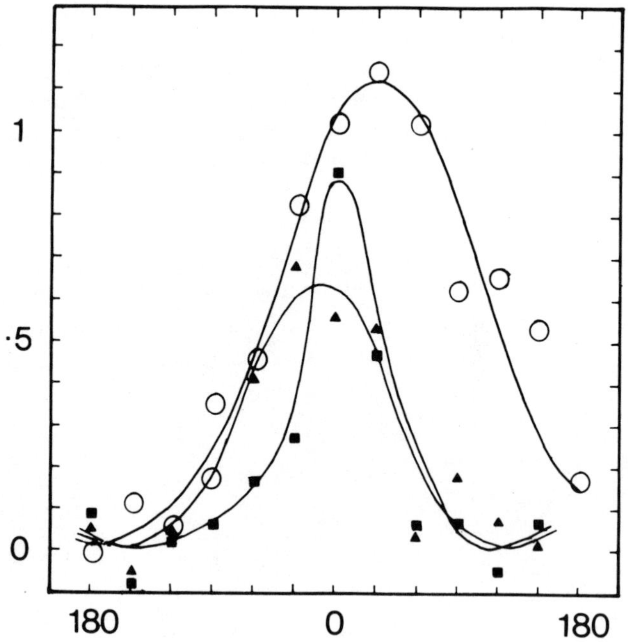

FIG. 7. Measurements of intra-ocular flare using the arrangement shown in Fig. 6. At 0° the illuminating light reaches the eye from a ventral direction; at 180° the light reaches the eye from a dorsal direction. The amount of flare is presented here in relative units (for a quantitative treatment see Lythgoe, in press, a). Open circles, Sand Goby; Closed squares, *Trachinus vipera*; Closed triangles, *Anthias* sp.

Perhaps the most difficult feature of the visual environment is the reduction of contrast. The light scattering and absorbing properties of the water cause the apparent contrast of an object to be rapidly reduced as the visual range increases. Intra-ocular flare will be particularly damaging in this regard since it acts to reduce apparent contrast. Contrast is defined as $C = ({}_{t}N - {}_{b}N)/{}_{b}N$ where ${}_{t}N$ is the object radiance and ${}_{b}N$ is the radiance of the background against which it is seen. Light resulting from intra-ocular flare (f) added to both ${}_{t}N$ and ${}_{b}N$ will obviously tend to reduce C. It is possible to calculate how much any known amount of flare will reduce visual range for horizontal paths of sight provided the narrow-beam attenuation coefficient of the water is known.

The actual amount of flare present in the eye can be calculated by integrating together the flare produced experimentally by light reaching the eye from the full 360° with the light known from Tyler & Preisendorfer's (1962) data to be actually present under water.

Details of the calculation procedure have been given elsewhere (Lythgoe, in press[a]). It is not possible to know the flare that would be present in the eye were the iridescent layer absent. An estimated value as a basis for comparison was obtained by calculating the mean value of flare produced by equal intensity light over 360° so that if the underwater spacelight radiance was equal in all directions, real and comparison values for flare would be the same.

Some calculated values are shown in Table 1 for the visible ranges of a black and a grey target assuming a beam attenuation coefficient of 1.73 (in this water a diver could see a large black object at a horizontal range of some 3.4 m) and a contrast perception threshold of 0.05. It is clear that flare could make a significant difference to the distance that objects can just be detected, especially if they present a low inherent visual contrast.

TABLE 1

	No flare	Comparison flare	Actual flare
Black target C = -1	260 cm	241 cm	256 cm
Grey target C = -.1	60 cm	41 cm	56 cm

Biological Significance of Iridescence

Regular stacks of thin lamellae whose dimensions are such as to produce interference of the reflected ray have the potential of serving at least four different functions.

1. Act as a wavelength-selective reflector and/or as a wavelength selective transmission filter.
2. Act as a wavelength unselective reflector and hence as a neutral density filter.
3. Act as an anti-reflexion layer.
4. Act as a polarising reflector and hence as a polarising filter.

Wavelength-Selective Reflector

In no fish so far examined has the iridescent reflexion been strong enough to cause obvious colouration of the cornea by transmitted light. It is probably safe to say that the cornea does not act as a significant colour filter, at least at angles of light near to the optic axis.

Iridescent reflexions do, however, give obvious colour to the cornea in the pupillary area and iridescence may very well serve to camouflage the pupil in some species. In others, such as the Sapphirine Gurnard, Trigla lucerna, the iridescence is so vivid that it seems reasonable to postulate some function in behavioural display.

The fishes where corneal iridescence may have some cryptic function seem generally to live their lives in close association with the bottom, sometimes indeed half buried in it. Thus the Plaice, Pleuronectes platessa, has a bronze coloured cornea when viewed at the surface, but underwater the colour matches more closely the body colour surrounding the eye, thus making it more difficult to detect. Amongst the Serranidae the Groupers, such as the Caribbean Red Hind, Epinephelus guttatus, and the Indo-Pacific Honeycomb Grouper, Epinephelus merra, have pinkish-bronze corneas when seen at the surface, but at even moderated depths the eye takes on a curiously "dead" look and the pupil loses its "brillant" black appearance (Lythgoe, in press[b]).

One of the best examples of iridescent camouflage for the pupil may be in the cryptic bottom-dwelling Crocodile fish, Platycephelus indicus. Here the cornea also appears pinkish-bronze (when photographed by white light) but the age is nearly invisible when viewed under water by natural light.

Wavelength Unselective Reflector

The silvery sides of many teleosts owe their reflecting properties to guanine crystal plates embedded in a cytoplasm matrix (Denton & Nichol, 1965; Denton & Land, 1971). Fish such as the Herring, Clupea harengus, do not look coloured, but silvery-white. The explanation is that each scale reflects about one-third of the spectrum, but where three scales, each reflecting a different third of the spectrum, overlap, a silvery reflexion results (Denton & Nicol, 1965). In other fish, such as the Roach, Rutilus rutilus, light reflected from a single scale is almost white even though analysis of the crystal spacings would suggest a red reflexion. In this case Land (1972) suggests that the white colour is probably caused by systematic variations in the spacing within each scale.

Something of the sort may well occur in the very strong reflexions at oblique angles of incidence in the cornea of the Lizard fish, Synodus indicus, and the Red Hind, Epinephelus guttatus (Lythgoe, in press[a]). In the Sand Goby, the oblique reflexion is pale blue in colour and there is no need to find an explanation for the absence of colour in these oblique reflexions.

Anti-Reflexion Layer

Losses in transmitted light caused by reflexion at the boundary between two layers of different refractive index can be suppressed if the light rays reflected from the front and back surface of the layer are out of phase and thus destructively interfere with each other. Such a layer is used to coat the surface of optical lenses and a similar structure could possibly be found at the surface of the cornea.

In fish the similarity in the refractive index of water and cytoplasm make it unlikely that such a surface coating would be useful. However, if the multilayer was present to serve some other purpose, such as protecting the eye from bright downwelling light, or camouflaging the pupil, consequent losses in the visually important light passing through the cornea at near normal incidence would be reduced.

Polarising Filters

The underwater spacelight is plane polarised due to small-particle scattering in the water. The spacelight is polarised down to all depths with the plane of maximum polarisation at right angles to the direction of the incident light. Thus near the surface the polarisation is at right-angles to the sun's rays and in deeper water, or when the sky is overcast, it is horizontal (see Jerlov, 1968, for a review).

Lythgoe & Hemmings (1967) have shown that the visible range of objects can be increased under water with the use of the appropriate plane-polarising faceplate. It is, however, unlikely that the teleost cornea is a sufficiently effective polariser to serve this function.

Forward & Waterman (1973) consider that some fishes may be sensitive to the plane of polarised light, which they use to orientate themselves and perhaps to navigate under water. They point out that a simple birefringent filter will not of itself tell the animal in what plane the light is orientated. In this context, it is worth noting that the fish that possess iridescent corneas are bottom-dwelling species (Lythgoe, in press [a & b]) that are not known to undertake long migrations; whereas migratory pelagic fishes do not possess iridescence.

CONCLUSIONS

Iridescent corneas are common amongst shallow-living marine teleost fishes that live in close association with the bottom. They have not been observed in pelagic fish from any depth except for the untypical case of the Sargassum fish, Histrio histrio (Lythgoe, in press[b]). The iridescent layer in the cornea is constructed in at least six morphologically distinct ways in different taxonomic groups of fishes, and it is reasonable to assume that the iridescent layer performs at least one important function in the visual life of fishes. The number of possible functions for these layers is large and include the enhancement or suppression of reflexion, birefringence, a colour filter, camouflage or display, and enhanced reflection of light arriving from some particular direction.

Observations by swimmers suggest that the iridescent cornea in some normally cryptic fish serves to obscure the blackness of the pupil and hence to enhance the camouflage of the fish as a whole. Silvery-sided pelagic fish which are thought to be very well camouflaged in open water do not show iridescence as might be expected if pupil camouflage was of great adaptive value in open water.

Corneal reflexion may also have the function of display in some species, and it seems likely that species will be found where corneal iridescence has some behavioural role. Such a role is suspected, but certainly not proved, in the blue iridescence of the Sapphirine Gurnard.

The iridescent multilayer could in theory act both as colour and polarising filters, but such functions seem unlikely since the effect are so small. The multilayer could act as one component of a system to recognise the plane of polarised light either in the underwater spacelight or in the sky. However, such a function is hypothetical with little or no evidence to support it.

The most likely function remains that suggested by Lythgoe (1971) and Locket (1972), namely, that the iridescent layer acts as a sophisticated sunshade that protects the eye from the very bright downwelling light, characteristic of the underwater environment.

It is postulated that this light, reaching the eye at grazing incidence, suffers multiple reflexions between the cornea and the lens and thus produces intraocular flare. This flare will degrade visual contrasts and consequently reduce the range that objects can be seen through the water.

The obliquely orientated multilayer within the cornea reflects much of this bright downwelling light out of the eye. At the same time, losses due to reflection at the multilayer of the visually important image-forming light are reduced because at angles of incidence near to the optic axis, the reflected rays mutually interfere and reflexions are suppressed.

REFERENCES

Bernard, G.D. & Miller, W.H. (1968) Interference filters in the corneas of Diptera. Inves. Ophthalmol. 7: 416 - 434.

Born, M. & Wolf, E. (1970) Principles of Optics, 4th Ed. Pergamon, Oxford

Cox, T.L., Farrel, R.A., Hart, R.W. & Langham, M.E. (1970) The transparency of the mammalian cornea. J. Physiol. (Lond.) 214 - 233.

Denton, E.J. & Land, M.F. (1971) Mechanism of reflexion in silvery layers of fish and cephalopods. Proc. R. Soc. B. 178: 43 - 61.

Denton, E.J. & Nicol, J.A.C. (1965) Reflexion of light by external surfaces of the Herring Clupea harengus. J. Mar. Biol. Assoc. U.K. 45: 711 - 738.

Forward, R.B. & Waterman, J.H. (1973) Evidence for E-vector and light intensity pattern discrimination by the Teleost Dermogenys. J. Comp. Physiol, 87: 189 - 202.

Huxley, A.F. (1968) A theoretical treatment of the reflexions of light by multilayer structures. J. Exp. Biol. 48: 227 - 245.

Jerlov, N.G. (1968) Optical Oceanography, Elsevier, New York.

Land, M.F. (1972) The physics and biology of animal reflectors. Prog. Biophys. Mol. Biol. 24: 75 - 106.

Locket, N.A. (1972) The reflecting structure in the iridescent cornea of the serranid teleost, Nemanthias carberryi. Proc. R. Soc. B. 182: 249 - 254.

Lythgoe, J.N. (1971) Iridescent corneas in fishes. Nature, Lond. 233: 205 - 207.

Lythgoe, J.N. The structure and function of iridescent corneas in teleost fishes. Proc. R. Soc. B. (in Press[a]).

Lythgoe, J.N. The ecology, function and phylogeny of iridescent multilayers in fish corneas. In: Light as an ecological factor II., a symposium of the British Ecological Society, edited by R. Bainbridge and G.C. Evans (in press[b]).

Lythgoe, J.N. & Hemmings, C.C. (1967) Polarized light and underwater vision. Nature, Lond., 213: 893 - 894.

Maurice, D.M. (1957) The structure and transparency of the cornea. J. Physiol. 136: 263 - 286.

Maurice, D.M. (1970) The transparency of the corneal stroma. Vision Res. 10: 107 - 108.

Muntz, W.R.A. (1973) Yellow filters and the absorption of light by the visual pigments of some Amazonian fishes. Vision Res. 13: 2235 - 2254.

Smith, R.W. (1969) The transparency of the corneal stroma. Vision Res. 9: 393 - 396.

Smith, R.W. (1970) The transparency of the corneal stroma. Vision Res. 10: 109 - 110.

REFRACTIVE STATE, ACCOMMODATION, AND RESOLVING POWER OF THE FISH EYE

H.O. SCHWASSMANN

Department of Zoology, University of Florida

Gainesville, Fla. 32611 U.S.A.

'Sblood, there is something in this more than natural,
if philosophy could find it out.
Hamlet, Act II, Sc. 2

INTRODUCTION

The refractive state of the eye is a measure of the difference between the focal distance of the dioptric elements and the distance at which the receptors are located. If parallel rays of light incident on the pupil, as they would be when they come from an object at long distance, are brought to a point focus in the plane of the light-sensitive outer segments of the photoreceptors, the eye is said to be emmetropic. If the rays from the distant object focus in front of the receptor layer, the eye is considered myopic, or near-sighted, and only a blur would exist at the photoreceptor level. This blurred image can be brought into sharp focus by reducing the object-to-eye distance, since the image plane of a near object lies at a longer distance from the lens than that of an object farther away.

Conditions opposite to myopia exist when parallel light rays reach the retina before coming to a point focus. In this case the eye is said to be hyperopic (hypermetropic), and bringing the object closer will only increase blurring of the image. Improvement can only be achieved by adding to the refractive power of the lens, or by increasing the lens-to-retina distance.

Accommodation is the ability to bring into focus at the receptor level the images of near as well as distant objects. One can distinguish between static and dynamic accommodation. In static accommodation the axial length of the eye varies for different directions of regard, as in the eyes of horses and rays with a ramp retina (Walls, 1942). Dynamic accommodation involves muscular contraction. The adjustment for sharp vision at different distances can be accomplished either by altering the focal power of the refractive elements, usually the lens, or by adjusting the lens-to-retina distance. The first possibility is realised in the case of the human eye in which active accommodation for near vision is done by contraction of the ciliary muscle. This releases the tension of the zonula, holding the lens, and the anterior face of the lens becomes more convex, thereby increasing its refractive power. Accommodation in the teleost eye utilises the second possibility which is principally like the adjustment of a photographic camera for different distances. The spherical lens of the fish eye can be moved towards the retina or away from it in accommodation.

Present knowledge concerning refractive state and accommodation mechanism of the teleost eye is based largely on results of extensive experiments by Beer (1894). A detailed review of earlier theories of accommodation in the eye of fishes was published recently (Munk, 1973). According to this literature survey, essentially correct mechanisms by lens retraction were suggested by von Haller in 1764, by Jurine in 1821, and by Dalrymple in 1838. Although Albrecht von Haller was the first one not only to describe the campanula, later recognised as the retractor muscle, but also to mention the possibility of its being a muscle, he mistook the choroidal gland for the accommodation muscle. According to Munk's review, Jurine apparently misinterpreted the suspensory ligament and the tendon of lens muscle insertion in the lens capsule to be the contractile elements. Dalrymple suggested that the lens is moven slightly inwards, in direction towards the retina; he also recognised the campanula Halleri as the accommodation muscle, not being aware that similar ideas were expressed a few years earlier by Wallace who, however, was incorrect about the actual mechanism (Munk, 1973).

There exists some confusion regarding the refractive state of the fish eye, and the experimental results and conclusions arrived at by Beer (1894) have sometimes been questioned. Beer preferred the use of an ophthalmoscope for refractive state measurements. Rather infrequently, he employed a skiascope which works similarly to the more modern retinoscope, but utilised reversal of direction of movement of the pupillary shadow on the fundus rather than that of the reflexion of a slit of light. Beer focussed on prominent structures at the level of the inner limiting membrane, parts of the falciform process or small blood vessels at the optic papilla. In the eyes of some fish, notably *Blennius* and *Scorpaena,* he was also able to measure the "true" refractive state by focussing on the cone outer segments. The difference in diopters between these two measurements corresponded to the difference calculated from the histogically verified thickness of the retina. Visible niveau differences at the inner limiting membrane near the optic papilla were correlated with measured changes in refractive state. With the few exceptions mentioned above, Beer's measurements were made by focusing on the vitreal-retinal boundary and later correction for retinal thickness measured from sections. He was concerned with refractive state determination in a "resting" state, in the non-accommodated eye, and his animals were injected with atropin or curare in order to relax the accommodation muscle. Considering the structures on which Beer focused,his measurements must have been made from an angle of 50 to 70 degrees from the forward direction. He obtained measurements indicating hyperopic conditions which, when corrected for retinal thickness, resulted in myopia of several diopters. Beer investigated accommodation by electrically stimulating the excised eye and observing the resulting lens movement in direction and magnitude. He correctly stated and illustrated that in many teleosts the lens moved in a temporal direction as well as away from the cornea. Another landmark study was the work by Tamura (1957) with marine teleosts. He utilised refractive state measurements made from many different directions in order to specify a true visual axis. He demonstrated a cocidence between the direction of accommodative lens movement and the line through retinal areas of greatest cell densities. Tamura dit not correct the measurements for the thickness of the retina, assuming, and also verifying experimentally with an excised piece of retina, that the reflexion of light came from the receptor cell layer. His refractive state measurements indicate appreciable hyperopia. Similar observations were made and a general hyperopic state of the fish eye was advocated by Baylor and Shaw (1962), affirming earlier notions of French workers (Rochon-Duvigneaud, 1943; Verrier, 1948). Like Tamura (1957), they insist that the reflecting surface is near the receptor cell layer. That this may not necessarily be the case and that the fish eye might not be hyperopic at all, was suggested by the observation that apparent hyperopia was progressively higher with decreasing eye diameter, as shown in a report by Glickstein and Millodot (1970). A recent study suggests that the retinoscopically determined hypermetropia may be related to lenticular chromatic aberration (Sivak, 1974).

REFRACTIVE STATE DETERMINATION BY ELECTROPHYSIOLOGICAL METHOD

A different technique for the objective measurement of refractive state utilises electrical responses of units in the optic tectum of fish (Meyer and Schwassman, 1970; Schwassmann and Meyer, 1971). Although the method permits monitoring the activity of only one, or a few, retinal ganglion cells at a time, it has the great advantage of not interfering with the optical properties of the eye and of using natural photic stimulation. It is necessary to immobilize the fish during the experiment. This is advantageous since it results in obtaining successive measurements in different visual directions which are comparable because of the absence of eye movements and accommodative changes.

The rationale and technique in this experimental approach was the following: The electrode in the subsurface of the mid-brain tectum records electrical activity that represents the responses of ganglion cells in the retina of the contralateral eye to photic stimulation of a very restricted area, the ganglion cell's receptive field. After isolating a single unit by electrode adjustment, its receptive field is marked on an illuminated screen at a long distance (2-3m) from the eye. Any stimulus of sufficient dimension and contrast that passes through the imaginary line between the eye and the receptive field will cause the unit to change its discharge pattern. Assuming the simple and common case of a unit with very little, or no spontaneous activity, moving a thin black wire across the receptive field results in a burst of action potentials. It appears as if the unit is able to "see" the wire. Every traverse through the unit's line of vision results in a response when it is done at close distance. Moving this stimulus farther away from the eye, one comes to a point beyond which there is no response. The distance at which responses fail can be designated as a "threshold", and the visual angle taken up by the wire's diameter at this distance is the "threshold visual angle". Employing a set of wire stimuli of different diameters, one can obtain a threshold visual angle for each wire. If one then plots these threshold angles in a graph against the distance from the eye at which this minimum angle was calculated, one obtains a curve the shape of which is an indication of the refractive state of the eye in this particular direction. Threshold angles, approximately constant, along the distance axis, with slightly elevated values very close to the eye, represent emmetropic conditions. Very low thresholds near the eye, but increasing with distance show myopia of differing degrees, while high thresholds throughout, but highest near the eye, should mean hyperopic conditions. Calibration threshold curves have been achieved by placing underwater lenses in front of the eye as close as possible to the cornea in order to reduce the "spectacle magnification" effect. The results of several studies, some yet to be published, utilising this method of tectal unitary recording, will be summarised briefly.

Accommodation in the eye of *Paralabrax* is accomplished by displacement of the lens along the antero-posterior axis. This "visual axis" passes through the centre of the temporal fovea. The retractor lentis muscle in these serranid basses is bipartite (Fig. 1). Electrical stimulation of the muscle by needle electrode and simultaneous photographing of lens margins showed the temporal displacement (Schwassmann and Meyer, 1971). Under moderate doses of a neuromuscular blocking drug, electrical activity was recorded by implanted electrodes directly from the retractor muscle. As shown in Fig. 2, spontaneous firing at a relatively steady frequency

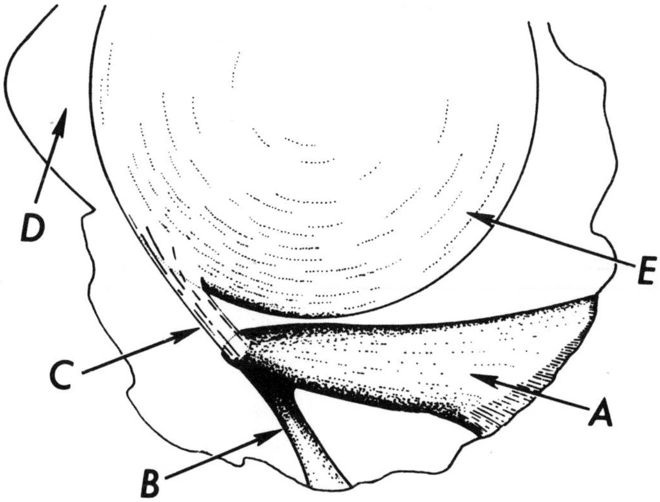

FIG. 1. Side view of the left eye of *Paralabrax* with the pupil opened ventrally to show the larger posterior (A) and the smaller anterior (B) head of the retractor lentis muscle inserting with a transparent ligament (C) anterior on the lens capsule. E: lens, D: anterior aphakic space formed by the triangular shape of the pupil.

indicates some intermediate tonus of the retractor muscle. An accommodative response occurs whenever a stimulus is moved through the anterior visual field, coinciding with the area of foveal vision. The response is a reduction in firing frequency of the units recorded, indicative of a lessening of retractor muscle tonus and resulting in rostrad lens movement which would bring near objects into focus on the temporal retina (Schwassmann and Meyer, unpublished). Threshold visual angle curves were determined for more than 200 tectal units in 30 of the foveate basses. Accommodative lens movements were prevented by doubling the concentration of the neuro-muscular blocking agent; the electromyogram now showed continued retractor muscle tonus without noticeable alteration in frequency during photic stimulation in the anterior visual field. The results of these experiments are evidence for near-emmetropic conditions of the eye in all of the directions measured. Tricaine anesthetic added to the respiratory water abolished the retractor tonus and the lens could be seen to move rostrally. Strong myopia in the forward direction and corresponding hyperopia in the temporal visual field resulted. This tricaine effect was reversed and the emmetropic conditions were restored by washing out the anesthetic. No change in refractive state could be measured lateral to the eye in the central visual field indicating that there is almost no lens movement in this

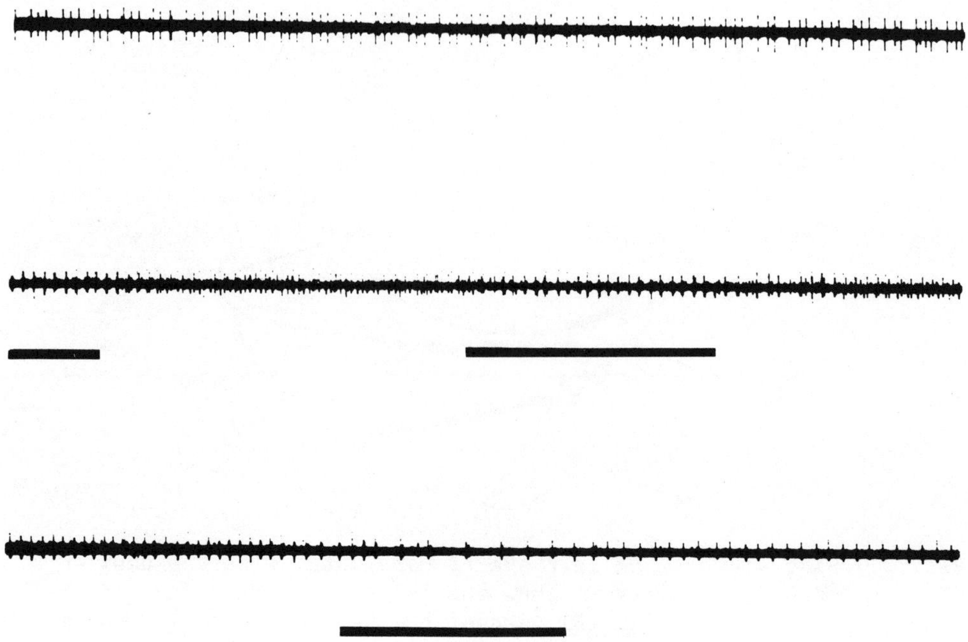

FIG. 2. Multi-unit records by implanted steel electrode from the posterior head of the retractor lentis muscle. Bars under the recordings indicate movement of a wire in front of the fish in the region of the foveal vision. The response to this stimulation is seen as a decreased rate of spike discharge.

direction (Meyer and Schwassmann, 1970; Schwassmann and Meyer, 1971).

During the past five years these studies have been extended to several other teleost species and similar refractive state conditions were found. Moreover, investigations of the refractive state of eyes in other vertebrates by similar electrophysiological methods during the last few years have resulted in identical evidence in favour of emmetropic conditions while measurements by retinoscopy indicated hyperopia. The notion advanced by Glickstein and Millodot (1970) that some reflexion of light in retinoscopy might occur at the vitreal-retinal border must be considered seriously.

It was of advantage to make simultaneous, or at least immediately successive determination of the refractive state in the same visual direction by our electrophysiological technique as well as by retinoscopy. These experiments were successfully done in 1971 with four specimens of the common opal eye (*Girella nigricans*). After establishing the distance thresholds for the series of line stimuli of one or more units, we measured the refractive error with the retinoscope in the same direction. The fish were sacrificed after measurements at three different angles, the eyes were fixed in Kolmer's, embedded in paraffin and cut serially in the horizontal plane. The thinnest section from the centre portion was used to mark the angles of previous measurements and the distances between inner and outer limiting membrane along these lines were determined under the microscope. The results (Schwassmann and Kennedy, in preparation) show emmetropic to slightly myopic conditions by electrophysiology but hypermetropic refractive errors from +2 to +7 diopters at the same angles. Testing for the possibility of reflexion by the inner limiting membrane was done by converting the two measures of focal length into diopters, the difference then indicating the hypermetropic errors to be expected. These were found to be about twice as great as those actually measured. The question of where the reflexion of the streak in retinoscopy takes place has not yet been answered, but there is another indication that it occurs at a plane vitread to the outer segments. While moving the streak along the horizontal meridian and measuring the refractive errors, the thin pigmented falciform process appeared at about 60 degrees. When moving the light across this point on the retina towards the anterior, the refractive error changed from +3 to about +6. Inspection of the sectioned eye showed clearly a corresponding niveau difference of the vitreous-retina boundary; the optic nerve fibre layer was thinner temporal to the falciform process.

LINE ACUITY AND GRATING RESOLUTION

Our electrophysiological technique was convenient for investigating differences in response properties of ganglion cells in different regions of the fish retina. The thin wire stimuli gave surprisingly low thresholds; values of less than 20 seconds of arc were frequently found not only in serranid basses but also in opal eye, blue gill, and others. No appreciable differences in line acuity thresholds were present when comparing foveal units with those elsewhere. Moreover, very recent work shows that line acuity at the ganglion cell level in goldfish is not greatly inferior to that in other fish with demonstrably better eyes, and single line thresholds of around one half minute were not uncommon.

Significant interspecies differences in threshold visual angles were encountered when the ganglion cells were investigated for their ability to respond to moving grating patterns. Ronchi rulings were projected on a screen on which the receptive field of the unit under study was marked. After focusing the grating pattern of the screen, the slide was moved up-and-down, or sideways when the stripes were vertically oriented. The locations of the projector and screen could be changed for altering the visual angle of the grating. Two kinds of responses were normally encountered: one threshold response occurring to every reversal in direction of slide motion; another response, at a larger angle, consisting of one or more spikes at the passing of each stripe across the receptive field centre. A sample of these data, comparing blue gill with goldfish, is shown in Fig. 3.

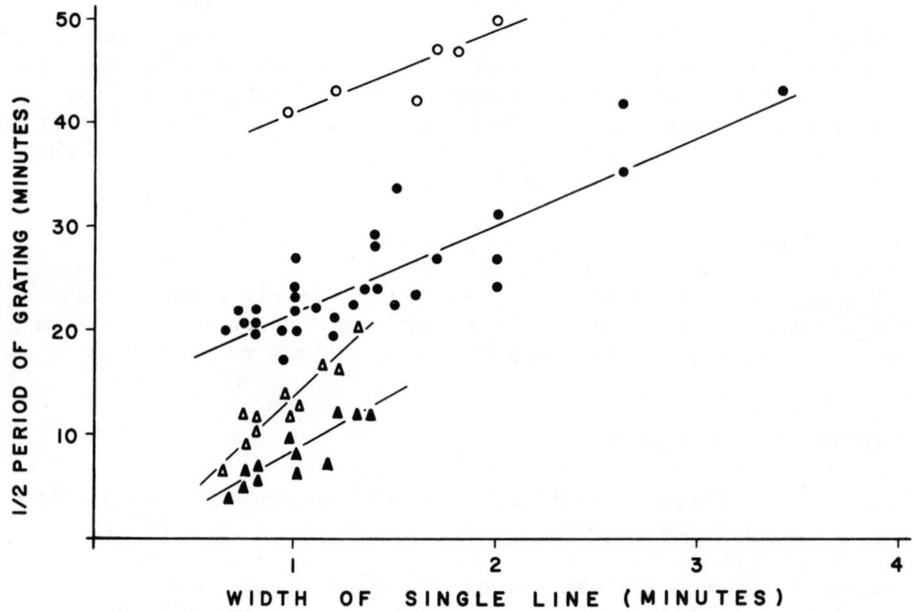

FIG. 3. Electrophysiologically determined threshold visual angles from a group of seven goldfish (circles) compared with the same data obtained with one blue gill (triangles). Size of the goldfish was 75 to 125 mm standard length, 6 to 9 mm eye diameter. The blue gill was 170 mm standard length, its eye measured 15 mm in diameter. Filled symbols represent minimum angles determined by the response to reversal of movement of the grating, open symbols show the values obtained when a response to the traverse of each bar of the grating pattern was used as criterion.

Each unit was tested for line and for grating response and grating threshold is plotted against single line threshold at the same distance from the eye. The clustering of data points of the four groups along the statistically fitted lines indicates a high degree of dependence of the two acuity measures; the correlation coefficients of the four samples range from 0.85 to 0.9. A useful parameter for comparing acuity measures in the two fish species is the ratio of line to grating acuity. When averaging the persent data, this ratio is 1:8 for the blue gill and 1:20 for the goldfish. Of paramount importance for arriving at a realistic value are the lowest values obtained since a certain proportion of the ganglion cells responding would be sufficient for detecting the pattern. The lowest grating resolution found in this series was four minutes for the blue gill and 17 minutes for the goldfish. This last value was also found as the minimum angle for grating resolution by behavioural method (Wilkinson, 1972).

REFERENCES

Baylor, E.R. (1967). Vision of Bermuda reef fishes. Nature (London) 214: 306-307.

Baylor, E.R. and Shaw, E. (1962). Refractive error and vision in fishes. Science 136: 157-158.

Beer, T. (1894). Die Accommodation des Fischauges. Pflügers Archiv. Gesamte Physiol. 58: 523-649.

Glickstein, M. and Millodot, M. (1970). Retinoscopy and eye size. Science 168: 605-606.

Meyer, D.L. and Schwassmann H.O. (1970). Electrophysiological method for determination of refractive state in fish eyes. Vision Res. 10: 1301-1303.

Munk, O. (1973). Early notions of dynamic accommodatory devices in teleosts. Vidensk. Medd. Dan Naturhist. Foren. KBH 136: 7-28.

Rochon-Duvigneaud, A. (1943). Les yeux et la vision des vertébrés. Masson, Paris.

Schwassmann, H.O. and Meyer, D.L. (1971). Refractive state and accommodation in the eye of three species of *Paralabrax* (Serranidae, Pisces). Vidensk. Medd. Dan Naturhist Foren KBH 134: 103-108.

Sivak, J.G. (1974). The refractive error of the fish eye. Vision Res. 14: 209-213.

Tamura, T. (1957). A study of visual perception in fish, especially on resolving power and accommodation. Bull. Jap. Soc. Sci. Fish. 22: 536-557.

Verrier, M.L. (1948). La vision des vertébrés et les théories de la vision. Ann. Biol. 24: 209-239.

Walls, G.L. (1942). The vertebrate eye and its adaptive radiation. Cranbrook Inst. Science, Bloomfield Hills, Mich., Reprint Hafner, New York 1967, 785 pp.

Wilkinson, F. (1972). A behavioural measure of grating acuity in the goldfish. Thesis, Dalhousie University, Halifax, N.S., Canada.

ACCOMMODATIVE MECHANISMS IN AQUATIC VERTEBRATES

Jacob G. SIVAK

School of Optometry, University of Waterloo

Waterloo, Ontario

O excellent device! Make a sop of him.
King Richard The Third, Act I, Sc. 4

INTRODUCTION

Accommodation in fishes involves two problems concerning interdependent and, yet at the same time, separate optical entities. These entities are: a) the actual mechanism of accommodation (i.e. the dynamic response which gives the eye the capacity to vary its focal power) and b) the refractive states of the eye (i.e. the degree to which the focal plane coincides or fails to coincide with the retina) which exist concurrently with a). Investigations of more than a century have gone a long way toward resolving the first problem. That is not to say that there do not exist any more difficulties and puzzles that are yet to be worked out. The second problem has resisted solution more tenaciously.

TELEOST ACCOMMODATION

The existence of an intraocular lens muscle in fishes and the possibility that it could move the crystalline lens was noted by von Haller (1764; cited by Munk, 1973), Wallace (1834), and Dalrymple (1838). A comprehensive study of accommodation by Beer (1894) led him to conclude that teleosts are myopic in the unaccommodated state and that they accommodate by moving the lens toward the retina by means of a retractor lentis muscle. This interpretation is appealing. It relates refractive state to accommodation in a tidy way. In the instance of myopia, a point somewhere between the eye and optical infinity is in conjugate focus with the retina. This eye is myopic for targets more distant than that point. Movement of the lens towards the retina brings those distant targets into focus on the retina, thereby reducing the myopia. The natural habitat would tend to favour myopia; for, with the usually restricted range of light, characteristic of water, only near objects would be visible and the most advantageous condition would be for the eye to be unaccommodated when such objects are in focus. The above reasoning cannot be applied to the refractive state of hyperopia since the conjugate focal point of the hyperopic eye lies behind the eye.

Beer relied principally on direct ophthalmoscopy to determine refractive status. He assumed that his measurements were based on a light path limited by the retinal internal limiting membrane and was obliged to compensate for an inherent bias toward hyperopia by an amount equivalent to the thickness of the retina. This would certainly be the case where the human (or mammalian) eye is concerned since the vessels coursing between the vitreous humor and the retina are the most obvious structures to be seen. The fish eye is a different matter. In most fishes the falciform process is the only vascular structure observed when viewing the fundus (Walls, 1942; Rochon-Duvigneaud, 1943). For the most part one sees a relatively empty field or, if the eye is large enough, the retinal mosaic produced by the geometric arrangement of retinal receptors. In either case it is difficult to specify where the focal plane lies.

While Beer's work gained widespread acceptance, some opposition did develop, due largely to the difficulty in establishing refractive state. As early as 1911, Sicherer measured the refractive state of a number of species of freshwater fishes, using the retinoscope exclusively, and reported a preponderance of hyperopia. Admittedly such findings are not very meaningful since accommodation was uncontrolled. However, any finding of significant hyperopia is an embarrassment to Beer's description of accommodation since movement of the lens toward the retina will only augment the hyperopia. This point was raised as recently as 1962 by Baylor and Shaw who noted that hyperopia is common when refractive state is measured retinoscopically (accommodation was uncontrolled in this study as well).

Verrier (1927, 1928, 1934, 1948), supported by Rochon-Duvigneaud (1943), provided the major opposition to Beer's work. Their criticism is based on the measurement of hyperopia and an inability to induce accommodative movements in the manner described by Beer. In 1927 Rochon-Duvigneaud and Verrier reported the existence of serous pouches between the posterior tunics of the teleost eye. Ciliary muscle fibres, having no direct or indirect articulation with the lens, have been observed in a number of teleosts (Rochon-Davigneaud 1943). These factors led Verrier (1948) to suggest that the true accommodative muscle is the ciliary muscle. She argued that hyperopia, present in the unaccommodated state, is reduced when this muscle contracts and produces an increase in the axial length of the eye. The presence of pouches provides room for the axial expansion.

DIRECTION OF ACCOMMODATIVE LENS MOVEMENT

Subsequent researchers have, without exception, agreed with the essentials of the accommodative mechanism which Beer demonstrated. Verrier's concept was resurrected by Baylor and Shaw (1962). However, their study dealt with the question of the validity of retinoscopic measurements of refractive error and not with accommodation per se. For the most part the more recent efforts have been concerned with establishing the direction of accommodative lens movements. Tamura (1957) and Tamura and Wisby (1963) demonstrated a correlation between the primary axis of accommodation, the location of high retinal receptor density and feeding behavior. A relationship between direction of accommodation and feeding behavior was also noted by Kimura and Tamura (1966) and Sivak (1973 and 1974c).

While these reports have all presented evidence as to the actual occurrence of lens movement there has been disagreement with regard to the primary axis of the lens motion. Tamura (1957), Tamura and Wisby (1963), Kimura and Tamura (1966), Meyer and Schwassmann (1970), and Schwassmann and Meyer (1971) emphasised lens movement in the pupil plane (rostral-caudal). In his study Beer noted that large lens movements close to the pupil plane were accompanied by smaller but significant lens movements along the pupil axis. Kimura and Tamura (1966) suggest that the lens motion along the pupil axis reported by Berr was a spurious finding. Nor did Meyer and Schwassmann (1970) and Schwassmann and Meyer (1971) record any lens motion along the pupil axis. However, Sivak (1973, 1974c) and Sivak and Howland (1973) reported results which are in closer agreement with those of Beer. These may be summarised as follows: In a few species the principle direction of lens movement is along the pupil axis (e.g. *Carassius auratus* and *Catostomus commersonni*). In the majority, the main accommodative movement occurs close to the pupil plane (e.g. *Perca flavescens* and *Amblo-plites rupestris rupestris*); but not directly along the pupil plane since a significant amount of lens motion is still found in the direction of the pupil axis.

The differences mentioned above may be due to species variation or to differences in methodology. The former possibility is strengthened by a report (Munk, 1971) concerning the existence of two lens muscles, one medially placed and the other laterally placed, in the eyes of a number of teleosts.

The methods which have been used to study accommodation are quite varied. Tamura (1957) used a modified optometer to measure refractive state before and after a fish was subjected to various treatments. These treatments included: severing of the optic nerve, laceration of the heart, injection of atropine and of curare. In every case the resulting change in refractive state indicated relaxation of the retractor lentis muscle. Tamura and Wisby (1963) recorded lens movements along the pupil plane by photographing lens position before and after specimens were sacrified. Kimura and Tamura (1966) recorded lens position photographically in enucleated eyes before and after electrical stimulation.

Schwassmann and Meyer (1971, see also Meyer and Schwassmann, 1970) developed an elegant method involving the measurement of visual acuity by way of electrophysiological recordings at the optic tectum. Lenses were then used to produce acuities indicating emmetropic conditions. Their method of inducing accommodative changes (tricaine methanesulfonate and Flaxedil) has been questioned (Sivak, 1974c) in view of the fact that Flaxedil may not affect the antonomic nervous system.

Meader (1936) demonstrated that the retractor lentis muscle is innervated by post-ganglionic parasympathetic nerve fibers. Sivak (1973 and 1974c) used pilocarpine and atropine (parasympathomimetic and parasympatholytic respectively) to induce accommodative changes. Simultaneous recording of heart rate changes support the contention regarding parasympathetic involvement in accommodation (Sivak, 1973). It is felt that these studies more closely approximate the method used by Beer than those mentioned earlier, since Beer's work involved both the stimulation and relaxation of the retractor lentis. A video tape study of the naturally occurring accommodative response in the northern rock-bass (Sivak and Howland 1973) agrees with the results obtained when accommodation is induced with pilocarpine and atropine. In a very recent report Somiya and Tamura (1973) describe accommodative responses which agree with the general description mentioned above (Sivak, 1973 and 1974 c; Sivak and Howland, 1973).

RECENT CONSIDERATIONS OF REFRACTIVE STATE

The question of refractive state has not been of paramount importance as far as the more recent studies of accommodation are concerned. Firstly, methods involving the photographic measurement of accommodation have avoided this issue. Secondly, the wide variation of refractive state with direction of regard, as first shown definitively by Tamura (1957), has added further confusion to this area.

Tamura, as well as Baylor and Shaw (1962) and Verrier (1928), considered that their measurements of refractive state were coincident with the assumption that the receptor layer of the retina is the focal plane of the eye. Meyer and Schwassmann (1970) pointed out that retinoscopic measurements of refractive state may be biased in the direction of hyperopia due to reflection from a retinal point between the lens and the receptor layer or pigmented epithelium. Glickstein and Millodot (1970) presented indirect evidence that this was true for the mammalian eye. It should be pointed out, however, that the authors cited by Meyer and Schwassmann used a variety of optical techniques, not just retinoscopy. Moreover, recent measures of mammalian refractive state (e.g. Massof and Chang, 1972; the rat eye) do not agree with the data cited by Glickstein and Millodot. Finally, a comparison of retinoscopic measurement with an anatomical determination of refractive state in goldfish (Sivak, 1974a) indicates that retinoscopic measurements coincide with calculations based on the distance between the lens and the receptor layer of the retina.

It has been suggested that the restricted spectral transmission of large bodies of water in combination with the existence of ocular chromatic aberration could produce a bias toward hyperopia (Sivak, 1974a). This possibility should be considered in any method used to measure refractive state and the spectral quality of the normal environment should be duplicated during measurement.

ACCOMMODATION IN NON-TELEOSTS

In comparison with the teleosts, accommodation in other groups of lower aquatic vertebrates has been sparsely investigated. In general, those reports which have appeared are concerned with establishing the presence or absence of an ability to move the lens, while the question of refractive state has been considered to be an extension of the teleost case.

CYCLOSTOMES

According to a number of authors, (Walls, 1942; Duke-Elder, 1958; Tansley, 1965), Franz (1932, 1934) described the accommodative mechanism of *Petromyzon fluviatilis* as follows: An extraocular muscle, the cornealis, originates at a point temporal to the orbital capsule and is inserted on the inner surface of the primary spectacle. Contraction of this muscle flattens the spectacle and cornea. The lens, which is in contact with the cornea, is pushed closer to the retina and the eye becomes less myopic or more hyperopic.

Unfortunately this description is an oversimplification, for, in addition to the above effect Franz noted two others. Strong contraction of all the extraocular muscles pulled the eye toward the cranium and produced an elongation of the optic axis. The refractive result is opposite to that just mentioned. In addition, Franz observed large refractive changes produced by lateral bending of the head. Refractive state was measured with a retinoscope and, in an apparent extrapolation from Beer's work, the results were adjusted to account for retinal thickness. The most common refractive state was that of moderate hyperopia, although by inducing the effects noted above (electrically) Franz reported that he could alter refractive state by as much as 100 diopters (plus to minus or vice versa). Attempts to produce similar changes in *Petromyzon marinus* have failed (Sivak and Woo, unpublished data).

ELASMOBRANCHS.

It has been reported (Franz, 1931) that elasmobranchs accommodate by moving the lens toward the cornea, and that this is achieved by contraction of an intraocular muscle, the protractor lentis or pseudo-campanule. Beer (1894), Verrier (1930), Somiya and Tamura (1973) and Sivak (1974b), however, were unable to induce refractive changes in this group. Furthermore, histological examination of the pseudo-campanule (Rochon-Duvigneaud, 1943; Wang, 1968) indicates the lack of contractile fibres. It should be noted however that Franz's study involved mainly rays, while the other reports are generally concerned with sharks.

CHONDROSTEANS AND HOLOSTEANS

A structure resembling the pseudo-campanule exists in the sturgeon. However, an attempt to induce its contraction was unsuccessful (Hess, 1912). A retractor lentis-like muscle has been demonstrated in holosteans (Rochon-Duvigneaud, 1943). This muscle is probably responsible for accommodative changes, characteristic of the teleost eye, that were observed in the case of *Amia calva* and *Lepisosteus osseus oxyurus* (Sivak, unpublished data).

DIRECTIONS OF FURTHER RESEARCH

The study of accommodation in fishes has been dominated by

investigations in which refractive changes or lens movements are measured in association with artificial treatments. Less attention has been paid to accommodation as it occurs naturally or to the behavioral importance of this response to the animal in question. With regard to the latter point it should be noted that Pumphrey (1961), and Charman and Tucker (1973) suggested that the depth of focus available to the fish eye may obviate the need for active accommodation. In line with these considerations, the studies which represent the probable directions for future research are those which have a) related accommodation to behavioral information (e.g. Tamura, 1957), b) examined the naturally occuring response (e.g. Sivak and Howland, 1973) and c) evaluated accommodation in terms of visual acuity (e.g. Schwassmann and Meyer, 1971).

With regard to the problem of determining refractive state, a quantitative study of the effect of restricted spectral ranges of aquatic habitats is needed. In addition, it would seem that further attention should be given to the identification of the reflecting surface upon which measurements of refractive state are based. One possible aspect which appears not to have been considered in the past is the effect of retinomotor responses on refractive state. It will be of interest to do so in view of Glickstein and Millodot's (1970) observations, and the fact that in teleosts, particularly the movement of the visual cells is considerable (Ali, this volume).

REFERENCES

Baylor, E.R. and Shaw, E. (1962). Refractive error and vision in fishes. Science 136: 157-158.

Beer, T. (1894). Die Accommodation des Fishauges. Pflügers Arch. Physiol. 58: 523-650.

Charmen, W.N. and Tucker, J. (1973). The optical system of the goldfish eye. Vision Res. 13: 1-8.

Dalrymple, J. (1838). Some acount of a peculiar structure in the eyes of fishes. Mag. Nat. Hist. N. S. 2: 136-141.

Duke-Elder, S. (1958). System of Ophthalmology. Vol. I. The eye in Evolution. Henry Kimpton, London.

Franz, V. (1931). Die Akkommodation des Selachierauges und seine Abblendrungsapparate, nebst Befunden an der Retina. Zool. Jahrb., Abt. Allegm. Zool., Physiol. Tiere 49: 323-462.

Franz, V. (1932). Auge und Akkomodation von Petromyzon (Lampreta) fluvialilis L. Zool. Jahrb., Abt. Allegm. Zool. Physiol. Tiere 52: 118-178.

Franz, V. (1934). Vergleichende Anatomie des Wirbeltierauges. In Handbuch der Vergleichende Anatomie der Wirbeltiere, edited by Bolk, Göppert, Kallius, Lubosch, Bd. 2, III Höhere Sinnesorgane, pp. 989-1292, Urban und Schwartzenberg, Berlin.

Glickstein, M. and Millodot, M. (1970). Retinoscopy and eye size. Science 168: 605-606.

Von Haller, A. (1794). Mémoire sur les yeux de quelques poissons. Hist. Acad. r. Sci. Paris 1762, Mém. Math. Phy. 76-95. (cited by Munk, O. (1973). Early notions of dynamic accommodatory devices in Teleosts. Vidensk. Medd. Dan. Naturhist. Foren. 136: 7-28.

Von Hess, C. (1912). Vergleichende Physiologie des Gesichtssinnes. In Handbuch der Vergleichenden Physiologie, edited by Winterstein, Bd. 4: pp. 1-290, Gustav Fisher, Jena.

Kimura, K. and Tamura, T. (1966). On the direction of the lens movement in the visual accommodation of teleostean eyes. Bull. Jap. Soc. Fish. 32: 112-116.

Massof, R.W. and Chang, F.W. (1972). A revision of the rat schematic eye. Vision Res. 12: 793-796.

Meader, R.G. (1936). Accommodation and its reflex pathways in the teleosts. Yale, J. Biol. Med. 8: 511-522.

Meyer, D.L. and Schwassmann, H.O. (1970). Electrophysiological methods for determination of refractive state in fish eyes. Vision Res. 10: 1301-1303.

Munk, O. (1971). On the occurrence of two lens muscles within each eye of some teleosts. Vidensk. Medd. Dans. Naturhist, Foren. 134: 7-19.

Munk, O. (1973). Early notions of dynamic accommodatory devices in teleosts. Vidensk. Medd. Dan. Naturhist. Foren. 136: 7-28.

Pumphrey, R.J. (1961). Concerning Vision. In: The Cell and the organism, edited by Ramsay and Wigglesworth, pp. 193-208. Cambridge University Press, London.

Rochon-Duvigneaud, A. (1943). Les Yeux et la Vision des Vertébrés, Masson, Paris.

Rochon-Duvigneaud, A. and Verrier, M.L. (1927). Sur l'existence des poches séreuses dans l'oeil et l'orbite des Téléosteens. C. r. hebd. Séances Acad. Sci. 184: 539-542.

Schwassmann, H.O. and Meyer, D.L. (1971). Refractive state and accommodation in the eye of three species of *Paralabrax* (Serranidae, Pisces). Vidensk. Medd. Dan. Naturhist. Foren. 134: 103-108.

Sicherer, O. (1911). Untersuchungen über die Refraktion der Augen der Süsswasserfische. Arch. Vergl. Ophthal. 1: 481-488.

Sivak, J.G. (1973). Interrelation of feeding behavior and accommodative lens movements in some species of North American freshwater fishes. J. Fish. Red. Board Can. 30: 1141-1146.

Sivak, J. and Howland, H.C. (1973). Accommodation in the northern rock bass (Ambloplites rupestris rupestris) in response to natural stimuli. Vision Res. 13: 2059-2064.

Sivak, J.G. (1974a). The refractive error of the fish eye. Vision Res. 14: 209-213.

Sivak, J.G. (1974b). Accommodation of the lemon shark eye (*Negaprion brevirostris*). Vision Res. 14: 215-216.

Sivak, J.G. (1974c). Accommodative lens movements in fishes: lateral vs. rostral-caudal motion. Abstract in Program of the 54th Annual Meeting of the American Society of Ichthyology and Herpetology, National Museum of Natural Sciences, Ottawa. (in press).

Somiya, H. and Tamura, T. (1973). Studies on the visual accommodation in fishes. Jap. J. Ichthyol. 20: 193-206.

Tamura, T. (1957). A study of visual perception in fish especially on resolving power and accommodation. Bull. Jap. Soc. Sci. Fish. 22: 536-557.

Tamura, T. and Wisby, W.J. (1963). The visual sense of pelagic fishes especially the visual axis and accommodation. Bull. Mar. Sci. Gulf Caribb. 13: 433-448.

Tansley, K. (1965). Vision in Vertebrates. Chapman and Hall, London.

Verrier, M.L. (1927). Sur la réfraction statique de l'oeil chez les poissons. C. R. Hebd. Seances Acad. Sci. 185: 1070-1072.

Verrier, M.L. (1928). Recherches sur les yeux et la vision des poissons. Bull. Biol. Fr. Belg. suppl. XI: pp. 1-222.

Verrier, M.L. (1930). Contribution à l'étude de la vision chez les selaciens. Annls. Sci. Nat. Zool. 13: 5-54.

Verrier, M.L. (1934). La réfraction de l'oeil des poissons. Bull. Soc. Zool. Fr. 59: 333-338.

Verrier, M.L. (1948). La vision des vertébrés et les théories de la vision. Année Biol. Ser. 3.24: 209-238.

Wallace, W.C. (1834). Discovery of a muscle in the eye of fishes. Am. J. Sci. Arts. 26: 394.

Walls, G.L. (1942). The Vertebrate eye and its adaptive radiation, Cranbrook Institute of Science, Bloomfield Hills, Michigan.

Wang, C.S.T. (1948). The Eye of fishes with special reference to pigment migration. Ph.D. dissertation, Cornell University.

TAPETA LUCIDA AND THE ORGANISATION OF VISUAL CELLS IN TELEOSTS

Edward ZYZNAR

Département de Biologie, Université de Montréal

Montréal, Qué. Canada

He holds him with his glittering eye.
Coleridge: The Ancient Mariner 1

The tapetum lucidum is often typified as a reflecting layer which returns light transmitted through the retina to the visual cells. In primitive fishes tapeta are smooth and mirror-like lying in the choroid. The tapeta of most teleosts are found in cells of the pigment epithelium modified with long processes which extend vitread and contain reflecting material. Considerable variation occurs in the structure of the processes and the organisation of the visual cells. This variation and its implications to vision will be herein discussed.

It is convenient to categorise retinal tapeta in accord with the organisation of the visual cells of the dark-adapted eye. The position of the rods relative to the tapetal processes determines, in part, how efficiently the reflected light is utilised and the resolution of images. The rods of a few teleosts form an unbroken layer vitread to the tapetum; in others, they are grouped among the epithelial processes. In several species the organisation is intermediate between the two types.

NON-GROUPED VISUAL CELLS

In some Malacosteidae, Sternoptychidae, Argentinidae and Ophidiidae (Brauer, 1908; Ali and Hanyu, 1963; Nicol, Thurston and Zyznar, in press) the epithelial processes are short and blunt as is characteristic of many deep-sea fishes. The surface of the tapetum is knobbed and almost continuous. The rods are situated in front of the tapetum and little or no interdigitation with the epithelial processes takes place. There is either a single tier of elongated rods or several regular tiers of rod outer segments (Fig. 1). Such an arrangement permits the tapetum to reflect light toward the rods a single time, thereby potentially doubling light absorption by the retina.

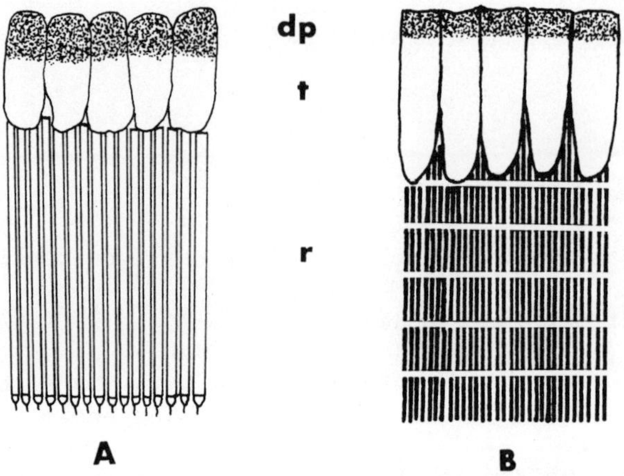

FIG. 1. Schematic representations of tapeta lucida and non-grouped receptors with (A) a single tier of long rods and (B) several regularly stacked tiers of short rods. dp. dark pigment; r, rods; t, tapetum lucidum.

The reflecting material has been studied in detail only in shallow water species of the Ophidiidae (Nicol et al., in press). It is comprised of numerous, minute, membrane-bound, lipid spheres containing mostly a long chain, polyunsaturated triglyceride. The spheres provide refractive heterogeneity within the tapetal processes and scatter light in all directions. The component scattered forward is the greatest; nevertheless, due to the high density of spheres and low absorption of light, multiple scattering results which eventually returns most of the light backward toward the rods. These tapeta are thence excellent diffuse reflectors.

Diffuse reflexion from these tapeta spreads light away from the centre of illumination increasing sensitivity, perhaps at some expense to acuity. The intensity, flux per unit solid angle, from a point on a good diffusing surface diminishes as the cosine of the angle, θ, to the normal. Although the brightness or luminosity of the surface is about the same in all directions (because the surface area intercepted by the pupils of the viewer over some solid angle, if projected orthogonally, increases as the cos θ), its luminance decreases with obliquity. A large part of the scattered light is, however, absorbed locally by the retina. The rods are able to channel some of the non-axial rays. Most of the energy of oblique rays which are not channelled is absorbed by the rods nearest to the point of reflexion on the tapetum as a result of self screening e.g. if a retina has a density 0.5, about 2X more light is absorbed by the first half of the retina than by the remaining part. The oblique rays striking the rods side-on are absorbed more efficiently than are axial rays as a consequence of linear dichroism (Liebman, 1962; Harosi, 1971) increasing self screening for the same optical distance through the rods. Finally, histological evidence indicates that retinas with tapeta are highly convergent. Where the lateral summation is large, scattered light is spread over fewer summated fields.

GROUPED VISUAL CELLS

The receptors of some clupeidae (*Dorosoma cepedianum*- personal observation), Evermannellidae, Mormyridae, Elopidae, Hiodontidae, Percidae, Engraulidae and Scopelarchidae are distributed entirely among the tapetal processes (Brauer, 1908; McEwan, 1938; Moore, 1944; O'Connell,1963; Locket, 1971; Nicol, Arnott and Best, 1973; Zyznar and Ali, in press), and in several of these families the visual cells are segregated into discrete groups optically isolated from each other by the reflecting material. Light can be reflected back and forth between the tapetal cells potentially more than doubling the amount absorbed by the visual pigment at any ambient intensity. Because the entire surface of the retina is not blanketed with rods, light insensitive areas exist at the loci of the tapetal processes. In such retinas, elaborate adaptations are found which seemingly alleviate this problem and augment the light directed toward the rods.

Well defined rod groups are present in retinas of the Evermannellidae, Mormyridae, Hiodontidae, Scopelarchidae and Elopidae. The groups are polygonal or round in tangential section and arranged in a regular hexagonal pattern. The myoids of the rods are broad, often tapering toward the outer segments (Fig. 2A). The tapetal processes are wedge shaped and fine near the external limiting membrane. Light may be collected over a wide angle with minimal interference from the tapetum, especially if the rod myoids channel light.

The cones of some Percidae (*Stizostedion*) contain lipid inclusions and function as light funnels which increase the amount of energy reaching the rods (Zyznar and Ali, in press). The cone ellipsoids are hypertrophied. There is a gradient of increasing lipid content from the periphery to the centre of each cell giving

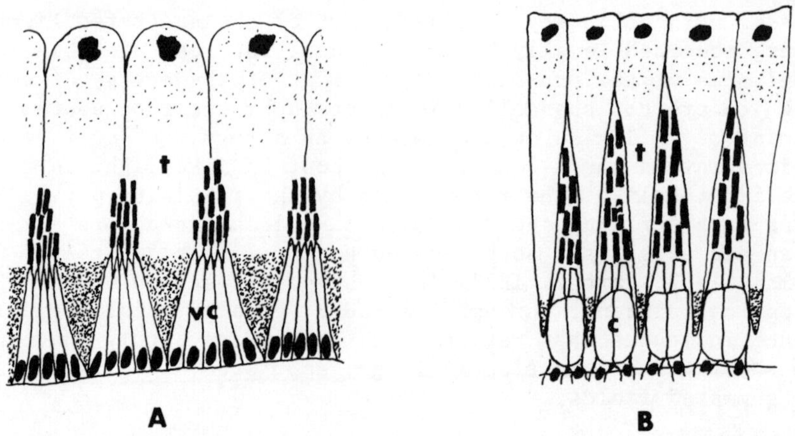

FIG. 2. Schematic representations of tapeta lucida and grouped receptors. The visual cells (A) of the Elopidae, Hiodontidae and Mormyridae have broad myoids which taper sclerad. Light can be collected by the outer segments with minimal interference by the tapetal processes. The cones of *Stizostedion* (B) are interposed between the rods and the incoming light. Light arriving at the rods must pass through the cones which function as lenses. Dark pigment in the tips of the epithelial processes delimits the angle of acceptance of each rod group. c, cones; r, rods; t, tapetum lucidum; vc, visual cells. Dark pigment is denoted by stippling and nuclei by black spheres.

rise to a gradient of refractive index becoming greater centrally. The vitread portions of the cones are surrounded by tapetal material. In the dark-adapted state, the cones are interposed between the rods and the incoming light, so the light reaching the rods must first pass through the cones (Fig. 2B). The cones increase the angle of acceptance of the rod groups.

The reflecting particles in the tapetal processes are distributed randomly in *Stizostedion* and *Dorosoma*, but are oriented and arranged in orderly rows in at least two species of the Scopelarchidae and Engraulidae (Locket, 1971; Nicol et al., 1973; Zyznar and Ali, in press). The processes of the latter two are filled with geometrically anisotropic granules or platelets oriented with their broad faces toward the rods. The value of this arrangement is presently uncertain except that the regular orientation probably introduces a specular element into the reflected light. The lamellar structure of the visual cell outer segments of *Anchoa mitchilli* interestingly parallels the length of the tapetal processes (Nicol et al., 1973).

Locket (1971) discussed the significance of rod groups to visual acuity. The rods of each group in *Scopelarchus* are bound by tight junctions between inner segments indicating possible electrical couplings. Whether or not true gap junctions between rods exist, reflected light will be absorbed by all of the rods about equally hindering imaging within a group. Resolution would then depend on the size of the rod group.

A final intriguing point concerns the disposition of dark pigment in the tips of the epithelial processes of the dark-adapted eyes of most of these families. Unlike the more usual occurrence in which the dark pigment retracts basally, the tips of the tapetal processes remain absorbing rather than reflecting at low light intensities. In this position the pigment limits the angle of acceptance of each group of rods by absorbing very oblique rays. The exposed fundus of *Stizostedion* appears to darken when observed obliquely.

INTERMEDIATE TYPES

The rods of the Sciaenidae (Arnott, Nicol and Querfeld, 1972), Pomadasyidae and Ephippidae (*Orthopristes chrysoptera* and *Chaetodipterus faber*, personal observation) are located among the tapetal processes and vitread to them in the dark-adapted eye. The rods among the processes are not segregated into discrete units nor are they completely enveloped by reflecting material. Light absorption is increased somewhat by multiple reflexions without forming insensitive areas on the retina.

REFERENCES

Ali, M.A. and Hanyu, I. (1963). A comparative study of retinal structure in some fishes from moderately deep waters of the western North Atlantic. Can. J. Zool. 41: 225-241.

Arnott, H.J., Nicol., J.A.C. and Querfeld, C.W. (1972). Tapeta lucida in the eyes of the seatrout (Sciaenidae). Proc. R. Soc. Lond. Ser. B. Biol. Sci. 180: 247-271.

Brauer, A. (1908). Die Tiefseefische. B. Augen, Wissenchaftliche Ergebnisse der deutschen Tiefsee. Expedition auf dem Dampfel "Valdivia" 1898-99, edited by C. Chun, Bd. 15: II. Anatomische Teil.

Harosi, F.I. (1971). Frog rhodopsin in situ: orientational and spectral changes in the chromatophores of isolated retinal rod cells. Ph.D. Dissertation. Johns Hopkins University.

Liebman, P.A. (1962). In situ microspectrophotometric studies on the pigments of single retinal rods. Biophys. J. 2: 161-178.

Locket, N.A. (1971). Retinal anatomy in some Scopelarchid deep-sea fishes. Proc. R. Soc. Lond. Ser. B. Biol. Sci. 178: 168-169.

McEwan, M.R. (1938). A comparison of the retina of the mormyrids with that of various other teleosts. Acta. Zool. 19: 427-465.

Moore, G.A. (1944). The retinae of two North American teleosts, with special reference to their tapeta lucida. J. Comp. Neurol. 80: 369-379.

Nicol, J.A.C., Arnott, H.J. and Best, A.C.G. (1973). Tapeta lucida in bony fishes (Actinopterygii): a survey. Can. J. Zool. 51: 69-81.

Nicol, J.A.C., Thurston, E.L. and Zyznar, E.S. The tapetum lucidum of cuskeels. Can. J. Zool. (In press)

O'Connell,C.P. (1963). The structure of the eye of *Sardinops caerulea*, *Engraulis mordax*, and four other pelagic marine teleosts. J. Morphol. 113: 287-319.

Zyznar, E.S. and Ali, M.A. An interpretative study of the organization of the visual cells and tapetum lucidum of *Stizostedion*, Can. J. Zool. (In press)

THEORETICAL CONSIDERATIONS ABOUT TAPETA LUCIDA

Edward ZYZNAR

Département de Biologie, Université de Montréal

Montréal, Qué. Canada

I have had a most rare vision.
A Midsummer's Night's dream, Act IV, Sc. 1

Many fishes employ a catoptric device, the tapetum lucidum, to increase light absorption by the retina. The ocular structure and habits of these fishes indicate that they are adapted to a mode of existence in dim light and that tapeta might aid vision under such circumstances by enhancing sensitivity. Unfortunately, little experimental evidence is available about the role of fish tapeta in vision, so even an indirect approach may contribute useful information.

A MODEL

It will be attempted in this discussion to bring out some attributes of tapeta by means of a simplified model of the tapetum as a reflecting surface lying immediately behind an absorbing layer, the retina. Visual stimulation is considered here to depend solely on the amount of light absorbed. Factors such as absorption and scattering by pre-retinal ocular elements are neglected. It is further assumed that light intensities are sufficiently low to cause only negligible bleaching. In this system, light is reflected specularly.

There is considerable variation in the density of visual pigments among fishes and in the light intensities they encounter. Pigment density determines the amount of light which is transmitted to the tapetum and the fraction of reflected light absorbed. Also, the tapeta of many fishes are occluded at higher intensities whereas the visual cells of others are permanently exposed to tapetal reflexions causing uncertainty about the function of this reflecting tissue at different intensities.

A totally reflecting tapetum effectively doubles the optical density (O.D.) of the retina, however, the amount of reflected light which is absorbed may best be determined by the absorptance, A_1, of the visual pigments where

$$A_1 = 1 - \frac{F}{F_o}.$$

F_o is the total flux incident on the rods and F is the flux transmitted. That part, A_T, of F_o absorbed by a retina with tapetum is the sum of the absorptance and the reflected light absorbed, A_R.

$$A_T = A_1 + A_R$$

and

$$A_R = RA_1\ (1-A_1)$$

$$= R\ (A_1 - A_1^2).$$

R is the reflectance coefficient of the tapetum. Differentiating, we find that

$$\frac{dA_R}{dA_1} = R\ (1-2A_1)$$

and a maximum occurs when $A_1 = 0.5$. A retina with an absorptance of 50% (O.D. = 0.3) absorbs the largest amount of reflected light; the total absorbed increases 1.5X. A_R is not a steep function of A_1 and changes, at best, by 4% between absorptances of 0.3 and 0.7 (Fig. 1). A_T almost doubles in the same interval (Fig. 2). The tapetum is most efficient at moderate retinal absorptances whereas A_T increases continuously with higher absorptance.

The above relationships are only approximations for diffusely reflecting tapeta because the obliquely scattered light follows a longer path through the retina and meets the visual cells laterally. Although the path length of obliquely reflected light through the rods is longer, the luminance of a diffusing tapetum falls off rapidly with increased obliquity minimizing the disparity between n specular and diffuse reflectors. Secondly, the rods exhibit dichroic properties such that unpolarised light which is incident upon them side-on is more efficiently absorbed than in the axial direction (Liebman, 1962; Harosi, 1971). Absorption of the scattered light

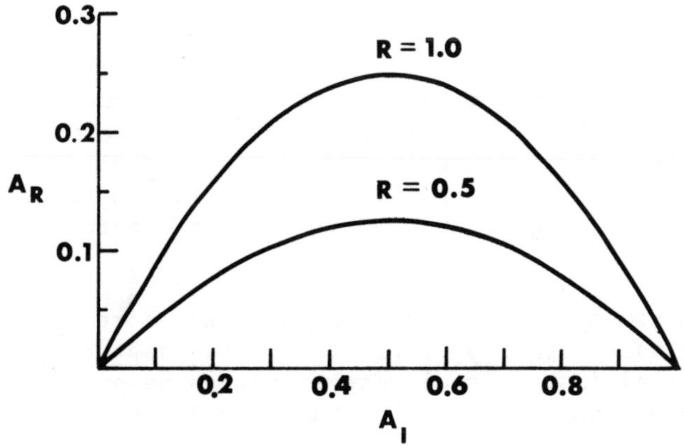

FIG. 1. The reflected light absorbed by the retina as a function of retinal absorptance. R is the reflectance coefficient of the tapetum.

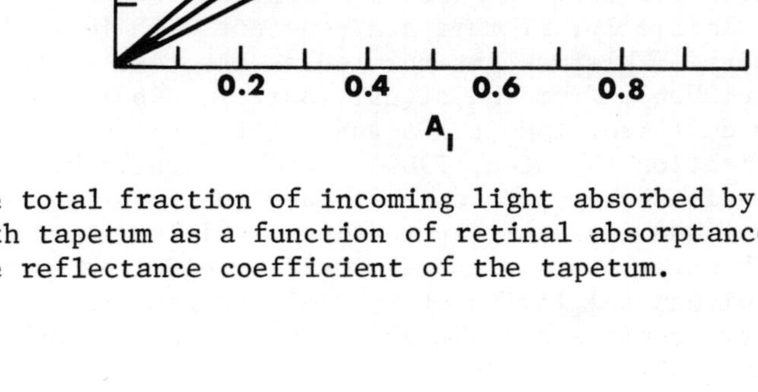

FIG. 2. The total fraction of incoming light absorbed by a retina with tapetum as a function of retinal absorptance. R is the reflectance coefficient of the tapetum.

is probably greater and the maximum value of A_R is shifted to absorptances slightly below 0.5.

Denton and Nicol (1964) found that retinas of fishes with tapeta do not absorb more light than those lacking tapeta, rather an equal amount of light is absorbed with less visual pigment. Several possible advantages of tapeta were enumerated by Denton and Nicol, but one is particularly striking. If the absolute threshold of vision depends on noise derived from thermal decomposition of visual pigment (Barlow, 1956), a contention which is by no means certain (Hubbard, 1958), tapeta associated with retinas containing less visual pigment would reduce noise and lower the threshold of vision. The expected gain is small, but greatest at moderate retinal absorptances.

Light incrementation by the tapetum must be considered together with the intrinsic capacity of the retina to adapt to greatly differing intensities. Visual adaptation can be accounted for by amplification mechanisms in the retina. This view is supported by the findings of Dowling (1960) who showed that a linear relationship exists between the amount of visual pigment bleached and the log threshold. The retina is responsive to a wide range of intensities which varies by several log units, but the tapetum raises light absorption at the retina by only a fraction of a log unit. The increase provided by the tapetum in the log number of photons absorbed is small; it is largest at low light levels and drops quickly at higher intensities approaching a limit of 1.0 asymptotically. If the signal without the tapetal increment is ample for vision, the addition of reflected light reduces somewhat the amount of amplification needed. When the signal is too weak to be clearly amplified, the tapetum is markedly of value increasing it to a level which can be further processed.

Tapeta probably do not aid in the discrimination of objects in the visual field at higher ambient intensities. Barlow (1965) measured increment thresholds (ΔI) as functions of the background illumination (I) under rigorously defined conditions. He was able to discern three distinct branches of response before rod saturation was reached. Initially, ΔI remained constant with increased background intensity. This was interpreted as the result of noise which was indistinguishable from the signal (Barlow, 1965) or which was qualitatively different than the signal, but affected the gain of visual amplification (Rushton, 1965). With slightly higher background, the retina behaved as does an ideal detector and seemed to be restricted by random quantum fluctuations. ΔI increased in proportion to I^{-2}, however, this response could be elicited only when the stimulus was within the limits of spatial and temporal summation. Finally, the eye conformed to Weber's law; ΔI and I rose at approximately the same rate. White tapeta increase light at each point on the retina by a constant factor irrespective of background intensity or variations within the visual field, thus the differential between an object and its background is unaltered by tapetal reflexions unless one or the other is below the absolute threshold of vision. When Weber's law is operative, visual discrimination is not improved and scattered light may degrade the image. In the rather restricted circumstances whereby the retina functions as an ideal detector, the tapetum can be beneficial because the increment threshold does not rise as rapidly as the background intensity. Lastly, tapeta can lower the absolute threshold by increasing the retinal light catch.

HIGHER SENSITIVITY AND VISION UNDERWATER

Higher sensitivity in an aqueous environment implies extended visual range when light intensities are low. Dimmer, more distant objects can be perceived and vision can take place at greater depths in the water column. Natural light is rapidly attenuated in water according to the inverse square law and Lambert's law. On the other hand, the extra sensitivity gained from tapeta is quite small. The distance, ΔD, by which the tapetum extends vision underwater relative to a similar retina without a tapetum can be estimated.

Firstly, it should be pointed out that threshold intensity and luminosity as functions of target distance also depend on the size of the target observed. Neither the threshold intensity nor luminosity of a large target of uniform luminance is influenced solely by target distance because the illumination of the image on the retina remains constant. The flux received by the pupil from the target decreases as the square of the distance, as does the area of the image projected on the retina (Walsh, 1965). The static relationship between threshold intensity and distance breaks down for small or very distant targets. By Ricco's law, the threshold intensity and subjective brightness of an object subtending a few minutes of arc is the product of the retinal illumination and the solid angle between the pupil and the light source (LeGrand, 1957). This effect occurs because the light stimulus is summated over small areas. Ricco's law was initially established for foveal and near foveal vision in humans, but applies equally well to the retinas of fishes. Complete spatial summation is found for targets up to 200' of arc in the dark-adapted goldfish (Hester, 1968). The effect of spatial summation becomes much less pronounced as larger patches of retina are involved and disappears when the solid angle subtended is about 15° (in humans).

With this in mind, let us examine a situation whereby a reflective object suspended in water is observed from above. The ratio of downwelling to upwelling light in natural water bodies is often very great (Tyler, 1960). At low light intensities, the background appears black and the image of the object is unobstructed by intervening scattered light. For a small object, the image of which can be projected onto the retina within a summated field, threshold depends on the product of intensity and area of retina illuminated, and varies as the inverse square of the target distance. Light losses resulting from absorption and scattering by a medium are described by Lambert's equation. If I_1 is the flux per area of retina with a tapetum illuminated by small object of uniform luminance, I, at a distance, D_1, then

$$I_1 = \frac{1}{D_1^{\,2}} \left[I \cdot e^{-\alpha D_1} \right]$$

and

$$\log \frac{I}{D_1^{\,2} I_1} = \frac{\alpha D_1}{2.3} .$$

α is the beam attenuation coefficient of the water.

The tapetum increases retinal illumination by some factor, t. Under optima. conditions t is about 1.5. The illumination of a similar retina without a tapetum is I_2 at a distance, D_2, nearer to the object. When I_2 is equal to tI_1 which just evokes a threshold response, the same amount of light is absorbed by both retinas, so

$$I_2 = tI_1 = \frac{I}{D_2^{\,2}} \left[I \cdot e^{-\alpha D_2} \right]$$

and

$$\log \frac{I}{D_2^{\,2} tI_1} = \frac{\alpha D_2}{2.3} .$$

By definition,

$$\Delta D = D_1 - D_2 \qquad D_1 > D_2,$$

$$= \frac{2.3}{\alpha} \left[\log \frac{I}{D_1^{\,2} I_1} - \log \frac{I}{D_2^{\,2} tI_1} \right]$$

$$= \frac{2.3}{\alpha} \left[\log t - \log \frac{D_1^{\,2}}{D_2^{\,2}} \right].$$

Values of D_1 and D_2 can be tabulated for the range of extinctions found in natural waters, but the order of magnitude of ΔD can be quickly determined. The ratio of D_1^2 to D_2^2 must be less than t, e.g. ΔD is less than 0.9 m or 1.8 m when t is 1.5 and the threshold distance, D_1, is 5.0 m or 10.0 m, respectively. In an aqueous environment attenuation takes place which depends on extinction by the water. ΔD varies as the reciprocal of the beam attenuation coefficient and is probably a fraction of a metre in turbid waters.

The luminosity of a large object is independent of the inverse square law because the image on the retina extends over several summated fields. This concept was incorporated in Duntley's (1963) equations describing the visibility of objects underwater and in Lythgoe's (1968) measurements. So, for large objects

$$\Delta D = \frac{2.3 \log t}{\alpha} .$$

Now ΔD has values between a fraction of one metre and several metres for the extinctions commonly found in natural waters. Hester (1968) suggested that in the limited visual range of an underwater environment, objects are more likely to subtend large angles even near the limits of visibility. The advantage gained from tapeta by extending the visual range in dim light is small, but not insignificant.

CONCLUSIONS

Although the arguments presented are often based on simplified conditions, they serve the useful purpose of delineating limitations and trends which otherwise are not obvious. By these arguments, one expects that (1) tapeta are most efficient when associated with retinas of moderate absorptances, but the total light catch is greatest at high absorptances, (2) at higher scotopic intensities, visual discrimination is not improved by tapeta because the differential between an object and its background is unchanged by tapetal reflexions unless one or the other is below the absolute threshold, (3) tapeta increase sensitivity most at low intensities near the absolute threshold of vision (4) the sensitivity gains from tapeta are small, (5) the distance by which the visual range is extended is small, and (6) depends on the size of the target.

Tapeta are certainly an elaborate approach to the problem of maximizing visual awareness of the surroundings in dim light. In view of the number of fishes that have independently evolved tapeta, a large premium must be placed on these and other adaptations which increase visual sensitivity where little light is available.

REFERENCES

Barlow, H.B. (1956). Retinal noise and absolute threshold. J. Opt. Soc. Am. 46: 634-639.

Barlow, H.B. (1965). Optic nerve impulses and Weber's law. Cold Spring Harbor Symposia on Quantitative Biology. XXX Sensory Receptors.

Denton, E.J. and Nicol, J.A.C. (1964). The chorioidal tapeta of some cartilaginous fishes (Chondrichthyes). J. Mar. Biol. Assoc. U.K. 44: 219-258.

Dowling, J.E. (1960). Chemistry of visual adaptation in the rat. Nature (London) 188: 114-118.

Duntley, S.W. (1963). Light in the sea. J. Opt. Soc. Am. 53: 214-233.

Harosi, F.I. (1971). Frog rhodopsin *in situ*: orientational and spectral changes in the chromophores of isolated retinal rod cells. Ph. D. Dissertation. Johns Hopkins University.

Hester, F.J. (1968). Visual contrast thresholds of goldfish (*Carassius auratus*). Vision Res. 8: 1315-1336.

Hubbard, R. (1958). Bleaching of rhodopsins by light and heat. Nature (London), 181: 1126.

LeGrand, Y. (1957). Light, colour and vision. Chapman and Hall, London.

Liebman, P.A. (1962). *In situ* microspectrophotometric studies on the pigments of single retinal rods. Biophys. J. 2: 161-178.

Lythgoe, J.N. (1968). Visual pigments and visual range underwater. Vision Res. 8: 997-1012.

Rushton, W.A.H. (1965). Bleached rhodopsin and visual adaptation. J. Physiol. 181: 645-655.

Tyler, J.E. (1960). Radiance distribution as a function of depth in an underwater environment. Bull. Scripps Inst. Oceanog. Univ. Calif. 7: 363-412.

Walsh, J.W.T. (1965). Photometry. Dover Publ., New York.

RETINOMOTOR RESPONSES

M. A. ALI

Département de Biologie, Université de Montréal

Montréal 101, Canada

An ill-favoured thing, sir, but mine own.
As You Like It, Act V, Sc. 4

INTRODUCTION

The term "Retinomotor responses" denotes the dispersion or concentration of the masking pigment within the retinal epithelial cells and the expansion and contraction of the rods and cones in response to changes in the ambient light (Fig. 1).

These were noted as early as 1856 by Müller. The variation in the thickness of the retinal epithelial pigment (REP) was noted by Morano (1872) but it was Czerny (1867) who first made a qualitative observation of the effect of light on REP. He noted that it was more difficult to separate it from the rest of the retina when the animal had been previously exposed to bright light. However, we may regard Boll (1877) and Kühne (1877) to be the real discoverers of REP migration. Working independently, they simultaneously found that the REP of the frog was considerably dispersed in the light-adapted animal and concentrated in the form of a band near the choroid in the dark-adapted one. Eight years later Engelmann (1885) reported the movement of cones in response to light and dark as observed by his graduate student van Genderen-Stort who himself described the phenomenon a year later (1886). A couple of years earlier Angelucci (1884) had observed the movement of the rods of the frog in response to light. Curiously enough what he and later Arcoleo (1890) claimed to have observed was a shortening of the rod in response to light, a notion contrary to the observations of Lederer (1908) and all the subsequent investigations

A great deal of work has been done by many workers on this phenomenon and the literature has been reviewed from time to time (Garten, 1907; Arey, 1915; Walls, 1942; Detwiler, 1943; Studnitz, 1952; Nicol, 1963; Blaxter, 1970; Ali, 1971). Of particular interest to us at the moment are the results of the work concerning the speed of these movements, the influence of environmental factors on the speed, the state of the REP, cones and rods under various environmental factors, the relationship of these responses to the retinal structure of the fish, its habitat and behaviour, the causation or mechanism of these responses and, finally, their significance to the fish. My colleague, Pierre Couillard, will deal mainly with the mechanism in his paper (this volume) while I propose to emphasise the significance of these changes. Before doing that I shall briefly go over some of the other aspects in order to better orientate what follows. Bearing in mind the theme of this Advanced Study Institute, I plan to stress what I consider to be some methods of studying this phenomenon.

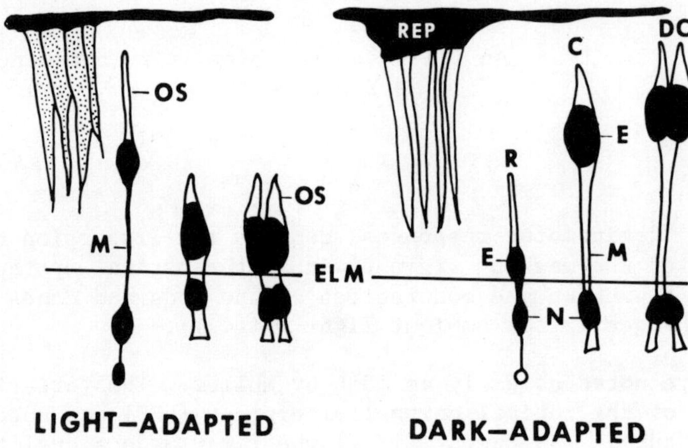

FIG. 1. Diagrammatic presentation of the position of REP, cones and rods in the light- and dark-adapted retinas of a salmonid (modified after Ali, 1958).

FIG. 2. Photomicrographs of transverse sections of the retina from the ventral regions of the dark- and light-adapted eyes of the Artic char, *Salvelinus alpinus*. The scale indicates 50 µm. REP, retinal epithelial pigment R, rod; C, cone; ELM, external limiting membrane; ENL, external nuclear layer; EPL, external plexiform layer; INL, internal nuclear layer; IPL, internal plexiform layer; GCL, ganglion cell layer; NFL, nerve fibre layer; ILM, internal limiting membrane.

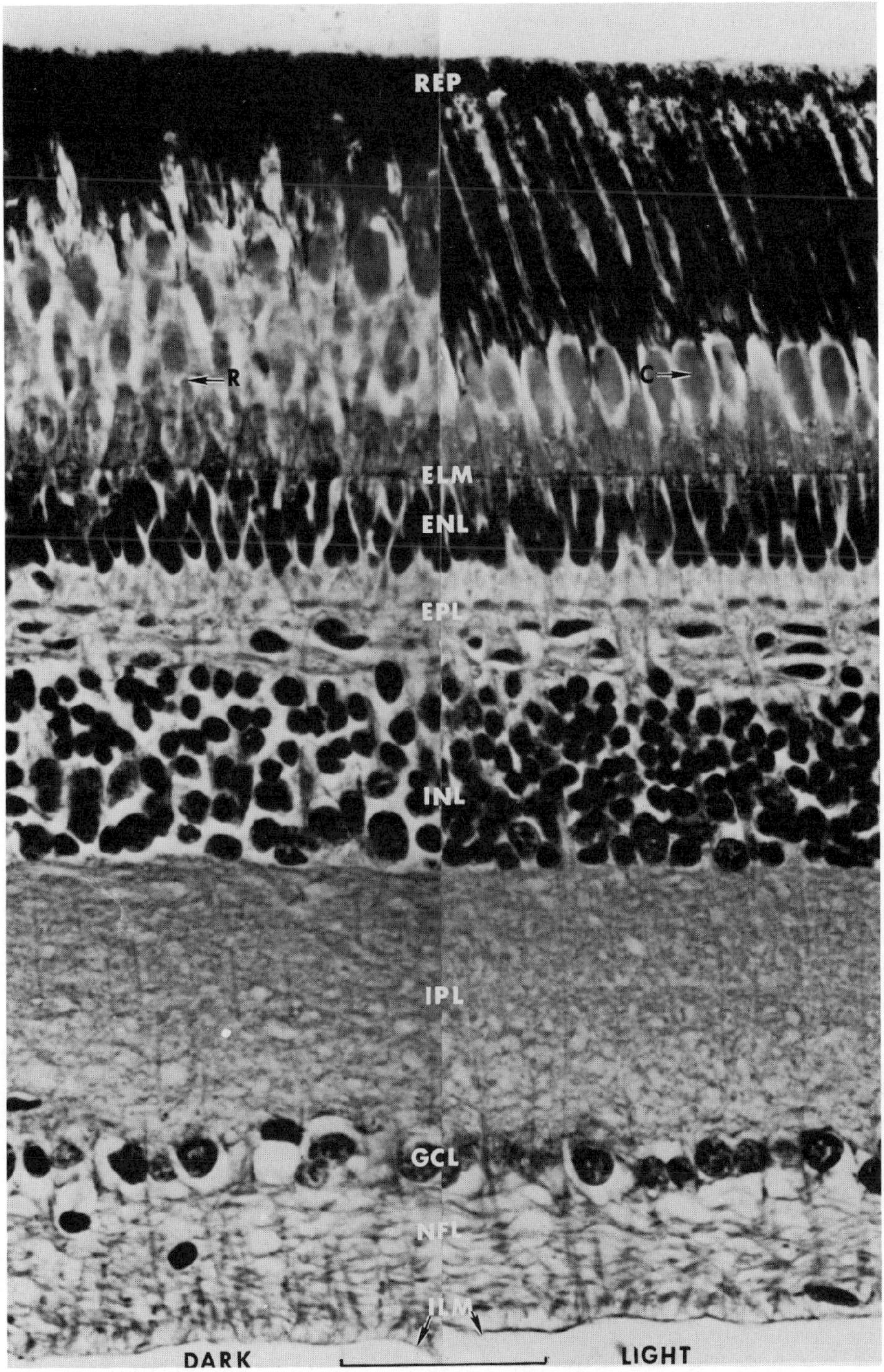
REP
R
C
ELM
ENL
EPL
INL
IPL
GCL
NFL
ILM
DARK
LIGHT

STATE OF THE ELEMENTS

The retinas of light- and dark-adapted fish, such as the trout, show the typical state (Fig. 2). In the light-adapted fish the REP is expanded, the cones contracted and the rods elongated, while in the dark-adapted one the REP is contracted, the cones are elongated and the rods contracted. The extent of expansion and elongation is variable depending on ambiant conditions. This is an aspect which has been extensively studied by a number of workers and the results have been reviewed (Ali, 1971). There are a few general conclusions which may be drawn from these investigations. 1. The role of the REP appears to be, at least in retinas of this type, to control the quantity of light impinging on the visual cells. Thus, when the intensity is greater the extent of expansion is greater, obviously up to a limit. The relationship between the intensity of light and extent of expansion of the REP follows the Weber-Fechner law. Rods and cones, on the other hand, seem to react generally in another way. That is, they are either completely contracted or completely elongated. Rarely are they seen in an intermediate position as the REP is (Fig. 3). One must discount here complex situations such as those of the double cones. These will be dealt with in a later section. 2. Temperature does have an influence on the state of these elements but its role is not clear (Fig. 4). The available evidence indicates, in general, that temperature has some effect in the dark but none in the light. Its effect is greater on the REP than on the rods and cones. Further, its influence varies from fish to fish. No attempt has been made to study the role of acclimation such as that done in the case of ERG (Hanyu and Ali, 1963; 1964; Ali and Kobayashi, 1967). It may be worthwhile combining such a study with the bleaching rates of the visual pigment using microspectrophotometry.

RHYTHMS

Diurnal rhythms in the positions of REP and visual cells of some fishes kept in constant darkness or continuous light have been found (Fig. 5). Some fishes such as the catfish (*Ictalurus nebulosus*) are very rhythmic while others, such as the Pacific salmon (*Oncorhynchus*) are not. Perhaps some aspects of behaviour are related to this characterisation. For example, the downstream migration of the juvenile Pacific salmon may be related to the ambiant light intensity (Fig. 15). Very rhythmic retinal elements in such a fish should hamper its responses to the environmental light conditions. There

FIG. 4. Graph showing the thicknesses of retinal epithelial pigment and cone layers of Atlantic salmon (*Salmo salar*) and Sockeye salmon (*Oncorhynchus nerka*) adapted to light or to dark for two hours at various temperatures (after Ali, 1971).

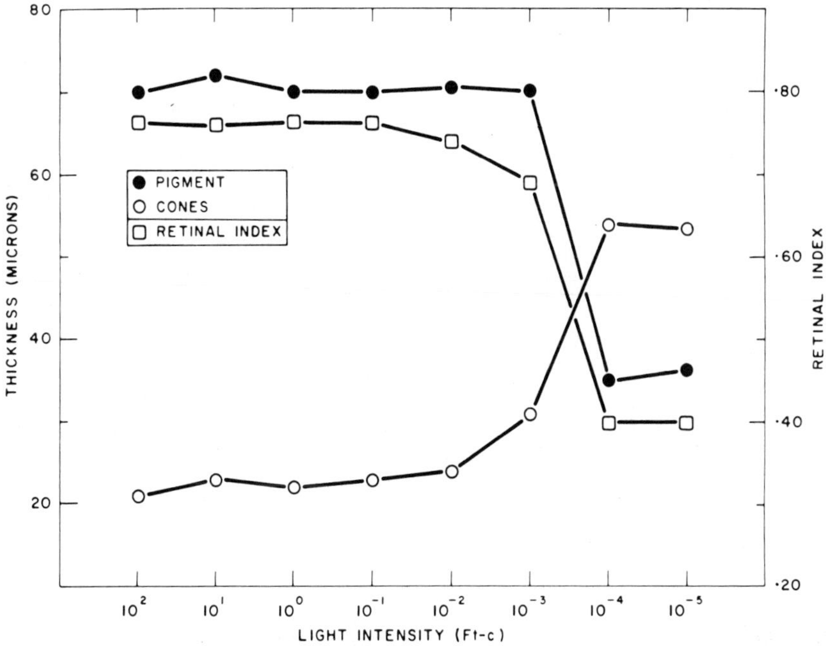

FIG. 3. Graph showing the thicknesses of the retinal epithelial pigment and cone layers in the eyes of Atlantic salmon (*Salmo salar*) exposed to various light intensities. The calculated retinal indices which may also be referred to as pigment indices are also given. This is calculated as given in Ali, Stevenson and Press (1961).

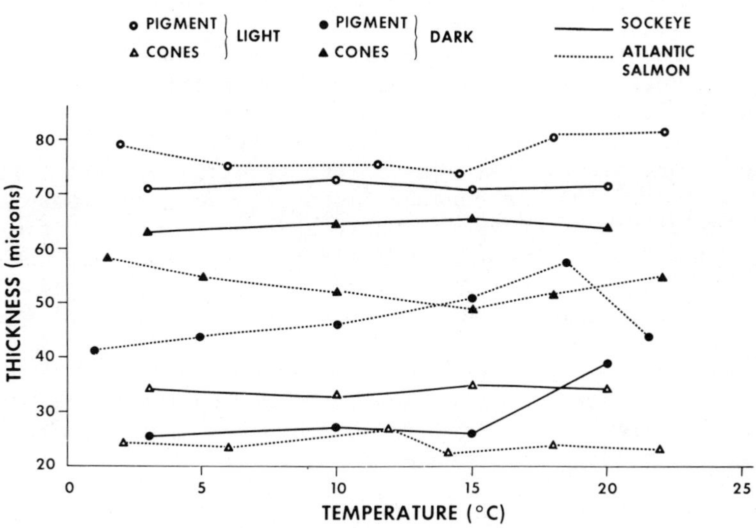

is also another factor. In general, neither the REP nor the visual cells show a rhythm for longer than a day in their positions in continuous light whereas the rhythm persists for a longer period in constant darkness. It also appears that both continuous light and darkness become a source of stress after a few days, usually three to five days. This is indicated by the erratic behaviour of the REP after a few days. It should be recalled that the REP is under the influence of Intermedin, among other things (Pickford and Atz, 1957). I made a few attempts over ten years ago to investigate if there was any relationship between metabolic rates and the behaviour of the REP (Ali, 1964e). There appeared to be a relationship. However, to obtain valid results it would be necessary to measure metabolic and activity rates simultaneously. That would have posed considerable technical problems in the case of fishes (Kramer and Ali, 1965). Usually, the REP tends to be more rhythmic than the visual cells except in some fishes such as the bluefish, *Pomatomus saltatrix* (Olla and Marchioni, 1968).

MONOCHROMATIC STIMULI

An interesting aspect, which has been the object of just a few investigations and deserves to be studied in greater detail, is the response of the REP, cones and rods to monochromatic light stimuli of equal energies (Pergens, 1899; Wunder, 1926; Honjo, 1935; Ali, 1961a) (Fig. 6).

This aspect has taken on a particular significance since the observations of Müller (1952; 1954) showing the presence of three different types of cones in the guppy, and those of Marks (1965) and Hárosi and MacNichol (1974) establishing the presence of three cone pigments in the goldfish, and the subsequent (Stell, Hárosi and Lightfoot, 1974) morphological identification of the three types (see also Stell, this volume). The investigations of Pergens (1899) and Wunder (1926) were merely on qualitative observations while Honjo (1935) and Ali (1961a) measured the thickness of the REP and the length of the cones. However, even the results of the latter serve only to give an idea of the spectrum visible to the fishes concerned. This is because light intensity was measured in terms of lux or foot-candles and the possible presence of the three types of cones and differences in their distribution were not taken into account. In an investigation, currently being carried out in my laboratory, we find that in the brook trout, (*Salvelinus fontinalis*) three types of cones may also be distinguished. The short, stubby barrel-like cones of the dorsal region of the eye are probably the blue cones, while the longer and slender cones of the ventral region are possibly the red cones. The area in-between has a mixture, and quite a few double cones, which appear to be made up of green and red members. We are in the process of finding out if

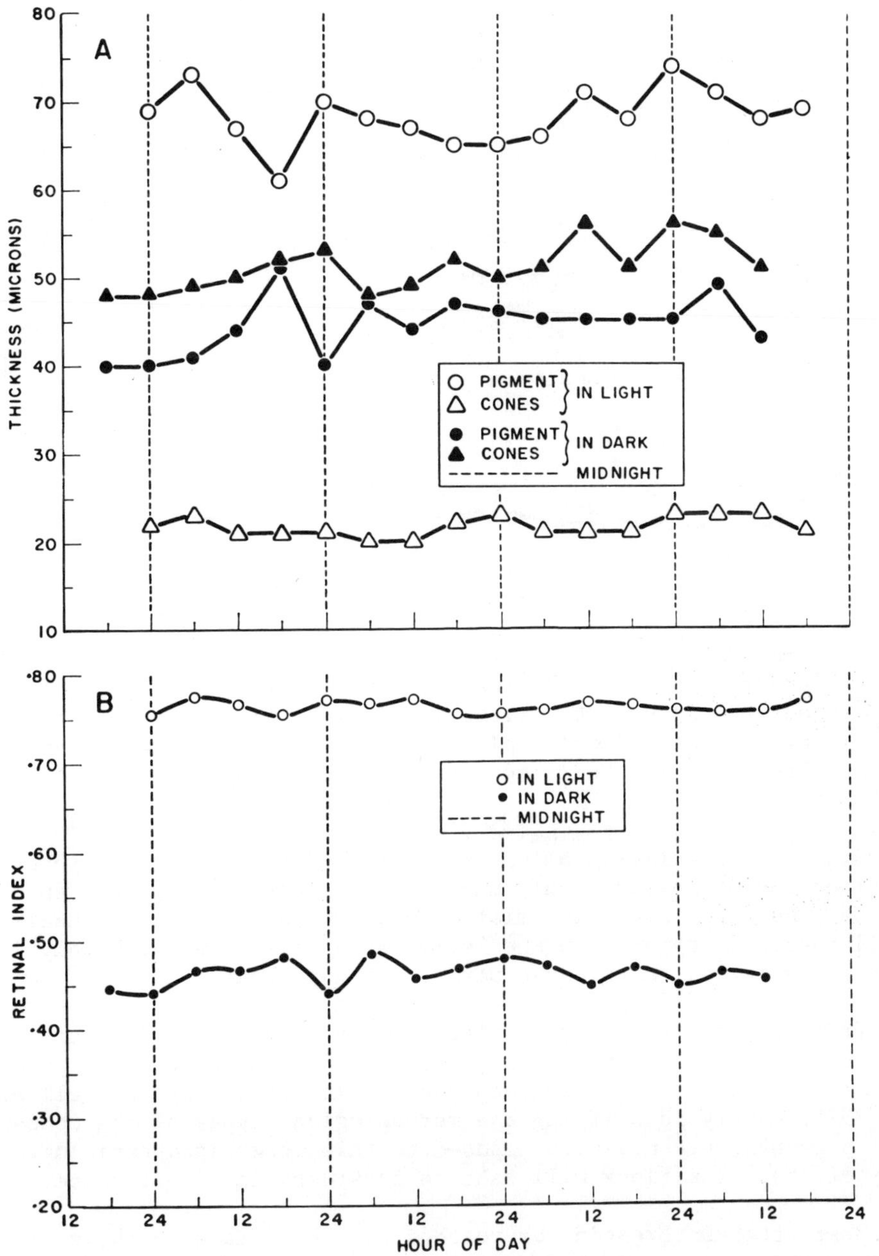

FIG. 5. A. Diagram showing the thickness of the REP and the length of the cones in Atlantic salmon (*Salmo salar*) maintained under constant light or darkness. In B, retinal (pigment) index values, calculated as per Ali, Stevenson and Press (1961) are given. (After Ali, 1961a).

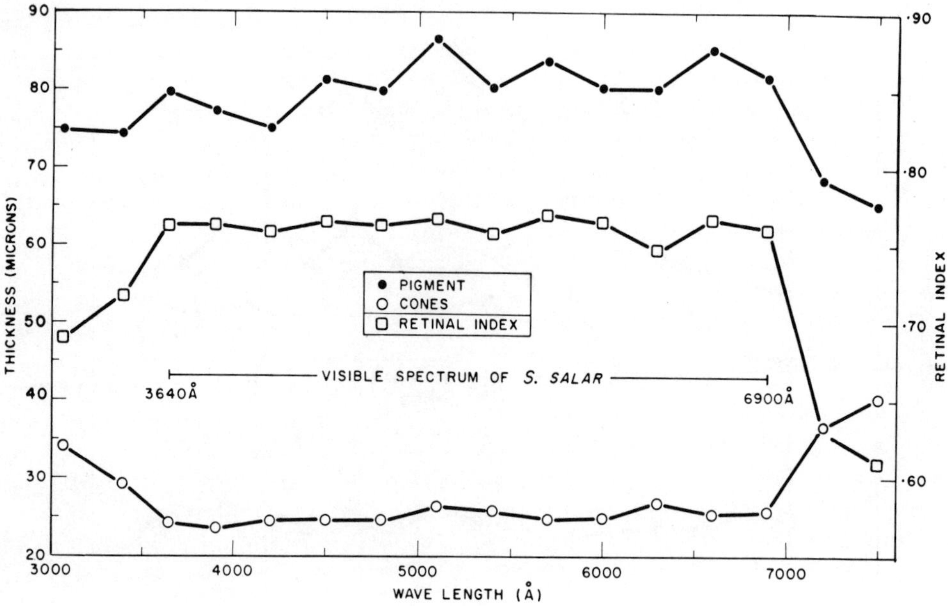

FIG. 6. Graph showing the state of retinal epithelial pigment, cones, and retinal index in the eyes of Atlantic salmon (*Salmo salar)* exposed to various wavelengths for one hour. Retinal or pigment index is calculated as per Ali, Stevenson and Press, 1961. Modified after Ali (1961a).

monochromatic stimulation of short and long durations brings about a contraction of one type only and not the other. It appears so far that the inner synaptic relationships may bring about a reaction of all the three types even when stimulus of one wavelength only is given if it is for an extended period, like an hour.

From an adaptive point of view this is understandable since the fish's environment practically never offers a narrow-band monochromatic stimulus. The best approach seems to be that of Stell et al. (1974) namely identifying the morphological types of the cones whose pigments were measured. Added to this, the histophysiological study of their reactions will take on considerable significance.

Some other interesting approaches have been made to the study of the influence of different wavelengths on the state of the cones and the REP. Weidemann (1966) studied the influence of different wavelengths of equal energy on the state of adaptation of the cones in the guppy, *Lebistes reticulatus*. He found that a five minute exposure to violet or red light brought about an almost complete contraction of the cones. The state of adaptation, according to him, was dependent on the intensity of light and the retinal region. The three types of cones were found to be similar in this respect except possibly for the reaction of one type (outer cones) to greenish light. He suggested that this exception may be due to the regional differences and to a decrease in intensity from the dorsal to the ventral region. Weidemann also measured the length of the cones in fishes subjected to darkness after initially adapting them to different wavelengths. Plotting the lengths of the cones against the wavelengths was inversely proportional to the absorption spectrum of porphyropsin. That is, the cones of fishes initially exposed to green were less relaxed than those of fishes exposed to other colours.

Statistical analysis showed a difference among the cones especially in the central part of the eye. Cone populations could be established after these findings. He also did a preliminary histochemical analysis of light-and dark-adapted retinas and found a basis for believing that different types of cones may exist. He could not however establish with certainty cone types with generally different spectral sensitivities. Ali and Crouzy (1968) found, that in the brook trout, at a wavelength of 520 nm, less energy was required to bring about the expansion of the REP in a previously dark-adapted fish (Fig. 7). Even though this value does not correspond exactly to the maximum of the absorption spectrum (510-515nm) of the scotopic pigment, it is fairly close. The discrepancy may be attributed to the fact that in this fish there is a mixture of rhodopsin and porphyropsin whose ratio varies with age and season (Beatty, 1966). It is also interesting that in the goldfish (*Carassius auratus*) the electroretinogram (ERG) threshold is lower at a wavelength of 520nm, which corresponds to the maximum absorption spectrum of its scotopic pigments (Crouzy and Ali, 1966). A correlation between the action spectrum of the expansion of the REP and the scotopic pigment is not surprising. This had even been confirmed in the case of the frog (Liebman, Carroll and Laties, 1969). However, it is surprising to observe a relationship between the response of the cones which are photopic elements, and the porphyropsin which is found in the rods. This aspect also merits further careful investigation.

1. Intensity. The progression of changes from one state to another is influenced by the intensity of the stimulus used and, in the case of adaptation to dark, the light intensity to which the fish had been previously exposed (Ali, 1962b).

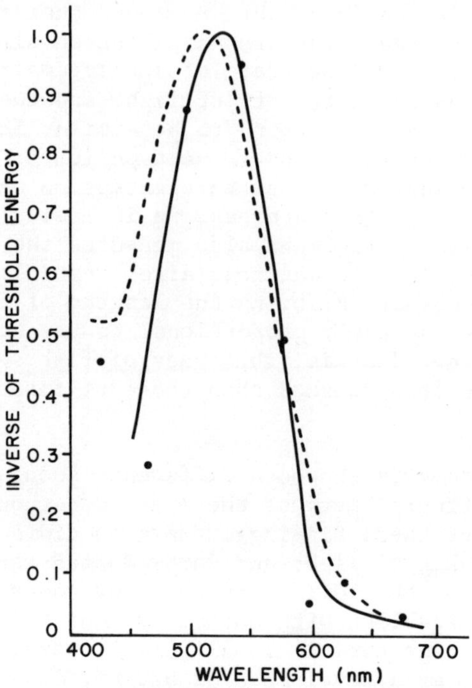

FIG. 7. The relative action spectrum of retinomotor responses (solid line) in the brook trout (*Salvelinus fontinalis*). The ordinates are proportional to the inverse of the threshold radiant energy per unit area arbitrarily taking the maximum value to be 1.0. The broken line represents the scotopic pigment absorption curve (Wald, 1941). After Ali and Crouzy (1968).

In general, the movement of the REP and the visual cells is more rapid in a fish exposed to a higher light intensity, than in a fish exposed to a lower intensity. In addition the REP is more expanded in the former (Fig. 8).

Conversely, these elements attain the dark-adapted state more slowly in a fish which had been previously exposed to a higher light intensity, than in one which had been exposed to a lower intensity. Thus differences among investigations with regard to the speed of these changes might often be attributed to the use of different light intensities.

2. Temperature. REP and cones attain the light-adapted state faster at a higher temperature than a lower one (Fig. 9). It also

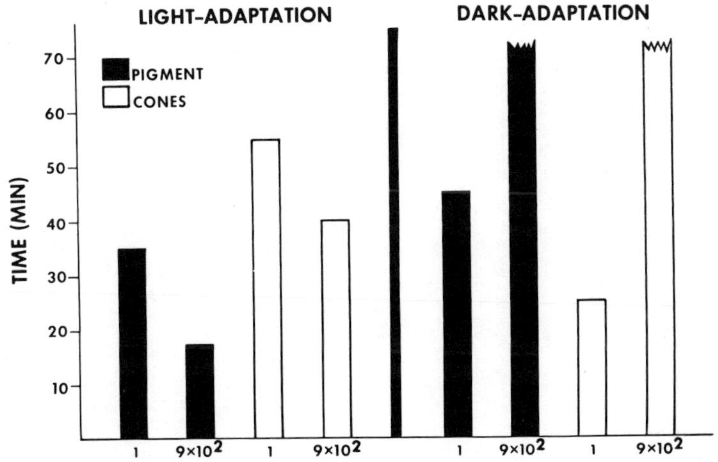

FIG. 8. Histogram showing the time taken by the retinal epithelial pigment and the cones to attain the light- and dark-adapted positions. In light-adaptation dark-adapted fish were exposed to high (9 x 10^2 ft-c) and low (1 ft-c) light intensities. In dark-adaptation fish previously exposed to high and low light intensities were subjected to darkness. Data from Ali (1962b).

takes them longer to attain the dark-adapted state at a lower temperature (Ali, 1964f) (Fig. 10).

3. Wavelengths. Wavelengths may also influence the speed of these reactions as has been suggested by the work of Honjo (1935) and Weidemann (1966). That is, when a fish is exposed to a given wavelength, for example green, its retina may attain the state of light-adaptation faster than that of another fish exposed to another colour. This aspect needs more study and might throw some light on the relationship among the different cones on the one hand and the connexion between the cone and the horizontal bipolar cells on the other. Retinas of fishes exposed to various wavelengths for different periods, along the line of experiments done to determine the speed of changes (Ali, 1959) should be examined for a careful study of this aspect. Findings may also indicate clearly if the three types of cones respond the same way or differently to stimuli of different wavelengths.

4. Age and taxonomic factors. Age and species differences also appear to influence the course and speed of these responses. This

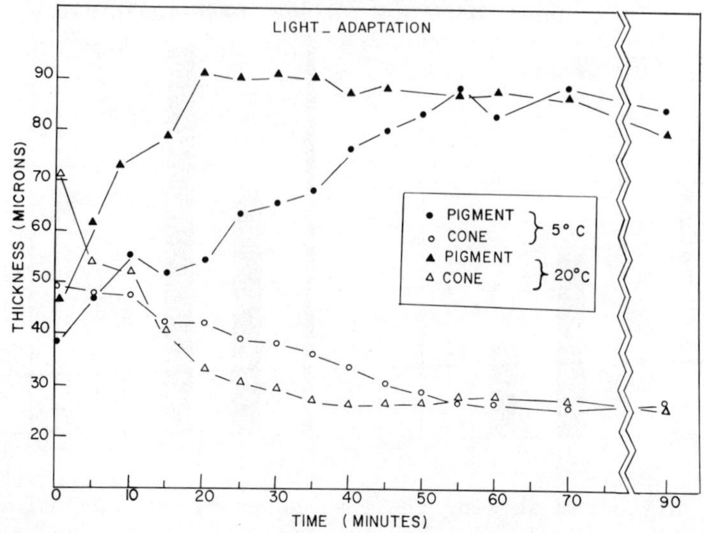

FIG. 9. Expansion of the REP and contraction of cones in Atlantic salmon smolts acclimated to two different temperatures and exposed to light after previous dark-adaptation. After Ali (1964f).

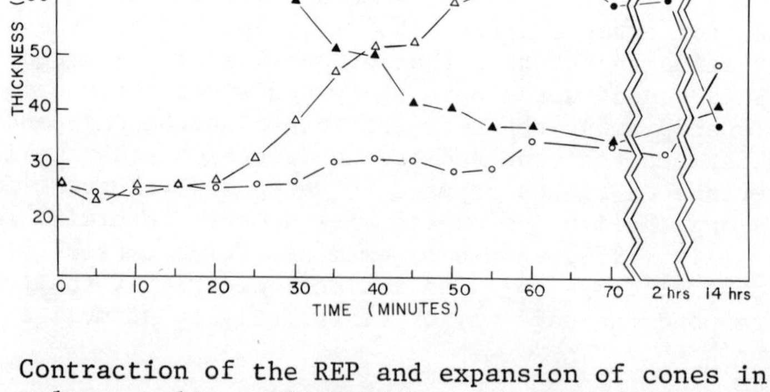

FIG. 10. Contraction of the REP and expansion of cones in Atlantic salmon smolts acclimated to two temperatures adapted to light and then subjected to darkness. After Ali (1964f).

has been studied in detail in the case of four species of the Pacific salmon (*Oncorhynchus*) and the different stages of the four species. The results show differences among the stages on the one hand and among the species on the other. In general, it takes slightly longer for the cones to attain the light-adapted state in the older (e.g. smolt) than in the younger (e.g. fry) stage. (Figs. 11,12) These differences may be related to the stage of development of the retina (Ali, 1963, 1964a) and the differences in the habitats and modes of life. To elaborate, the retina is not fully formed until the Pacific salmon fry emerges from the gravel and, after that, as the animal gets longer and bigger and its eye gets larger, the retina becomes thinner which would have a bearing on the responses of the retinal elements. Further, the pink and chum salmon fry commence migrating towards the sea while the sockeye and coho fry do not. The preferences of these different stages regarding light are also different (Ali, 1962c). It would be interesting to continue such studies also with marine fish stages along the lines Blaxter (1970) has done.

RELATIONSHIP WITH HABITAT

Fishes inhabit a marine, freshwater or brackish water environment. Some are capable of moving from one to another, rapidly or slowly (euryhaline) while others are restricted to one (Stenohaline). Some migrating fishes move from one to the other during their life time. Fish such as the salmon which are hatched in freshwater and move to the sea to become adults and return to freshwater to spawn are anadromous while fish such as the eel which do the opposite are catadromous. In the marine environment one could find fishes inhabiting the littoral zone, the continental shelf, the pelagic zone or deep waters. Some fishes are capable of living in very deep waters while some others live in moderately deep (300 to 600 metres) waters. Some migrate vertically during a 24 hours period, going down during the day and coming up to less deeper waters during dusk and night. Even in shallower waters there are surface living fish and bottom living ones. As far as photic qualities are concerned, there are clear waters, muddy waters and even black and milky waters as in the Amazon basin.

As for habits, one finds nocturnal or arythmic fishes. There are no known diurnal fish. Most of those which are referred to as nocturnal are essentially those which live in muddy or deep waters and consequently they are adapted to function in a dim milieu. These have rod-rich or pure rod retinas while the arythmic fishes, capable of functioning in bright as well as dim environments have mixed retinas. In this respect fishes differ from the amniotes which have distinctly nocturnal, diurnal or arythmic forms as far as activity patterns are concerned. It is interesting to examine retinal struc-

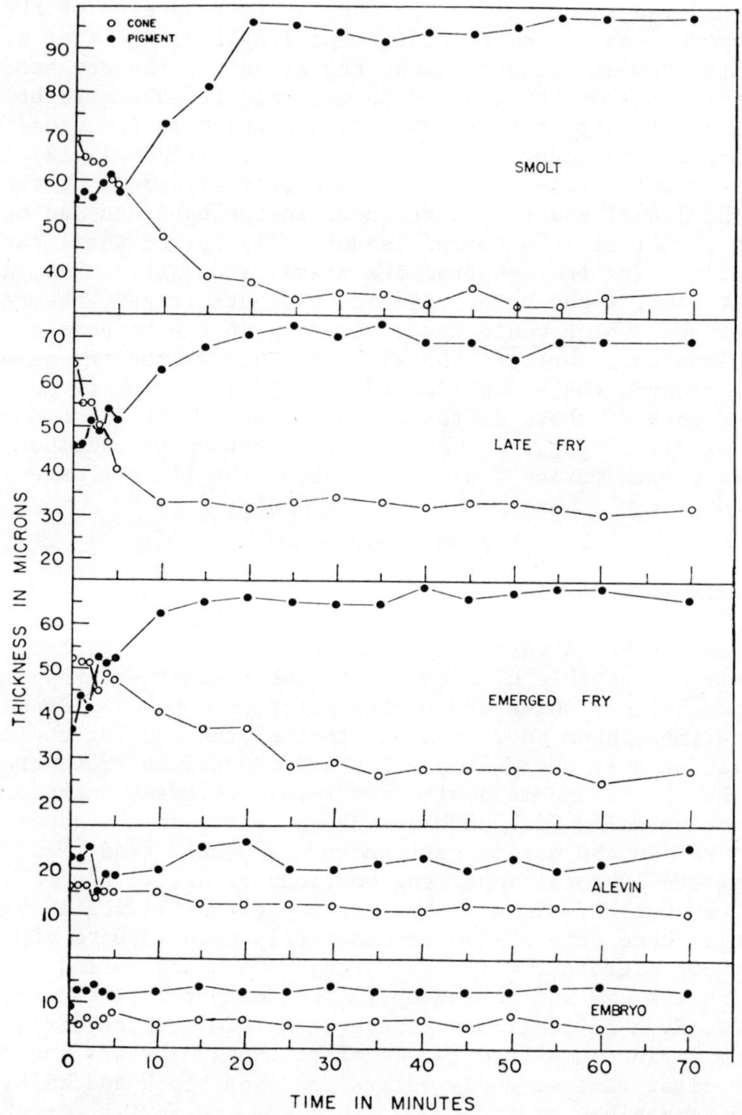

FIG. 11. Graph showing the expansion of the pigment and contraction of cones when dark-adapted coho salmon (*Oncorhynchus kisutch*) of various stages are exposed to light of 400 ft-c. After Ali (1959).

ture and the occurrence of retinomotor responses in relation to habitat and habits. The relationship is nearly perfect in the case of the fishes in general and teleosts in particular. The nocturnal, deep-water or muddy water fishes have a very sparse REP which is

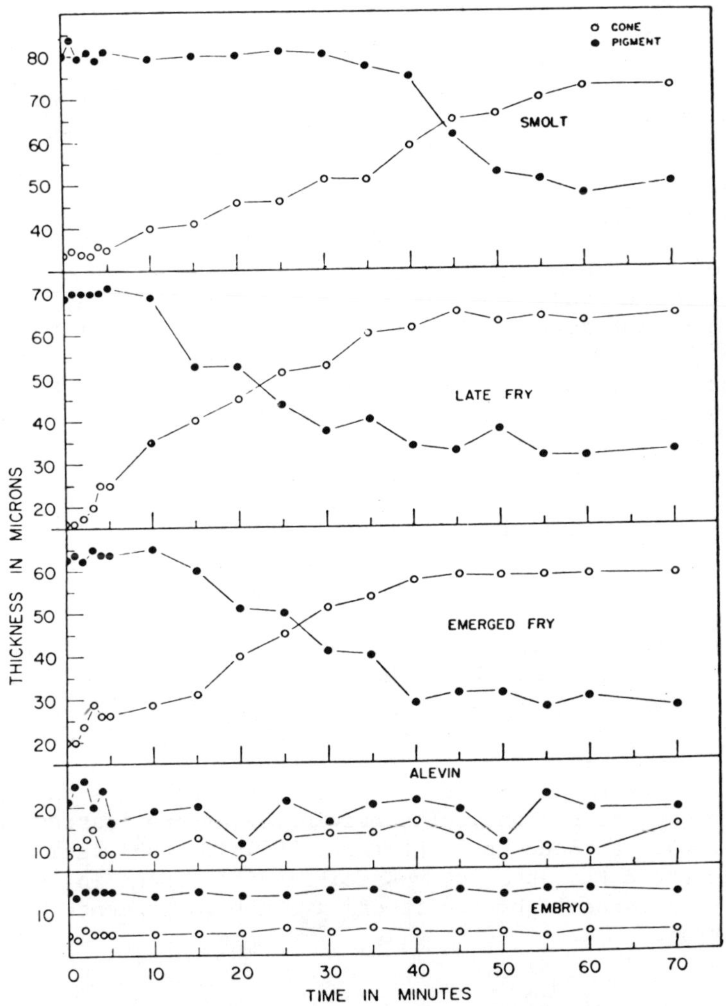

FIG. 12. Graph showing the contraction of pigment and expansion of cones when light-adapted coho salmon (*Oncorhynchus kisutch*) of various stages are exposed to dark. After Ali (1959).

usually incapable of movement. When the retinas of these fishes contain reflecting material, the distribution of the REP, coupled with the movement of the rods and the light guiding function of the cone ellipsoids, serves to aid sensitivity (Zyznar and Ali, 1975; Zyznar, this volume). In these fishes one may encounter pure rod retinas or rod-rich retinas with a few large cones. The REP is sparser in the pure rod retinas and usually totally incapable of any movement.

The arythmic fishes with mixed retinas have REP, rods and cones capable of movement. These three parts are generally well developed. In this group of fishes one finds a great deal of variation. Some are almost diurnal with a good proportion of cones while some others are almost entirely nocturnal with an abundance of rods and reflecting material in their epithelial layer. Those inhabiting turbid waters resemble the latter. There are forms intermediate in varying degrees. These arythmic fish show the most marked retinomotor responses. The dispersion of the REP may, in some forms, increase the width of the pigmented layer by as much as five times while the length of the rods and cones may vary much more as a result of their expansion and contraction. Many publications have described these relations (Munk, 1959; 1964; 1966; Engström 1963a; 1963b; Engström and Rosstorp, 1963; Ali and Hanyu, 1963; Ali, Anctil and Mohideen, 1968; Ali and Anctil, 1968; Anctil, 1969; Ali, Anctil and Raymond, 1973).

Some investigations have also demonstrated a relationship among habitat, retinal structure and vascularisation (Hanyu, 1962; Ali et al., 1968; Anctil, 1968; 1969). In general, an active, arythmic fish with a mixed retina capable of retinomotor responses possesses a richer vascularisation than an inactive fish dwelling in turbid-water or deep-sea with a rod rich retina containing few cones or no cones at all. It has also been shown that the eyes removed from an active fish living in clear water, such as the salmon (Ali, 1962a), do not survive for an hour even in oxygenated Ringer's (Fig. 18) while those from less active forms such as goldfish and catfish (Figs. 19,20) which are capable of putting up with less oxygenated waters, are capable not only of surviving after enucleation but respond for over an hour to light (Ali, 1964b). Kobayashi (1957) rightly suggested that the inter-specific differences in retinal reactions to diverse light conditions are the consequence of ecological adaptation.

RELATIONSHIP WITH BEHAVIOUR

A very clear relationship exists between behaviour and retinomotor responses in fishes (Ali, 1961b). Certain aspects of behaviour lend themselves particularly well for establishing such a correlation. These are: 1) Dorsal light action, 2) phototactism, 3) formation and maintenance of schools, 4) feeding and 5) migration.

1. Dorsal light reaction. When a laterally flattened fish such as an angel-fish is kept in a dark tank and a light shone from one side, it tilts its body sideways in order to have the light shining on its dorsal part. The angle of its tilt is in direct proportion to the intensity of the light. This reflex or reaction is called Lichtrückenreflex or dorsal light reaction (Holst, 1935).

A good correlation exists between the retinomotor responses and this reaction in *Pterophyllum eimeckei* and *Gymnocorymbus ternetzi* (Braemer, 1957). The intensity required to make the fish tilt is also the intensity required to trigger the contraction of the cones. Further work using a combination of this technique with electrophysiology and histology should yield some valuable information on sensitivity, acuity and colour vision on the one hand and the ERG, histophysiological and behavioural thresholds on the other.

2. Phototactism. There is a difference between the larvae of fresh-water fishes and those of marine fishes. In general, the egg of a fish such as trout or salmon, which is anadromous, is big and yolk-rich while that of a marine fish like the herring is small with very little yolk. Obviously, this difference is reflected in their subsequent development. The salmonid egg is laid and remains on river beds, between pebbles and rocks while the herring egg is pelagic and forms part of the plankton. The incubation period of the former is much longer than that of the latter. The larvae also obviously differ in size. The salmonid alevins are relatively big and possess a yolk-sac while the herring larvae are very small. The former could afford to remain for a few weeks at the bottom hidden between the stones while the latter, due to the absence of yolk, must feed themselves. The former are photonegative while the latter are not. While the retina in both is not capable of retinomotor responses, that of the herring is better developed (Ali, 1959; Blaxter, 1968; Blaxter and Jones, 1967; Blaxter and Staines, 1970). The fresh-water salmonid alevin's retina contains very few visual cells, and these appear to be cones while the marine herring larva has more visual cells although these also appear to be cones. In either case the REP does not respond to light. Since the inner retinal layers of the herring larva are better formed and since it is capable of catching its food, it may be assumed that its retina is almost functional. On the other hand, the retina of the newly hatched alevin is not well formed and it shows no reaction to light except avoidance of bright light which could as well be due to its pineal organ. As it gets older, the salmonid alevin becomes less photonegative and this corresponds to the development of the rods and cones and the absorption of the yolk-sac (Fig. 13). The development of a retina with all its layers in their proper proportions, appearance of clearly identifiable rods and cones capable (together with the REP) of retinomotor responses, complete resorption of the yolk and, positive phototactism transform the alevin into a fry. The fry is fully capable of retinomotor responses and full visual acuity as seen in feeding experiments (Fig. 14). In the case of the marine larva, the metamorphosis coincides with the appearance of rods and the capacity for retinomotor responses (Blaxter, 1969; 1970). It is also the period when the juvenile fish becomes more active and, instead of forming a part of the passive plankton, becomes independent and active, capable of swimming on its own.

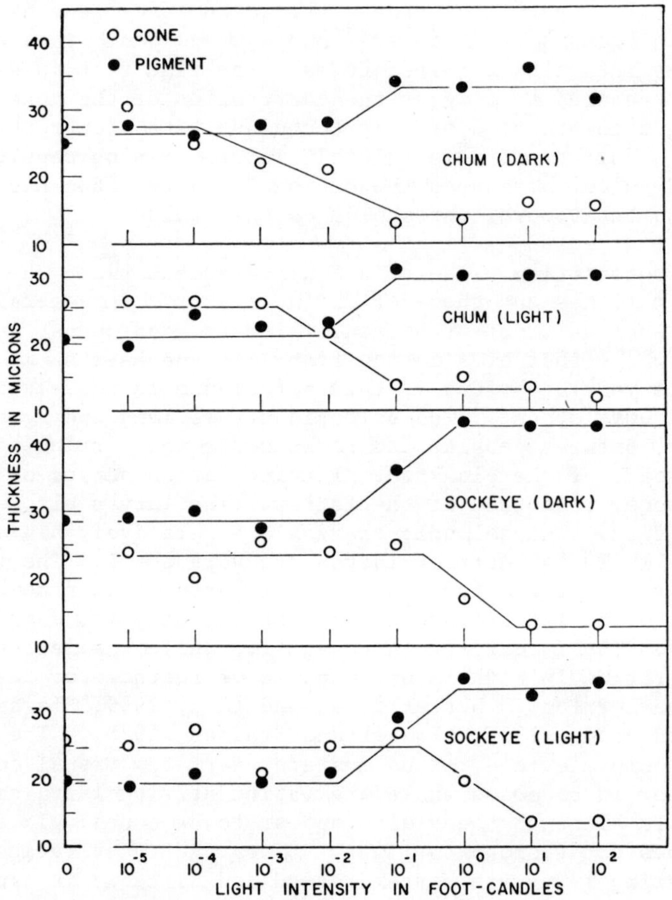

FIG. 13. Graph showing the thicknesses of pigment and cones of sockeye (*Oncorhynchus keta*) alevins in various light intensities. (Light) and (Dark) refer to the fish that were kept under light or dark, respectively, for 3 days before exposure to the various light intensities mentioned. After Ali (1959).

3. Schooling. That vision is the primary sense involved in schooling is certain (Keenleyside, 1955; Loukashkin and Grant, 1959). It has also been shown that although vision is the main factor in the formation and maintenance of schools, full acuity is not essential (Ali, 1959). So long as the light intensity is above the rod threshold the fish are able to form a school. The school disbands when the intensity falls below this level. At such intensities even big objects such as rocks are not distinguishable. No direct

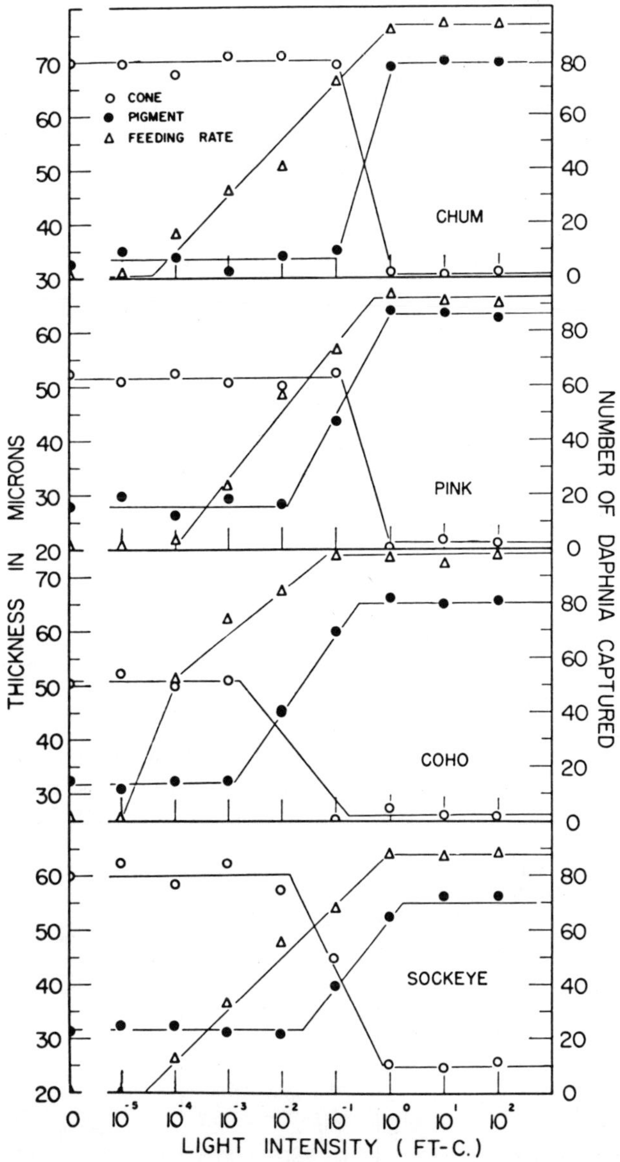

FIG. 14. Graph showing the thicknesses of pigment and cones and the feeding rates of late fry of the four species of Pacific salmon (genus *Oncorhynchus*) under various light intensities. After Ali (1959).

relationship between schooling and the position of the REP, cones and rods has been established. When dark-adapted fish are exposed to light of a minimum intensity, they form a school in five to ten minutes (Fig. 15). The results obtained with *Oncorhynchus*

(Ali, 1959), are essentially similar to those obtained with *Astyanax mexicanus* (John and Haut, 1964). The formation of schools at night due to illumination has been used in fishery in the case of *Lateolabrax japonicus* and *Cyprinus carpio* (Tamura, 1957) and *Osmerus eperlanus* (Girsa, 1967). Kobayashi (1957) has shown a relationship between this aspect and the retinomotor responses in *Trachurus*.

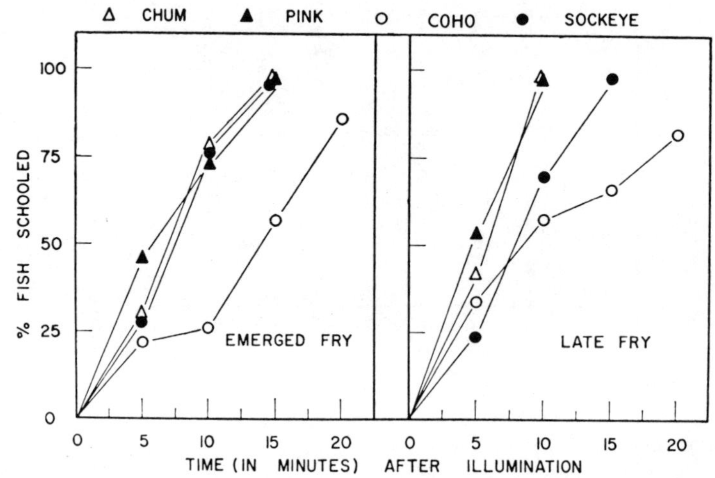

FIG. 15A. The rate of schooling of dark-adapted emerged fry and late fry of the four species of Pacific salmon (genus *Oncorhynchus*) on exposure to light of 400 ft-c.

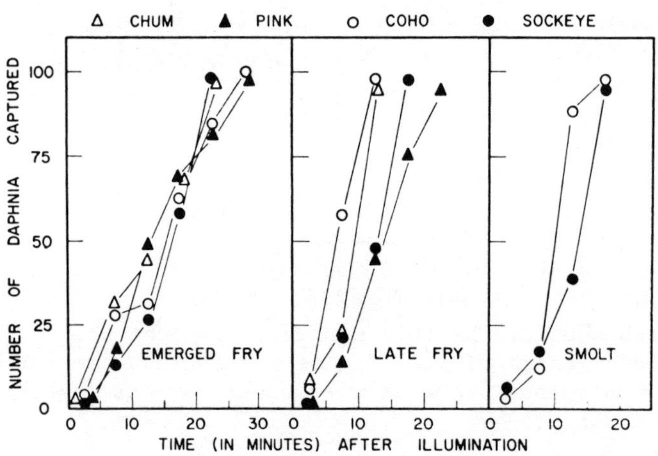

FIG. 15B. The rate of feeding of conditioned, dark-adapted, emerged fry, late fry, and smolts of the four species of Pacific salmon (genus *Oncorhynchus*) on exposure to light of 400 ft-c. After Ali (1959).

4. Feeding. In *Oncorhynchus* (Ali, 1959; 1961b) and *Pleuronectes* (Blaxter, 1968) feeding seems to depend on vision. Various stages of four species of Pacific salmon were conditioned to feed on *Daphnia* and the relationship between the number of prey captured in a given time and the state of adaptation of the REP and cones was studied. These follow the Weber-Fechner law at intensities between the rod and cone thresholds. The salmon fry and juveniles depend on vision to locate and capture their prey. They are unable to capture any except by chance, when the light intensity is below the rod threshold, or in darkness. The ability to capture prey is at its maximum when the light intensity is at or above the cone threshold. At intensities between rod and cone thresholds, this ability diminishes in direct proportion to the light intensity (Fig. 16).

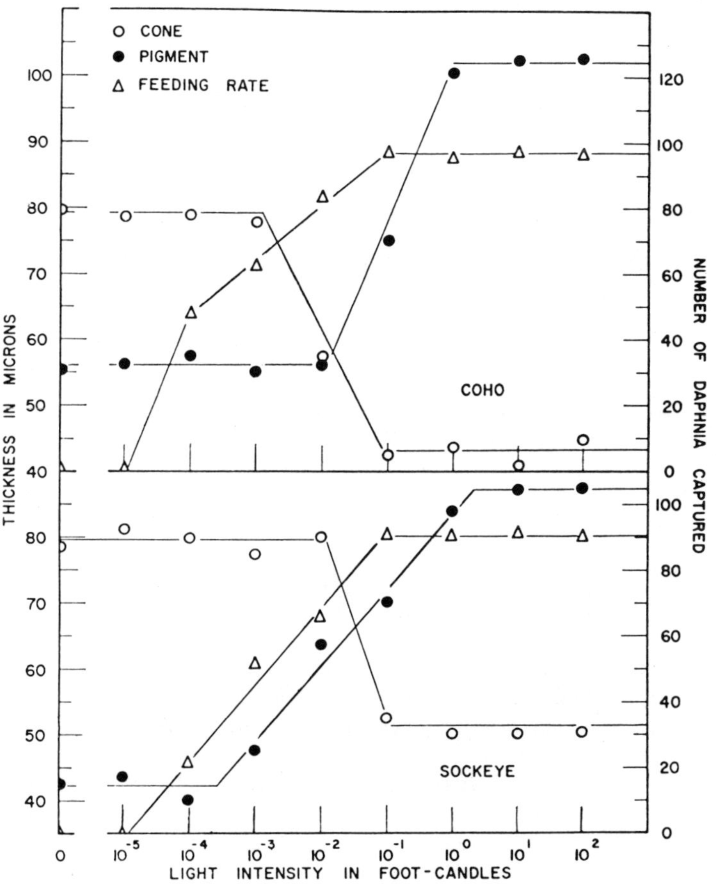

FIG. 16. Graph showing the thicknesses of pigment and cones and the feeding rates of sockeye (*Oncorhynchus nerka*) and coho (*Oncorhynchus kisutch*) smolts under various light intensities. After Ali (1959).

The method of capture also varies. At lower intensities the fish remains at the bottom and locates its prey by its silhouette. This change of mode indicates that scotopic vision has taken over with a loss of acuity. As the intensity further decreases, the difference between the prey and the background diminishes and is completely lacking when the light intensity falls below the rod threshold. At this stage the fish becomes incapable of locating its prey and feeding ceases. In the *Pleuronectes* larva (Blaxter, 1968), sensitivity increases with age and a metamorphosis it is about one to two log units greater, which is about the same as in the Pacific salmon (Ali, 1959). In the young *Pleuronectes* feeding on a small scale continues even in darkness presumably by the use of some other sense.

5. Migration. Three types of migration may be considered. 1. seaward migration, with the current, 2. upstream migration and, 3. vertical migration in lakes and oceans.

The downstream migration of the juvenile pink salmon (*Oncorhynchus gorbuscha)* may be triggered by vision (Brett and Ali, 1958; Ali and Hoar, 1959). As mentioned above, when the fry emerge from gravel, their vision is fully developed, their visual acuity is good and they are capable of feeding. They are able to maintain their position in the stream, against the current, by using some small objects as points of reference. When the light intensity diminishes rapidly at dusk, there is a period during which the REP and visual cells are in a semi-adapted state because of the retinomotor responses not keeping pace with the rate of decline of the light intensity (Fig. 17a). All fish are not affected in the same way since some areas of the stream are less exposed than others. The capacity of the fish to orient itself with reference to a position is reduced or lost. It then rises to the surface and swims actively with the current or is swept by it. The juveniles of sockeye (*O. nerka)* and coho (*O. kitsutch)* have a cone threshold lower than that of the other species of Pacific salmon. The speed of movements of their REP and cones is also higher. Their downstream migration occurs later at dusk since it commences when the light intensity falls below the cone threshold and brings about a state of semi-blindness until the retina is completely dark-adapted. When the REP and visual cells are completely dark-adapted, the fish is able to perceive large objects and maintain its position in the stream. Then migration stops or is greatly slowed down until dawn when the procedure is repeated but in the opposite direction and more rapidly since the speed of changes in light is faster (Fig. 17b). The longer the time to adapt the longer the period of migration.

The case of the coho salmon (*O. kisutch)* is interesting because of its greater sensitivity. If one compares the migratory

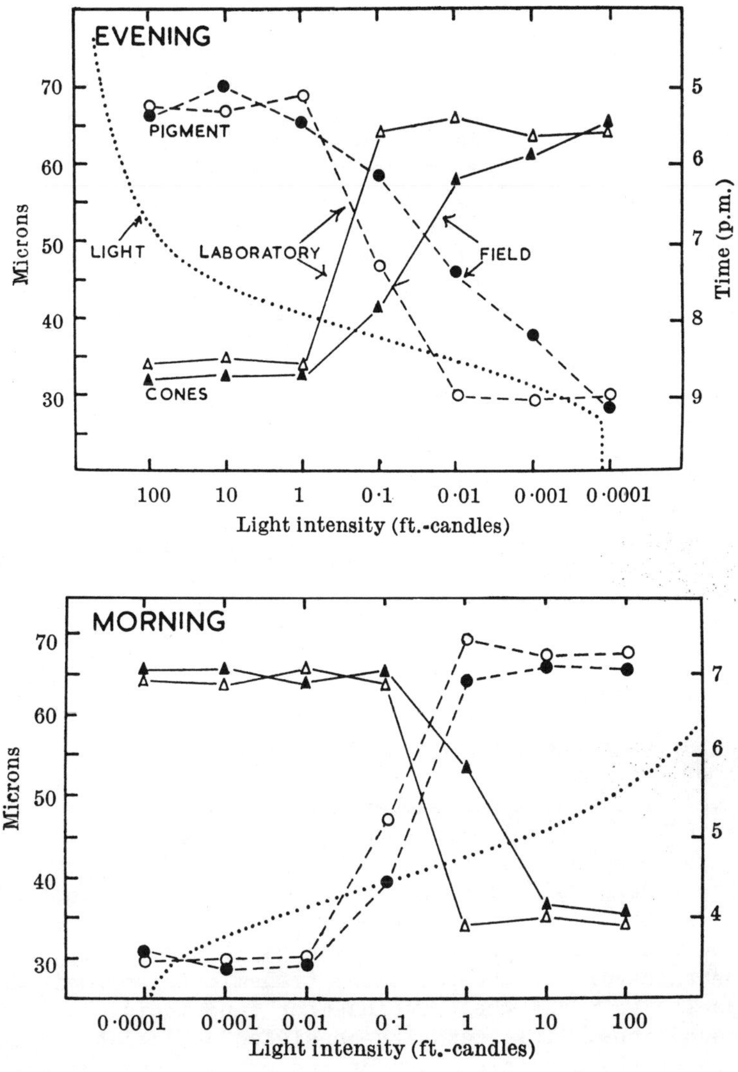

FIG. 17. Thicknesses of the pigment and cone layers of pink salmon (*Oncorhynchus gorbuscha*) retinae when fully adapted to a series of light intensities in the laboratory and at the same intensities under conditions of rapidly changing morning and evening light in the field. After Ali and Hoar (1959).

behaviour of the five species of Pacific salmon occuring in Canadian waters one finds the coho is the closest to the genus *Salmo* (Atlantic salmon and trouts). In addition, its life in freshwater resembles more that of the Atlantic salmon than that of the other Pacific salmon species.

The Atlantic salmon yearling lives in fairly deep fresh-water bodies. It does not migrate at dusk. The smolts in the little Codroy river in Newfoundland feed on small organisms and are in turn eaten by eels. They move around within the estuary during the day. The cone threshold is lower in the Atlantic salmon than in the Pacific salmon species of all ages (compare Figs. 3 and 16) (Ali, 1959; 1961a). To this difference may be attributed, partially at least, the lack of downstream migration at dusk. They could also feed for longer periods and avoid predators with higher cone thresholds. It is probable that because of its influence on the state of the retinal elements, temperature may play a role by modifying acuity and sensitivity and thus indirectly governing migration.

PHYLOGENY OF THESE RESPONSES

There is a considerable degree of uniformity within the class Pisces as far as the occurrence and nature of the retinomotor responses are concerned, except for the variations attributable to differences in habitats (Ali and Wagner, this volume). These variations seem to be an adaptive radiation in respect to the retinomotor responses.

As far as the distribution of these responses in the different classes of vertebrates is concerned it appears that they may be fairly well related to a phylogenetic tree (Ali, 1971). Paucity of information concerning vertebrates does not permit well founded generalisations on this aspect. The mobility of the choroidal pigment, the absence of retinomotor responses and the contractility of the iris in the elasmobranchs are of phylogenetic interest. Presumably the ancestors of the elasmobranchs (Placoderms), which were also the ancestors of the teleosts, transmitted to their descendents the capacity to adapt their retinas for functioning in different light conditions. This capacity consisted of variations in the distribution of the screening pigment and/or movements of the visual cells - movements which are usually related to the movement of the pigment. A third factor could have been varying the diameter of the pupil. The elasmobranchs reduce the quantity of light impinging on the visual cells by masking the crystals of the reflecting material in the choroid. It exploits the available dim light by exposing the crystals. At least one teleostean, the big-eye (*Priacanthus arenatus)* is known to have a similar mechanism (Nicol and Zyznar, 1973). The elasmobranchs are also capable of slow pupillary movement. Alternatively it may be argued that the retinomotor responses have not evolved in the cyclostomes and elasmobranchs because their visual cells are not clearly differentiated into rods and cones as we know them in the teleosteans and mammals. Obviously, further detailed work, both structural and physiological, along the lines undertaken by Stell and by Gruber (this volume) on the elasmobranch retina, is necessary to arrive at valid conclusions.

From these potential capacities transmitted by the Placoderms the teleosteans seem to have retained the retinomotor responses and evolved them further. Some teleosts such as the flat fishes have also somewhat developed the pupillary response. By another line, passing via the Crossopterygians these capacities seem to have been transmitted to the amphibians.

ONTOGENY OF THESE RESPONSES

As has been mentioned above (phototactism) there is an ontogenetic parallel between the structural and functional development of the retinal elements in the Pacific salmon (Ali, 1959). This is also true in the case of the Atlantic salmon (Ali, 1963) and several species of marine fishes (Blaxter and Jones, 1967; Blaxter, 1968; Blaxter and Staines, 1970). Retinal development in the guppy (Müller, 1952) and the brown trout (Lyall, 1957) has been studied. Although these investigations did not deal with the ontogeny of the retinomotor responses specifically, it is clear from the details presented that in these two fishes also there is a parallel between structural and functional development of the retina. In the salmonids, neither the embryo nor the alevin possesses differentiated rods and the retina appears altogether different from that of an older stage. It is incapable of responding to light. As the various layers of the retina develop and attain normal proportions (Figs. 11, 12) the retinomotor responses become evident. The morphometric relationships also vary from one stage to another (Ali, 1963) and in specimens of the same age (Ali, Copes and Stevenson, 1961). As the young fish grows the retina and its layers do not increase in thickness, as a function of length, eye diameter etc., but tend to decrease due to a stretching of the retina, which also brings about a greater dispersion of the components such as visual cells and the ganglion cells (Lyall, 1957; Ali, 1964a). As mentioned earlier, in many marine fishes the larvae also have an incompletely developed retina which lacks rods and is incapable of retinomotor responses (Blaxter and Jones, 1967; Blaxter, 1968; Blaxter and Staines, 1970). Retinomotor responses appear later concomitantly with the fuller development of the retina and the metamorphosis of the larvae (Ali and Wagner, this volume).

EFFECTOR MECHANISMS

How are these movements brought about? Since their discovery a great deal has been written about them and quite a few explantions proposed. However, as Arey (1915) commented "Many such explanations reveal the resourcefulness of the human mind rather than the ingenuity of nature". But imaginative theories do stimulate thinking and serve a purpose by bringing the possible processes and interrelations to the surface. The various hypotheses proposed have not been fully satisfactory because of the variation in the responses to light among vertebrates. Hormonal and nervous factors have often been suggested (Arey, 1915; 1916; 1919; 1928; Vilter, 1942; 1946; Studnitz, 1952). The suggestion that pH is an important factor (Studnitz, 1937), has received support from some recent work Hanawa, 1956; Hanawa, Tanaka and Nakano, 1959; Hongo, Hara, Sekoguti, Hara, Kato and Oki, 1958). However, none of these investigations has really discovered the mechanism or mechanisms underlying the phenomenon. How do the pigment granules move within the epithelial cell processes? How do the visual cells elongate and contract? Why do the rods and cones react oppositely to the same stimulus? These are the questions which need to be answered. The most logical way is to tackle the matter step by step in an analytical way. First, the responses of the REP and the visual cells should be studied together. After that attempts should be made to ascertain their inter-relationships. The subsequent step would be to inhibit one or the other of these. Such studies have to be undertaken employing histological, photometric, biochemical, histochemical and electron-microscopical techniques. I shall discuss here the results of some investigations using the histological and photometric techniques and in another paper Pierre Couillard (this volume) will discuss the others which deal more directly with contractility.

1. Histophysiology. Since the discovery of the retinomotor responses many workers have used enucleated eyes in their investigations of this phenomenon. While the results are generally in agreement in the case of the frog they are not so in the case of the fishes! A review of the results of investigations with the fishes' eyes failed to give me a coherent picture with regard to the capacities of their enucleated eyes to respond to photic stimulation (Ali, 1964b). Garten (1907) thought that there was a migration of the REP and cones in the enucleated eyes of *Leuciscus rutilus* after they were placed in the fixative but Chiarini (1904) and Wunder (1926) thought there was not. The REP and cones in the enucleated eyes of *Ameiurus* (*Ictalurus*) *nebulosus* were observed to respond to light but not to dark, while those of *Abramis crysoleucas*, *Fundulus heteroclitus* and *Carassius auratus* responded neither to light nor to darkness (Arey, 1916). There was no response in the eyes of *Gobio fluviatilis*, *Gasterosteus aculeatus*, *G. pungitius*, *Tinca vulgaris* and *Barbus fluviatilis* left in hemisections of the heads (Wunder, 1926). Bayliss, Lythgoe

and Tansley (1936) observed no responses to light in some dark-adapted, decapitated marine fishes. Wigger, (1937) found that the enucleated eyes of *Carassius auratus, Perca fluviatilis* and *Scardinius erythrophthalmus* manifested retinomotor responses to both light and darkness. However, he found that the retinas in the enucleated eyes did not assume the completely dark-adapted state. He ascribed this to the accumulation of the products of bleaching of the visual purple in the retina due to the slowing down of the regeneration process. Further, enucleated eyes, usually of frogs, have been the subject of many electro-physiological investigations, (Müller-Limmroth, 1959) and a survey of this work reveals that the results are not always in agreement. Further it was of interest to know to what extent the interruption of blood and nerve supply to the eye had affected these results. It has been found (Hanyu and Ali, 1963; 1964) that though the eyes *in situ* of goldfish and salmon are capable of responding for even an hour, the enucleated eyes of goldfish are able to respond to flickering light for only 10 to 15 minutes while those of the salmon failed to respond even for a few minutes after enucleation. In view of this, carefully controlled experiments were undertaken with fishes from different habitats. In the Atlantic salmon, a very active and demanding fish, which lives in clear and well oxygenated waters and whose metabolic demands are quite high, the enucleated eye is incapable of retinomotor responses even in well aerated Ringer's (Fig. 18). The cones are in a relaxed state in the eyes exposed to light as well as to darkness and the pigment granules in an intermediate state. It would appear that the cells are not alive. On the other hand in fishes which are less demanding from the point of view of metabolism etc., the retina in the enucleated eyes retains the capacity for retinomotor responses. These responses are not identical to those of these eyes *in situ* but they are closer in the case of the bullhead (Fig. 19) than in the case of the goldfish (Fig. 20). It may be recalled that the catfish is a less active fish living in stagnant waters which are not well oxygenated.

The capacity of the visual cells to move albeit not as much as in the eyes *in situ*, shows that these are direct responses to light. It has also been shown that exposure of some part of the body other than the eye does not bring about a movement of the retinal elements (Fujita, 1911; Arey, 1916; Ali, 1964d).

Since the retinomotor responses are not complete even in the eyes enucleated from the catfish, it would appear that humoral and nervous interactions may exist. This has been investigated by experimenting with unilateral illumination. The unexposed eye is not in a perfect state of dark-adaptation showing that these interactions indeed may exist (Fig. 21). The REP of the non-exposed eye responds somewhat while the cones practically do not which suggests that hormonal interaction may be significant since the REP is to a

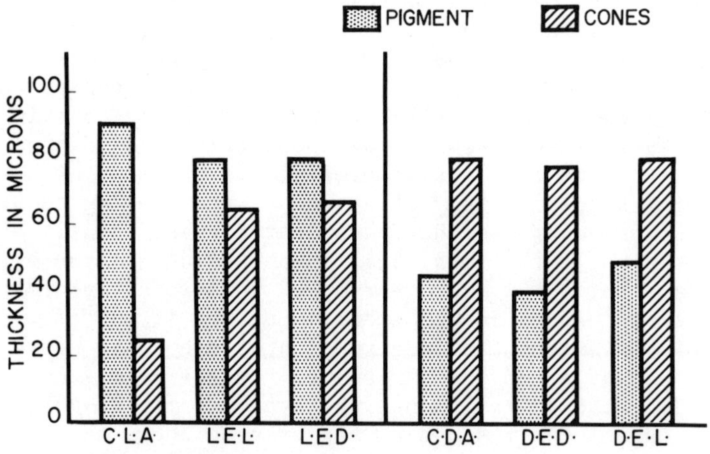

FIG. 18. Histogram showing the mean thicknesses of the retinal epithelial pigment and cone layers from the retinas of Atlantic salmon (*Salmo salar*) used in the experiments: C.L.A. - Control light-adapted; L.E.L. - Eye enucleated from a light-adapted fish and left in light; L.E.D. - Eye enucleated from a light-adapted fish and subjected to darkness; C.D.A. - Control dark-adapted. D.E.D. - Eye enucleated from a dark-adapted fish and left in darkness; D.E.L. - Eye enucleated from dark-adapted fish and exposed to light. After Ali (1962a).

greater extent under hormonal control than are the visual cells which are first order neurons. The results of various investigations are generally in agreement on this aspect (Scharrer, 1930; Wigger, 1937; Vilter and Thibault, 1948; Ali, 1964c).

Peyraud (1965; 1966) suggests a retinal compensatory phenomenon in unilateral illuminations. He thinks that central relay mechanisms participate in retinomotor responses. By studying the effects of unilateral or bilateral cutting of the optic nerve, it was proposed (Bimes, Guilhem, Peyraud and Serfaty, 1966a; 1966b) that the movement of the REP is governed primarily by the contra-lateral nerve reflex and secondarily by a humoral mechanism. They felt that the covered eye emits a humoral message to the exposed one. Bimes et al. (1966b) showed this by placing a light-adapted, enucleated eye of the carp in a Ringer's solution in which a dark-adapted eye had been kept

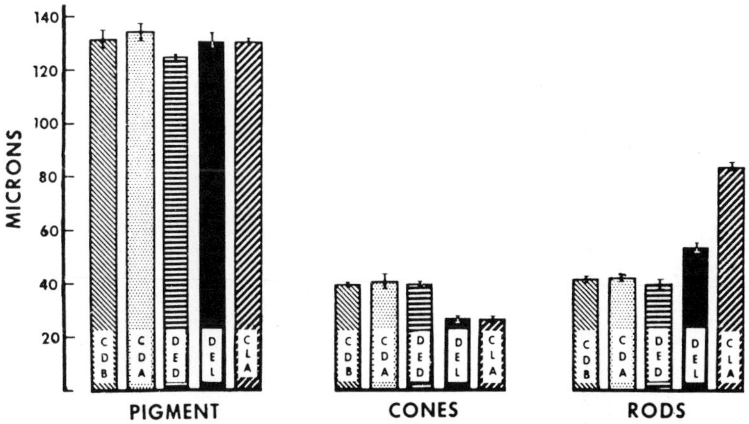

FIG. 19. Histogram showing the thickness of the retinal epithelial pigment layer and the lengths of cones and rods (from the external limiting membrane) in the eyes of brown bullheads (*Ictalurus nebulosus*) sampled under various conditions. C.D.B.: Control dark-adapted sampled before the commencement of the experiment. C.D.S.: Control dark-adapted sampled after the experiment. D.E.D.: Enucleated from dark-adapted fish and left in dark for 1 1/2hr. D.E.L.: Enucleated from dark-adapted fish and exposed to light for 1 1/2hr. C.L.A.: Control light-adapted. The vertical bars indicate standard errors. After Ali (1964b).

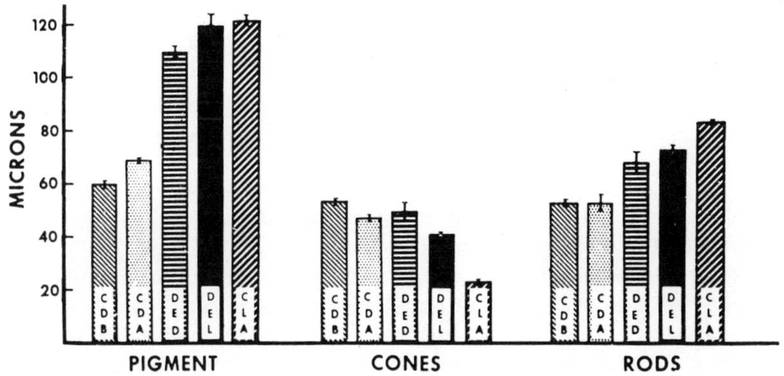

FIG. 20. Histogram showing the thickness of retinal epithelial pigment layer and the lengths of cones and rods (from the external limiting membrane) in the eyes of goldfish (*Carassius auratus*) sampled under various conditions. Abbreviations as in Fig. 19. The vertical bars represent standard errors. After Ali (1964b).

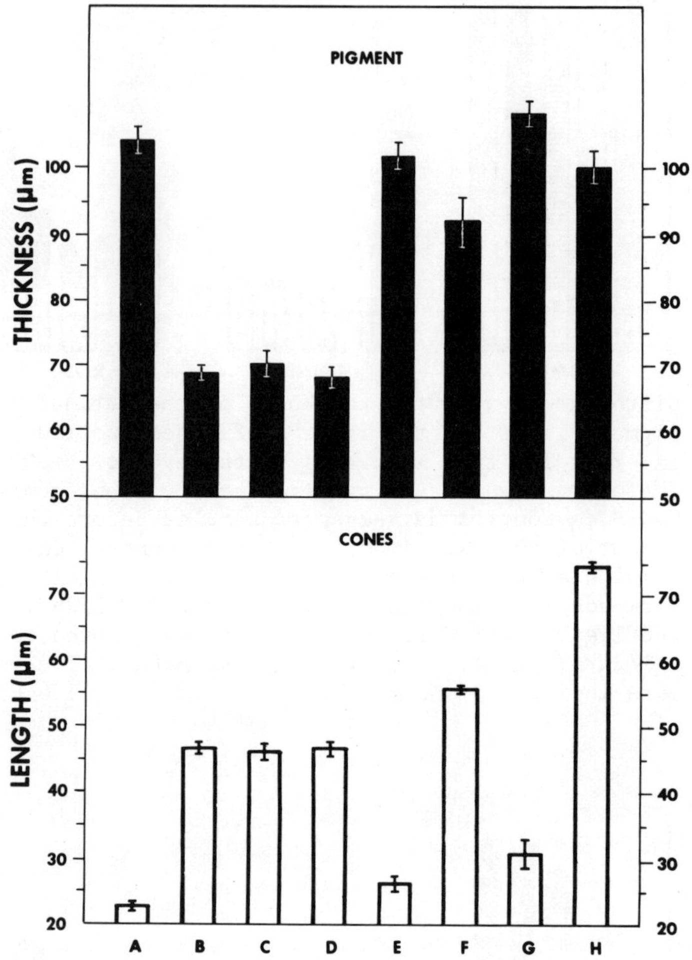

FIG. 21. Histogram showing the mean thicknesses and lengths of the pigment layer and cones in the retina of goldfish (*Carassius auratus)*. The vertical bars indicate standard errors of the means. A, control light-adapted. B, control dark-adapted. C, light-adapted fish subjected to the experimental procedure except for the exposure of the eye. E, exposed eye of previously light-adapted fish. F, unexposed eye of the previously light-adapted fish. G, exposed eye of the previously dark-adapted fish. H. unexposed eye of the previously dark-adapted fish.

previously. Apparently this influence of the dark-adapted eye diminished after having had a growing effect which increased during the first 30 min. In this aspect also, carefully controlled further work with a number of species from different habitats is warranted.

2. Photometry. One of the characteristic features of any light induced phenomenon is the spectral response curve or action spectrum. It is also useful to know the energy threshold of a response. A knowledge of this value permits the comparison of a photoresponse with others on an absolute scale of universal significance. With this aim in mind one could try to obtain quantal thresholds and action spectra of retinal elements which are capable of retinomotor responses. Ali and Crouzy (1968) undertook such a study and showed that the action spectrum of REP expansion is fairly similar to that of the scotopic pigment's absorption spectrum in the brook trout (Fig. 7). The maximum of this action spectrum corresponds to a radiant energy per unit area (B_{max}) of about two microwatts per steradian per square metre. Assuming that the rod pigment triggers the expansion of the REP, one may apply the values obtained in another investigation (Crouzy and Ali, 1965) to the absorption of the ocular media and to the absorption of the brook trout's scotopic pigment *in situ*. From the value for B_{max} and assuming the visual field of the trout to be 2π steradians it can be calculated that the REP expansion is triggered by an absorption of about 7000 quanta per second per square degree of the field by the rod pigment. It is also interesting that in two elasmobranchs (*Mustelus mustelus* and *Scyliorhinus canicula)* the action spectrum for pupillary contraction corresponds to the absorption curve of rhodopsin (Kuchnow, 1970). It seems then that the isomerisation of the visual pigment molecule triggers off other chemical, electrical and mechanical phenomena (Ali, 1972). The question arises as to how the absorption of light by the rod pigment triggers off the movement of the REP. Is the rod pigment also related to the contraction of the REP? Is it also related to the movement of the cones and rods? Weidemann's (1966) observations, referred to earlier, may be recalled. What is the relation between these movements and the cone pigments? Three types of cone are present in some fishes (Müller, 1952; 1954; Weidemann, 1966; Eberle, 1968) and three cone pigments have been recorded from the goldfish (Marks, 1965; Harosi, this volume).

Weidemann's (1966) observation raises an interesting point. Why is the elongation of the cones related to the rod pigment's absorption spectrum? Since interaction between rods and cones and inhibition at the bipolar cell level may be ruled out in the case of the goldfish (Stell, 1967; 1972), two suggestions may be made:

1. Since Weidemann used a fairly low intensity (170 metre-candles), the REP may not have been fully dispersed. This is known to happen as has been mentioned above (Ali, 1962b). Thus, green (521 nm) being the wavelength to which the rods are most sensitive,

a greater quantity of their pigment would have been decomposed by this wavelength than by the others. When the light was turned off, the reaction of the rods in the retinas of fishes exposed to green would have been slower which would have slowed down the elongation of the cones. A parallel between rod and cone movements is known (Ali, 1964b).

2. The cone pigments of goldfish have a maximum around 455 nm, 530 nm and 625 nm. The mean of these is 537 nm and the wavelength closest to 521 nm among those used by Weidemann (1966). It is probable, of course, that the cone pigments are not the same in the guppy just as variations occur in the case of rod pigments (Dartnall and Lythgoe, 1965). Even if they were the same, their proportions probably will not be the same. In any case when the entire eye is exposed to one wavelength, the total response would reflect the mean of the responses of all the photoreceptors as happens in the case of the absorption curves of scotopic pigment mixtures (Crouzy and Ali, 1965; Ali, Hubbard and Wald, 1970) or the spectral sensitivity (Muntz and Northmore, 1970; Northmore and Muntz, 1970). Thus if the entire cone population is more sensitive to 521 nm than to the other wavelengths used, the effect of 521 nm will be greater, as if its intensity were relatively greater. It has been mentioned elsewhere that the effect of a higher light intensity lasts longer (after it is turned off) on the expansion of cones than a lower light intensity (Ali, 1962b) (Fig. 6).

Even in the case of the correlation between the dispersion of the REP and the absorption curve of the rod pigment an explanation may be possible. When a fish, such as the trout, is in darkness, its rod outer segments are not masked by the REP. If these are exposed to increasing intensities of light, at a given point the energy will be too great for the rods and this may be about the same as the cone thresholds. It is at this intensity that the transition from scotopic to photopic vision takes place. So it is around this energy that the REP commences to disperse and if the rods are more sensitive to 500 nm or to 520 nm than to 450 nm or to 650 nm it will require less energy at 500 nm or 520 nm to trigger the movement of the REP or the contraction of the cones. Therefore, the end result will be a seeming correlation between the rod pigment absorption spectrum, their movement, the dispersion of the REP and, even the contraction of the cones. It will be worth while to undertake detailed comparative studies of these aspects using fishes having various mixtures of scotopic and photopic pigments.

SIGNIFICANCE TO VISION

The significance of these responses from the point of view of habitat, behaviour, etc. has already been discussed and here I shall try to examine their significance to some aspects of the visual process and scrutinise the theories which have been proposed by various authors.

It should be interesting to see what, if any, relationship exists between the retinomotor and electroretinographic responses; for, the significance of the latter to the visual process cannot be overemphasised (Brown, 1968). In at least one investigation with fish (Gramoni and Ali, 1970) two points of similarities have been found. In both responses the intensity of the pre-adapting light influences the rate of subsequent dark-adaptation (Mühlmann, 1967). Both commence later and take a longer period for completion, the stronger the pre-adapting light has been. Secondly, in both types of response, light-adaptation is more rapid than dark-adaptation. Apart from these two general similarities no other direct correlation between ERG responses and retinomotor responses has been established. This is also an aspect which merits further investigation.

The most comprehensive explanation of the significance of the REP migration is one of the first (Garten, 1907) given. Garten thought that the function of the expanded REP was to absorb the light escaping from the visual cells by refraction. Were it not for the optical isolation of the cells by the REP a great deal of light would be scattered in all directions due to the large ellipsoids and oil droplets (in amphibians, reptiles and birds) and would stimulate neighbouring rods and cones. He also argued that in the case of long slender rods total internal reflexion prevented this dispersion. Results of some recent investigations with deep-sea fish supports this view. In fact, in pure rod retinas whose rods are slender (e.g. deep-sea fishes) REP is sparse. Histophysiological and electrophysiological results also provide much evidence in support of this theory. For example, the electroretinographic flicker fusion frequency is higher in the albino brook trout than in the pigmented one (Ali and Kobayashi, 1968a, b). Further, the latencies of the ERG components are shorter and the amplitudes higher in the albino than in the pigmented fish. It is clear that the lack of melanin in the former increases the amount of light available to the visual cells and produces a dazzling effect at higher light intensities. It has also been noted that in fishes with a mixed retina and in which there is reflecting material in the retinal epithelium, the REP masks the reflecting material when the light is bright (Ali and Anctil, 1968; and unpublished observations). It is of interest in this connexion that even in some diurnal lizards and birds, which have pure cone retinas, the REP insulates the cones from each other in the light-adapted state (Walls, 1942; Ali, 1960). In the albino rat, which has an almost pure rod retina, the minimum separable angle is twice that in its pigmented relative although the REP in the latter is incapable of dispersion in light. Garten (1907) also found that in the bream (*Abramis brama*) cones did not elongate in dim light. This happens in some other fishes such as *Stizostedion* also and calls for another explanation (Zyznar and Ali, 1975, and Zyznar, this volume).

Walls (1928) suggested that the retinomotor and pupillary responses were mutually exclusive except in birds. According to this view the pupillary reaction is a more efficient and more advanced mechanism while the retinomotor responses are primitive. This is not entirely acceptable since pupillary responses also occur in some fishes (e.g. elasmobranchs), and the bird's eye is not all that weird as he had thought. Walls (1942) thus favoured the hypothesis of Exner and Januschke (1906) which is that the retinomotor changes are basically a mechanism for adapting the eye to scotopic and photopic vision. They also discussed the role of the tapetum and this aspect is dealt with by Nicol and by Zyznar (this volume). Studnitz (1952) felt the retinomotor responses were a means of fully exploiting the available space in the retina. Thus in the dark the cones move away sclerad, away from the external limiting membrane, leaving the space to the rods which now place themselves directly in the path of the incoming light without its having to traverse the cone ellipsoids and outer segments first. In the light, the rods elongate and have their outer segments masked by the dispersed REP while the cones, due to the contraction of their myoids, are situated near the external limiting membrane. This exploitation of the available space may have a significance in the understanding of the mechanism of the movements of the rods and cones, (Couillard, this volume).

The masking of the rods by the REP and the consequent protection of their pigment from bleaching by bright light is in fact found to be true. The relation between the thickness of the REP and the light intensity has been discussed above. The lack of this screening pigment changes the form of the photopic spectral sensitivity curve of the albino trout as compared to that of its pigmented relative (Kobayashi and Ali, 1971). This may be at least partly due to the rods not being masked in the light-adapted state. The fact that the REP is capable of migration only after the appearance of rods during development (Ali, 1959; Blaxter and Staines, 1970) adds further support to this hypothesis. The relationship between rhodopsin concentration and the position of the REP in frogs (Bäck, Donner and Reuter, 1965) exposed to light also shows that the REP reduces the effect of light by about a third. Essentially the same observations have been made in the case of the squid (Hagins and Liebman, 1962) and moth and wasp (Hoglund and Struwe, 1970; Hoglund, Langer, Struwe and Thorrell, 1970).

It is of historical interest to summarise some of the other theories which had been proposed. Kühne (1878) suggested that the decomposition of the REP by light initiated the sensation of vision. Boll (1881) and Gad (1894) had the same thought. This is hard to accept since there is no evidence that the REP is decomposed by light. Ayres and Kühne (1878) related the REP to the regeneration of the visual purple. They considered the retinal epithelium to be a gland which secreted the visual pigment. This also obviously is not acceptable. Kolmer (1909) held essentially a similar view. Their ideas may have been inspired by the observation of the bleaching of some outer segment particles found in the retinal epithelium either due to breakage or phagocytosis. Later (1879) Kühne found a parallel between the time for regeneration of the visual pigment and the time for REP contraction. Modern work (Bäck, Donner and Reuter, 1965) supports this possibility.

Chiarini (1904) thought that the function of the dispersed REP was nutritive and that it compensated for the losses by rods and cones. This may be indirectly valid but more work on the epithelial and visual cell relationship is needed. Herzog (1905) felt that the migration of the REP stored up energy and made available "dark heat waves" which accelerated the chemical processes in the visual cells! He also thought that the retinomotor responses provided evidence supporting the duplexity theory by adapting the eye for scotopic and photopic vision. This is essentially what Exner and Januschke (1906) also thought.

The wide disagreements and disparities seen in the mass of theoretical consideration surveyed in these pages, indicate that the greatly varied retinal structure, habitats and behaviours are too complex for formulation of any all inclusive theory.

REFERENCES

Ali, M.A. (1958). The ocular structure, retinomotor and photo-behavioural responses of juvenile Pacific salmon. Ph. D. Thesis, University of British Columbia, 102 p.

Ali, M.A. (1959). The ocular structure, retinomotor and photo-behavioral responses of juvenile Pacific salmon. Can. J. Zool. 37: 965-996.

Ali, M.A. (1960). The effect of temperature on the juvenile Sockeye salmon retina. Can. J. Zool. 28: 169-171.

Ali, M.A. (1961a). Histophysiological studies on the juvenile Atlantic salmon (*Salmo salar*) retina. II. Responses to light intensities, wavelengths, temperatures, and continuous light or dark. Can. J. Zool. 39: 511-526.

Ali, M.A. (1961b). The correlation on photobehavioural and retino-motor responses in the Pacific salmon. In: Progress in Photobiology, Proc. 3rd Intl. Congr. Photobiol., edited by Christensen, B.C. and Buchmann, B., Elsevier, Amsterdam, 415-417.

Ali, M.A. (1962a). Retinal responses in enucleated eyes of Atlantic salmon (*Salmo salar*). Rev. Can. Biol. 21: 7-15.

Ali, M.A. (1962b). Influence of light intensity on retinal adaptation in Atlantic salmon (*Salmo salar*) yearlings. Can.J.Zool. 40: 561-570.
Ali, M.A. (1962c). Light preference of control, light- or dark-adapted sockeye salmon (*Oncorhynchus nerka*) juveniles. Am. Zool. 2: 385.
Ali, M.A. (1963). Correlation of some retinal and morphological measurements from the Atlantic Salmon (*Salmo salar*). Growth 27: 57-76.
Ali, M.A. (1964a). Stretching of retina during growth of salmon (*Salmo salar*). Growth 28: 83-89.
Ali, M.A. (1964b). Retinomotor responses in the enucleated eyes of the brown bullhead (*Ictalurus nebulosus*) and the goldfish (*Carassius auratus*). Rev.Can.Biol. 23: 55-66.
Ali, M.A. (1964c). Retinomotor responses of the goldfish (*Carrassius auratus*) to unilateral photic stimulation. Rev.Can.Biol. 23: 45-53.
Ali, M.A. (1964d). Retinomotor responses and skin illumination in trout (*Salmo irideus*) and goldfish (*Carassius auratus*). Am. Zool. 4: 419.
Ali, M.A. (1964e). Diurnal rhythm in the rates of oxygen consumption, locomotor and feeding activity of yearling Atlantic salmon (*Salmo salar*) under various light conditions. Proc.Indian Acad. Sci.60: 249-263.
Ali, M.A. (1964f). Über den Einfluss der Temperatur auf die Geschwindigkeit der retinomotorischen Reaktionen des Lachses (*Salmo salar*). Naturwissenschaften 51: 471.
Ali, M.A. (1964g). Retinomotor responses and skin illumination in trout (*Salmo irideus*) and goldfish (*Carrassius auratus*). Am. Zool. 4: 419.
Ali, M.A. (1971). Les réponses rétinomotrices: caractères et mécanismes. Vision Res. 11: 1225-1288.
Ali, M.A. (1972). Action spectra of retinomotor and pupillary responses. Vision Res. 12: 1199.
Ali, M.A. et Anctil, M. (1968). Corrélation entre la structure rétinienne et l'habitat chez *Stizostedion vitreum vitreum* et *S. canadense*. J. Fish Res. Board Can. 25: 2001-2003.
Ali, M.A., Anctil, M. et Mohideen, H.M. (1968). Structure rétinienne et la vascularisation intraoculaire chez quelques poissons marins de la région de Gaspé. Can.J.Zool. 46: 729-745.
Ali, M.A., Copes, P. and Stevenson, W.R. (1961). Correlation of morphological and intra-ocular measurements in the Atlantic salmon (*Salmo salar*) yearling. J. Fish Res. Board Can. 18: 259-272.
Ali, M.A., M. Anctil et N. Raymond (1973) La rétine de quelques poissons marins du littoral brésilien. Revista Biol. 9: 101-114.
Ali, M.A. and Crouzy, R. (1968). Action spectrum and quantal thresholds of retinomotor responses in the brook trout, *Salvelinus fontinalis* Mitchill). Z.vergl.Physiol. 59: 86-89.

Ali, M.A. and Hanyu, I. (1963). A comparative study of retinal structure in some fishes from moderately deep waters of the Western North Atlantic. Can.J.Zool. 41: 225-241.

Ali, M.A. and Hoar, W.S. (1959). Retinal responses of pink salmon associated with its downstream migration. Nature 184: 106-107.

Ali, M.A., Hubbard, R. and Wald, G. (1970). Scotopic visual pigments in the albino brook trout, *Salvelinus fontinalis*. Rev. Can. Biol. 29: 153-156.

Ali, M.A. and H. Kobayashi (1967). Temperature: influence on the electroretinogram-flicker fusion frequency of the sunfish (*Lepomis gibbosus* L.). Rev. Can. Biol. 26: 341-345.

Ali, M.A. and Kobayashi, H. (1968a). Electroretinogram of albino and pigmented brook trout *Salvelinus fontinalis* (Mitchill). Rev. Can. Biol. 27: 145-161.

Ali, M.A. and Kobayashi, H. (1968b). Electroretinogram-flicker fusion frequency in albino trout. Experientia 24: 454.

Ali, M.A., Stevenson, W.R. and Press, J.S. (1961). Histophysiological studies on the juvenile Atlantic salmon (*Salmo salar*) retina. I. Rates of light- and dark-adaptation. Can.J.Zool. 39: 123-128.

Anctil, M. (1968). Intraocular vascular supply in some marine teleosts. Rev.Can.Biol. 27: 347-355.

Anctil, M. (1969). Structure de la rétine chez quelques téléostéens marins du plateau continental. J.Fish.Res.Board Canada 26: 597-628.

Angelucci, A. (1884). Una nuova theoria sulla visione. Acad.Med.d. Roma. sed. 14 Juglio. (from Detwiler, 1943).

Arcoleo, E. (1890). Osservazioni sperimentali sugli elementi contrattili della retina negli animali a sangue freddo. Ann. d'Ottalmologie 19: Fasc 3 and 4, p. 253-262.

Arey, L.B. (1915). The occurrence and significance of photochemical changes in the vertebrate retina. An historical survey. J. Comp. Neurol. 25: 535-554.

Arey, B. (1916). The movements in the visual cells and retinal pigment of the lower vertebrates. J. Comp. Neurol. 26: 121-201.

Arey, L.B. (1919). A retinal mechanism of efficient vision. J.Comp. Neurol. 30: 343-353.

Arey, L.B. (1928). Visual cells and retinal pigment. In: Special cytology, Sec. 25, edited by Cowdry. v. 2, p. 887-926. Hoeber, New York.

Ayres, W. and W. Kühne (1878). Ueber Regeneration des Sehpurpurs beim Säugethiere. Untersuch. aus d. Physiol. Inst. d. Univ. Heidelberg 2: 215-240.

Bäck, I, Donner, K.O. and Reuter, T. (1965). The screening effect of the pigment epithelium on the retinal rods in the frog. Vision Res. 5: 101-111.

Bayliss, L.E., Lythgoe, R.J. and Tansley, K. (1936). Some new forms of visual purple found in sea fishes with a note on the visual cells of origin. Proc.R.Soc.Lond.Ser.B. 120: 95-113.

Beatty, D.D. (1966). A study of the succession of visual pigments in Pacific salmon (*Oncorhynchus*). Can. J. Zool. 44: 429-455.

Bimes, C., Guilhem, A., Peyraud, C. et Serfaty, A. (1966a). Mise en évidence d'un mécanisme humoral dans la dynamique rétinienne d'un poisson téléostéen: la carpe commune (*Cyprinus carpio L.*) C.R. Séances Soc. Biol. 160: 2470-2473.

Bimes, C., Guilhem, A., Peyraud, C. et Serfaty, A. (1966b). Mise en évidence chez un poisson téléostéen: la carpe (*Cyprinus carpio L.*) de l'influence d'un broyat de rétine adaptée à l'obscurité sur la dynamique, in vitro, des photorécepteurs d'un oeil exposé à la lumière. C.R. Séances Acad. Sc. 263: 668-670

Blaxter, J.H.S. and Jones, M.P. (1967). The development of the retina and retinomotor responses in the Herring. J. Mar. Biol. Ass. U.K. 47: 677-697.

Blaxter, J.H.S. (1968). Light intensity, vision and feeding in young plaice. J. Exp. Mar. Biol. Ecol. 2: 293-307.

Blaxter, J.H.S. (1969). Development: eggs and larvae. In: Fish Physiology, V. 3, Chap. 4, edited by Hoar, W.S. and Randall, D.J. Academic Press, New York, p. 177-252.

Blaxter, J.H.S. (1970). Light: Fishes. In: Marine ecology, Vol. 1, Part 1 (Environmental factors). Edited by Kinne, O., Wiley-Interscience, London, p. 213-320.

Blaxter, J.H.S. and Staines, M. (1970). Pure-cone retinae and retinomotor responses in larval teleosts. J. Mar. Biol. Ass. U.K. 50: 449-460.

Boll, F. (1877). Zur Anatomie und Physiologie der Retina. Arch. Anat. Physiol. 4: 783-787.

Boll, F. (1881). Thesen und Hypothesen zur Licht und Farbenempfindung. Arch. Anat. Physiol.: 5-1.

Braemer, W. (1957). Verhaltensphysiologische Untersuchungen am optischen Apparat bei Fischen. Z. vergl. Physiol. 39: 374-398.

Brett, J.R. and Ali, M.A. (1958). Some observations on the structure and photomechanical responses of the Pacific salmon retina. J. Fish. Res. Board Canada 15: 815-829.

Brown, K.T. (1968). The electroretinogram: its components and their origins. Vision Res. 8: 633-677.

Chiarini, P. (1904). Changements morphologiques que l'on observe dans la rétine des vertébrés par l'action de la lumière et de l'obscurité. Arch. Ital. Biol. 42: 303-322.

Crouzy, R. and Ali, M.A. (1965). Données photométriques spectrales sur l'oeil de la truite albinos et sur son pigment scotopique. C.R. Acad. Sc. Paris. 261: 4509-4512.

Crouzy, R. et Ali, M.A. (1966). Relation entre la sensibilité électrorétinographique et le spectre d'absorption du pigment visuel scotopique chez le poisson rouge. Nombre minimum de quanta absorbés. Bull. Mus. Nat. Hist. Naturelle, 2e série, 38: 730-743.

Czerny, V. (1867). Ueber Blendung der Netzhaut durch Sonnenlicht. Stzungen Akad. Wiss. Wien, Math. Naturwiss. Kl., 56: 409-428.

Dartnall, H.J.A. and Lythgoe, J.N. (1965). The clustering of fish visual pigments around discrete spectral positions and its bearing on chemical structure. In: "Colour Vision", edited by de Reuck, A.V.S. and Knight, J. Churchill, London, p. 3-26.

Detwiler, S.R. (1943). Vertebrate photoreceptors. MacMillan, New York.

Eberle, H. (1968). Zapfenbau, Zapfenlange und chromatische Aberration im Auge von *Lebistes reticulatus* Peters (Guppy). Zool. Jb. Physiol. 74: 121-154.

Engelmann, T.W. (1885). Über Bewegungen der Zapfen und Pigmentzellen der Netzhaut unter dem Einfluss des Lichtes und des Nervensystems. Arch. ges. Physiol. 35: 498-508.

Engström, K. (1963a). Studies on teleostean visual cells. Ake Nyblom & Co. Stockholm. 16 p.

Engström, K. (1963b). Cone types and cone arrangements in teleost retinae. Acta Zool. 44: 179 243.

Engström, K. and Rosstorp, E. (1963). Photomechanical responses in different cone types of *Leuciscus rutilus*. Acta Zool. 19: 145-160.

Exner, S. and Januschke, H. (1906). Die Stäbchenwanderung im Auge von *Abramis brama* bei Lichtveränderungen. Ber. d.K.K. Akad. d. Wiss. zu Wien, Math.-Natur. Kl. 115: 269-280.

Fujita, H. (1911). Pigmentbewegung und Zapfenkontraktion im Dunkelauge des Frosches bei Einwirkung verschiedener Reize. Arch. vergl. Ophthalmol. 2: 164-179.

Gad, J. (1894). Der Energieumsatz in der Retina. Arch. Anat. Physiol. (Physiol. Abt.), p. 491-501.

Garten, S. (1907). Die Veränderungen der Netzhaut durch Licht. Graefe-Saemisch Handbuch der gesamten Augenheilkunde, Leipzig, V. 2, chapt. 12, p. 250-380.

Genderen-Stort, A.G.H. van (1886). Über Form und Ortsveränderung der Elemente in der Sehzellenschicht nach Beleuchtung. Bericht 18. Versamm. deutsche Ophthalmol. Gesell. Heidelberg, p. 43-49.

Girsa, I.I. (1967). Specific retinomotor reaction features of the retina of the eye when fish are being lured by light (Russian). Vopr. Ikhtiol. 7: 540-546.

Gramoni, R. and Ali, M.A. (1970). Changes in the ERG components of the landlocked salmon (*Salmo salar ouananiche*, McCarthy) during dark- and light-adaptation. Rev. Can. Biol. 29: 35-47.

Hagins, W.A. and Liebman, P.A. (1962). Light-induced pigment migration in the squid retina. Biol. Bull. 123- 498.

Hanawa, I. (1956). Photo-metabolism of the isolated outer-segments of rods. Jap. J. Physiol. 6: 218-225.

Hanawa, I., Tanaka, S. and Nakano, H. (1959). Free amino acids in the choroid and effects of some metabolites on the retinal pigment-migration. Jap. J. Physiol. 9: 56-62.

Hanyu, I. (1962). Intraocular vascularization in some fishes. Can. J. Zool. 40: 87-106.

Hanyu, I. and Ali, M.A. (1962). Intra-sub-specific variation in retinal structure in *Sebastes marinus mentella*. Nature, 196: 554-556.
Hanyu, I. and Ali, M.A. (1963). Flicker fusion frequency of electroretinogram in light-adapted goldfish at various temperatures. Science 140: (3567): 662-663.
Hanyu, I. and Ali, M.A. (1964). Electroretinogram and its flicker fusion frequency at different temperatures in light-adapted salmon (*Salmo salar*). J. Cell. Comp. Physiol. 63: 309-322.
Hárosi, F.I. and MacNichol, E.F. (1974). Visual pigments of goldfish cones. Spectral properties and dichroism. J. Gen. Phys. 63: 279-304.
Herzog, H. (1905). Experimentelle Untersuchungen zur Physiologie der Bewegungsvorgänge in der Netzhaut. Arch. Anat. Physiol. (Physiol. Abt.) 5/6: 413-464.
Hoglund, G. and Struwe, G. (1970). Pigment migration and spectral sensitivity in the compound eye of moths. Z. vergl. Physiol. 67: 229-237.
Hoglund, G., Langer, H., Struwe, G. and Thorell, B. (1970). Spectral absorption by screening pigment granules in the compound eyes of a moth and a wasp. Z. vergl. Physiol. 67: 238-242.
Holst, E. von (1935). Über den Lichtrückenreflex bei Fischen. Pubbl. Staz. Zool. Napoli 15: 143-158.
Hongo, I., Hara, T., Sekoguti, Y., Hara, R., Kato, M. and Oki, I. (1958). Studies on the mechanism of the retinal photoreception by the use of phosphorus-32. Proc. 1st. UNESCO Intl. Conf. Sc. Res. 4: 90-99.
Hongo, I. (1935). Die Wirkung monochromatischen Lichtes auf die motorischen Elemente der Knochenfischnetzhaut. Z. vergl. Physiol. 22: 293-297.
John, K.R. and Haut, M. (1964). Retinomotor cycles and correlated behavior in the teleost *Astyanax mexicanus* (Fillipi). J. Fish. Res. Board Can. 21: 591-595.
Keenleyside, M.H.A. (1955). Some aspects of the schooling behaviour of fish. Behaviour 8: 183-248.
Kobayashi, H. (1957). Notes on retinomotor phenomena in some fishes under various light conditions. J. Shimonoseki Coll. Fish. 7: 169-177.
Kobayashi, H. and Ali, M.A. (1971). Electroretinographic determination of spectral sensitivity in albino and pigmented brook trout (*Salvelinus fontinalis*, Mitchell). Can. J. Physiol. Pharmacol. 49: 1030-1037.
Kolmer, W. (1909). Über einen sekretartigen Bestandteil der Stäbchen-Zapfenschicht der Wirbeltierretina. Vorläufige Mitteilung. Arch. f.d.ges.Physiol. 129: 35-45.
Kramer, E. and Ali, M.A. (1965). A new method of recording fish activity continuously over long periods. Am. Zool. 5: (2) 232.

Kuchnow, K.P. (1970). Threshold and action spectrum of the elasmobranch pupillary response. Vision Res. 10: 955-964.

Kühne, W. (1877). Über den Sehpurpur. Untersuch. Physiol. Inst. Univ. Heidelberg, 1: 235-337.

Kühne, W. (1878). Fortgesetzte Untersuchungen über die Retina und die Pigmente des Auges. Untersuch. Physiol. Inst. Univ. Heidelberg, 2: 89

Kühne, W. (1879). Chemische Vorgänge in der Netzhaut. Hermann's Handb. der Physiologie, Leipzig 3: 235- 337.

Lederer, R. (1908). Veränderungen an den Stäbchen der Froschnetzhaut unter Einwirkung von Licht und Dunkelheit. Centralbl. Physiol. 22(24): 762-764.

Liebman, P.A., Carroll, S. and Laties, A. (1969). Spectral sensitivity of retinal screening pigment migration in the frog. Vision Res. 9: 377-384.

Loukashkin, A.S. and Grant, N. (1959). Behavior and reactions of the Pacific sardine, *Sardinops caerula* (Girard), under the influence of white and colored lights and darkness. Proc. Calif. Acad. Sci. 29: 509-548.

Lyall, A.H. (1957). The growth of the trout retina. Q.J. Microsc. Sci. 98: 101-110.

Marks, W.B. (1965). Visual pigments of single goldfish cones. J. Physiol. 178: 14-32.

Morano, F. (1872). Die Pigmentschicht der Netzhaut. Arch. Mikros. Anat. 8: 4-81.

Mühlmann, D. (1967). Untersuchungen zum Ablauf der Dunkeladaptation bei Fischen. II. ERG beim Flussbarsch (*Perca fluviatilis*, L). Z. vergl. Physiol. 55: 134-144.

Müller, H. (1856). Anatomisch-physiologische Untersuchungen über die Retina bei Menschen und Wirbelthieren. Z. Wiss. Zool. 8: 1-122.

Müller, H. (1952). Bau und Wachstum der Netzhaut des Guppy (*Lebistes reticulatus*). Zool. Jahrb. Physiol. 63: 276-324.

Müller, H. (1954). Die Dundeladaptation beim Guppy (*Lebistes reticulatus* P.). Z. vergl. Physiol. 37: 1-18.

Müller-Limroth, W. (1959). Electrophysiology des Gesichtssinnes. Springer-Verlag, Berlin, p. 331.

Munk, O. (1959). Scientific results of the Danish deep-sea expedition round the world 1950-52. Galathea Report 3: 79-86.

Munk, O. (1964a). The eyes of three benthic deep-sea fishes caught at great depths. Galathea Report 7: 137- 149.

Munk, O. (1964b). The eyes of some ceratioid fishes. Dana Report 61.

Munk, O. (1966). Ocular anatomy of some deep-sea teleosts. Dana Report 70.

Munk, O. (1969). On the visual cells of some primitive fishes with particular regard to the classification of rods and cones. Vidensk. Meddr. Dansk.Naturh.Foren 132: 25-30.

Muntz, W.R.A. and Northmore, D.P.M. (1970). Vision and visual pigments in a fish, *Scardinus erythrophthalmus* (The rudd). Vision Res. 10: 281-298.

Nicol, J.A.C. (1963). Some aspects of photoreception and vision in fishes. Adv. Mar. Biol. 1: 171-208.

Nicol, J.A.C. and Zyznar, E. (1973). The Tapetum lucidum in the eye of the big-eye *Priacanthus arenatus* Cuvier. J. Fish. Biol. 5: 519-522.

Northmore, D.P.M. and Muntz, W.R.A. (1970). Electroretinogram determinations of spectral sensitivity in a teleost fish adapted to different daylengths. Vision Res. 10: 799-816.

Olla, L. and Marchioni, W. (1968). Rhythmic movements of cones in the retina of Bluefish, *Pomatomus saltatrix*, held in constant darkness. Biol. Bull. 135: 530-536.

Pergens, E. (1899). Über Vorgänge in der Netzhaut bei farbiger Beleuchtung gleicher Intensität. Z. Augenheilk. 2: 125-141.

Peyraud, C.M. (1965). Recherches sur la régulation des mouvements respiratoires chez quelques téléostéens: Analyse d'un réflexe opto-respiratoire. Thèse, Fac. des Sciences, Université de Toulouse, p. 258.

Peyraud, C.M. (1966). Contribution à l'étude de la dynamique des pigments et des photorécepteurs rétiniens chez la carpe (*Cyprinus carpio* L.). C.R. Acad. Sc. Paris 263: 65-67.

Pickford, G.E. and Atz, J.W. (1957). The physiology of the pituitary gland of fishes. New York Zoological Society, N.Y.

Scharrer, E. (1930). Über Hell - und Dunkelstellung im Fischauge bei einseitiger Belichtung. Z. vergl. Physiol. 11: 104-106.

Stell, W.K. (1967). The structure and relationships of horizontal cells and photoreceptor-bipolar synaptic complexes in the goldfish retina. Am. J. Anat. 120: 401-424.

Stell, W.K. (1972). The structure and morphologic relations of rods and cones in the retina of the spiny dogfish *Squalus*. Comp. Biochem. Physiol. 42A: 141-151.

Stell, W.K., Hárosi, F.I. and Lightfoot, D.O. (1974). Cones and the color-coding of cone-horizontal cell contacts in the goldfish retina. The Association for Research in Vision and Ophthalmology. Abst. 3, p. 47.

Studnitz, G. von (1937). Die retinale Säurebildung. Pflüg. Arch. ges Physiol. 238: 802-813.

Studnitz, G. von (1952). Mikroskopische Untersuchungen an der Netzhaut des Wirbeltierauges. Mikrokosmos 42: 58-62.

Studnitz, G. von (1952). Physiologie des Sehens. 2nd ed. Akademische Verlagsgesellschaft, Leipzig, p. 493.

Tamura, T. (1957). On the relation between the intensity of illumination and the shifting of cones in the fish retina. Bull. Jap. Soc. Sci. Fish. 22: 742-746.

Vilter, V. (1942). Régulation sympathico-hypothysaire de migrations pigmentaires dans la rétine de l'anguille (*Anguilla vulgaris* L.). Arch. Phys. Biol. 16(Suppl): 6-8.

Vilter, V. (1946). Déterminisme intrinsèque de migrations pigmentaires dans la rétine. C.R. Soc. Biol. Paris 140: 277-278.

Vilter, V. and Thibault, C. (1948). Etude du mécanisme photoadaptif du pigment rétinien mélanique chez la carpe par les éclairements unilatéraux. C.R. Séances Soc. Biol. 142: 290-292.
Wald, G. (1941). The visual systems of euryhaline fishes. J. Gen. Physiol. 25: 235-245.
Walls, G.L. (1928). An experimental study of the retina of the brook lamprey, *Entosphenus appendix* (De Kay). J. Comp. Neurol. 46: 465-473.
Walls, G.L. (1942). The vertebrate eye and its adaptive radiation. Cranbrook Institute of Science Bull. 19, 785 p. Bloomfield Hills, Michigan.
Weidemann, H.L. (1966). Der Einfluss quantengleicher Farblichter auf die Retinomotorik dreier Zapfentypen in der Netzhaut des Guppy (*Lebistes reticulatus* Peters). Diss. Mathematisch-Naturwissenschaftliche Fakultät der Georg-August-Universität Göttingen, Allemagne.
Wigger, H. (1937). Versuche zur Kausalanalyse der retinomotorischen Erscheinungen. Pflüg. Arch. ges Physiol. 239: 215-246.
Wunder, W. (1926). Die Bedeutung des Adaptationszustandes für das Verhalten der Sehelemente und des Pigmentes in der Netzhaut von Knochenfischen. Z. vergl. Physiol. 3: 595-614.
Zyznar, E.S. and Ali, M.A. (1974). An interpretative study of the organization of the visual cells in the tapetum lucidum of *Stizostedion*. Can. J. Zool. (in press)

APPROACHES TO THE STUDY OF CONTRACTILITY IN THE RODS AND CONES

P. COUILLARD

Département de Biologie, Université de Montréal

C.P. 6128, Montréal 101, Canada

"Cones and rods in serried ranks,
Stand sentinel in retinal flanks;
Back and forth they move to light,
And finny life is gay and bright."
P.K. Menon

In the present lecture, I shall try to give a necessarily complicated answer to a simple question: Is there an active contractile mechanism in the rods and cones of the animals like fish, which demonstrate retinomotor response? What is the available evidence for such a mechanism and how should we look for additional evidence?

As a preamble, we have to make a rapid survey of the known mechanisms of cellular motility and of the various experimental approaches in this field, keeping in mind their possible applications to our problem.

The highly differenciated system of striated muscle has long been the classical model of protoplasmic contractility and has served as a reference point for techniques and theories in this field. In the sarcomere, the chief contractile proteins, globulins, soluble at high ionic strength, are assembled into an interdigitating, highly ordered array of thick and thin filaments, each comprising several hundreds of molecular subunits. (For general reference, see Bendall, 1969 and Carlson & Wilkie, 1974).

Myosin is easy to characterise; in the pure, non-aggregated state, it is a highly asymetrical molecule, a 1500Å long fibre with a 100Å globular head. (Molecular wt. 460,000). Myosin is an enzyme, an ATPase, needing reduced SH-groups for its activity. In the purified state, it is activated by calcium ions with two optimum ranges of pH: below 6.4 and above 9.2. Histochemical ATPase assays are run at pH 9.0 in the presence of calcium at the light microscope (Padykula and Hermann, 1955) (Guth & Samala, 1970) and at the electro-microscope level (Tice and Barnett, 1962). Myosin is extremely sensitive to denaturation (heat, drying, pressure, heavy metals). At low ionic strengths it associates into filamentous 1.5μ aggregates, analagous to the thick filaments of the sarcomere.

Actin is a globular protein (molecular weight 42000.) devoid of any known enzymatic activity. It can associate into long, helical, twin-chain filaments analogous to the thin filaments of the sarcomere.

Actin and myosin can interact to form a complex, actomyosin, in which myosin molecules attach by the head to the actin filament in a three-dimensional herring-bone pattern. This affinity for myosin is considered one of the most fundamental and characteristic properties of actin. The so-called "decoration" of actin filaments by myosin (or by the heavy, shorter subfragment of the myosin molecule) can easily be detected in negatively stained electron microscope preparations. (Ishikawa, Bischoff and Holtzer, 1969). This reaction has made possible the detection of "Actin like molecules" in various types of non muscular cells, for example, *Acanthamoeba*, (Pollard, Shelton, Weihing and Korn, 1970).

In combination with actin, myosin becomes a Mg-activated ATPase with optimum pH of 7.2. The presence of minor proteins, tropomyosin and troponin, within the actin filament in the sarcomere allows this Mg-ATPase to be controlled by minute concentration of calcium (10^{-5}-10^{-8}M) liberated upon stimulation by vesicles of the endoplasmic reticulum.

Amino acid composition can also be used to characterise actin-like contractile proteins (Mohri, 1968). Similarities in pattern are found, not only between muscle actins from insects and mammals, but also with the various tubulins, which are the monomeric units of microtubules. Bacterial flagellin and myosin stand well apart from actins in this respect.

Of course, characterising a suspected actin-like protein by amino acid pattern requires the preparation of important amounts of this protein in pure form. This severely restricts the types of cells to which the technique can be applied.

Glycerinated models provide a widely applicable technique in the study of cell contractility. As developed by Szent Gyorgyi in the early 1950's it consists in extracting muscle fibres in the cold with a solvent of low ionic strength (glycerol 50%). This leaves intact the contractile protein framework of the cell within a completely permeabilised membrane. These models can then be reactivated (made to contract) by the simple addition of ATP and ions. Thus the characteristics of the actual contractile event, ion and pH requirements, immediate energy source, inhibitors etc. can be studied in the absence of the metabolic machinery and membrane control mechanisms.

The glycerinated model technique and its variants have been extended to a variety of cell types, ranging from protozoa (Simard-Duquesne & Couillard, 1962) to dividing fibroblasts in tissue culture (Hoffmann-Berling, 1954). (For a recent review, see Arronet, 1973). This technique constitutes a very physiological approach to problems of cell movements of many types.

Actin-like microfilaments, of 40 Å diameter and indefinite length have been found associated with various types of cell movements such as cytokinesis (Schroeder, 1968, Szollosi, 1970), protoplasmic streaming (Nagai & Rebhun, 1966), amoeboid movement (Pollard & Ito, 1970) and morphogenetic cell movements (Wessels, et al., 1971). It is generally recognized that bundles or meshwork of such filaments may represent a form of protoplasmic contractile machinery, although the mechanism of their action is not yet known.

Homology with muscle is difficult to establish since, except in rare cases, (Gicquaud & Couillard, 1971) myosin-like large filaments cannot be found associated with the fine filaments.

Bacterial flagella are a class by themselves of filamentous motile structures. They also are assembled from a single type of 45 Å protein subunit ("Flagellin") but its amino-acid composition has nothing in common with that of actin (Mohri, 1968). Furthermore current theories about flagellar motion in Bacteria consider these organelles as rigid structures affixed to actively rotating disk-shaped bases in the cell wall (Berg & Anderson, 1973).

Cytoplasmic contractile or tonic structures composed of microfilaments, can sometimes be visualised at the light microscope level. The *TPA technique* (Tannic acid, phosphomolybdic acid, amidoblack) was first used by Puchtler (1956) and by Leblond (Leblond, Puchtler and Clermont, 1960) to reveal such things as glandular cell webbs, terminal bars, and also smooth and striated muscles in mammalian tissues. We were later successful in adapting the TPA technique for the demonstration of fibrillar contractile structure in various protozoa (Latendresse & Couillard, 1971; Latendresse, 1972).

The very broad specificity of the TPA technique makes it particularly suitable for locating contractile cytoplasmic structures, its only handicap being that fixation procedure differs from tissue to tissue and must be established empirically.

The cytochalasins are a class of heterocyclic compounds extracted from fungi. They were first described in 1964 as potent inhibitors of cell movements in tissue cultures at concentrations as low as 0.5 μg/ml. Six types of cytochalasins had been described by 1972 (Carter, 1972).

Further biological activities of cytochalasins were discovered, the most notable being their action on cytokinesis. In general, they came to be considered as specific inhibitors of microfilament-generated cell movements (Wessels et al., 1971). More precisely, it can be said that, when a cellular movement is inhibited by cytochalasin, microfilaments are likely to be found in a strategic location for effecting this movement and, in many cases, the cytochalasin treatment has altered the morphology of these microfilaments to some extent.

However, the situation is complicated by known cases in which cytochalasin inhibition of movement has not affected the microfilaments (Blueming, 1971), and by a general cytotoxic effect of cytochalasins at higher concentration (Carter, 1972).

With the discovery of suitable fixation procedures, notably those involving glutaraldehyde, microtubules have come to be considered as an extremely widespread category of cytoplasmic organelles. Some microtubules are certainly involved in the maintenance of cell shape but their role in cytoplasmic flow and cell movements is no less important. Here are some examples of microtubule-generated movements:

Ciliary and flagellar movements
Axostyle movements in trichomonads (Grimstone & Cleveland, 1965)
Saltatory movements (Freed & Lebowitz, 1970)
Axoplasmic flow (Samson, 1971)
Chromosome anaphase movements
Melanophore pigment migration (Green, 1968).

When they move cytoplasm within cells, they can act either singly or in loose association, but when cell deformation is implied, they are generally found in geometric arrays involving a great number of elements which often are related by intertubular bridges (McIntoch, 1974) as in cilia, axostyle, ciliate Km fibres (Huang and Pitelka, 1973) or heliozoan axopodia.

Microtubules are straight and inextensible. They cannot change length except by the removal or addition of subunits and the only way for them to achieve active, rapid change in length is by sliding along one another.

In some cases, intertubular bridges (for example, the dynein "arms" of ciliary peripheral doublets) have been shown to be the sites of energy-liberating ATPases.

Microtubules may look alike in electron micrographs but they have proved to be quite heterogeneous as to their susceptibility to solvents, to physical factors such as temperature and pressure, and to specific disrupting agents like colchicine and vimblastine (Behnke and Forer, 1967). All these influences can be considered as affecting the equilibrium between a cellular pool of precursor subunits (tubulin) and the formed microtubules.

Thus it seems that actin-like globular subunits can assemble either as filaments (1 strand or 2 strand) or as single or double microtubules. It is tempting, though it is a little too early now, to suppose that we could be dealing with different states of association of a limited number of types of molecular elements.

It often occurs that both microtubules and microfilaments are involved in movement within the same cell. A particularly interesting case is that of the spirotrich ciliate, *Stentor*, which possesses a cortical array of longitudinally-oriented microfibrillar bundles (myonemes) and of microtubular ribbons (Km fibres) (Bannister & Tatchell, 1968).

These two sets of structural elements are believed to play antagonistic roles in cell deformation. The myonemes (microfilaments) being responsible for the rapid shortening of the trumpet-shaped cell (to 20% of the stretched length in milliseconds) and the Km fibres (microtubules) for the slow extension back to stretched length (Huang & Pitelka, 1973).

This system presents some unique features:

1. Microtubular extension of the Km ribbons (by active sliding of constituent microtubules along one another), represents the first case in which a cellular contractile system has been postulated to do work in the extension phase of its cycle.

2. Shortening of the myonemes (40 Å microfilaments) involves their reversible transformation into tubular structures (100-200 Å with 40-50 Å walls), presumably by a process of helical coiling of the constituent actin-like globular subunits. (Huang & Pitelka, 1973).

These concepts open new perspectives in our thinking about cell motility. First, they raise the possibility that work can be done in the extension phase of the contraction cycle. Second, they describe a mechanism whereby a single microfilament can shorten actively, thus doing away with the necessity of always supposing two-component interaction in protoplasmic contractile systems.

With this background, the problem of demonstrating a contractile system in the myoids of rods and cones can be approached in many different ways, even if methods such as the extraction and characterisation of contractile proteins have to be ruled out a priori.

Let us look first at histochemical techniques:

We have applied the TPA procedure, using 3 different fixatives to retinal sections of fishes (*Amia* and *Salvelinus*), frog (*Rana pipiens*) and rat (*Rattus norvegicus*) (Ali, Latendresse & Couillard, 1971). In all the twelve preparations, the myoid turned out to be the least TPA-stainable region of rods as well as cones.

At the ultrastructural level, Mg-activated, pH 7.2 ATPase has been found only in the outer segment and in the mitochondria of the ellipsoid of frog's visual cells (Scarpelli & Craig, 1963).

We intend to look for myosin-like ATPase at the light microscope level in the myoid and to adapt a technique of glycerination to visual cells. For both, we need living, isolated visual cells in sizeable quantities.

Techniques of visual cell isolation from the retina of fishes and amphibians have been perfected in various laboratories including our own. (Couillard & Ali, 1970; Drujan & Svaetichin, 1972; Lam, 1972; Anctil, Ali & Couillard, 1973). All rely on "general-purpose" proteolytic enzymes such as Papain, Pronase or Trypsin, or more specialised ones like Elastase, Collagenase or Hyaluronidase, together with the Calcium-chelating agent EDTA, used singly or in various sequences and combinations. Treatment of the dissected-out retina can be preceded by perfusion of the rear chamber of the eye *in situ* after removal of the vitreous. Perfusion of the intact eye via the circulatory system should perhaps also be attempted.

These procedures are excellent if only small amounts of isolated cells are needed, for microscopy, microelectrode work or microspectrophotometry. But we have yet to devise a means for obtaining isolated visual cells in sufficient quantities for such low-yield procedures as glycerination experiments.

At the ultrastructural level, relatively little deliberate search for contractile structures (microtubules or microfilaments) has been done on the myoids of fishes or amphibians. In osmium tetroxide-fixed retina of labrid fishes Engström (1963) described 200 Å "tubuli" which run lengthwise inside the myoids of rods and cones continuous with the endoplasmic reticulum elements of the ellipsoid. Yound & Droz (1968) have shown the red rod myoid of frog to be rich in endoplasmic reticulum and Golgi bodies. Logically enough, they also showed, by autoradiography, that this myoid is the primary site of amino-acid incorporation into protein. No mention is made of microtubules or microfilaments in their Osmium tetroxide-fixed material.

A more recent study (Ali, 1971), on fish and toad retinas, mentions a few longitudinally-oriented microtubules in the myoid after glutaraldehyde fixation.

As with all negative results, considerably more work must be done, with appropriate preparative techniques, before we can conclude unequivocally that the myoid is poor in microtubules and lacking in microfilaments.

It is not enough that there should be an active contractile process in the myoid; we need, in addition, some kind of recovery mechanism. In other words, though it is easy to imagine the visual cells being pulled out of the pigment epithelial processes by the postulated contraction of the myoid, it is difficult to claim that the same myoid can push those cells back to where they came from during the reverse adaptation. One thing is certain, the myoid is far too poor in microtubules to effect active extension as presumed for the Km fibres of *Stentor* by Huang & Pitelka (1973).

One possibility we have been considering for some time, is the presence of a series elastic element connected to the outer segment or to the ellipsoid of the visual cells. This element would stretched by the myoid contraction and would elastically pull the visual cell back into place during the relaxation phase of the myoid cycle. In search of such an elastic element, we are in the process of applying to retinal sections various stains for elastin. So far, the classical technique of Weigert has yielded negative results. We are now adapting to our material the spirit blue technique of Owen (1959).

A more indirect approach to the same problem would involve the long-term effects of lathyrogenic agents on the retinomotor response. These drugs, of which Penicillamine and Beta-amino-propionitrile are typical representatives, interfere with the cross-linking of connective tissue fibres in mammals (Selye, 1957; Tanzer, 1965).

What does a lathyric fish look like? How well does its retina adapt to variations of the level of illumination? These remain, for the moment, purely hypothetical questions.

Preliminary experiments by V. Virabhadrachari, in Prof. M.A. Ali's laboratory, have shown us that treating fish with penicillamine is a tricky business, most often likely to kill the fish prematurely.

Results from electron microscopy are, again, not encouraging. In a recent scanning electron microscope survey of the frog retina, Steinbert (1973) shows a tenuous reticular network of amorphous material surrounding the outer segments of rods and cones, but no more. It may be important to note that the pigment epithelium had been stripped away from these preparations.

Thus, evidence for any active contractile mechanism of structure in the myoids of the visual cells of fishes and amphibia is scanty, and, for the most part, negative.

At this stage, it is tempting to take a second look at our problem from another aspect. Let us consider first the speed of the retinomotor responses. Ali's (1971) histological measurements show that the light-dark adaptation in fishes is nearly complete in some 30 minutes at room temperature. Taking into account the changes of cell lengths involved, we obtain a rate of movement of 1-2 μm/min. This requires little energy and, conceivably, little structural differenciation, so the search for overt or rapid manifestations of a contractile mechanism under experimental conditions is bound to be difficult.

Going one step further, is it possible to consider alternatives to an active contractile process in the myoid? For example, could the movements of the visual cells be passive? Could the driving force for these movements be the melanophore-like active expansion and retraction of the epithelial pigment? If we assume the retina to be a closed system, volume-wise, it would be logical to expect that any active redistribution of matter (pigment granules) within it should be compensated for by an equivalent, passive, relocalisation of other matter (visual cells). Thus, in the course of the retinomotor response, the visual cells would be either crowded inwards or outwards by shifts of pigment granules, and, this passive response would be in opposite directions for rods and for cones depending on their respective shape, size and position in the retina.

Considered in this perspective, the myoid becomes an "elastoid", a part of the visual cell capable of elastic changes in length in response to the passive movements of the ellipsoid-outer segment complex.

Such a passive mechanism of visual cell movement during the retinomotor responses would eliminate a number of difficulties:

1. It would insure the observed synchronization of pigment, rod and cone movements during the response.

2. We would no more have to suppose, in the myoids of rods and cones respectively, a contractile mechanism responding in opposite ways to the same external stimulus.

3. More generally speaking, we would not have to suppose any contractile mechanism in any myoid; myoids of animal groups with retinomotor responses need not differ, functionnally, from those of other groups.

4. The evolutionary acquisition of the myoid, in spite of its passive role in retinomotor response, is amply justified by its other essential functions as the site of protein synthesis (Yound and Droz, 1968), as an optical wave-guide (Miller and Snyder, 1972) ... Still, it will not be justifiable to say that the myoid plays no role in retinomotor responses; the elastic recoil of the stretched myoid helps to pull back the outer part of the visual cell during the "contraction" phase of the response.

5. Finally, the evolutionary acquisition or loss of retinomotor responses would simply imply acquiring or losing the capacity for active pigment movements in response to the light regime. It is interesting to note that pigment mobility is always and only associated with the retinomotor responses.

This opens new research perspectives:

1. A closer consideration of the kinetics of pigment, rods and cones displacement in the course of the retinomotor response, in search of the implied synchronisation of such movements.

   In this respect, Ali's (1971) histological analysis shows cone and pigment movements to be mirror images or rod movements. Also Engstrom (1963), working on Labrid fishes, states that, in the course of dark adaptation, pigment tubes and cones always move together and do not change their positions relative to each other. He goes on to describe very intimate contact of the outer (vitreal) processes of pigment cells with the ellipsoids of rods and cones, particularly with the "accessory outer segment" of these cells.

2. Research on the action of colchicine and vimblastine on retinomotor responses acquires new significance. These drugs inhibit melanophore pigment movements (Schliva and Bereiter-Hahn, 1973). They should have no effect on myoids if it is true that these are poor in mocrotubules. Work on the effects of colchicine and cytochalasine on fish retinomotor response is in progress in Prof. M.A. Ali's laboratory.

REFERENCES

Ali, M.A. (1971). Les réponses rétinomotrices, caractères et mécanismes. Vision Res. 11: 1225-1288.

Ali, M.A., Latendresse, D. & Couillard, P. (1971). Colorabilité au TPA de la rétine de divers vertébrés. Vision Res. 11: 1023-25.

Anctil, M., Ali, M.A. & Couillard, P. (1973). Isolated retinal cells of some lower vertebrates. Rev. Can. Biol. 32: 107-19.

Arronet, N.I. (1973). Motile muscle and cell models. Consultants Bureau, N.Y. Translation of 1972, Russian edition.

Bannister, L.H. & Tatchell, E.C. (1968). Contractility and the fibre systems of *Stentor coeruleus*. J. Cell. Sci. 3: 295.

Behnke, O. & Forer, A. (1967). Evidence for four classes of microtubules in individual cells. J. Cell. Sci. 2: 169-92.

Bendall, J.R. (1969). Muscles, molecules and movement. Heinemann, London.

Berg, H.C. & Anderson, R.A. (1973). Bacteria swim by rotating their flagellar filaments. Nature 245: 380-82.

Bluemink, J.G. (1971). Effects of cytochalasin B on surface contractility and cell junction formation during egg cleavage in *Xenopus laevis*. Cytobiology 3: 176-87.

Carlson, F.D. & Wilkie, D.R. (1974). Muscle Physiology. Prentice Hall, Englewood Cliffs, N.J.

Carter, S.B. (1972). Les cytochalasines, outil de recherche en cytologie. Endeavour 31: 73-82.

Couillard, P. & Ali, M.A. (1970). Sur une technique d'isolement des cellules visuelles chez les vertébrés à sang froid. An. ACFAS Suppl. 37: 6.

Drujan, B.D. & Svaetichin, G. (1972). Characterization of different classes of isolated retinal cells. Vision Res. 12: 1777-84.

Engström, K. (1963). Structure, organization and ultrastructure of the visual cells in the teleost family Labridae. Acta Zool. 44: 1-41.

Freed, J.J. & Ledowitz, M.M. (1970). The association of a class of saltatory movements with microtubules in cultured cells. J. Cell. Biol. 45: 334-54.

Gicquaud, C.R. & Couillard, P. (1972). Préservation des mouvements dans le cytoplasme démembrané d'*Amoeba proteus*. II Mise en évidence de filaments de type myosine dans les préparations. Cytobiologie 5: 139-45.

Green, L. (1968). Mechanism of movements of granules in melanocytes of *Fundulus heteroclitus*. Proc. Nat. Acad. Sci. 59: 1179-86.

Grimstone, A.V. & Cleveland, L.R. (1965). The fine structure and function of the contractile axostyle of certain flagellates. J. Cell. Biol. 24: 387-400.

Guth, L. & Samala, F.J. (1970). Procedure for the histochemical demonstration of Actomyosin ATTase. Exp. Neurol. 28: 365-67.

Hoffmann-Berling, H. (1954). Die Bedeutung des Adenosintriphosphat für die Zell und Kernteilungsbewegungen inder Anaphase. Biochim. Biophys. Acta 15: 226.

Huang, B. & Pitelka, D.R. (1973). The contractile process in the ciliate *Stentor coeruleus*. I The role of microtubules and microfilaments. J. Cell. Biol. 57: 704-28.

Huxley, H.E. (1969). The mechanism of muscular contraction. Science 164: 1356.

Ishikawa, H., Bischoff, R. & Holtzer, H. (1969). Formation of arrowhead complexes with heavy meromyosin in a variety of cell types. J. Cell. Biol. 43: 312.

Latendresse, D. (1972). Adaptation d'une technique de coloration à l'amidoblack (TPA) à l'étude des systèmes fibrillaires chez *Stentor coeruleus*. Thèse de maîtrise Sc., Département des Sciences biologiques, Université de Montréal.

Latendresse, D. & Couillard, P. (1971). Fibrillar structures in protozoa stained by the TPA techniques. Proc. Can. Fed. Biol. Soc. 13: 626.

Leblond, C.P., Puchtler, H. & Clermont, Y. (1960). Structures corresponding to terminal bars and terminal web in many types of cells. Nature 186: 784-88.

McIntosh, J.R. (1974). Bridges between microtubules. J. Cell Biol. 61: 166-87.

Miller, W.H. & Snyder, A.M. (1972). Optical functions of myoids. Vision Res. 12: 1841-48.

Mohri, H. (1968). Amino acid composition of tubulin constituting microtubules of sperm flagella. Nature 217: 1053-54.

Nagai, R. & Rebhun, L.I. (1966). Cytoplasmic microfilaments in streaming Nitella cells. J. Ultrastruct. Res. 14: 571-89.

Nilsson, S.E.J. (1964). An electron microscopic classification of the retinal receptors of the leopard frog (*Rana pipiens*). J. Ultrastruct. Res. 10: 390-416.

Owen, G. (1959). A new method for staining connective tissue fibres with a note on Liang's method for nerve fibres. Q. J. Microsc. Sci. 100: 421-24.

Padykula, H.A. & Herman, E. (1955). Factors affecting the activity of adenosine triphosphatase and other phosphatases as measured by histochemical techniques. J. Histochem. Cytochem. 3: 161-95.

Poglazov, B.F. (1962). A myosin-like protein in the brain tissue. Bull. Exp. Biol. Med. (USSR) 52: 1030-33.

Pollard, T.D. & Ito, S. (1970). Cytoplasmic filaments of *Amoeba proteus* I Role of filaments in consistency changes and movement. J. Cell. Biol. 46: 267-89.

Pollard, T.D., Shelton, E., Weihing, R.R. and Korn, E.D. (1970). Ultrastructural characterization of F-actin isolated from *Acanthamoeba castellanii* and identification of cytoplasmic filaments as F-actin by reaction with rabbit heavy meromyosin. J. Mol. Biol. 50: 91-97.

Puchtler, H. (1956). Histochemical analysis of terminal bars. J. Histochem. Cytochem. 4: 439-40.

Samson, F.E. (1971). Mechanism of axoplasmic transport. J. Neurobiol. 2: 347-60.

Scarpelli, D.G. & Craig, E.L. (1963). The fine localization of nucleoside triphosphatase activity in the retina of the frog. J. Cell.Biol. 17: 279-88.

Schliwa, M. & Bereiter-Hahn, J. (1973). Pigment movements in fish melanophores morphological and physiological studies III The effects of colchicine and vimblastine. Z. Zellforsch 147: 127-48.

Schroeder, T.E. (1968). Cytokinesis: filaments in the cleavage furrow. Exp. Cell. Res. 53: 272-318.

Selye, H. (1957). Lathyrism. Rev. Can. Biol. 16: 1-82.

Simard-Duquesne, N. & Couillard, P. (1962). Ameboid movement. I Reactivation of glycerinated models of *Amoeba proteus* with adenosine triphosphate. Exp. Cell. Res. 28: 85-91.

Sleigh, M. (1973). The biology of protozoa, P. 47. Edward Arnold, London.

Steinberg, R.H. (1973). Scanning electron microscopy of the bull frog's retina and pigment epithelium. Z. Zellforsch 143: 451-63.

Szollosi, D. (1970). Cortical cytoplasmic filaments of cleaving eggs A structural element corresponding to the contractile ring. J. Cell. Biol. 44: 192-209.

Tanzer, M.L. (1965). Experimental Lathyrism. Int. Rev. Connect. Tissue Res. 3: 91-112.

Tice, L.M. & Barrnett, R.J. (1962). Fine structural localization of ATPases activity in heart muscle myofibrils. J. Cell. Biol. 15: 401-16.

Weis-Fogh, T. & Amos, W.B. (1972). Evidence for a new mechanism of cell motility. Nature 236: 301-04.

Wessels, N.K., Spooner, B.S., Ash, J.F., Bradley, M.O., Luduena, M.A., Taylor, E.L., Wrenn, J.T. & Yamada, K.M. (1971). Microfilaments in cellular and developmental processes. Science 171: 135-143.

Young, R.W. & Droz, B. (1968). The renewal of protein in retinal rods and cones. J. Cell. Biol. 39: 169-84.

DISTRIBUTION AND DEVELOPMENT OF RETINOMOTOR RESPONSES

M. A. ALI & H. -J. WAGNER, Dépt. de Biol. Univ. de Montréal, Canada, and Lehrstuhl für Morphologie & Anatomie, Universität Regensburg, Germany

He's got'em on the list - he's got'em on the list;
And they'll none of 'em be missed.
W.S. Gilbert: The Mikado, Sc. 1

INTRODUCTION

Retinomotor responses as we understand them are almost completely restricted to the teleosts among fishes (see also Ali, this volume). Outside this group they occur only in the holostean, *Amia*, which is closely related to the teleosts. In *Amia* only the epithelial pigment and the rods move and even these responses are low in amplitude. Among the teleosts retinomotor responses occur in a large number of species belonging to a wide variety of orders, families and genera and occupying different habitats.

The following two tables also give various types of information concerning the fish, its retina, and its habitat. The classification followed is that of Greenwood, Rosen, Weitzmann and Myers (1966). The sources for both published and un-published informations are given. In some cases even if a particular paper did not deal specifically with retinomotor responses, if the evidence presented showed beyond a reasonable doubt that the fish was capable of these responses it has been so listed. For example, if the photomicrograph of the retina of a fish sampled at dusk showed the elements to be in an intermediate state and if a closely related fish was shown to be capable of these responses then the fish sampled at dusk was considered also to be capable of them.

DISTRIBUTION

Retinomotor responses occur in active as well as inactive species, in both fresh-water and marine habitats, provided that the fish lives in water shallow or clear enough to permit the perception of diel light regimes. These responses occur in retinas with both photoreceptor types, rods and cones except in two cases which need to be confirmed (Ophidium, Careproctus). It does not follow that both photoreceptor types undergo positional changes in response to light changes.

In a few species belonging to unrelated groups (Amia calva Mallotus villosus, Synodontis spp., Gadus ogac, Stizostedion vitreum and S. canadense), the cones do not migrate in response to light and darkness (i. e. remain in the **vitread position**), whereas the rods are responsive. The reverse is never true, i. e. when the rods do not migrate, neither do the cones. Although we cannot explain definitely as to why the cones do not respond, recent evidence (Zyznar and Ali, 1975) suggests the possibility of the cone ellipsoids functioning as light guides in some of these fishes. It would appear that the displacement of the cones is not only not critical for scotopic sensitivity but in fact aids, together with the tapetum the detection of the feeble light intensity by the rods. Cone migration may take place (as it does in most fishes capable of retinomotor responses) for accommodatory reasons, to permit the location of the rods directly in the path of the incoming light quanta.

When the responses involve only the epithelial pigment, as in the cave fish, Anoptichthys, this is due to the degenerative state of visual cells.

Some fragmentary, observations indicate (Wang, 1968) that in the dark, the cone myoid (located vitread to the external limiting membrane) of the nurse shark retina elongates in the vitread direction, which is opposite to the direction of the cone movements in the teleosts. The cone myoids of teleosts are mostly sclerad to the external limiting membrane. The position of the cone nucleus, and not the ellipsoid and outer segment of the nurse shark is affected by this phenomenon. Whether this may be considered a retinomotor response technically remains to be decided on the basis of further work with the elasmobranchs. This may have some significance in our understanding of the homology of the elasmobranchs and teleosts "cones".

Table 1. Survey of the occurrence of retinomotor responses among fishes. The arrangement of the teleostean species follows the recent classification of Greenwood et al. (1966).
C cones; DC double cones only; M mixed retina (rods and cones); P retinal epithelial pigment; QC quadruple cones; R rods; TC triple cones; + positive response; - no response; ? doubtful.

Classification	Retina	Retinomotor response P	C	R	Author (s)	Comments
Subclass Actinopterygii						
Holostei						
Amiidae						
1. _Amia calva_	M	+	-	+	Ali & Anctil (1973)	Freshwater, nocturnal, sluggish
Teleostei						
Elopomorpha						
Anguilliformes						
Anguillidae						
2. _Anguilla anguilla_	M	+	+	+	Chèze (1974)	Catadromous, active swimmer
Clupeomorpha						
Clupeiformes						
Clupeoidei						
Clupeidae						
3. _Sardinops caerulea_	M	+	+	+	O' Connell (1963)	Marine pelagic, active
4. _Alosa sapidissima_	M	+	+	+	O' Connell (1963)	plancton feeder
5. _Clupea harengus_	M	+	+	+	Blaxter & Jones (1967)	Coastal pelagic, schooling fish, plancton eater

Classification	Retina	Retinomotor response P	C	R	Author (s)	Comments
Engraulidae						
6. *Engraulis encrasicholus*	M	+	+	+	Protasov et al. (1960)	Marine
7. *Engraulis mordax*	M, DC	+	+	+	O' Connell (1963)	pelagic,
8. *Anchoa compressa*	M, DC	+	+	+	O' Connell (1963)	retinal tapetum
9. *Anchoa mitchilli*	M	+	+	+	Zyznar (1973)	lucidum
Protacanthopterygii						
Salmoniformes						
Salmonoidei						
Salmonidae						
10. *Oncorhynchus gorbuscha*	M	+	+	+	Ali (1959)	Anadromous, active swimmer
11. *O. keta*	M	+	+	+	Ali (1959)	Anadromous, active swimmer
12. *O. kisutch*	M	+	+	+	Ali (1959)	Anadromous, active swimmer
13. *O. nerka*	M	+	+	+	Ali (1959)	Anadromous, active swimmer
14. *O. tshawytscha*	M	+	+	+	Ali (1959)	Anadromous, active swimmer
15. *Salmo trutta fario*	M	+	+	+	Wunder (1926a) Verrier (1928)	Freshwater, gravelly bottom, active
16. *Salmo irideus*	M	+	+	+	Chèze (1974)	territorial fishes
17. *Salmo salar*	M	+	+	+	Ali et al. (1961)	Anadromous, active swimmer
18. *Salvelinus fontinalis*	M	-	+	+	Ali (1964c)	Deep freshwater moderately active,
x *S. namaycush*						albino
19. *S. alpinus*	M	+	+	+	Ali (1965)	Euryhaline,
20. *S. fontinalis*	M	+	+	+	Wang (1968)	active
21. *Coregonus spp.*	M	+	+	+	Wunder (1926b)	Freshwater, shallow or deep waters, active

Classification	Retina	Retinomotor response P	C	R	Author (s)	Comments
Osmeridae						
22. Mallotus villosus	M, DC	+	-	+	Protasov (1964) Anctil (1969)	Marine, deep shelf waters, schooling fish
23. Osmerus eperlanus	M	+	+	+	Girsa (1967)	Euryhaline, shallow waters, schooling fish
Esocoidei						
Esocidae						
24. Esox lucius	M	+	+	+	Wunder (1926a) Engström (1963b)	Freshwater, slowly moving fish, predator
Ostariophysi						
Cypriniformes						
Characoidei						
Characidae						
25. Gymnocorymbus ternetzi	M	+	+	+	Braemer (1957)	Freshwater, tropical
26. Anoptichthys jordani	M	+	-	-	Kähling (1961)	Freshwater, cave-dwelling
27. Astyanax mexicanus	M	+	+	+	John & Haut (1964) John & Kaminester (1969)	Freshwater, schooling fish
28. Serrasalmus marginatus	M	+	+	+	Ali & Raymond (1972)	Freshwater, tropical schooling fish, predator
Cyprinoidei						
Cyprinidae						
29. Abramis brama	M, TC	+	+	+	Exner & Januschke (1905, 1906)	Freshwater, bottom dwelling fish

Classification	Retina	Retinomotor response P	C	R	Author (s)	Comments
30. Phoxinus laevis	M, TC, QC	+	+	+	von Frisch (1924) Wunder (1926a) Engström (1963 b)	Freshwater, schooling fish
31. Gobio gobio (fluviatilis)	M	+	+	+	Wunder (1926a)	Freshwater, bottom dwelling, schooling fish
32. Carassius auratus	M	+	+	+	Arey (1916) Kobayashi (1957) Ali (1964a) John et al. (1967) Chèze (1974)	Freshwater, over calm, muddy waters, vegetarian
33. Rutilus rutilus	M, TC	+	+	+	Wunder (1926a) Engström & Rosstorp (1963)	Freshwater, sluggish predator
34. Cyprinus carpio	M, TC	+	+	+	Peyraud (1966) Chèze (1974)	Life habits as in Carassius auratus
35. Exoglossum maxillingua	M	+	+	+	Wang (1968)	Life habits as in Carassius auratus
36. Notemigonus crysoleucas	M	+	+	+	Arey (1916) Wang (1968)	Freshwater, acitve
37. Notropis cornutus	M	+	+	+	Wang (1968)	Freshwater, active
38. Notropis rubellus	M	+	+	+	Wang (1968)	Freshwater, active
39. Rhinichthys atratulus	M	+	+	+	Wang (1968)	Freshwater, active
Catostomidae						
40. Hypentelium nigricans	M	+	+	+	Wang (1968)	Freshwater, scavenger

Classification	Retina	Retinomotor response P	C	R	Author (s)	Comments
Cobitidae						
41. Cobitis barbatula	M	+	+?	+?	Wunder (1926a)	Freshwater,
42. Cobitis taenia	M	+	+?	+?	Wunder (1926a)	bottom-living,
43. Misgurnus anguillicaudatus	M	+	+	+	Kobayashi (1957)	burrow in sand or mud
Siluriformes						
Ictaluridae						
44. Ictalurus nebulosus	M	+	+	+	Arey (1916) Wunder (1926a) Arey & Mundt (1941) Ali (1964b) Arnott et al. (1974)	Freshwater, sluggish, saprotrophic, retinal tapetum lucidum
Mochokidae						
45. Synodontis nigriventris	M	+	-	+	Wagner (1970)	Freshwater, crepuscular,
46. S. notatus	M	+	-	+	Wagner (1970)	grouped in large shoals
47. Rhamdia sebae	M	+	+	+	Ali (unpublished)	Marine, tropical predator
Ariidae						
48. Arius felis	M	+	+	+	Zyznar (1973) Arnott et al. (1974)	Marine, crepuscular, turbid waters, retinal tapetum lucidum

Classification	Retina	Retinomotor response P	C	R	Author (s)	Comments
Loricariidae						
49. Plecostomus plecostomus	M	+	+	+	Wagner (unpublished)	Tropical freshwater fish; crepuscular; pupil of falciform shape in the light becomes round during darkness
Paracanthopterygii						
Gadiformes						
Gadoidei						
Gadidae						
50. Gadus morrhua	M	+	+	+	Engström (1961) Ali & Hanyu (1963) Protasov (1964)	Marine, shelf waters, diurnal
51. Gadus ogac	M	+	-	+	Ali et al. (1968)	Marine, shelf waters, diurnal
52. Lota lota	M, TC	+	+	+	Wunder (1926a) Engström (1961)	Freshwater, predator
53. Pollachius virens	M	+	?	?	Ali et al. (1968)	Deep shelf waters, active
Ophidioidei						
Ophidiidae						
54. Lepophidium graellsi	M	+	+	+	Zyznar (1973)	Marine, benthic,
55. Ophidium barbatum	R	+	-	+	Pora et al. (1963)	retinal tapetum
56. O. welshi	M	+	+	+	Zyznar (1973)	lucidum

Classification	Retina	Retinomotor response P	C	R	Author (s)	Comments
Zoarcoidei						
Zoarcidae						
57. Lycodes lavalei	M, TC	+	-	+?	Ali & Hanyu (1963) Anctil (1969)	Marine, shelf waters, benthic, sluggish
58. L. vahlii maculatus	M, TC	+	-	+?	Ali & Hanyu (1963) Engström (1963b)	Marine, shelf waters, benthic, sluggish
Acanthopterygii						
Atherinomorpha						
Exocoetoidei						
Exocoetidae						
59. Exocoetus volitans	M	+	+	+	Anctil & Ali (1970)	Epipelagic,
60. Fodiator acutus	M	+	+	+	Anctil & Ali (1970)	swift swimmers
61. Hyporhamphus unifasciatus	M	+	+	+	Ali et al. (1973)	Epipelagic, active
Cyprinodontoidei						
Cyprinodontidae						
62. Fundulus heteroclitus	M	+	+	+	Arey (1916)	Euryhaline, active
Poecilidae						
63. Lebistes reticulatus	M	+	+	+	Müller (1954) Weidemann (1966)	Freshwater, feed on insect larvae
Atherinoidei						
Atherinidae						

Classification	Retina	Retinomotor response P	C	R	Author (s)	Comments
64. Atherina mochonpontica	M	+	+	+	Protasov et al. (1960)	Marine, active
Percomorpha						
Gasterosteiformes						
Gasterosteoidei						
Gasterosteidae						
65. Eucalia inconstans	M	+	+	+	Wang (1968)	Freshwater, active
66. Gasterosteus aculeatus	M	+	+	+	Wunder (1926) Ali et al. (1968) Wagner (1972)	Euryhaline, active
Syngnathoidei						
Syngnathidae						
67. Hippocampus hippocampus	M	+	?	?	Pora et al. (1963)	Marine shorefish, swim in vertical position, eat crustaceans and larvae
Scorpaeniformes						
Scorpaenoidei						
Scorpaenidae						
68. Sebastes marinus	M, DC	+	+	+	Ali & Hanyu (1963)	Deep shelf water, benthic or semi-pelagic, moderately active

Classification	Retina	Retinomotor response P	C	R	Author (s)	Comments
Cottoidei						
Cottidae						
69. Triglops murrayi	M	+	+	+	Ali & Hanyu (1963)	Shelf, benthic, sluggish fish
Agonidae						
70. Aspidophoroides monopterygius	M	+	+	+	Ali & Hanyu (1963)	Shelf, benthic, sluggish fish
Cyclopteridae						
71. Careproctus sp.	R	+	-	?	Ali et al. (1968)	Deep shelf and oceanic waters, rocky bottoms, sedentary
Perciformes						
Percoidei						
Serranidae						
72. Lateolabrax japonicus	M	+	+	+	Tamura (1957)	Marine shore fish, diurnal
Centrarchidae						
73. Lepomis macrochirus	M	+	+	+	Engström (1963b) John & Gring (1967)	Calm fresh waters, feed on
74. Lepomis gibbosus	M	+	+	+	Wang (1968)	plancton
Priacanthidae						
75. Priacanthus arenatus	M	?	?	?	Zyznar (1973)	Tropical, marine fish, noc-turnal retinal tapetum lucidum

Classification	Retina	Retinomotor response P	C	R	Author (s)	Comments
Percidae						
76. Acerina cernua	M	+	+	+	Wunder (1926a) Engström (1963b)	Freshwater, moderately active, retinal tapetum lucidum
77. Perca fluviatilis	M	+	+	+	Wunder (1926a) Engström (1963b)	Freshwater, diurnal
78. Etheostoma olmstedi	M	+	+	+	Wang (1968)	Freshwater, dwells under stones
79. Stizostedion sandra	M	+	+	+	Wunder (1930)	Turbid freshwaters, retinal tapetum lucidum
80. Stizostedion vitreum	M (DC)	+	-	+	Ali & Anctil (1968)	Clear and turbid waters as well, crepuscular fish, retinal tapetum
81. S. canadense	M(DC)	+	-	+	Zyznar & Ali (1975)	Mainly turbid waters, retinal tapetum lucidum
Pomatomidae						
82. Pomatomus saltatrix	M	+	+	+	Olla & Marchioni (1968)	Marine, inshore, fast-moving, schooling fish, voracious
Carangidae						
83. Trachurus japonicus	M	+	+	+	Kobayashi (1957)	Coastal
84. T. symmetricus	M	+	+	+	O'Connel (1963) Pora et al. (1963)	pelagic, active

Classification	Retina	Retinomotor response P	C	R	Author (s)	Comments
Lutjanidae						
85. Lutjanus synagris	M	+	+	+	Ali et al. (1973)	Marine shore fish, nocturnal
Pomadasyidae						
86. Orthopristis chrysoptera	M	+	+	+	Zyznar (1973)	Tropical, marine fishes, retinal tapetum lucidum
Sparidae						
87. Spargus annularis	M	+	+	+	Protasov et al. (1960)	Marine shore fish, nocturnal
88. Archosargus rhomboidalis	M	+	+	+	Ali et al. (1973)	Marine shore fish, nocturnal
89. Oligonlitos saurus	M	+	+	+	Ali et al. (1973)	Marine shore fish, nocturnal
Sciaenidae						
90. Cynoscion arenarius	M	+	+	+	Arnott et al. (1970, 1972)	Turbid shore waters, retinal tapetum lucidum
91. C. nebulosus	M	+	+	+	Arnott et al. (1970)	
Ephippidae						
92. Chaetodipterus faber	M	+	+	+	Ali et al. (1973) Zyznar (1973)	Turbid shore waters, retinal tapetum lucidum
Cichlidae						
93. Pterophyllum eimeckei	M	+	+	+	Braemer (1957) Wagner (1972)	Freshwater, quiet fish, use vision in feeding
94. Nannacara anomala	M	+	+	+	Kuenzer & Wagner (1969)	

Classification	Retina	Retinomotor response P	C	R	Author (s)	Comments
95. N. taenia	M	+	+	+	Wagner (unpublished)	Freshwater, quiet fish, use vision in feeding
96. Symphysodon aequifasciata	M	+	+	+	Wagner (unpublished)	Freshwater, quiet fish, use vision in feeding
97. Apistogramma ramirezi	M	+	+	+	Wagner(unpublished)	Freshwater, quiet fish, use vision in feeding
98. A. reitzigi	M	+	+	+	Wagner(unpublished)	Freshwater, quiet fish, use vision in feeding
99. Cichlasoma nigrofasciatum	M	+	+	+	Wagner(unpublished)	Freshwater, quiet fish, use vision in feeding
100. Haplochromis multicolor	M	+	+	+	Wagner(unpublished)	Freshwater, quiet fish, use vision in feeding
101. H. burtoni	M	+	+	+	Rohrschneider (pers. comm.)	Freshwater, quiet fish, use vision in feeding
102. Hemichromis bimaculatus	M	+	+	+	Wagner(unpublished)	Freshwater, quiet fish, use vision in feeding
Scaridae						
103. Cryptotomus roseus	M	+	+	+	Ali et al. (1973)	Tropical marine fish of intermediary depth, diurnal
Mugiloidei						
Mugilidae						
104. Mugil curema	M	+	+	+	Ali et al. (1973)	Marine shorefish, active

Classification	Retina	Retinomotor response P	C	R	Author (s)	Comments
Labroidei						
Labridae						
105. Labrus ossifagus	M	+	+	+	Engström (1963a)	Marine shorefish, rely
106. Symphodus melops	M	+	+	+	Chèze (1974)	on vision for feeding
107. Pseudolabrus miles	M	+	+	+	Fineran & Nicol (1974)	Marine inshore fish,
108. Pseudolabrus celidotus	M	+	+	+	Fineran & Nicol (1974)	diurnal, live in moderately
109. Pseudolabrus pittensis	M	+	+	+	Fineran & Nicol (1974)	turbid waters
Blennioidei						
Anarhichadidae						
110. Anarhichus lupus	M	+	+	+	Protasov (1964)	Marine, benthic in shelf waters, sluggish
Stichaeidae						
111. Ulvaria subbifurcata	M	+	-?	-?	Ali et al. (1968)	Marine, coastal shelf waters, active
Gobioidei						
Gobiidae						
112. Gobius sp.	M	+	+	-?	Pora et al. (1963)	Freshwater, active

Classification	Retina	Retinomotor response P	C	R	Author (s)	Comments
Scombroidei						
Scombridae						
113. Scomber japonicus (Pneumotophorus diego)	M	+	+	+	O'Connell (1963)	Marine, epipelagic,
114. Scomber scombrus	M	+	+	+	Engström (1963b)	active
Anabantidae						
115. Betta splendens	M	+	+	+	Wagner(unpublished)	Freshwater, diurnal fish, use vision in feeding
Pleuronectiformes						
Pleuronectoidei						
Pleuronectidae						
116. Hippoglossoides platessoides	M, TC	+	?	?	Protasov (1964) Ali et al. (1968)	Marine, benthic,
117. Pleuronectes platessa	M, TC	+	+	+	Protasov (1964) Nicol (1965) Blaxter (1968)	sandy bottoms, sluggish, undertake vertical
118. Microstomus kitt	M, TC, QC	+	+	+	Nicol (1965)	migrations of moderate scale
Soleoidei						
Soleidae						
119. Solea solea	M, TC	+	+	+	Nicol (1961)	Life habits as in Pleuronectidae

Classification	Retina	Retinomotor response P	C	R	Author (s)	Comments
Brachiopterygii						
Polypteriformes						
Polypteridae						
120. *Calamoichthys calabricus*	M	-	-	-	Pfeiffer (1968)	Freshwater,
121. *Polypterus delhezi*	M	-	-	-	Pfeiffer (1968)	sluggish
Dipnoi						
Lepidosireniformes						
Lepidosirenidae						
122. *Lepidosiren paradoxa*	M	-	-	-	Ali & Anctil (1973)	Freshwater, sluggish
Protopteridae						
123. *Protopterus dolloi*	M	-	-	-	Pfeiffer (1968)	Freshwater, sluggish

DEVELOPMENT

Ontogenetic parallels of both structural and functional development of the retinal elements have been observed in teleosts as well as in amphibians (Saxen, 1954). There is general agreement that in the larval eye there is only one class of receptor cell, which, in most cases, has been described as a cone. During this stage, the pigment epithelium is contracted, the pigment granules being located in a narrow band along the basement membrane. Neither visual cells nor pigment epithelial cells appear to be capable of showing any retinomotor responses. Only after a delay of variable length typical, fully differentiated rods can be visualised. At the same time, retinomotor responses appear for the first time, involving rods, cones and the pigment epithelium.

On the basis of these observations, it may be suggested that the duplex system of the retina begins to function after rods and retinomotor responses are present. Observations of behaviour of the fry during twilight period (Ali, 1959; Wagner, 1974) would support this hypothesis.

If the basic features of the ontogeny of the fish retina seem to hold for all species, there are, on the other hand, marked differences in the developmental and behavioural patterns between young marine and freshwater fish larvae. The latter are characterised by a big yolk sac and slowly developing eyes; during this stage they are nourished by the yolk, resting on the bottom between the gravel and exhibiting a strongly photonegative behaviour. This is true for solitarily growing salmonids and for cichlid fry which are guarded by the parents. Marine fish larvae have on the other hand no yolk sac but well developed eyes, which appear to be functional because the larvae feed actively on plancton, relying on vision to a great extent. Contrary to the freshwater larvae they show a photopositive reaction.

Other types of behaviour, such as schooling, feeding, and migration can also be related to the developmental stage of the retina. In every case, however, the kind of response which is enabled by the growing capacity of the eye must be considered in the context of the specific ecological conditions of the species.

Table 2. List of ontogeny of retinomotor responses including some details concerning retinal structure and behaviour of the species studied. P retinal pigment epithelium; C cones; R rods; DC double cones; + positive response; O negative response

Species	Stade	Retina	Retinomotor response P	C	R	Comments
Anguilla anguilla (Blaxter & Staines, 1970)	leptocephalus	only R	0	0	0	
Clupea harengus (Blaxter & Jones, 1967)	larvae	only single C	0	0	0	Active, feed on plancton vertical migration
	metamorphosis	DC and R appear	+	+	+	Threshold 1 - 0.01 lux
Sardina pilchardus (Blaxter & Staines, 1970)	larvae	no R	0	0	0	
Oncorhynchus sp. (Pacific salmon) O. kisutch O. keta O. nerka O. gorbuscha O. tshawytscha (Ali, 1959)	embryos before hatching	P and C permanently contracted no R	0	0	0	In the egg
	alvins	some R	$\pm$	$\pm$	0	Photonegative, very mobile, active
			Weak responses to light; exception: O. gorbuscha			

Species	Stade	Retina	Retinomotor response P	C	R	Comments
Oncorhynchus sp. *O. kisutch* *O. keta* *O. nerka* *O. gorbuscha* *O. tshawytscha*	emerged fry	R visible; external nuclear layer broadening; no diurnal rhythm of P and C under constant light	+	+	+	Swim actively; *O. keta* and *O. gorbuscha* migrate downstream
	late fry		Enhancing gradually; P shows a period of latency before contracting at darkness, except for *O. gorbuscha*			Swim actively; *O. nerka* and *O. kisutch* migrate from the river to the lake
(Ali, 1959)	smolt (yearlings)					*O. nerka* and *O. kisutch* migrate towards the sea
Salmo salar (Atlantic salmon) (Ali, 1963; Blaxter & Staines, 1970)	embryos	Density of ganglion cells and C decreases (C diminish less because they take more time to develop) during extension of the retina (growth of the retina is slower than growth of the eye)	0	0	0	
	yearlings		+	+	+	Migrate towards the sea

Species	Stade	Retina	Retinomotor response P	C	R	Comments
Microstomus kitt (Blaxter & Staines, 1970)	larvae	no R	0	0	0	
	metamorphosis	R appear	0	0	0	
Gasterosteus aculeatus (Blaxter & Staines, 1970)	larvae	no R	0	0	0	
	metamorphosis	R appear	+	+	+	Threshold 10 - 0.1 lux for migration of P
Haplochromis burtoni (Rohrschneider, in preparation)	young alvins	no R	0	0	0	Mouth breeder, larvae still in the mouth of the mother
	emerging fry	R anlagen without outer segments	0	0	0	Leaving the mouth, schooling
	late fry	R with outer segments	+	+	+	

Species	Stade	Retina	Retinomotor response P	C	R	Comments
Nannacara anomala (Wagner, 1974)	embryos	C, R, P undiffe-rentiated	0	0	0	In the egg
	alvins	no R	±	0	0	Wriggling larvae, photonegative
	fry	R appear after onset of free swimming	+	+	+	Begin to swim free 5 days after hatching; active, photopositive, schooling; inactive during twilight; 12 days after onsert of free swimming activity extends to twilight period
Pomatoschistus minutus (Blaxter & Staines, 1970)	larvae	no R	0	0	0	
Pleuronectes platessa (Blaxter, 1968)	young larvae	no R	0	0	0	Threshold of light intensity for feeding is 1 - 0.01 lux
	metamorphosis	R seem to be mixed	+	+	+	Threshold at 0.01 lux

Species	Stade	Retina	Retinomotor response P	C	R	Comments
Solea solea (Blaxter & Staines, 1970)	larvae	no R	0	0	0	
	metamorphosis	R appear	+	+	+	Threshold 10 - 0.1 lux for migration of P

ACKNOWLEDGEMENTS

We thank Michel Anctil for help provided in 1973 for updating the information which had been gathered until then by one of us (MAA). We are also grateful to Prof. H.J. Autrum for permitting one of us (HJW) to make use of his library's facilities.

REFERENCES

Ali, M. A. (1959). The ocular structure, retinomotor and photobehavioural responses of juvenile Pacific salmon. Can. J. Zool 37: 965-996.

Ali, M.A. (1961). Histophysiological studies on the juvenile Atlantic salmon (Salmo salar) retina. II. Responses to light or dark. Can. J. Zool. 39: 511-526.

Ali, M. A. (1963). Correlation of some retinal and morphological measurements from the Atlantic salmon (Salmo salar). Growth 27: 57-76.

Ali, M. A. (1964a). Retinomotor responses of the goldfish (Carassius auratus) to unilateral photic stimulation. Rev. Can. Biol. 23: 45-53.

Ali, M. A. (1964b). Retinomotor responses in the enucleated eyes of the brown bullhead (Ictalurus nebulosus) and the goldfish (Carassius auratus). Rev. Can. Biol. 23: 55-66.

Ali, M. A. (1964c). Retina of the albino splake (Salvelinus fontinalis x S. namaycush). Can. J. Zool. 42: 1158-1160.

Ali, M. A. (1965). Retinal structure in the Arctic char (Salvelinus alpinus L.). J. Fish. Res. Bd. Can. 22: 221-223.

Ali, M. A., and Anctil, M. (1968). Corrélation entre la structure rétinienne et l'habitat chez Stizostedion vitreum vitreum et S. canadense. J. Fish. Res. Bd. Can. 25: 2001-2003.

Ali, M. A., and Anctil, M. (1974). Retinomotor responses and isolated photoreceptors in Amia calva (Holostei, Amiidae). Copeia 2: 379-386.

Ali, M. A., and Anctil, M. (1973). Retina of the South American Lungfish. Lepidosiren paradoxa Fitzinger. Can. J. Zool. 51: 696-972.

Ali, M. A., Anctil, M., and Mohideen, H. M. (1968). Structure rétinienne et la vascularisation intraoculaire chez quelques poissons marins de la région de Gaspé. Can. J. Zool. 46: 729-745.

Ali, M. A. , and Hanyu, I. (1963). A comparative study of retinal structure in some fishes from moderately deep waters of the Western North Atlantic. Can. J. Zool. 41: 225-241.

Ali, M. A., and Raymond, N. (1972). La rétine de piranha (Serrasalmus marginatus). Revista de Biologia 8: 27-32.

Ali, M. A., Stevenson, W. R., and Press, J. S. (1961). Histophysiological studies on the juvenile Atlantic salmon (Salmo salar) retina. I. Rates of light- and dark-adaptation. Can. J. Zool. 39: 123-128.

Anctil, M. (1969). Structure de la rétine chez quelques téléostéens marins du plateau continental. J. Fish. Res. Bd. Can. 26: 597-628.

Anctil, M., and Ali, M. A. (1970). Retina of Exoceotus volitans and Fodiator acutus (Pisces: Exocoetidae). Copeia 1: 43-48.

Arey, L. B. (1916). The movements in the visual cells and retinal pigment of the lower vertebrates. J. comp. Neurol. 26: 121-201.

Arey, L. B., and Mundt, G. H. (1941). A persistent diurnal rhythm in visual cones. Anat. Rec. (Suppl.) 79: 5-11.

Arnott, H. J., Maciolek, N. J., and Nicol, J.A.C. (1970). Retinal tapetum lucidum: A novel reflecting system in the eye of teleosts. Science,(Washington), 169 : 478-480.

Arnott, H. J., Nicol, J.A. C., and Querfeld, C. W. (1972). Tapeta lucida in the eyes of the seatrout (Sciaenidae). Proc. R. Soc. Lond. B. 180: 247-271.

Arnott, H.J., Best, A.C.G., Ito, S., and Nicol, J.A.C. (1974). Studies on the eyes of catfish with special reference to the tapetum lucidum. Proc. Roy. Soc. Lond. B. 186 : 13-36.

Blaxter, J.H.S. (1968). Light intensity, vision and feeding in young plaice. J. exp. mar. Biol. Ecol. 2: 293-307.

Blaxter, J.H.S., and Jones, M.P. (1967). The development of the retina and retinomotor responses in the herring. J. mar. biol. Ass. U. K. 47: 677-697.

Blaxter, J.H.S., and Staines, M. (1970). Pure-cone retinae and retinomotor responses in larval teleosts. J. mar. biol. Ass. U. K. 50: 449-460.

Braemer, W. (1957). Verhaltensphysiologische Untersuchungen am optischen Apparat bei Fischen. Z. vergl. Physiol. 39: 374-398.

Chèze, G. (1974). L'épiphyse des poissons téléostéens. Techniques d'approche. Recherches morphologiques et expérimentales. Thèse D. ès Sciences Naturelles. Université Paul Sabatier de Toulouse. pp. 352.

Engström, K. (1961). Cone types and cone arrangement in the retina of some gadids. Acta Zool. 17: 227-243.

Engström, K. (1963a). Structure, organization and ultrastructure of the visual cells in the teleost family Labridae. Acta Zool. 19: 1-41.

Engström, K. (1963b). Cone types and cone arrangements in teleost retinae. Acta Zool. 19: 179-243.

Engström, K., and Rosstorp, E. (1963). Photomechanical responses in different cone types of Leuciscus rutilus. Acta Zool 19: 145-160.

Exner, S., and Januschke, H. (1905). Über Verschiebung der Tapetummasse in Chorioidealepithel unter dem Einfluß des Lichtes. Zentralbl. Physiol. 19: 327-328.

Exner, S., and Januschke, H. (1906). Die Stäbchenwanderung im Auge von Abramis brama bei Lichtveränderungen. Sitz. - Ber. Akad. Wissensch. Wien, Math. -Nat. Kl. 115: 243-244.

Fineran, B.A., and Nicol, J.A.C. (1974). Studies on the eye of New Zealand parrotfishes (Labridae). Proc. R. Soc. Lond. B. 186: 217-247.

Frisch, v. K. (1924). Sinnesphysiologie der Wasserthiere. Verh. Dt. Zool. Ges. 29: 34.

Girsa, I.I. (1967). Specific retinomotor reaction features of the retina of the eye when fish are being lured by light. Vop. Ikhtiol. 7: 540-546.

Greenwood, P.H., Rosen, D.E., Weitzman, S.H., and Myers, G.S. (1966). Phyletic studies of teleostean fishes, with a provisional classification of living forms. Bull. Am. Mus. Nat. Hist. 131: 341-455.

John, K.R., and Haut, M. (1964). Retinomotor cycles and correlated behavior in the teleost Astyanax mexicanus (Fillipi). J. Fish. Res. Bd. Can. 21: 591-595.

John, K.R., Segall, M., and Zawatzky, L. (1967). Retinomotor rhythms in the goldfish, Carassius auratus. Biol. Bull. mar. biol. Lab., Woods Hole 132: 200-210.

John, K. R., and Gring, D.M. (1968). Retinomotor rhythms in the bluegill, Lepomis macrochirus. J. Fish. Res. Bd. Can. 25: 373-381.

John, K. R., and Kaminester, C.H. (1969). Further studies on retinomotor rhythms in the teleost Astyanax mexicanus. Physiol. Zool 42: 60-70.

Kobayashi, H. (1957). Notes on retinomotor phenomena in some fishes under various light conditions. J. Shimonoseki Coll. Fish 7: 169-177.

Kuenzer, P., and Wagner, H.-J. (1969). Bau und Anordnung der Sehzellen und Horizontalen in der Retina von Nannacara anomala (Cichlidae, Teleostei). Z. Morph. Tiere, 65: 209-224.

Müller, H. (1954). Die Dunkeladaptation beim Guppy (Lebistes reticulatus P.) Z. vergl. Physiol. 37: 1-18.

Nicol, J.A.C. (1961). Photomechanical changes in the eyes of fishes. I. Retinomotor changes in Solea solea. J. mar. biol. Ass. U. K. 41: 695-698.

Nicol, J.A.C. (1965). Retinomotor changes in flatfishes. J. Fish. Res. Bd. Can. 22: 513-520.

O'Connell, C.P. (1963). The structure of the eye of Sardinops caerulea, Engraulis mordax, and four other pelagic marine teleosts. J. Morph. 113: 287-329.

Olla, B.L., and Marchioni, W.W. (1968). Rhythmic movements of cones in the retina of Bluefish, Pomatomus saltatrix, held in constant darkness. Biol. Bull. mar. biol. Lab., Woods Hole 135: 530-536.

Peyraud, C.M. (1966). Contribution à l'étude de la dynamique des pigments et des photorécepteurs rétiniens chez la carpe (Caprinus carpio L.). C. r. Acad. Sci. Paris 263: 65-67.

Pfeiffer, W. (1968). Retina und Retinomotorik der Dipnoi und Brachiopterygii. Z. Zellforsch. 89: 62-72.

Pora, E.A., Wittenberg, C., Rusdea, D., and Dragos, M. (1963). Observations on retinomotor phenomena among some fishes of the Black Sea. Stud. Cercet. Biol. 14: 299-304.

Protasov, V.R. (1964). Some features of the vision of fishes. Inst. Morfologii Zhivotnykh, Akad. Nauk. SSSR pp. 29-48. Translation Mar. Lab. Aberdeen, No. 949.

Protasov, V.R., Altukhov, Y.P., and Kovaleva, N.D. (1960). Morphological and functional features of the transition from day vision to twilight vision in some Black Sea fish. Dokl. Akad. Nauk. SSSR 134: 195-198. Translation in Dokl. biol. sci. sect. 134: 694-696 (1961).

Saxén, L. (1954). The development of the visual cells. Embryological and physiological investigations on Amphibia. Ann. acad. sc. Fenn. Ser. A., IV Biologica 23: 1-93.

Tamura, T. (1957). On the relation between the intensity of illumination and the shifting of cones in the fish retina. Bull. Jap. Soc. Sci. Fish 22: 742-746.

Verrier, M.L. (1928). Recherches sur les yeux et la vision des poissons. Suppl. au Bull. biologique de France et de Belgique, Suppl. XI.

Wagner, H. -J. (1970). Der Bau der Retina und der multiplen optischen Pupille bei zwei Synodontis-Arten (Teleostei, Siluroidae). Z. Morph. Tiere 68: 69-82.

Wagner, H. -J. (1972). Vergleichende Untersuchungen über das Muster der Sehzellen und Horizontalen in der Teleostier-Retina (Pisces). Z. Morph. Tiere 72: 77-130.

Wagner, H. -J. (1974). Die Entwicklung der Nezthaut von Nannacara anomala (Cichlidae, Teleostei) mit besonderer Berücksichtigung regionaler Differenzierungsunterschiede. Z. Morph. Tiere 79: 113-131.

Wang, C.S.J. (1968). The Eye of Fishes with Special References to Pigment Migration, 291 p. Ph. D. Thesis, Cornell University.

Weidemann, H. L. (1966). Der Einfluß quantengleicher Farblichter auf die Retinomotorik dreier Zapfentypen in der Netzhaut des Guppy (Lebistes reticulatus Peters). 49 p. Thesis, Mathematisch-Naturwissenschaftliche Fakultät der Georg-August-Universität Göttingen.

Wunder, W. (1926b). Über den Bau der Netzhaut von Süßwasserfischen, die in großer Tiefe leben. Z. vergl. Physiol. 4: 22-36.

Wunder, W. (1930). Bau und Funktion der Netzhaut beim Zander (Lucioperca = Stizostedion sandra) und einigen anderen im Balatonsee häufigen Fischarten. Z. vergl. Physiol. 11: 749.

Zyznar, E. S. (1973). A study of the tapeta lucida of eight teleost fishes. Ph. D. thesis. University of Texas. Austin, Texas. pp. 148.

Zyznar, E. S., and Ali, M.A. (1975). An interpretative study of the organization of the visual cells and tapetum lucidum of Stizostedion. Can. J. Zool. (in press).

HORIZONTAL EYE MOVEMENTS IN FISH

S.S. EASTER, Jr.

Department of Zoology, University of Michigan

Ann Arbor, Michigan 48104, U.S.A.

> See my lips tremble, and my eyeballs roll
> Pope, Eloisa to Abelard

INTRODUCTION

The enormous group of animals included under the label *Pisces* defies generalisation in most respects. A herpetologist friend of mine once described the teleosts as "a bottomless pit" and to be avoided for that reason. Nevertheless, I have attempted to study eye movements in fish, realising at the outset that what is true for the goldfish, my choice, may not be true for another. Indeed, we know this to be so already, for Lanchester and Mark (1969) have described eye movements in a foveated reef fish, and the fixation behaviour evident in this very visual fish is never seen in goldfish, an animal much more dependent on chemical than visual stimuli. Why, then, study the goldfish? The main reason is that our knowledge of this species' visual system is quite detailed, and becoming more so. Therefore it seemed an appropriate organism in which to interpret visually-cued behaviour. Also the eye movements are a very objective and easily measured behaviour.

There are two main points to be inferred from the work described below. First, the smooth eye movements made by goldfish are predominantly controlled by visual stimuli. Secondly, saccadic eye movements, although influenced by visual stimuli, are not sensitive to the position of the stimuli.

In what follows, I shall first describe experiments on restrained animals, then go to freely swimming fish, and return to the restrained state. In all cases, the measurements will be restricted to the horizontal plane. I shall deal first with the experiments leading to the two statements made in the preceding paragraph, then go on to the subject of what parts of the brain are involved in eye movements, and finish with an observation on their ontogeny.

OBSERVATIONS

Unanesthetised animals were restrained by clamping them between two sponge-lined plates, inside a transparent, water filled cylinder. I developed an optical gadget which samples the orientation of the eyes sixty times per second and gives out a pair of analog voltages linear with the angular orientation of each eye in the horizontal plane. This machine has already been described (Easter, 1971; 1972).

When a fish is restrained in this way, it spontaneously moves its eyes from side to side (Johnstone and Mark, 1969; Hermann and Constantine, 1971), in steps, as shown in the first part of Fig. 1. This goes on in light or dark. When the animal is surrounded by a visible striped drum, and the drum rotates about its vertical axis, the eyes respond with optokinetic nystagmus (OKN) as illustrated by Fig. 1 (Easter, 1972). Note that the wide range of positions assumed by the eyes in the absence of movement is very much diminished once the movement begins. When the drum rotates counterclockwise, the eyes shift toward an extreme clockwise position, and vice-versa. The nystagmus consists of opposed smooth and saccadic eye movements, and the saccades are the larger, with the result that the eyes shift toward the direction from which the movement is originating. It might be guessed that this represents an effort to centre the eyes on the site of origin of the movement, which is of course impossible inside the drum, since the movement originates nowhere, and exists everywhere. But this hypothesis was ruled out by limiting the fish's field of view to very small (45 degrees of horizontal subtense) windows through which the drum was viewed. It turned out that when the window was visible to only one eye, and situated either in the front or rear of that eye's field of view, the nystagmus was always the same. For instance, when the window was in the anterior field of the left eye and invisible to the right, the positions of the eyes during nystagmus still depended only on the direction of rotation. When the drum rotated clockwise in the anterior field, the left eye shifted to a counterclockwise position, with the result that the movement which prompted the nystagmus was seen only out of the corner of the eye, rather than through the centre. Analogous results were obtained for other positions of the window, which led to the conclusion that saccadic eye movements are independent of the position, and depend only on the direction of movement, of visual stimuli.

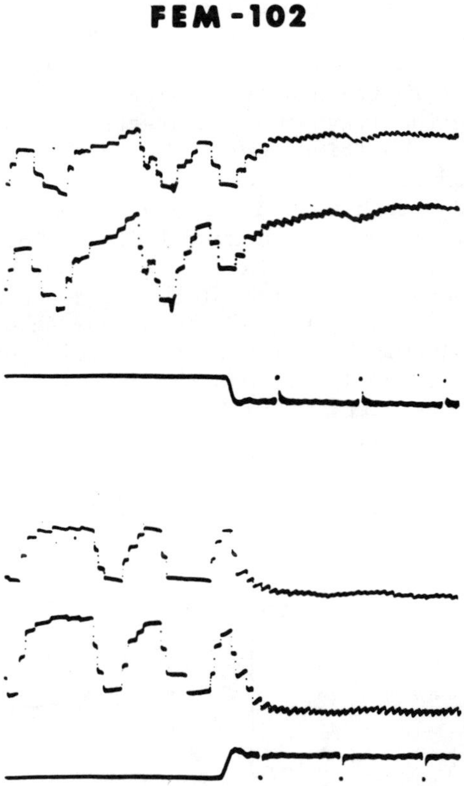

FIG. 1. OKN and ocular position. In both records, the left eye is above, right eye below, drum velocity lowest. Clockwise rotation, viewed from above, is downward in all traces. (From Easter, 1972).

The smooth movements, or slow phase of OKN, are also tailored to the drum movement. In general, the pursuit velocity is a fraction of the drum velocity; this fraction, or the velocity gain, can be as high as 0.9. It has been shown in other animals (Horridge, 1966; Collewijn, 1972) that the smooth puisuit system is servomechanistically controlled. Servomechanisms typically have gain which is smaller in the closed than open loop condition. I have investigated the goldfish smooth pursuit system to see if the same result obtains. The experiment was done as follows. One eye was blinded, either, by retrobulbar crush of the optic nerve or by masking the cornea. The effectiveness of the blinding was confirmed by establishing that the animal failed to make pursuit eye movements when a moving target was presented to this eye alone. The other eye was sighted, but restrained by reining its corneal stalk by a thin thread. The effectiveness of the restraint was confirmed by monitoring this eye's position during the experiment. When a moving target was presented to the sighted eye under these conditions, the blind eye responded with nystagmus, the slow phase velocity of which was often much greater than the drum velocity. The results of one such experiment are illustrated in Fig. 2, in which the velocity gain is 6. This number ranged from 1 to 10, depending on the individual fish, and on the subtense of the window through which the moving pattern was seen. The bigger the fraction of the field occupied by the target, the larger was the open loop gain. This high open loop gain is consistent with the smooth pursuit system operating servomechanistically.

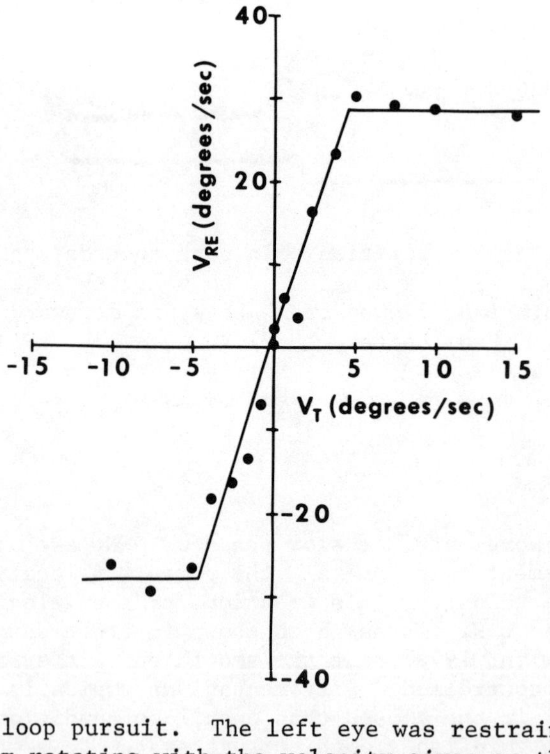

FIG. 2. Open loop pursuit. The left eye was restrained and viewed a drum rotating with the velocity given on the abscissa. The right eye was blind and responded with pursuit movements of velocity given on the ordinate.

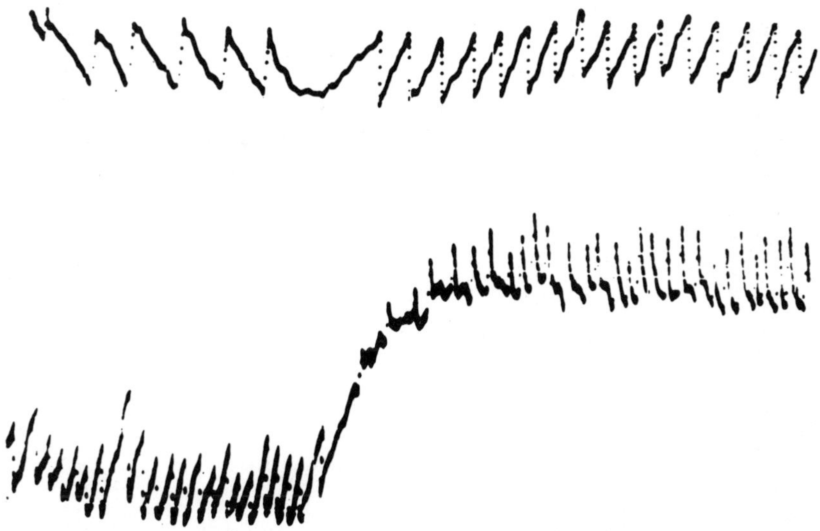

FIG. 3. Passively evoked OKN. The left eye (above) was moved nystagmoidally by the experimenter, in view of a stationary striped drum. The right eye (below) was blind, and responded to the contralateral retinal image velocity.

This question was investigated further by passively moving the sighted eye (by pulling on the rein) in the presence of a visible, stationary drum. The results of this experiment are shown in Fig.3. The passive movements of the left eye are shown in the upper trace; the right eye's response is shown below. Note that the direction of the slow phases are reversed, as expected, since clockwise rotation of the left eye in a stationary field induces retinal image movement which corresponds to counterclockwise rotation of the field about a stationary eye. Note also that the slow phase velocity of the right eye exceeded that of the left; i.e., the velocity gain was greater than unity. When this experiment was reproduced in the dark, or with both eyes blinded, the right eye did not respond. Thus the nystagmus depended on visual cues alone, and not proprioception.

The conclusions begged the question of the degree of involvement of the visual system in compensatory eye movements made by freely swimming fish. We investigated this by filming them, sighted and blind, from above (Easter, Johns and Heckenlively, 1974; Easter and Johns, 1974). Stalks were attached to their eyes, and the angles of the stalks and the animal's head were obtained from a frame-by-frame analysis of the film. As any observer of goldfish can confirm, the animals are nearly constantly in motion, and

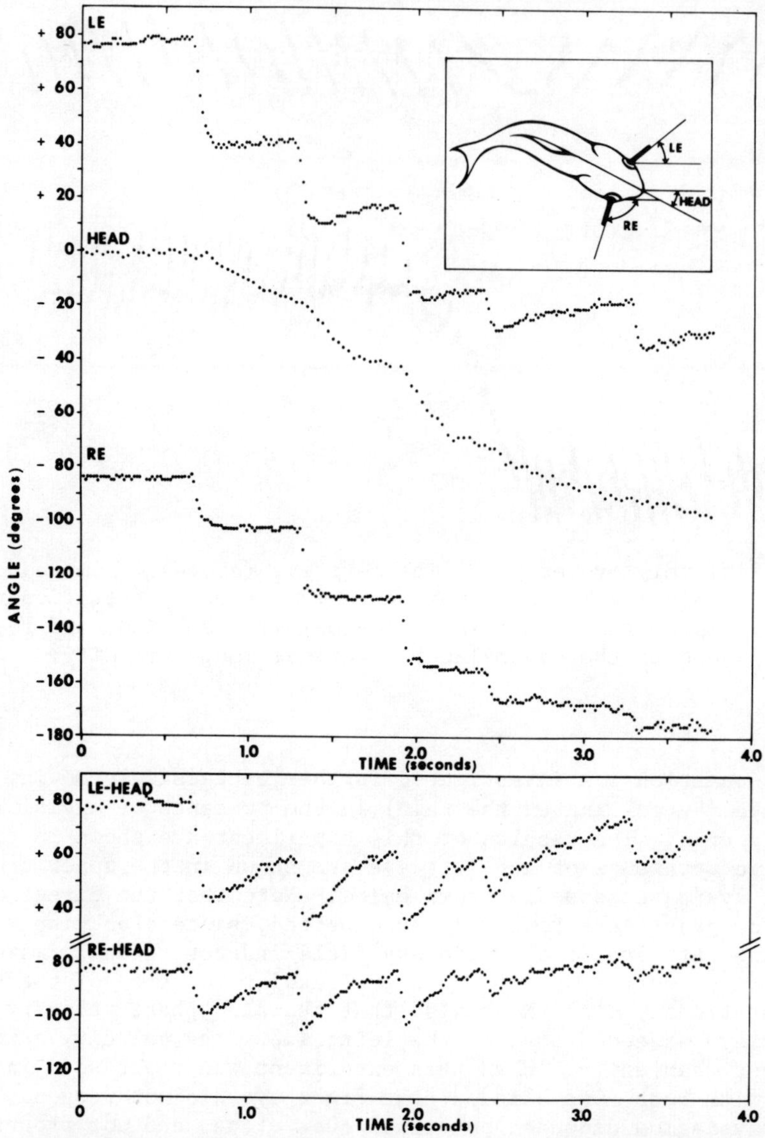

FIG. 4. The upper three traces were measured as shown in the inset. The lowest two traces represent eye positions relative to the head. (From Easter, Johns, and Heckenlively, 1974).

Fig. 4 illustrates the coordination of eyes and head. As the record begins, the animal was swimming in from the left with its stalks roughly symmetrically aligned relative to the head. Just before turning to the right, it flicked both eyes in that direction. Notice that the saccades are quite unequal, just as had been observed in restrained animals (Easter, 1971). After completion of the first pair of saccades, the head continued to turn, but the eyes' orientations remained relatively constant, indicated by the nearly flat plateaux. This implies, of course, that the eyes were rotating relative to the head with an angular velocity opposite and roughly equal to that of the head. As the turn continued, more saccades and counter-rotations occurred until the turn was completed.

The accuracy of these compensatory movements was investigated quantatively by measuring the amounts of rotation by head and eyes between saccades. A numerical estimate of the degree of compensation is given by the "compensation factor" (CF), defined as:

$$CF = \frac{\Delta E - \Delta H}{\Delta H},$$

in which ΔE and ΔH are the intersaccadic angular displacements of eyes and head, respectively. The numerator of the fraction gives the ocular rotation relative to the head, while the denominator gives that of the head alone. When CF = -1, compensation was perfect. We computed CF's for a large number of turns and found that compensation was, in fact, nearly perfect, on the average:

$$CF \pm 2\ S.E.M. = -0.95 \pm 0.10.$$

We sought next to learn what inputs were important to the execution of these compensatory eye movements (Easter and Johns, 1974). We did so by depriving the animals of either vision, horizontal canals, or both, and measured the CF's exactly as we had done with the normals. The results are summarised by:

CF (canal lesioned) = -0.80
CF (blind) = -0.50
CF (blind, canal lesioned) = -0.41.

The contribution of vision to the CF was obtained by:

CF (normal) - CF (blind) = -0.45
CF (canal lesioned) - CF (blind, canal lesioned) = -.39.

The two differences are equal, within the experimental error. The contribution from the canals was obtained by:

CF (normal) - CF (canal lesioned) = -0.15
CF (blind) - CF (blind, canal lesioned) = -0.09.

Again, these two differences are essentially identical, within experimental error.

There were two surprises here. Frist was the fact that vision contributed so much more significantly than the horizontal canals. Second was the fact that with neither canals nor vision functional, the animals still compensated. This surprised us because we had imagined that the canals and vision would wholly control eye movements, and we had expected to find a CF of zero. However, control experiments confirmed that the canals were indeed inoperative, and blinding was always confirmed by the animal's inability to make pursuit eye movements while restrained inside the rotating striped drum. Although we tried, we were never able to identify the other sensory input responsible for the remaining compensation.

Having thus established that the visual input was the major determinant of ocular compensation, we then set out to determine what structures in the brain were involved in OKN. We looked first to the retinal projection sites. Sharma (1972) and Northcutt (unpublished observations) had found seven such areas, all contralateral, the major one of which is of course the tectum. Although no careful estimates have been made of the relative numbers of fibres going to these centres, a rough comparison of the volumes of the nuclei suggests that well over 95% of retinal afferents must terminate in the tectum. We (Easter, Northcutt and Landreth, 1974) assessed the role of the tectum by removing one eye and the ipsilateral tectum. This left the animal with one tectum intact, but no eye to innervate it, and one eye without a tectum to connect with. Following the surgery, we waited two weeks, long enough for the animal to recover from the trauma, but too short to allow innervation of the ipsilateral tectum (Sharma, 1973). We found that eye movements, including optokinetic nystagmus, were quite normal. Evidently, pursuit eye movements do not depend on information sent to the tectum. We checked the surgery histologically, and confirmed that the entire tectum had been removed. In addition, we injected the eye with tritiated proline (Neale, Neale, and Agranoff, 1972), and made radioautographs of the slides. These showed that all the contralateral target nuclei were radioactive, but none of the ipsilateral ones were, indicating that no post-surgical sprouting into the ipsilateral side had occurred. Careful examination of these slides also revealed the existence of yet another site of retinal projections, this one quite caudal in the mesencephalic tegmentum. It may well be homologous with the basal optic nucleus, shown to be essential to optokinetic body movements made by frogs (Lazar, 1972).

The conclusion that the tectum is not involved in pursuit eye movements is in keeping with similar experiments on adult frogs (Lazar, 1972), tadpoles (Mark and Feldman, 1972) and monkeys (Pasik and Pasik, 1964). It was surprising, nonetheless, in view of the numerical dominance of tectum-bound retinal fibres.

Lesions in the brain stems of otherwise normal goldfish successfully destroyed the ability to make pursuit eye movements, even in the presence of an intact tectum. These were checked histologically, and we concluded that either the pretectal nuclei or the putative basal optic nucleus, mentioned above, must include the visual centre or centres essential to this function. Projections from these nuclei are not known, so we are at a loss to carry the analysis further. Perhaps the projections go directly to the ocular motor nuclei themselves, or perhaps there are other nuclei interposed.

In an effort more completely to determine what structures are involved in pursuit eye movements, we (Easter and Ferguson, in preparation) have made other lesions; namely, we have ablated the telencephalon and the cerebellum. The absence of the former had no effect, as we expected, but we had anticipated, on the basis of Collewijn's (1970) work on rabbits and that of Westheimer and Blair (1973) on monkeys, that cerebellectomy would affect the execution of eye movements. Our expectation was not fulfilled. Saccades were found to have the normal time course and range of velocities. Optokinetic nystagmus was also unaffected. The only abnormality observed in restrained fish was a spontaneous nystagmus when the fish were blinded, observed in three of the four animals. We filmed the fish swimming freely, and here again, behaviour was quite normal. Apart from some small locomotory problems, these fish were indistinguishable from normal. They made the typical sort of graceful turns, accompanied by binocular saccades and very accurate compensatory smooth movements. Even when blinded, the freely swimming behaviour was quite normal, although the compensation factors vadied a bit more than usually.

We conclude that pursuit eye movements are controlled exclusively in the brain stem, and involve no nuclei anterior to the diencephalon.

Finally, a word about the ontogeny of eye movements. We (Easter and Johns, unpublished observations) have investigated this problem in the Japanese medaka, *Oryzias latipes*. This particular fish is widely used in embryological instruction because it has such a clear egg shell that it is possible to view organogenesis in vivo. The eyes develop very early in ovo and their movements can be observed at low power under a dissecting microscope. The extreme transparency of the shell allows the embryo to see outside, and we have attempted to evoke pursuit eye movements by rotating a cylindrical striped drum several centimetres in diameter around the animal. We found that on the seventh day of development, the animals began to make spontaneous eye movements, and within a few hours after that, it was possible to evoke pursuit eye movements. Hatching occurred after ten days. I know of no other experiments aimed at determining when eye movements begin; but I doubt if any animal could be beginning the movements much earlier!

CONCLUSION

To summarise, fish develop very early in life the ability to stabilise the retinal image of the environment by compensatory eye movements. These movements depend critically on non-tectal visual signals, and to a lesser degree on non-visual signals. The neuronal machinery controlling eye movement seems to be restricted to the brain stem posterior to the telencephalon.

ACKNOWLEDGEMENT

The work described in this paper was supported by grant EY-00168, from the National Eye Institute of the U.S. Public Health Service, a special research grant from the L.S.A. Deans, and Project Grant 494 from the Michigan Memorial Phoenix Project to myself, grants MH-12506 from the National Institute of Mental Health and GB-32409 from the National Science Foundation to B.W. Agranoff, and grant N-508417 from the National Institute of Neurological Diseases and Stroke of the U.S. Public Health Service to R.G. Northcutt.

REFERENCES

Collewijn, H. (1970). Dysmetria of fast phase of optokinetic nystagmus in cerebellectomized rabbits. Exp. Neurol. 28: 144-154.

Collewijn, H. (1972). Ocular stability in variable visual feedback conditions in the rabbit. Brain Res. 36: 47-57.

Easter, S.S. (1971). Spontaneous eye movements in restrained goldfish. Vision Res. 11: 333-342.

Easter, S.S. (1972). Pursuit eye movements in goldfish (*Carassius auratus)*. Vision Res. 12: 673-688.

Easter, S.S., R.G. Northcutt, and G.E. Landreth (1974). Optokinetic nystagmus and the goldfish brain. Abst. Soc. Neurosc. 4.

Easter, S.S., D.R. Heckenlively, and P.R. Johns (1974). Horizontal compensatory eye movements in goldfish (*Carassius auratus*). I. The normal animal. J. Comp. Physiol. 92: 23-36.

Easter, S.S. and P.R. Johns (1974). Horizontal compensatory eye movements in goldfish (*Carassius auratus*). II. A comparison of normal and deafferented animals. J. Comp. Physiol. 92: 37-58.

Hermann, H.T. and M.M. Constantine (1971). Eye movements in the goldfish. Vision Res. 11: 313-332.

Horridge, G.A. (1966). Study of a system, as illustrated by the optokinetic response. In: Nervous and hormonal mechanisms of integration, Symp. S.E.B. 20, 179-197.

Johnstone, J.R. and R.F. Mark (1969). Evidence for efference copy for eye movements in fish. Comp. Biochem. Physiol. 30: 931-939.

Lanchester, B.S. and R.F. Mark (1969). Eye and body movements of fish tracking moving objects. Aust. J. Exp. Biol. Med. Sci. 47: 16.

Lazar, G. (1972). Role of the accessory optic system in the optokinetic nystagmus of the frog. Brain Behav. Evol. 5: 443-460.

Mark, R.F. and J. Feldman (1972). Binocular interaction in the development of optokinetic reflexes in tadpoles of *Xenopus laevis*. Invest. Opthalmol. 11: 402-410.

Neale, J.H., E.A. Neale, and B.W. Agranoff (1972). Radioautography of the optic tectum of the goldfish after intraocular injection [^3H] proline. Science 176: 407-410.

Pasik, R. and T. Pasik (1964). Oculomotor functions in monkeys with lesions of the cerebrum and superior colliculi. In: The oculomotor system, edited by M. Bender, Harper and Row: New York, pp. 40-80.

Sharma, S.C. (1972). The retinal projections in the goldfish: an experimental study. Brain Res. 39: 213-223.

Sharma, S.C. (1973). Anomalous retinal projection after removal of contralateral optic tectum in adult goldfish. Exp. Neurol. 41: 661-669.

Westheimer, G. and S.M. Blair (1973). Oculomotor defects in cerebellectomized monkeys. Invest. Ophthalmol. 12: 618-621.

5. COMPARATIVE AND DEVELOPMENTAL ASPECTS

DEVELOPMENT OF THE OPTIC TECTUM IN BROWN TROUT

S.C. SHARMA

Department of Ophthalmology, New York Medical College*

New York, New York 10029

For here comes the trout that must be caught with tickling.
Twelfth Night, Act II, Sc. 5

The complexity of cellular characteristics and their interrelationships in the teleostean optic tectum is quite apparent from Dr. Venegas' review (this volume). In order to accomplish the early events of neurogenesis, certain morphological and cellular changes occur in the presumptive optic tectum and these changes probably cause the transformation of the "matrix zone" (proliferative ventricular cells) into post-mitotic neurons.

The present study was undertaken to investigate neural differentiation, migration and growth of the optic tectum in the brown trout (Salmo fario). Fertilised eggs were kept in running water at 8-10° C in the laboratory. Hatching commenced at approximately day 71 after spawning.

In an attempt to unravel the initial stages of differentiation of the developing fish optic tectum, three well-defined matrix zones, or areas of the neural epithelium lining the tectal ventricle, were found (Kirsch, 1960; Richter & Kranz, 1970; and Schmatola and Erdmann, 1973). These areas were labelled as a) dorsal matrix zone in the medial tectum, b) basal matrix zone in the ventral tectum, and c) caudal matrix zone in the caudal tectum. Our data on

* This work was supported by Grant GB 43506 from the National Science Foundation.

developing Salmo fario confirmed the presence of such proliferative zones. We also observed that by about day 30 after spawning, there was enhanced mitotic activity in the caudal tectum as compared to the rostral tectum. However, this trend reversed itself by about day 35, and mitotic activity in the optic tectum was rarely observed after day 46.

Schmatola and Erdmann (1973) studied the effects of optic innervation in the embryonic teleost tectum and showed that the absence of the eye had no effect on the proliferation and distribution of mitotic activity. However, cellular migration and differentiation in the marginal zones was increased by optic innervation.

We observed retinal fibres in Salmo entering the rostral optic tectum around day 35, but it was not until day 46 that the first sign of the outer marginal layer appeared on the rostrolateral side of the optic lobes. A day-to-day study of the growth of the outer layers of the optic tectum was conducted.

Ramón y Cajal (1960) proposed a schema for the early differentiation of neurons in the nervous system which essentially consisted of a) germinative cells of His, b) apolar or polygonal cells, c) bipolar cells, d) unipolar cells and e) multipolar cells. The presence of germinative cells of His has not yet been confirmed, although apolar cells have been shown to exist in chick central nervous system (Sechrist, 1969). The presence of the other types of cells has been confirmed recently (Hinds and Hinds, 1974). Numerous other versions of neuronal differentiation have also been proposed (Morest, 1970; Rakic, 1972).

In the fish tectum, the scheme of differentiation is not fully understood, but some observations can be made. Post-mitotic migratory neuroblasts in the early stages of development (day 40-day 60) are oriented toward the marginal zones. Most differentiated neuroblasts in the area of greater cell division have one or more protrusions extending outward on their processes. On the rostral side of the tectum, some neurons have already started migrating outwardly. In front of the zone of cell division, close to the surface of the tectum, are a few cells which are in the late metaphase (Fig. 1). These cells may be the predecessors of cells in the stratum opticum layer. In Golgi preparations, the youngest cells to be impregnated are found near the tectal surface and are small with a single process pointing outwardly.

In the lateral tectum cell differentiation is well advanced in Golgi preparations. Dendritic branches have started arborising, some with beaded side branches. Presence of these cells is well marked at day 54.

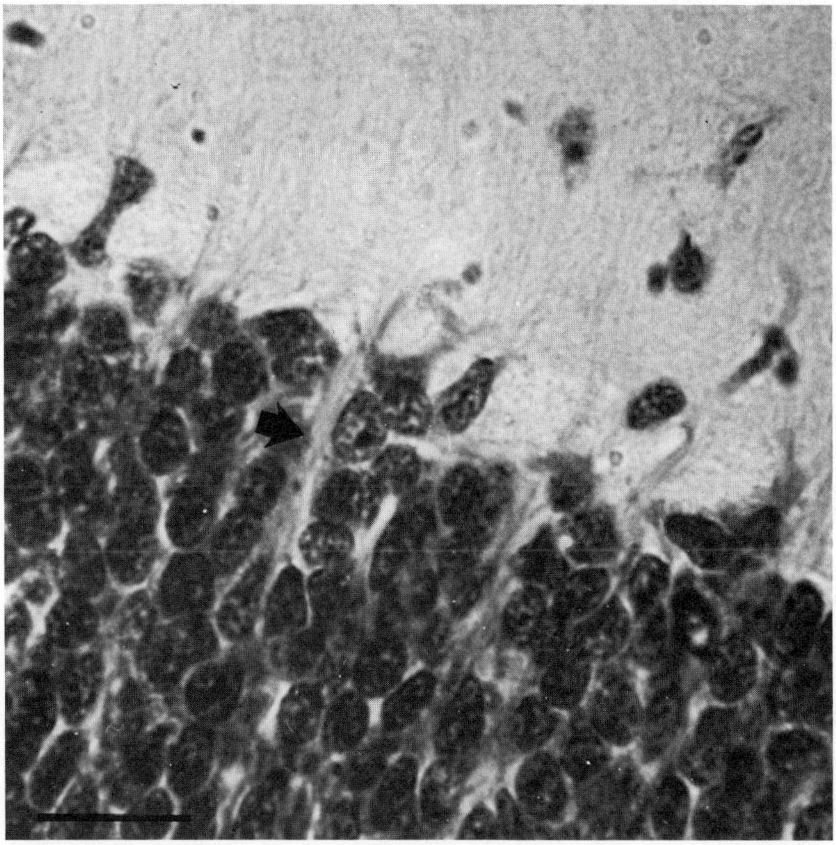

Fig. 1 Photomicrograph of the transverse section of the developing optic tectum at day 45 after spawning. Notice the presence of numerous thin fibres and fibre bundles (arrow) which are extensions of the newly differentiating underlying cells. At top left, notice a cell in the late metaphase. Iron Haematoxylin stain. Scale: 20 um.

Ganglionic cells are first seen around day 55 in Golgi preparations on the lateral tectum. At this stage they are single with one or two laterally oriented dendrites.

A few days before hatching (day 62), cell population in the outer layer has increased, although stratification is not very clear. The first well-marked layer is the stratum opticum. There is a significant increase in the number of cells in the outer zones by the day of hatching.

The tectum now becomes rather complicated. The differentiation of tectal neurons continues, and in the outer layers it results in increased growth of dendritic processes and their ramifications.

Six days after hatching, stratum album centrale is clearly demarcated.

During the early period of differentiation, the inner gray peri-ventricular zone is extensive in size as compared to the outer marginal white zone. As differentiation proceeds and cells migrate to the outer zones, the total thickness of the outer white zone increases proportionally. By day 37 after hatching the outer white zone is nearly equal in thickness to the inner gray peri-ventricular zone. Cell migrations were rarely observed at this stage. There is a progressive decrease in the thickness of the gray peri-ventricular zone and a significant increase in the outer zones. (Fig. 2) These changes in the thickness of the tectum and that of the outer white and inner gray zones were measured from photographs with identical enlargements of transverse sections made at the level of the fourth cranial nerve. From 27 to 37 days after hatching, the gray cellular zone decreases in thickness. At this time cell death is commonly observed in this area, and this is the probable cause of the decrease in thickness. From day 37 after hatching the white zones grow rapidly. The relative cell counts in different stages of development were made in the gray peri-ventricular zone. The packing density of cells in this zone does not differ considerably.

Mention should be made here on the mode of migration of tectal neurons. Freshly prepared slides of the developing optic tectum studied with phase contrast microscopy, revealed that the tectum is composed of many fine processes lying roughly parallel. These processes originate from ovoid-cells and often contain an oval cytoplasmic mass (Fig. 3a). In the later stages of differentiation (Fig. 3b) when migratory cells are reaching the outer white zone, the migratory cell is still attached to its parent cell which has started extending dendritic branches. It is my belief that when the migratory cell has reached a certain level in the tectum, it detaches itself from its parent cell after having formed connections with the surroundings.

Regarding visual functional development, the earliest opto-kinetic responses in trout embryos were observed a few days after hatching. A definite visual response to moving stripes was noticed around day 37 after hatching, the same stage in development in which the outer white zone was equal to the inner gray peri-ventricular zone in the optic tectum.

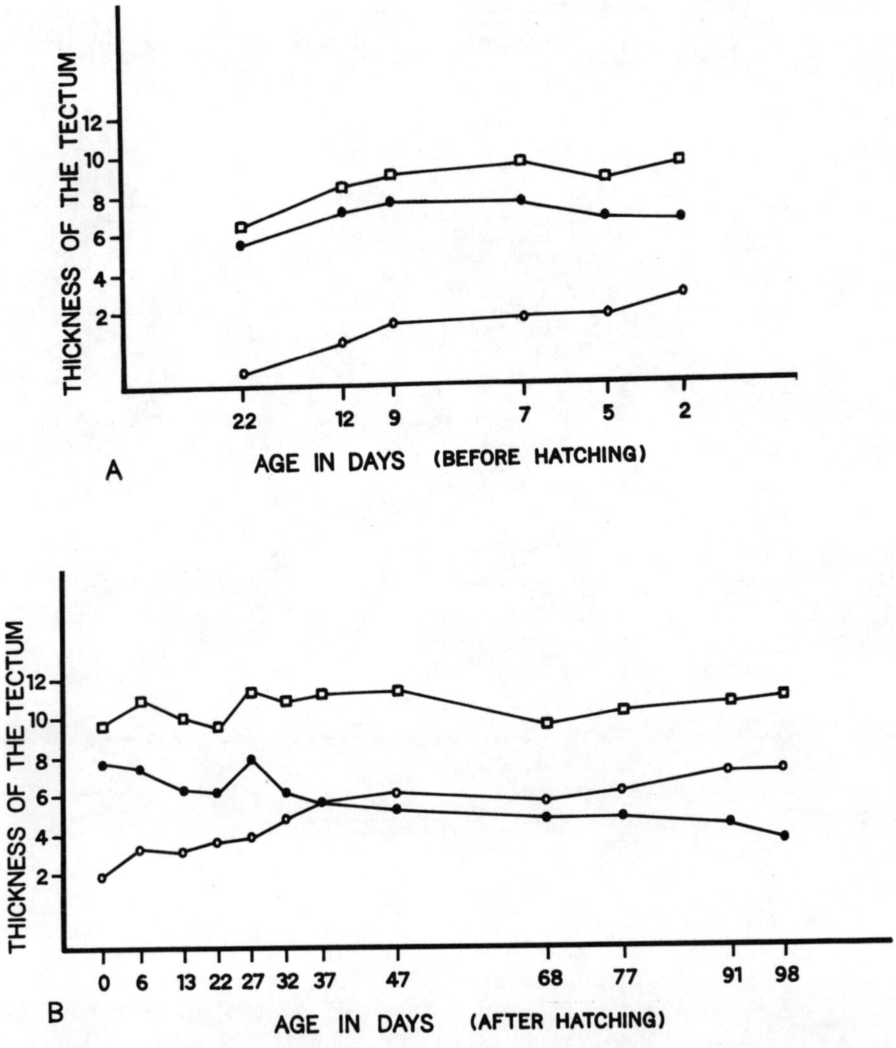

Fig. 2 Graph showing the rate of growth of the optic tectum in relation to the growth of white and gray zones: a) before hatching, b) after hatching. Open squares represent total width of the tectum; solid circles - width of the inner gray peri-ventricular zone; open circles - width of the outer white zone.

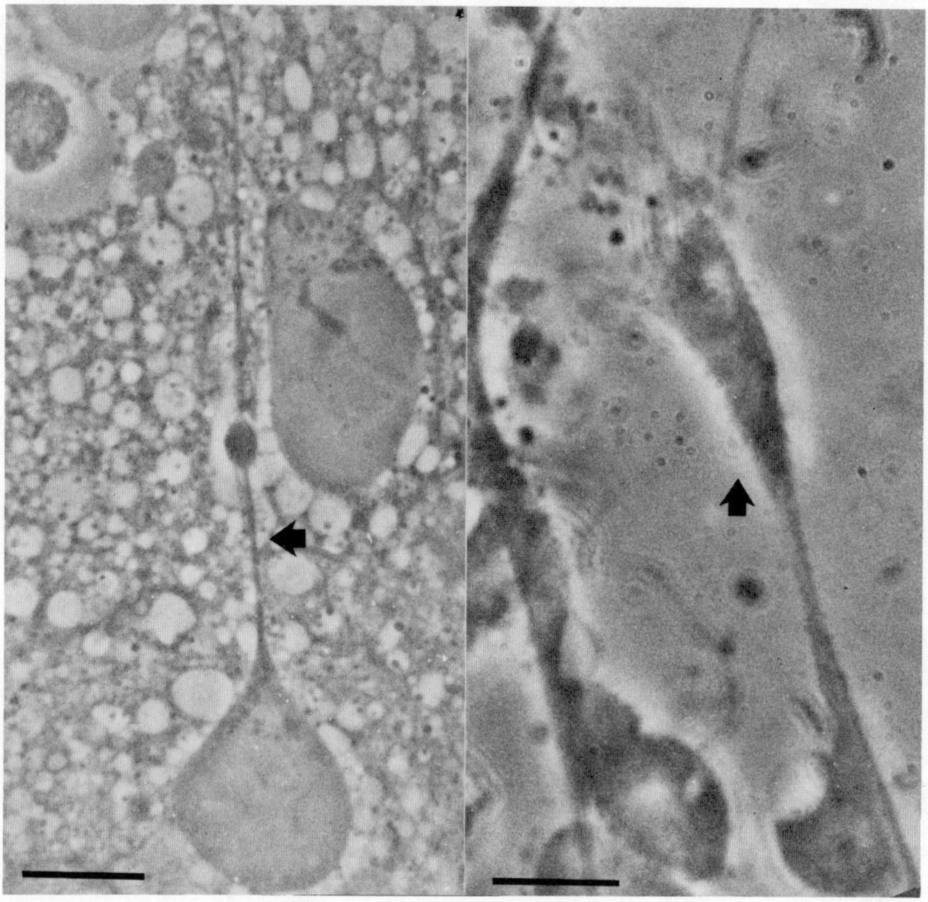

A B

Fig. 3 Fresh preparations of the optic tectum. Phase contrast photographs. A. 10 days before hatching. An ovoid cell body with an elongated process (arrow). Note the oval mass in the process. Scale: 10 um. B. 10 days after hatching. Notice the dendritic formation of the migratory neuroblast (arrow) still attached to the parent cell at the bottom. Scale: 20 um.

This preliminary report on the development of the optic tectum of Salmo fully recognises that information provided herein is far from sufficient. However, this study offers several bases for potential investigation. Exploration of the early neural differentiation in the tectum and correlation of anatomical findings with electrophysiological work on the developing visual system should certainly extend our understanding of the complexity of this heretofore unexplored area of fish vision. It is further suggested that due to their rather lengthy embryonic and larval life spans, trout provides an ideal experimental animal for further studies on the developing visual system in fish.

REFERENCES

Hinds, J.W. and Hinds P.L. (1974) Early ganglion cell differentiation in the mouse retina: an E.M. analysis utilizing serial sections. Dev. Biol. 37: 381-416.

Kirsche W. (1960) Zür Frage der Regeneration des Mittelhirnes der Teleostei. Verh. Anat. Ges. 56: 259-270.

Morest, D.K. (1970) A study of neurogenesis in the forebrain of young opposum. Z. Anat. Entwicklungsgesch 130: 265-305.

Rackic, P. (1972) Mode of cell migration to the superficial layers of foetal monkey neocortex. J. Comp. Neurol. 145: 61-84.

Richter, W. and Kranz, D. (1970) Die Abhängigkeit der DNS - Synthesis in den Matrixzonen des Mesencephalous vom Lebensalter der Versuchstiere (Lebistes reticulatus - Teleostei) Autoradiographische Unterschungen. Z. Mikrosk. Anat. Forsch. (Lerpg) 82. 76-92.

Ramón y Cajal, S. (1960) Studies on vertebrate neurogenesis (revised and translated by L. Guth). Thomas. Springfield, Ill.

Schmatola, E. and Erdmann G. (1973) Influence of retinotectal innervation on cell proliferation and cell migration in the embryonic teleost tectum. J. Embryol. Exp. Morph. 29: 697-712.

Sechrist J.W. (1969) Neurocytogenesis, neurofibrils, neurofilaments and terminal mitotic cycle. Am. J. Anat. 124: 117-134.

GENETIC PATHS LEADING TO BLINDNESS IN *ASTYANAX MEXICANUS*

Perihan SADOGLU

Div. of Biological & Medical Sciences, Brown University

Providence, Rhode Island 02912

Thine eyes, sweet lady, have infected mine.
King Richard The Third, Act 1, Scene 2

INTRODUCTION

The nature of the genetic controls of the eye and its development is still a largely unexplored field. Embryologists postulate that lens development is initiated by a process of inductive interaction of the developing eye cup and the surface ectoderm. Little light has been shed, however, on the action of genetic controls in this process, and particularly on the degree of independence of the controls of the various eye components.

What work has been done in this area has been done on mutations (discovered in dogs, rats and mice) that cause either lens abnormalities (reviewed by Varnum and Stevens, 1968) or optic degeneration as a whole (Parry et al., 1953, Bourne et al., 1938). These studies suggest specific genetic controls for these abnormalities and often point to extensive pleitrophic effects of the mutant genes. Because of the isolated nature of these mutations, however, the relations of the various genes involved in the eye remain undetermined.

The blind cave varieties of *Astyanax mexicanus* provide a unique opportunity for the study of genetic controls since they alone of the blind cave animals can breed with their normal eyed, surface ancestor (R_t) to produce fertile offspring. Their blindness results from a natural collection of mutations which affect the eye, and the crosses with sighted forms allow a dissection of the eye into its genetic components. Since I first succeeded in crossing these fish with the R_t ancestor nearly twenty years ago, work with them, though still in its infancy, has produced some interesting results, and considerable conflict among investigators.

QUANTITATIVE WORK

Sadoglu (1957) postulated that different genes control the size of the lens and the size of the eye as a whole. Quantitative studies based on the eye and lens diameters of seven-day-old larvae of parental forms and of first (F_1H), second (F_2H), and backcross hybrids, indicated a polygenic system controlling both the size of the lens and the size of the eye, but the persistence of bimodal distribution in some of the crosses indicated that one or two size, or structural genes, might be segregating on the polygenic background. In addition, a higher coefficient of variation for the lens, as shown in the chart below, though it does not prove independence of eye size genes and lens size genes, does strongly suggest that fewer genes operate in determining the lens size.

TABLE I

Coefficients of Variation (C V) for the relative Eye and Lens Diameters as Percentage of total Body Length in F_2H Generation Crosses between Chica Cave ♀ and R_t ♂ fish. (N = 1068)

	MEANS ± S.E.	C V
EYE	7.24 ± .016	.070
LENS	1.88 ± .018	.318

Pfeiffer (1967), however, finding a significant correlation between eye and lens size in adult F_2H fish, proposed that the genes responsible for the size of the eye also determine the development and size of the lens. He concluded that two additive pairs of genes, segregating independently, provide sufficient explanation for the multiplicity of eye sizes he observed. Similarly, Wilkens (1971) stressed the impossibility of separating single genes into those responsible for the lens and those for the eye-cup, preferring to speak only of eye genes in general rather than of lens genes, eyeball genes, etc. Although he also reported a biomodal distribution of eye size, he interpreted this to be the result of partial dominance of certain eye genes, or the influence of genic balance. He concluded that the eye gene system is organised hierarchically: "the manner of its manifestation therefore is less dependent on which than on how many genes are present".

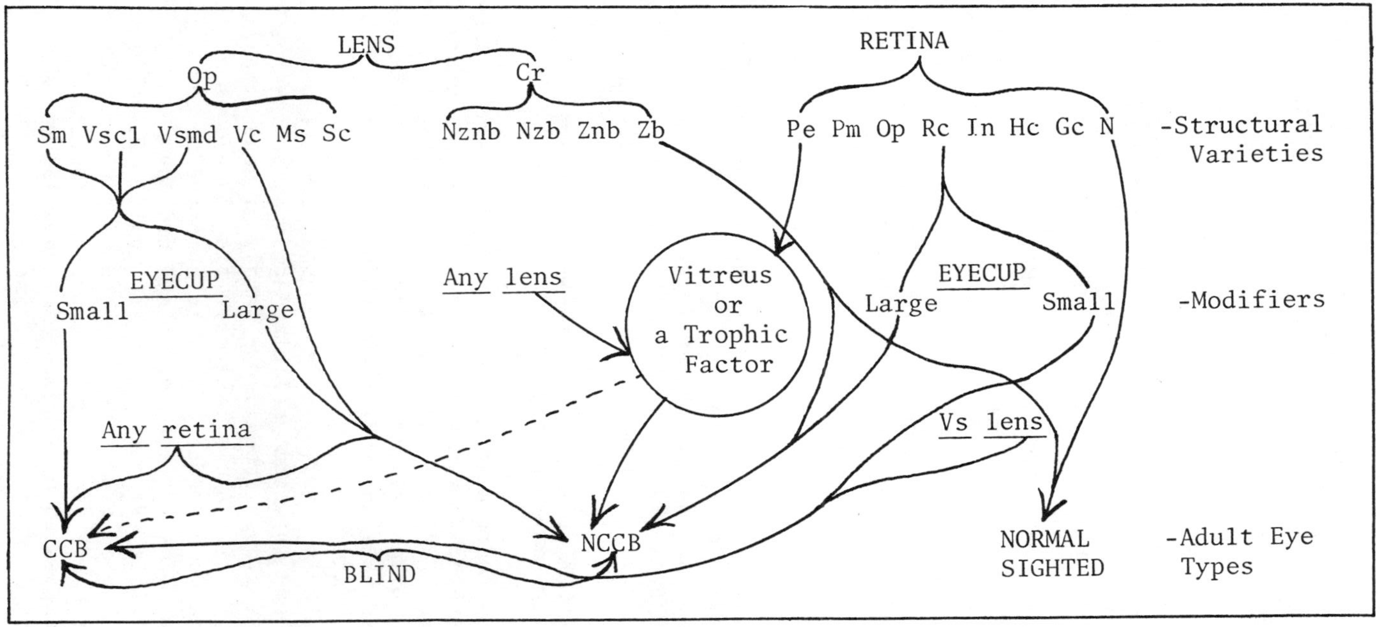

Fig. 1. A chart of structural varieties in the larval eye of *Astyanax* and their observed relationship to adult eye types. Pathways for the adult types for some of the structural types are not shown since they have not yet been isolated for observation of their resultant adult eye types, though the author feels that the structural defects preclude a normal eye. The broken line from the circled, postulated trophic factor to CCB indicates some uncertainty about this pathway. Normal eyes have a normal (N) retina and a zoned and brittle lens (Zb).

The variety of the above interpretations and their incompatibility clearly indicate that statistical work alone fails to provide an unambiguous picture of the genetic controls operating in the eye of *Astyanax*.

QUALITATIVE WORK

Fortunately quantitative variations are the least important of the observable differences in *Astyanax* populations. Numerous structural variations also appear, especially in F_2H crosses.

Cahn (1958) studied the optic development of *Astyanax* from fertilization up to the fifth larval day. She found that the lenses of the surface fish showed, 24 hours after fertilization, two zones of lens nuclei, an inner and an outer. The inner cells appeared to elongate and arrange themselves concentrically into the primary fibre cells while the outer zone cells formed into lens epithelium.

In cave populations, normal lens fibres never developed. Cahn postulated that this might be due to inadequate inductive ability of the small eye buds of the cave fish to induce normal lenses. Retinal tissue in the cave fish is well differentiated by the fifth larval day in the central region of the eye, but in the dorsal and ventral areas it has coalesced into an undifferentiated mass. Following the theory of Chase and Chase (1941) in their studies on anophthalmic mice, Cahn postulated that the primary effect of mutant genes in *Astyanax* was to retard the growth of the optic vesicle, with retinal degeneration and inadequate lens differentiation as secondary effects of inadequate optic inductive ability.

In my own studies, carried out on seven-day-old larvae, I have found, as did Cahn, that cave fish never develop a lens comparable to that in normal eyed R_t fish. At this stage the R_t lens is crystalline (Cr) and has three distinct zones (see Fig. 2). The cave fish on the other hand have an apparently opaque (Op) lens. Histological studies show that these lenses are either a solid mass (Sm) or, more usually, vesicular (Vs). Vs lenses usually have a residue within the lumen and may show some cellular activity on the posterior side of the vesicle (vesicular modified - Vsmd), or none at all (vesicular classic - Vscl). Almost all cave fish have eyes at this stage, though they are much reduced in size compared to those of the Rt fish, and later sink into the orbit and are covered by layers of skin leaving no visible eyes at all in the adult. This classic cave blindness (CCB) breeds true, even in crosses between various isolated cave populations. Studies on the sexually mature fish show that the lens has disappeared or remains as a capsule, while the retina shows differing degrees of degeneration.

F_1H hybrids between Rt and cave populations have a normal, functional eye, though it is smaller than that of the surface parent. These eyes remain at the surface, and function through out the life of the fish. This indicates that a single normal allele from each eye gene will suffice to produce a structurally normal eye, while eye size depends on the nature of all alleles.

F_2H hybrids show an enormous variety of combinations of optic structures involving especially the lens and the retina. They sort themselves nicely, however, into a set of structures for the retina and another for the lens which segregate independently.

In addition to the normal Cr lens which is zoned and brittle and the Op types found in the pure cave forms, three additional Cr and three more Op types appear. The retina, which is mostly normal in both Rt and cave populations shows seven additional structural types in the F_2H populations. As is shown in the chart, Figure 1, some of these abnormalities definitely lead to blindness with eye cup size acting as a modifier determining whether the blind eye in the adult will be completely sunken (CCB) when the eyecup is small or remain at or near the surface (non-classic cave blindness -NCCB when the eyecup is large. In Figures 2 through 9, a few of the larval variations for lens and retina are shown. It is important to note that the structural varieties of retina and lens manifest themselves independently of each other. This structural independence strongly suggests that there are separate genetic controls for the eye and for the lens. This hypothesis finds strong support in the few genetic studies that I have been able to do with these fish, though much of the potential genetic work remains incomplete owing to the difficulty of raising certain crosses.

Among the structural types, two types of lens and one type of retina have been genetically isolated. Fish with Vs lenses, which comprise 18 to 26% of the F_2H offspring, produce 100% Vs lenses when crossed, though some of the offspring of parents with Vscl lenses will have Vsmd lenses. The Vs characteristic seems to involve a degeneration of inner zone cells in the developing lens placode, and these crossing results indicate that a single recessive gene or supergene controls the expression of this degeneration. Another, but rarer type of lens, which appears in all F_2H crosses - the Vacuoler (Vc), also breeds true to opacity, producing no normal or Cr lenses. The offspring of Vc to Vc crosses are 81% Vc with the remainder, another Op type, scrambled, with large unorganised granular cells (though this may be a stage in the development of Vc lenses). This also indicates a recessive gene or supergene in control of this lens type. The nature of the genetic controls of the other types of lenses has not yet been determined, and they may be the result of certain genetic recombination or developmental retardation.

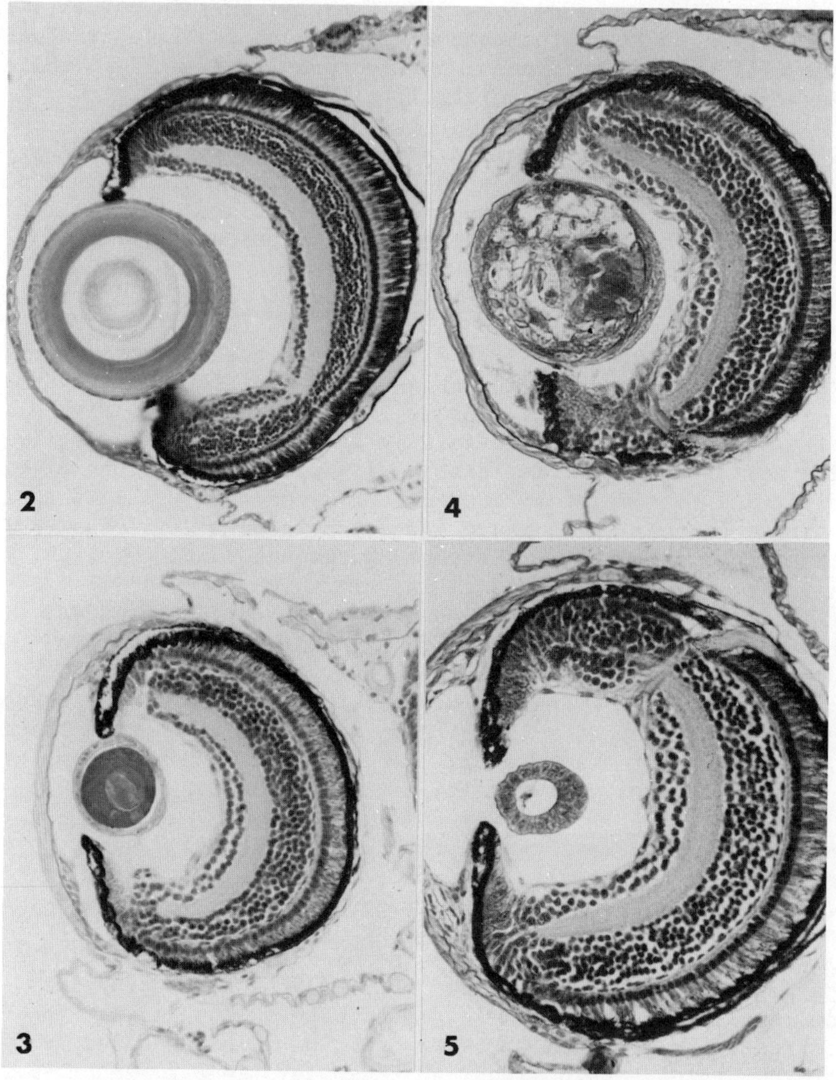

Fig. 2-5. Various lens types with normal (N) retina.
Fig. 2. From a surface fish, a normal Cr lens, zoned and brittle (Zb). (X218).
From hybrids, all (X300).
Fig. 3. Another Cr lens without zonation but brittle (Nzb).
Fig. 4. An Op, Vacuoler (Vc) lens.
Fig. 5. An Op, Vesicular classic (Vscl) lens, with no cellular activity appearing in the vesicle.

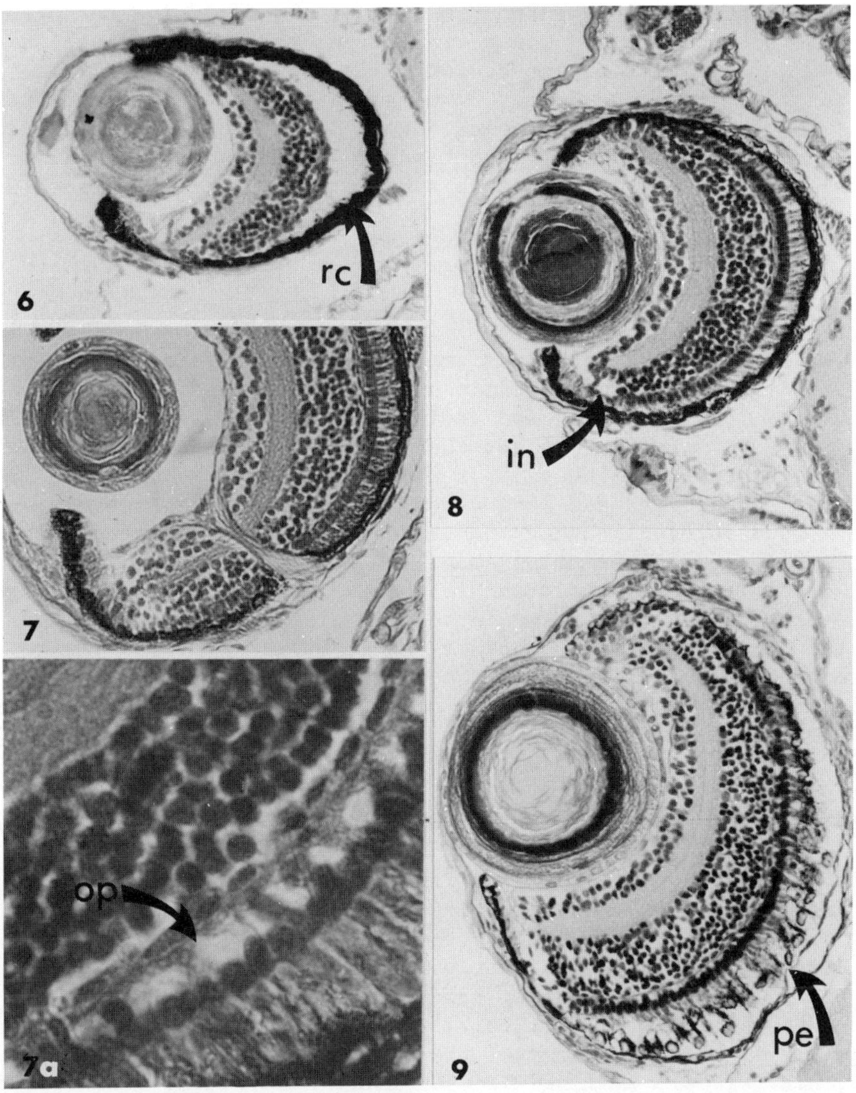

Fig. 6-9. Various retinal types, all appearing with normal lens, all X300.
Fig. 6. Retina Rc, rodless and coneless, lacking all photoreceptors.
Fig. 7. Retina Op, degenerating areas in the outer plexiform layer.
Fig. 7a. Enlargement of Fig. 7 (X1170).
Fig. 8. Retina In, gaps in the inner nuclear layer of the lip of the eye.
Fig. 9. Retina Pe. pigment epithelial cells are not contiguous and have reduced pigment. This defect always appears without any vitreus.

Similar uncertainty surrounds most of the retinal structures that have been observed. One retinal type, rodless and coneless (Rc), which lacks all photoreceptor cells, however, breeds true, this indicates that the defect is an autosomal recessive characteristic, though the number of genes involved has not been determined. In relation to the argument of optic insufficiency for lens induction, it is interesting to note that this Rc retina appears with both normal and vesicular lenses.

CONCLUSION

The discovery of individual genetic controls for some of the retina and lens types of *Astyanax* hybrids, and especially the observation that lens variation and retinal variation manifest themselves independently, strongly suggest that separate genetic controls act in the development of the various parts of the eye. There are most certainly some retina and lens genes segregating independently, perhaps on a polygenic background, and they can be isolated. This does not preclude interaction among the eye structures. The theory of induction is still important in the explanation of the ordering of eye development, but it would seem to affect only whether and when the lens will develop and not how it will. Similarly, size controls of both lens and eye, though still not clear, must take a subsidiary, modifying role in eye development to the effects of major structural genes.

REFERENCES

Bourne, M.C., Campbell, D.A. and Tansley, K. (1938). Hereditary degeneration of the rat retina. Br. J. Ophthal. 22: 613-623.

Cahn, P.H. (1958). Comparative optic development in *Astyanax mexicanus* and in two of its blind cave derivatives. Bull. Am. Mus. Nat. Hist. 93: 501-531.

Chase, H.B. and Chase E.B. (1941). Studies on an anophthalmic strain of mice. I. Embryology of the eye region J. Morphol. 68: 279-301.

Parry, H.B., Tansley, K. and Thomson L.C. (1953). The electroretinogram of the dog. J. Physiol. 120: 28-40.

Pfeiffer, W. (1967). Die Korrelation von Korperlänge, Augen-, Linsen-, und Pupillengrösse bei Hybriden aus Astyanax, Anoptichthys (Characidae, Pisces). Wilhelm Roux Arch. Entwicklungsmech. Org. 158: 218-245.

Sadoglu, P. (1957). Mendelian inheritance in the hybrids between the Mexican blind cave fish and their overground ancestors. Verh. Dtsch. Zool. Ges. 432-439.

Varnum, D.S. and Stevens, L.C. (1968). Aphakia, a New Mutation in the Mouse. J. Hered. 59: 147-150.

Wilkens, H. (1971). Genetic interpretation of regressive evolutionary process: studies on hybrid eyes of two Astyanax cave populations (characidae, pisces). Evolution 25: 530-544.

THE EYES OF LARVAL FISH

J.H.S. Blaxter

Dunstaffnage Marine Research Laboratory, P.O. Box 3

Oban, Argyll, PA34 4AD, Scotland

> These things seem small and undistinguishable.
> A Midsummer Night's Dream. Act IV, Sc. 1

INTRODUCTION

Fish may be oviparous or live-bearing. The incubation time for the eggs is usually short (except in some species like the salmonids). The eggs float in the water or are attached to the bottom. The larvae of many species of teleost are small, transparent and delicate and lack many of the anatomical features found in the adult. The gut and kidney are relatively simple, the blood is colourless and acellular, the gills are without filaments, the skin lacks scales and pigment and some of the median fins are absent. These structures develop later. In particular, the larvae often undergo a metamorphosis when the blood turns pink and the scales form, the larvae taking on the adult form. In flatfish this metamorphosis is more elaborate, since one eye migrates and the fish come to lie permanently on one or other side. As an example of such development some stages of growth in herring and plaice are shown in Fig. 1.

The sensory system is also often incomplete, judging from the little work which has been done on it in fish larvae. The lateral line becomes more complex with age and in the early larvae the mechanoreceptors often consist only of naked neuromast organs. Some species hatch without functional eyes and rely on the mechanoreceptors to avoid predation during the yolk-sac stage. By the end of the yolk-sac stage the eyes are functioning and most species of larvae probably feed mainly by sight (see Blaxter 1968, 1969).

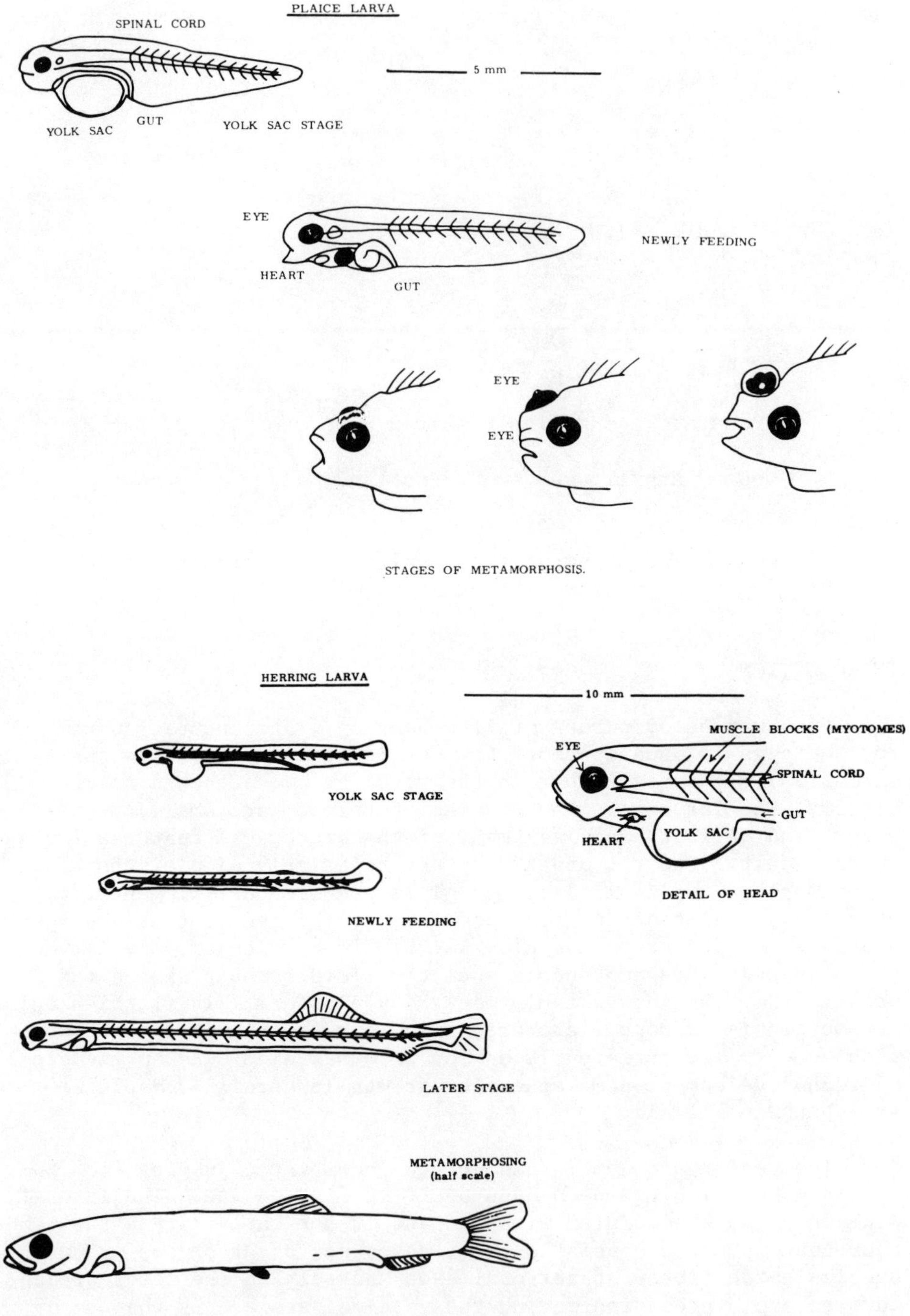

FIG. 1. Stages of herring and plaice larvae.

ANATOMY

Ali (1959) reported that the eyes of the young stages of *Oncorhynchus* spp. contained no rods in the retina. This has since been investigated in other species from different families of teleost by Blaxter and Jones (1967), Blaxter (1968) and Blaxter and Staines (1970); see Table 1. Generally the eyes were sectioned at 5 μm with the lens intact and stained with haematoxylin and eosin for light microscopy. In addition herring eyes were mounted in Epon, sectioned at 50 nm and stained with uranyl acetate for electron microscopy.

The nature of the visual cells was investigated by making counts, at a minimum of four positions in the retina of each larva, of the distinct ellipsoidal structures in the visual cell layer, the visual cell nuclei in the outer nuclear layer and the nuclei in the bipolar layer. A 1:1 ratio between the ellipsoids and the visual cell nuclei would indicate that only one type of visual cell was present. (In the duplex retina of the adult the visual cell nuclei greatly outnumber the ellipsoids, which represent the cones, the rods being too small and numerous to count.) The number of cells in the bipolar layer may indicate the degree of summation and the stage of morphogenesis in the retina.

The size and density of the visual cells and ratios between the various retinal layers are shown in Table 1. Some of the photomicrographs taken with a Wild Micro-Photoautomat using KB14 film (20ASA) and a x50 oil immersion objective are shown in Figs 2 and 3. Young pilchard, haddock, lemon sole and perch (Fig. 2A, C, E and Fig. 3A) had a 1:1 ratio between the distinct ellipsoids and the nuclei of the outer nuclear layer, indicating a pure-cone retina with no rods present. This was confirmed by the counts shown in Table 1, both for these four species and for the herring, goby, stickleback, plaice and sole.

It would be very difficult to resolve the rods by light microscope even if they existed. Apart from the indirect evidence of the ratios, the electronmicroscope can be used to show the presence of smaller elements. Work on the herring (Fig. 4) confirmed that only one type of visual cell was present in the larval retina.

The retinae of two deep water larvae, the leptocephalus stage of the eel (Fig. 3E) and a macrurid (Fig. 3D), showed a homogeneous pattern of rod-like structures very similar to those of the adult deep sea fish (Munk, 1966).

During later development of the pure-cone retinae there appeared, at least in the herring (Blaxter and Jones, 1967), to be a progressive increase in the number of cones and a decrease in

TABLE 1. SIZE, DENSITY AND RATIOS OF CELL TYPES IN THE RETINA (from Blaxter and Staines, 1970) (Bipolar layer nuclei are bipolar + amacrine cells)

Species	Length (mm)	Stage	Mean ellipsoid diameter (μ)	Eliipsoid density (no/100 μ)	Ratios: Cones	Ratios: Visual cell nuclei	Ratios: Bipolar layer nuclei
Pilchard	6-7	Larva	1.4	49	1	1.0	1.6+0.6
Herring	13	Larva	1.9	38	1	1.0	1.6+1.0
	29-31	Metamorph.	4.5	18	1	2.5	2.1+1.7
Salmon	14-17	Larva	not recognizable			not recognizable	
	23-24	Metamorph.	2.8	30	1	2.6	5.2+3.4
Leptocephalus	80	Larva	2.0 (rods)	45	1 (rods)	1.7	1.2+1.2
Macrurid	17	Larva	2.0 (rods)	50	1 (rods)	1.0	1.7+1.0
Haddock	4	Larva	1.3	50	1	1.0	1.6+1.0
Nile perch	6	Larva	2.8	32	1	1.3	3.5+1.0
	12	Metamorph.	3.3	24	1	3.0	4.7+1.0
Sand goby	3-4	Larva	1.2	56	1	1.0	1.7+0.8
Stickleback	7-8	Larva	2.6	31	1	1.1	3.2+2.0
	14-18	Metamorph.	3.0	27	1	2.5	3.5+2.2
Plaice	5-6	Larva	2.1	40	1	1.0	1.8+1.3
	14-16	Metamorph.	3.6	20	1	2.6	4.0+2.4
Lemon sole	5	Larva	1.6	46	1	1.0	1.6+0.9
	9-11	Metamorph.	2.4	37	1	1.1	2.0+1.0
Sole	5	Larva	1.3	47	1	1.0	1.5+0.9
	10-12	Metamorph	1.7	43	1	1.9	1.5+1.2

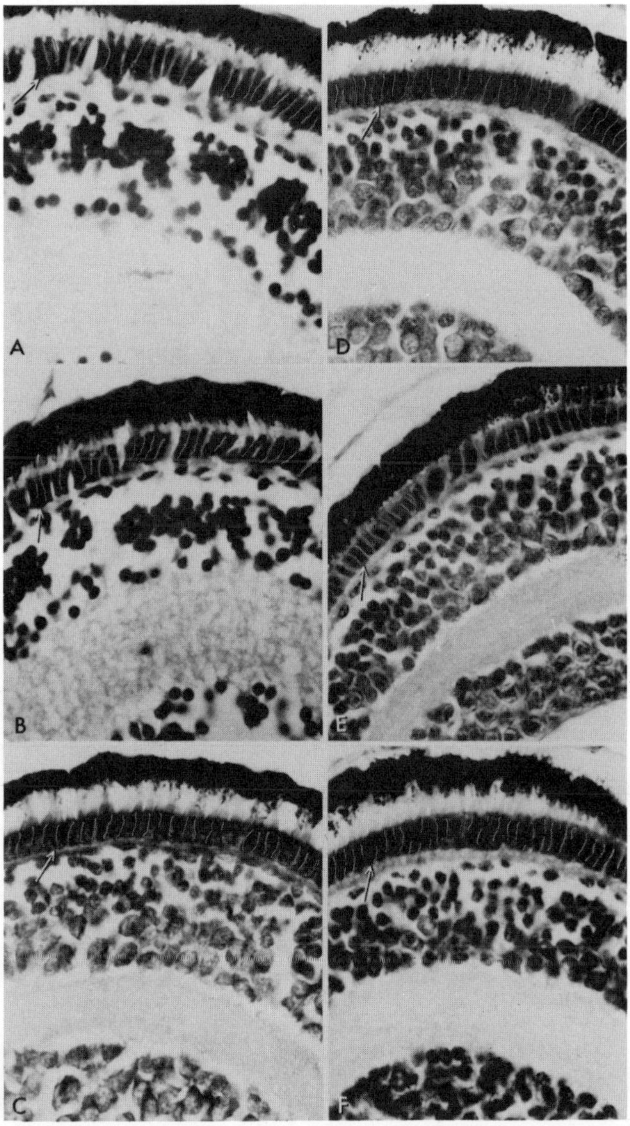

FIG. 2. Retinae of larvae (body length in brackets) showing pure-cone structure and lack of retinal masking pigment migration; each print is about 55 μ wide (from Blaxter and Staines, 1970).

A. pilchard (6-7 mm) light-adapted
B. pilchard (6-7 mm) dark-adapted
C. haddock (4 mm) light-adapted
D. haddock (4 mm) dark-adapted
E. lemon sole (5 mm) light-adapted
F. lemon sole (5 mm) dark-adapted

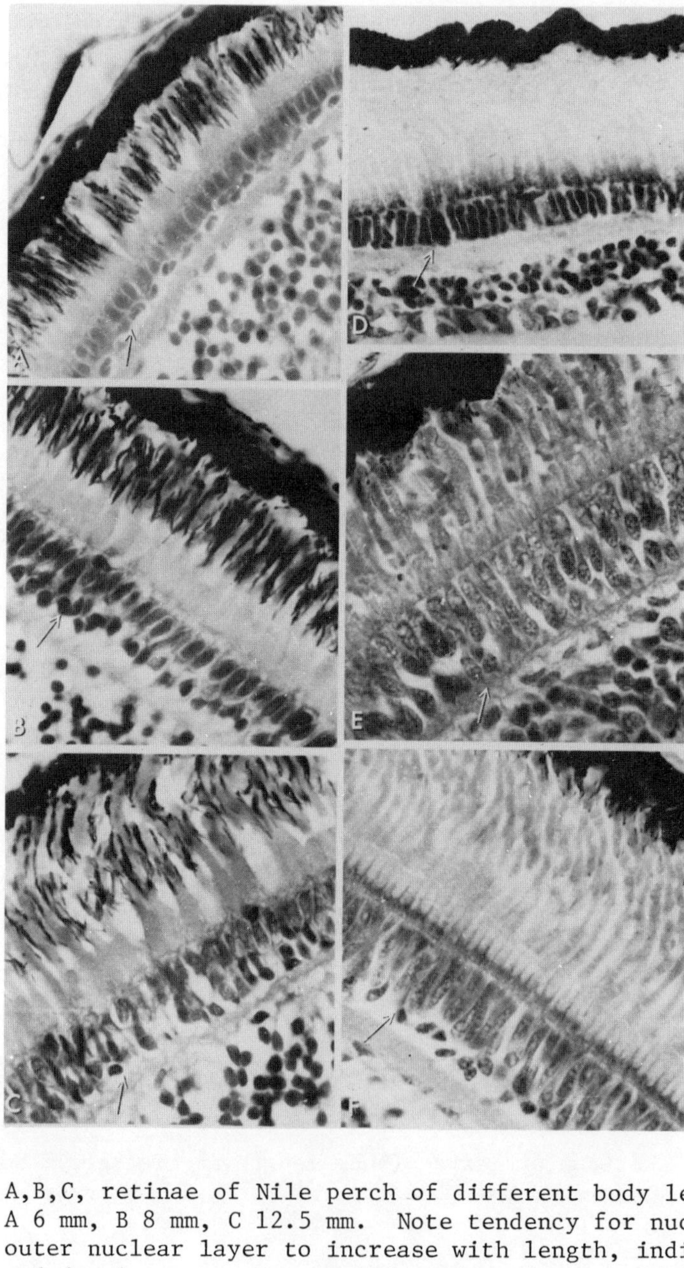

Fig. 3. A,B,C, retinae of Nile perch of different body length A 6 mm, B 8 mm, C 12.5 mm. Note tendency for nuclei in outer nuclear layer to increase with length, indicating rod development. D pure-rod retina of macrurid larva (17 mm). E, F light-adapted (E) and dark-adapted (F) retina of leptocephalus showing pure rod retina. Each print about 55 μ wide (from Blaxter and Staines, 1970).

their density (Figs. 5 and 6). Mitoses are only found peripherally. The double cones of the adult-herring retina also appear first at metamorphosis.

The appearance of the rods can best be observed by calculating the ratio of countable visual cells (which will be cones) to the visual cell nuclei in the outer nuclear layer. As shown in Table 1, in metamorphosing herring, salmon, perch (see also Fig. 3B, C), stickleback, plaice, sole and lemon sole the ratio changed from 1:1 to 1:2 or more, except in the lemon sole, which was at an early stage of metamorphosis. The best series of data were obtained for herring (Fig. 7) where the 1:1 ratio first changed at metamorphosis, eventually reaching 1:15 or more in juvenile and adult fish.

The origin of the rods has been discussed by Lyall (1957 a, b) and Blaxter and Jones (1967). Lyall (1957a) suggested that the rod nuclei migrated from the bipolar layer and also (1957b) that the rods might develop from single cones, finding transitional elements to support this view. The additional nuclei in the outer nuclear layer of the fish larvae, which seem to indicate the development of rods, were small, darkly staining and often with processes pointing towards the external limiting membrane (Fig. 8A). Transitional elements were found in the herring (Fig. 8B), in particular, but only in the pure-cone area. This appears to cast doubt on Lyall's (1957b) hypothesis. The presence of mitoses mainly in the periphery may indicate that this is the site of all visual cell proliferation.

It seems to be generally true, then, that fish with duplex retinae in the adult stage have larvae with pure-cone retinae, the rods developing at metamorphosis. How is this incomplete development reflected in other functions of the eye such as dark adaptation, and in physiological functions such as acuity, spectral sensitivity and visual thresholds?

LIGHT AND DARK ADAPTATION

In the fully-developed duplex retina light-adaptation is characterized by an inward (towards the lens) movement of the masking pigment, extension of the rod myoids (so that the rods are invested by the pigment) and contraction of the cone myoids (Fig. 9). In the dark-adapted retina, on the other hand, the masking pigment is withdrawn to the periphery of the eye, the cone myoids are extended and the rod myoids contracted. The simplest way to represent the state of light- or dark-adaptation of the eye is to measure the pigment index (Ali, 1959) which is the ratio of the thickness of the pigment layer to the thickness of the visual cell layer. The index is thus independent of the way in which the eye is sectioned.

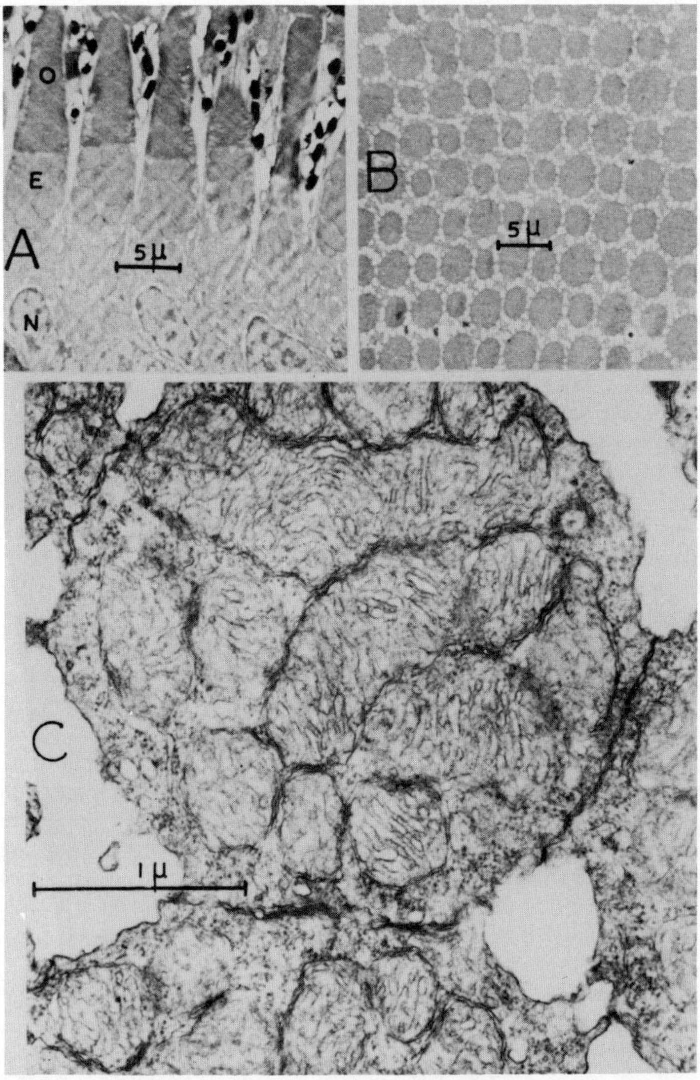

FIG. 4. Electron micrographs of visual cells in the herring retina. A longitudinal section, B transverse section, C transverse section through ellipsoid region showing mitochondria. Note that only one type of cell is present (Blaxter and Jones, 1967).

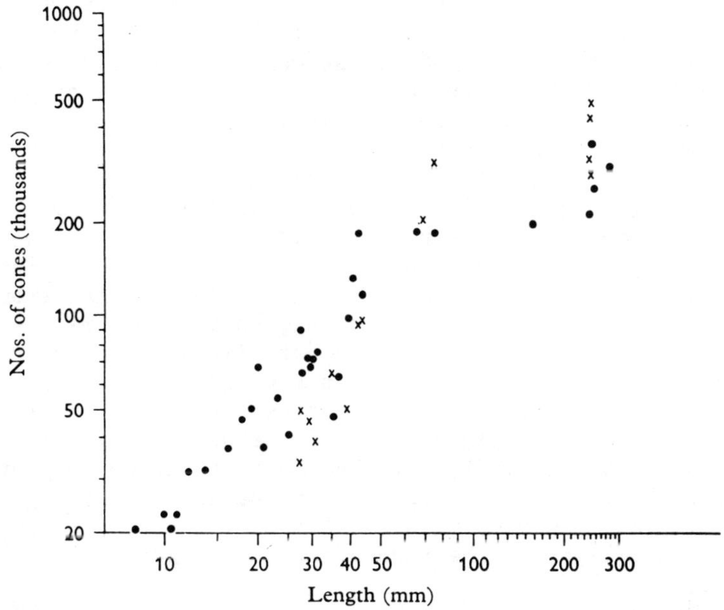

FIG. 5. Numbers of cones estimated at different body lengths in herring (Blaxter and Jones, 1967).

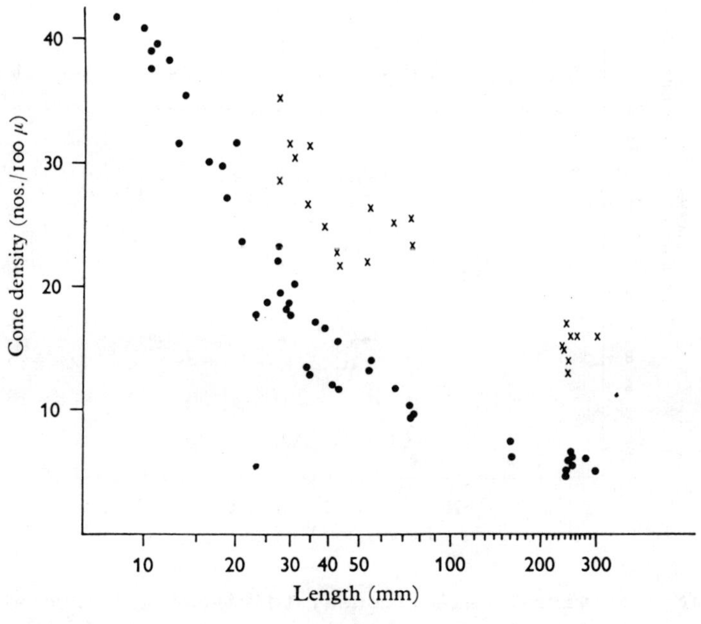

FIG. 6. Density of cones at different body lengths in herring (Blaxter and Jones, 1967).

Where sufficient live material was available larvae were sampled at intervals over the dusk period, the light intensity being monitored by a photometer. The larvae were immediately anaesthetised and fixed and later the index was measured at a number of points in the retina after sectioning and staining. In some species larvae which had been light or dark-adapted for at least 1 hr. were sampled. The most complete series of measurements have been made by Blaxter and Jones (1967). An example of their results is shown in Fig. 10 where it can be seen that the pigment did not migrate in the larval stage of herring and, in fact, only started to migrate at metamorphosis when the rods appeared (Fig. 11). The results from other species (Fig. 12) confirmed this finding, as did comparisons of light-adapted and dark-adapted retinae of pilchard larvae as shown in Fig. 2A and B, of haddock larvae in Fig. 2C and D, of lemon sole larvae in Fig. 2E and F; the leptocephalus larva with a pure-rod retina showed no masking pigment changes between the light- and dark-adapted states (Fig. 3E, F). The lack of pigment migration in the pure-cone larval retina thus seems to be established, providing some evidence that the masking pigment does, indeed, protect the rods from bright light in the fish after metamorphosis.

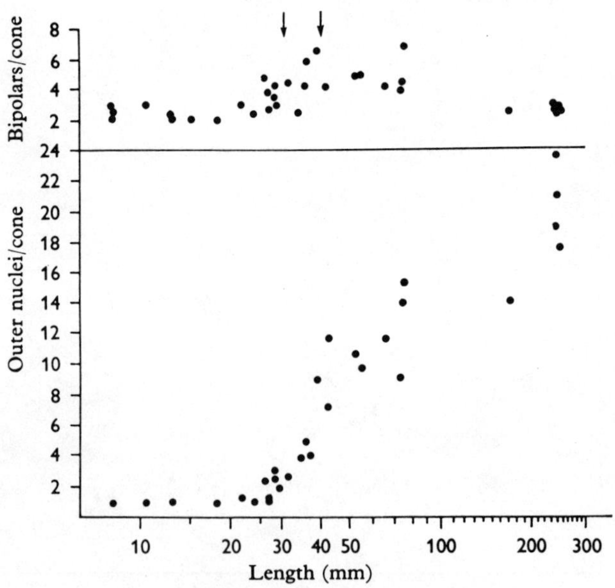

FIG. 7. Ratio of visual cells (cones) to visual cell nuclei in the outer nuclear layer in herring, showing rapid increase of rods, as judged from their nuclei, at metamorphosis, (arrows) (Blaxter and Jones, 1967).

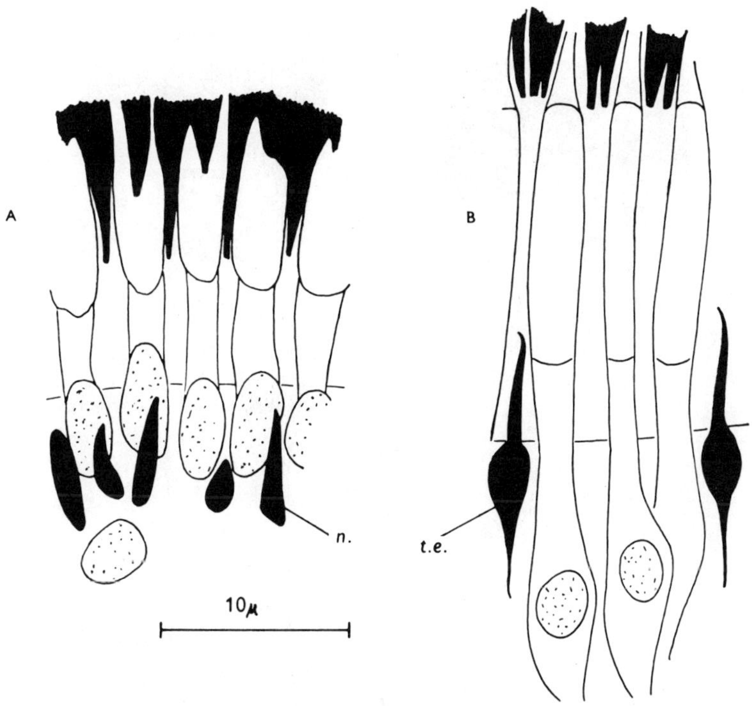

FIG. 8. A. Additional nuclei in outer nuclear layer of herring.
B. Possible "transitional elements" in herring retina (Blaxter and Jones, 1967).

ACUITY

It is possible to calculate acuity (α) in minutes of minimal separable angle from Tamura and Wisby's (1963) formula:

$$\text{Sin}\ \alpha = (^1/_f)\left[(0.1(1 + 0.25) \times 2)/\sqrt{n}\right]$$

where f is the focal length, which can be obtained from the radius of the lens, and n is the density of visual cells. Using some of the data in Figs. 5 and 6, the change in acuity with length of the body was calculated and is shown in Fig. 13 for herring. The herring has a specialised pure-cone area temporalis on the main visual axis. Acuity was better in this area and here and in the rest of the retina acuity improved with body length, a feature which was due to the increasing focal length of the lens rather than the decreasing density of cones as the retina grew.

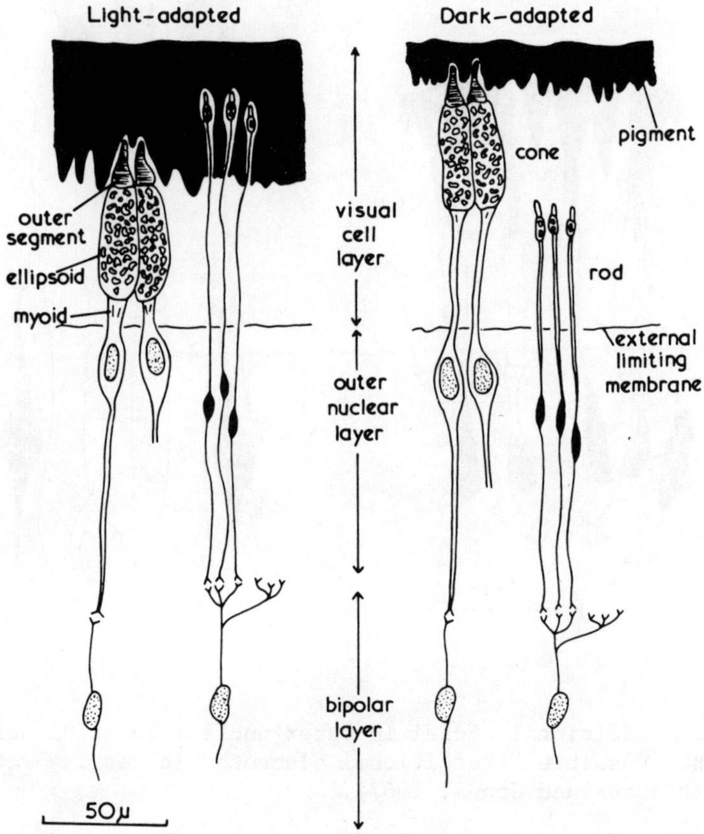

FIG. 9. Diagram of fish retina in the light-adapted and dark-adapted state.

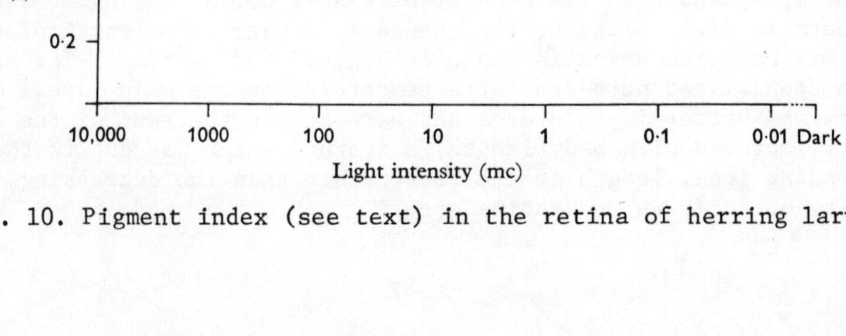

FIG. 10. Pigment index (see text) in the retina of herring larvae.

SPECTRAL SENSITIVITY

Teleosts with a duplex retina show a Purkinje shift (see Blaxter, 1970); i.e., there is a change of maximum sensitivity in the eye towards the blue end of the spectrum in the dark. One might expect that the larval retina would shown a cone-type of spectral sensitivity curve and no Purkinje shift.

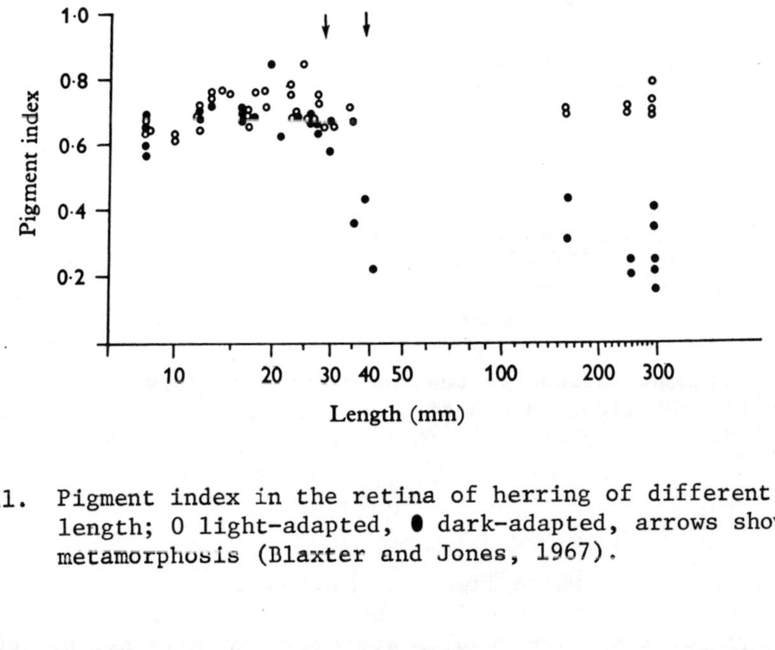

FIG. 11. Pigment index in the retina of herring of different body length; 0 light-adapted, ● dark-adapted, arrows show metamorphosis (Blaxter and Jones, 1967).

Only Blaxter (1968, 1969) has measured the spectral sensitivity of fish larvae-behaviourally, by establishing at different wavelengths the threshold for negative phototaxis. The results for herring, plaice and sole larvae of different age are shown in Fig. 14. The results were somewhat equivocal in that they showed a very broad plateau-like curve which may well, in fact, be typical of spectral sensitivity curves as determined by behavioural techniques which may also give a variety of action spectra depending on the task to be performed (see Muntz and Northmore, 1970). There were signs of the involvement of different cone populations with the maximum sensitivity well away from the 500 nm region more typical of rhodopsin (Lythgoe, 1966).

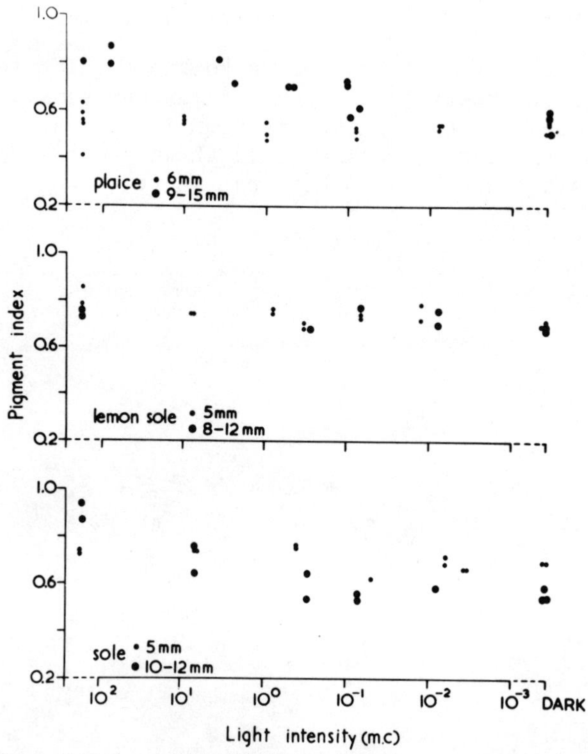

FIG. 12. Pigment indices in the retina of various species at different light intensities showing the movement of pigment in older larvae only (Blaxter and Staines, 1970).

VISUAL THRESHOLDS

The threshold to white light was also determined by measuring the light intensity at the point where there was extinction of negative phototaxis. The results are shown for herring, plaice and sole larvae of different age in Fig. 15. There was a general increase in sensitivity with age, the threshold being surprisingly low for what is thought to be a pure-cone retina, although it is difficult to compare thresholds for different species and ages, as a variety of techniques and sources of light stimuli have been used by other sensory physiologists (see Blaxter, 1968).

DISCUSSION

It is of great interest that fish larvae of many species can live for weeks or months with visual equipment apparently inferior

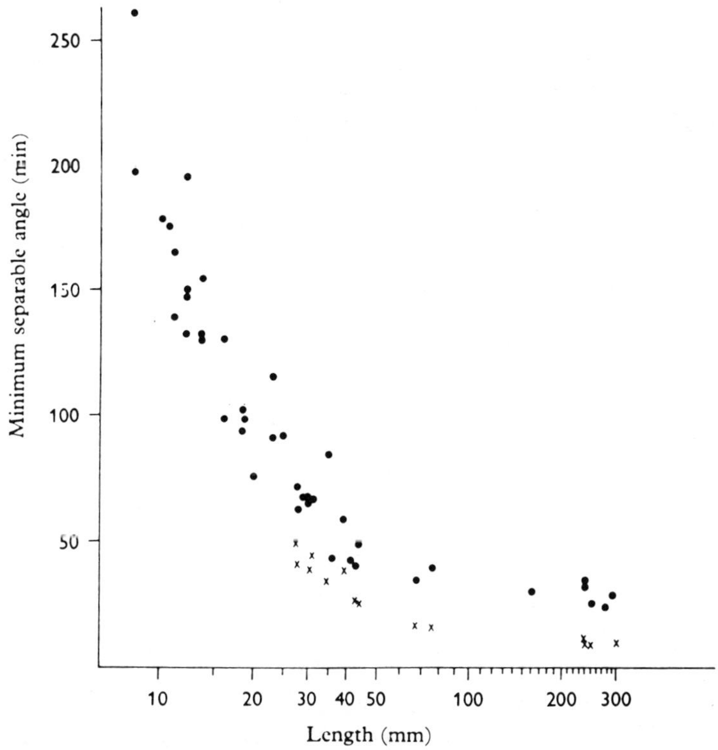

FIG. 13. Changes in acuity calculated for herring of different body length peripheral retina, x *area temporalis* (Blaxter and Jones, 1967).

to that available to the adults. The pure-cone retina, in species other than those from the deep sea, means that the larvae cannot, presumably, perceive the same wide range of intensity of light as the adult. The low cone threshold in the larvae may, however, be a larval adaptation to a rod-free eye. The acuity is poorer in the young larvae but this is determined by the cones. One may question what value rod vision has for the older stages. Rod vision is generally more active in the peripheral retina and is especially involved in movement perception. One must conclude that the larvae are somewhat more vulnerable in poor light, especially to predatory attack, and this may be one reason why fish larvae are produced in such large numbers - an adaptation to the high loss rate in the early stages of their life history.

The other points of interest which arise from developmental studies are:

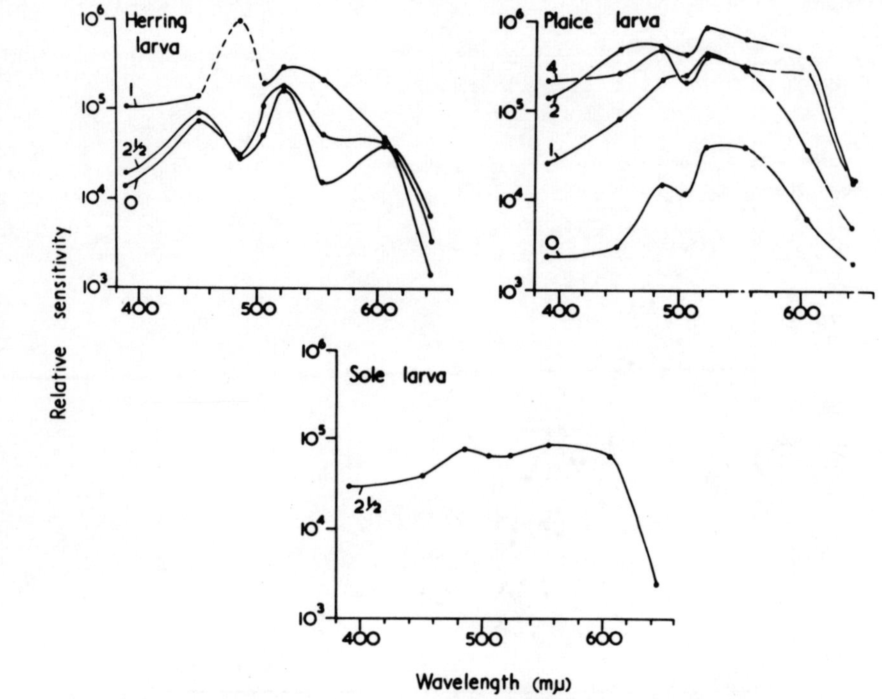

FIG. 14. Spectral sensitivity of herring, plaice and sole larvae of different age (in weeks from hatching, by each curve) (from Blaxter 1968, 1969).

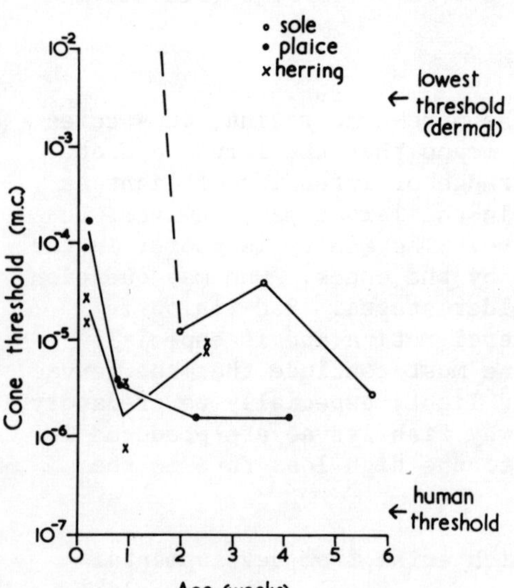

FIG. 15. Light intensity thresholds for phototaxis in herring, plaice and sole larvae (from Blaxter 1968, 1969).

(1) From where do the rods originate; mitoses only seem to occur at the periphery of the retina? Could the rod precursors migrate from the periphery or do they originate in the bipolar layer?

(2) What it the relationship between the development of the eye and the optic tectum? How does the optic nerve develop?

REFERENCES

Ali, M.A. (1959). The ocular structure, retinomotor and photobehavioural responses of juvenile Pacific salmon. Can. J. Zool. 37: 965-996.

Blaxter, J.H.S. (1968). Visual thresholds and spectral sensitivity of herring larvae. J. Exp. Biol. 48: 39-53.

Blaxter, J.H.S. (1969). Visual thresholds and spectral sensitivity of flatfish larvae. J. Exp. Biol. 51: 221-230.

Blaxter, J.H.S. (1970). 2. Light. 2.3. Animals. 2.32. Fishes. In: Marine Ecology, Vol. 1, Pt. 1, pp. 213-320, edited by O. Kinne Wiley Interscience, London.

Blaxter, J.H.S. and Jones, M.P. (1967). The development of the retina and retinomotor responses in the herring. J. Mar. Biol. Ass. U.K. 47: 677-697.

Blaxter, J.H.S. and Staines, M. (1970). Pure-cone retinae and retinomotor responses in larval teleosts. J. Mar. Biol. Ass. U.K. 50: 449-460.

Lyall, A.H. (1957a). The growth of the trout retina. Q.J. Microsc. Sci. 98: 101-10.

Lyall, A.H. (1957b). Cone arrangements in teleost retinae. Q.J. Microsc. Sci. 98: 189-201.

Lythgoe, J.N. (1966). Visual pigments and underwater vision. Br. Ecol. Soc. Symp. 6: 375-391.

Muntz, W.R.A. and Northmore, D.P.M. (1970). Vision and visual pigments in a fish, *Scardininus erythrophthalmus* (the rudd). Vision Res. 10: 281-298.

Tamura, T. and Wisby, W.J. (1963). The visual sense of pelagic fishes especially the visual axis and accomodation. Bull. Mar. Sci. Gulf Caribb. 13: 433-448.

EXTRARETINAL PHOTOSENSITIVITY IN FISH LARVAE

W. WALES, University of Stirling and Dunstaffnage
Marine Research Laboratory, P.O. Box 3
Oban, Argyll, PA34 4AD, Scotland

By gar, de herring is no dead so as I vill kill him.
The Merry Wives of Windsor, Act II, Sc. 3

The presentation of a paper on extraretinal photoreception may at first appear incongruous in an Advanced Study Institute on vision in fish. Extraretinal photoreceptors are, however, far more important to the life of fishes than the relatively small bulk of published research on this subject may reflect. Thus the inclusion of this contribution may serve to remind us that, although no other photoreceptor can approach the complexity and sophistication of the eye, there are other photosensitive organs, or cells, of functional importance in fish.

Eyes have readily lent themselves to study as they are discrete organs, externally located and hence accessible. The precise location or structure is known for only a few extraretinal photoreceptors and the bulk of our knowledge has resulted from observation of the responses of eyeless fish to light. Many species of naturally blind fish are known to respond to light, mainly by photonegative responses (Steven 1963). Both experimentally blinded minnows (Jones 1956) and trout (Erickson 1972) respond to changes in light intensity.

The term "dermal light sense" used widely to describe extraretinal photoreception (Blaxter, 1970; Millott, 1968; Steven, 1963) is misleading in that it necessarily implies the receptors are located in the skin. This is true of some cases such as the ammocoete larva of the brook lamprey, *Lampetra planeri*, which has dermal photoreceptors located in the tail region (Stevens, 1951). Information from these photoreceptors is conducted to the central nervous system via the lateral line nerve (Young, 1935). Photoreceptive regions also occur in the brains of many vertebrates (Menaker, 1971). The pineal and adjacent regions of the mid brain have been shown to be light-sensitive in several fish species (Steven, 1963), and in invertebrates individual neurones of the central nervous system are photosensitive (Bullock & Horridge, 1965). It may be that in some fish, the extraretinal photoreceptors are relatively morphologically unspecialized groups of neurones.

OBSERVATIONS ON FISH LARVAE

Fish larvae are transparent animals with their nervous system unshielded from the light. They are, therefore, interesting animals in which to investigate extraretinal photosensitivity as any photoreceptors will be readily stimulated.

My observations were made on two species of fish larvae readily available at Oban, the herring, *Clupea harengus*, and the plaice *Pleuronectes platessa*. The larvae were reared from laboratory fertilized eggs provided by Dr. J.H.S. Blaxter. Experiments were performed on larvae at various stages from hatching to metamorphosis. The eyes were removed from the larvae anaesthetised in 1:20,000 MS 222 using a pair of hair loops. The loops were placed over the eye and pulled in opposite directions to produce a scissors-like action. The larvae were handled by the eyes only, thus preventing further injury. After eye removal the animals were maintained in one third strength sea water containing antibiotics for two days before returning to sea water. Observations were made on the larvae from two to fourteen days after the operation.

Two responses of herring and plaice have been shown by earlier workers to be evoked by changes in light intensity. These are phototaxis and vertical migration. Herring larvae respond to a horizontal light beam with a positive phototaxis at intensities above 10^0 to 10^{-1} m.c. and a negative phototaxis at light intensities below this down to a lower threshold of 10^{-5} to 10^{-6} m.c. (Blaxter, 1968). Plaice larvae were not examined at light intensities higher than 10^0 m.c. but exhibit a negative phototaxis over a similar range of light intensities (Blaxter, 1969).

Blind herring and plaice in the later larval stages were tested for phototactic responses using a method similar to that of Blaxter (1968). The dark-adapted larvae, in a rectangular perspex dish, were placed in a light-proof box illuminated by a tungsten light source through an aperture in the box. After a period of ten minutes the box was opened in a darkened room and the position of the larvae photographed instantaneously. The phototactic response was then measured by counting the larvae present in equal sized regions at the two ends of the dish and subtracting one from the other to give a measure of the response. Preliminary observations indicate that although the control larvae respond normally, the blind larvae exhibit no distinct positive or negative phototactic movements within the range of light intensities tested (10^2 to 10^{-6} m.c.)

Vertical migration is an interesting phenomenon of unknown function which has been demonstrated to occur naturally and over a considerable range of depths (see Blaxter, 1970 and this volume). The evidence for vertical migration of fish larvae in nature is, however, conflicting due to the inadequecy of the methods used to detect migrations over small distances. Blaxter (1973) has shown that under laboratory conditions both herring and older plaice larvae will migrate in a vertical column of water during changes in natural light intensity at dusk and dawn. He also demonstrated that this vertical migration can be produced at other times of the day by reducing the intensity of an artificial source with neutral density filters. The migration is clearly elicited by decreasing light and apparently has no endogenous rhythmical component (Blaxter, unpublished observation).

Blinded larvae of herring and plaice were tested for vertical migration with the same technique. The animals were placed in a cylindrical column with a water depth of 117 cm having an activity monitoring device, which consisted of a pair of matched thermistors, in a balanced bridge circuit, located 2 cm from the water surface. Thus the degree of surface activity at different light intensities could be recorded on a pen recorder. The water column was placed in a light-tight box illuminated from above through a single aperture with a tungsten light source which produced an intensity of 10^2 m.c. at the water surface. The light intensity was reduced by a series of neutral density filters in logarithmetic steps from 10^2 to 10^{-7} m.c. The filters were changed automatically every 15 minutes to give a regular pattern of reducing light intensity.

As shown in figure 1, blind herring larvae respond to decreasing light intensity by an upward migration similar to the control animals. They do not exhibit the early peak at higher light intensities suggesting that the U shaped curve of the control animals may result from the interaction of inputs from retinal and extraretinal photoreceptors. The peak of activity at low light intensity is clearly due to extraretinal photosensitivity and is light intensity dependent as it can be switched off in blind fish by returning to the original high light intensity.

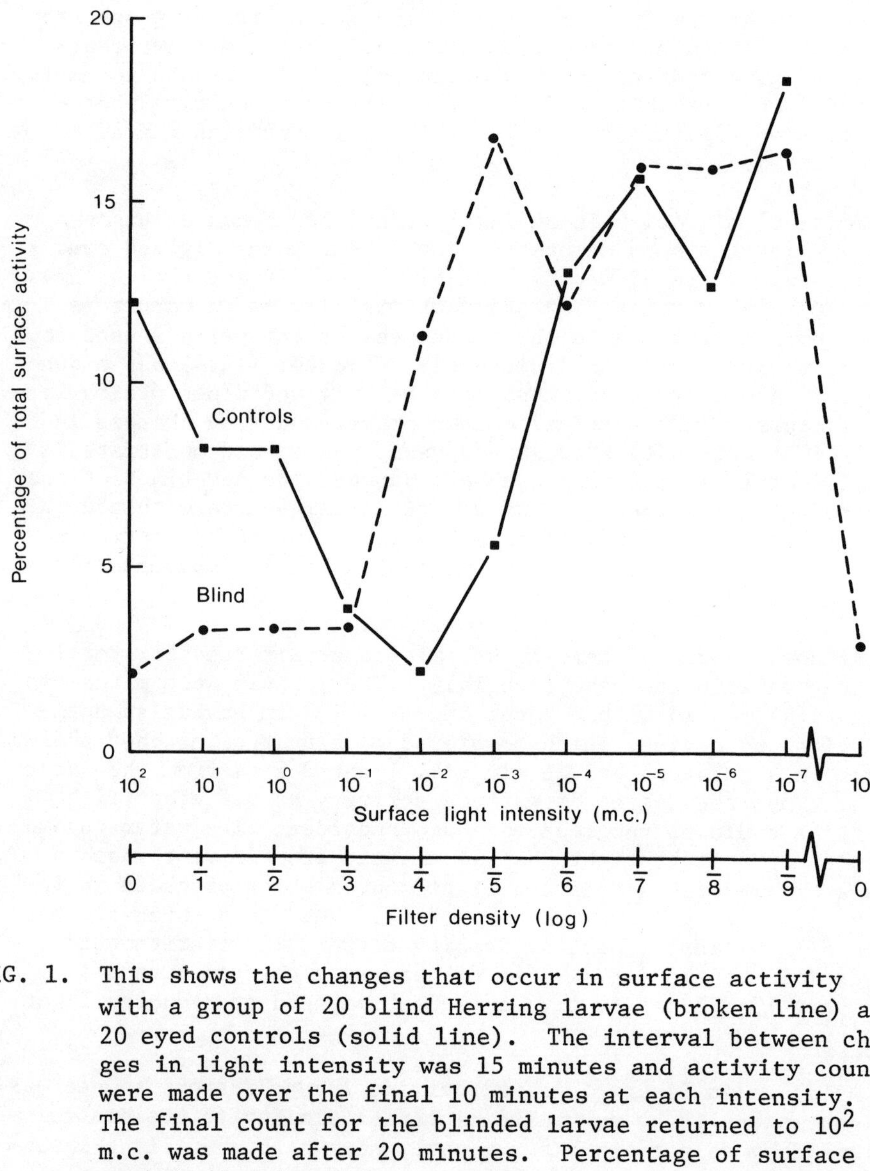

FIG. 1. This shows the changes that occur in surface activity with a group of 20 blind Herring larvae (broken line) and 20 eyed controls (solid line). The interval between changes in light intensity was 15 minutes and activity counts were made over the final 10 minutes at each intensity. The final count for the blinded larvae returned to 10^2 m.c. was made after 20 minutes. Percentage of surface activity has been plotted rather than actual activity counts to compensate for differences in total activity between experimental groups of larvae.

DISCUSSION

Although my results clearly indicate the presence of extraretinal photoreceptors in fish larvae nothing can be said at this stage regarding the location or structure of the receptors involved. With larger herring larvae it may prove possible to devise experiments which will provide information on this point.

The response of blind herring larvae is similar to that of blind minnows (Jones, 1956) in that they both show increased activity during periods of low light intensity. Comparison of the behaviour of the blinded animals and of the normal eyed controls indicates that they also exhibit this pattern of activity with some modifications and hence we can conclude the extraretinal light sense may play a role in the life of these fish in nature. Clearly the retinal and extraretinal photosensory inputs interact to produce the observed behaviour in eyed animals, and the larvae of herring and plaice offer an interesting preparation in which to study this interaction.

ACKNOWLEDGEMENTS

I am indebted to Dr. J.H.S. Blaxter for introducing me to a fascinating experimental animal, the fish larva, and for the advice he has given to me. I also thank Dr. Blaxter and Dr. R.N. Gibson for proof reading the manuscript.

REFERENCES

Blaxter, J.H.S. (1968) Visual thresholds and spectral sensitivity of herring larvae. J. Exp. Biol. 48: 39-53.

Blaxter, J.H.S. (1969) Visual thresholds and spectral sensitivity of flatfish larvae. J. Exp. Biol. 51: 221-230.

Blaxter, J.H.S. (1970) Light, animals, fishes. In: Marine biology I, 213-320, edited by O. Kinne. Wiley, New York.

Blaxter, J.H.S. (1973) Monitoring the vertical movements and light responses of herring and plaice larvae. J. Mar. Biol. Ass. U.K. 53: 635-647.

Bullock, T.H. and Horridge, G.A. (1965) Structure and function in the nervous systems of invertebrates. Freeman. London.

Erikson, L.O. (1972) Tagesperiodik geblendeter bachsaiblinge. Naturwissenschaften 59: 219-220.

Jones, F.R.H. (1956) The behaviour of minnows in relation to light intensity. J. Exp. Biol. 33: 271-281.

Menaker, M. (1971) Synchronization with the photic environment via extraretinal receptors in the avian brain. In: Biochronometry, edited by M. Menaker. National Academy of Sciences, Washington.

Millot, N. (1958) The dermal light sense. In: Invertebrate receptors, edited by J.D. Carthy and G.E. Newell, p. 1-36. Academic Press, New York.

Steven, D.M. (1951) Sensory cells and pigment distribution in the tail of the ammocoete. Q.J. Microsc. Sci. 92: 233-234.

Steven, D.M. (1963) The dermal light sense. Biol. Rev. (Camb) 38: 204-240.

Young, J.Z. (1935) The photoreceptors of lampreys. 1. Light sensitive fibres in the lateral line nerves. J. Exp. Biol. 12: 229-238.

RETINAL GROWTH IN ADULT GOLDFISH

Pamela R. JOHNS & S.S. EASTER, Jr.

Department of Zoology, University of Michigan

Ann Arbor, Michigan 48104, U.S.A.

For the Eye altering alters all.
William Blake, "The Mental Traveller"

INTRODUCTION

Most studies of growth and development have been done on embryonic or larval animals, and with good reason. The dramatic increase in size, rapid changes in morphology, and sudden emergence of function in the embryo are much more exciting from a developmental perspective than the relatively static adult state. A separate but related problem is how to grow while functioning, and that is the problem which adult organisms must solve. We are concerned here with growth of the nervous system, in which function is dependent on a precise, orderly set of connexions between neurons. During growth, how are new nerve cells integrated into the system without disturbing the ongoing activity?

Growth of the vertebrate retina by cell addition occurs post-embryonically in many species. Recent work using ^{3}H-thymidine radioautography has shown cell proliferation and migration during larval and even post-metamorphic development in amphibians (Hollyfield, 1968, Straznicky and Gaze, 1971). Wilson (1971) has shown an increase in numbers of optic nerve fibres and retinal ganglion cells by direct counts in post-metamorphic *Xenopus laevis*. Teleost fishes also appear to add cells to the retina during post-embryonic growth. Müller (1952) and Blaxter and Jones (1967) have described mitotic figures at the margins of the retina in adult fish, which suggests that new cells are being added. Ali (personal communication) has shown increasing numbers of optic nerve fibres in Atlantic salmon (*Salmo salar*) up to the yearling stage. Lyall (1957) in an elegant geometrical study of the cone mosaic in adult trout (*Salmo trutta*) demonstrated that photoreceptors are added during growth. These studies indicate that cell proliferation does occur in the adult fish retina and may result in increased numbers of retinal neurons. Simply observing mitotic figures does not allow one to conclude that the number of cells is increasing because cell death may be occurring simultaneously. Similarly, increased numbers of optic nerve fibres may be due to additional centrifugal axons (Witkovsky, 1971) and not to new retinal ganglion cells. The present work is an attempt to quantify the addition of retinal ganglion cells in adult goldfish, *Carassius auratus*, by direct cell counts in the retina, thus avoiding the problems mentioned above.

METHODS AND RESULTS

The eyes from ten goldfish ranging in body length from 4 cm to 20 cm tip-to-tip were fixed in Bouin's, embedded in paraffin and sectioned radially at 5 or 10 μm through various meridians. The tissue was stained with cresyl violet or hematoxylin and eosin. The diameter of the enucleated eye was measured with a hand micrometer. The length of the retina, defined as the distance in millimetres from border to border along a path through the retina in a radial section through a maximum diameter of the eye, was measured with an ocular micrometer.

We were concerned in this study only with the retinal ganglion cells, and the first problem was how to identify them in radial sections. We first categorised all the cells in the retina in the region between the inner plexiform layer and the vitreal surface. This region contains the ganglion cell layer, the optic fibre layer, or axons of the ganglion cells, and the blood vessels. Approximately 25% of the cells were recognisable ganglion cells. They were characterised by a large, pale, spherical nucleus from 5 to 9 μm in diameter. The size of the ganglion cell nuclei correlated roughly with the size of the fish. The ganglion cells were arranged in a layer a single cell thick and were often separated from one another by long empty spaces. Of the remaining cells in the region of interest, 40% were glial cells and the rest were red blood cells and associated mesodermal elements.

To be certain of our identification of the ganglion cells we sectioned the optic nerve and observed the chromatolytic response. This consists of specific morphological changes in the soma of a neuron in response to axonal injury, and for goldfish ganglion cells includes cellular and nuclear hypertrophy, cytoplasmic basophilia, an increase in size and number of nucleoli, and nuclear eccentricity (Murray and Grafstein, 1969). We did not see any cells which we would identify as ganglion cells which were not chromatolytic, nor did we see chromatolytic cells in any other retinal layer. We were therefore confident that we could reliably identify and count ganglion cells in the adult goldfish retina.

The total number of ganglion cells in the retinas of fish of varying sizes were determined from the cell density and the surface area of the retina. Ganglion cell nuclei were counted in an equatorial section through the centre of the eye, corrected for split cells using Abercrombie's formula (1946), and the cell density determined. This method of determining density is valid because of the homogeneity of ganglion cells in the goldfish retina (Schellart, 1973). From 200 to 400 cell nuclei were counted in each retina, and we estimate the percent error in cell density determinations due to the size of the sample to be $1/\sqrt{N}$ or 5-7%. The area of the retina was calculated from the measured retinal length.

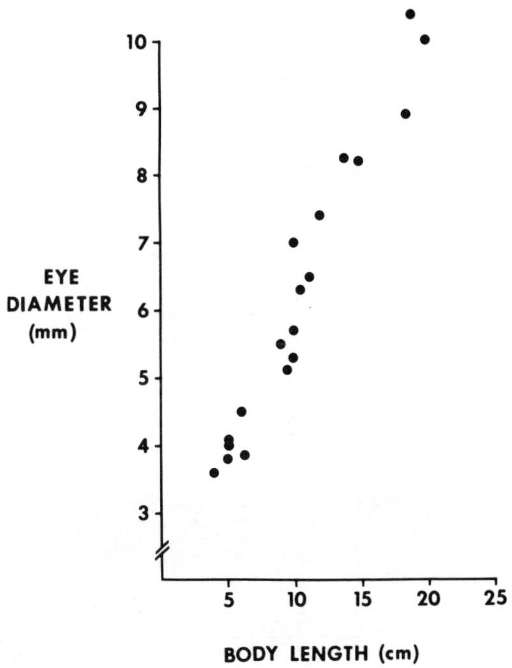

Fig. 1. The diameter of the eye as a function of body length. Each point represents the average of the diameters of the right and left eyes of each fish, measured with a hand micrometer after enucleation. The data include the ten goldfish used in other parts of this study as well as fish not used here.

As shown in Fig. 1, the diameter of the eye is a linear function of the body length for fish of the size studied. The retinal length increased from 3.3 mm in the smallest fish (4 cm body length) to 9.8 mm in the largest fish (20 cm). Likewise, the retinal area increased from 7 mm^2 to 62 mm^2. In contrast, the ganglion cell density decreased from 6000 cells/mm^2 in the smallest fish to 2000 cells/mm^2 in the largest. This latter value is in good agreement with Schellart (1973) who found an average of 2600 cells/mm^2 in tangential retinal sections from goldfish 18 to 21 cm body length. The decrease in density with retinal area agrees with the observations by Ali (1964).

Fig. 2 is a graph of the total number of ganglion cells versus body length. Each point represents one retina. The number of cells in the largest retina is two and one-half times that in the smallest, 120,000 and 45,000 cells, respectively. We conclude that growth of the adult goldfish retina, with respect to ganglion cells, is due to a combination of both increased cell number as shown in Fig. 2 and spreading out of existing cells as shown by the decrease in cell density.

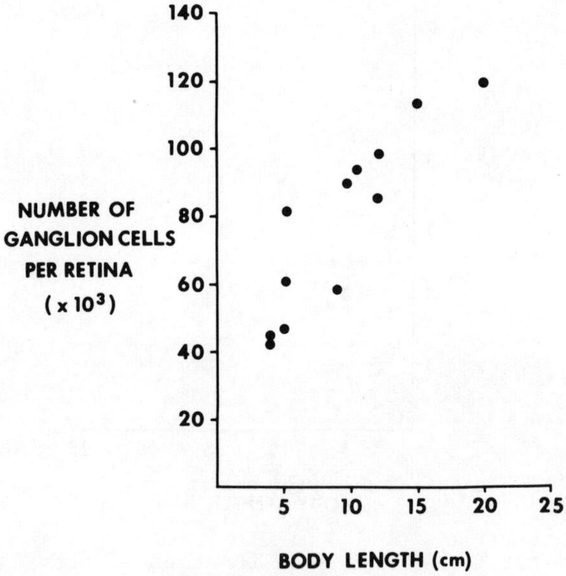

Fig. 2. The total number of ganglion cells in adult goldfish retina versus the body length of the fish. Each point represents one retina.

Because the rate of growth in fish is variable depending on numerous environmental factors, we do not know the age of these fish. However, the parameters measured (e.g. retinal area and ganglion cell number) are monotonic functions of the body length of the fish regardless of age. We do have preliminary data on the rate of growth of four fish. All four were 4 cm in length at time t = 0. Two of the fish were killed immediately and retinal ganglion cells counted in one eye. One fish was left to grow for one month, during which time it increased in length by one cm. The last fish grew for five months and increased 6.5 cm to a final length of 10.5 cm. Figure 3 shows the change in total number of ganglion cells through time; the rate of cell addition is 300 ganglion cells/day. Since the other types of retinal cells are probably also increasing in number, the total rate of cell proliferation in the adult goldfish retina is undoubtedly even larger.

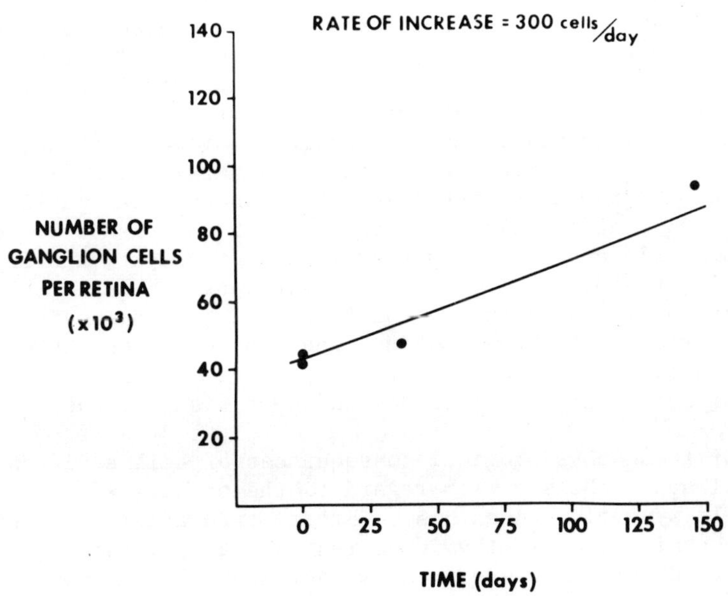

Fig. 3. Rate of ganglion cell addition. Each point represents the number of ganglion cells in the retina of a fish allowed to grow for the specified length of time. All four fish were 4 cm in body length at t = 0.

DISCUSSION

This dramatic increase in numbers of retinal ganglion cells has two obvious physiological implications for ganglion cell receptive fields. The first has to do with field size. In our laboratory, measurements of the acceptance angle of the pupil of the intact fish in water have shown that the extent of the visual field in goldfish is $185^0 \pm 5^0$ regardless of the size of the fish (L. Baumann, unpublished). But the large fish have many more ganglion cells dividing up that 185^0 than do the smaller fish. If we assume that the ganglion cells maintain a receptive field of constant visual angle, for example 10°, then the area of the retina from which that cell is receiving information must increase as the eye grows, because 10° visual angle in a retina 4 mm in length corresponds to a retinal distance of 216 μm, while that same 10° represents a distance of 540 μm in a retina 10 mm in length. Since the area of the receptive field is related to the extent of the dendritic arborisation of the ganglion cell (Brown, 1965), as the eye grows the dendritic arborisation of the ganglion cell must increase and probably forms new synapses as well. The alternative is that, if the ganglion cell dendritic tree remains constant in size at a diameter of 216 mm, then as the eye grows the visual angle subtended must decrease from 10° to 4°.

A second interesting implication of the continued addition of ganglion cells has to do with the region of the visual world to which a given cell is responding. If we assume, for example, that growth of the adult retina continues in the pattern found for embryonic and larval animals, that is, addition of cells to the periphery, then the cells which were in the peripheral retina in a small eye would become gradually displaced toward the centre and likewise their receptive fields would move centrally in the visual field as the eye grows. To ensure the appropriate motor output, the information ascending the optic nerve from a given ganglion cell must be coded for its position in the visual field. This implies that the central connexions of these "moving" ganglion cells may have to be re-wired. Again this suggests that synaptic connexions may be altered during normal growth processes in adult fish retinas.

Both of these physiological consequences of cell addition have more general applicability with regard to the problem of hard-wired versus labile synaptic connexions. Experiments are in progress using the techniques of ^{3}H-thymidine radioautography and electrophysiology to determine where the new ganglion cells in adult goldfish come from and where they go and the changes in physiological properties and synaptic connexions that ensue.

In summary we would like to emphasise the dramatic rate of increase in ganglion cell numbers in adult goldfish retina, 300 cells/

day, and to suggest that the physiological implications of growth processes in functional adult nervous systems make this an interesting area for further study.

REFERENCES

Albercrombie, M. (1946). Estimation of nuclear population from microtome sections. Anat. Rec. 94: 239-247.

Ali, M.A. (1964). Stretching of the retina during growth of salmon (*Salmo salar*). Growth 28: 83-89.

Blaxter, J.H.S. and Jones, M.P. (1967). The development of the retina and retinomotor response in the herring. J. Mar. Biol. Ass. U.K. 47: 677-697.

Brown, J.E. (1965). Dendritic fields of retinal ganglion cells of the rat. J. Neurophysiol. 28: 1091-1100.

Hollyfield, J.G. (1968). Differential addition of cells to the retina of *Rana pipiens* tadpole. Devel. Biol. 18: 163-179.

Lyall, A.H. (1957). The growth of the trout retina. Q. J. Microsc. Sci. 98: 101-110.

Müller, H. (1952). Ban und Wachstum der Netzhaut des Guppy (*Lesbistes reticulatus*). Zool. Jahrb. 63: 275-324.

Murray, M. and Grafstein, B. (1969). Changes in the morphology and amino acid incorporation or regenerating goldfish optic neurons. Exp. Neurol. 23: 544-560.

Schellart, N.A.M. (1973). Dynamics and statistics of photopic ganglion cell response in isolated goldfish retina. Thesis, University of Amsterdam.

Straznicky, K. and Gaze, R.M. (1971). The growth of the retina in *Xenopus laevis*: an autoradiographic study. J. Embryol. Exp. Morphol. 26: 67-79.

Wilson, M.A. (1971). Optic nerve fiber counts and retinal ganglion cell counts during development of *Xenopus laevis* (Daudin). Q. J. Exp. Physiol. 56: 83-91.

Witkovsky, P. (1971). Synapses made by myelinated fibers running to teleost and elasmobranch retinas. J. Comp. Neurol. 142: 205-221.

A UNIQUE LACTATE DEHYDROGENASE ISOZYME IN THE TELEOST RETINA*

Gregory S. WHITT

Department of Genetics and Development

University of Illinois, Urbana, Illinois, U.S.A.

I could a tale unfold whose lightest word
Would harrow up thy soul, freeze thy young blood,
Make thy two eyes, like stars, start from their spheres.
Hamlet, Act I, Sc. 5

The main research of this laboratory over the past few years has focused on the developmental and biochemical genetics of fishes. One of the most interesting gene products investigated is the enzyme lactate dehydrogenase (LDH), particularly the isozyme of LDH found in the retina of many groups of teleosts (Markert and Faulhaber, 1965). This isozyme may play an important role in visual metabolism (Whitt, 1970; Whitt and Booth, 1970; Whitt et al., 1973).

The LDH of all vertebrates exists in multiple forms or isozymes (Markert and Møller, 1959). The two types of LDH subunits commom to all vertebrates, A and B (which differ in their amino acid sequence), combine randomly to form tetrameric isozymes which are readily resolved by gel electrophoresis.

Different physiological roles have been proposed for these isozymes based in part upon their different tissue specificities, and in part upon their different enzymatic properties. The A_4 isozyme is found predominantly in tissues subject to periods of transient anaerobiosis, and it possesses kinetic properties suited

* Supported by National Science Foundation Grants G.B. 16425 and GB 43995

for the conversion of pyruvate to lactate when there are high substrate concentrations, as during anaerobic glycolysis in skeletal muscle. In contrast, the B_4 isozyme is relatively predominant in more oxygenated tissues and has kinetic properties suited for functioning at the lower substrate concentrations typical of aerobic tissues such as the heart muscle.

The isozyme patterns of fish are more complex than those of other vertebrates, but are also more interesting. In Fig. 1 are shown the LDH isozymes of a representative teleost, the mackerel (*Scomber scombrus*). In this fish, as in many others, fewer than five isozymes are formed due to a restriction of A and B subunit assembly (Markert and Faulhaber, 1965). In addition to the usual A and B subunit synthesis in most tissues, a unique isozyme is synthesised in the eye. This C_4 isozyme possesses a very distinctive anodal electrophoretic mobility.

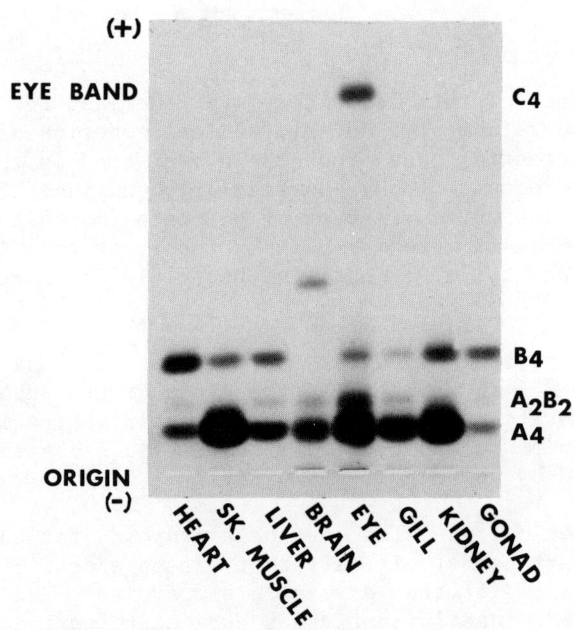

FIG. 1. Lactate dehydrogenase isozymes of the tissues of the mackerel (*Scomber scombrus*). Note the highly anodal retinal-specific LDH C_4 isozyme.

Genetic crosses made with species in the Centrarchidae, using electrophoretic variants of the C_4 isozyme as markers, indicate that the eye band is encoded in a third LDH locus which is not sex linked (Whitt et al., 1971).

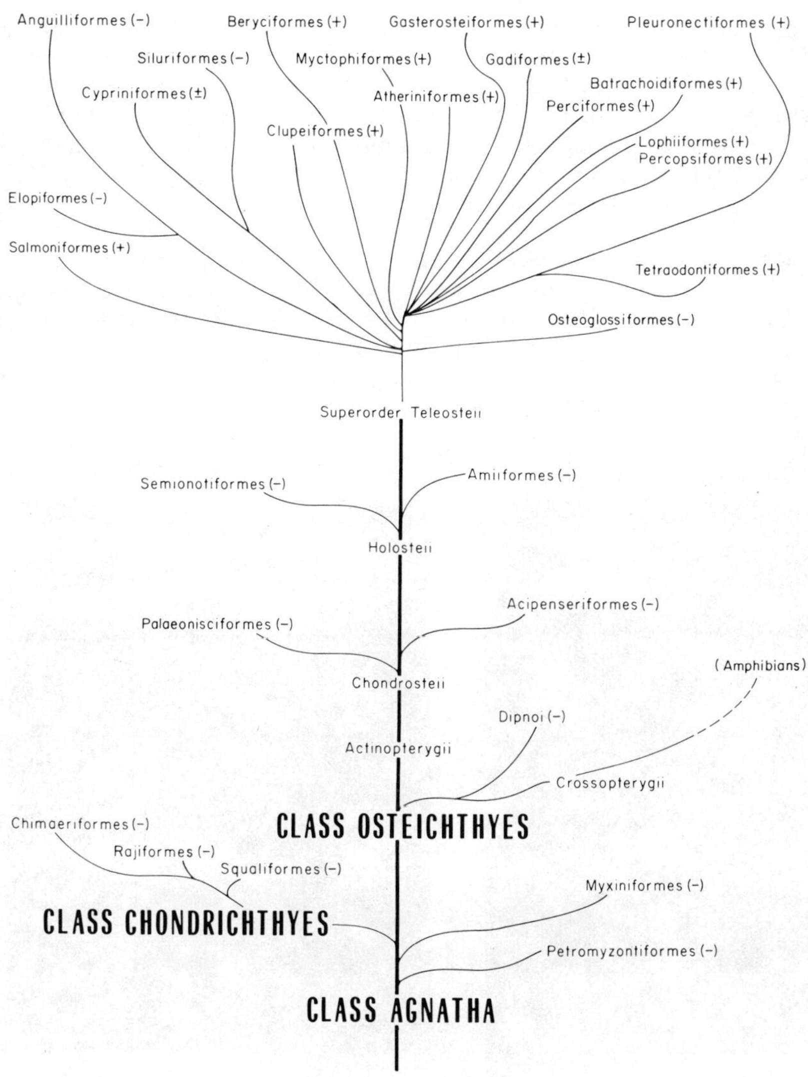

FIG. 2. Distribution of the retinal-specific LDH C_4 isozyme in the fishes. ± indicates some families in an order possess the eye band, other families do not.

The retinal-specific LDH has distinctive physical properties. In addition to its very large net negative charge, the C_4 LDH is considerably more resistant to denaturation by heat or urea than the other isozymes (Whitt, 1970). Kinetic, physical, and immunochemical analyses indicate that the LDH C_4 isozyme is much more closely related to the LDH B_4 than to the A_4 isozyme (Whitt, 1969; 1970).

After determining the genetic basis and some of the physical properties of the LDH C_4 isozyme, a phylogenetic survey of fishes was made to determine which groups possess this retinal-specific LDH (Shaklee et al., 1973; Whitt et al., 1973; 1974). As shown in Fig. 2 none of the primitive fishes possess this isozyme, whereas with the exception of the more primitive orders most of the teleosts do.

Since the retina of many species of fishes has the highest level of C subunit synthesis we employed histochemical procedures to determine which differentiated cells of the retina possessed the C gene function. Frozen sections of the swordtail (*Xiphophorus helleri*) retina were made in a cryostat and then incubated in a staining solution similar to that used for staining gels. However,

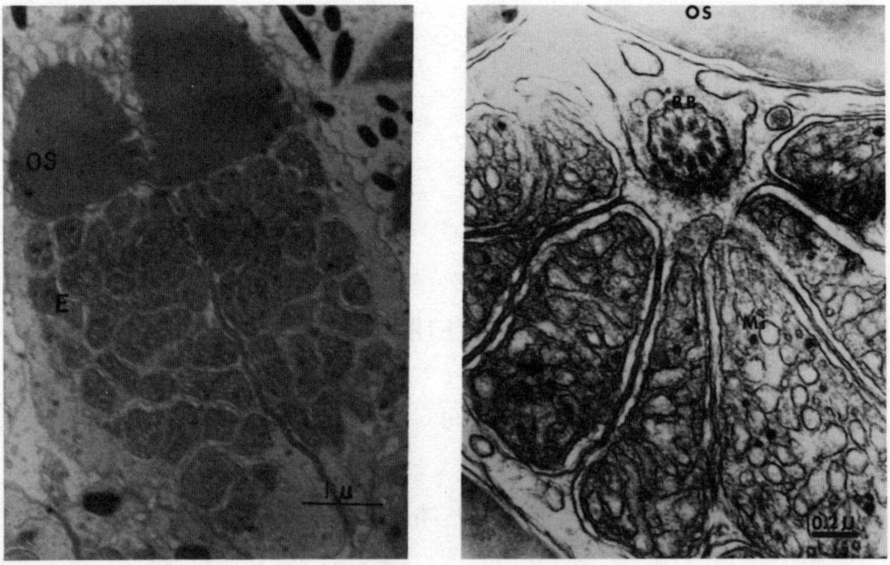

FIG. 3. (Left) Electron micrograph of a pair of twin cones from the adult green sunfish; OS = outer segment, E = ellipsoid. (Right) High magnification electron micrograph of a cone cell ellipsoid. The mitochondria (Mi) are arranged in a radial pattern around the basal body (BB).

one additional component (1.8 M urea) was added. This concentration of urea inactivated all the LDH isozymes except the highly stable LDH C_4, C_4 isozyme activity was thereby demonstrated to be restricted to the inner segment region of the photoreceptor cells, particularly the ellipsoid region (Whitt and Booth, 1970, 1970; Whitt et al., 1973).

Electron microscopy was used to examine the fine structure of the photoreceptor cells (Whitt et al., 1973; Miller and Whitt, 1974). The ultrastructure of cones from an adult green sunfish is shown in Fig. 3 to the left. The ellipsoid region of this cell possesses large numbers of mitochondria. The ellipsoid region is shown at higher magnification to the right in Fig. 3. The observed concentration of mitochondria in this same region where the retinal-specific LDH is localised is very interesting because the sperm specific LDH isozyme of birds and mammals may also be associated with the mitochondria (Machado de Domenech et al., 1972).

In order to determine whether this unique retinal isozyme has

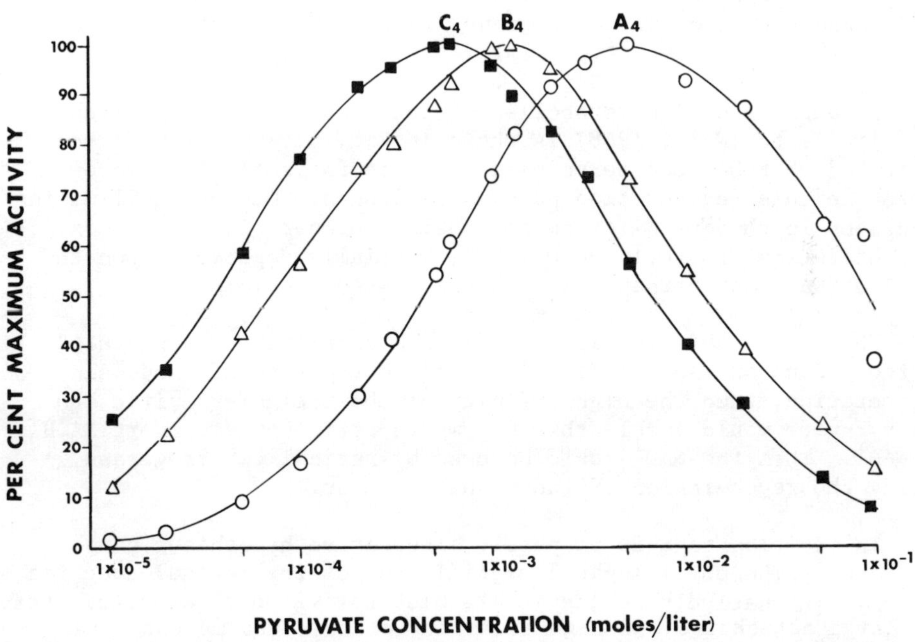

FIG. 4. Pyruvate optima of the homotetrameric isozymes of the mackerel (*Scomber scombrus*). The kinetic properties of LDH C_4 are more like the LDH B_4 than the LDH A_4.

kinetic properties tailored for this particular metabolic environment, the three homopolymeric isozymes (A_4, B_4, and C_4) from the mackerel (*Scomber scombrus*) eye were isolated and purified and their kinetic properties examined (Whitt, 1970). Fig. 4 illustrates the pyruvate concentration optima of these isozymes. The C_4 isozymee has the lowest substrate concentration optimum, Km, and the greatest susceptibility to substrate inhibition. Thus the eye specific LDH behaves more like the B_4 isozyme found in aerobic tissues than the A_4 isozyme predominant in more anaerobic tissues. The kinetic properties of the retinal isozyme are consistent with its localisation in the more aerobic region of the retinal cell near the mitochondria.

The cellular specificity of C gene expression and the kinetic behaviour of the C_4 isozyme suggest that this enzyme may play an important role in the metabolism of the photoreceptor cell and perhaps interacts with the visual cycle itself. Since the retinal-specific LDH has a higher affinity for substrate than the other LDH isozymes, it appears especially suited for cells with a more aerobic metabolism. As pointed out by Wittenberg and Wittenberg (1974), the retinal oxygen tensions of some fish species are very high. Since most of the carbohydrate metabolism of the photoreceptor cell is directed toward the regeneration of rhodopsin (or porphyropsin) it has been proposed that the retinal-specific LDH may play a role in this regeneration of visual pigments in some groups of teleosts (Whitt, 1970; Whitt et al., 1973).

An outline of the metabolism of the visual cycle of vertebrates (modified after Wald, 1968) is shown in Fig. 5. Wald (1968) has indicated that NAD can serve as a coenzyme for retinol dehydrogenase in this regeneration pathway in fishes. One of the limiting processes in rhodopsin (or porphyropsin) regeneration is the oxidation of retinol to retinene which is partially dependent upon the level of NAD, and perhaps also the $NAD^+/NADH_2^+$ ratio.

Where does the retinal-specific LDH come into this picture? If the amount of NAD^+ becomes limiting to the rate of rhodopsin regeneration, then the high affinity of the retinal-specific LDH for pyruvate would enable this enzyme to efficiently convert $NADH_2^+$ to NAD^+. Then the NAD^+ could be used by retinol dehydrogenase during the regeneration of the visual pigment.

In addition to this proposal, alternative hypotheses can be generated. The physiological significance of the retinal-specific LDH may alternatively reside in its distinctive physical attributes. The large net charge and high stability of this enzyme could be important in its subcellular localisation or its resistance to degradation. In any event, the restriction of this isozyme to the highly differentiated photoreceptor cells certainly suggests that

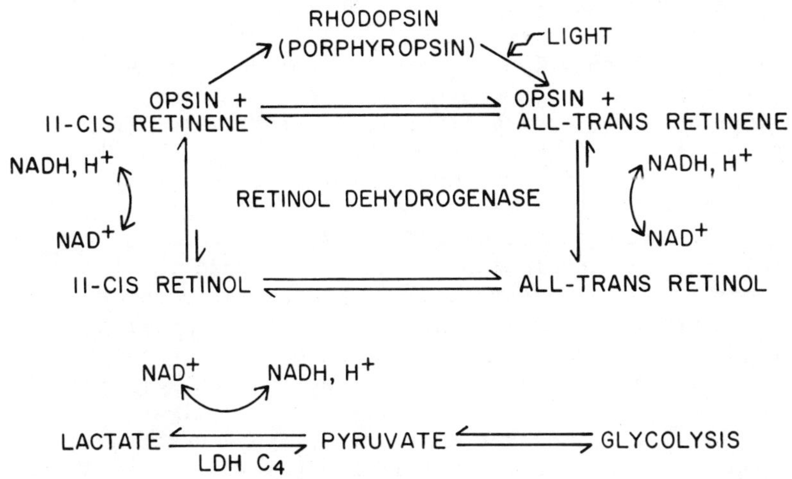

FIG. 5. (Above) Outline of the metabolism of the teleost visual cycle. (Below) The reaction catalysed by the retinal-specific LDH C_4 isozyme. The LDH C_4 may play a role in regulating NAD^+ / $NADH_2^+$ ratio necessary for visual pigment regeneration.

it plays a very specialised role--a role unable to be fulfilled by the other LDHs.

If the retinal LDH isozyme is important to visual metabolism, its appearance during embryogenesis should be tightly coupled to the morphological and functional differentiation of the retina. In 1965 Nakano and Whiteley, working with the medaka (*Oryzias latipes*), observed that the retinal-specific LDH appeared abruptly at the time of hatching. These researchers attributed the sudden appearance of the C_4 isozyme to the rapid change in metabolism at hatching. Specifically, they proposed that the LDH C gene activation was caused by the change from the anaerobic conditions of the egg to the aerobic conditions after hatching.

In order to test the generality of this hypothesis, the ontogeny of the LDH isozymes was analysed in a different teleost species, *Fundulus heteroclitus* (Whitt, 1970). In this fish the C_4 isozyme was first detected at least 24 hours prior to hatching. Therefore, the LDH C gene had to be activated at least 24 hours prior to hatching. It was concluded therefore that the act of hatching is not an obligatory stimulus for C gene activation (Whitt, 1970). Two other laboratories (Hitzeroth et al., 1968; Goldberg et al., 1969) working independently with trout observed a different pattern of C gene expression. In these species the retinal-specific isozyme first appeared several weeks after hatching.

In order to determine why the C_4 isozyme was synthesised after hatching in these species, a study of the earlier literature on the morphological development of trout and salmon was initiated. Ali's work on salmonid eye development demonstrated that the larvae do not develop a mature visual apparatus, i.e. rods, retinomotor response, and visual acuity until two to three weeks after hatching (Ali, 1959). This timing of functional differentiation of the retina correlated fairly well with the time of appearance of the retinal-specific LDH in these salmonids. More recent analyses at the light microscope level by Kunz (1971) indicate that the C_4 appears at the time of photoreceptor cell differentiation in the guppy (*Lebistes*). In addition, Nakano and Hasegawa (1971) working with the medaka, have suggested that morphological differentiation may precede the appearance of the eye band by two days in this fish.

In the light of the data obtained from light microscopy, it appeared reasonable to postulate that C locus expression is tightly coupled with the differentiation of the photoreceptor cells. This hypothesis was tested by following LDH isozyme synthesis during the embryogenesis of the green sunfish and by monitoring the structural differentiation of the photoreceptor cells (Whitt et al., 1973; Miller and Whitt, 1974).

The progression of LDH isozymes during the early development of the sunfish at 24°C is displayed in Fig. 6. At the time of hatching the eyes are poorly developed with no pigmentation. Two days after hatching the eyes are faintly pigmented. By three days post-hatch the eye is heavily pigmented. By five days after hatching the eye appears completely differentiated on the basis of external morphology. No C polypeptide synthesis is detected until five days after hatching when the B_3C_1 heteropolymer is first observed. The retinal LDH C_4 isozymee is first detected at six days after hatching.

Two stages of sunfish development were analysed by electron microscopy (Miller and Whitt, 1974): five days post-hatch (the time when C subunit synthesis is first detected) and three days post-hatch (the time prior to both C_4 synthesis and completion of external morphological differentiation).

There are a number of important differences between the structure of the retina in the larval fish at three days post-hatch and in the adult. The micrograph of the three day post-hatch retina reveals that the ellipsoid regions of the photoreceptor cells have reduced

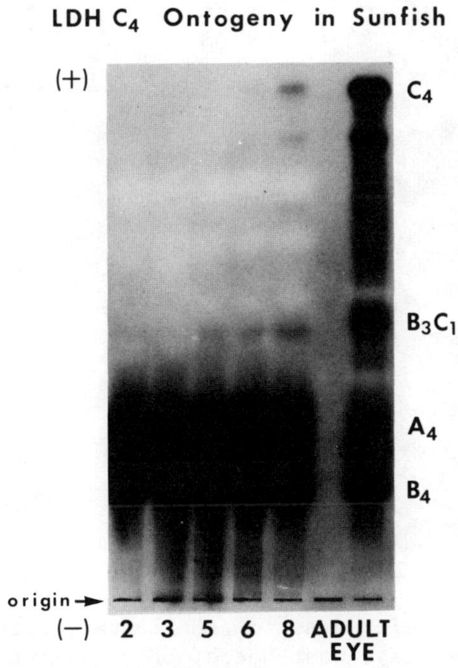

FIG. 6. Synthesis of the C_4 isozyme of lactate dehydrogenase during development of the larval sunfish. The numbers below the slots indicate the number of days after hatching. LDH C subunit synthesis is first detected as the B_3C_1 isozyme five days post-hatch.

numbers of mitochondria (Fig. 7, left). Most photoreceptor cells do not possess outer segments at this stage. The few outer segments which are present appear as small triangles. The nuclei are very much longer at this stage than those seen in the adult.

A number of changes have occurred by the fifth day post-hatch (Fig. 7, right), the time when C subunit synthesis is first detected. Since the previous stage, there has been a dramatic increase in the number of photoreceptor cells with outer segments. Electron micrographs of the five day post-hatch retina reveal a much closer arrangement of cells with larger outer segments. Furthermore, the receptor cells now possess well-differentiated ellipsoids with tightly clustered mitochondria. In fact, cells of the central region of the five day post-hatch retina appear very similar to those of the adult retina.

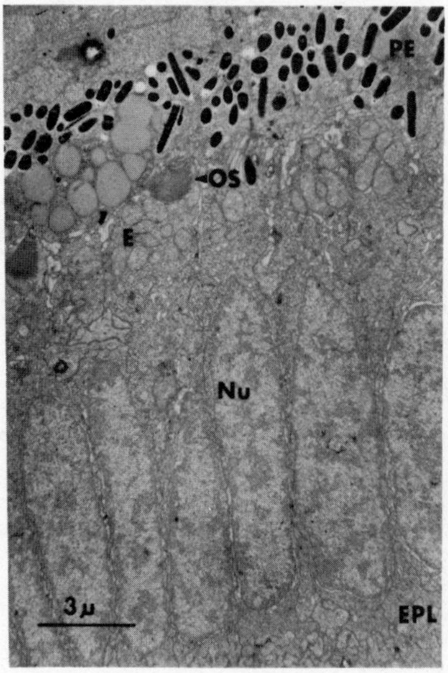

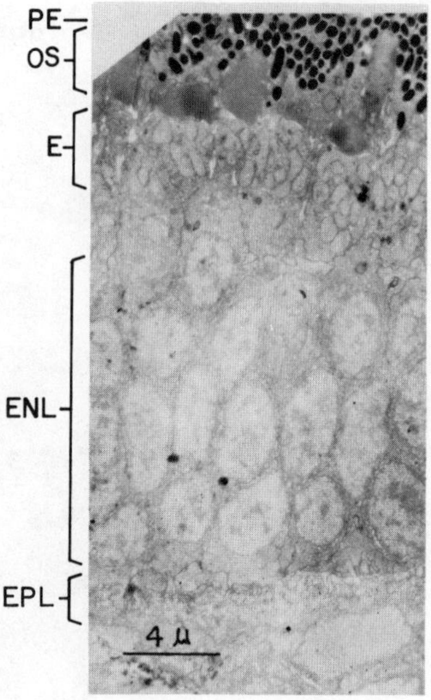

FIG. 7. (Left) Electron micrograph of the larval green sunfish retina at three days post-hatch; PE = pigment epithelium, OS = outer segment, E = ellipsoid, NU = nucleus, and EPL = external plexiform layer. (Right) Electron micrograph of the larval green sunfish retina at five days post-hatch; ENL = external nuclear layer.

In conclusion, the multigenic isozyme system of lactate dehydrogenase is an excellent model system for studying the specificity of gene function during cellular differentiation. One of these LDH gene products--the C_4 isozyme--is specifically localised in the highly differentiated photoreceptor cell of many teleosts and may be intimately involved in visual metabolism.

REFERENCES

Ali, M.A. (1959). The ocular structure, retinomotor and photobehavioural responses of juvenile Pacific salmon. Can. J. Zool. 37: 965-996.

Goldberg, E., Cuerrier, J.P. and Ward, J.C. (1969). Lactate dehydrogenase ontogeny, paternal gene activation, and tetramer assembly in embryos of brook trout, lake trout, and their hybrids. Biochem. Genet. 2: 335-350.

Hitzeroth, H., Klose, J., Ohno, S., and Wolf, U. (1968). Asynchronous activation of parental alleles at the tissue-specific gene loci observed on hybrid trout during early development. Biochem. Genet. 1: 287-300.

Kunz, Y. (1971). Distribution of lactate dehydrogenase (and its E-isozymes) in the developing and adult retina of the guppy (*Lebistes reticulatus*). Rev. Suisse Zool. 78: 761-776.

Machado de Domenech, E., Domenech, C.E., Aoki, A. and Blanco, A. 1972). Association of the testicular lactate dehydrogenase isozyme with a special type of mitochondria. Biol. Reprod. 6: 136-147.

Markert, C.L., and Faulhaber, I. (1965). Lactate dehydrogenase isozyme patterns of fish. J. Exp. Zool. 159: 319-332.

Markert, C.L. and Møller, F. (1959). Multiple forms of enzymes: Tissue, ontogenetic, and species specific patterns. Proc. Natl. Acad. Sci. U.S.A. 45: 753-763.

Miller, E.T. and Whitt, G.S. (1974). Lactate dehydrogenase isozyme synthesis and cellular differentiation in the teleost retina. In: Isozymes, Vol. 3. Edited by C.L. Markert. Academic Press. (In press).

Nakano, E. and Whiteley, A.H. (1965). Differentiation of multiple molecular forms of four dehydrogenases in the teleost, *Osyzias latipes*, studied by disc electrophoresis. J. Exp. Zool. 159: 167-179.

Nakano, E. and Hasegawa, M. (1971). Differentiation of the retina and retinal lactate dehydrogenase isoenzymes in the teleost, *Oryzias latipes*. Dev., Growth and Diff. 13: 351-357.

Shaklee, J.B., Kepes, K.L. and Whitt, G.S. (1973). Specialized lactate dehydrogenase isozymes: The molecular and genetic basis for the unique eye and liver LDHs of teleost fishes. J. Exp. Zool. 185: 217-240.

Wald, G. (1968). Molecular basis of visual excitation. Science 162: 230-239.

Whitt, G.S. (1969). Homology of lactate dehydrogenase genes: E gene function in the teleost nervous system. Science 166: 1156-1158.

Whitt, G.S. (1970). Developmental genetics of the lactate dehydrogenase isozymes of fish. J. Exp. Zool. 175: 1-35.

Whitt, G.S. and Booth, G.M. (1970). Localization of lactate dehydrogenase activity in the cells of the fish (*Xiphophorus helleri*) eye. J. Exp. Zool. 174: 215-224.

Whitt, G.S., Childers, W.F., and Wheat, T.E. (1971). The inheritance of tissue-specific lactate dehydrogenase isozymes in interspecific bass (*Micropterus*) hybrids. Biochem. Genet. 5: 257-273.

Whitt, G.S., Miller, E.T., and Shaklee, J.B. (1973). Developmental and biochemical genetics of lactate dehydrogenase isozymes in fishes, p. 243-276. In: Genetics and mutagenesis of fish Edited by J.H. Schröder. Springer-Verlag, New York, 356 pp.

Whitt, G.S., Shaklee, J.B. and Markert, C.L. (1974). Evolution of the lactate dehydrogenase isozymes of fishes. In: Isozymes Vol. 4. Edited by C.L. Markert. Academic Press. (In press).

Wittenberg, J.B. and Wittenberg, B.A. (1974). The choroid rete mirabile of the fish eye. 1. Oxygen secretion and structure: comparison with the swimbladder rete mirabile. Biol. Bull. 145: 116-136.

LANDOLT'S CLUB IN SOME PRIMITIVE FISHES

N.A. LOCKET

Institute of Ophthalmology, University of London

Judd Street, London SC1H9QS, U.K.

The actors are at hand; and, by their show,
You shall know all that you are like to know.
Midsummer Night's Dream, Act V, Sc. 1

In 1871, Landolt described a cell process in the retina of certain urodele amphibians which lay between the nuclei of visual cells and extended as far as the outer junctional zone. He thought that these club shaped processes were not of neural origin, but that they were in direct communication with the 'supporting' tissue of the retina. Subsequent investigations, notably those of Ranvier (1882) and Dogiel (1883), showed that the clubs arose from bipolar cells.

Not all bipolar cells give rise to Landolt's clubs. Ramón y Cajal (1893) described them in the frog, a lizard, *Lacerta viridis*, and the chicken. He did not find them in the three species of teleost fish he had studied, nor in sheep, cattle and dogs, though he quoted Dogiel as finding a possible club in mammals. In those species which had the clubs Ramón y Cajal found them arising from the dendritic tuft of the smaller, cone bipolars, but not from that of the larger, rod bipolars. The cone bipolars in the fovea of chameleons and birds, however, lacked the clubs.

Recently Stell (1972) has reviewed the previous work on the clubs, and states that they have been seen in all vertebrate groups except mammals, nocturnal birds and teleost fishes. Witkovsky and Stell (1973) found that in the selachian, *Mustelus canis*, bipolars of certain types only bear the clubs.

Landolt's clubs have not been extensively studied by electron microscopy. Hendrickson (1966) gave a detailed account of those in the newt *Diemictylus v. viridescens*; Locket described the unusual relations of the clubs in the African lungfish, *Protopterus aethiopicus* (1970), and those in the coelacanth *Latimeria chalumnae* (1973).

Retinal bipolar cells typically have their nucleus in the inner nuclear layer. A vitread process extends to the inner synaptic layer where it arborises, and a sclerad process extends to the outer synaptic layer. This sclerad process bears dendrites which make contact with the synaptic pedicles of the visual cells. The details of these contacts vary with species and with the type of bipolar cell, and will not be further considered here.

In certain bipolar cells of lower vertebrates the nuclei are not located in the inner nuclear layer. These cells were described in sturgeons by Dogiel (1883) and in selachians by Neumayer (1897). Ramon y Cajal (1893) described and figured them in frog and lizard, and called them displaced bipolar cells. These cells differ from the conventional bipolar cells in that their nuclei are situated sclerad to the outer synaptic layer, and thus amongst the nuclei of the visual cells. Displaced bipolars and visual cells are easily distinguished from each other in Golgi preparations, but in conventionally stained paraffin sections their nuclei may appear similar and so be confused. This is a potential source of error in retinal cell counts. If all the nuclei in the outer nuclear layer are assumed to belong to visual cells when some are in fact those of displaced bipolars, then the visual cell count will be higher, and the bipolar count lower by the number of displaced bipolars wrongly identified.

Both conventional and displaced bipolar cells may bear Landolt's clubs. These are easily identified in Golgi preparations, when they are seen as a process, often stout, arising from the dentritic tree of a conventional bipolar cell in the outer synaptic layer, and extending between the outer nuclei towards the outer junctional zone. The club may continue the line of the main sclerad process, but commonly does not do so, arising from one of the horizontal dendrites. The club of displaced bipolars commonly arises from the perikaryon, sclerad to the nucleus, and the cell soma may look pearshaped. The dendritic tree arises from the vitread aspect of the perikaryon, and the main vitread process arises amongst these dendrites. The arrangements are seen in Fig. 3 and diagrammatically in Fig. 1.

The retinae of primitive fishes have been described by optical, and in part by electron microscopy. *Protopterus* was first considered by Schiefferdecker (1886), and later by Hosch (1904), Grynfeltt (1911), Walls (1942), Rochon-Duvigneaud (1943) and Munk (1964). Locket (1970) described Landolt's clubs in *P. aethiopicus*

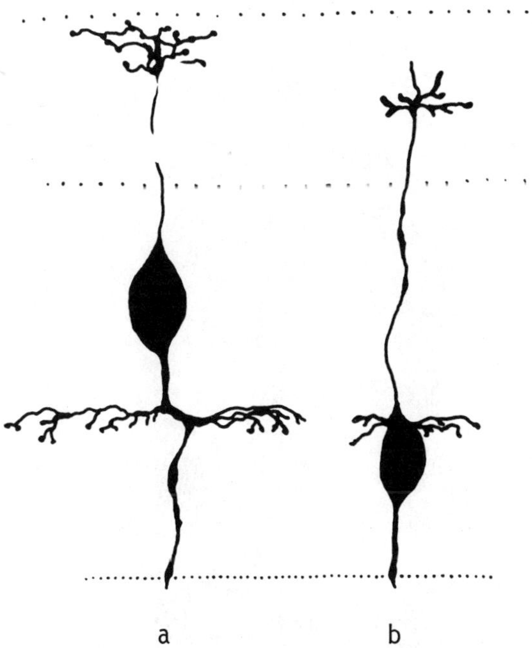

FIG. 1. Conventional (a) and displaced (b) bipolar cells of the frog. Both cells bear a Landolt's club; that of the conventional cell arises from the dendritic tree in the outer plexiform layer, and that of the displaced cell from the perikaryon. The clubs extend to the outer junctional zone. After Ramón y Cajal.

and Leure-duPree (1973) studied the same species by electron microscopy. The earlier authors did not report the clubs, which are hard to see in paraffin embedded material. They are well shown by electron microscopy, and once recognised are easily identified in 1μm plastic sections.

In *Protopterus*, Fig. 2, the clubs have an unusual disposition; they arise from the bipolar dendrites as in other examples, but they extend sclerally far beyond the outer junctional zone. In their course vitread to that zone they appear in tangential sections as smooth rounded profiles about 2-3μm in diameter, mostly surrounded by radial fibre, but in some places in contact with visual cells. At the outer junctional zone they show the junctional complexes with adjacent cells which make up the 'membrane', and sclerad to this they continue as expanded club shaped endings which extend more deeply into the pigment epithelium cells than even the rod outer segments.

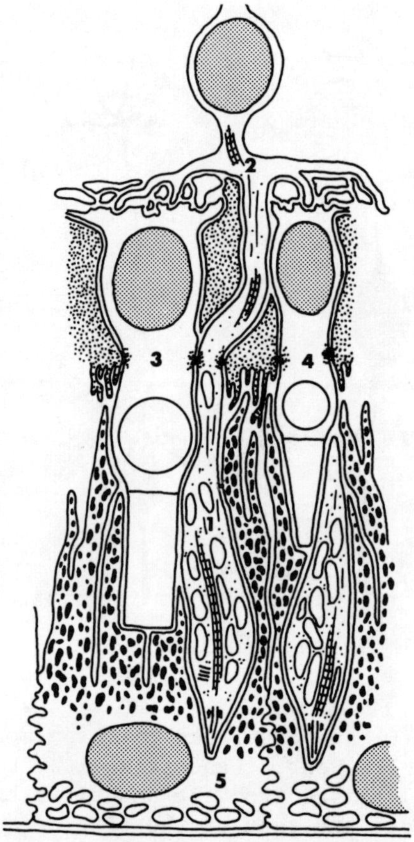

FIG. 2. Diagram of the arrangement of Landolt's clubs in *Protopterus aethiopicus*. The clubs, 1, arise from the dentritic tree of a conventional bipolar cell, 2. They are related to rods, 3, and cones, 4, and largely surrounded by radial fibre, stippled. The clubs extend through the outer junctional zone, and are inserted into the pigment epithelium cells, 5, reaching to the level of the nuclei. The clubs bear an apical cilium, from the basal body of which striated rootlets extend along the club stalk at least as far as the outer synaptic layer.

Electron microscopy shows that the club shaped end contains numerous mitochondria, mostly aligned along the axis of the process. These are surrounded by pale cytoplasm, containing granules about 25nm diameter which have the appearance of glycogen granules.

Situated in the expanded part are two centrioles. One of these forms the basal body of a 9+0 cilium, and from it extend striated ciliary rootlets, cross-banded with a 68nm periodicity. These rootlets, which appear as a patch of punctate densities in clubs cut transversely, extend at least as far as the outer synaptic layer. Similar densities seen in processes sectioned amongst the bipolar nuclei probably represent rootlets which have extended through the club stalk to the main sclerad process of the bipolar cell.

Some of the mitochondria of the club are elongated and lie in close association with the striated rootlets, but others have a more random distribution. Though the clubs and the pigment epithelium cells come into close apposition the junctions between them are not marked by any special features.

Ali and Anctil (1973) have reported the presence of similar long Landolt's clubs in paraffin sections of *Lepidosiren paradoxa*, the South American lungfish, and the present author confirms this finding from plastic embedded material. Though not studied in detail, these clubs appear similar to those in *Protopterus*. The retinae of the two forms are very similar, though that of *Lepidosiren* is less well differentiated.

The Australian lungfish, *Neoceratodus forsteri*, appears less closely related to the previous two than they are to each other, and the retinal structure is also somewhat different. *Neoceratodus* shares with the other lungfishes, and with urodele amphibians, the possession of very large cells. The visual cells, described by Schiefferdecker (1886) and by Munk (1969) include cones with oil droplets and rods without. My observation shows that conventional and displaced bipolar cells are present, and that the nuclei of the displaced variety are indistinguishable from those of visual cells in paraffin preparations. They are more clearly seen in plastic sections, and the two types are clearly differentiated in Golgi preparations (Fig. 3). Both types of bipolar cell give rise to Landolt's clubs, which extend as far as the outer junctional zone, but do not come into relationship with the pigment epithelium cells (Fig. 4). The clubs terminate in small varicosities, just projecting into the ventricular space, amongst which a single short cilium is present. In some examples the plasma membrane delimiting the varicose ending of the club is deficient in places; this may be artifactual. The portion of the club just vitread to the outer junctional zone is expanded, and this region contains mitochondria, striated ciliary rootlets and microtubules, like the expanded clubs in *Protopterus*. The cells in *Neoceratodus* are large enough for the clubs to be distinguished easily in flat sections, and they are seen to be surrounded by radial fibre in almost every case.

The retina of the surviving coelacanth, *Latimeria chalumnae*, was studied from paraffin embedded material by Millot and Carasso (1955) and Millot and Anthony (1965). Locket (1973) described suitably fixed and embedded material by optical and electron microscopy. The bipolar cells in this fish are almost entirely of

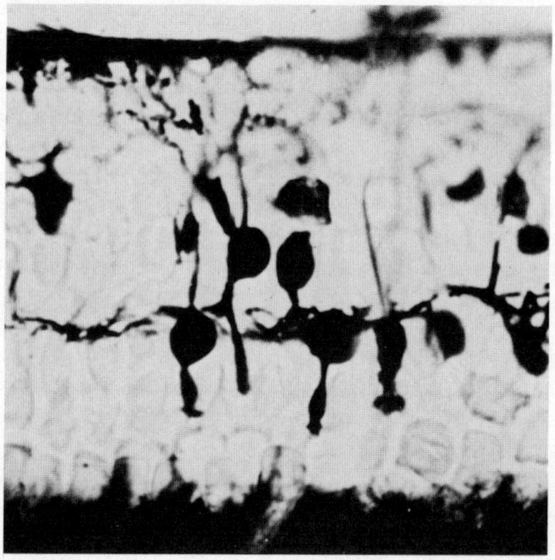

FIG. 3. Bipolar cells in *Neoceratodus forsteri*. Golgi impregnation. Examples of conventional and displaced bipolars are present, both types bearing Landolt's clubs which extend to the outer junctional zone.

the displaced variety, and they bear Landolt's clubs with certain unusual features (Fig. 5). The more vitread portions of the clubs resemble those of displaced bipolars in *Neoceratodus*, arising as a broad extension of the perikaryon towards the outer junctional zone. As in the lungfishes, they contain mitochondria, microtubules, granules resembling glycogen and the striated rootlets of a cilium.

The tips of the clubs project through the outer junctional zone, and form a swollen ending. This ending typically contains a few small mitochondria, and a single mitochondrion about three times the size of the others. The short cilium, which projects from one side of the end swelling, originates from one of the two centrioles. Commonly associated with these are several dense bodies, and the swollen ending may contain a number of vesicles. These vesicles mainly appear clear, though some show a small punctate central density, and at 60nm diameter are larger than the 40nm vesicles present in the visual cell synapses. Some clubs contain these vesicles, some contain none, and in a few cases the plasma membrane of the ending appears disrupted, and vesicles are present within and outside the ending.

The features of Landolt's clubs described do not define their function, but they permit speculation. The clubs do not occur on all bipolars. Mammals, nocturnal birds and teleosts lack them, and

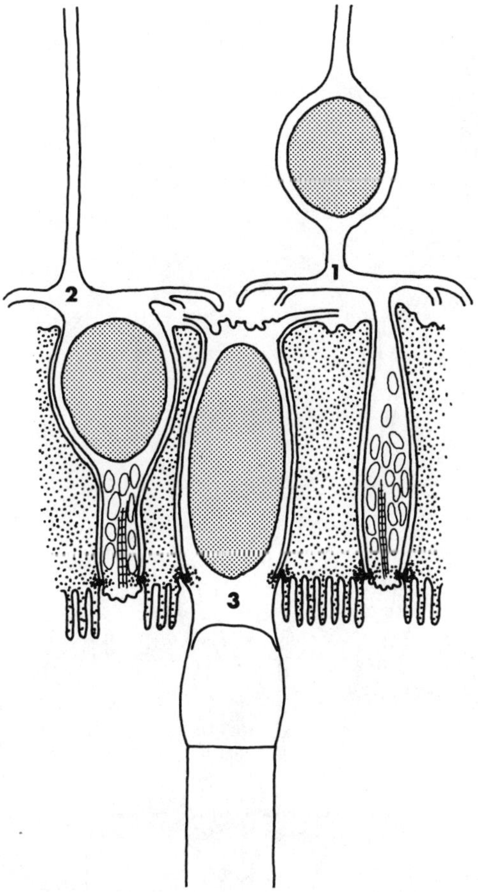

FIG. 4. Diagram of Landolt's clubs in *Neoceratodus forsteri*. Clubs from a conventional, 1, and a displaced, 2, bipolar arc shown, surrounded by radial fibre, stippled. The clubs contain mitochondria and striated ciliary rootlets, and end in varicosities just sclerad to the outer junctional zone. A rod is shown between the clubs, 3.

in *Mustelus canis* some bipolars have them and some not. (Witkovsky and Stell, 1973). Whatever their function, if it is an essential one, it must be subserved by some other agency in those bipolars which lack the clubs. Stell (1972) observes that the groups which have the clubs do not have typical oligosynaptic rod spherules. Both rods and cones in lungfishes have complex pedicles, but the rod synapses in *Latimeria* are simple spherules except for the possession of a single basal filament (Locket, 1973).

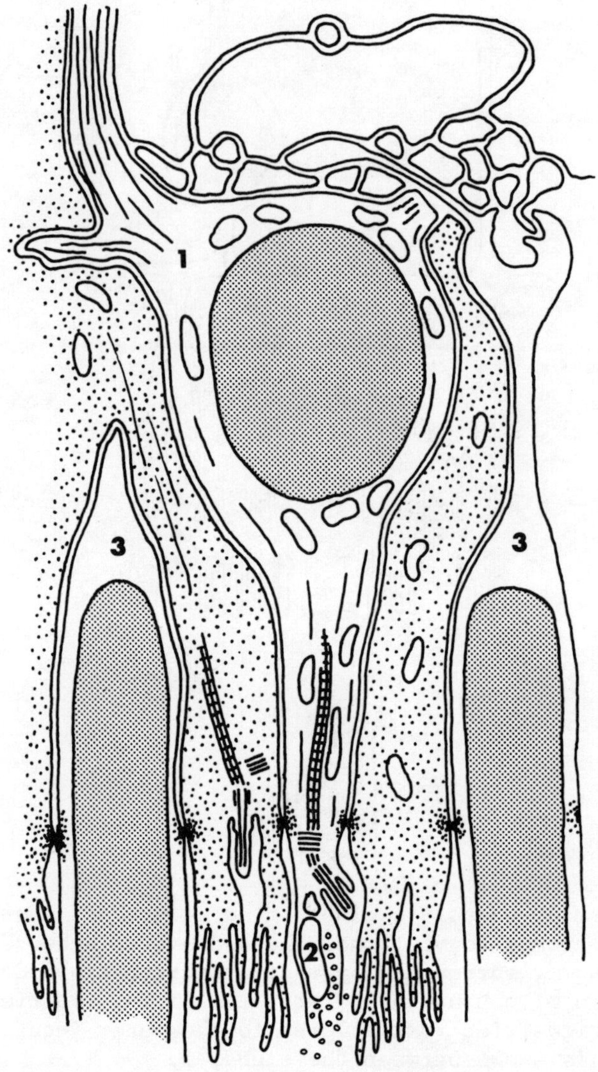

FIG. 5. Diagram of Landolt's club in *Latimeria chalumnae*. In *Latimeria* most of the bipolar cells are displaced, 1. The club extends sclerally, and ends in the ventricular space in an expansion. This contains a few small and one large mitochondria, 2, and two centrioles, of which one is the basal body of a short cilium. The expansion may contain vesicles; in some cases these are present inside and outside the expansion, the membrane of which is disrupted, as shown here. Rods, 3. Radial fibre stippled.

The presence of an aggregation of mitochondria and the 9+0 cilium in the clubs with prominent rootlets, calls to mind the inner segment of the receptors, and it may be that the clubs are receptors of some sort themselves. The photoreceptor cells in the fishes considered do not have prominent rootlets, and do bear an outer segment, of which there is no trace in the clubs. The clubs might respond to some stimulus other than light; there is no evidence to support this suggestion, nor on the possible nature of such a stimulus.

The presence of packed mitochondria suggests that the clubs may have a metabolic function. They represent a site at which the bipolar cytoplasm comes into close contact with the ventricular space, and in *Protopterus* with the cytoplasm of the pigment epithelium as well. They are thus a potential pathway for metabolites passing between bipolars and the pigment cells, in one or both directions.

The vesicles sometimes present in the clubs of *Latimeria* have an appearance consistent with an intermittent release of some vesicle-borne substance from the clubs to the ventricular space. The existence and nature of such a substance remain conjectural.

Stell (1972) concluded that the function of Landolt's clubs is totally obscure, a situation which remains unchanged today.

REFERENCES

Ali, M.A. and Anctil, M. (1973). Retina of the South American lungfish, *Lepidosiren paradoxa* Fitzinger. Can. J. Zool. 51: 969-972.

Dogiel, A. (1883). Die Retina der Ganoïden. Arch. Mikrosk. Anat. Entw Mech. 22: 419-472.

Grynfeltt, G. (1911). Etudes anatomiques et histologiques sur l'oeil de *Protopterus annectens*. Bull. Mens. Acad. Sci. Lett. Montpellier, 210.

Hendrickson, A. (1966). Landolt's club in the amphibian retina. A Golgi and electron microscope study. Invest. Ophthal. 5: 484-496.

Hosch, F. (1904). Das Sehorgan von *Protopterus annectens*. Arch. Mikrosk. Anat. EntwMech. 64: 99-110.

Landolt, E. (1871). Beiträge zur Anatomie der Retina von Frosch, Salamander und Triton. Arch. Mikrosk. Anat. EntwMech. 7: 81

Leure-duPree, A.E. (1973). Electron microscopic observations of the retina of the African lungfish *Protopterus aethiopicus*. Anat. Rec. 175: 370-371.

Locket, N.A. (1970). Landolt's club in the retina of the African lungfish, *Protopterus aethiopicus*, Heckel. Vision Res. 10: 299-306.

Locket, N.A. (1973). Retinal structure in *Latimeria chalumnae*. Phil. Trans. R. Soc. Lond. B. 266: 493-521.

Millot, J. and Anthony, J. (1965). Anatomie de *Latimeria chalumnae*. II. Système nerveux et organes des sens. Centre National de la Recherche Scientifique, Paris.

Millot, J. and Carasso, N. (1955). Note préliminaire sur l'oeil de *Latimeria chalumnae* (Crossoptérygien coelacanthidé). C.r. Hebd. Séances Acad. Sci. 241:576-577.

Munk, O.(1964). The eye of *Calamoichthys calabaricus* Smith 1865 (Polypteridae, Pisces) compared with the eye of other fishes. Vidensk. Medd. Dan. Naturhist Foren., KBH 127: 113-126.

Munk, O. (1969). On the visual cells of some primitive fishes with particular regard to the classification of rods and cones. Vidensk. Medd. Dan. Naturhist Foren., KBH 132: 25-30.

Neumayer, L. (1897). Der feinere Bau der Selachier-Retina. Arch. Mikrosk. Anat. EntwMech. 48: 83-111.

Ramon y Cajal, S. (1893). La Rétine des Vertébrés. Cellule., 9: 131-225.

Ranvier, L. (1882). Traité technique d'histologie. F. Savy, Paris.

Rochon-Duvigneaud, A. (1943). Les yeux et la vision des vertébrés. Masson, Paris.

Schiefferdecker, P. (1886). Studien zur vergleichenden Histologie der Retina. Arch. Mikrosk. Anat. EntwMech., 28: 305-396

Stell, W.K. (1972). The morphological organisation of the vertebrate retina. In: Handbook of Sensory Physiology. VII/2, edited by M.G.F. Fuortes, Springer, Berlin.

Walls, G.L. (1942). The vertebrate eye and its adaptive radiation. Cranbrook Inst. Sci., Michigan.

Witkovsky, P. and Stell, W.K. (1973). Retinal structure in the smooth dogfish, *Mustelus canis*: Light microscopy of bipolar cells. J. Comp. Neurol. 148: 47-59.

VISUAL PIGMENTS: PHYLOGENY AND ECOLOGY

M. A. ALI & H. -J. WAGNER, Dépt. de Biol. Univ. de Montréal, Canada, andLehrstuhl für Morphologie & Anatomie, Universität Regensburg, Germany

Draw near,
And list what with our council we have done.
King Richard the Second, Act. I, Sc. 3

INTRODUCTION

Although John Lythgoe's (1972) list of vertebrate visual pigments has been extremely useful and appreciated it was felt that a repertoire of visual pigments in fishes should also be given in this volume, for various reasons. First, it seemed that a volume of this sort will not be complete without such information. Secondly, considerable information quality and quantity wise has appeared since the material for Lythgoe's list was gathered. Also, we wanted to give some information on the habitat of the various families and follow a more modern classification. The work was undertaken with the naive feeling that it will essentially be a reorganisation of Lythgoe's list but experience proved it to be otherwise. Almost as many more species have been added and the integration of new information, re-organisation of the classification and, compilation of ecological information proved to be formidable and time consuming but worthwile tasks. We did not also set a limit as Lythgoe did. Any available information known or provided to us has been included. Thus, information obtained by the Carr-Price method (Ali and Heumann, 1970, 1972) and by microspectrophotometry of single cells (Hárosi, 1973) as well as retinal patches (Muntz, 1973) has been integrated. The information concerning visual pigments tends to be labile in the sense that as techniques improve, previously obtained information appears less exact. Thus, for example Salmo salar was known to have a mixture of $P503_1$,and $P527_2$ pigments (Munz and Beatty, 1965) but Beatty (personal communication) has lately found that they are P505, and $P531_2$. Such has also been the case for Anguilla anguilla, Salmo trutta, etc. Similarly, measurements made with pigment in the intact retina differ from those obtained not only with extracts and suspensions but also from single cells, side-on, (Dartnall & Lythgoe, 1965; Hárosi, 1971; Bowmaker, 1973). We do not expect much comparative information covering a large number of species to come out in the next few years as it did in the late sixties and early seventies (Schwanzara, 1967; Munz & McFarland, 1973; Beatty, 1973). But we do expect that many of the maxima as we know them now will be displaced by a few nanometres, as in the case of the salmon, brown trout and eel, due to the development of improved techniques.

A word about the inclusion of information obtained by the Carr-Price method. Since the aim was to give information about as many species as possible we felt that information if a fish had single or mixed pigments, is in itself interesting. This information is provided by this method. Since exact maxima and spectra are not necessarily immutable, the information on the type of vitamin A only would appear acceptable in cases where no other information about the visual pigment (s) is available.

ACKNOWLEDGEMENTS

We thank Dr. J. L. Lythgoe for the advice he gave when we decided to undertake this task; Drs. D. D. Beatty and F. I. Hárosi for the published and unpublished information which they brought to our attention. We appreciate Professor H. J. Autrum's kindness in allowing the use of his library.

List of visual pigments in teleost fishes. Classification according to Greenwood et al. (1966); B bottom living; C crepuscular; De deep water; Di diurnal; F freshwater; I intermediate depth; M marine; N nocturnal; S surface; T tropical.

Classification	Pigment type and or $\lambda_{max.}$ (nm) A_1	A_2	Ratio	Author	Ecological Comments
Teleostei					
Elopomorpha					
Anguilliformes					
Anguilloidei					
Anguillidae					
1. *Anguilla*					
anguilla	487	523		47	catadromous,
	502			48	active swimmer
2. *A. rostrata*	482	523		11	yellow stage
	501				metamorphosis
Muraenidae					T, M, B
3. *Gymnothorax*					cavernicolous
undulatus	495.4			42	
4. *Pridonophis*					
equatorialis	+			1	
Congridae					M, predator
5. *Conger*					
marginatus	486.8			42	
6. *C. conger*	487			49	
Clupeomorpha					
Clupeiformes					
Clupeoidei					
Clupeidae					M, pelagic,
7. *Opisthonema*					schooling,
libertate	+			1	plancton feeder
8. *Lile stolifera*	+			2	
9. *Dorosoma*					
cepedianum	500	521		14	
10. *Alosa*					
pseudoharengus	504	528		45	
11. *Clupea sprattus*	500			26	

Classification	Pigment type and or λ max. (nm) A1	A2	Ratio	Author	Ecological Comments
Engraulidae					M, pelagic, plancton feeder
12. Anchoa compressa	508			36	
13. A. sp.	+			2	
Osteoglossomorpha					
Osteoglossiformes					
Osteoglossoidei					
Osteoglossidae					
14. Osteoglossum bicirrhosum	503	527		45	
Notopteroidei					
Hiodontidae					T, Di, F
15. Hiodon alosoides		535		9	
Notopteridae					M, C
16. Xenomystus nigri	503	527		45	
Protacanthopterygii					
Salmoniformes					
Salmonoidei					
Salmonidae					Anadromous, active swimmers
17. Oncorhynchus gorbuscha	503	527		40	
18. O. keta	503	527		40	
19. O. kisutch	503	527		40	
20. O. nerka	503	527		40	
21. O. tshawytscha	503	527		40	
22. O. masou	502	527		9	
23. Salmo clarkii	503	527		40	
24. S. gairdneri	503	527		40	
	504	526		17	
25. S. salar	503	527		40	
	505	531		10	
26. S. trutta	503	527		40	
	503	527		17	
	508	534		6	

Classification	Pigment type and or $\lambda_{max.}$ (nm) A_1	A_2	Ratio	Author	Ecological Comments
27. Salvelinus fontinalis	503	527		40	
(albino)	500	518		17	
28. S. malma	503	527		40	
29. S. namaycush	512	(545)		40	
	511	540		20	
30. S. alpinus	509			40	
	508	+		18	
31. S. perisii	508	+		18	
32. S. willughbii	508	+		18	
33. Thymallus thymallus	-	527		15	
34. T. arcticus	502	529		9	
35. Coregonus clupeoides		536		16	
36. C. pollan	-	536		16	
37. C. clupeoides stigmaticus	510	+		16	
38. C. clupeoides pennantii	520			15	
39. C. vandesius gracilior	510	+		16	
40. C. wartmannii	510	+		18	
41. C. macrophthalmus	510	+		18	
42. C. sp.	509	536		18	
43. C. artedii	510	+		20	
	509			9	
44. C. clupéaformis	509			9	
Osmeridae					M, De
45. Osmerus eperlanus		543		16	shelf waters, schooling
46. O. eperlanus mordax	512	541		20	
	514 –	542		21	ratio depending on habitat

Classification	Pigment type and or $\lambda_{max.}$ (nm) A_1	A_2	Ratio	Author	Ecological Comments
Argentinoidei					
Argentinidae					M, bathypelagic
47. Searsia koefoedi	477			26	
48. Nansenia groenlandica	491			26	
49. Bathylagus microphthalmus	467			26	
50. B. wesethi	478 501			35	
Esocoidei					
Esocidae					F, slowly moving, predator
51. Esox lucius		533		23	
Umbridae					F, predator
52. Umbra limi		526		14	
Stomiatoidei					
Sternoptychidae					M, bathypelagic
53. Sternoptyx obscura	485			35	
54. Argyropelecus affinis	478			35	
Malacosteidei					
Malacosteidae					M, bathypelagic, bioluminescent
55. Aristostomias scintillans	526	551		44	
Alepocephaloidei					
Alepocephalidae					M, bathypelagic
56. Xenodermichthys copei	479			26	
Myctophoidei					
Synodontidae					T, M, pelagic; Di
57. Synodus lucioceps	+			1	

Classification	Pigment type and or λ max. (nm) A1	A2	Ratio	Author	Ecological Comments
58. S. evermanni	+			2	
59. Synodus scituliceps	+			2	
60. S. sechurae	+			2	
Myctophidae					M, De, N, vertical migrations
61. Lampanyctus mexicanus	490			35	
62. L. sp.	485			26	
Ostariophysi					
Cypriniformes					
Characoidei					
Characidae					T, F, Di, often schooling
63. Arnoldichthys spilopterus	503	527		45	
64. Colossoma nigripinnis	503	527		45	
65. Copeina arnoldi	503			45	
66. Ctenobrycon spilurus	503	527		45	
67. Exodon paradoxus	503	527		45	
68. Gymnocorymbus ternetzi	503	527		45	
69. Hemigrammus rhodostomus	503	527		45	
70. Hyphessobrycon sp.	503	527		45	
71. Metynnis schreitmuelleri		530		45	
72. Micralestes interruptus	503	527		45	
73. Moenkhausia oligolepis	503	527		45	
74. Rooseveltiella nattereri		527		45	
75. Hoplias malabaricus	519.8	+	26.2% A_1	43	

Classification	Pigment type and or $\lambda_{max.}$ (nm) A1	A2	Ratio	Author	Ecological Comments
76. Acestrorhynchus falcatus	511.3	+	66.1% A_1	43	
77. Serrasalmus rhombens		527		43	
78. Moenkhausia lepidura	513.2	+	56.1% A_1	43	
79. Leporinus fasciatus		526.2	2.9% A_1	43	
80. Bryconops affinis		527		43	
Gasteropelecidae					T, F, Di, often schooling
81. Gasteropelecus levis	503	527		45	
Anostomidae					T, F, Di, often schooling
82. Abramites microcephalus	503	527		45	
83. Anostomus anostomus	503	527		45	
84. Prochilodus insignis	503	527		45	
	511.2	+	66.2% A_1	43	
Hemiodontidae					T, F; often schooling
85. Hemiodus semitaeniatus	503	527		45	
Cyprinoidei					
Cyprinidae					F; calm waters, mostly vegetarian
86. Ptychocheilus oregonensis	506			45	
87. Rutilus rutilus		535		25	

Classification	Pigment type and or $\lambda_{max.}$ (nm) A_1	A_2	Ratio	Author	Ecological Comments
88. Scardinius erytrhophthalmus	507	535		19	
89. Gobio gobio		535		25	
90. Tinca tinca auratus		533		25	
91. T. tinca	467	533		25	
		620		30	(principal cones)
92. Pimephales notatus	505	535		14	
93. Notropis cornutus frontalis	505	536		14	
94. Richardsonius balteatus balteatus	506	531		4	ratio depending on photic environment
95. Notemigonus crysoleucas boscii	502	529		14	
	502	527		5	ratio depending on temperature
96. N. crysoleucas auratus	504	532		14	
	504	530		5	ratio depending on temperature
97. Hybopsis biguttata	505	535		14	
98. Alburnus lucidus	510	533		24	
		550			
99. Cyprinus carpio		523		22	
100. Carassius carassius		523		25	
101. C. auratus		522		45	
		455		32	(cone pigments)
		530, 527		30	
		625, 625			
		575			dark adapted

Classification	Pigment type and or λ max. (nm) A1	A2	Ratio	Author	Ecological Comments
102. Balantiocheilus melanopterus		522		45	
103. Barbus conchonius	502			45	
104. B. everetti	500			45	
105. B. fasciolatus	500	522		45	
106. B. schwanenfeldi	500	522		45	
107. Brachydanio rerio	500			45	
108. Chela mouhoti	501			45	
109. Danio malabaricus	502			45	
110. Morulius chrysophekadion		522		45	
111. Rasbora lateristrata elegans	501			45	
Catostomidae					F, sluggish, vegetarians
112. Catostomus commersonii		524		14	
Siluriformes					
Ictaluridae					F, N, B, sluggish
113. Ictalurus nebulosus marmoratus		534		14	
114. I. punctatus		620		30	(principal cones)
		543			(accessory cones)
115. I. melas		543		30	(rods & cones)
Bagridae					T, F, C, B
116. Mystus tengera		535		45	
Mochokidae					T, F, C
117. Synodontis nigriventis		533		45	

Classification	Pigment type and or $\lambda_{max.}$(nm) A$_1$	A$_2$	Ratio	Author	Ecological Comments
Pimelodidae					T, F, C, B
118. Pimelodella gracilis		527		31	
Loricariidae					T, F, N sluggish
119. Plecostomus plecostomus		533		8	
Ariidae					T, F, N
120. Arius felis		527		6	sluggish
Paracanthopterygii					
Batrachoidiformes					
Batrachoididae					M, B
121. Porichthys marginatus	+			1	
122. P. analis ?	+			2	
123. P. notatus	496			9	
124. Opsanus tau	499			9	
Percopsiformes					
Percopsoidei					
Percopsidae					F, predator
125. Percopsis omiscomaycus		529		9	
Lophiiformes					
Antennarioidei					
Ogcocephalidae					T, M, pelagic
126. Zalieutes elater	+			1	
Gadiformes					
Gadoidei					
Gadidae					M, B, or pelagic in continental shelf
127. Gadus macrocephalus	498			8	
128. G. morrhua	499			26	
129. G. minutus	493			26	
130. G. merlangus	500			26	

Classification	Pigment type and or $\lambda_{max.}$ (nm) A1	A2	Ratio	Author	Ecological Comments
131. G. aeglifinus	494			26	
132. Microgadus proximus	499			8	
133. Theragra chalcogrammus	498			8	
134. Antimora rostrata	485			26	
135. Lota lota	523	527		8	
Atherinomorpha					
Atheriniformes					
Exocoetioidei					
Exocoetidae					M, N, S, epi-pelagic
136. Cypselurus simus	475.1			42	
137. Parexocoetus brachypterus	499.3			42	
138. P. mento	498.6			42	
Belonidae					M, C, S
139. Strongylura incisa	500.5			42	
Hemiramphidae					T, M, epipelagic
140. Hemiramphus saltator	+			2	
Scomberesocidae					M, Di
141. Cololabis adocetus	497.4			42	
Cyprinodontoidei					
Cyprinodontidae					Euryhaline, Di
142. Aplocheilus lineatus	501			45	
143. Jordanella floridae	500	522		45	
144. Cyprinodon variegatus	506			14	

Classification	Pigment type and or $\lambda_{max.}$ (nm) A1	A2	Ratio	Author	Ecological Comments
Poeciliidae					T, F, Di
145. Belonesox belizanus	498	521		14	
146. Poecilia latipinna	502			45	
147. P. reticulata	500			45	
148. P. sphenops	502			45	
149. Xiphophorus helleri	500			45	
150. X. maculatus	500			45	
151. X. variatus	501			45	
Anablepidae					T, F, Di, S
152. Anableps sp.	506			45	
Atherinoidei					
Atherinidae					M, active, schooling
153. Atherinops affinis	508			36	
154. Melanotaenia nigrans	501			45	
Acanthopterygii					
Beryciformes					
Stefanoberycoidei					
Melamphaeidae					M, deep sea fishes
155. Melamphäes bispinosus	488			35	
Berycoidei					
Holocentridae					T, M, N, I
156. Holocentrus diadema	490.3			42	
157. H. lacteoguttatus	494.7			42	
158. H. spinifer	490.3			42	
159. H. tiere	489.2			42	
160. H. violaceus	498.4			42	
161. H. xantherythrus	485.4			42	
162. H. laevis	502.2			42	

Classification	Pigment type and or $\lambda_{max.}$ (nm) A1	A2	Ratio	Author	Ecological Comments
163. H. sammara	501.6			42	
164. H. scythrops	480.7			42	
165. Myripristis argyromus	491.8			42	
166. M. berndti	492.5			42	
167. M. multiradiatus	491.2			42	
Zeiformes					
Zeidae					M, active, predators
168. Zeus faber	492			26	
Caproidae					M, I, B
169. Capros aper	493			26	
Gasterosteiformes					
Gasterosteidei					
Gasterosteidae					F, Di, active also coastal or brackish waters
170. Gasterosteus aculeatus	501	522		34	
Aulostomoidei					
Aulostomidae					M, C, I
171. Aulostomus chinensis	490.4			42	
Scorpaeniformes					
Scorpaenoidei					
Scorpaenidae					M, N, I
172. Dendrochirus brachypterus	493.9	+		42	
173. Pterois antennata	491.2			42	
174. P. radiata	494			42	
175. Scorpaenoides parvipinnis	495.8			42	
176. Scorpaenopsis diabolus	494.4	+		42	
177. Scorpaena guttata	500			34	

Classification	Pigment type and or $\lambda_{max.}$ (nm) A1	A2	Ratio	Author	Ecological Comments
178. S. mystes	+			1	
179. Pentius sp.	+			1	
180. Sebastes caurinus	497			9	
181. S. maliger	496.5			9	
Triglidae					M, N, B
182. Trigla gurnardus	491 511			26	
183. T. cuculus	492			26	
184. Prionotus punctatus	+			2	
185. P. stephanophrys	+			2	
Hexagrammoidei					
Hexagrammidae					M, S - I, coastal waters active swimmers
186. Hexagrammos decagrammus	498.5			9	
187. Ophiodon elongatus	501			9	
Anoplopomatidae					M, S - I, coastal waters active swimmers
188. Anoploma fimbria	491			9	
Cottoidei					
Cottidae					F or M, C, B, bottom dwelling
189. Cottus sp.	511			36	
190. C. bubalis	503			26	
191. C. ricei	503			18	
192. Leptocottus armatus	511			36	
193. Myxocephalus quadricornis thompsonii	508	515		20	
194. M. polyacanthocephalus	502			9	

Classification	Pigment type and or $\lambda_{max.}$ (nm) A_1	A_2	Ratio	Author	Ecological Comments
195. Enophrys bison	501			9	
196. Dasycottus setiger	492			9	
197. Chitonotus pugetensis	496			9	
198. Blepsias cirrhosus	502.5			9	
Agonidae					M, S-I, B
199. Agonus acipenserinus	499			9	
Perciformes					
Percoidei					
Centropomidae					T, F, Di
200. Chanda ranga	500			45	schooling
201. Lates niloticus	499	520		31	
Serranidae					T, M, C, I
202. Anyperodon leucogrammicus	495.9			42	
203. Epinephalus fuscoguttatus	493.7	+		42	
204. E. kohleri	493.3			42	
205. E. macrospilus	496.5			42	
206. E. merra	496.2	+		42	
207. E. microdon	496.1	+		42	
208. E. socialis	496.1			42	
209. E. tauvina	496.1			42	
210. E. niveatus	+			1	
211. E. labriformis	+			2	
212. E. guaza	494 501			26	
213. E. alexandrinus	494 499			26	
214. Diplectrum macropoma	+			1	
215. D. pacificum	+			2	

Classification	Pigment type and or $\lambda_{max.}$ (nm) A1	A2	Ratio	Author	Ecological Comments
216. D. sciurus	+			2	
217. Mycteroperca rosacea	+			1	
218. M. jordani	+			2	
219. M. prionura	+			2	
220. Paralabrax maculato-fasciatus	+			1	
221. P. auro-guttatus	+			1	
222. Anthias ? sp.	+			1	
223. Cephalopholis acanthistius?	+			2	
224. Serranus fasciatus	+			2	
225. S. cabrilla	493			26	
226. Petrometopon panamense	+	+	4:1	2	
227. Centropristis striata	496			9	
228. Paracentro-pristis scriba	492 498			26 26	
229. Morone labrax	502	534		26	
Grammistidae					T, M, C, I
230. Grammistes sexlineatus	494.4			42	
Theraponidae					T, M, N, S
231. Eutherapon theraps	497			42	
Centrarchidae					F, Di
232. Pomoxis nigro-maculatus		525 527		39 45	
233. P. annularis		525		37	
234. Micropterus salmoides floridanus		525		14	

Classification	Pigment type and or $\lambda_{max.}$ (nm) A_1	A_2	Ratio	Author	Ecological Comments
235. Lepomis microlophus		525		14	
		620		30	(principal cone)
		537			(accessory cone)
236. L. macrochirus purpurescens		525		14	
237. Chaenobryttus coronarius		524		14	
Priacanthidae					T, M, N, S
238. Priacanthus cruentatus		496.4		42	
Apogonidae					T, M, N, I
239. Apogon brachygrammus (adults)		493.9		42	
(larvae)	481.6			42	
240. A. maculiferus	484.2			42	
241. A. menesemus	494.9			42	
242. A. snyderi	490			42	
243. Lovamia novemfasciata	491.4			42	
244. Sphaeramia orbicularis	495.4			42	
Echeneidae					T, M, S
245. Echeneis naucrates	497.2			42	
Branchiostegidae					T, M, pelagic
246. Caulolatilus princeps ?	+			2	
Carangidae					T, M, C, I
247. Carangoides ajax	496.6			42	
248. Caranx mate	496.1			42	
249. C. sexfasciatus	496.8			42	

Classification	Pigment type and or $\lambda_{max.}$ (nm) A$_1$	A$_2$	Ratio	Author	Ecological Comments
250. C. trachurus	498			26	
251. C. hippos	+			1	
252. Decapterus pinnulatus	494.8			42	
253. Elagatis bipinnulatus	500.1			42	
254. Gnathanodon speciosus	496.1			42	
255. Trachurops crumenoph-thalmus	499.8			42	
256. Trachurus symmetricus	495			34	
257. Trachinotus rhodopus	+			1	
Coryphaenidae					M, Di, S
258. Coryphaena hippurus	+	+	2:1	1	
	500.6			41	
Centracanthidae					
259. Merolepis maena	501	522		26	
260. M. chryselis	512			26	
	501				
Hoplopagridae					T, M
261. Hoplopagrus guntheri	+			1	
Lutjanidae					M, N, I
262. Lutjanus fulvus	498			1	
263. L. gibbus	493.6			42	
264. L. mono-stigmus	496.6			42	
265. L. argenti-ventris	+			2	
266. L. novem-fasciatus	+			2	

Classification	Pigment type and or $\lambda_{max.}$ (nm) A$_1$	A$_2$	Ratio	Author	Ecological Comments
Nemipteridae					T, M, Di, S
267. Pentapus caninus	498.1			42	
Gerridae					T, M, B
268. Gerres cinereus	+	+	2:1	1	
269. Eucinostomus ? sp.	+			2	
Pomadasyidae					T, M, I predators
270. Haemulon sexfasciatus	+			1	
271. H. steindachneri	+			1	
272. Pomadasys branicki	+			1	
273. Anisotremus interruptus	+			2	
274. Orthopristis reddingi	+			2	
275. O. chrysoptera	500			9	
Lethrinidae					T, M, N, I
276. Lethrinus kallopterus	498			42	
277. L. miniatus	494.3			42	
Sparidae					M, coastal, often schooling
278. Spondyliosoma cantharus	498 518			26	
279. Sarpa salpa	500			26	
280. Pagellus erythrinus	493 499			26 26	
281. Oblada melanura	500			26	
282. O. sp.	585			26	
283. Lagodon rhomboides	500			9	

Classification	Pigment type and or $\lambda_{max.}$(nm) A1	A2	Ratio	Author	Ecological Comments
284. Stenotomus caprinus	496.5			9	
285. Calamus brachysomus	+			2	
Sciaenidae					M, B coastal waters
286. Leiostomus xanthurus	499			9	
287. Micropogon undulatus	496.5			9	
288. Seriphus politus	504			36	
289. Johnnius umbra (= Corvina (Sciaena) nigra)	500			26	
290. Cynoscion parvipinnis	+			2	
291. C. regalis	496			9	
Mullidae					T, M, Di, N, I
292. Mulloidichthys samoensis	484.4			42	
293. Parupeneus cyclostomus	492.8			42	
294. P. multifasciatus	491.1			42	
295. P. porphyreus	491.8			42	
296. Upeneus arge	495.6			42	
297. Pseudopeneus grandisquamis	+			2	
Monodactylidae					T, M, S schooling
298. Monodactylus argenteus	502			45	
Pempheridae					T, M, N, S
299. Pempheris oualensis	498.8			42	

Classification	Pigment type and or $\lambda_{max.}$ (nm) A1	A1	Ratio	Author	Ecological Comments
Kyphosidae					T, M, S
300. Kyphosus cinerascens	497.5			42	
301. K. vaigiensis	497.8			42	
302. K. ? sp.	+			2	
303. Microcanthus strigatus	493.9			42	
Girellidae					T, M
304. Girella simplicidens	+			2	
Ephippidae					T, M, N, S
305. Platax platax	496.5			42	
306. P. teira	497			42	
307. Chaetodipterus zonatus	+			1	
Scatophagidae					T, M, N, S
308. Scatophagus argus	500			45	
	496.4			42	
Chaetodontidae					T, M, Di, I
309. Centropyge flavissimus	491			42	
310. Chaetodon auriga	493.8			42	
311. C. citrinellus	495.6			42	
312. C. corallicola	491			42	
313. C. ephippium	494.2			42	
314. C. falcula	493.3			42	
315. C. fremblii	490.4			42	
316. C. kleini	490.6			42	
317. C. lineolatus	493.2			42	
318. C. lunula	494.4			42	
319. C. melanotus	493.4			42	
320. C. mertensii	489.7			42	
321. C. miliaris	492			42	
322. C. ornatissimus	490.2			42	

Classification	Pigment type and or $\lambda_{max.}$ (nm) A_1	A_2	Ratio	Author	Ecological Comments
323. C. rafflesi	494.1			42	
324. C. semeion	492.6			42	
325. C. triangulum	493			42	
326. C. trifasciatus	491.7			42	
327. C. unimaculatus	491			42	
328. C. vagabundus	493.9			42	
329. Chaetodontoplus mesoleucas	491.7			42	
330. Heniochus acuminatus	490.3			42	
331. H. varius	492.6			42	
332. Holacanthus arcuatus	479.7			42	
333. H. ciliaris	+			1	
334. Megaprotodon strigangulus	493.5			42	
335. Pomacanthus zonipectus	+			2	
Embiotocidae					M, shore fishes
336. Hyperprosopon argenteum	506			36	
337. Embiotoca jacksoni	506			36	
Cichlidae					T, F, Di mostly predators
338. Aequidens portalegrensis	500	522		45	
339. A. tetramerus	505	522.3	83.7% A_1	43	
340. Apistogramma sp.	500			45	
341. Astronotus ocellatus	500	522		45	
		522.3		43	
342. Cichlasoma meeki	500	522		45	
343. C. festivum		522.2		43	
344. Pelmatochromis guentheri	500	522		45	

Classification	Pigment type and or $\lambda_{max.}$ (nm) A_1	A_2	Ratio	Author	Ecological Comments
345. P. pulcher	500	522		45	
346. Pterophyllum eimeckei	500	522		45	
347. Symphysodon aequifasciata	497	516		45	
348. Tilapia nilotica	503	527		45	
349. T. zilli		519		31	
350. T. leucosticta		519		31	
351. T. variabilis		520		31	
352. T. melanopleura	499	520		31	
353. Haplochromis melanopleura	499	520		31	
354. Acarichthys heckelii	509.3		60.2% A_1	43	
355. Geophagus jurupari	510.9		50.2% A_1	43	
356. G. surinamensis	502.1		100.0% A_1	43	
357. Geophagus wavrini ?	514.1		33.5% A_1	43	
358. Cichla ocellaris	512.2		43.2% A_1	43	
359. Uaru amphiacanthoides	511.5		46.9% A_1	43	
360. Petenia spectabilis	502.7		96.5% A_1	43	
361. Acaronia nassa	507.9		68.2% A_1	43	
362. Etroplus maculatus	495	515		46	freshwater seawater
363. Acathochromis polyacanthus	492.2			42	

Classification	Pigment type and or $\lambda_{max.}$ (nm) A_1	A_2	Ratio	Author	Ecological Comments
364. Chromis caeruleus	494.3			42	
365. C. dimidiatus	482.7			42	
366. C. verater	480			42	
367. C. punctipinnis	495			34	
Pomacentridae					T, M, Di, I
368. Abudefduf abdominalis	496.3			42	
369. A. filifer	494.8			42	
370. A. saxatilis	494.7			42	
	+			1	
371. A. septemfasciatus	496.9			42	
372. A. sordidus	496.7			42	
373. Amphiprion bicinctus	490.8			42	
374. A. melanopus	493			42	
375. Dascyllus albisella	491.2			42	
376. D. auranus	495.7			42	
377. D. reticulatus	493.6			42	
378. D. trimaculatus	494.4			42	
379. Pomacentrus nigricans	495.9			42	
380. P. notophthalmus	492.5			42	
381. P. pavo	495.8			42	
382. P. taeniurus	492.7			42	
383. P. vaiuli	491.8			42	
384. Premnas biaculeatus	494.2			42	
385. Microspathodon dorsalis	+			1	
Cirrhitidae					T, M, N
386. Cirrhitus rivulatus	+			2	
387. C. pinnulatus	496.7			42	

Classification	Pigment type and or $\lambda_{max.}$ (nm) A$_1$	A$_2$	Ratio	Author	Ecological Comments
388. Paracirrhites arcatus	494.8			42	
Mugiloidei					
Mugilidae					M, shore fishes
389. Mugil cephalus	499.1			42	
	499	522		14	
390. M. curema	499	522		14	
Sphyraenoidei					
Sphyraenidae					T, M, C, S
391. Sphyraena barracuda	497.6			42	
392. S. helleri	498.1			42	
Labroidei					
Labridae					M, Di, S or I
393. Anampses diadematus	501			42	
394. Bodianus bilunulatus	480			42	
395. Cheilinus diagrammus	489.1			42	
396. C. rhodochrous	489.3			42	
397. C. trilobatus	492.7			42	
398. C. undulatus	484.6			42	
399. Choerodon zosteropherus	494.9			42	
400. Coris aygula	493.8			42	
401. Halichoeres centriquadrus	504.2			42	
402. Thalassoma ballieni	503			42	
403. T. duperrey	506.4			42	
404. Pimelometopon pulchrum	497 520			37	
405. Labrus mixtus	486 527			26	
406. L. merula	492	510		48	

Classification	Pigment type and or λ max. (nm) A1	A2	Ratio	Author	Ecological Comments
407. *Crenilabrus festivus*		513		48	
408. *C. massa*		513		48	
409. *Paralabrax clathratus*	500 520			48	
410. *Decodon sp.*	+			2	
Scaridae					M, Di, I
411. *Calotomus spinidens*	483			42	
412. *Scaridea zonarcha*	483.4			42	
413. *Scarus aeruginosus*	484			42	
414. *S. chlorodon*	484.1			42	
415. *S. dubius*	483.6			42	
416. *S. fasciatus*	483.9			42	
417. *S. forsteri*	484.4			42	
418. *S. ghobban*	484.6 +			42 2	
419. *S. niger*	484.1			42	
420. *S. oviceps*	484			42	
421. *Sparisoma cretense*	485	520		26	
Blennioidei					
Blenniidae					M, Di, S cavernicolous
422. *Cirripectus variolosus*	496.1			42	
Tripterygiidae					M, Di, B shore fishes
423. *Axoclinus carminalis*	500			34	
Callionymoidei					
Callionymidae					M, Di shore fishes
424. *Callionymus lyra*	485 511			26	

Classification	Pigment type and or λ max. (nm) A1	A2	Ratio	Author	Ecological Comments
Goibioidei					
Gobiidae					M, B, shore fishes
425. Gillichthys mirabilis	512			33	
426. Eucyclogobius newberryi	512			36	
427. Clevelandia ios	512			36	
428. Coryphopterus nicholsii	500			36	
Acanthuroidei					
Acanthuridae					T, M, Di, I
429. Acanthurus gahhm	429.6			42	
430. A. glaucopareius	491.9			42	
431. A. guttatus	492.8			42	
432. A. leucosternum	491			42	
433. A. lineatus	492.9			42	
434. A. mata	492.5			42	
435. A. olivaceus	492.3			42	
436. A. triostegus triostegus	497.2			42	
437. A. triostegus sandvicensis	496.7			42	
438. A. xanthopterus	492.6			42	
439. Callicanthus lituratus	492.6			42	
440. Ctenochaetus striatus	491.1			42	
441. C. strigosus	491.2			42	
442. Naso brevirostris	491.6			42	
443. N. hexacanthus	490.3			42	
444. N. lituratus	492.3			42	
445. N. unicornis	492			42	
446. Zanclus cornutus	492			42	

Classification	Pigment type and or $\lambda_{max.}$ (nm) A1	A2	Ratio	Author	Ecological Comments
447. Zebrasoma flavescens	490.5			42	
448. Z. scopas	491.1			42	
449. Z. veliferum	492.5			42	
450. Xesurus punctatus	+			1	
451. Prionurus punctatus	+			2	
Gobiidae					T, M, pelagic
452. Bollmannia sp.	+			2	
Siganidae					T, M, Di
453. Siganus fuscescens	495.5			42	
454. S. puellus	495.6			42	
Scombroidei					
Scombridae					M, Di
455. Acanthocybium solandri	482			42	
456. Euthunnus yaito	483.1			42	
457. Katsuwonus pelamis	483.1			42	
458. Neothunnus macropterus	484.1			42	
459. Scomberomorus commersoni	484.6			42	
460. Pneumatophorus japonicus diego	491			34	
461. Scomber scombrus	521 487			26	
462. S. japonicus	491			38	
463. Scomberomorus concolor	485			34	
464. S. sierra	486			38	
465. Sarda chiliensis	488 +			38 1	

Classification	Pigment type and or λ (nm) max. A1	A2	Ratio	Author	Ecological Comments
Istiophoridae					T, M, Di
466. Istiompax marlina	486			42	
467. Makaira ampla	484.4			42	
Anabantoidei					
Anabantidae					T, F, Di
468. Helostoma temmincki	500	522		45	
489. Macropodus opercularis concolor	500	522		45	
Pleuronectiformes					
Bothidae					M, I, B
490. Rhombus maximus	499			26	
	517				
491. R. laevis	497			26	
492. Arnoglossus	486			26	
megastoma	493			26	
Pleuronectidae					M, I, B
493. Pleuronectes	492			26	
microcephalus	501			26	
494. P. limanda	499			26	
	506			26	
495. P. flesus	507			26	
496. Hippoglossoides elassodon	499			9	
497. Isopsetta isolepis	497			9	
498. Microstomus pacificus	492			9	
499. Perophrys vetulus	497			9	
500. Platichthys stellatus	508			9	

Classification	Pigment type and or λ max. (nm) A1	A2	Ratio	Author	Ecological Comments
Soleidae					M, I, B
501. Solea vulgaris	501			26	
502. Achirus mazatlanus	+			2	
Cynoglossidae					T, M, B
503. Symphurus sp.	+			2	
Tetraodontiformes					
Balistoidei					
Balistidae					T, M, Di, B
504. Alutera scripta	491.5			42	
505. Balistapus undulatus	501.9			42	
506. Oxymonacanthus longirostris	493.3			42	
507. Pervagor spilosoma	491.5			42	
508. Rhinecanthus aculeatus	493.7			42	
509. R. verrucosus	488.6			42	
510. Sufflamen verres	+			2	
511. Monacanthus hispidus	495 496			9	
Tetraodontoidei					
Tetraodontidae					M, I
512. Arothron hispidus	495.1			42	
513. Sphaeroides sechurae	+			2	
514. S. annulatus	500	522		36	
515. S. maculatus	498			9	
516. Tetraodon schontedeni	503	527		45	

Classification	Pigment type and or λ_{max} (nm) A₁	A₂	Ratio	Author	Ecological Comments
Diodontidae					T, M, N
517. Diodon holocanthus	494.2			42	
	+			2	
518. Chilomycterus californiensis	+			1	

REFERENCES

1. Ali, M. A., and Heumann, W. R. (1970). Distribution of vitamins A_1 and A_2 in the retinas of some marine fishes from the Gulf of California. Vision Res. 10, 1307-1310.
2. Ali, M. A., and Heumann, W. R, (1972). Distribution of vitamins A_1 and A_2 in the retinas of some marine fishes from the Gulf on California II. Vision Res. 12, 2157 - 2159.
3. Ali, M. A., Hubbard, R., and Wald, G. (1970). Scotopic visual pigments in the albino brook trout *Salvelinus fontinalis*. Rev. Can. Biol. 29, 153-156.
4. Allen, D. M. (1971). Photic control of the proportions of two visual pigments in a fish. Vision Res. 11, 1077-1112.
5. Allen, D. M., and McFarland, W. N. (1973). The effect of temperature on rhopsin-porphyropsin ratios in a fish. Vision Res. 13, 1303-1309.
6. Allen, D. M., MacFarland, W. N., Munz, F. W., and Poston, H. A. (1973). Changes in the visual pigments of trout. Can. J. Zool. 51, 901-914.
7. Arnott, H.-J., Best, A. C. G , Ito, S., and Nicol, J. A. C. (1974). Studies on the eyes of catfishes with special reference to the tapetum lucidum. Proc. Roy. Soc. Lond. B. 186, 13-36.
8. Beatty, D. D. (1969). Visual pigments of the burbot *Lota lota* and seasonal changes in their relative proportions. Vision Res. 9, 1173-1183.
9. Beatty, D. D. (1973). Visual pigments of several species of teleost fishes. Vision Res. 13, 989-992.
10. Beatty, D. D. (1974). unpublished observations.
11. Beatty, D. D. (1975). Visual pigments of the American eel, *Anguilla rostrata*. Vision Res. (in press).
12. Bowmaker, J. K. (1972). Kundt's rule: the spectral absorbance of visual pigments in situ and in solution. Vision Res. 12, 529-548.
13. Bowmaker, J. K. (1973). Spectral sensitivity and visual pigment absorbance. Vision Res. 13, 783-791.
14. Bridges, C. D. B. (1964). Periodicity of absorption properties in the pigments based on vitamin A_2 from fish retinae. Nature (Lond.) 208, 303-304.
15. Bridges, C. D. B. (1965). Absorption properties, interconversions and environmental adaptation of pigments from fish photoreceptors. Cold Spr. Harb. Symp. quant. Biol. 30, 317-334.

16. Bridges, C. D. B. (1967). Spectroscopic properties of porphyropsins. Vision Res. 7, 349-369.

17. Bridges, C. D. B. (1972). The Rhodopsin-Porphyropsin Visual System. In: Handbook of Sensory Physiology Vol. VII/1: Photochemistry of Vision, Ed. H. J. A. Dartnall, pp. 417-480, Berlin - Heidelberg - New York: Springer.

18. Bridges, C. D. B., and Yoshikami, S. (1970a). Distribution and evolution of visual pigments in Salmonid fishes. Vision Res. 10, 609-626.

19. Bridges, C. D. B., and Yoshikami, S. (1970b). The rhodopsin-porphyropsin system in freshwater fishes. -I. Effects of age and photic environment. Vision Res. 10, 1315-1332.

20. Bridges, C. D. B., and Delisle, C. E. (1974a). Brief observations concerning the visual pigments of some selected fishes from Lake Heney, Québec, a relict of glacial Lake Gatineau. Vision Res. 14, 187-193.

21. Bridges, C. D. B., and Delisle, C. E. (1974b). Postglacial evolution of the visual pigments of the smelt, Osmerus eperlanus mordax. Vision Res. 14, 345-356.

22. Crescitelli, F., and Dartnall, H. J. A. (1954). A photosensitive pigment of the carp retina. J. Physiol. (Lond.) 125, 607-627.

23. Dartnall, H. J. A. (1952). Visual pigment 467, a photosensitive pigment present in tench retinae. J. Physiol. (Lond.) 116, 257-289.

24. Dartnall, H. J. A. (1955). Visual pigments of the bleak (Alburnus lucidus). J. Physiol. (Lond.) 128, 131-156.

25. Dartnall, H. J. A. (1962). The photobiology of visual processes. In: The Eye, Vol. 2, (Ed. H. Davson). pp. 523-533, New York - London, Academic Press.

26. Dartnall, H. J. A., and Lythgoe, J. N. (1965). The spectral clustering of visual pigments. Vision Res. 5, 81-100.

27. Greenwood, P. H., Rosen, D. E., Weitzman, S. H., and Myers, G. S. (1966). Phyletic studies of teleostean fishes, with a provisional classification of living forms. Bull. Am. Mus. Nat. Hist. 131, 341-455.

28. Hárosi, F. J. (1971). Frog rhodopsin in situ: Orientational and spectral changes in the chromophores of isolated retinal rod cells. Ph. D. Thesis, John Hopkins University, Baltimore, Md. 21218, 122 pp.

29. Hárosi, F. J., McNichol, E. F. (1974). Visual pigments of goldfish cones. Spectral properties and dichroism. J. gen. Physiol. 63, 279-304.

30. Liebman, P. A. (1973). Microspectrophotometry of Visual Receptors. In: Biochemistry and Physiology of Visual Pigments, Ed. H. Langer, pp. 299-305, Berlin-Heidelberg-New York: Springer.
31. Lythgoe, J. N. (1972). List of Vertebrate Visual Pigments. In: Handbook of Sensory Physiology. Vol. VII/1. Photochemistry of Vision. Ed. H. J. A. Dartnall, pp. 604-624. Berlin-Heidelberg-New York : Springer.
32. Marks, W. B. (1965). Visual pigments of single goldfish cones. J. Physiol. (Lond.) 178, 14-32.
33. Munz, F. W. (1956). A new photosensitive pigment of the euryhaline teleost Gillichthys mirabilis. J. gen. Physiol. 40, 233-249.
34. Munz, F. W. (1957). The photosensitive retinal pigments of marine and euryhaline teleost fishes. Ph. D. Thesis University of California, L. A.
35. Munz, F. W. (1958a). Photosensitive pigments from the retinae of certain deep-sea fishes. J. Physiol. (Lond.) 140, 220-235.
36. Munz, F. W. (1958b). The photosensitive retinal pigments of fishes from relatively turbid coastal waters. J. gen. Physiol. 42, 445-459.
37. Munz, F. W. (1958c). Retinal pigments of a labrid fish. Nature (Lond.) 181, 1012-1013.
38. Munz, F. W. (1964). The visual pigments of epipelagic and rocky-shore fishes. Vision Res. 4, 441-454.
39. Munz, F. W. (1965). Adaptation of visual pigments to the photic environment. Ciba foundation symposium. Colour vision. Physiology and Experimental Psychology, pp. 27-45. London: J. & A. Churchill.
40. Munz, F. W., and Beatty, D. D. (1965). A critical analysis of the visual pigments of salmon and trout. Vision Res. 5, 1-17.
41. Munz, F. W., and McFarland, W. N. (1965). A suggested hereditary mechanism for visual pigments of chars (Salvelinus sp.) . Nature (Lond.) 206, 955-956.
42. Munz, F. W., and McFarland, W. N. (1973). The significance of spectral position in the rhodopsin of tropical marine fishes. Vision Res. 13, 1829-1874.
43. Muntz, W. R. A. (1973). Yellow filters and the absorption of light by the visual pigments of some Amazonian fishes. Vision Res. 13, 2235-2254.

44. O'Day, W. T., and Fernandez, H. R. (1974). Aristostomias scintillans (Malacosteidae): A deep-sea fish with visual pigments apparently adapted to its own bioluminescence. Vision Res. 14, 545-550.
45. Schwanzara, S. A. (1967). The visual pigments of freshwater fishes. Vision Res. 7, 121-148.
46. Virabhadrachari, V., Krishnamoorthy, R. V., and Parvatheswararao (1967). Visual pigments in a tropical freshwater fish Etoplus maculatus (Teleostei). J. Exp. Biol. 47, 307-311.
47. Wald, G. (1958). The significance of vertebrate metamorphosis. Science 128, 1481-1490.
48. Wald, G. (1960). The distribution and evolution of visual systems. In: Comparative Biochemistry, Vol. 1, New York - London: Academic Press.
49. Walker, M. A. (1956). Homogeneity tests on visual pigment solutions from two sea fish. J. Physiol. (Lond.) 133, 56.

COMPARATIVE ANALYSIS OF THE PATTERNS OF RECEPTOR AND HORIZONTAL CELLS IN TELEOST FISHES

H.-J. WAGNER, Lehrstuhl für Morphologie und Anatomie der

Universität Regensburg, Universitätstr. 31

D-84 Regensburg, W. Germany

I doubt it not; and all these woes shall serve
For sweet discourses in our time to come.
Romeo and Juliet, Act III, Sc. 5

INTRODUCTION

Many day active fish with an acute vision have their visual cells arranged in very regular, mosaic-like patterns. This has been known for a long time and described by many authors, some of whom pointed to the apparent similarity to the arrangement of receptor cells in the insect eye. In both cases, the highly organised orientation of the visual cells were taken as a prerequisite for the functional capacity of the eyes. The real significance of the mosaic of receptor cells, however, remained a matter of speculation. More recently, it was found that the regular arrangment was not restricted to the visual cells but could be traced further down to the underlying neural layers. The first indications of this were found in labrids (Engström, 1963) and mugilids (Selvin de Testa, 1966) and concerned the parallels in the orientation of cones and horizontal cells. The aim of the present investigation is to examine whether the finding has more general applicability in teleost fish, and to discuss its possible consequence for the processing of primary visual information.

MATERIAL AND METHODS:

Eyes of 40 diurnal species of 21 different families were embedded in paraffine, sectioned both vertically and tangentially (thickness 5 μm), and stained after Foot-Masson. For the classification of the horizontal cell types, retinas of *Nannacara anomala* were impregnated with silver according to the rapid Golgi method. From the latter preparations, 40 μm sections were cut and examined in the light microscope. In this communication, the findings of only a few species are presented which are, however, typical for the observations in the remaining specimens. For further details see Wagner (1972).

RESULTS:

Basically, two kinds of mosaics can be distinguished at the level of visual cells, independently of the systematic classification of the species. They differ only in the orientation of the double cones. In the first case, these are arranged in such a way that the clefts separating the two members of a double cone run parallel to each other. As the distance between the double cones is fairly constant they form a row pattern. In the second type, the clefts of two adjacent double cones are at right angles giving rise to a zig-zag arrangement of double cones. Where such rows meet the double cones come into square patterns. These two basic kinds of mosaic may consist either of equal double cones as in *Scomber* (row) or *Zoarces* (square), or of unequal double cones as in *Brachydanio* (row) or *Poecilia* (square) (see fig. 1, 5). In addition, one or two kinds of single cones may be present in both the patterns. They are distinguished either by their length (long and short single cones) or by their position within the mosaic. In the square pattern long single cones are situated in the centre of the units, i.e. at the point where the clefts between the double cones of a square meet (central single cones, *Nannacara, Zoarces*), whereas the short single cones lie at the corners of the squares (additional single cones; *Nannostomus*).

In vertical sections, the horizontal cells occupy consecutive strata at the vitread border of the outer plexiform layer. The most sclerad layer consists of a cell type with a cuboid perikaryon and numerous short processes, all running in the direction of the receptor pedicles. It corresponds well to the brush-like cell of Cajal (1894), with the only difference that a horizontal "axon" could not be demonstrated (fig. 2). In the intermediate layer of horizontal cells, two different cell types are found. Both are characterised by radiating processes, which are horizontally orientated for 10-15 μm before bending towards the plexiform layer. They differ, however, in the shape of the horizontal parts of the processes. In the first type, they are very slender and branched ("stellate" horizontal cell) (Fig. 2, 3A), whereas in the second, they are stout and almost unbranched ("lobated" horizontal cells). A third layer of vitread horizontal cells is not encountered in every species (see

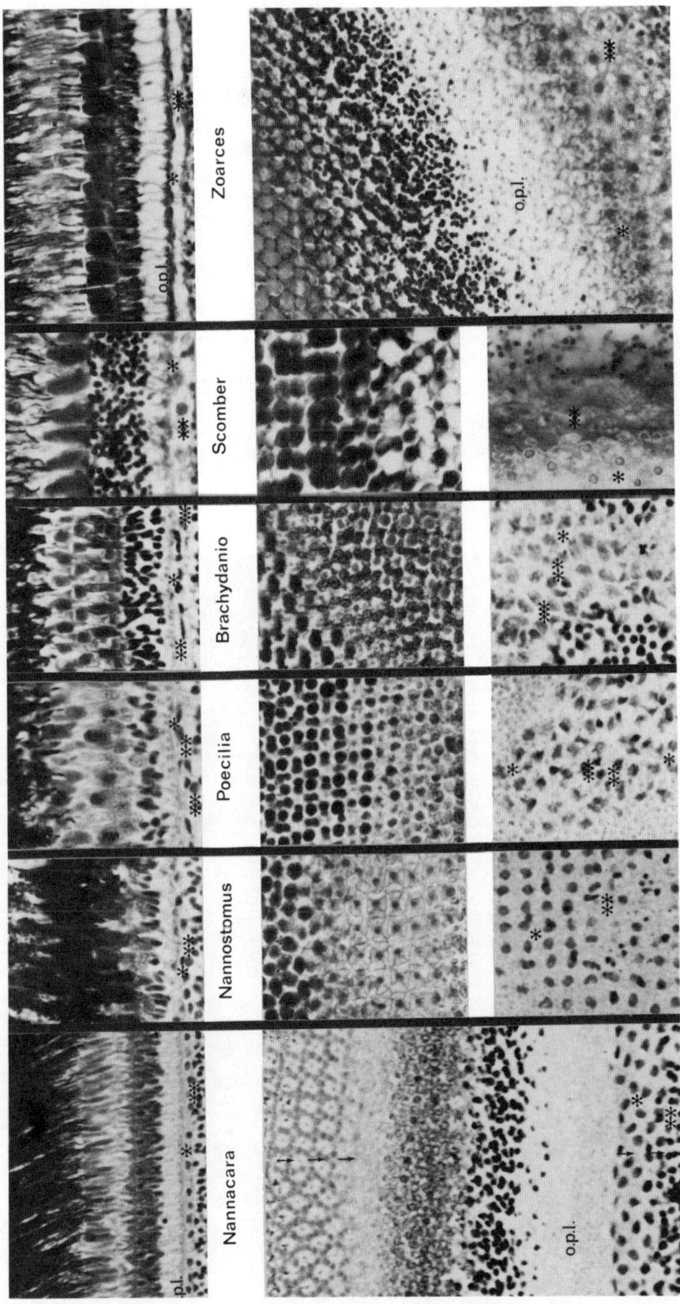

FIG. 1. Vertical and tangential sections of receptor and horizontal cells of six different species. Upper part: vertical sections; middle: tangential sections at the level of the receptor ellipsoids; lower part: tangential sections of the horizontal cell layers. Note the close spatial relationship between cone mosaic and horizontal cell mosaic. In *Poecilia* and *Brachydanio*, the pattern of horizontal cells is obscured by the fact, that the three layers are cut obliquely. o.p. l. outer plexiform layer; the asteriks mark the types of horizontal cells; see also fig. 2, 3 and 5; x 320.

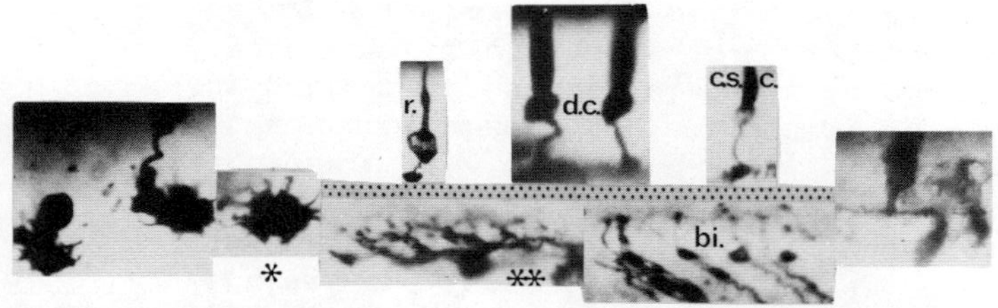

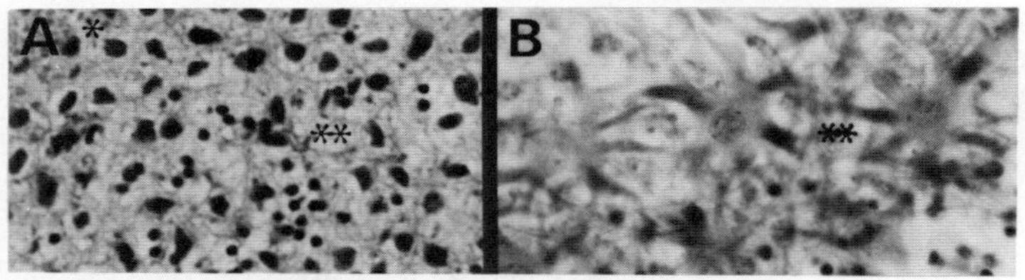

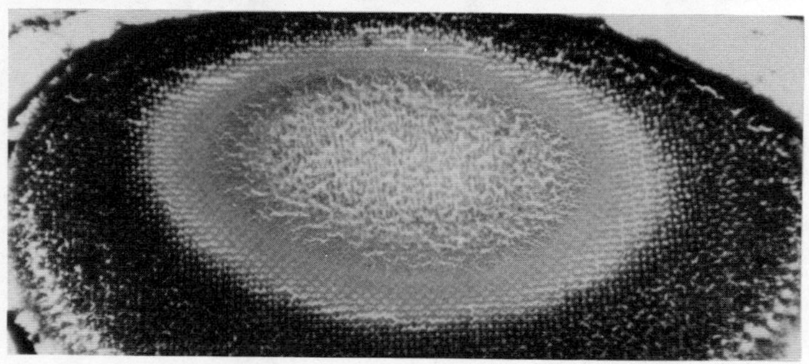

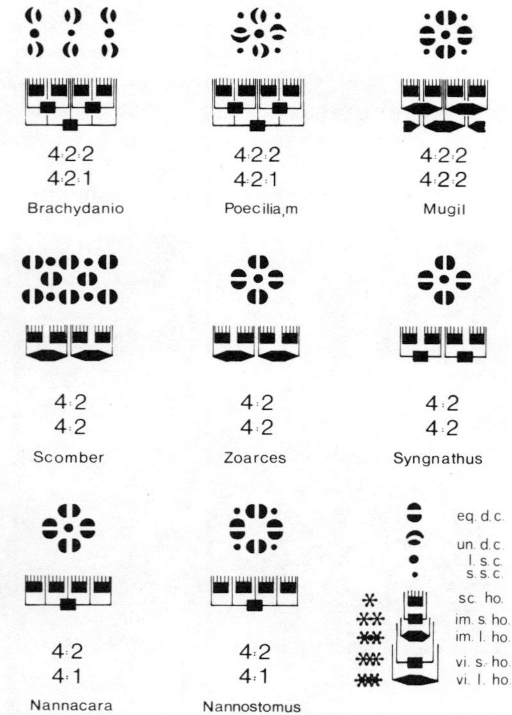

FIG. 2. (upper left) Silver-impregnated receptors and neurones of the outer plexiform layer (o.p.l.) pf *Nannacara anomala*. Note the sclerad (*) and the intermediate "stellate" (**) horizontal cells. r. rod; d.c. double cones; c.s.c. central single cone; bi. bipolar cell dendrites. x 85,000.

FIG. 3. (middle left) Intermediate horizontal cells after Foot-Masson staining. x 640.

FIG. 3. A. *Monocirrhus polyacanthus;* stellate type (~~XX~~)
B. *Perca fluviatilis;* lobated type (~~XX~~).

FIG. 4. (lower left) Tangential summary of the receptor ellipsoids of *Nannacara anomala* near the ora serrata. In the upper part, double cones are arranged in a row pattern which is changed into a square mosaic with increasing distance from the ora serrata. x 210.

FIG. 5. (right) Schematic summary of the quantitative relationship between cones and horizontal cells: 1. The number of cone types equals the number of horizontal cell types. 2. The ratio of double cones to single cones (upper figures) is in most cases identical to the ratio of sclerad to intermediate and vitread horizontal cells (lower figures). eq.d.c. equal double cone; uneq.d.c. unequal double cone; l.s.c. long single cone; s.s.c. short single cone; sc.ho. sclerad horizontal cell; im.s.ho. intermediate stellate horizontal cell; im.l.ho. intermediate lobated horizontal cell; vi.s.ho. vitread stellate horizontal cell; vi.l.ho. vitread lobated horizontal cell.

below). If it is present, the cell types are similar to those of intermediate layer.

In all the species studied, the tangential sections of the retinas revealed a striking accordance in the patterns of receptor and horizontal cells (fig. 1). It is most obvious for the rows of single cones and horizontal cells of the sclerad layer, which run in parallel directions. This is true for both the row and the square patterns. Quantitative evaluation shows that the distance between two single cones of the same type equals the distance between two sclerad horizontal cell nuclei. A closer examination of the outer plexiform layer (fig. 1, *Nannacara, Zoarces)*, where the cone mosaic is also seen at the level of the receptor terminals, shows that the nucleus of a sclerad horizontal cell takes the place of a central single cone. The close spatial relationship is further emphasised by the fact that dislocations in the cone patterns can be followed as far as to the level of the horizontal cells (fig. 1, *Nannacara,* arrows). -The horizontal cells of the intermediate and vitread layers are found to fit into this pattern only if they are of the "lobated" cell type. The distance of the nuclei of these cells is about twice the distance between two single cones of the same type.

A comparison of the number of layers of horizontal cells with the number of cone types in different species, reveals another parallel. Species with double cones and one type of single cone have only two layers of horizontal cells, whereas species with an additional single cone have three layers of horizontal cells (fig.

1, 5). In *Poecilia*, this can be demonstrated in different regions within the same retina. In most parts there are three cone types and three types of horizontal cells; in the ventromedial region, however, where the additional single cone is missing, only two layers of horizontal cells are found.

As to the ratio of double cones to single cones, the figures are 4: 2 for species with only one single cone and 4:2:2 for species with three cone types. The ratio of the numbers of horizontal cell types between the different layers was determined by cell counts within a distance of 200 μm in several consecutive sections. Apart from minor variations, the figures exhibit a basic similarity. The ratio is about 4 (sclerad):2(intermediate):2 (vitread) horizontal cells (fig. 5) which is in good agreement with that of the cones types.

DISCUSSION

Occurence and distribution of cone types and cone patterns have been studied and reviewed by Engström (1963) and Anctil (1969) (see both authors for further literature). There is general agreement concerning the classification of the basic types of cone mosaics, i.e. square and row patterns with one or two kinds of single cones. In a detailed analysis, Anctil (1969) tried to establish additional kinds of patterns which are intermediate between the main types. It can, however, be shown that an ontogenetic relationship exists between row and square patterns. Fig. 4 demonstrates parallel rows of double cones in the zone of additional growth near the ora serrata; with increasing distance from this region, they are reorientated, forming zig zag rows, two of which finally give rise to a pure square pattern which is found all over the fundus of the eye.

As for the phylogenetic relationship of the cone mosaics, it has been assumed that row patterns are more primitive than square patterns. There are,however, numerous exceptions rendering any developmental approach rather difficult. It would appear that ecological factors are far more important in the determination of the type of cone mosaic than taxonomical position. All studies corroborate the fact that highly organised cone mosaics are only found in day-active pelagic fish for which vision is the most important of their senses, whereas, in bottom-dwelling species which are less visually orientated the cone patterns tend to disintegrate. It is therefore concluded that cone patterns may provide a structural basis for a high resolution both spatially and temporally. Furthermore, it has been suggested that they may be involved in the perception of polarised light (Watermann et al., 1972).

Regular arrangements of visual cells not only concern the cones but can also be found in relation to the rods. In a few cases, these are integrated into the cone mosaic (Wagner, 1972). In the pure rod retina of the deep-sea fish *Scopelarchus*, Locket (1970) reported that the visual cells lie in groups which, in turn, are distributed in a hexagonal lattice. One of the causes for this arrangement may be seen in the interaction with the tapetum lucidum; on the other hand, a very dense packing of elements, for example inner and outer segments of visual cells) also results in a regular array. Thus, it cannot be excluded that the mosaic-like distribution of cones may be due to a packing effect, too. However, this hypothesis would not explain the different types of cone patterns.

The morphological parallels between cones and horizontal cells described in this communication occur irrespectively of the type of cone mosaic. This is easily explained by the ontogenetic connection between row and square patterns (fig. 4). In the course of the transformation the position of the single cones does not change; therefore, it seems that not the double cones but the single cones determine the parallelism between the horizontal cells and the cones.

The functional importance of this close relationship may be seen in two aspects of vision: enhancement of contrast and color vision.

The role of the horizontal cells in the processing of color information in the goldfish has been demonstrated by Tomita and Stell (both this volume) with electrophysiological and morphological techniques. The numerical parallels between the number of cone types and the number of layers of horizontal cells on the one hand, and the similarities of the respective relative ratios on the other hand (fig. 5) seem to be in accordance with the findings of these authors. In the present investigation the identification of cone types and of cell types within a layer of horizontal cells of most species was, however, not based on Golgi studies which give direct evidence of the correlation between different types of cells by establishing their connections. Therefore, the existence of additional types of cones or of horizontal cells (such as rod horizontal cells) cannot be excluded.

The enhancement of contrast is of great importance for the shape detection in underwater vision. This is mainly achieved by lateral inhibition. As horizontal cells are able to inhibit the activity of visual cells (Baylor et al., 1971) they may provide a possible site for this mechanism. In retinae with parallelism of cones and horizontal cells, one may imagine the following process: an excitatory impulse originating from e.g. a square unit of cones, reaching the underlying horizontal cell may be propagated by gap junctions (Stell, 1972) to neighbouring cells of the same layer from which an inhibitory influence might be exerted on the overlying cone squares. Thus, only the central unit of receptor cells would be able to transmit impulses to bipolar cells.

REFERENCES

Anctil, M. (1969). Structure de la rétine chez quelques téléostéens marins du plateau continental. J. Fish. Res. Board Can. 26: 597-628.

Baylor, D.A., M.G.F. Fuortes and P.M. O'Bryan (1971). Receptive fields of cones in the retina of the turtle. J. Physiol. Lond 214: 265-294.

Cajal, R.S. (1894). Die Retina der Wirbeltiere. Bergmann, Wiesbaden.

Engström, K. (1963). Structure, organisation and ultrastructure of visual cells in the teleost family labridae. Acta. Zool. (Stockh.) 44: 1-41.

Locket, N.A. (1970). Deep-see fish retinas. Br. Med. Bull. 26: 107-111.

Selvin de Testa, A. (1966). Morphological studies of the horizontal and amacrine cells of the teleost retina. Vision Res. 6: 51-59.

Stell, W.K. (1972). The Morphological Organisation of the Vertebrate Retina. In: Physiology of Photoreceptor Organs, edited by M. C.F. Fuortes, p. 112-213, Springer, New York.

Wagner, H.-J. (1972). Vergleichende Untersuchungen über das Muster der Sehzellen und Horizontalen in der Teleostier-Retina (Pisces). Z. Morphol. Tiere, 72: 77-130.

Waterman, T.H. and Forward, R.B. Jr. (1972). Field Demonstration of Polarotaxis in the Fish *Zenarchopterus*. J. Exp. Zool. 180: 33-54.

DUPLEX VISION IN THE ELASMOBRANCHS: HISTOLOGICAL, ELECTROPHYSIOLOGICAL AND PSYCHOPHYSICAL EVIDENCE*

SAMUEL H. GRUBER

Rosenstiel School of Marine and Atmospheric Science

University of Miami, Virginia Key, Florida 33149

The Buddhas...regard all of the perceptions of sight...
as having no true nature of existence within themselves.
The Surangama Sutra

INTRODUCTION

Over the past century, biologists recognising the unique phylogenetic position of sharks have studied these animals to gain insight into the evolutionary history of vertebrates. Early in their development, sharks left the mainstream of vertebrate evolution and today retain, in many ways, primitive structural and physiological attributes (Shaeffer, 1967). Thus systematic studies of vision in sharks offer the potential of providing evolutionary information about the basic visual function in all vertebrates. Unfortunately, a fund of misinformation concerning both the structure and function of the visual system and visual behaviour in the elasmobranchs has become fixed in the biological literature. Typically, sharks have been described as nocturnal scavengers with a pure rod retina and crude visual system, probably relying on sensory modalities other than vision for survival (Bigelow & Schroeder, 1948; Verrier, 1930; Walls, 1942). Such descriptions stem no doubt from conclusions based on incomplete histological data, anecdotal behavioural evidence and pure speculation. Certain

* Contribution from the Rosenstiel School of Marine and Atmospheric Science. This research was entirely supported by contract # 0014-67-A-0201-0013 between the Oceanic Biology Branch (Code 488) Office of Naval Research and the University of Miami.

occular adaptations such as an occlusable tapetum (Kuchnow and Gilbert, 1967), mobile pupil (Young, 1933) and duplex retina (Gruber, Hamasaki and Bridges, 1963) suggest the possibility of day vision. Actually, many sharks are active during daylight and spend much of their lives in the well-illuminated epipelagic zone. Beyond phylogenetics, the elasmobranchs are perhaps the last well-known class of vertebrates whose behaviour has not been systematically investigated (except see Myrberg and Gruber, 1974). If we hope to understand the drive states influencing the activities of sharks, reliable data on the range of stimuli affecting their visual systems will have to be obtained. These data must be supplied by laboratory investigations. Through such investigations, we have attempted to provide normative data on some aspects of vision in sharks. Our studies involve histological descriptions, electrophysiological analyses and psychophysical measurements. The rationale underlying the research is taken from the Duplicity theory of vision (Schultze, 1866). From our results, a quantitative profile of vision in the lemon shark, Negaprion brevirostris (Poey), has emerged which differs radically from the early predictions and speculations of visual capability in elasmobranchs. We have shown that the lemon shark possesses a functional and extremely sensitive visual system capable of analysing spatio-temporal photic information under both diurnal and nocturnal conditions and is probably capable of analysing wavelength variations of light as well.

HISTOLOGY

Our basic findings concern the types of photoreceptors occurring in the retina of elasmobranchs. Using standard histological techniques we have examined more than twenty species of sharks and rays and have reported the presence of anatomically duplex retinas in all species (Gruber, et. al., 1963, Hamasaki and Gruber, 1965; and Gruber, et al. in press).

All retinas were amply provided with visual cells whose morphological characteristics agreed with the classical description of the cone. As Stell (1973) has stated, "...The differences between rods and cones were so obvious that one wonders why they were overlooked by previous investigators." We have found rod-cone ratios in certain species exceeding 5:1 at selected areas of the retina: the average was about ten rods to each cone observed. Our evidence strongly suggests that the typical elasmobranch possesses a morphologically duplex retina. We have provided data from nine families of elasmobranchs and the literature now contains examples of duplex retinas from all modern orders.

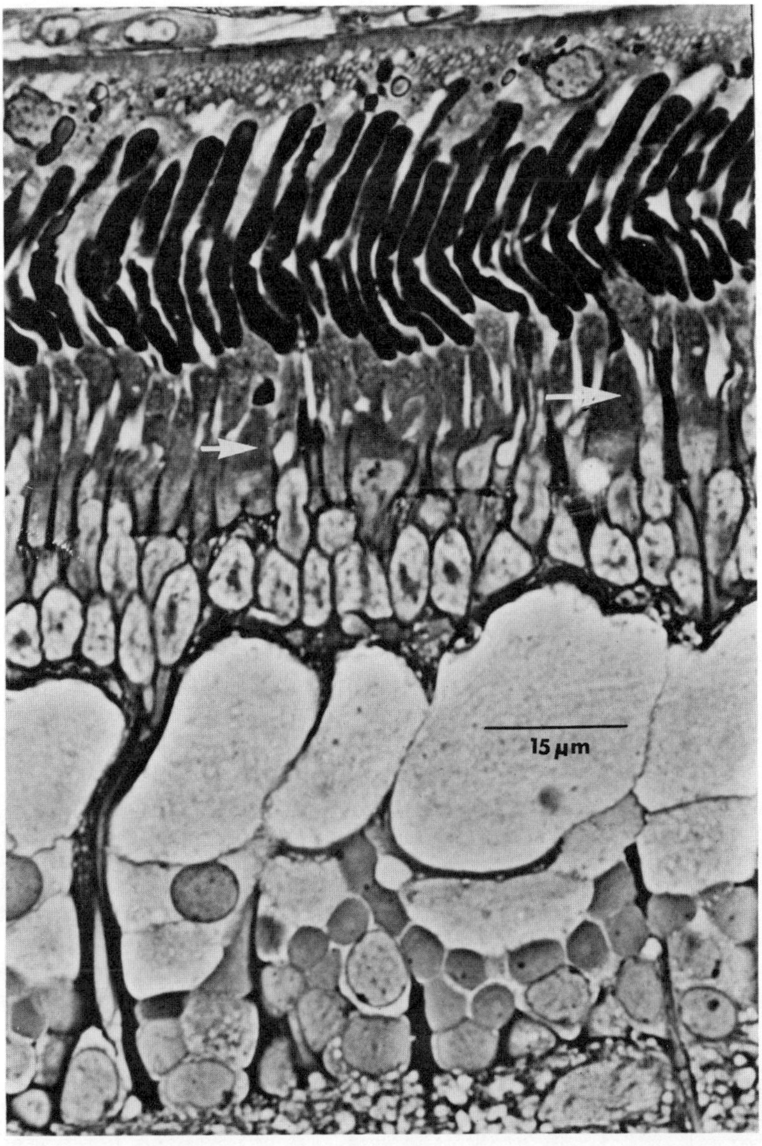

Fig. 1 Retina of the Mako shark, Isurus oxyrinchus Raf. Arrows indicate cone photoreceptors. Gluteraldehyde fixed, phenylene-diamine para (Pease, 1964), stained tissue was embedded in plastic and cut at 2 um.

VISUAL STUDIES

Duplicity theory assigns separate visual roles to the two classes of photoreceptors found in the vertebrate retina. The assigning of a particular function to a specific type of receptor rests on tenuous grounds, but the presence of rods and cones in a retina still predicts duplex retinal properties (Cohen, 1973). It was this prediction, especially regarding photopic (daylight) visual mechanisms, that we wished to test. Beyond this, we wanted to establish the range of vision and visual capability in terms of normative data as a standard against which results from experimental procedures on the shark visual system could be compared.

The subject in all studies was the lemon shark, Negaprion brevirostris. This animal is particularly suited for a variety of experimental procedures.

The procedures used in these studies include both electrophysiological and psychophysical techniques. Because of the relative familiarity of recording ERGs, procedural details, first worked out for the lemon shark by Hamasaki and Bridges (1965), will be omitted.

However, true respondent conditioning has infrequently been reported in aquatic animals. It therefore seems useful to give some details of our present training methods. Fig. 2 shows the manner in which sharks were restrained during testing and training and also the electro-optical apparatus. Subjects were conditioned in a tank containing 200 litres of running sea-water. A stimulus field was projected on a viewing screen placed 80 centimetres in front of the tank. An optical programmer automatically controlled the type of stimulus projected, number of trials, duration of the conditioned stimulus (CS-light), duration of the unconditioned stimulus (US-shock), intertrial interval and CS-US interval. All responses were monitored using an infra-red sensitive video system (VTR).

Naive, healthy sharks were individually adapted to the apparatus for 10 minutes and given 100 acquisition trials with a random intertrial interval of 25, 30 or 35 seconds. Conditioning consisted of pairing a flash of white light with an electric shock delivered as an electric field between two electrodes placed on either side of S's head. The intensity of the electric field was varied throughout training to maintain a uniform extension of the nictitating membrane as the unconditioned stimulus (US). Responses were analysed by playing back the video record of each session in stop-frame mode. A conditioned response (CR) to the photic stimulus was defined as an extension of the nictitating membrane occurring within 400 msec (i.e. before the 25th video field) after the onset of light. Re-

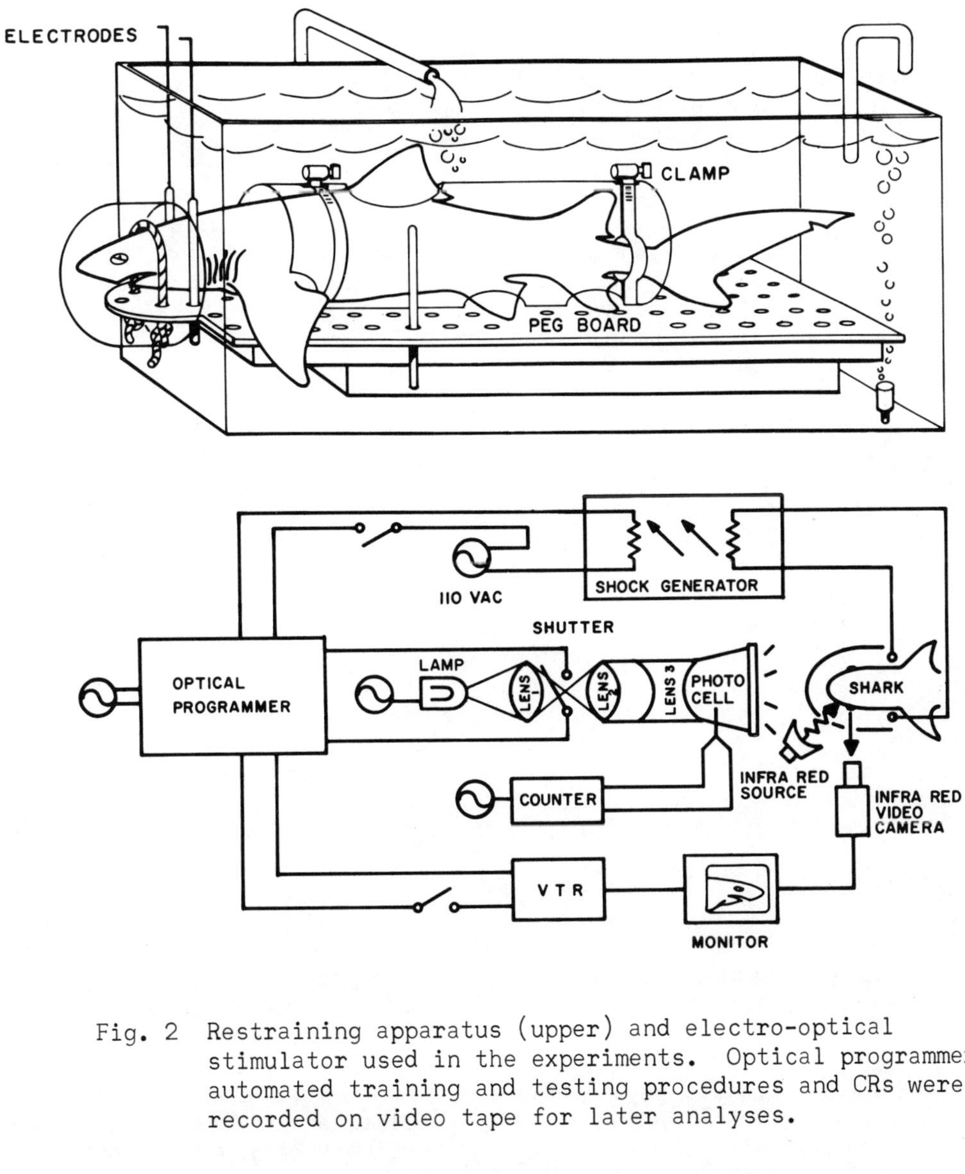

Fig. 2 Restraining apparatus (upper) and electro-optical stimulator used in the experiments. Optical programmer automated training and testing procedures and CRs were recorded on video tape for later analyses.

sponses <u>after</u> 400 msec were evoked by the shock. Reliable CRs occurred on the first session and quickly stabilised at 98% by the second half of the second session. Appropriate controls were made to insure that the CR was to the light alone and not confounded with unconditioned, pseudoconditioned, sensitised or other non-associative responses (Gruber and Schneiderman, in press). The general training procedure was modified as stimulus variables in the experiments dictated. In each experiment, control studies for the particular stimulus dimension were made. For example, in the flicker study described below, time-random, non-associative decreases in stimulus luminance were given throughout training and testing to insure that S attended to flicker rather than the decrease in stimulus luminance accompanying flicker.

A response-adjusting or discrete-trial titration schedule was used to evaluate threshold conditions. Thus, the stimulus was decreased until the response waned, then increased, and in this manner tracked back and forth around a specific value about 30 times for each threshold measurement. The relative merits of this psychophysical method have been described in detail by Rosenberger (1970).

Duplicity theory implies a sensitivity mechanism associated with the retinal rods and a chromatic, acuity mechanism associated with the cones each operating through a transitional range of light intensities (Walls, 1942). Depending on which system was in operation, differences in visual function might be expected. Indeed, when graphs of visual parameters are plotted, a characteristic "kink" often marks the changeover from rod- to cone-control of the process being measured. Our experiments were designed to reveal such "kinks" and other indirect evidence for cone function. Thus, we have determined, both by classical conditioning and electrophysiological techniques, the rate and range of dark-adaptation, the relation between stimulus illumination and critical frequency of flicker-fusion and the light- and dark-adapted spectral sensitivity of the lemon shark.

DARK-ADAPTATION

Details of this experiment can be found in Gruber (1966, 1967) and Hamasaki and Bridges (1965). The basic results, shown in fig. 3, demonstrate that the lemon shark is capable of extensive, slow dark-adaptation. The shark's final sensitivity exceeded that of two human observers measured with the same adaptometer under similar test conditions. The physiological and behavioural data were in good agreement which enhanced the reliability of the results and indicated that higher visual centres exert very little influence over dark-adaptation. As in man (Craik and Vernon, 1941), the process of dark-adaptation in the shark is apparently controlled by physiological changes within the retina. Incidental to the dark-adaptation measurements, we determined that pupillary dilation accounts for about one log of the six log unit increase in sensitivity accompanying dark-adaptation. The curve of dark-adaptation is smooth, and thus does not provide evidence for a photopic mechanism. However alternative explanations for these data have been given by Gruber (1967).

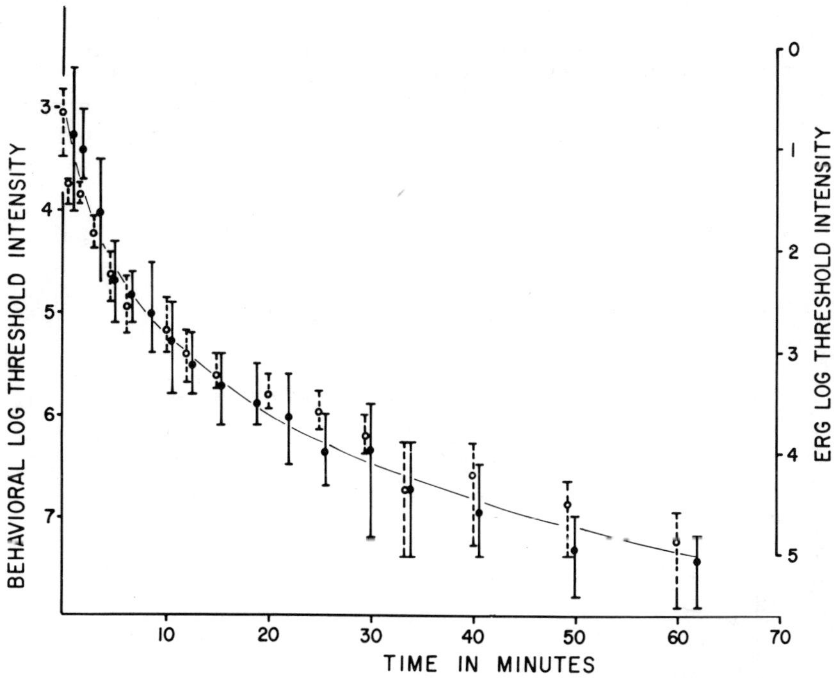

Fig. 3 Course of dark-adaptation in the lemon shark. Left ordinate represents the psychophysical measures. Solid conventions are mean psychophysical thresholds and bars represent the range of variation for 5 animals over 11 dark-adaptation runs. Right ordinate represents ERG data originally given in Hamasaki and Bridges (1965). The curves were aligned at 60 min. Open conventions are mean ERG thresholds and dotted bars give variation for dark-adaptation runs from 8 eyes.

CRITICAL FREQUENCY OF FLICKER-FUSION (CFF)

Details of this experiment can be found in Gruber (1969) and Gruber and Hamasaki (in prep.). Results are shown in fig. 4. Again, the behavioural and physiological data are in good agreement. The most important feature of these data is that the curves can be easily resolved into two intensity-dependent portions. Both curves ran parallel throughout their course and both have high and low intensity branches. The behavioural data are invariably higher at "corresponding" intensity values. Such differences between electrophysiological and behavioural flicker data are often reported. For example, Bernhard (1940) found the same situation for human flicker data. Van der Tweel (1961) has provided a general discussion of the differences between psychophysical and ERG determinations of CFF in man and animals. Both ERG and behavioural flicker-intensity curves for the shark fit the pattern established in dozens of experiments for animals with mixed retinas.

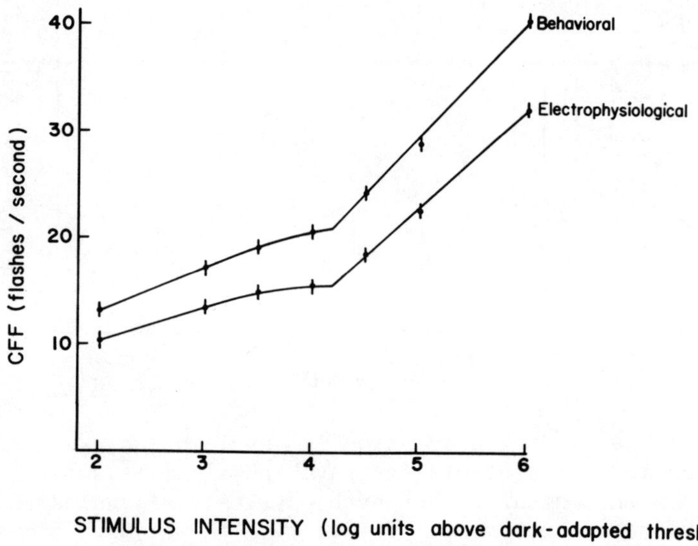

Fig. 4 Relation between CFF and stimulus intensity in the lemon shark. Upper curve represents mean (± one std. error) of the behavioural CFF values from 5 animals. Lower is the same for ERG data from 8 eyes. Curves were fitted by eye after statistical resolution into 2 branches.

Thus, it seems reasonable to conclude that the lemon shark possesses a functional photopic mechanism presumably associated with the retinal cones.

SPECTRAL SENSITIVITY

Details of this experiment can be found in Gruber (1969) and Gruber (1973). The spectral-sensitivity curves are shown in fig. 5. The major feature of these data is that a small (Purkinje) shift occurred upon moderate light adaptation. The point of peak sensitivity moved about 20nm toward the red. We have found some interesting and presently unexplained differences between the behavioural and physiological spectral-sensitivity data. For example, the behaviourally derived scotopic curve of spectral-sensitivity approximates the Dartnall (1953) nomogram for visual pigment 500 and also matches Bridges' unpublished observations on extracts from the retina of lemon sharks. The corresponding ERG curve peaked at 520nm, a 20nm departure from the expected position. Because of these differences and because our chromatic stimulus field was not adequate for intense photopic adaptation, we are continuing this work. However, these results represent further evidence that the visual system in the lemon shark is functionally duplex.

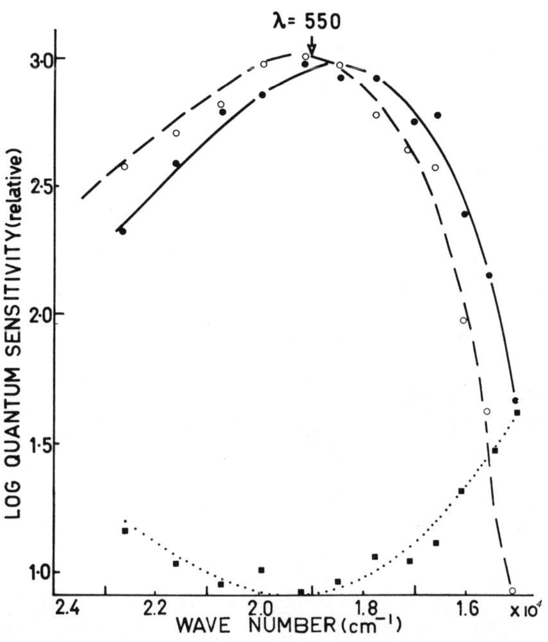

Fig. 5 Spectral-sensitivity of the light- and dark-adapted lemon shark retina. Open circles are scotopic measures and closed circles photopic measures of sensitivity. The two curves were aligned at their maximum sensitivity points. Closed squares show the difference between photopic and scotopic data. Results are from ERG recordings of 12 eyes. Abscissa is given in wave numbers to facilitate comparison with visual pigment curves.

COLOUR VISION

Discrimination between wavelength variations of two stimuli under photopic conditions is an important feature of the cone system. Accordingly, we devised experiments to determine whether lemon sharks could be trained to distinguish between colours. First, we empirically determined the test stimuli by finding two colours which differed greatly in wavelength, but whose energy relations were very close at the threshold of the CR. Given the conditions of our optical stimulor, blue (460nm) and red (650nm) were consistantly within 0.1 log at threshold for four light-adapted sharks. Several sharks were then trained to respond when the blue (460nm) stimulus, used as an adapting field, was silently switched to the red (test) stimulus. Blue-to-blue catch trials were randomly interspersed among the blue-to-red training trials. While the blue-to-red trials elicited responses in 70% of the trials, the catch trials remained below 30%. Thus, a fair degree of discrimination was achieved by the subjects. To test for hue discrimination, a series of trials were arranged with the blue adapting field lowered by 40% relative to the brightest red or blue obtainable from the optical stimulator. Recall that at photopic levels the blue and red had about the same measured (subjective) threshold brightness to the shark.

Results of these tests are shown and explained in table I. Complete details of this work are given in Gruber (1969). These data are presented with several reservations since results refer to the behaviour of one subject only. Next, a statistical test was required to demonstrate the role of hue in the discriminations and finally, the total number of trials was relatively low for such an experiment. The major problem in this work was subject mortality. Nevertheless, this conservative interpretation in no way invalidates the observation that hue was a factor in establishing the discriminative behaviour shown in table I.

CONCLUDING REMARKS

Our primary objective was to find out if sharks possessed a duplex retina and, if so, whether both scotopic and photopic mechanisms were present as predicted by the presence of retinal cones. We also wished to learn something about the visual capability of these animals in terms of sensitivity, spectral limits of vision and temporal resolution. The first problem concerned choosing a suitable method for studying the visual system of a non-verbal animals with low "intelligence". The ERG seemed to be adaptable to elasmobranchs and in fact, Kobyashi (1962) had completed such an analysis of the retinas of both elasmobranchs and teleosts. However, demonstrating that the retina responds electrically under certain

Table I Results of colour vision test in which lemon sharks were conditioned to respond when a blue adapting field was switched to a red stimulus field. Details are given in the text.

Test	Response To Stimulus YES	NO	Chi-Square	Signif.	Comment
dim blue→bright red vs. bright blue→bright red	33 47	14 56	7.83	P<0.01	Test compares effect of color and brightness to color alone. Result indicates both factors together provide a stronger stimulus than the factor of color alone.
dim blue→bright blue vs. bright blue→bright blue	45 19	42 35	3.65	N.S.*	Test compares the effect of intensity to trials where neither color or intensity change. x^2 value does not detect a difference between the two situations.
bright blue→bright red vs. bright blue→bright blue	47 19	56 35	1.58	N.S.*	Test compares the effect of color in the same way that above test compared intensity alone. x^2 value again does not detect a difference between the two situations.
dim blue→bright red vs. dim blue→bright blue	33 45	14 42	4.28	P<0.05	Test compares the effect of both factors to intensity alone in the same way that test one compared the combined effect to color alone. Results illustrate that the combination of color and intensity provided a stronger stimulus than intensity alone.

*not significant

The conclusion is that color is exerting a synergistic effect on the discrimination in combination with intensity and therefore the subject must have detected color differences.

stimulating conditions is quite different from specifying what the intact, unanesthetised animal actually sees. So, from the standpoint of visual capability, we had to investigate some relevant aspect of the shark's behaviour. We wanted to ascertain if the animal detected a visual stimulus. Moreover, we had to demonstrate the validity of the resultant answer. Thus, the question of validity assumes a central position in animal psychophysics and must be dealt with if the final data are to have any meaning.

We choose the method of classical conditioning because of certain inherent advantages not found in other types of conditioning. For example, the CR is quickly and easily elicited and under control of the experimentor at all times. Motivation plays little role in establishing the CR and the length of sessions is not limited by changes in motivation. Difficulties in defining the CR (Stebbins, 1970) have been eliminated in this study by developing an objective criterion for assessing time relations between onset of the visual stimulus and movement of nictitating membrane. However, the major advantage of this conditioning technique is that the animal is spatially fixed so that stimulus conditions are nearly identical from trial to trial. This contrasts with the usual training methods in which the stimulus constantly changes as the fish swims toward it.

The problems of validity and reliability of the behavioural data were assessed in several ways. First, we evaluated the conditioning paradigm through a series of control experiments (Gruber and Schneiderman, in press). Results clearly established that the lemon shark can be rapidly and reliably conditioned to a visual stimulus and that CRs were stable over time and were dependant on the visual stimulus only. To assess the reliability of the adaptometer, human subjects were used in place of sharks and dark-adaptation measured. The results were consistent in all respects with the well-known duplex curve of dark-adaptation reported in the literature (see Graham, 1965). Thus, the opto-mechanical apparatus appeared adequate for the psychophysical measurements. Another indication of reliability was that the range of variation in threshold between and within sharks amounted to about one log unit which agreed quite well with the variability of human subjects (Pirenne, 1962). The final test of validity concerned use of an entirely different methodology (ERG) for determining the same visual parameters. When psychophysical and electrophysiological measures are taken from the same animal, it not only enhances the validity of both sets of data, but comparison of results may yield information unobtainable from either method alone. For example, the agreement between the two sets of dark-adaptation data strongly suggests that retinal processing controls this function.

The finding of retinal cones, combined with the agreement between ERG and behavioural results, establishes that the lemon shark possesses both retinal and central photopic mechanism. The relation between receptor type, visual pigment and the Purkinje shift has been adequately covered by Graham (1965). The evidence indicates that the photopic curve represents the combined spectral range and sensitivity of the cone system(s). Similarly, the high intensity, fast-flicker branch of the CFF log I curve represents cone activity (Landis, 1953). Thus, the predictions of the Duplicity theory have been reasonably well satisfied by our data. It should be noted that recent evidence suggests the simplistic relations of Duplicity are in need of re-examination as is the classification of vertebrate photoreceptors (Pedler, 1965; Underwood, 1968; Cohen, 1972). However, the broad implications of Duplicity are probably correct and widely accepted, especially for human vision (Graham, 1965).

The measurements presented here also quantitatively establish visual capability of this animal. However, speculation on the significance of the data with regard to the natural visual behaviour of the lemon shark is premature. Toward this end, we have carried out ethological investigations on a sympatric species to try, among other things, to gain insight into the role that vision plays in the social behaviour of sharks (Myrberg and Gruber, 1974). Again, speculation is premature. Most desirable in this regard would be a more direct method of assessing the role of vision in field behaviour. However, appropriate methodology has not been forthcoming.

Visual capabilities of marine fishes are relatively unknown. This is true especially for elasmobranch fishes. Existing data have been obtained from anatomical and, more recently, electrophysiological studies. The data presented here represent the first systematic, multi-disciplinary investigation of visual function in the elasmobranchs and argue that the lemon shark possesses a highly functional visual system which operates both in daylight and at night. Behavioural observations indicate that lemon sharks are active during daylight hours (Stark, 1968) and may be efficient predators capturing a specific variety of prey. This differs from the role of nocturnal scavengers and opportunistic feeders which has been placed upon the shark. If Stark's (1968) fragmentary observations turn out to be correct, the visual system of the lemon shark should be adequate to provide it with the visual information needed by such a predator operating under a wide variety of light levels. However, only through behavioural, ecological and natural history studies on sharks will one be confident in defining the role of these animals in the biosphere.

Finally, the future of these studies will take the following directions: Laboratory investigations will centre on the role of colour in the lemon shark's vision. By operant conditioning, sharks will be tested in a Y-maze on a color discrimination task. Baseline studies have already been completed. By classical conditioning with a 3-channel, 5-kw xenon optical system, experiments including chromatic adaptation, heterochromatic mixtures and high intensity photopic spectral-sensitivity will be accomplished. The electrophysiological studies will shift emphases to the role of higher centres in the visual mechanisms of sharks. Ultimately, all these data must be related to the behaviour and ecology of sharks. This is the final objective of our studies.

REFERENCES

Bernhard, C.G. (1940) Contributions to the neurophysiology of the optic pathway. Acta Physiol. Scand., 1: (Suppl 1) 1-94.

Bigelow, H.B. and Schroeder, W.C. (1948) Fishes of the Western North Atlantic Part 1. Lancelets, cyclostomes and sharks. Memoir no. 1, Sears Foundation for Marine Research. Yale University, New Haven.

Cohen, A.I. (1972) Rods and Cones. In: Physiology of Photoreceptor Organs edited by M.G.E. Fourtes. Springer Verlag, Berlin.

Craik, K.T.W. and Vernon, M.D. (1941) The nature of dark-adaptation. Brit. J. Psychol. 32: 62-81.

Dartnall, H.J.A. (1952) The interpretation of spectral sensitivity curves. Br. Med. Bull. 9: 24-30.

Graham, C.H. (1965) Vision and Visual Perception. New York.

Gruber, S.H. (1966) A technique for Producing Respondent Conditioning in the Lemon Shark, *Negaprion brevirostris* (Poey), and its Application to Dark-adaptation Studies. M.S. Thesis, University of Miami, Coral Gables, Florida.

Gruber, S.H. (1967) A behavioural measurement of dark-adaptation in the lemon shark, *Negaprion brevirostris*. In: Sharks, Skates and Rays, edited by Johns Hopkins Press, Baltimore.

Gruber, S.H. (1969) The Physiology of Vision in the Lemon Shark, *Negaprion brevirostris* (Poey): A Behavioural Analysis. Ph.D. Diss. University of Miami, Coral Gables, Florida.

Gruber, S.H. (1973) Purkinje shift in the lemon shark: Behavioural and electrophysiological findings. Publ. abstracts ARVO Spring Meeting 1973.

Gruber, S.H. and Schneiderman, N. Classical conditioning of the nictitating membrane response in the lemon shark (*Negaprion brevirostris*). J. Animal Behav. Conditioning. (in press).

Gruber, S.H., Gulley, R. and Brandon, J. Photoreceptors in seven elasmobranch species. Bull. Mar. Sci. (in press).

Gruber, S.H., Hamasaki, D.I. and Bridges, C.D. (1963) Cones in the retina of the lemon shark (*Negaprion brevirostris*). Vision Res. 3: 397-399.

Hamasaki, D.I. and Bridges, C.D. (1965) Properties of the electroretinogram in three elasmobranch species. Vision Res. 5: 483-496.

Hamasaki, D.I. and Gruber, S.H. (1965) The photoreceptors of the nurse shark, Ginglymostoma cirratum, and the stingray, Dasyatis sayi. Bull, Mar. Sci. 15: 1051-1059.

Kobayashi, H. (1962) A comparative study on electroretinogram in fish, with special references to ecological aspects. J. Shimonoseki Coll. Fish. 11: 17-148.

Kuchnow, K.P. and Gilbert, P.W. (1967) Preliminary in vivo studies on pupillary and tapetal pigment responses in the lemon shark, Negaprion brevirostris. In: Sharks, Skates and Rays, edited by P.W. Gilbert et al. Johns Hopkins Press, Baltimore, 634 pp.

Landis, C. (1953) An annotated bibliography of flicker-fusion phenomena, covering the period 1740-1952. Armed Forces National Research Council Vision Committee Secretariat. Ann Arbor, Univ. Michigan.

Myrberg, A.A. and Gruber, S.H. (1974) The behaviour of the bonnet head shark, Sphyrna tiburo. Copeia 3: 358-374.

Pease, D. (1964) Histological Techniques for Electron Microscopy. Academic Press, New York.

Pedler, C.H.M. (1965) Rods and cones - a fresh approach. In: Color Vision edited by A.V.S. de Reuk. Little Brown, Boston.

Pirenne, M.H. (1962) Dark-adaptation and night vision. In: The Eye Vol. II. The Visual Process edited by H. Davson, Academic Press, New York.

Rosenberger, P.B. (1970) Response-adjusting stimulus intensity. In: Animal Psychophysics, edited by W.C. Stebbins. Appleton-Century-Crofts, New York.

Schultze, M. (1866) Zur Anatomie und Physiologie der Retina. Arch. Mikr. Anat. 2: 175-286. (Quoted by Pirenne, 1962).

Schaeffer, B. (1967) Comments on elasmobranch evolution. In: Sharks, Skates and Rays edited by P.W. Gilbert et al. Johns Hopkins Press, Baltimore.

Stark, W.A. (1968) Life history and ecology of the lemon shark, Negaprion brevirostris (Poey), in Florida Bay. In: Current Investigations in the United States dealing with the Elasmobranch Fishes. Washington, A.I.B.S. Shark Research Panel.

Stebbins, W.C. (1970) Studies of hearing and hearing loss in the monkey. In: Animal Psychophysics edited by W.C. Stebbins. Appleton-Century-Crofts, New York.

Stell, W.K. (1972) The structure and morphologic relations of rods and cones in the retina of the spring dogfish, Squalus. Comp. Biochem. Physiol. 42A: 141-151.

Underwood, G. (1968) Some suggestions concerning vertebrate visual cells. Vision Res. 8: 483-488.

van der Tweel, L.H. (1961) Some problems in vision regarded with respect to linearity and frequency response. Ann. N.Y. Acad. Sci. 89: 829-856.

Verrier, M.L. (1930) Contribution à l'étude de la vision chez les sélachians. Ann. Sci. Nat. 13: 52-63.
Walls, G.L. (1942) The Vertebrate Eye and Its Adaptive Radiation. Bull. No. 19: Cranbrook Inst. of Science, Bloomfield Hills, Michigan.
Young, J.Z. (1933) Comparative studies of the physiology of the iris. I. Selachians. Proc. R. Soc. 112B: 228-241.

6. ENVIRONMENTAL ASPECTS

ASSESSING THE FITNESS OF VISUAL PIGMENTS FOR THEIR PHOTIC ENVIRONMENTS

H.J.A. DARTNALL

MRC Vision Unit, University of Sussex

Falmer, Brighton, BN1 9QG, U.K.

> The eye may be said to owe its existence to light, which calls forth, as it were, a sense that is akin to itself; the eye, in short, is formed with reference to light, to be fit for the action of light; the light it contains corresponding with the light without.
>
> Goethe

In the dictionary sense a pigment is a substance that is coloured to the human eye. Colour is so striking a sensation that it was once thought to be an attribute of the "coloured" substance itself. The appearance of an object, however, depends not only on the inherent properties of the object, and the sensorium of the organism that perceives it, but on the nature of the light that illuminates it. In a certain respect, however, we can relate the colour of a substance to a physical attribute, and that is by its absorption spectrum. This is a mathematical concept expressing the variation with wavelength of the ability to absorb light. Apart from white, grey and black pigments, which absorb all wavelengths equally (the white not at all, the grey a constant fraction, and the black completely) pigments absorb light selectively and characteristically.

It would be logical to extend the meaning of the word pigment to include substances that absorb in any part of the electromagnetic spectrum. Such an extension, however, would bring in most, if not all, substances. The apparently transparent atmosphere of the earth, for example, has several strong absorption bands in both the ultraviolet and infra-red. It is, perhaps, better, therefore, to keep to the popular definition of a pigment as that which absorbs visible radiation. But we should avoid the anthropocentric restriction of what is visible, and remember that some organisms are more sensitive than us to the near ultraviolet or the near infrared.

The electromagnetic spectrum extends over an enormous range of frequencies - at least from 10^4 to 10^{22} hertz. And yet virtually all the photochemical reactions important to living organisms are activated by the narrow band of frequencies between the near ultraviolet and the near infra-red, less than a single order.

The reason for this is provided by the quantum theory. The size of a quantum, that is the amount of energy it contains, is given by the frequency in hertz multiplied by Planck's constant h, which has the value 6.62517×10^{-27} erg seconds. From the long radio waves to the near infra-red the frequencies are low, and the quanta of correspondingly low energy. The near infra-red, the visible, and the near ultraviolet have medium frequencies and energies, and the far ultraviolet, X-rays and beyond, progressively higher frequencies and energies.

When radiation reacts with matter it does so in multiples of quanta. In its passage through matter a particular quantum is either absorbed by a molecule or it is not. There are no half measures. The quanta of the far ultraviolet, X-rays and above, when absorbed by matter - particularly the complex molecules of living matter - can be destructive. The quanta of the far infra-red and below, on the other hand, have such low energies that they affect only the intra-atomic vibrations and rotations of molecules - motions that are soon shared, and appear on a macro scale as a mere rise in temperature.

The quanta of the middle part of the spectrum, however, i.e. between the near infra-red and the ultraviolet, have energies sufficient to cause electronic excitation of the molecules that absorb them. In such excited states, the molecules have quite different properties from those in their ground states, and reactions that would not otherwise occur become possible.

This band of radiation is copiously emitted by the sun, which has a surface temperature of about 6000°C and an emission spectrum roughly the same as a black body of this temperature. The peak of emission, when expressed in the usual units, is at 480 nm in tbe blue-green. If the surface temperature of the sun were only 3000°C then the peak emission would occur at 1000 nm in the infra-red, and the intensity of the total radiation would be one sixteenth of its present value. Conversely, if its temperature were 12,000°C the emission peak would move to 250 nm in the ultraviolet and the total radiation (which would include much more of the lethal short wave radiation) would be sixteen times the present value.

The fact that our sun has a surface temperature of 6000°C, that the Earth is the right distance from it, and that our atmosphere and waters are transparent, have all contributed to the generation and evolution of life.

The radiation that is absorbed by a pigment molecule represents an acquisition of energy to that molecule, and this may lead to its chemical change. The radiation that is not absorbed is reflected or transmitted; in other words, it is returned to the environment.

In most organically produced pigments, the colour serves a useful purpose to the organism. It can be employed to camouflage, or conversely to advertise, the presence of an animal; or to prevent harmful or unwanted radiation from reaching a sensitive tissue.

An apparent exception to the rule of usefulness is haemoglobin, the colour of which is perhaps an accidental property of a molecular structure conditioned by other requirements. Even so, use is made of its colour in human lips, the cock's comb, etc., and we might not pay so much attention to our wounds if our blood were colourless.

The pigments that I wish particularly to discuss, however, are not those involved in camouflage, advertisement or radiation screening. These we might regard as inactive; inactive, that is, in the sense that, in general, no use is made by the organism of the radiant energy they absorb. On the contrary I wish to consider the active pigments, that is those that do utilise this absorbed energy. The prime examples are the chlorophylls, which use it to promote photosynthesis of sugars in plants and bacteria; and the visual pigments, which use it to trigger the processes of vision.

It seems reasonable to expect a close spectral correspondence between the absorptive capacities of these active pigments and the radiation available in the environment. This is what I mean by the "fitness" of pigments. I use the word fitness, with its Darwinian undertones, deliberately to suggest that one would also expect such pigments to evolve in harmony with their environments.

Fifteen years ago Wald (1959) made a revealing comparison between the absorbance spectra of the chlorophylls and the spectrum of sunlight. His comparison is reproduced as Fig. 1, and in his remarks Wald wrote "chlorophylls, the pigment universally used in photosynthesis, have absorption properties that seem just the opposite of what is wanted in a photosynthetic pigment. The energy of sunlight as it reaches the surface of the earth forms a broad maximum in the blue-green to green region of the spectrum, falling off at both shorter and longer wavelengths. Yet it is precisely in the blue-green and green, where the energy of sunlight is maximal, that the chlorophylls absorb light most poorly; this, indeed, is the reason for their green colour. Where the absorption by chlorophyll is maximal - in widely separated bands in the violet and red - the energy of sunlight has fallen off considerably". Wald concluded, laconically, that "after perhaps two billion years of selection, involving a process whose efficiency is more important than that of any other process on earth, this seems an extraordinarily poor performance". Wald then went on to consider what other desirable properties in the chlorophyll molecule might have denied the possibility of a better match between absorptive properties and the sun's spectrum.

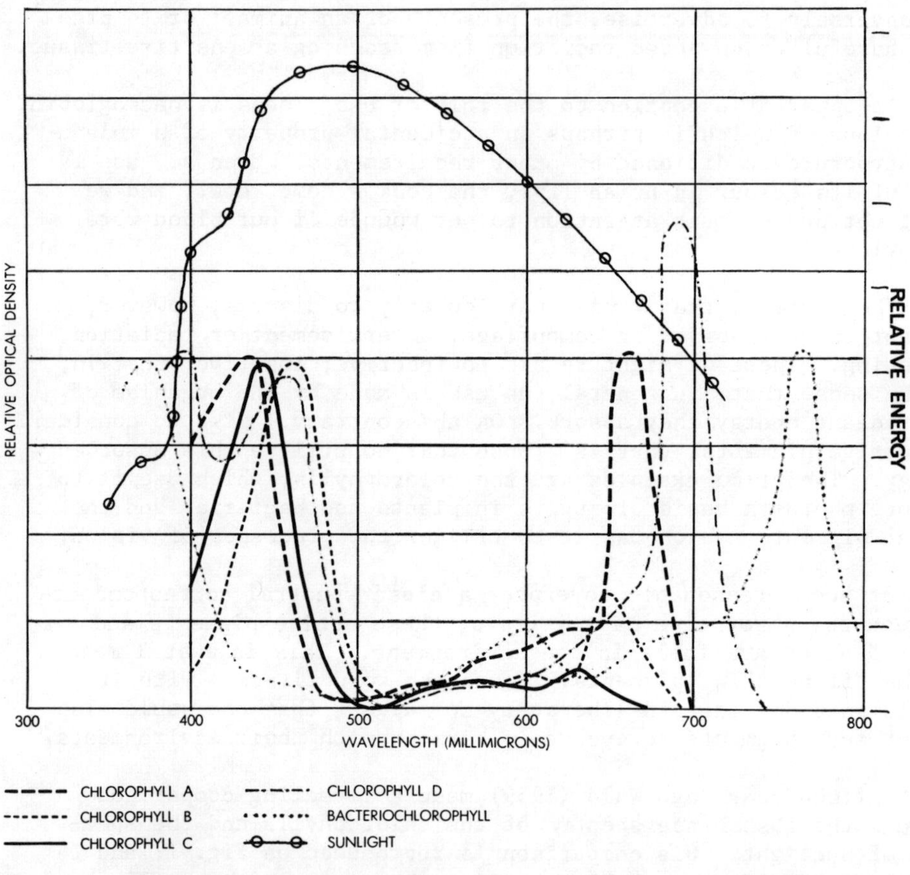

FIG. 1. Comparison between the energy spectrum of sunlight and the absorbance spectra of various chlorophylls (After Wald, 1959).

The "window" in the middle of the absorbance spectra of the chlorophylls clearly means that they are not such efficient absorbers of radiation as they would be without it. Wald's penetrating analysis of the possible reasons for their peculiar spectra is compelling. But there is another aspect to be considered as well. It is this. Given that the spectra of the chlorophylls have to be the way they are, are they positioned in the spectrum to make the best possible use of the available solar radiation?

We can put the matter in a more general way. If one wants to assess the performance of an active pigment is it valid to make a direct "eyeball" comparison between a spectral energy distribution curve and the absorbance spectrum of the pigment? To answer this question we need to consider carefully the dimensions and units of the functions compared.

Since radiation is a flow of energy an appropriate ordinate scale for the sun's emission spectrum in Fig. 1 is units of energy emitted in unit time (i.e. power) measured in, say, ergs per second or watts. In a continuous emission spectrum of finite intensity, however, one cannot measure the power at a single, precise wavelength for it is infinitesimally small. One has to make the measurement for a small wavelength interval, $\Delta\lambda$, within which an element ΔP of the total power, P, is contained. Thus the proper ordinate for the solar emission spectrum is ΔP per $\Delta\lambda$, i.e. $\Delta P/\Delta\lambda$ or, rather, the differential $dP/d\lambda$ to which this approaches as $\Delta\lambda$ tends to zero. The normal units in which this, the spectral irradiance at wavelength λ is expressed, are watts per square metre per micrometre. The power emitted over a finite range of wavelengths between λ_1 and λ_2 can be obtained by integration, i.e. by evaluating

$$\int_{\lambda_1}^{\lambda_2} (dP/d\lambda)\, d\lambda$$

There are other ways, however, in which the spectral irradiance can be presented. In photochemical reactions the rate of change of the active pigment is given by the product of the quantum efficiency and the rate of absorption of quanta:

$$\frac{-dn}{dt} = \gamma J_t$$

where n molecules of pigment remain at time t, J_t is the quantal flux absorbed by them, and γ is the quantum efficiency. The flux absorbed is related to the flux incident, I, by the expression

$$J_t = I(1-10^{-D_t})$$

where D_t is the optical density of the pigment at time t.

Absorbed quanta are all equally effective, irrespective of

their spectral origin. Thus in making a comparison between pigment absorption spectra and spectral irradiance it is appropriate to present the latter in quantal units. The energy content E of a quantum is given by

$$E = h\nu = hc/\lambda$$

Planck's constant h and the velocity of light c are both constant and so the conversion of spectral irradiance from energy to quantal units can be carried out in a relative way by multiplying the units of power by the wavelength. The new ordinate values $dN/d\lambda$ are then expressed in photons per second per unit area per unit wavelength interval.

Yet another way of presenting the spectral irradiance function is to express the photon flux for a constant frequency interval. The frequency, ν, and the wavelength λ are related by the expression

$$\nu = c/\lambda$$

Whence by differentiation,

$$d\nu = -c\lambda^{-2}d\lambda$$

Thus in relative terms the presentation of spectral irradiance in photon flux per unit frequency interval can be obtained by multiplying the ordinates of the photon flux per unit wavelength interval by the square of the wavelength. Alternatively, one can multiply the ordinates of the original data ($dP/d\lambda$) by the cube of the wavelength.

These three ways of presenting an irradiance spectrum yield functions of such different appearance that it is almost difficult to believe that they are based on identical data. They are shown for the solar spectrum outside the atmosphere in Fig. 2. The $dP/d\lambda$ basis gives a curve maximal at about 21,000 wave numbers (476 nm), the $dN/d\lambda$ basis one maximal at about 17,000 wave numbers (588 nm) and the $dN/d\nu$ basis one with a broad maximum at about 6,800 wave numbers (1500 nm), well into the infra red.

Which of these ways of presenting a spectral irradiance function is most appropriate for graphic comparison with the absorbance spectra of photosensitive pigments? An ordinate scale of $dP/d\lambda$ is clearly unsuitable, for it is not the energy absorbed by a pigment that determines the amount of photochemical change; it is the number of quanta. (For the same reason a fourth method of presentation - not shown in Fig. 2 - namely, an ordinate scale of $dP/d\nu$, is also inappropriate).

This leaves the choice between $dN/d\lambda$ and $dN/d\nu$. On both bases the irradiance is correctly expressed in quantal flux - per unit wavelength interval in the former and unit frequency interval in the latter. At first sight both would seem equally suitable provided the absorbance spectra with which they are to be compared are plotted correspondingly in wavelengths or frequencies. But there is a good reason for preferring dN/dν. It is that on a frequency

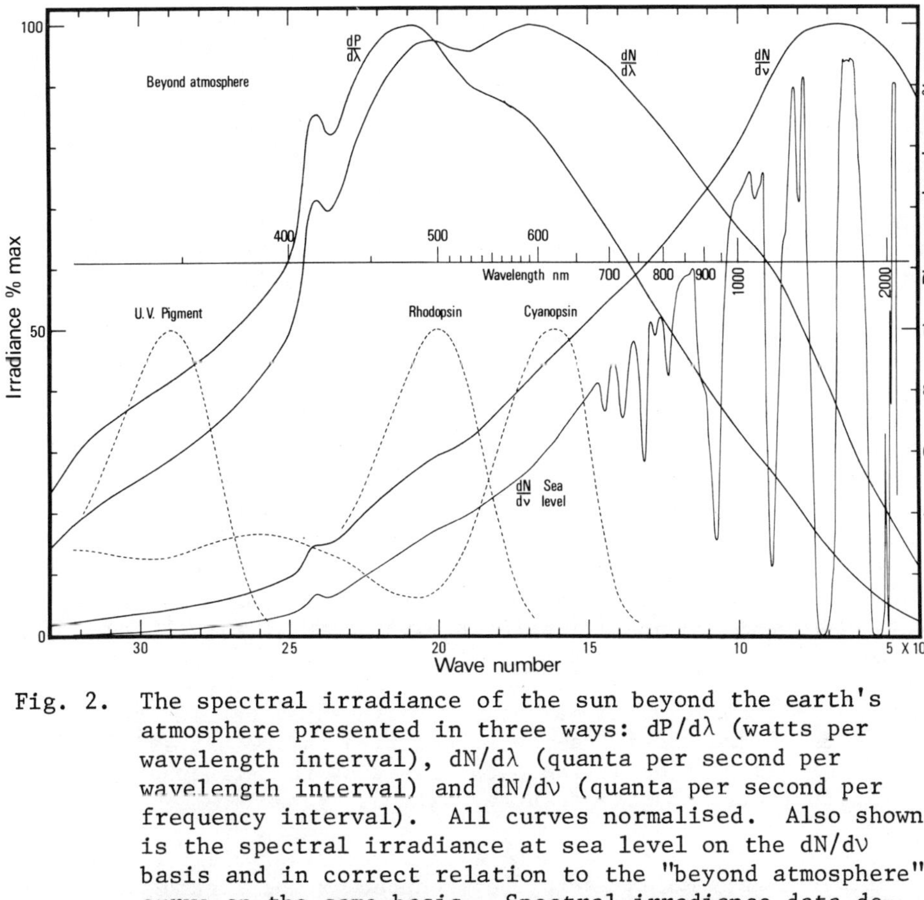

Fig. 2. The spectral irradiance of the sun beyond the earth's atmosphere presented in three ways: dP/dλ (watts per wavelength interval), dN/dλ (quanta per second per wavelength interval) and dN/dν (quanta per second per frequency interval). All curves normalised. Also shown is the spectral irradiance at sea level on the dN/dν basis and in correct relation to the "beyond atmosphere" curve on the same basis. Spectral irradiance data derived from Moon (1940). The absorbance spectra of rhodopsin and two "extreme" visual pigments are also shown. Cyanopsin is drawn according to the data of Wald et al. (1953) The "UV pigment" is the 345 nm pigment reported by Hamdorf et al. (1973) and this and the rhodopsin curve are drawn according to the nomogram for A_1-based pigments (see Wyszecki & Stiles, 1967).

scale the absorbance spectra of the visual pigments have very similar shapes or, rather, that there are two shape patterns, one for the A_1 pigments and the other for A_2 pigments. Consequently, when one wants to show graphically what λ_{max} pigment would be the best quantum catcher in a given environment, the representation of irradiance as dN/dν allows one to do this with least complication.

For this reason I suggest that irradiance spectra would be better presented on a dN/dν basis. Conversion from other bases is simple arithmethic and, in an appendix to this paper, I have recalculated Moon's (1940) data for the solar irradiance at sea level from his dP/dλ units (watts per square metre per micrometre) into dN/dν units (quanta per second per square millimetre per 10^{13} hertz).

Shown also in Fig. 2 are the absorbance spectra of three visual pigments; first, cyanopsin (Wald, Brown and Smith, 1953), the red-sensitive, A_2-based pigment having maximum absorbance at 620 nm (16,500 wave numbers), second, a typical rhodopsin with maximum at 500 nm (20,000 wave numbers) and third, the ultraviolet-sensitive pigment with maximum at 345 nm (29,000 wave numbers) found in the insect *Ascalaphus macaronius*, (Hamdorf, Paulsen & Schwemer, 1973).

Cyanopsin and the UV pigment represent extreme spectral positions for known visual pigments. Thus all visual pigments have spectra much displaced from the irradiance maximum of the solar spectrum, which is at 1500 nm in the infra-red.

The solar irradiance spectrum so far considered is that for outside the earth's atmosphere. This smooth spectrum was chosen for clearly bringing out the effects of expressing irradiance in different ways. But for comparison with the absorbance spectra of animal and plant pigments we should start with the spectrum of solar irradiance at sea-level. This is also shown in Fig. 2, on the dN/dν basis.

The sea-level curve is markedly less intense, particularly in the ultraviolet and the infra-red. Reduction in the ultraviolet is mainly due to absorption by the ozone layer of the atmosphere and, in the infra-red, by water vapour as well as by ozone.

The loss of infra-red at sea-level is considerable. Moreover, such infra-red as remains is further reduced by the liquid water present in animal and plant humours and tissues. Additionally, those organisms living in water are virtually denied the infra-red completely, as we shall see later. These factors largely explain why the spectra of natural photosensitive pigments are not centred beyond about 700 nm in the spectrum. Another possible reason is that the infra-red, with its low-energy quanta, is primarily the region of rotation-vibration spectra. A pigment that could be electronically excited by such feeble quanta might be chemically unstable.

Between about 450 and 700 nanometres (22,000-14,000 wave numbers) the solar irradiance spectrum is roughly linear and increases by a factor of rather less than 4.

The chlorophyll spectra with their blue-and red-absorbing peaks at 450 nm and 650-750 nm respectively appear to be optimally placed. An advance of the whole spectrum towards the infra-red, though it would improve the quantum-catching power in the "blue" peak, would seriously diminish it in the "red" one.

As regards the visual pigments, all but a few have their peak absorbance well within this 450-700 nm range. An interesting exception is the ultraviolet-sensitive pigment with maximum at 345 nm (29,000 wave numbers) found in the insect *Ascalaphus macaronius* (Hamdorf et al, 1973). The function of this pigment is clearly to sensitise insect vision to the differences in ultraviolet reflectivity between various parts of flowers and thus reveal the nectar guides. The irradiance of the solar spectrum in the ultraviolet at sea-level is so small compared with the total spectrum, that a rhodopsin with λ_{max} at 500 nm, for example, would be useless for mediating contrast between the reflecting and non-reflecting parts. Indeed to human vision there is no contrast and the different parts have the same appearance. By photography in ultraviolet light alone, however, the contrast can be shown.

These considerations suggest that on the land during daytime there is ample radiation from the sun to operate any visual mechanism from the near ultraviolet to the near infra-red. This view is supported by the low effective densities of visual pigments in cone cells, which may waste more than 90% of the light incident on the retina. Maximising the quantum catch, therefore, is not a major factor in determining the spectral locations of the visual pigments of colour vision. It is, for example, just as important that the 345 nm pigment of the insect shall lack sensitivity to most of the sun's spectrum as that it shall be sensitive to the ultraviolet. Otherwise the UV-response would be swamped. To account for the spectral positions of visual pigments in animals with colour vision it seems we should especially look for sensitivity to the reflected radiations from surfaces that are important to the welfare of the animals (food colour, predator, prey and mate coloration, and so on).

Under the full moon at night the spectral irradiance is some 6-7 log units lower than that from full sunlight (Richardson, 1969). The moon, acting as a mirror, gives a spectral irradiance basically similar to the sun's but relatively somewhat richer in the longer waves of the visible spectrum. Under dim conditions there may not be enough light to activate the mechanisms of colour vision. In those animals that possess a duplex retina the scotopic mechanism then takes over.

Some compensation for diminished environmental light can be obtained by increasing the fraction of it that is absorbed by visual pigment. A visual pigment with an optical density of 0.05 will absorb 11½% of incident light of that wavelength, and one of density 1.0 will absorb 90%. Clearly, however, there is here a case of diminishing returns, and little more than a log unit of compensation can be clawed back in this way. Nevertheless, the fact that the effective densities of rod pigments are generally much higher than those of cone pigments and that nocturnal animals have tapeta to reflect the light back through the retina again suggests two things: that the scotopic visual mechanism is short of light and that the animals have evolved to make maximum use of what there is.

This being so it is the more surprising that the scotopic pigments of land animals are not more advantageously placed in the spectrum than they are. In humans the change from photopic to scotopic vision results in the Purkinje shift of maximum sensitivity from 560 to 500 nm. The solar irradiance, however, increases linearly with increasing wavelength by a factor of about 4 in the visible spectrum (Fig. 2), and so this shift is in the wrong direction for improved sensitivity, particularly since moonlight (and starlight) is "redder" than sunlight. If, for example, our rhodopsin had λ_{max} at 620 instead of 500 nm the quantum catch would improve by 100%. Admittedly a doubled catch is not much compared with the improvement that can be obtained by increasing the pigment density; but it is more than a perfect tapetum can achieve.

It might be argued that we are not nocturnal animals and that our 500 nm rhodopsin is a case of adaptation to the Chappuis effect during the special conditions of twilight to which Munz and McFarland (1973) have recently drawn attention in their paper on tropical marine fishes. Against this is the fact that 500 nm rhodopsins have been found in truly nocturnal species as well as in the large majority of land animals.

No satisfactory explanation has been given for the paradoxical Purkinje shift. Lythgoe (1972) has speculated "that the almost universal possession of rhodopsin of λ_{max} near 500 nm in land vertebrates provides a means under scotopic conditions for distinguishing a brightness contrast between different species of green plants, which have spectral radiances dominated by the optical properties of chlorophyll".

The photic environment of aquatic animals is greatly modified depending on the light scattering and absorptive properties of the water and the depth of habitat. Even land-dwelling animals lose much of the infra-red and ultraviolet by absorption in the lens and humours of the eye though some, for example the frog, have media that are especially transparent to the near ultraviolet. Pure water appears to have maximum transmission in the region of 425 nm and these absorptive characteristics are approached in Crater Lake, Oregon (Smith & Tyler, 1967)but most natural waters have maximum transmission between 470 nm in the blue and 590 nm in the yellow, depending on the amount of yellow substance they contain. Jerlov (1968) has classified marine waters into a number of optical classes. The oceanic waters fall into three main types denoted by the roman numerals I, II and III, and the coastal waters into five main types denoted by arabic numerals 1, 3, 5, 7 and 9. The spectral transmittance of the clearest coastal water, arabic 1, is identical with that for the oceanic water, roman III, between 700 and 500 nm, but is much less at shorter wavelengths because of the presence of yellow substance, typical of coastal waters.

Jerlov's data for the spectral transmittance of the various water types are given for a path length of one metre. According to Jerlov (1968), however, the values can be used for calculating the irradiance at greater depths - even to the limit of the photic zone - without introducing more than small errors due to optical non-homogeneity. If the transmittance per metre at wavelength λ is T_λ then the proportion of incident light transmitted through a thickness of x metres is T_λ^x. The irradiance at x metres is then obtained by multiplying T_λ^x by the solar irradiance at sea level for that wavelength.

The spectral irradiance curves for a depth of 10 metres have been calculated in this way for a selection of Jerlov waters, and are shown in Fig. 3. The irradiance ordinate in this figure is scaled logarithmically to accommodate clearly the large range of values. The data for solar irradiance, both outside the atmosphere and at sea level, are the same as in Fig. 2.

Fig. 3 brings out two points about underwater photic environment. First, its spectral quality is very different in the different water types. Second, its intensity is likewise extremely variable. Thus the irradiance loss at 10 metres in different waters can range (at the wavelengths of maximum transmittance) from 0.07 log unit in the case of the clearest oceanic water JI to 2.5 log units in the case of the densest coastal water J9. Moreover this difference is for a mere 10 metres of depth and will progress in proportion for every further 10 metres.

All water types act similarly as sharp cut-off filters in the infra-red but the near ultraviolet significantly penetrates the surface layers of oceanic waters, though not of the coastal ones.

It is instructive to calculate the irradiance spectra at greater depths, down to the limit of the photic zone, and this has been done for three Jerlov waters in Fig. 4. The spectra for oceanic JI water are given in the middle of Fig. 4 for depths of 10, 50, 100 and 300

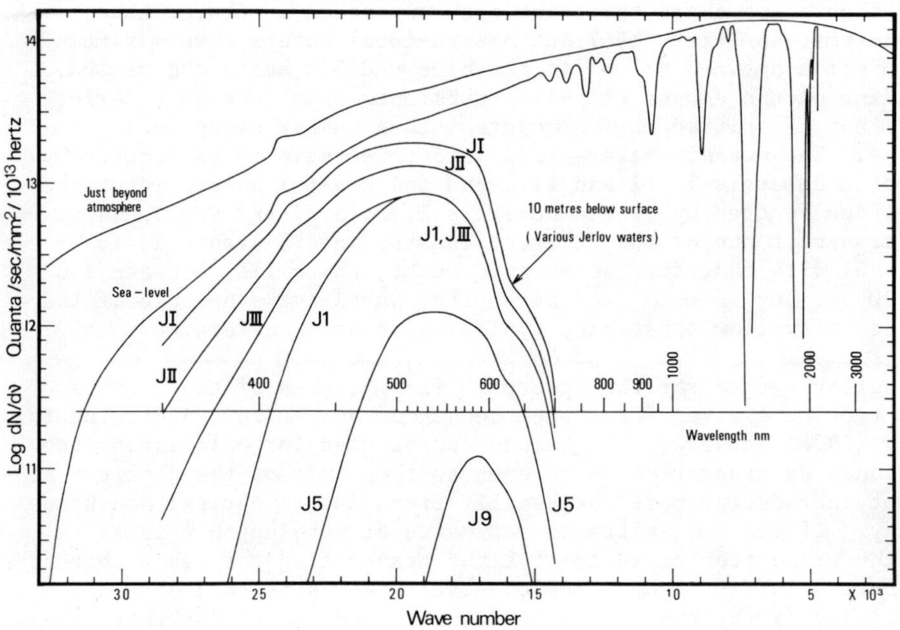

Fig. 3. Spectral irradiances of solar origin: outside the earth's atmosphere; at sea level; and 10 metres below the surfaces of various Jerlov-type waters as indicated. JI, JII and JIII are oceanic water types and J1, J5 and J9 are coastal water types. All irradiances expressed on the dN/dν basis (quanta per unit time per unit area per unit frequency interval).

metres and then at 300-metre intervals down to 1500 metres. Each spectrum is drawn to extend over only 2 log units of irradiance. For example, at 50 metres the irradiance drops 2 log units below maximum at c. 350 nm on the short-wave side and at c. 600 nm on the long-wave side. Similarly, at 600 metres depth the limits are at c. 430 nm and c. 500 nm respectively. By drawing the interrupted lines through the short and long wave limits respectively one gets a picture of the shrinking of the available spectrum with increasing depth.

On either side of the JI spectra are spectra similarly drawn for oceanic water JII and coastal water J9. In all cases the monochromatising effect of passage through water is evident, tending to 465 - 470 nm for JI and JII waters and to 580 - 590 nm for the coastal water.

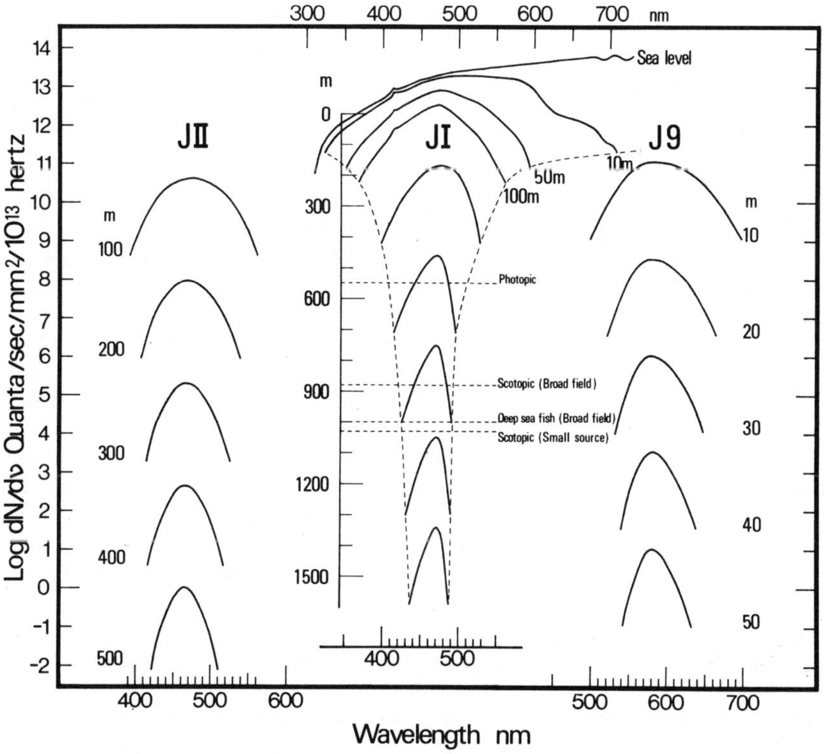

Fig. 4. The spectral irradiances at various depths in metres, as indicated, of two oceanic waters (types JI and JII) and a coastal water (type J9). All curves are drawn in absolute relation on the lgo dN/dν axis. Consequently, the limits of vision indicated by the interrupted lines drawn to the depth scale for JI water are applicable also to the other waters by horizontal extension.

Deep-living fishes are commonly found to have rhodopsins of λ_{max} in the 470 - 485 nm range and seem to be well adapted to their photic environments. Coastal and estuarine fishes, on the other hand, seem less well adapted (on a sensitivity basis) for they generally have pigments in the 500 - 510 nm range.

In Fig. 4 a vertical depth scale in metres is shown in register with the spectra of JI water. The limits for photopic and scotopic vision are marked off at points along this scale. These values have been taken from Clarke & Denton (1962) for clearest ocean water having k = 0.033. According to Jerlov the percentage of surface irradiance (465 nm) that penetrates JI water falls to

1% at 143 metres and consequently the value of k is given by 1/143 $\log_e$ 100 or 0.032. Thus Clarke & Denton's "clearest ocean water" is equivalent to JI water. In moonlight, according to Clarke & Denton, the limits of vision in this water shift towards the surface by about 400 metres.

The spectra for the other waters JII and J9 have been drawn at regular depth intervals such as to yield curves of roughly similar irradiance levels to those for the JI series. In the case of JII water this interval is 100 metres and in J9 water is only 10 metres. The limits of vision indicated for JI water can consequently be applied (with suitable depth adjustment) to the others. Thus the broad field scotopic limit, at 880 metres in JI water, would be reached at about 250 metres in JII water and at about 30 metres in J9 water. Similarly, the photopic limit at 550 metres in JI would be 150 metres in JII and 20 metres in J9.

Thus merely by swimming upwards a distance of 10 metres a J9 fish can pass from a scotopic to a photopic environment. For a JI fish this would entail a vertical migration of 300 metres and would involve as well the adaptation to changes in pressure and temperature.

The relatively slow rate of change of spectral irradiance with depth in clear oceanic waters may be one reason why deep-living fishes have visual pigments that are so well adapted on a sensitivity basis. Not only is the light of low intensity, putting a premium on quantum-catching power, but the normal vertical excursions of such fishes would result in little change to their photic environment, either as regards irradiance level or spectral composition.

Coastal fishes, on the other hand, living in optically dense waters, must experience considerable changes in irradiance level and spectral composition as they approach or leave the surface. In their case one might expect their visual apparatus to be a compromise between sensitivity and other performance requirements such as contrast perception (Lythgoe, 1966) and rapid adaptation.

In the clearest oceanic water the limit of solar-based scotopic vision is reached at about 1000 metres for human observers (Fig. 4). Even deep-sea fishes with their large-aperture eyes and "chrysopsins" (rhodopsins of λ_{max} 472 - 485 nm) of high optical density probably do not push this limit down more than another 100 metres or so. And yet fishes and other organisms with well-developed visual apparatus are found at very much greater depths than this.

Vision in these regions where the "ether" is void of solar radiation is entirely dependent on bioluminescence. Much of this appears to be of a blue to green colour (for review, see Nicol, 1962) though there are exceptions. This is somewhat surprising for, in the clearest water at any rate, long-wave light can penetrate several metres (Fig. 4). It is gratifying, therefore, to note the reports by Denton, Gilpin-Brown & Wright in 1970 and, more recently, by O'Day & Fernandez (1974) of mesopelagic and deep-sea fishes that emit red light and possess visual pigments suitable for its detection. It would indeed be curious if, in an empty ether, advantage were not taken of operating on "secret" wave-bands. I expect these are the first of many such examples.

I have tried to give a rough sketch of some of the problems in relating visual pigments and light environments. However, apart from the classic case of the deep-sea fishes, the Sensitivity Hypothesis that the function of a visual pigment is to absorb the greatest number of quanta, does not seem to have led us very far. But basically, I still believe this viewpoint is correct. The fault lies, I think, in our ignorance of what various regions of the spectrum are important to different animals.

Only plants look at the sun; animals don't. They, except in the "silhouette" situation (and in bioluminescence), look at the solar light that is reflected from surfaces important to their well-being - berries against foliage, the differential reflectivity of petals to UV light, the pattern of a mate. I think that if we knew much more than we do about the reflectivity of biologically important surfaces, then the Sensitivity Hypothesis might come into its own - at photopic levels as well as scotopic ones.

ACKNOWLEDGMENT

I record my thanks for the useful discussions I have had by correspondence with Dr. W.S. Stiles, OBE, FRS.

REFERENCES

Clarke, G.L. and Denton, E.J. (1962). Light and animal Life. In: The sea, Vol. 1, Physical oceanography, edited by M.N. Hill. Interscience, New York.

Denton, E.J., Gilpin-Brown, J.B. and Wright, P.G. (1970). On the "filters" in the photophores of mesopelagic fish and on a fish emitting red light and especially sensitive to red light. J. Physiol. (London) 208: 72-73.

Hamdorf, K., Paulsen, R. and Schwemer, J. (1973). Photoregeneration and sensitivity control of photoreceptors of invertebrates. In: Biochemistry and physiology of visual pigments, edited by H. Langer. Springer-Verlag, New York.

Jerlov, N.G. (1968). Optical oceanography. Elsevier, New York.

Lythgoe, J.N. (1966). Visual pigments and underwater vision. In: Light as an ecological factor, edited by R. Bainbridge, G.C. Evans and O. Rackham. Blackwell, Oxford.
Lythgoe, J.N. (1972). The adaptation of visual pigments to the photic environment. In: Handbook of sensory physiology, Vol. VII/1, edited by H.J.A. Dartnall. Springer-Verlag, New York.
Moon, P., (1940). Proposed standard solar radiation curves for engineering use. J. Franklin Inst. 230: 583-617.
Munz, F.W. and McFarland, W.N. (1973). The significance of spectral position in the rhodopsins of tropical marine fishes. Vision Res. 13: 1829-1874.
Nicol, J.A.C. (1962). Animal luminescence. Adv. Comp. Physiol. Biochem. 217-273.
O'Day, W.T. and Fernandez, H.R. (1974). *Aristostomias scintillans* (Malacosteidae): A deep-sea fish with visual pigments apparently adapted to its own bioluminescence. Vision Res. 14: 545-550.
Richardson, E.A. (1969). Contrast enhancement imaging devices by selection of imput photosurface spectral response. Adv. Electronics and Electron Phys. 28B, 661-675.
Smith, R.C. and Tyler, J.E. (1967). Optical properties of clear natural water. J. Opt. Soc. Amer. 57: 589-595.
Tyler, J.E. and Smith, R.C. (1970). Measurements of spectral irradiance underwater. Gordon and Breach, New York.
Wald, G. (1959). Life and light. Sci. Amer. (Oct) 92-108.
Wald, G., Brown, P.K. and Smith, P.S. (1953). Cyanopsin, a new pigment of cone vision. Science 118: 505-508.
Wyszecki, G. and Stiles, W.S. (1967). Colour science. Concepts and methods, quantitative data and formulas. Wiley, New York.

APPENDIX

The conversion of irradiance data from dP/dλ units to dN/dν units

As explained in the text, the conversion of irradiance values given in units of energy per unit time per unit area per unit wavelength interval into numbers of quanta per unit time per unit area per unit frequency interval, can be carried out in relative terms by multiplying the former values by the cube of the wavelength.

Moon's data (1940)

Moon (1940) gives values for the solar spectral irradiance, both outside the atmosphere and at sea level under various conditions, in the units watts per square metre per micron (micrometre). After multiplying each value, as given, by λ^3 (in micrometres) to yield relative values, the absolute values for the irradiance dN/dν expressed in numbers of quanta per second per square millimetre per 10^{13} hertz can be obtained by multiplying the relative values by 1.67942×10^{11}. In calculating the value of this constant multiplier the following constants were used:

Velocity of light	2.99793×10^8 m/sec
Planck's constant	6.62517×10^{-27} erg.sec

Table I gives the solar irradiance at sea level in dN/dν units listed versus wavelength from 295 nm up to 1000 nm. These values are a straightforward conversion of Moon's data (his Table III) as described in the previous paragraph.

Table II also gives the solar irradiance, but at regular 400 wave-number intervals from 33400 to 9800 (299.4 nm to 1020.4 nm). These values have been derived from those of Table I by linear interpolation (33,400 to 15,000) or graphical interpolation (15,400 to 9800) and are only representative of, not quantitatively faithful to, Moon's values.

If the irradiance values in Tables I or II are required for any other frequency interval than 10^{13} hertz they can be adjusted by simple proportion. The frequency interval of 10^{13} hertz is equivalent to 333.5635 wave numbers. Thus, for example, if irradiance values were required to refer to an interval of 400 wave numbers the tabled values would be multiplied by 1.199172 ($400 \div 333.5635$). This can be done by adding 0.0789 to the logarithmic values in the Tables.

Other data

In converting the dP/dλ of other authors, adjustment of the value of the constant multiplier (1.67942×10^{11}) will be necessary if different units from Moon's (watts/m^2/μm) have been used. For example, Smith and Tyler (1967) have expressed the downwelling irradiance in Crater Lake, Oregon, at various depths in units of microwatts/cm^2/nm. To convert their data to quanta/sec/mm^2/10^{13}H, therefore, first multiply each value individually by λ^3 (in μm) as before and then convert to absolute units by the constant multiplier 1.67942×10^{12} (1 μw/cm^2/nm = 10w/m^2/μm).

The conversion of irradiance data from dN/dλ units to dN/dν units

Some authors have expressed irradiance in quanta per unit time per unit area per unit wavelength interval ($dN/d\lambda$). As explained in the text the conversion to unit frequency interval can be carried out in relative terms by multiplying each datum by the square of the relevant wavelength. Thus Tyler & Smith (1970) and Munz & McFarland (1973) have expressed irradiances in quanta per second per square centimetre per nanometre ($dN/d\lambda$). To convert their values to $dN/d\nu$ in quanta per second per square millimetre per 10^{13} hertz they should each first be multiplied by the square of the relevant wavelength (in micrometres) and then by the constant multiplier 0.3335635.

TABLE I. The solar irradiance ($dN/d\nu$) at sea level at regular wavelength intervals. Data converted from Moon's (1940) values of $dP/d\lambda$ for normal incidence and air mass = 2 (his Table III).

Wavelength (nm)	log $dN/d\nu$ (quanta/sec /mm^2/10^{13}H)	Wavelength (nm)	log $dN/d\nu$ (quanta/sec /mm^2/10^{13}H)
295	5.9547	320	11.4730
296	10.5607	330	11.7850
297	7.1013	340	11.9986
298	10.6421	350	12.1315
299	8.1912	360	12.2614
		370	12.3754
300	8.5650	380	12.4908
301	8.9088	390	12.5971
302	9.1992		
303	9.4804	400	12.7034
304	8.7382	410	12.8908
		420	12.9600
305	9.9591	430	13.0215
306	10.1432	440	13.1150
307	10.3046		
308	10.4769	450	13.1874
309	10.6183	460	13.2469
		470	13.2976
310	10.7406	480	13.3419
311	10.8465	490	13.3785
312	10.9432		
313	11.0340	500	13.4066
314	11.1208	510	13.4292
		520	13.4520
315	11.1972	530	13.4728
316	11.2658	540	13.5008
317	11.3282		
318	11.3847	550	13.5218
319	11.4311	560	13.5423

TABLE I (continued)

Wavelength (nm)	log dN/dν (quanta/sec /mm^2/10^{13}H)
570	13.5639
580	13.5829
590	13.6025
600	13.6267
610	13.6486
620	13.6687
630	13.6936
640	13.7137
650	13.7332
660	13.7505
670	13.7678
680	13.7830
690	13.7320
700	13.8050
710	13.8083
720	13.7173
730	13.7997
740	13.8503
750	13.7884
760	13.6204
770	13.8705
780	13.8590
790	13.8832
800	13.8674
810	13.7945
820	13.8702
830	13.9184
840	13.9315
850	13.9372
860	13.9387
870	13.9457
880	13.8468
890	13.7868
900	13.7691
910	13.6763
920	13.5281
930	13.3585
940	13.5886
950	13.8458
960	13.9384
970	13.9869
980	14.0084
990	14.0203
1000	14.0245

TABLE II. The solar irradiance (dN/dν) at sea level at regular frequency (wave-number) intervals. Values derived from Table I by linear interpolation (between 33,400 and 15000 wave-numbers) or graphical interpolation (between 14,600 and 9,800 wave-numbers).

Wave number (reciprocal wavelength in cm.)	Wavelength (nm)	log dN/dν (quanta/sec /mm²/10¹³H)
33,400	299.4	8.3412
33,000	303.0	9.4882
32,600	306.7	10.2640
32,200	310.6	10.7998
31,800	314.5	11.1564
31,400	318.5	11.4066
31,000	322.6	11.5535
30,600	326.8	11.6851
30,200	331.1	11.8090
29,800	335.6	11.9040
29,400	340.1	12.0004
29,000	344.8	12.0628
28,600	349.7	12.1269
28,200	354.6	12.1914
27,800	359.7	12.2577
27,400	365.0	12.3180
27,000	370.4	12.3797
26,600	375.9	12.4440
26,200	381.7	12.5087
25,800	387.6	12.5716
25,400	393.7	12.6364
25,000	400.0	12.7034
24,600	406.5	12.8253
24,200	413.2	12.9131
23,800	420.2	12.9610
23,400	427.4	13.0052
23,000	434.8	13.0662
22,600	442.5	13.1329
22,200	450.5	13.1901
21,800	458.7	13.2393
21,400	467.3	13.2839
21,000	476.2	13.3250
20,600	485.4	13.3618
20,200	495.0	13.3927
19,800	505.1	13.4180
19,400	515.5	13.4417
19,000	526.3	13.4651

TABLE II (continued)

Wave number (reciprocal wavelength in cm.)	Wavelength (nm)	log dN/dν (quanta/sec /mm²/10^{13}H)
18,600	537.6	13.4942
18,200	549.5	13.5206
17,800	561.8	13.5462
17,400	574.7	13.5729
17,000	588.2	13.5990
16,600	602.4	13.6320
16,200	617.3	13.6632
15,800	632.9	13.6995
15,400	649.4	13.7319
15,000	666.7	13.7620
14,600	684.9	13.752
14,200	704.2	13.826
13,800	724.6	13.732
13,400	746.3	13.833
13,000	769.2	13.865
12,600	793.7	13.874
12,200	819.7	13.865
11,800	847.5	13.939
11,400	877.2	13.882
11,000	909.1	13.692
10,600	943.4	13.704
10,200	980.4	14.009
9,800	1020.4	14.037

VISUAL PIGMENTS AND THE ENVIRONMENT

W.R.A. MUNTZ

Lab. of Exptl. Psychology, University of Sussex

Brighton, BN1 9QG, England

> More light and light, - more dark and dark our woes!
>
> Romeo and Juliet, Act III, Sc. 5

The fishes live in a more varied light environment than any other group of vertebrates. Thus many species inhabit the clear waters of the open ocean, occupying both the well lit surface levels and the great depths where no daylight penetrates. Others inhabit the relatively turbid and muddy waters of rivers and coastal regions, and yet others may leave the water at times and use their eyes in air. These environmental differences not only apply to different species, but also often to a given species at different periods of its life. A number of fishes, for example, migrate between the sea and freshwater during their breeding cycles, and thus expose themselves to very different light environments, and the quantity and quality of the light in many bodies of water show marked seasonal variations due to factors such as the growth of plankton. Such different environments must present widely different visual problems to their inhabitants.

The fishes also show a very great diversity in the structure and physiology of their visual systems. This diversity is reflected in their visual pigments, of which a much greater variety have been found in fishes than any other vertebrate group. Thus pigments have been extracted with absorption maxima ranging from 473 nm (Dartnall, 1972) to 551 nm (O'Day and Fernandez, 1974), many fishes have more than one extractable visual pigment, and the pigments may change in response to various factors during an individual animal's lifetime (see eg Crescitelli, 1972, for a review). It seems clear that this diversity of pigments is related to the diversity of the environment, and reflects the different visual tasks the animals have to face. While in some cases, such as the deep-sea fishes, the way in which the pigments are adapted to the environment is fairly clear, in other cases no satisfactory explanation has yet been advanced. The purpose of this paper is to consider some of the problems involved in studying the adaptation of visual pigments to the environment, and to review some of the possible benefits accruing from the different pigments different fishes possess.

METHODS

Information is now available on the visual pigments of over 250 species of fish. These data have been obtained by the several methods discussed below, each of which has both advantages and disadvantages (Liebman, 1972). From the point of view of comparative work one of the main problems is that interesting species are often only found in remote places, where adequate laboratory facilities are lacking. Since the visual pigments may change rapidly under conditions of captivity, this means that either some method of preserving material for later analysis is necessary, or measurements must be made with simple techniques that do not require elaborate facilities.

The first and most common method of studying visual pigments is by means of extracts, for which the techniques are now very well known and need not be described here (eg Dartnall, 1957). The chief advantage of the technique is that large quantities of pigment may be obtained, allowing various chemical analyses to be performed. Where more than one pigment is present in the extract they can often be separately characterised by the technique of partial bleaching, and if the bleaching is carried out under appropriate conditions the position of the photoproduct maximum shows whether the pigments are based on vitamin A_1 or vitamin A_2. The technique does, however, require adequate laboratory facilities, and cannot be used under field conditions. One common method of avoiding this problem is to freeze the retinas at the collection site for later analysis, but this is not always completely satisfactory because keeping the material frozen during transport can be difficult to arrange, and defective specimens cannot be replaced.

A second technique that has been used involves measuring the spectral absorbance of small areas of excised retina. This method has been called macrospectrophotometry by Leibman (1972), in order to distinguish it from the technique of microspectrophotometry, in which much smaller measuring spots are used to obtain information on the absorbance of single receptor outer segments. In many cases the absorbance of the intact retina is almost entirely due to the visual pigment, which can therefore be adequately characterised by such measurements. The technique has been used successfully on several occasions in comparative work, and indeed the first convincing correlation between pigments and the environment, namely the shift in the spectral absorbance of deep-sea fish pigments to shorter wavelengths, was obtained by this method (Denton and Warren, 1956, 1957). The disadvantages of the method include the difficulty of allowing for the effects of any non-visual pigments the retina may contain, possible effects of scattering by the retinal elements, and the fact that the chemical nature of the pigment (including whether it is based on vitamin A_1 or vitamin A_2) cannot be determined with any certainty. The method offers, however, several compensating advantages. It is rapid, and the apparatus can be made simple, robust, and portable, so that the method can be used under field conditions (eg Muntz, 1973). In species with more than one pigment the distribution over the retina may not be uniform, and this can be readily detected by macrospectrophotometry (Denton et al., 1971). The *in situ* optical density can be measured, and where more than one pigment is present the method measures the effective absorption of the mixture, which is the datum that should be compared to the light environment. Finally, many fishes possess corneas and lenses that are a vivid yellow in colour (see below), and their characteristics, which help to determine the effective absorbance of the visual pigments, can also be measured with the same apparatus.

A final technique is that of microspectrophotometry, which measures the spectral absorbance of single receptors. Its advantages have been described in detail by Liebman (1972) and Harosi (this volume). In particular, the method allows the pigments of individual, anatomically recognisable, receptors to be determined, and it is also the only technique at present available that routinely allows the characterisation of cone pigments as well as rod pigments. The technique has, however, so far been little used in comparative work (but see Svaetichin et al., 1960). The apparatus is complex, and so far no satisfactory method has been developed that would allow material to be collected in the field and preserved for later analysis. Once this problem has been solved, however, it seems certain that microspectrophotometry will make a very substantial contribution to our knowledge in this area.

In order to make meaningful comparisons between the visual pigments and the environment it is not only necessary to know the pigments' spectral absorbance, but also the relationship between this and the animal's visual behaviour. In the case of man there is essentially perfect agreement between the spectral absorbance of the extractable visual pigment of the eye and the scotopic spectral sensitivity curve determined behaviourally (eg Crescitelli and Dartnall, 1953; Wald and Brown, 1958), and it has often been suggested that in other animals too the scotopic spectral sensitivity can be predicted from a knowledge of their extractable pigments. The situation with the fishes is, however, more complicated than it is in man. In the first place many fishes have more than one extractable pigment. In such cases it is not certain to what degree the different pigments will contribute to the animal's spectral sensitivity: they could, for example, be segregated into different receptors with different sensitivities in such a way that their contributions to sensitivity are not in proportion to their concentrations in retinal extracts. An extreme case would be if one of the extractable pigments was a cone pigment, and so did not contribute to scotopic sensitivity at all. Munz and McFarland (1973), in their study of the visual pigments of coral fishes, suggested that this was the case in many of their samples, and they consequently correlated environmental characteristics with the absorbance of the major pigment in the extracts alone.

In contrast to the situation in man, the visual pigments of fishes usually occur at relatively high optical densities. Thus densities (measured in the intact retina) of 0.6 are common, and in deep-sea species they may exceed 1.0 (Denton, 1959; Denton and Warren, 1957; Muntz, 1973); the density may also vary with retinal position (Denton et al., 1971). It is well known that the form of the spectral absorbance curve depends quite strongly on the optical density, becoming broader as the density increases, and with densities as high as those found in fishes the broadening will be marked (Dartnall, 1957).

The effect of coloured lenses, corneas, and tapeta is another problem that is more serious in fishes than in man. The effects of yellow lenses and corneas have been discussed in Muntz (1973). Fig. 1 shows an extreme example: it can be seen that short wavelength light is totally excluded from the retina in this species, and that the effective spectral absorbance curve of the retina is both narrowed and substantially shifted towards longer wavelengths. The effects of coloured tapeta are discussed by Nicol (this volume).

Relating visual pigments to the environment can clearly only be carried out if effects such as those described in the preceeding paragraphs are allowed for. This can best be done by determining spectral sensitivity behaviourally, and comparing the results with the properties of the visual pigments. Fig. 2 shows the scotopic spectral sensitivity of the cyprinid fish Scardinius erythrophthalmus (the rudd), determined in a 2-choice discrimination situation (further details of the methods may be found in my other paper in this volume). This species has two extractable visual pigments in the retina, $VP507_1$ and $VP535_2$ (Bridges and Yoshikami, 1970); since the relative proportion of these pigments is known to depend on daylength three separate determinations were made under each of three different daylengths. The results showed that the spectral sensitivity was affected by daylength, which presumably means that in this species both the extractable visual pigments are involved in scotopic vision.

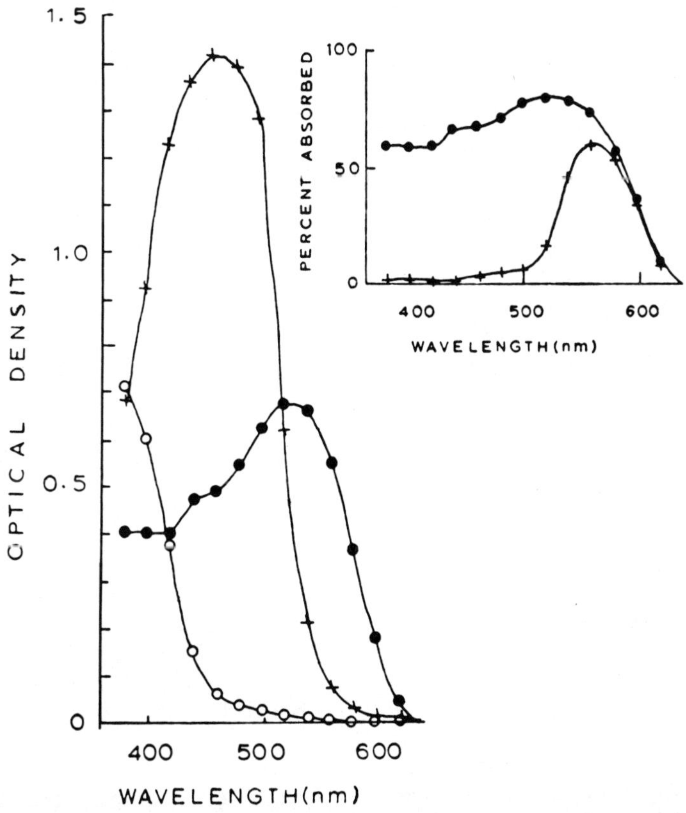

FIG. 1. Spectral absorbance of cornea, lens, and retina of *Astronotus ocellatus*. In the main part of the figure the crosses show the absorption of the cornea, the empty circles that of the lens, and the filled circles that of the retina. The inset shows the percentage of the incident light that would be absorbed by the visual pigment alone (filled circles), and by the visual pigment in life when the lens and cornea have the characteristics shown in the main part of the figure (crosses).

The spectral sensitivity curve also showed a breadth that was compatible with the visual pigments having an optical density of 0.64, which is a value that had actually been obtained macrospectrophotometrically for the rudd (Denton et al., 1971). The scotopic behavioural data that Yager (1968) obtained with goldfish suggest that a similar broadening of the scotopic spectral sensitivity curve occurs in this species also.

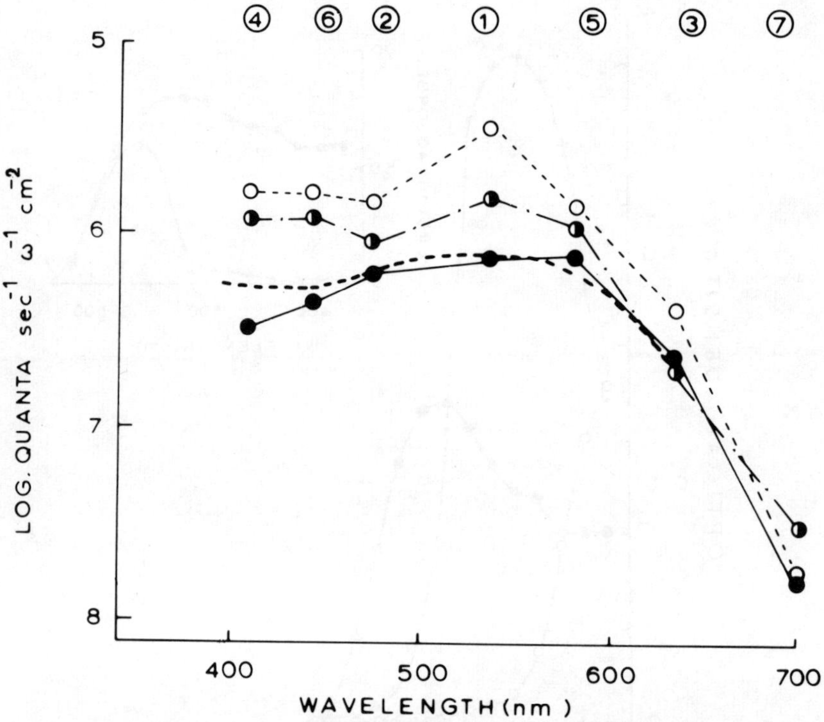

FIG. 2. Dark-adapted spectral sensitivity curves for four fish when kept under 18 hr (empty circles and dashed line), 12 hr (half-filled circles and long-dashed line), and 4 hr (filled circles and continuous line) daylengths. The heavy dashed line shows the spectral absorbance of a point on the retina of rudd C from Denton et al. (1971). The figures in circles show the order in which the different stimuli were used.

Finally, it should be emphasised that pigment extracts and macrospectrophotometry seldom provide any information about cone pigments (Dartnall 1960). Behavioural data also show that photopic vision is much more complex than scotopic vision, and seldom bears any obvious relationship to the rod pigments (eg my other paper). This means that, as Munz and McFarland (1973) point out, data on visual pigments should be compared to the light that is available in water at twilight and during the night. This presents a real problem, since it is difficult to measure the very low light levels prevailing at these times.

ADAPTATION OF THE VISUAL PIGMENTS TO THE ENVIRONMENT

A number of hypotheses have been put forward on how the visual pigments are adapted to the environment. Perhaps the most convincing of these is that the visual pigments are adapted to the spectral quality of the ambient light in such a way as to catch the greatest possible number of quanta. This notion was put forward by Clarke (1936) and by Bayliss et al. (1936), who independently suggested that the spectral sensitivity of deep-sea fishes might be shifted towards short wavelengths in order to make them more sensitive to the blue (approx 470-480 nm) light that penetrates deep oceanic waters; and was later confirmed by Denton and Warren (1956, 1957) using the technique of macrospectrophotometry, and by Munz (1957, 1958) using conventional pigment extracts. The data showed that the maxima of the visual pigment absorption curves of deep-sea fishes were shifted about 20 nm towards short wavelengths as compared to those of shallow water and esturine forms, which agrees well with expectations based on the spectral composition of light in deep water. The sensitivity hypothesis as applied to deep-water forms has been described in some detail elsewhere (Dartnall, this volume). In the present context it is worth emphasising that a deep-water shift to short wavelengths has now been described for the visual pigments of a wide variety of groups, including teleost fishes (Denton and Warren,1956, 1957; Munz, 1957, 1958; Wald et al., 1957), cartilagenous fishes (Denton and Shaw, 1956), pinnepids (Lythgoe and Dartnall, 1971), and whales (McFarland, 1971). Fig. 3 shows macrospectrophotometric data from the retinas of a variety of cichlid fishes caught at different depths in Lake Malawi, a deep and clear freshwater lake that has probably been completely isolated for two million years. Here again as the depth increases there is a clear shift towards short wavelengths in the spectral position at which the pigments absorb. The variety of animals and locations for which the effect is found strongly suggests that it has an ecological rather than a phylogenetic basis.

Since the spectral quality of the light reaching a fish depends on the light source as well as the water's transmission characteristics, visual pigments may on occasion not match the spectral characteristics of the water, but nevertheless still be well adapted according to the sensitivity hypothesis. A striking example where this may be true is in certain fishes which have both photophores that emit red light and red sensitive visual pigments. This phenomenon was first reported by Denton et al., (1971) for the mesopelagic genus *Pachystomias*, in which the photophores close to the eyes were seen to emit red light and for which the maximum of the retinal absorption curve, measured macrospectrophotometrically, was at about 575 nm. Similar findings were obtained by O'Day and Fernandez (1974) for the deep-sea bioluminescent fish *Aristostomias*

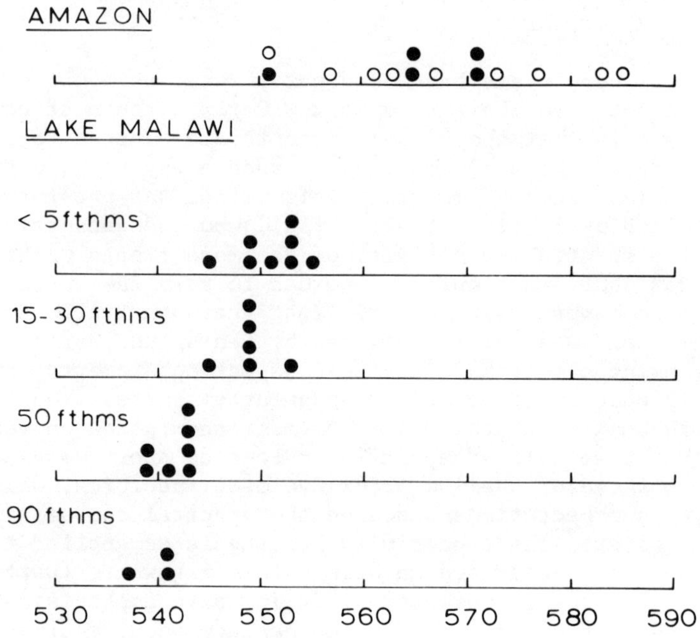

FIG. 3. Summary of macrospectrophotometric data for cichlids from the Amazon and from different depths (in fathoms) in Lake Malawi. The horizontal axis gives the wavelength on the long-wavelength side of the spectral absorbance curve at which the optical density has fallen to 50% of its maximum value (the λ 50), and each point represents a different specimen. Empty sumbols show specimens for which the lens of cornea were visibly yellow in colour.

scintillans using pigment extracts: in this case two pigments with λ max at 526 and 551 nm were found, and here again some of the photophores emit red light. Another possible example is provided by Munz and McFarland (1973), who have emphasised the fact the ambient light becomes markedly bluer at twilight. These authors suggest that the pigments of many coral fishes absorb at unexpectadly short wavelengths because they are adapted to vision at twilight, which is a critical time on coral reefs during which predation is particularly high.

The sensitivity hypothesis in intuitively plausible and appears to work well in many situations. Nevertheless on its own it cannot account for the distribution of all the different fish visual pigments that are known. For example, the waters of the English

Channel are considerably greener than those of the Mediterranean, but this is not reflected in any difference in the λ max of the pigments of fishes from the two locations (Dartnall and Lythgoe, 1965). Similarly, most bodies of fresh-water are green or yellow-green in colour, and although fresh-water fishes do possess visual pigments with rather longer λ max than sea fishes, the shift towards long wavelengths is less than would be expected on the basis of the sensitivity hypothesis. Some of the limitations of the hypothesis in these respects have been discussed in detail by, among others, Lythgoe (1972).

A further complication is that many fresh-water fishes have two extractable visual pigments, one based on vitamin A_1 and the other on vitamin A_2, with the latter absorbing at considerably longer wavelengths than the former. In paired-pigment species of this sort the proportion of the two pigments may often be altered by a variety of environmental factors (see eg Beatty, this volume). In the rudd (*Scardinius erythrophthalmus*), for example, increasing the daylength increases the proportion of the A_1- based pigment (Dartnall et al., 1961; Bridges and Yoshikami, 1970; Muntz and Northmore, 1972), so that the animals should become relatively more sensitive to short wavelengths during the summer. There is also more of the A_1-based pigment in the dorsal part of the retina (Muntz and Northmore, 1971; Denton et al., 1971), so that they should be more sensitive to short wavelength light for downward lines of sight. Such data as are available, however, show that in relatively productive bodies of fresh-water, similar to those inhabited by rudd, the spectral distribution of the ambient light shifts towards long wavelengths in the summer (eg Vollenweider, 1961 for Lake Maggiore) and that the up-welling light is redder than the downwelling light (eg Tyler and Priesendorfer, 1970 for San Vicente Reservoir). Since visual pigment determinations have not been made on fishes from either of these bodies of water this comparison is in principle insecure, but light measurements that are at present being carried out in a lake actually containing rudd are confirming both these trends, and pigment extracts on fish caught at regular intervals in the same lake are showing that as the water becomes redder in the spring, the pigments come to absorb at shorter wavelengths (Muntz and Wainwright, unpublished data). From the point of view of the sensitivity hypothesis, therefore, the visual pigments of the rudd are strikingly maladaptive.

Disparities of this type led Lythgoe (1966, 1968) to suggest that considering sensitivity alone is misleading. Underwater objects are generally perceived because they are slightly brighter or darker than the water background; and the task of the eye in such circumstances is not merely to catch as many quanta as possible, but to detect the object by its contrast with the background. A pigment that matches the background is not necessarily optimal for detection tasks of this type. This arises because the selective absorption of water reduces the bandwidth of light passing through it, and the further the passage of light through the water, the more monochromatic it will become (Tyler, 1959). Consider now the case of a grey or white object that is brighter than the water background and is being viewed in shallow water along a horizontal sight path. When the object is nearby not only will it be brighter than the background, it will also reflect into the eye a broader band of wavelengths since it is reflecting light that has come from the surface and therefore travelled on average a shorter distance through the water than the background light. Although therefore the object will radiate more light than the background at all wavelengths, the relative amount by which it radiates more light will be least at the wavelength at which the water spacelight radiance is maximal, and will increase progressively as we move away from this wavelength. The ratio of the light absorbed from the object and the background (its contrast) will therefore be greater for an offset pigment (one in which the λ max does not coincide with the wavelength of maximum transmission of the water) than for a matched pigment. Using Duntley's (1963) equations and various assumptions about the eye, Lythgoe has calculated the increase in visual range that would result from this improvement in contrast.

As Lythgoe points out, offset pigments would only improve contrast in rather special circumstances, namely the detection of objects brighter than the water background viewed at short ranges in shallow water. In all other cases (objects darker than the background, and both bright and dark objects in deep water or at longer distances) the advantage disappears and matched pigments become more efficient. The contrast sensitivity hypothesis therefore implies that this one particular class of stimuli is especially important in the life of fishes. At present we do not know if this is so.

One striking finding about the visual pigments is that in paired-pigment species retinal extracts from different individuals caught in the same locality always show a very considerable variation in the proportion of the A_1- and A_2-based pigment (eg Bridges, 1964). This variability usually persists even under laboratory conditions: as just one example, after 147 days of constant conditions, the proportion of A_2-based pigment in a group of twenty rudd varied between 15% and 55% (Bridges and Yoshikami 1970). In a similar way, the visual pigments of different species from the same body of water may vary considerably, even though they are probably subjected to similar light environments. It therefore seems likely that the variability in itself may confer some selective advantage. This idea was first put forward in (1964) by Bridges, who suggested that paired-pigments are particularly common in schooling species, with the result that the spectral sensitivity of the school will be broader than that of the individuals.

It seems likely that the variability of fish visual pigments is highest in bodies of water that are themselves very variable in the quantity and spectral quality of the light. An example of this is given in Fig. 3, which summarises macrospectrophotometric data that have been obtained on a variety of cichlid fishes. The different species from Lake Malawi show little variability, while those from the Amazon show a wide range in the location of their spectral absorbance curves, probably largely due to different proportions of A_1- and A_2-based pigments. In the Amazon different individuals of of the same species also often showed different spectral absorbance curves (which did not occur in Lake Malawi), and the effective range will be considerably greater than that shown in the figure since many species possessed yellow lenses and corneas.

Lake Malawi is a body of water in which conditions are fairly constant. It is very transparent throughout the year: twelve Secchi disc readings taken in 1961, for example, varied between 11.2 and 22m, with a mean of 16.1m (D.H. Eccles, personal communication). Under these circumstances the sensitivity hypothesis appears to provide a good description of the results: light conditions are constant and the fishes match their pigments to them.

In the Amazon, however, conditions are very variable. Thus Secchi disc readings range from about 4m or more for clear water to about 15cm for white water (Giesler et al., 1971), and these different types of water are often in communication with each other so that fish may pass between them. Considerable variations also occur with time. For example, the colour of the water in a tributary near Manaus varied between 250 and 65 platinum units over an eight month period during which measurements were made (Santos et al., 1971), and these changes in colour were presumably accompanied by equivalent changes in intensity and transparency. In variable situations such as these it may be impossible for the fishes to develop pigments matching the water. The environmental changes may occur too rapidly for physiological adaptations (such as changing the proportion of A_1- and A_2-based pigments) to take place, and the fish may also encounter very different environments through swimming relatively short distances. The most profitable strategy may then be for different individuals to have different pigments, so that some individuals will always be adequately adapted. In Levins' (1968) terminology, the Amazon river system is coarse-grained from a visual point of view, and Lake Malawi is fine-grained. The former type of environment is best fitted by a series of specialised forms in proportions depending on the frequency of the different environmental patches, while the latter type is best fitted by a single phenotype specialised for the most frequently encountered set of conditions.

The different hypotheses on adaptation that have been discussed are of course not mutually incompatible, and they may all be true under appropriate circumstances. Some problems also remain very obscure, such as the functional significance of the pigment changes that occur in paired-pigments species in response to environmental factors. It is likely that further progress will depend largely on obtaining more information on the visual tasks that actually face fishes in their normal environments. This will require many more field measurements of the light in different bodies of water, and also behavioural experiments carried out under natural conditions.

REFERENCES

Bayliss, L.E., Lythgoe, R.J., and Tansley, K. (1936). Some new forms of visual purple found in sea fishes, with a note on the visual cells of origin. Proc. R. Soc. B., 816: 95-113.

Bridges, C.D.B. (1964). Variations of visual pigment amongst individuals of an American minnow, *Notemegonus crysoleucas boseii*. Vision Res. 4: 233-239.

Bridges, C.D.B. and Yoshikami, S. (1970). The rhodopsin-porphyropsin system in freshwater fishes - 1. Effects of age and photic environment. Vision Res. 10: 1315-1332.

Clarke, G.L. (1936). On the depth at which fishes can see. Ecology 17: 452-456.

Crescitelli, F. (1972). The visual cells and visual pigments of the vertebrate eye. In: Handbook of Sensory Physiology, Vol. VII/I, Photochemistry of Vision, edited by H.J.A. Dartnall, Springer-Verlag, Berlin.

Crescitelli, F., and Dartnall, H.J.A. (1953). Human visual purple. Nature (Lond). 172: 195-196.

Dartnall, H.J.A. (1957). The Visual pigments. Methuen, London.

Dartnall, H.J.A. (1960). Visual pigments of colour vision. In: Mechanisms of Colour Discrimination, edited by Y. Galifret, Pergamon, Oxford.

Dartnall, H.J.A. (1973). Visual pigment of the coelacanth. Nature (Lond). 239: 341-342.

Dartnall, H.J.A., Lander, M.R. and Munz, F.W. (1961). Periodic changes in the visual pigment of a fish. In: Progress in Photobiology, edited by B.C. Christensen and B. Buchmann, Elsevier, Amsterdam.

Dartnall, H.J.A. and Lythgoe, J.N. (1965). The spectral clustering of visual pigments. Vision Res. 5: 81-100.

Denton, E.J. (1959). The contributions of the orientated photosensitive and other molecules to the absorption of whole retina. Proc. R. Soc. B. 150: 78-94.

Denton, E.J., Gilpin-Brown, J.B., and Wright, P.G. (1970). On the "filters" in the photophores of mesopelagic fish and on a fish omitting red light and especially sensitive to red light. J. Physiol. (Lond). 208: 72P-73P.

Denton, E.J., Muntz, W.R.A. and Northmore, D.P.M. (1971). The distribution of visual pigment within the retina in two teleosts. J. Mar. Biol. Ass. U.K. 51: 905-915.

Denton, E.J. and Shaw, T.I. (1956). The visual pigments of some deep-sea clasmobranchs. J. Mar. Biol. Ass. U.K. 43: 65-70.

Denton, E.J., and Warren, F.J. (1956). Visual pigments of deep-sea fish. Nature (Lond). 178: 1059.

Denton, E.J., and Warren, F.J. (1957). The photosensitive pigments in the retinae of deep-sea fish. J. Mar. Biol. Ass. U.K. 36: 651-662.

Duntley, S.Q. (1963). Light in the sea. J. Opt. Soc. Amer. 53: 214-233.

Giesler, R., Knoppel, H.A., and Sioli, H. (1971). Ökologie der Süsswasserfische Amazoniens.Stand und Zukunftsaufgaben der Forschung. Naturwissenschaften 58: 303-311.

Levins, R. (1968). Evolution in changing environments. Princetown Univ. Press: Princetown.

Liebman, P.A. (1972). Microspectrophotometry of photoreceptors. In: Handbook of Sensory Physiology, Vol. VII/I, edited by H.J.A. Dartnall, Springer-Verlag, Berlin.

Lythgoe, J.N. (1966). Visual pigments and underwater vision. In: Light as an ecological factor, edited by R. Bainbridge, G.C. Evans, and O. Rackham. Blackwell, Oxford.

Lythgoe, J.N. (1968). Visual pigments and visual range underwater. Vision Res. 8: 997-1012.

Lythgoe, J.N. (1972). The adaptation of visual pigments to the photic environment. In: Handbook of Sensory Physiology, Vol. VII/I. Photochemistry of vision, edited by H.J.A. Dartnall, Springer-Verlag, Berlin.

Lythgoe, J.N., and Dartnall, H.J.A. (1971). A deep-sea rhodopsin in a mammal. Nature (Lond). 227: 955-956.

McFarland, W.N. (1971). Cetacean visual pigments. Vision Res. 11: 1065-1076.

Muntz, W.R.A. (1973). Yellow filters and the absorption of light by the visual pigments of some Amazonian fishes. Vision Res. 13: 2235-2254.

Muntz, W.R.A., and Northmore, D.P.M., (1971). Visual pigments from different parts of the retina in rudd and trout. Vision Res. 11: 551-562.

Muntz, W.R.A., and Northmore, D.P.M., (1973). Scotopic spectral sensitivity in a teleost fish (*Scardinius erythrophthalmus*) adapted to different daylengths. Vision Res. 13: 245-252.

Munz, F.W. (1957). Photosensitive pigments from retinas of deep-sea fishes. Science. 125: 1142-1143.

Munz, F.W. (1958). Photosensitive pigments from the retinae of certain deep-sea fishes. J. Physiol. (Lond). 140: 220-225.

Munz, F.W., and McFarland, W.N. (1973). The significance of spectral position in the rhodopsins of tropical marine fishes. Vision Res. 13: 1829-1874.

O'Day, W.T., and Fernandez, H.R. (1974). *Aristostominas scintillans* (Malacosteidae): a deep-sea fish with visual pigments apparently adapted to its own bioluminescence. Vision Res. 14: 545-550.

Santos, V. de M., Santos, A., and Brinkman, W.L.F. (1971). A composicão quimica do Rio Preto da Eva - Amazonia. Estudo preliminar. Ciencia y Cultura 23: 643-646.

Svaetichin, G., Negishi, K., and Fatehchand, R. (1960). Cellular mechanisms of a young - Hering visual system. In: Mechanisms of Colour Discrimination, edited by Y. Galefret, Pergamon, Oxford.

Tyler, J.E. (1959). Natural water as a monochromotor. Limnol. Oceanogr. 4: 102-105.

Tyler, J.E., and Preisendorfer, R.W. (1970). Measurements of spectral irradiance underwater. Gordon and Breach, New York.

Vollenweider, R.A. (1961). Photometric studies on inland waters: 1. Relations existing in the spectral extinction of light in water. Mem. Inst. Ital. Idrobiol. 13: 87-113.

Wald, G., Brown, P.K., and Brown, P.S. (1957). Visual pigments and depths of habitat of marine fishes. Nature (Lond). 180: 969-971.

Wald, G., and Brown, P.K.,(1958). Human rhodopsin. Science 127: 222-226.

Yager, D. (1968). Behavioral measures of the spectral sensitivity of the dark-adapted goldfish. Nature (Lond). 220: 1052-1053.

STUDIES ON THE EYES OF FISHES: STRUCTURE AND ULTRASTRUCTURE

J.A.C. NICOL

Marine Science Institute, University of Texas

Port Aransas, Texas 78373, U.S.A.

One of the prettiest touches of all and
that which angled for mine eyes, caught
the water though not the fish...
The Winter's Tale, Act V, Sc. 2

The outer part of the retina is bounded by the pigment epithelium derived from the outer wall of the optic cup. The epithelium is usually pigmented, containing melanin. Its basal membrane forms the innermost lamella of the glassy or Bruch's membrane, which also contains a thicker, loose fibrous layer, and an outer thin lamella which is the basal lamina of the choriocapillaris. There is usually an intricate meshwork of narrow spaces between the plasmalemma and the glassy membrane, spaces which are formed by foldings of the basal plasmalemma and which extend as clefts between the bases of adjacent cells as far as the desmosomes or tight junctions. The extent to which the basal plasmalemma is infolded varies greatly; it is, for example, greatly folded in stingrays. Walls of the endothelial cells of the choriocapillaris facing the pigment epithelium are pierced by capillary pores (fenestrated endothelial cells) (Arnott et al., 1972).

THE CHORIOID

The chorioid has a basis of collagenous fibres and contains a variable population of melanocytes, which are especially numerous on the upper or dorsal surface of the eyeball. Its outer region contains a stratum argentea, reduced or failing dorsally. Owing to the translucent character of the sclera, pigment and argentea are often visible to the exterior. Chorioids of *Amia* and many teleosts contain one or two retia mirabilia, the chorioidal and lentiform glands, which are interposed in the vascular supply to the choriocapillaris and the falciform body. The efferent supply to the chorioidal gland comes from the pseudobranch (see Beatty, this volume), which appears to be concerned with salt exchange; the lentiform body is supplied by the retinal artery (Barnett, 1951). Retia in fishes are organs of oxygen secretion, and the chorioidal rete seems actively to secrete oxygen in the fish's eye. Investigations have demonstrated the existence of high oxygen tensions in the vitreous humour of many species; tensions ranged from 250 to 1000 mm Hg or more (about 1.5 to 5 times that in the blood) in benthic and pelagic species, all of which had well-developed chorioidal glands. Fishes lacking or having but small retia had low levels of oxygen in the vitreous (Wittenberg and Wittenberg, 1962, 1974; Wittenberg and Haedrich, 1974; Randall, 1970).

There is a tapetum lucidum in the inner chorioid (immediately within the choriocapillaris) of some species. Additionally, the suprachorioid is sometimes the site of fat storage and contains substantial pockets of adipose cells. A peculiar feature of chimaerids is the presence of a large blood gland within the orbit; its sinusoidal spaces are loaded with eosinophiles.

The stratum argentea is a layer of closely packed crystal-containing cells, organised in depth. Cells contain, as their main cytoplasmic feature, many stacks of thin platelets; within each stack the platelets are regularly arranged above each in parallel; however, the orientations of the stacks with reference to the scleral surface vary greatly. Crystals of the argentea cells are raphidial hexagons; the system consists of regularly spaced quarter-wavelength films, to which reference will be made later; and light-reflexion is the result of constructive interference. Because of the variable orientation of the platelet-stacks, it is to be expected that reflexion would be diffuse.

It is generally believed that the stratum argentea is of functional importance in the larval fish during pelagic life. Like the argentea which lines the abdominal cavity it forms an opaque refulgent covering for a dark body, thereby rendering the fish less conspicuous. In the adult, Walls (1942) suggested that it may merely be residual and concealed. However, when the eyeball rotates in its socket, some of the outer surface covered by argentea is transiently revealed, and the skin over the orbit is often sufficiently translucent to allow the argentea to show through. It has been suggested that the argentea of the chorioid, like that of the swim bladder, may be a means of drastically reducing tissue permeability to gases (Denton et al., 1972).

Ocular melanins are sometimes distinguished from skin melanins (Lillie, 1965). Albinism involves the skin and ocular pigment (albino *Corydoras*); but there are white varieties of fish which have melanin in the eye (golden severan *Cichlasoma severum*). Other substances found in the chorioid and pigment epithelium are: zinc, 15.1 and 23.8 mg g^{-1} dry weight, trout and perch; riboflavin 0.9 mg g^{-1} dry weight, gar; pteridines, isoxanthopterin, sepiapterin, ichthyopterin, 2-amino-6-hydroxypteridine-8carboxylic acid; and thiamine, 0.053 and 0.276 mg g^{-1} dry weight, *Decapterus* and *Leiognathus*. The last is perhaps more concentrated in the retina (Fukuda and Hirami, 1955; Pirie and Van Heyningen, 1956; Christomanos and Pavlopulu, 1967; Nicol and Van Baalen, 1968; Nicol and Arnott, 1972). Several fishes have bright yellow chorioids, e.g. rock cod *Ruboralge* and toadfish *Opsanus*. The pigment, yellow and fluorescent, is intermingled with the stratum argentea; although lying behind the pigment epithelium, it suffices to cause eye shine (Nicol, Arnott and Best, 1973).

THE PIGMENT EPITHELIAL CELL

The pigment epithelium, a single sheet of cells, has a surprising variety of roles and produces a notable variety of substances. Adjacent cells near their bases are connected by junctional complexes consisting of three components. Distally, adjacent cells are closely apposed but are separated at intervals by narrow gaps; this region is the gap junction. Inside the gap junction the cells are encircled by prominent zonulae adherentes. Around each zonula the cytoplasm contains a circumferential band of fine filaments. The extracellular space of the junctional complex is occasionally obliterated by local fusion of apposed membranes, forming the zonula occludens. This complex, in conjunction with the folding of the basal plasmalemma, points to a role in active transport. The cells of the pigment epithelium are electrically coupled; the high resistance across the epithelium and the blood-retina barrier encountered there are linked with the sealing of intercellular spaces by zonulae occludentes. When a retinal circulation is wanting, as in selachians, most exchange and retinal nourishment must occur across the outer membrane. Evidence for enhanced activity is afforded by the concentration there of cell mitochondria (Moyer, 1969; Arnott et al., 1972; Hudspeth and Lee, 1973).

The pigment epithelium is concerned with the storage of vitamin A and the regeneration of rhodopsin (Baumann, 1972). The cells contain abundant smooth ER as tubules, phagosomes which are engaged in dissolution of lamellae of rod outer segments as part of the rod outer segment renewal process, and lipid vesicles. In some species there are so-called myeloid bodies, which are lozenge-shaped stacks of smooth lamellae located in the basal region of the cell and connected with the smooth ER and the nuclear membrane, e.g. cusk-eels;

of unknown function, they are, perhaps, concerned with vitamin A or lipid metabolism. The wealth of dictyosomes in pigment epithelial cells of the gar is noteworthy. We may notice, also, the unusual variety of included bodies and organelles encountered in the epithelia of fishes: melanosomes, crystals of purine and pteridine; spherules of lipid and melanoid substances. Finally, there are the curious red cylinders or micellae found in the cell processes of wrasses (Nicol and Van Baalen, 1968; Moyer, 1969; Young and Bok, 1969; Zyznar and Nicol, 1973; Nicol and Arnott, 1973; Nicol, Arnot and Best, 1973; Arnott et al., 1972, 1974; Fineran and Nicol, 1974).

Retinal pigment (melanin) is generally absent from the pigment epithelium of selachians, at least over the tapetal region. In many bony fishes it is migratory, advancing from the cell base into the cell process when the eye is illuminated; in some others (e.g. wrasses) it is confined to the basal region of the cell. It is not unusual to find two kinds of melanosomes present, spherical and rod-shaped (or bacillary), the latter lying more distally (Wunder, 1925 & 1926). When a retinal tapetum is present, the expanded pigment frequently occludes the reflector by covering the reflecting particles. Finally, in at least two fishes, the retinal pigment is confined to the cell processes and is static, viz. in the sturgeon and the bigeye; and in the latter it is limited to the ventro-temporal fundus (Nicol, 1969; Nicol and Zyznar, 1973). These fishes have chorioidal tapeta lucida and light is channeled down the photoreceptors between the pigmented processes (Figs. 1, 2). By this arrangement the latter can absorb reflected light which deviates grossly from the axis of the incident rays.

PHOTORECEPTOR LAYER

All or nearly all fishes have rods; they may be large and arranged in a single row, e.g. catfishes; or small, exceedingly numerous and distributed in depth, e.g. plaice (O'Connell, 1963; Nicol, 1965b). Most deep-sea fishes have pure rod retinae, or retinae with only sparse cones; but one deep-sea species has been reported that has an almost pure cone retina, viz. *Omosudis* (Munk, 1965; Locket, 1971b). It was once believed that selachians had only rods, but cones have now been distinguished in sundry species; indeed they are distinct and abundant in eyes of stingrays (Gruber, this volume; Gruber et al., 1963; Stell, 1972).

In neritic species when rods are small and numerous they are generally staggered irregularly through the width of the photoreceptor layer and are reached by filamentous myoids of varying lengths. A few coastal teleosts have rods arranged in regular rows, the ellipsoids being in alignment, e.g. *Priacanthus*. The latter animals, it may be observed, do not exhibit retinomotor activity. Most deep-sea

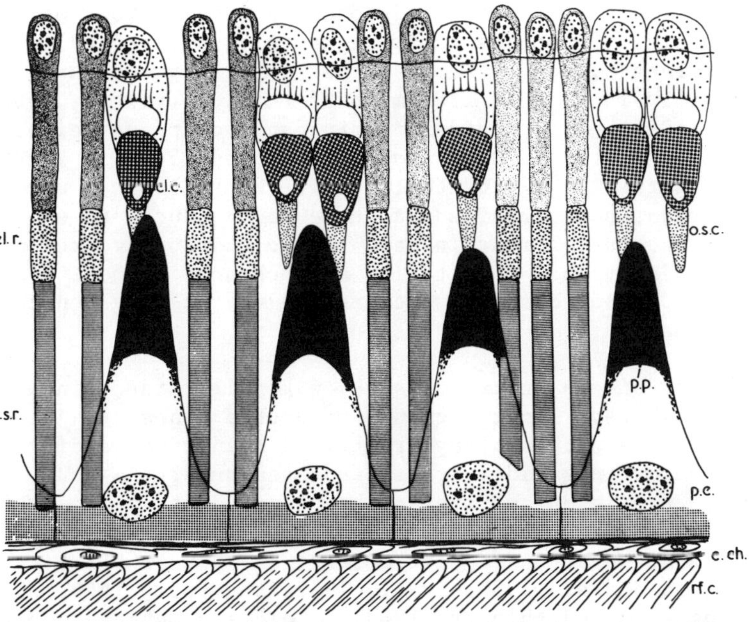

Fig. 1. Retina of the sturgeon. Legend: c.ch., choriocapillaris; el.c., el.r., ellipsoids of cone and rod; o.s.c., outer segment of cone; o.s.r., outer segment of rod; p.e., pigment epithelium cell; p.p., process of same; rf.c., reflecting cell. X 740

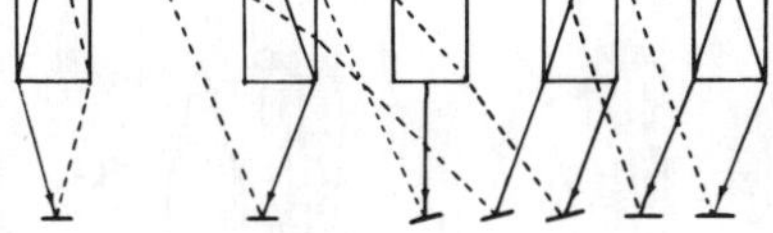

Fig. 2. Tapetum lucidum of the sturgeon. Diagram intended to illustrate the path of light rays channelled by the rods and reflected back by the platelets. Extreme cases are shown of rays emerging in the central axis of the rods and from the periphery of the rods, and reflected from platelets lying normal to the axis of the rods and tilted at 12° to the axis. R, rod; Pp, pigmented process of pigment epithelium cell. (From Nicol, 1969, modified.)

teleosts have pure rod retinae disposed in regular rows or tiers, from two to six. Sometimes outer segments of superposed rows differ markedly in length, e.g. *Diretmus*. Rod outer segments are sometimes extraordinarily long, e.g. 150 µm in *Sternoptyx* (Vilter, 1953; Munk, 1965, 1966a, b). Retinal densities in deep-sea teleosts are often very high, e.g. 1.08 in *Argyropelecus* and 1.21 in *Searsia* (Denton and Warren, 1957; Denton, 1959). Munk (1966a), after considering the various theories that have been adduced to explain the regular arrangement of rods in deep-sea teleosts, favoured one according to which visual sensitivity is enchanced without loss of acuity when superposed rods are connected to one ganglion cell in vertical summation.

Cones are single in many fishes, e.g. siluroids, single and multiple in others. Inner segments of double cones are joined; when the two members are equal they are termed twin cones. Single and twin cones occur in gadids; triple and even quadruple cones in flatfishes. Cyprinids and wrasses have, in addition to single cones, short and long single cones (Fig. 3). Frequently, the cones but not the rods are arranged in regular mosaics in part or all of the eye (Eigenmann and Shafer, 1900; Lyall, 1957; Engström, 1960, 1961, 1963a,b; Engström and Ahlbert, 1963). Goldfish retinae contain three types of cones, each having a single kind of visual pigment with absorption maxima at 455, 530 and 625 nm. Among the twin cones red-green and blue-green pairs were observed (Marks, 1965; Harosi, this volume). The role of oil droplets as short wavelength cut-off filters in turtle retinae has been demonstrated by Baylor and Hodgkin (1973).

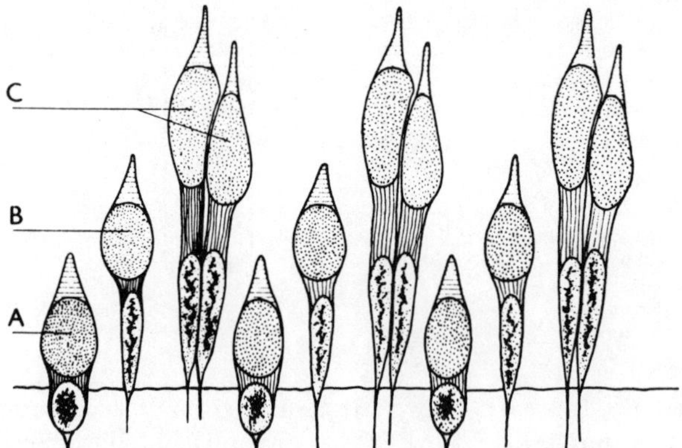

Fig. 3. Diagram of cone types of the roach *Leuciscus rutilus*. short single; B, long single; C, unequal double cone (From Engström, 1960).

Oil droplets are reported in a few fishes: colourless ones in sturgeon, coloured ones in cyprinodonts, and coloured ones, red and yellow, in the South American lungfish (Berger, 1966; Bowein and Hollenberg, 1973; Nicol, Arnott and Best, 1973 and Ali and Anctil, 1973). The peculiar micelles in processes of the pigment epithelial cells of wrasses contain a red pigment that cuts off short wavelengths (< 600 nm), and they seem to serve as a colour filter for long single cones, and as a screen for the rods when elongated (Engström, 1963a; Fineran and Nicol, 1974).

A peculiar accessory outer segment (which, perhaps, could be better termed an accessory sac) is found on some photoreceptors: examples are found on single cones of catfish, rods and cones of gadids (Engström, 1961; Arnott et al., 1974). It extends from the outer portion of the middle segment and lies alongside the main outer segment. In the catfish retina both accessory sac and main outer segment arise from the ciliary shaft which is subtended by a basal body and a second centriole; the former is filled with fine filamentous material. Also, cones are sometimes invested by a grille of processes which extend upwards from the ellipsoid to invest the outer segment like a calyx. There are about 20 of them around a catfish cone. Folds on the myoids, especially of double cones, form extensive areas of contact with the microvillous processes of Müller cells.

TAPETA LUCIDA

The two types of tapeta lucida of fishes are the chorioidal and the retinal, which are specular and diffuse reflectors, respectively.

Elasmobranch tapeta lucida: A chorioidal tapetum lucidum occurs in most selachians and in chimaerids. It is a nearly continuous layer, several cells deep, of flattened oval cells. The orientation of the innermost cells changes in a regular manner from the centre of the eye to the periphery. Whereas in the central fundus they lie parallel to the retina, proceeding away from the centre they become increasingly oblique until, approaching the ora terminalis, they are steeply inclined (Fig. 4) (Denton and Nicol, 1964).

The thickness of a reflecting cell is about one fifth the length, the nucleus lies centrally, and the cytoplasm is replete with more or less regularly arranged flat crystals which lie parallel to each other and to the long axis of the cell. Transverse to its flat surfaces a cell contains about 5 to 15 superposed crystals. There are many crystals in tandem along the length of the cell, and they are staggered and overlap with each other in a random manner. Each crystal is closely invested by a thin membrane forming a crystal

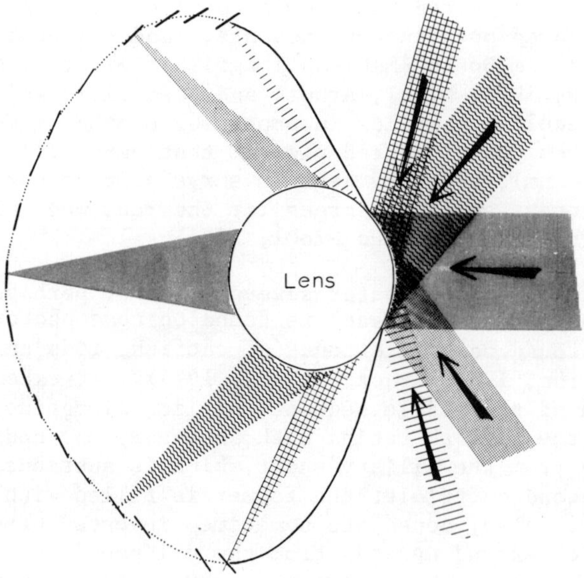

Fig. 4. Orientation of reflecting platelets in the eye of the blue shark, and the direction of light rays which can reach them. Vertical section. (After Denton and Nicol, 1964.)

chamber, which is enclosed in turn within a crystal sac delimited by a rather undulatory unit membrane (Best and Nicol, 1967).

CHORIOIDAL TAPETA OF OTHER FISHES

Chorioidal tapeta also occur in several groups of actinopterygians, viz. bichirs (*Polypterus*), sturgeons (*Acipenser*), glass-eyes (*Priacanthus*) and coelacanths (*Latimeria*) (Walls, 1942; Rochon-Duvigneaud, 1943, 1958; Lenoble and Le Grand, 1954; Denton and Nicol, 1965; Nicol, 1969; Nicol and Zyznar, 1973; Locket, 1974). Like that of elasmobranchs, the tapetum of the sturgeon also is composed of a layer of closely packed reflecting cells, up to ten in depth (Fig. 5). The cells change orientation from centre to periphery of the eye, from being parallel to the retina to being highly inclined (Fig. 6). A rare instance among teleosts of a chorioidal tapetum is found in big eyes or glass eyes (*Priacanthus*). It is a specular reflector organised like that of the selachian system: cells are stacked about four deep, and towards the periphery of the eye they become steeply tilted. Each cell contains one or two stacks of regularly arranged thin crystals lying parallel to each other. The crystals, although hexagons like those of the sturgeon, are peculiar in this respect: they have thick centres and become thin around the periphery.

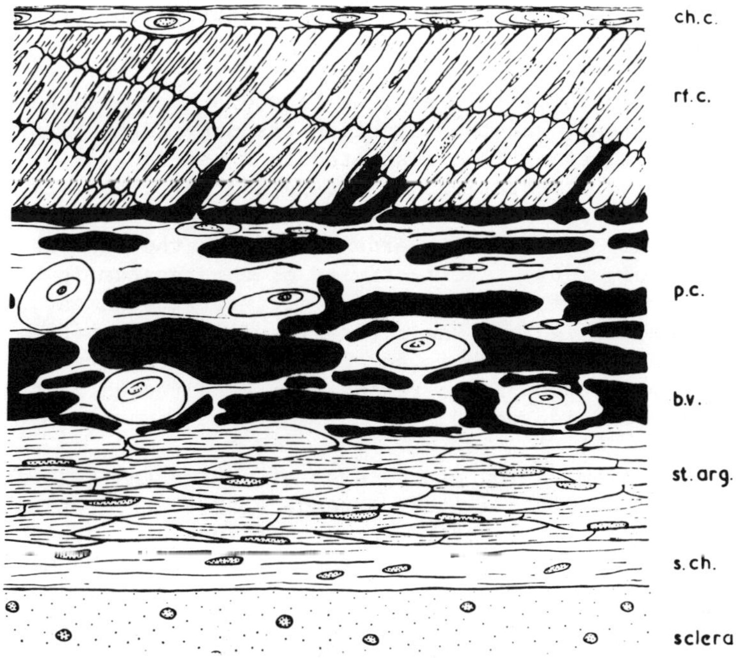

Fig. 5. Semidiagrammatic representation of a section through the chorioid of the sturgeon, towards the periphery of the eye. X 360. Legend: b.v., blood vessels among melanocytes, p.c.; ch.c., choriocapillaris; rf.c., oriented reflecting cells of the tapetum; st. arg., stratum argentea; s.ch., suprachorioid.

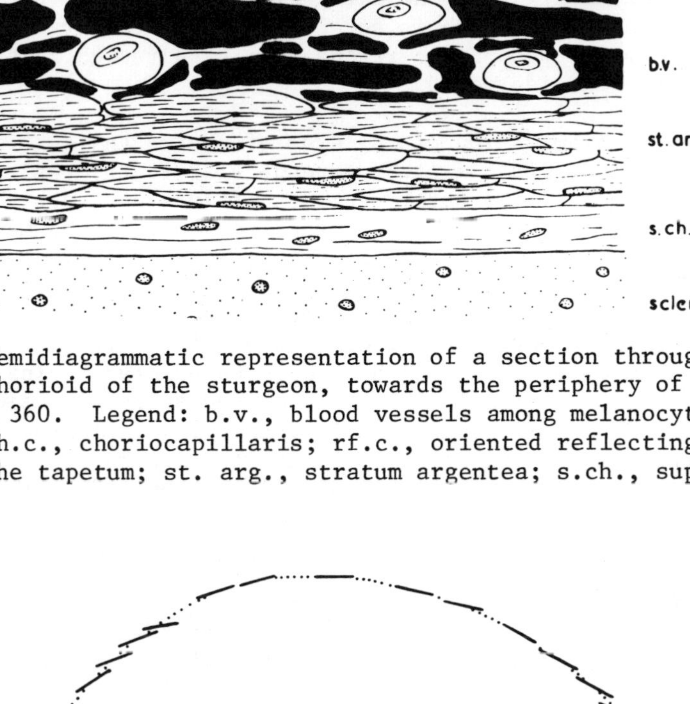

Fig. 6. Orientation of reflecting cells in the eye of the sturgeon. Vertical section (from Nicol, 1969).

THE SIGNIFICANCE OF CHANGING ORIENTATION OF REFLECTING CELLS IN THE CHORIOIDAL TAPETUM

Refraction in the fish's eye, we know, is accomplished almost entirely by the lens, and the direction taken by light rays from the cornea to the retina is illustrated in a section of a shark's eye (Fig. 4). The orientations of the reflecting platelets, parallel in the central and highly inclined in the peripheral retina, are such that they lie roughly perpendicular to the incident light (Denton and Nicol, 1964). Reflectivity of the tapetum is highly specular and spread of light is $\simeq 5^{o}$. If, for the moment, we neglect refraction of light in the retina it will be seen (Fig. 4) that most of the light which is not absorbed or scattered during its double passage through the retina will be directed back through the lens to the outside world.

Histologic examination of the eyes of sharks and dogfishes shows that the rods are oriented perpendicular to the retinal surface in all regions. The outer segments of rods have a higher refractive index (1.41, Sidman, 1957) than the surrounding medium (1.336, Le Grand, 1952), and the light which enters a rod close to its optical axis is totally internally reflected; additionally, some channelling effect probably takes place in the inner segment (Miller and Snyder, 1972). Internal reflectivities of the rod outer segment for various angles of incidence have been calculated, using Fresnel's equations (Denton and Nicol, 1964). When the angle of incidence is greater than 70° the light is totally internally reflected, whereas if it is less than 65° it is almost completely refracted through a very small angle. It has been calculated that over 80% of the light entering the rod will strike its internal surface at angles between 20° and 30°.

The focal length of the crystalline lens, viz. 2.6, is given by Matthiessen's ratio, focal length/radius of the lens (Pumphrey, 1961), and the angle that a lens having this focal length makes with the retina is about 45°. In the centre of the eye the maximal angle between light striking the rods and the rod axis is 22°. These approximations lead to the conclusion that in the centre of a shark's eye most of the light entering the rod outer segments is channelled along them, and that at the periphery of the eye the light passes almost directly through the layer of rods with no appreciable change of direction. In a small region of the eye the effect is complicated: light is in part reflected and channelled through the rods, and in part passes through them (Denton and Nicol, 1964). Now, although there is little doubt that the rods of dogfishes and sharks lie perpendicular to the retinal surface, it is by no means certain that this is the invariable arrangement, and an examination of the alignment of photoreceptors at the periphery of eyes of benthic fishes, such as rays, is desirable.

METHODS OF MEASURING SPECTRUM REFLECXION

Real values of spectrum reflexion of tissues are required in order to understand and evaluate the performance of reflecting systems. Limiting ourselves to tapeta lucida we find that five methods have on occasion been utilised.

One. A diffusing opal is evenly illuminated and a piece of tissue is placed opposite and approximately perpendicular to it. The opal scatters evenly according to the cosine law, and its surface looks equally bright from whatever direction it is viewed. When the tissue contains platelets inclined at various angles to its surface, each platelet reflects light from a different part of the opal. Light reflected from the preparation is collected by a microscope optically coupled to a photomultiplier. The microscope objective is move tow and from the opal and the tissue, and reflectivity of the tissue is determined as a fraction of light scattered from the opal to the objective. Nominal monochromatic light is provided by coloured or interference filters (Denton and Nicol, 1962, 1964).

Two. The tissue is illuminated from the circular mirror of a Leitz Ultropak attachment, and light reflected from the preparation and collected by the microscope objective is measured with a microphotometric attachment (Leitz MPV). Light is provided by a monochromator.

Three. The tissue is mounted in an integrating sphere and is illuminated with a small circle of light from a scanning monochromator. The tissue rests on a movable base, and its position is altered until its reflexion is cast upon the inner wall of the sphere a short distance from the entrance port of the light beam. Light is collected by a photomultiplier shielded by a screen from light directly reflected from the tissue, and monochromatic light is provided by a scanning monochromator.

Four. A coverslip, inclined at an angle of 23.5^{o} to the horizontal, casts light from a horizontal beam upon the specimen. Light reflected from the specimen enters a microscope, and the image is focused on a photoelectric sensor. Interference filters are used to give monochromatic light, and parallel measurements are made of some reference standard (Land, 1966).

Five. A piece of tissue is mounted on a freshly-prepared surface of magnesium oxide over which a cover glass is placed. It is illuminated at 45° by a collimated beam passing through an interference filter, and light is collected at 45° to the normal to the preparation by a microscope objective and detected by a photomultiplier. Reflectance is calculated from consecutive measurements made of the tissue and the standard (Dartnall et al., 1965).

SPECTRUM REFLEXION

A fresh tapetum of a dogfish or shark appears bright and silvery and it has, when viewed obliquely, a slight bluish tint. Reflexion occurs throughout the visible spectrum, and the reflectivity curves show broad bands centred in the blue or blue-green or, in the case of the stingaree, the yellow-green (Fig. 7), i.e. the reflected light is highly unsaturated (Denton and Nicol, 1964; Nicol, 1965a). Reflexion measured in an integrating sphere shows steeper curves and narrower spectrum bands (Fig. 7). The divergent results can perhaps be explained by differences in methods. When a specular reflector is illuminated by a large diffusing opal, it receives light through a large solid angle, and the reflected light will be shifted towards the blue as the angle of incidence increases. In the integrating sphere, however, the incident light is restricted to a narrow beam

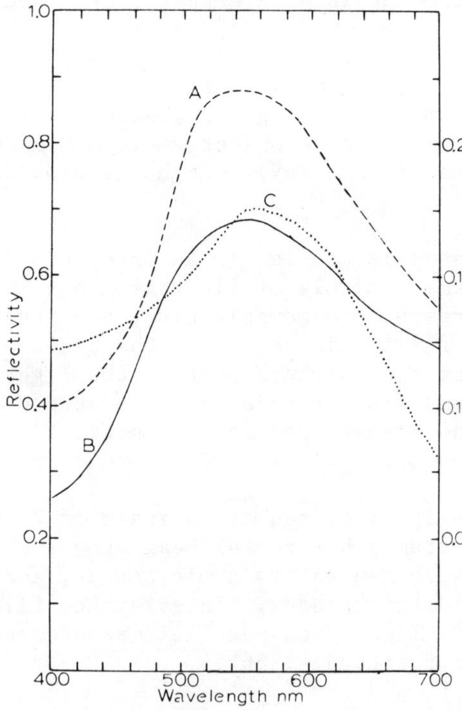

Fig. 7. Reflexion from the tapetum lucidum of the stingaree *Dasyatis sabina*. Curves: A, measured in an integrating sphere; B, with Leitz Ultropak microscope; mean of 10 measurements; C, unilateral illumination from a diffusing opal (ordinate to the right, fraction of front surface mirror, for curve B).

close to the normal to the tissue. Reflexions of tapeta of the spur dog and the stingaree at λ_{max} are > 0.85 (Fig. 7).

Oriented reflecting cells containing stacks of films are of widespread occurrence in the animal kingdom. Besides chorioidal tapeta lucida I have instanced the stratum argentea of the chorioid, and we encounter analogous systems in the silvery skin of fishes (Denton and Nicol, 1966; Arnott and Nicol, 1970; Denton and Land, 1971), and the reflector of photophores (Strum, 1969; Denton, 1970). The ultrastructure and reflecting mechanism have been worked out most fully in the eye of the scallop (Land, 1965, 1966). This reflector consists of 30 to 40 layers of thin guanine crystals, refractive index 1.80, separated by thin layers of low refractive index, and both layers have optical thicknesses of approximately one-quarter wavelength. Spacing and thicknesses cause constructive interference of rays of blue green light reflected from the surfaces of the films. A regular multilayered system containing many films (Fig. 8) can achieve very high reflectivity in a restricted spectrum band; in the case of the scallop's argentea there is a broad band centred at about 500-550 nm.

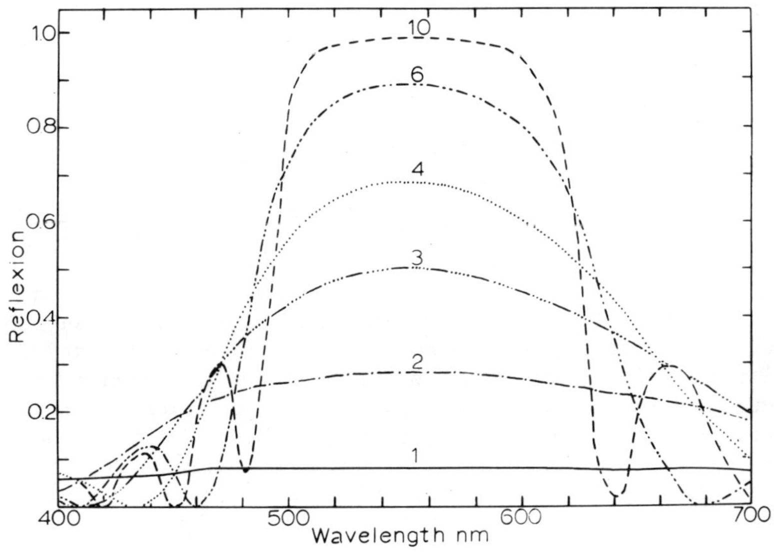

Fig. 8. Reflexion from one to nineteen $\frac{1}{4}$ wavelength films (1 to 10 thin crystals) having thicknesses set to achieve maximal reflexion at 550 nm.

Very regular systems that lend themselves to analysis are found in the skin of the ratfish (*Hydrolagus*) and the tapetum of the big eye (*Priacanthus*) (Arnott and Nicol, 1970; Nicol and Zyznar, 1973). In both animals, the cells contain one or two stacks of thin crystals disposed with remarkable regularity, and their characteristics serve to promote constructure interference of orange light centred at about 580-620 nm. High reflectivity in such a system is achieved not only by reflexion from the films in a stack, but also by reflexion from the stacks of superposed cells. Operating on the same principle but of greater complexity is the tapetum of selachians. In the reflecting cells the spacing is more variable: the system can be analysed into one consisting of sets of 3 to 5 one-quarter wavelength films separated by wider intervals. High reflectivity results from the summation of reflexion from each set of films, both within one cell and within superposed cells (up to 60 films may be involved). Variable spacing causes the spectrum band to be widened; the surface is a mosaic of spots of unsaturated colours; and the overall effect of blue or green sheen in the resultant summation of the light of innumerable small coloured areas (Fig. 7) (Best and Nicol, 1967).

DIFFUSE REFLECTORS OF TELEOSTS

Retinal tapeta of teleosts reflect well across the visible spectrum, e.g. the lipid tapeta of weakfishes; or maximally in particular spectral regions, e.g. the orange tapeta of gars. Some values for diffuse reflectance are: sand trout 0.52 at 500 nm; sea catfish 0.62 > 620 nm; gar 0.75 > 650 nm (Fig. 9) (Arnott et al., 1972, 1974; Wang and Nicol, 1974).

A variety of organic compounds - five have now been distinguished or described - are utilised as reflecting materials by teleosts (Zyznar and Nicol, 1973). Reflecting pigments of the gar and catfish have been referred to as melanoid pigments (Nicol, Arnott and Best, 1973) and my colleague Dr. Shosuke Ito has been successful in isolating, characterizing and synthesizing that of the catfish. He finds that it is an oligomer, most likely a tetramer, of 5,6-dihydroxyindole-2-carboxylic acid (Fig. 10), loosely bound to an adenosyl nucleoside; the monomers are probably joined by C-C bands at positions 4 and 7 (Ito and Nicol, 1974). Although the two components, the adenosyl nucleoside and the indole oligomer, are individually soluble in an aqueous medium, they form an insoluble precipitate at physiological pH (7.2), a property possibly not inconsistent with the inert condition of the pigment within tapetal spheres in vivo. This pigment probably is the tapetal reflecting material common to siluroids and many characins.

The physical operation of diffuse tapetal reflectors is understood in general, but further studies are needed for an understand-

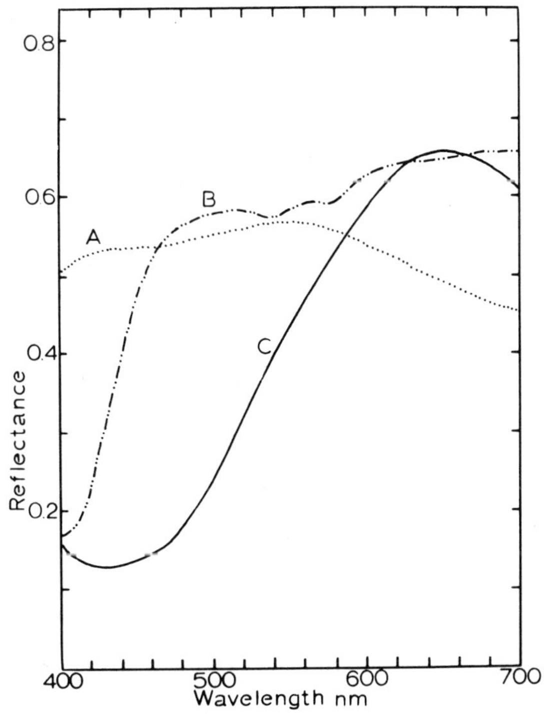

Fig. 9. Reflectance curves for the diffusely reflecting tapeta of some teleosts. A, golden croaker *Micropogon undulatus*; B B, sea catfish *Arius felis*; C, alligator gar *Lepisosteus spatula*.

ing of the performance of particular systems. Most contain small spheres acting as light-scatterers. We may glance at one type, however, which is made up of stacks of minute flat crystals. In the anchovy crystals are regularly arranged above one another, 5 to 14 deep. The characteristics and spacing are such that each stack constitutes a pile of one-quarter wavelength films, reflecting light by constructive interference. Although the surface of each tier might be expected to act as a specular reflector, the general impression, in macroscopic view, is of a matt white diffusing surface. A very interesting organisation of oriented reflecting platelets that surround groups of rods in the posterior retina of *Scopelarchus*, a deep-sea fish, has been described by Locket (1971a). The systems revealed in these fishes by observations on ultrastructure and reflectivity should make for very interesting research.

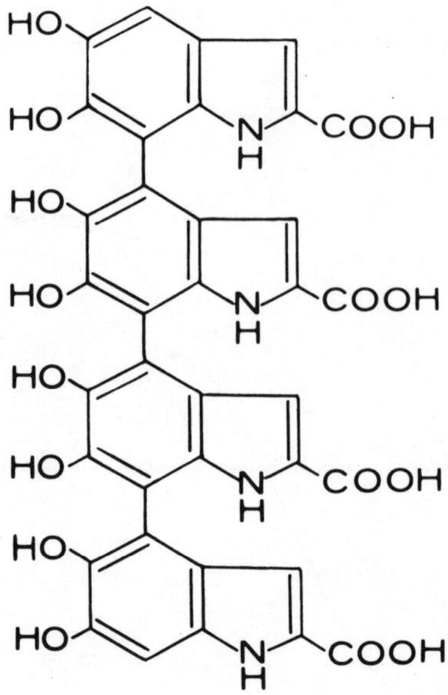

Fig. 10. Tapetal pigment of the sea catfish. The suggested formula is for the chromophore, tetramer of 5,6-dihydroxyindole-2-carboxylic acid.

TAPETAL PIGMENT AND PIGMENT MIGRATION

The pigment epithelium of a selachian is usually not pigmented and the chorioid outside the tapetum has a rich population of melanocytes. At the base of the tapetum is a single layer of stout melanophores whose processes extend inwards between the reflecting cells (Fig. 11). Franz (1905), on the basis of some observations which he made on a spur dog *Squalus acanthias* and a tope *Galeorhinus galeus*, came to the conclusion that the pigmented processes of these melanophores were mobile, and by moving inwards could conceal the tapetum when the eye was illuminated. This process of tapetal occlusion, although regarded as general among elasmobranchs (Franz, 1913; Walls, 1942; Rochon-Duvigneaud, 1943), does not occur in rough hounds *Scyliorhinus* and in rays *Raja* (Nicol, 1961; Denton and Nicol, 1964). When the eye is illuminated, pigment granules (melanosomes) move from the basal cell body into the cell processes, and in so doing they come to cover the reflecting surfaces.

The rate of pigment migration has been investigated by histologic, photographic and photometric methods. In one, the eye of a fish and a reference standard have been photographed simultaneously, densities of the pupillary area and of the standard on the film measured, and changes in relative intensities of light emerging

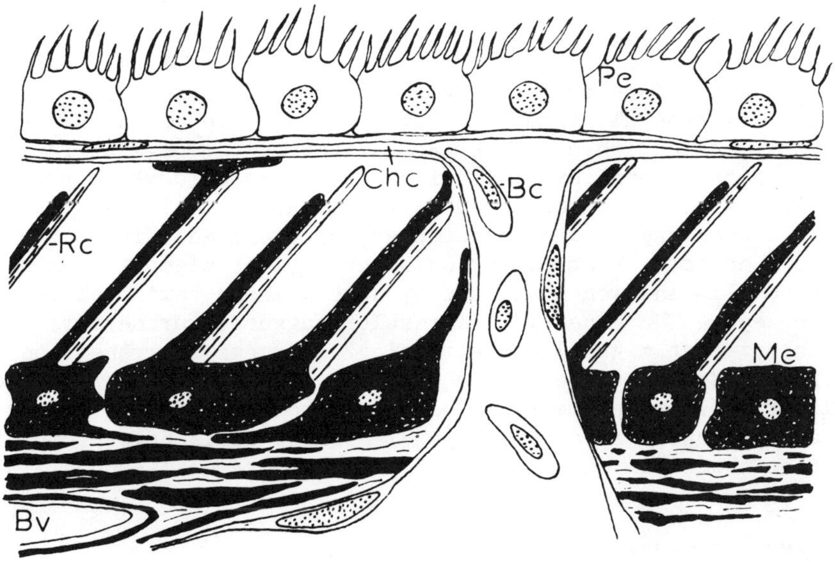

Fig. 11. Section through the tapetal region of the smoothhound *Mustelus levis*. Legend: Bc, blood cell, Bv, blood vessel; Chc, choriocapillaris; Me, melanophore; Pe, pigment epithelium; Rc, reflecting cell. (After Franz, 1913.)

from the pupil estimated (Kuchnow and Gilbert, 1967). In a second, surviving tapeta from which the retina had been removed were used, and reflexion was measured at successive intervals (Denton and Nicol, 1964; Nicol, 1965a). In a third, fish were sacrificed at selected intervals in light or darkness, the tapeta were fixed and light reflexion was measured; or fixed eyes were sectioned and a melanophore index was used to express the degree of pigment expansion (Nicol, 1965a).

A dark-adapted tapetum of a selachian has a reflectivity of 85% or more at λ_{max}, and a darkened one, 5% or less.

Pigment migration is a slow process. In the spur dog the tapetal pigment retreats from the surface of the tapetum in vitro completely in about $2\frac{1}{2}$ h, and occludes the tapetum in the same time; 80% of the tapetal surface is occluded within 1 h (Nicol, 1965a). In *Carcharhinus* the response in vivo occupies 1 h (Kuchnow, 1969).

The mechanism regulating pigment migration is unknown. Movement of pigment in one eye occurs independently of that in the other and normal movement of the pigment is not affected by severing nervous pathways, or by removal of the pineal and pituitary glands. The experiments indicate that the response, i.e. pigment migration, is not a reflex necessitating input of photosensory information into the brain since it occurs when optic, central and autonomic pathways are interrupted. Neither do the usual hormonal centres appear to be involved. The tapetal pigment migrates normally in excised eyes that are oxygenated: when the retina is removed the pigment remains extended (light-adapted eyes) or migrates outwards (dark-adapted eyes). It has also been observed that pigment migration sometimes takes place when the retina is detached and replaced (Nicol, 1965a). It seems, therefore, that migration of chorioidal pigment is a direct response to illumination, mediated by some process taking place within the eye. The experimental results suggest, furthermore, that for normal pigment migration (retraction) to occur on transition from light to darkness the retina must be present, and that the retina either inhibits pigment expansion or promotes pigment retraction. Nerve terminations seen on the tapetal pigment cells have prompted speculation about nervous control of pigment movement (Kuchnow,1970).

SIGNIFICANCE OF TAPETA LUCIDA

The optical densities of photosensitive pigments at λ_{max} in retinae of coastal selachians lie between 0.15 and 0.30 whereas in teleosts from the same waters that do not have tapeta the densities lie between 0.40 and 0.60. Comparable relations obtain when deep-sea selachians and teleosts are examined. When light reflected from the tapetum is taken into consideration the fraction of light which is absorbed is approximately the same in both kinds of fish. In a given environment, therefore, the reflector of the selachian achieves approximately the same absorption with about half as much photosensitive pigment.

Two suggestions have been made concerning the possible advantages of a reflecting tapetum and low density of visual pigment (Denton and Nicol, 1964). According to one, it is a device for improving the signal to noise ratio in the retina. Another starts from the premise that the regeneration of rhodopsin follows an exponential curve, and the increase of sensitivity during dark-adaptation depends on the absolute amount of rhodopsin unregenerated at a given time. The shark eye, compared with another containing twice as much rhodopsin, might regain its sensitivity more quickly (Denton and Nicol, 1964). However, several studies show that sensitivity is regained only slowly in the shark (Gruber, 1967).

LIGHT PENETRATION AND VISION

Following the invention of the barrier layer cell, Atkins and Poole (1933) made use of it to determine spectrum transmission of light in the English Channel. They discovered that green light was transmitted most effectively, blue somewhat less so, while red was quickly absorbed. Since then light transmission has been measured in many oceanic regions and Jerlov (1968) has presented a tabulation of water types. In clearest oceanic water (Jerlov's type I) blue λ_{max} 475 nm, penetrates furthest; transmissivity is 98% m^{-1} (extinction coefficient $\underline{k}$ 0.02). Clearest coastal water (Jerlov's type I) transmits maximally at 500 nm, 89% m^{-1} ($\underline{k}$ 0.12); turbid coastal water (Jerlov's type 9), at 575 nm, 56% m^{-1} ($\underline{k}$ 0.57).

At the sea surface the greatest intensity of daylight (sunlight plus skylight) which has passed through air mass 1 occurs at 490 nm. The sea acts as a monochromator, and in coastal areas with increasing depth maximum energy of submarine light shifts to 575 nm and beyond (Fig. 12). The displacement is due to scattering by particles, absorption by yellow substance (Gelbstoff) and absorption by chlorophyll (which has a peak at 675 nm) (Clarke, 1969).

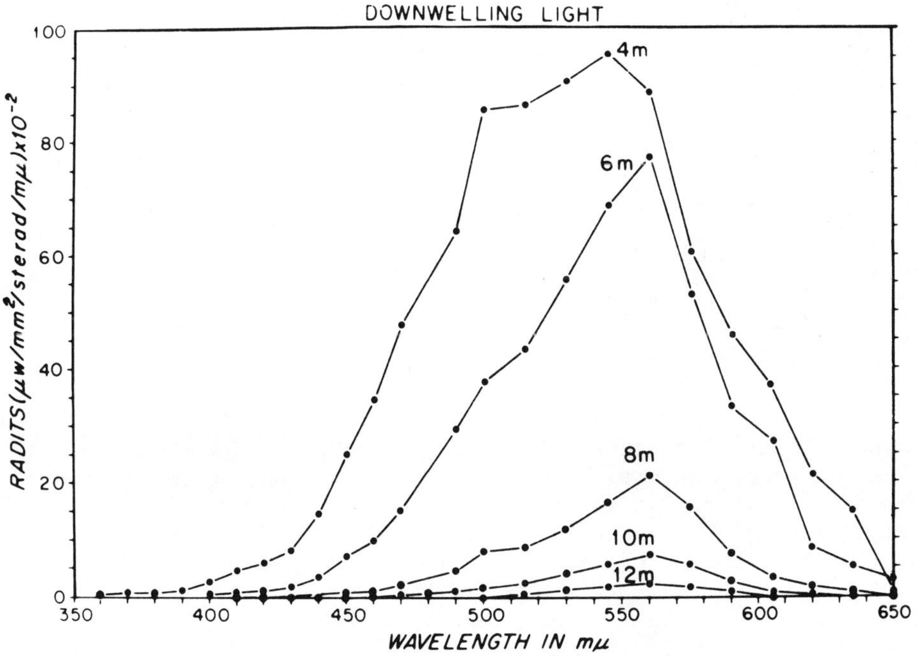

Fig. 12. Spectrum of down-welling light in Woods Hole Harbor (from Clarke, 1969, with permission).

The Gulf of Mexico has a wide continental shelf extending out to 100-200 km; and the northern Gulf, containing a large delta formation, receives the massive discharge of the Mississippi and many smaller rivers. River discharge and wind action, augmented by dredging activities, cause the inshore waters to be turbid most of the time. Jones et al. (1965) measured light transmission at a series of stations from nearshore to 46 km offshore. The instrument was a submarine photometer equipped with a barrier layer cell which has a spectrum response more or less corresponding to the photopic visibility curve. Their results show the great changes of light transmission which take place seasonally in this region because of meteorological and oceanographic factors. A series of spectrum irradiance curves obtained in this region (depth 17.6m) shows a λ_{max} at 585 nm (<u>k</u> 0.6).

A long chain of barrier islands extends around the northern Gulf, and the bays and estuaries behind them are shallow and generally moderately to exceedingly turbid. They receive the silt load of rivers, the turbidity is increased by wind and by man's activities, and the rich waters support crops of microorganisms which selectively absorb certain wavelengths. Transmissions range from 76 to 5.6% m^{-1} (<u>k</u> 0.28-2.9); and spectrum irradiance at several localities shows peaks at 575-580 nm (Lane, 1967, Arnott & al.,1974).

Data for fresh waters are more meagre, apart from Secchi disc readings. Crater lake, Oregon, noted for its clarity, has a transmission of 94% m^{-1} (<u>k</u> 0.06). Fairly clear lakes and rivers transmit green light preferentially, 74-9% m^{-1} (Clarke, 1954; Tyler and Smith, 1967). South Texas rivers can become very turbid during spate and, as a consequence of the heavy mud load, transmission shifts to long wavelengths (>575 nm) and can drop to <0.001% m^{-1}. Again, natural waters draining bogs and swamps are stained brown by decaying organic matter and minerals, short wavelengths are rapidly absorbed, and at a short distance below the surface only orange and red light are present (Birge and Juday, 1931; Clarke, 1939).

Because of forward scattering of background light into the path from the target, the distance at which objects can be seen is much reduced in turbid waters. Average underwater visibility in clear streams is 9 m, and as low as 0.5 m in turbid reservoirs and rivers (Carlander et al., 1963). It has been observed that some fishes living in very turbid waters have cloudy corneas. The conditions limiting visual acuity and sensitivity must differ greatly over the seasons, from one stream to another, and from species to species, and evaluation from available data is difficult. Nevertheless, we may suspect that image degradation or loss of contrast may often be the limiting factor for visual resolution in turbid water, and not retinal inadequacy, when visibility is reduced to very short distances, sometimes of a few decimetres. Vision, of course, is only one of the senses employed by fishes in their daily round. They utilise chemical and auditory information, and input from the lateral line. For example, marine catfish at night by calling to each other are able to maintain schools (Tavolga, 1960); and electric fish *Gymnarchus*, that live in turbid waters, emit electric pulses that inform them of changes of electric fields in their vicinity and therefore help them to orientate (Lissmann, 1961). We should remember that the visual sense is one in a broad spectrum of sensory modalites that are severally and conjointly utilised at different times and for different activities.

Estimates have been made from time to time of the depths at which fish can see (Clarke, 1936). A fish that has the same sensitivity as man can just detect background light of 10^{-8} μW cm^{-2}, and at an intensity of one order of magnitude greater it would begin to distinguish objects against background. In moderately turbid waters (k 0.5) scotopic vision by day would be possible down to 50 m and by night (starlight) down to 13 m. In very turbid rivers (k 2) vision would fail at 15 m during the day and 4 m at night. Fishes remaining at the surface or frequenting the shallows of such rivers experience some amazing changes of light intensities comparable to those encountered by researchers who stumble with red torches in and out of dark rooms. Many fishes living in turbid or stained waters have tapeta that selectively reflect long wavelengths, e.g. cats, and spectrum sensitivity at the red end of the spectrum is increased (Fig. 13) (Wang and Nicol, 1974).

The rivers of the southern United States, draining across the coastal plains, are generally turbid. In the Mississippi River, according to Higgins (1932), 90% of the light is absorbed in the first 25 mm of water, and in the Missouri River at Boonville, the light is reduced to 10^{-6} of surface intensity in 84 mm (Ellis, 1936). Of the mighty river Mark Twain writes, "The river annually empties four hundred and six million tons of mud into the Gulf of Mexico... This mud, solidified, would make a mass a mile square and forty-one feet high". And he goes on to explain --

"Here was a thing which had not changed; a score of years had not affected this water's mullato complexion in the least, a score of centuries would succeed no better, perhaps. It comes out of the turbulent, band-carving Missouri, and every tumblerful of it holds nearly an acre of land in solution. I got this fact from the bishop of the diocese. If you will let your glass stand half an hour, you can separate the land from the water as easy as genesis; and then you will find them both good: the one good to eat, the other good to drink. The land is very nourishing, the water is thoroughly wholesome. The one appeases hunger; the other thirst. But the natives do not take them separately, but together, as Nature mixed

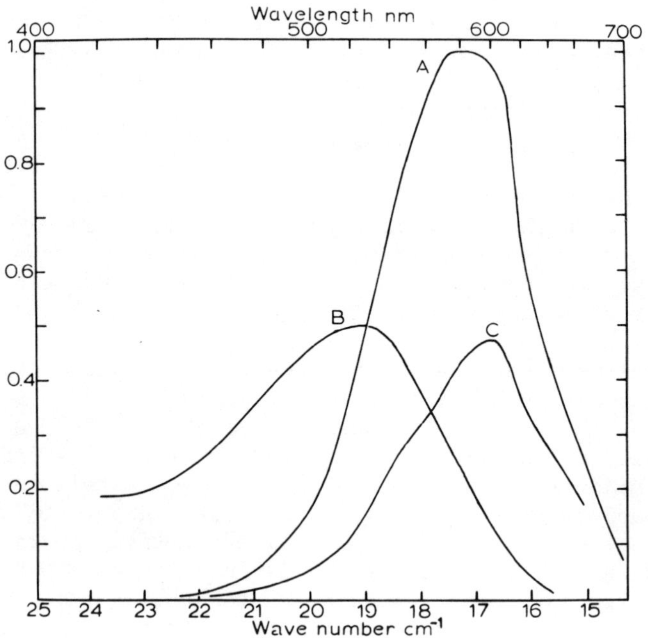

Fig. 13. Spectrum curves to show: A, submarine irradiance; B, absorption of visual pigment; and C, reflexion of light transmitted through the retina by the tapetum of the sea catfish *Arius felis*.

them. When they find an inch of mud in the bottom of a glass, they stir it up, and then take the draught as they would gruel. It is difficult for a stranger to get used to this batter, but once used to it he will prefer it to water. This is really the case. It is good for steamboating, and good to drink; but it is worthless for all other purposes, except baptizing." (Life on the Mississippi).

That the rivers have become much more turbid in recent historical times seems indisputable. The introduction of intensive agriculture by white settlers increased run-off and loss of topsoil (Miller, 1961; Cole, 1963); historical evidence that streams were clearer, and the composition of indigenous fish species different in early colonial days, has been adduced by Teal and Teal (1969). It is well known to ichthyologists that some fishes are far more tolerant than others to turbid conditions, e.g. white crappie and catfish (Coker, 1954). The determining factors are obscure, they involve higher temperatures, respiratory demand, change of food supply, and even vision dependent upon water clarity. It has been suggested that the river systems of the American southwest have been relatively stable during the post-Pleistocene until the arrival of settlers from the United States (Martin et al., 1961; Miller, 1961).

However, it is reasonable to suppose that the American Indian, when creating his civilisations on the basis of intensive agriculture, had already altered the condition of streams centuries before the Conquistadores arrived on the scene; they had engendered arroyos and, like the Pueblo people, having exhausted the land, had perforce been compelled to move to fresh fields. The rivers, as we know them, bear the imprint of man's activities, and the fishes they contain are the ones that are suited to, or were pre-adapted to, such conditions. It is not given to us to know what were the original or natural conditions of those waters, and of the conditions of underwater vision enjoyed by their original occupants. Perhaps, one of the last great river systems of the world where such studies can still be carried out is the Amazon basin, and that, too, is about to experience a great change ordained by man.

ACKNOWLEDGEMENT

Some of the research reported in this communication was supported by the National Eye Institute of the U.S. Public Health Service.

REFERENCES

Ali, M.A. and Anctil, M. (1973). Retina of the South American lungfish, *Lepidosiren paradoxa* Fitzinger. Can. J. Zool. 51: 969-972.

Arnott, H.J. and Nicol, J.A.C. (1970). Reflection of ratfish skin (*Hydrolagus coliei*). Can. J. Zool. 48: 137-151.

Arnott, H.J., Nicol, J.A.C. and Querfeld, C.W. (1972). Tapeta lucida in the eyes of the seatrout (*Sciaenidae*). Proc. R. Soc. Lond. Ser. B.180: 247-271.

Arnott, H.J., Best, A.C.G., Ito, S. and Nicol, J.A.C. (1974). Studies on the eyes of catfishes with special reference to the tapetum lucidum. Proc. Roy. Soc. Lond. Ser. B.186: 13-36.

Atkins, W.R.G. and Poole, H.H. (1933). The photo-electric measurement of penetration of light of various wavelengths into the sea and the physiological bearing of the results. Phil. Trans. Roy. Soc. Lond. Ser. B. 222: 129-164.

Barnett, C.H. (1951). The structure and function of the choroidal gland of the teleostean fish. J. Anat. 85: 113-119.

Baumann, C. (1972). The regeneration and renewal of visual pigment in vertebrates. In: Handbook of Sensory Physiology, Vol. VII/1, Photochemistry of vision, edited by H.J.A. Dartnall, p. 395-416, Springer-Verlag, Berlin.

Baylor, D.A. and Hodgkin, A.L. (1973). Detection and resolution of visual stimuli by turtle photoreceptors. J. Physiol., (Lond.) 234, 163-198.

Berger, E.R. (1966). On the mitochondrial origin of oil drops in the retinal double cone inner segments. J. Ultrastruct. Res. 14: 143-157.

Best, A.C.G. and Nicol, J.A.C. (1967). Reflecting cells of the elasmobranch tapetum lucidum. Contr. Mar. Sci. 12: 172-201.

Birge, E.A. and Juday, C. (1931). A third report on solar radiation and inland lakes. Trans. Wis. Acad. Sci. Arts Lett. 26:383-425.

Borwein, B. and Hollenberg, M.J. (1973). The photoreceptors of the "four-eyed" fish, *Anableps Anableps*. L.J. Morph. 140: 405-441.

Carlander, K.D., Campbell, R.S. and Irwin, W.H. (1963). Mid-continental states. In Limology in North America, edited by D.G. Frey, p. 317-348. Univ. Wisconsin Press, Madison.

Christomanos, A.A. and Pavlopulu, C. (1967). A comparative chromatographic study of the presence of pterins in the eyes of invertebrates and of lower and higher vertebrates. Praktika Akad. Ahtenon 42: 405-428.

Clarke, G.L. (1936). On the depth at which fish can see. Ecology 17: 452-456.

Clarke, G.L. (1939). The utilization of solar energy by aquatic organisms. Problems in Lake Biology. Am. Assoc. Adv. Sci. Publ. 10: 27-38.

Clarke, G.L. (1954). Elements of Ecology, Wiley, New York.

Clarke, G.L. (1969). The significance of spectral changes in light scattered by the sea. In Remote Sensing in Ecology, edited by P. Johnson, p. 164-172, Univ. Georgia Press, Ahtens.

Coker, E.C. (1954). Streams, Lakes, Ponds. Univ. North Carolina Press, Chapel Hill.

Cole, G.A. (1963). The American southwest and middle America. In:Limnology in North America, edited by D.G. Frey, p. 317-348, Univ. Wisconsin Press, Madison.

Dartnall, H.J.A., Arden, G.B., Ikeda, H., Luck, C.P., Rosenberg, M.E., Pedler, C.M.H. and Tansley, K. (1965). Anatomical, electrophysiological and pigmentary aspects of vision in the bush baby: an interpretative study. Vision Res. 5: 399-424.

Denton, E.J. (1959). The contributions of the oriented photosensitive and other molecules to the absorption of whole retina. Proc. R. Soc. Lond. Ser. B. 150: 78-94.

Denton, E.J. (1970). On the organization of reflecting surfaces in some marine animals. Phil. Trans. Roy. Soc. Lond. Ser. B. 258: 285-313.

Denton, E.J. and Land, M.F. (1971). Mechanism of reflexion in silvery layers of fish and cephalopods. Proc. Roy. Soc. Lond. Ser. B. 178: 43-61

Denton, E.J. and Nicol, J.A.C. (1962). Why fishes have silvery sides; and a method of measuring reflectivity. J. Physiol. (Lond.) 165: 13-15P.

Denton, E.J. and Nicol, J.A.C. (1964). The chorioidal tapeta of some cartilaginous fishes (*Chondrichthyes*). J. Mar. Biol. Assoc. U.K. 44: 219-258.

Denton, E.J. and Nicol, J.A.C. (1965). Direct measurements on the orientation of the reflecting surfaces in the tapetum of *Squalus acanthias*, and some observations on the tapetum of *Acipenser sturio*. J. Mar. Biol. Assoc. U.K. 45: 739-742.

Denton, E.J. and Nicol, J.A.C. (1966). A survey of reflectivity in silvery teleosts. J. Mar. Biol. Assoc. U.K. 46: 685-722.

Denton, E.J. and Warren, F.J. (1957). The photosensitive pigments in the retinae of deep-sea fish. J. Mar. Biol. Assoc. U.K. 36: 651-662.

Denton, E.J., Liddicoat, J.D. and Taylor, D.W. (1972). The permeability to gases of the swimbladder of the conger eel (*Conger conger*). J. Mar. Biol. Assoc. U.K. 52: 727-746.

Eigenmann, C.H. and Shafer, G.D. (1900). The mosaic of single and twin cones in the retina of fishes. Am. Nat. 34: 109-118.

Ellis, M.M. (1936). Erosion silt as a factor in aquatic environments. Ecology 17: 29-42.

Engström, K. (1960). Cone types and cone arrangement in the retina of some cyprinids. Acta. Zool. (Stockh.)41: 277-295.

Engström, K. (1961). Cone types and cone arrangement in the retina of some gadids. Acta. Zool. (Stockh.)42: 227-243.

Engström, K. (1963a). Structure, organization and ultrastructure of the visual cells in the teleost family Labridae. Acta. Zool. (Stockh.)44: 1-41.

Engström, K. (1963b). Studies on Teleostean Visual Cells, Ake Nyblom & Co., Stockholm.

Engström, K. and Ahlbert, Q.B. (1963). Cone types and cone arrangements in the retina of some flatfishes. Acta. Zool. (Stockh.) 44: 119-129.

Fineran, B.A. and Nicol, J.A.C. (1974). Studies on the eyes of New Zealand parrot-fishes (Labridae). Proc. R. Soc. Lond. Ser. B. 186: 217-247.

Franz, V. (1905). Zur Anatomie, Histologie und funktionellen Gestaltung des Selachierauges. Jean. Z. Naturw. 40: 697-840.

Franz, V. (1913). Sehorgan (Choroidea, Selachier). In: Lehrbuch vergl. mikr. Anat. Wirbeltiere, edited by A. Oppel, Teil 7: p. 166-169, Fischer, Jena.

Fukuda, S. and Yoshiko Hirami (1955). Vitamin B_1 in foods. I. Vitamin B_1 in eyeballs of fishes from province of Tosa. Nippon Nogei Kagaku Kaishi 29: 247-251.

Gruber, S.H. (1967). A behavioral measurement of dark adaptation in the lemon shark, *Negaprion brevirostris*. In Sharks, Skates, and Rays, edited by P.W. Gilbert, R.F. Mathewson and D.P. Rall, p. 479-490. Johns Hopkins Press, Baltimore.

Gruber, S.H., Hamasaki, D.H. and Bridges, C.D.B. (1963). Cones in the retina of the lemon shark (*Negaprion brevirostris*). Vision Res. 3: 397-399.

Higgins, E. (1932). Progress in biological inquiries. Appendix III. Rept. U.S. Fish Comm. 1931, p. 441-569.
Hudspeth, A.J. and Yee, A.G. (1973). The intercellular junctional complexes of retinal pigment epithelia. Invest. Ophthalmol. 12: 354-365.
Ito, S. and Nicol, J.A.C. (1974). Isolation of oligomers of 5,6-dihydroxyindole-2-carboxylic acid from the tapetum lucidum of the catfish. Biochem. J. (In press).
Jerlov, N.G. (1968). Optical Oceanography, Elsevier, Amsterdam.
Jones, R.S., Copeland, B.J. and Hoese, H.D. (1965). A study of the hydrography of inshore waters in the western Gulf of Mexico off Port Aransas, Texas. Contr. Mar. Sci. 10: 22-32.
Kuchnow, K.P. (1969). Tapetal pigment response of elasmobranchs. Vision Res. 9: 849-854.
Kuchnow, K.P. (1970). Pigment migration in the tapetum lucidum of the elasmobranch eye: evidence for a nervous mechanism. Vision Res. 10: 825-828.
Kuchnow, K.P. and Gilbert, P.W. (1967). Preliminary in vivo studies on pupillary and tapetal pigment responses in the lemon shark, *Negaprion brevirostris*. In Sharks, Skates, and Rays, edited by P.A. Gilbert, R.F. Mathewson and D.P. Rall, p. 465-477, Johns Hopkins Press, Baltimore.
Land, M.F. (1965). Image formation by a concave reflector in the eye of the scallop, *Pecten maximus*. J. Physiol., (Lond.) 179: 138-153.
Land, M.F. (1966). A multi-layer interference filter in the eye of the scallop, *Pecten maximus*. J. Exp. Biol. 45, 433-447.
Lane, E.D. (1967). A study of the Atlantic midshipman, *Porichthys porosissimus*, in the vicinity of Port Aransas, Texas. Contr. Mar. Sci. 12: 1-53.
Le Grand, Y. (1952). Optique physiologique. T. 1. La dioptrique de l'oeil et la correction. 2e ed. Editions de la "Revue d'Optique", Paris.
Lenoble, J. and Le Grand, Y. (1954). Le tapis de l'oeil du Coelacanthe (*Latimeria anjouanae* (Smith). Bull. Mus. Natl. Hist. Nat. 26: 460-463.
Lillie, R.D. (1965). Histopathologic Technic and Practical Histochemistry, McGraw-Hill, New York.
Lissmann, H.W. (1961). Zoology, locomotory adaptations and the problem of electric fish. In: The Cell and the Organism. Essays presented to Sir James Gray, edited by J.A. Ramsay and V.B. Wigglesworth, p. 301-307, Univ. Cambridge Press.
Locket, N.A. (1971a). Retinal anatomy in some scopelarchid deep-sea fishes. Proc. Roy. Soc. Lond. Ser. B. 178: 161-184.
Locket, N.A. (1971b). Retinal structure in *Platytroctes apus*, a deep-sea fish with a pure rod retina. J. Mar. Biol. Assoc. U.K. 51: 79-91.

Locket, N.A. (1974). The choroidal tapetum of *Latimeria chalumae*. Proc. Roy. Soc. Lond. Ser. B. 186: 281-290.

Lyall, A.H. (1957). Cone arrangements in teleost retinae. Q.J. Microsc. Sci. 98: 189-201.

Marks, W.B. (1965). Visual pigments of single goldfish cones. J. Physiol., (Lond.) 178: 14-32.

Martin, P.S., Schoenwetter, J. and Arms, B.C. (1961). Southwestern polynology and prehistory. The last 10,000 years. Geochromology Laboratories, Univ. Arizona, Tucson.

Miller, R.R. (1961). Man and the changing fish fauna of the American southwest. Mich. Acad. Sci. 46: 365-404.

Miller, W.H. and Snyder, A.W. (1972). Optical function of myoids. Vision Res. 12: 1841-1848.

Moyer, F.H. (1969). Development, structure and function of the retinal pigmented epithelium. In: The retina. Morphology, Function and Clinical Characteristics, edited by B.R. Straatsma, M.D. Hall, R.A. Allen and F. Crescitelli, p. 1-30, Univ. Calif. Press, Berkeley.

Munk, O. (1965). *Omosudis lowei* Günther, 1887, a bathypelagic deep-sea fish with an almost pure cone retina. Vidensk. Medd. Dan. Naturhist Foren. KBH 128: 341-356.

Munk, O. (1966a). Ocular anatomy of some deep-sea teleosts. Dan. Rep. No. 70: 63 p.

Munk, O. (1966b). On the retina of *Diretmus argenteus* Johnson, 1863 (Diretmidae, Pisces). Vidensk. Medd. Dan. Naturhist Foren. KBH 129: 73-80.

Nicol, J.A.C. (1961). The tapetum in *Scyliorhinus canicula*. J. Mar. Biol. Assoc. U.K. 41: 271-277.

Nicol, J.A.C. (1965a). Migration of chorioidal tapetal pigment in the spur dog *Squalus acanthias*. J. Mar. Biol. Assoc. U.K. 45: 405-427.

Nicol, J.A.C. (1965b). Retinomotor changes in flatfishes. J. Fish. Res. Board Can. 22: 513-520.

Nicol, J.A.C. (1969). The tapetum lucidum of the sturgeon. Contr. Mar. Sci. 14: 5-18.

Nicol, J.A.C. and Arnott, H.J. (1972). Riboflavin in the eyes of gars (*Lepisosteidae*). Can. J. Zool. 50: 1211-1214.

Nicol, J.A.C. and Arnott, H.J. (1973). Studies on the eyes of gars (*Lepisosteidae*) with special reference to the tapetum lucidum. Can J. Zool. 51: 501-508.

Nicol, J.A.C. and Van Baalen, C. (1968). Studies on the reflecting layers of fishes. Contr. Mar. Sci. 13: 65-88.

Nicol, J.A.C. and Zyznar, E.S. 1973. The tapetum lucidum in the eye of the big-eye *Priacanthus arenatus* Cuvier, J. Fish. Biol. 5: 519-522.

Nicol, J.A.C., Arnott, H.J. and Best, A.C.G. (1973). Tapeta lucida in bony fishes (Actinopterygii). Can. J. Zool. 51: 69-81.

O'Connell, C.P. (1963). The structure of the eye of *Sardinops caerulea*, *Engraulis mordox*, and four other pelagic marine teleosts. J. Morphol. 113: 287-329.
Pirie, A. and van Heyningen, R. (1956). Biochemistry of the Eye. Blackwell, Oxford.
Pumphrey, R.J. (1961). Concerning vision. In: The Cell and the Organism. Essays presented to Sir James Gray, edited by J.A. Ramsay and V.B. Wigglesworth, p. 193-207, Cambridge Univ. Press.
Randall, D.J. (1970). Gas exchange in fish. In: Fish physiology, edited by W.S. Hoar and D.J. Randall, Vol. IV, p. 253-292. Academic Press, New York.
Rochon-Duvigneaud, A. (1943). Les yeux et la vision des Vertébrés. Masson, Paris.
Rochon-Duvigneaud, A. (1958). L'oeil et la vision. In: Les yeux et la vision, traité de Zoologie, edited by P.P. Grasse, 3(fasc. 2), p. 1099-1142. Masson, Paris.
Sidman, R.L. (1957). The structure and concentration of solids in photoreceptor cells studied by refractometry and interference microscopy. J. Biophys. Biochem. Cytol. 3: 15-30.
Stell, W.K. (1972). The structure and morphologic relations of rods and cones in the retina of the spiny dogfish, *Squalus*. Comp. Biochem. Physiol. 42 (2A): 141-151.
Strum, J. (1969). Fine structure of the dermal luminescent organs, photophores, in the fish *Porichthys notatus*. Anat. Res. 164: 433-462.
Tavolga, W.N. (1960). Sound production and underwater communication in fishes. In Animal sound and communication, edited by W.E. Lanyon and W.N. Tavolga, p. 93-136. Publs. Am. Inst. Biol. Sci. No. 7.
Teal, J. and Teal, M. (1969). Life and Death of a Salt Marsh, Little Brown, Boston.
Tyler, J.E. and Smith, R.C. (1967). Spectroradiometric characteristics of natural light underwater. J. Opt. Soc. Am. 57: 595-601.
Vilter, V. (1953). Existence d'une rétine à plusieurs mosaiques photoréceptrices, chez un poisson abyssal bathypélagique, *Bathylagus benedicti*. C.R. Séances Soc. Biol. 147: 1937-1939.
Walls, G.L. (1942). The Vertebrate Eye and its Adaptive Radiation. Bull. No. 19, Cranbrook Inst. Sci., Bloomfield Hills, Mich.
Wang, R.T. and Nicol, J.A.C. (1974). The tapetum lucidum of gars *Lepisosteidae*) and its role as a reflector. Can. J. Zool. (In press).
Wittenberg, J.B. and Wittenberg, B.A. (1962). Active secretion of oxygen into the eye of a fish. Nature, Lond. 194: 106-107.

Wittenberg, J.B. and Wittenberg, B.A. (1974). The choroid rete mirabile of the fish eye I. Oxygen secretion and structure: comparison with the swimbladder rete mirabilis. Biol. Bull. (Woods Hole) 146: 116-136.

Wittenberg, J.B. and Haedrich, R.L. (1974). The choroid rete mirabile of the fish eye II. Distribution and relation to the pseudobranch and to the swimbladder rete mirabile. Biol. Bull. (Woods Hole) 146: 137-156.

Wunder, W. (1925). Physiologische und vergleichend-anatomische Untersuchungen an der Knochenfischnetzhaut. Z. vergl. Physiol. 3: 1-61

Wunder, W. (1926). Die Bedeutung des Adaptationszustandes für das Verhalten der Sehelemente und des Pigmentes in der Netzhaut von Knochenfischen. Z. vergl. Physiol. 3: 595-614.

Young, R.W. and Bok, D. (1969). Participation of the retinal pigment epithelium in the rod outer segment renewal process. J. Cell. Biol. 42: 392-403.

Zyznar, E.S. and Nicol, J.A.C. (1973). Reflecting materials in the eyes of three teleosts *Orthopristes chrysopterus*, *Dorosoma cepedianum* and *Anchoa mitchilli*. Proc. R. Soc. Lond. Ser. B. 184: 15-27.

RETINAL SPECIALISATIONS FOR AQUATIC VISION: THEORY AND FACTS

S.S. EASTER, Jr.

Department of Zoology, University of Michigan

Ann Arbor, Michigan 48104, U.S.A.

> There is occasions and causes why and wherefore in all things.
>
> Henry V, Act V, Sc. 1

INTRODUCTION

Most of the investigators who, over the past twenty-five years, have worked on the fish retina, have chosen this tissue for reasons unrelated to the fact that the fish is an aquatic organism, and as such, must cope with the severe problems of optical image transmission under water. Some of the problems are well understood, or at least well described, and I suggest that the physiological features which distinguish the retinas of aquatic vertebrates can be interpreted as evolutionary solutions of these problems. Taken as a general statement, this is probably unobjectionable, since nearly all biologists accept the theory of evolution via natural selection. It is when the assertion is made more specific that objections will arise. What are the features which distinguish aquatic retinas from those of land-dwelling vertebrates? What are the important optical problems underwater? What follows is one attempt to answer these questions and to tie together two previously disparate fields; namely, retinal physiology and underwater optics. At the outset, I acknowledge my debt to Dr. J.N. Lythgoe, whose papers (Lythgoe, 1966, 1968) stimulated my interest.

In the paragraphs below, I shall summarise the features which seem to me to be unique to the retina of fish, in particular, the cyprinids. Then I shall relate them to the problems of underwater vision, and try to convince the reader that my hypothesis might be valid. Finally, I shall outline some experiments to test the hypothesis. The experiments have not yet been carried out.

UNUSUAL FEATURES OF THE FISH RETINA

1. The red-sensitive mechanism is dominant in several unexpected ways, outlined below.

All goldfish retinal neurons which receive cone-derived inputs receive from the red-sensitive mechanism. This includes ganglion, amacrine, bipolar, and horizontal cells (Spekreijse, Wagner and Wolbarsht, 1972; Kaneko, 1973). Some of these receive other inputs as well, but the red-sensitive contribution is always present, and sometimes the only one present.

The red-sensitive mechanism often masks others. Frequently, an opponent-colour ganglion cell will seem to have only one underlying spectral mechanism. Invariably, the apparent input peaks in the red, and contributions from blue or green channels can be unmasked only by chromatic adaptation (Daw, 1968; Spekreijse et al., 1972). Even when two or more spectral mechanisms are apparent at treshold, the red one takes over at higher intensities, as Bicking (1965) noted.

The spatial extent of the central red-sensitive receptive field of a retinal ganglion cell is always narrower than its shorter wavelength antagonist (Spekreijse et al., 1972).

The red cones feed the same synaptic pathways as the rods. In any ganglion cell, the sign of the red-sensitive central response under photopic conditions is always the same as the scotopic central response (Raynauld, 1972; Beauchamp and Daw, 1972). This contrasts with colour-coded cells in other animals such as the cat (Pearlman and Daw, 1970) and the monkey (Wiesel and Hubel, 1966), where there is no such association between long-wavelength cone and rod. The anatomical basis of this linkage in the rudd has been described (Scholes and Morris, 1973).

2. Opponent-colour ganglion cells are very numerous. Daw found that 68% of goldfish ganglion cells are colour-coded. Hammond (1968) has described numerous chromatic units in the marine plaice, although he gave no percentages. In contrast, Hubel and Wiesel (1960) found that only 3% of monkey ganglion cells were colour-coded. Michael (1966b) found 19% in the ground squirrel. There have been no reports of colour-coded ganglion cells in the cat retina, but Pearlman and Daw (1970) found 3% of the lateral geniculate cells to be so organised. Reuter (1969) has described rare chromatic ganglion cells in the larval and adult frog, *Rana temporaria*, but gave no estimate of percentage.

3. Receptive field centers of ganglion cells are unusually large.

Jacobson and Gaze (1964) were the first to record single retinal units in the intact goldfish. They found that receptive field diameters varied from a low of 10 degrees to a high of 60. These compare with a range of 0.06-2.0 degrees in the spider monkey (Hubel and Wiesel, 1960), 0.5-4.0 degrees in the cat (Wiesel, 1960), and 0.5-4.5 degrees in the ground squirrel (Michael, 1966a; 1966b). The unusually large fields could have been due to poor focus of the targets, as Jacobson and Gaze dit not refract the eyes. My own measurements (unpublished) on 15cm goldfish ("comets") showed that one degree of arc corresponded to a distance of 45-50 μm on the retina. Therefore a 10 degree field would subtend 450-500 μm, and such dimensions are not uncommonly observed in isolated retinas with very sharp optics (Wagner, MacNichol and Wolbarsht, 1963). It must be concluded that the receptive field centres of goldfish ganglion cells subtend very large visual angles.

AQUATIC OPTICS AND EVOLUTIONARY ADAPTATIONS

In this section, I treat three features of underwater optics and speculate on the resultant evolutionary adaptations.

1. Underwater light scatter severely reduces contrast, and promotes development of dominant red-sensitivity and opponent-colour processes.

The role of light scatter in contrast reduction has been treated by Duntley (1962) and Lythgoe (1966, 1968), and I shall only summarise their analysis. First, natural bodies of water scatter light much more intensively than air, making the medium luminous. This veiling luminance, the aquatic space light, reduces contrast by adding equally to rays passing through it which originated on dark or light surfaces. Second, the aquatic space light is coloured, with most of its energy at the intermediate (blue-green) wavelengths, relatively less at the long and short wavelengths. Third, the conclusion is that transmission of optical contrast is better through spectral windows where the aquatic space light is weak; namely, the long and short wavelengths.

I suggest that the red-sensitivity of aquatic animals is a result of this; but why not become violet-sensitive? The answer probably lies in the transmission properties of the ocular media, which selectively attenuate short wavelengths near the aquatic transmission window (Kennedy and Milkman, 1956; Burkhardt, 1966). It may be that this is obligatory in proteinaceous dioptrics.

A second problem raised by the proposition that aquatic animals see primarily through the long wavelengths is that there is not much red light available underwater. Although the apparent contrast may be arbitrarily close to the intrinsic value, the image can not be transmitted if there are insufficient numbers of quanta. Are there enough?

In Fig. 1, I have replotted the data of Tyler and Smith (1970) obtained in the Gulf Stream. The downwelling irradiance is the number of photons passing from above through a one cm^2 surface during one second over a bandwidth of one cm^{-1}. This is the diffuse illumination which would provide most of the radiant energy incident upon a target. It is evident that there is much more blue-green than red light, but the question originally asked was, is there enough of the red? It seems to me that the answer must be, yes, even in this particular case, because at 15 m depth, the photon density at 600 nm is only down 1.8 log units from the peak value at that depth, and down only about 2.7 log units from the value at the surface. We know that visual systems readily adapt over many log units of ambient intensity. All we demand here is that the red-sensitive

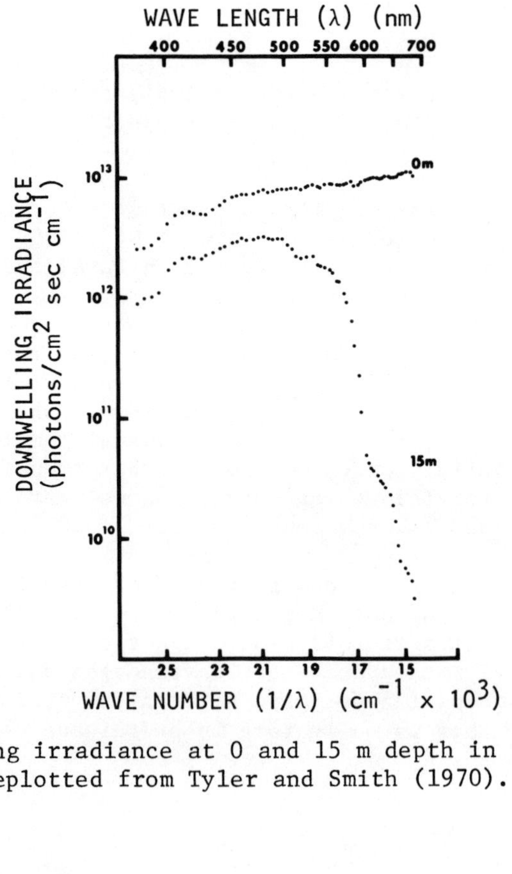

Fig. 1. Downwelling irradiance at 0 and 15 m depth in the Gulf Stream, replotted from Tyler and Smith (1970).

system be able to adapt over three log units; if it can, it can stay functional even in the blue-green underwater world.

Following this line of logic, one can account for the very numerous opponent-colour cells also, for it will be recalled that the optimal spectral filtre for our problem peaked at a wavelength where the contrast was high, and absorbed negligibly where the contrast was low. Visual pigments do not by themselves satisfy both of these requirements, because their extinction spectra are quite wide (e.g. 3625-4832 cm^{-1} in goldfish, Hárosi and MacNichol, 1974). Dartnall (1953) has described their shape. On the long wavelength side of the peak, they drop arbitrarily close to zero, but on the short wavelength side, they never drop below 20-30% of the maximum extinction coefficient. Therefore, a visual pigment with long λ_{max} may well absorb more quanta of shorter wavelengths where the extinction coefficient is low but the incident spectral energy high. For instance, a visual pigment with its λ_{max} at 600 nm, exposed to the downwelling irradiance illustrated in Figure 1 (15m) would absorb 13 times as many quanta at 479 nm as it would in a bin of comparable width centred at 600 nm. If the visual pigment itself were the only element in the spectral filtration, then it would not be very effective, for most quanta would be caught at wavelengths where the contrast was low.

The way around this difficulty is, of course, to narrow the spectral width (Legault and Riordan, 1964). One way to do this is to insert short wavelength "cut off filtres" (such as coloured oil droplets, Walls, 1942) in front of the photopigment. The other approach is to narrow the spectrum after the quanta have been absorbed, by subtracting the outputs of, say, a red- and a green-sensitive mechanism. There is abundant evidence for this, and the resulting spectral sensitivies of carp ganglion cells, for instance, are very much narrower than the absorption spectra of their cones (Witkovsky, 1965).

In summary, I suggest that the problem of light scatter under water has been responsible for the development of the most prominent features of photopic aquatic retinas: long-wavelength dominance and spectral opponent mechanisms. It is suggested further that multiple pigment vision first evolved in this context, and later was adapted to colour vision as we know it.

2. Refractive inhomogeneities in water decrease spatial resolution and thereby promote evolution of large receptive fields.

Refractive inhomogeneities bend light randomly, and in so doing, destroy fine detail of an image projected through the medium. The major refractive inhomogeneities in water are small transparent organisms.

It is generally thought (Pirenne, 1962) that in man, visual acuity is limited by the intercone spacing. Certainly it is true that at retinal sites where the cones are smaller and more closely spaced, the acuity is greater. The ganglion cell receptive fields also roughly follow the granularity of the receptor mosaic (e.g., Wiesel, 1960) and therefore the acuity. Thus, it seems that receptive field sizes have some relation to the resolving power of the visual system, and I suggest that the very large fields observed in fish are large because visual resolution is limited by the aquatic medium.

3. The shape of the radiance vector field is asymmetric and has promoted an asymetric retinal distribution of On-and-Off centre receptive fields.

Finally, there is a third fact of aquatic optics to which no physiological data now relate, but its predictions are fairly straightforward. The cloud which envelops an organism underwater is not uniformly radiant in all directions. Looking upward, the observer sees large radiance, and as the angle of elevation of gaze drops, the radiance decreases to reach a minimum directly below. Even when scanning the vertical at 180 degrees away from the sun, the radiance at a depth of 4 metres, in open ocean, on a clear sunny day, was 67 - fold greater straight up than straight down (Tyler and Preisendorfer, 1962).

The implication of such a radiance field for vision is that objects above the observer will almost always be sensed as dark against a light background, while objects below the observer will have the reversed contrast. This is very different from the case in air, where the radiance vector field is more nearly spherical owing to the high reflectivity of earth, rocks, and vegetation. (It must be pointed out that the radiant field in open ocean is quite different from that found in a reef, for instance, where the water is clear and shallow and the bottom highly reflective, much like the conditions on land.)

The asymmetrical radiance has certainly influenced the organisation of reflective surfaces in fishes, particularly oceanic fishes, which are characteristically much darker on their dorsal surface than ventrally, presumably as an aid against detection from above (Denton, 1971). One Pacific species has gone one step further, and developed a luminescent swim bladder which emits a diffuse radiation downward through a transparent ventral surface. This emission substitutes for the light that the animal's shadow interrupts. Emission of this light occurs only when the animal is illuminated (Hastings, 1971).

Observations of this sort indicate that evolutionary adaptations to the radiant field have already occurred, and it is likely that the visual system has also been affected. It is predicted that in pelagic fishes, the ventral retina should be dominated by dimming detectors (off centre cells), and the dorsal retina by on centre cells, the two types most suited to the detection of the radiance changes most often experienced.

EXPERIMENTS

Most of these speculations are verifiable experimentally, and I have begun to do so. I am using a comparative approach.

The four-eyed minnow, *Anableps*, is the principal animal for this purpose. I suggest that the dorsal (aquatic) retina should show more red-dominance, spectral opponency, and larger receptive fields than the ventral (aerial) retina. Certainly the two differ morphologically. The dorsal retina has very sparse ganglion cells, a thin inner nuclear layer, larger horizontal cells, and large cones, like most fish retinas. The ventral retina, on the other hand, has smaller cones and horizontal cells, and considerably thicker inner nuclear and ganglion cell layers (Borwein and Hollenberg, 1973).

An inter-animal comparison of retinal structure and function is to be made between closely related species, such as turtles, one terrestrial and the other aquatic. Larval aquatic and adult terrestrial forms of the same animal, such as an amuran, should also be compared, but this is technically more difficult.

These kinds of comparisons are the sort that could support or reject my speculation, but at the moment, I can only say, as Mark Twain did in the preface to his autobiography, "Some of it may even be true."

ACKNOWLEDGMENT

This work was supported by grant EY-00168, form the National Eye Institute of the U.S. Public Health Service.

REFERENCES

Beauchamp, R.D. and Daw, N.W. (1972). Rod and cone input to single goldfish optic nerve fibers. Vision Res. 12: 1201-1212.

Bicking, L.A. (1965). Some quantitative studies on retinal ganglion cells. Ph.D. Thesis, The Johns Hopkins University, Baltimore, Md.

Burkhardt, D.A. (1966). The goldfish electroretinogram: relation between photopic spectral sensitivity function and cone absorption spectra. Vision Res. 6: 517-532

Borwein, B. and M.R. Hollenberg (1973). The photoreceptors of the "four-eyed" fish, *Anableps anableps L.* J. Morph. 140: 405-442.

Dartnall, H.J.A. (1953). The interpretation of spectral sensitivity curves. Br. Med. Bull. 9: 24-30.

Daw, N.W. (1968). Colour-coded ganglion cells in the goldfish retina: extension of their receptive fields by means of new stimuli. J. Physiol. 197: 567-592.

Duntley, S.Q. (1962). Underwater visibility. In: The Sea Vol. 1 edited by M.N. Hill. Interscience: New York, 452-455.

Hammond, P. (1968). Spectral properties of dark-adapted retinal ganglion cells in the plaice *(Pleuronectes platessa*, L.). J. Physiol. 195: 535-556.

Harosi, F.I. and E.F. MacNichol (1974). Visual pigments of goldfish cones: Spectral properties and dichroism. J. Gen. Physiol. 63: 279-304.

Hastings, J.W. (1971). Light to hide by: ventral luminescence to camouflage the silhouette. Science 173: 1016-1017.

Hubel, D.H. and Wiesel, T.N. (1960). Receptive fields of optic nerve fibers in the spider monkey. J. Physiol. 154: 572-580.

Jacobson, M. and Gaze, R.M. (1964). Types of visual response from single units in the optic tectum and optic nerve of the goldfish. Quart. J. Exptl. Physiol. 49: 199-209.

Kaneko, A. (1973). Receptive field organization of bipolar and amacrine cells in the goldfish retina. J. Physiol. 235: 133-153.

Kennedy, D. and Milkman, R.D. (1956). Selective light absorption by the lenses of lower vertebrates and its influence on spectral sensitivity. Biol. Bull. 111: 375-386.

Legault, R.R. and Riordan, J.F. (1964). An algorithm for optimizing a spectral filter. App. Optics 3: 735-739.

Lythgoe, J.N. (1968). Visual pigments and underwater vision. In Light as an ecological factor, edited by R. Bainbridge, G.C. Evans and O. Rackham. Br. Ecol. Soc. Symp. 6: 375-391.

Lythgoe, J.N. (1968). Visual pigments and visual range underwater. Vision Res. 8: 997-1011.

Michael, C.R. (1966a). Receptive fields of directionally selective units in the optic nerve of the ground squirrel. Science 152: 1092-1095.

Michael, C.R. (1966b). Receptive fields of opponent color units in the optic nerve of the ground squirrel. Science 152: 1095-1097.

Pearlman, A.L. and Daw, N.W. (1970). Opponent color cells in the cat lateral geniculate nucleus. Science 167: 84-86.

Pirenne, M.H. (1962). Visual acuity. In: The Eye, Vol. 2 edited by H. Davson. Academic Press: New York, 175-196.

Raynauld, J.P. (1972). Goldfish retina: sign of the rod-input in opponent color ganglion cells. Science 177: 84-85.

Scholes, J. and Morris, Jacqueline (1973). Receptor-bipolar connectivity patterns in fish retina. Nature 241: 52-54.

Spekreijse, H., Wagner, H.G. and Wolbarsht, M.L. (1972). Spectral and spatial coding of ganglion cell responses in goldfish retina. J. Neurophysiol. 35: 73-86.

Tyler, J.E. and Smith, R.C. (1970). Measurement of Spectral Irradiance Underwater. Gordon and Breach: New York.

Tyler, J.E. and Preisendorfer, R.W. (1962). Transmission of energy within the sea. In: The Sea, Vol. 1 edited by M.N. Hill. Interscience: New York, 397-451.

Wagner, H.G., MacNichol, E.F. and Wolbarsht, M.L. (1963). Functional basis for on-center and off-center receptive fields in the retina. J. Opt. Soc. Am. 53: 66-70.

Walls, G.L. (1942). The Vertebrate Eye and its Adaptive Radiation. Cranbrook Inst. Sci: Bloomfield Hills, Michigan.

Wiesel, T.N. (1960). Receptive fields of ganglion cells in the cat's retina. J. Physiol. 153: 583-594.

Wiesel, T.N. and Hubel, D.H. (1966). Spatial and chromatic interactions in the lateral geniculate body of the Rhesus monkey. J. Neurophysiol. 29: 1115-1156.

Witkovsky, P. (1965). The spectral sensitivity of retinal ganglion cells in the carp. Vision Res. 5: 603-614.

PROBLEMS OF SEEING COLOURS UNDER WATER

J.N. LYTHGOE

MRC Vision Unit, University of Sussex

Falmer, Brighton, BN1 9QG. U.K.

> How should they, when wise men have no eyes?
>
> Romeo and Juliet, Act III, Sc. 3

INTRODUCTION

It is probably impossible to think about the perception of colour by fishes without being influenced by our own colour sense. Up to a point this can be helpful since fish visual pigments are not very different from our own. However, our habits of thought about colour perception, and indeed our language itself, have been moulded by our experience on land. This can be misleading when we consider human underwater vision and probably much more so where fishes are concerned.

There is no doubt that fishes can perceive the difference between two radiances by virtue of hue alone - in other words, they have colour vision. Exactly how different species perceive these radiance differences are considered in other papers in this volume.

Ultimately it may be possible to predict the visual performance of individual fish species by some sort of line element equation similar to the Stiles-Helmholtz Line Element for man, which is described below. Although such a mathematical description of the visual performance of fishes is a long way off, some generalisation about colour vision under water can be made with reasonable confidence. This is possible because the spectral absorption characteristics dominate the conditions for vision under water and restrict the wavelengths available for vision to a waveband much narrower than that encountered on land.

To the human eye the colour of natural water can vary between the deep blue of the open ocean to the reddish-brown of some land-locked waters. In spectral terms this represents a wavelength of maximum transparency anywhere from about 420 nm for very clear water to 600 nm and longer for highly coloured lakes and rivers (Fig. 1).

The Colour of Natural Water

There are four major materials present in natural water that influence the colour of the ambient light.

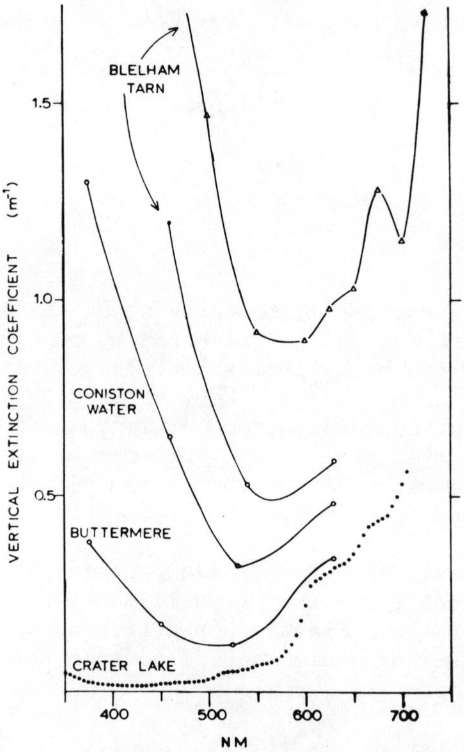

FIG. 1. The vertical extinction coefficient of various bodies of natural water. Note how the absorption in the blue is more variable than in the red. Also reduced transparency is generally associated with maximum transmission at longer wavelength. (From Talling, 1970).

1. The spectral absorption of pure water - The purest known water - Crater Lake in Oregon - is most transparent between 400 and 450 nm. It absorbs strongly in the ultraviolet and in the red with a very sharp increase in absorption at about 575 nm (Smith & Tyler, 1967).

2. "Gelbstoffe" or "Yellow substance" - These are soluble humus-like substances that are the stable products of organic decay (Kalle, 1938, 1966) chiefly, perhaps, from chlorophyll (Fogg & Boalch, 1958). They may be either of marine origin, in which case they are chiefly yellow melanoidines, or are washed down the rivers from the land, in which case they are brown pheno-humic acids (Shapiro, 1957). These yellow substances absorb strongly at short wavelengths and are relatively transparent at longer wavelengths. Yellow substances are present in nearly all natural waters and since they absorb strongly in the blue, and the water itself absorbs in the red, the wavelengths of maximum transparency tend to be located in the middle of the spectrum, typically around 475 nm for the waters of the open ocean and 530 - 550 nm for inshore waters. In extreme cases there is sufficient yellow substance to colour the water yellow or brown.

3. Phytoplankton - Phytoplankton contain large amounts of chlorophyll and the associated carotenoid pigments (Halldal, 1974). In very productive waters the phytoplankton may be sufficiently dense for the photosynthetic pigments to be a major factor in the spectral absorption of natural water (Tyler & Smith, 1967). Phytoplankton will be less important in deep water since they can only grow at depths where the ambient light is of the order of 1% of the surface light.

The absorption characteristics of phytoplankton is, of course, influenced by all the pigments present and the variation in relative spectral transmission is large. However, there seems always to be an absorption peak around 670 - 680 nm and a less well defined absorption band between 400 and 500 nm (Halldal, 1974).

4. Suspended matter - This may be particles of sand, silt or plankton. In brightly lit water with abundant phytoplankton, oxygen bubbles floating in the water are present and near the surface in rough water air bubbles are present in the water column. All these particles are large compared to the wavelength of light and the light they scatter is more or less equal at all wavelengths.

There appears to be no fundamental differences in the colour of fresh and salt water. The dissolved salts in sea water have little or no effect on its absorptive properties (Clarke & James, 1939; Sullivan, 1963).

Factors Affecting Visibility

The analysis of the factors affecting the visibility and appearance of underwater objects falls naturally into four stages. First there is the passage of light from the surface to the object in question; second, the passage of the reflected light from the object to the eye; third, the passage of light through the optic media and its absorption by the visual pigment; and fourth, the physiological events in the retina and brain.

1. Surface to object

The quantum distribution of typical daylight is given by Dartnall in this volume. Once it has traversed the surface layers of the water, it is selectively absorbed and scattered by the water so that in any particular direction it becomes less intense (except very near the surface) and more restricted in bandwidth as the depth increases. As the depth increases an increasing proportion of the light is scattered and eventually almost all the light is diffuse. More light is scattered in the forward direction than back towards the surface and is included in any measure of the downwelling light.

The rate that this light is attenuated is described by K, the diffuse attenuation coefficient, or perhaps more accurately, the broad beam attenuation coefficient

$$I_z = I \exp{-KZ} \qquad -1$$

where I is the radiance falling upon a surface in the water and I_z is the radiance falling upon a similar surface at depth Z below it.

Light that reaches the observer directly from above has travelled a shorter path than light reaching the eye from other directions. Since the bandwidth of light is progressively reduced as the path length increases, downwelling light has a broader spectral distribution curve and will be more intense than will light from other directions (Tyler & Smith, 1970). This non-uniform distribution of under water light has important consequences for underwater vision. An example is in the apparent colour of grey objects in blue water. In an experiment carried out at 30 m in the Mediterranean (Lythgoe, 1971) a series of grey targets from black to white, each with a flat spectral reflectance curve but with different reflectances, were displayed to a diver at the same depth so that they were seen in a horizontal direction against the water background. As expected the white and black targets were visible at the greatest distance. It was the dark grey targets that most nearly matched the water background in brightness and were visible at the shortest range. However, these matched targets still remained visible because they appeared pink in contrast to the blue of the water background. The downwelling light is 10 - 100 times brighter than sidewelling light (see Jerlov, 1968, for a review). Light reflected from the object mostly originates from the relatively broad-band downwelling light, whereas the object is seen against the much more monochromatic background light. The eye becomes adapted to the blue background light and the grey target appears slightly reddish in colour.

2. Object to eye

Some of the ambient light that reaches the object is reflected in the direction of the eye. In practice it is difficult to calculate the quantity of this reflected light since the irradiance from different directions varies both in intensity and spectral distribution. Tyler & Preisendorfer (1962) have shown that even on an overcast day sidewelling light may be only 0.1 or 0.01 the intensity of the downwelling light.

The image-forming light reflected from the object in the direction of the eye suffers a loss in intensity from absorption. It suffers a further loss due to light being scattered out of its path by the particulate matter in the water. The appropriate measure of attenuation in this case is α, the narrow beam attenuation coefficient and like K is strongly wavelength dependent.

This exponential attenuation of image-forming light does not by itself explain why distant objects become invisible under water. A black object under water is visible as a silhouette against the spacelight. The black object has zero radiance and this cannot be reduced further. Nevertheless, black objects are visible at an infinite range under water, but their apparent radiance increases to equal that of the background spacelight. This increase in radiance is derived by light scattered from other directions into the eye. It is also this scattered light that constitues the background spacelight (Fig. 2). The change in the apparent spectral radiance, ${}_tN_r$, of the object at distance r can thus be written

$$ {}_tN_r = {}_tN_o\,(\exp{-\alpha r}) + {}_bN(1-\exp{-\alpha r}) \qquad - 2 $$

where N_o and ${}_bN$ are the radiances from the object and the background respectively, and r is the range between the eye and the object. This is Koschmeider's classical equation for radiance transfer through turbid media. Duntley (1962, 1963) has adapted and expanded the equation for different directions of light under water. The equation shown here is for the simplest case when the path of light is horizontal.

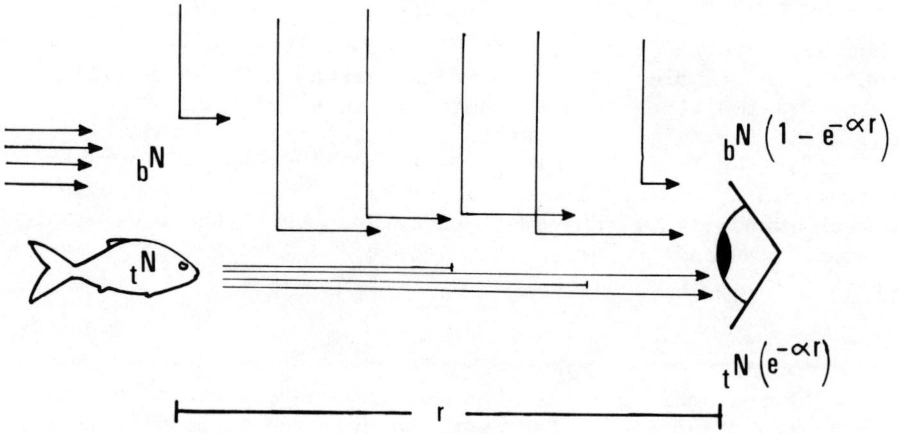

$$_tN_r = {_tN_o}\left(e^{-\alpha r}\right) + {_bN}\left(1 - e^{-\alpha r}\right)$$

FIG. 2. The main factors that influence the visibility of objects seen along a horizontal path of sight under water. From Lythgoe, 1969).

In Fig. 3 are shown the way that the relative spectral radiances from two objects, one grey, the other white, become transformed into the radiance of the water background spacelight. It will be seen that at each end of the spectrum where absorption is large, the apparent object radiance becomes indistinguishable from the water background radiance before wavelengths where the absorption is less rapid.

3. Intra-ocular media

In calculating the light actually reaching the outer segments of the rods and cones, account has to be taken of the absorption of the ocular media. The number of quanta actually absorbed in each type of outer segment can then be calculated (Lythgoe, 1968) provided the optical density of pigment in the outer segments is known.

4. The perception of objects

If an object is to be seen its apparent radiance must differ in some way from its background. Where only one receptor is

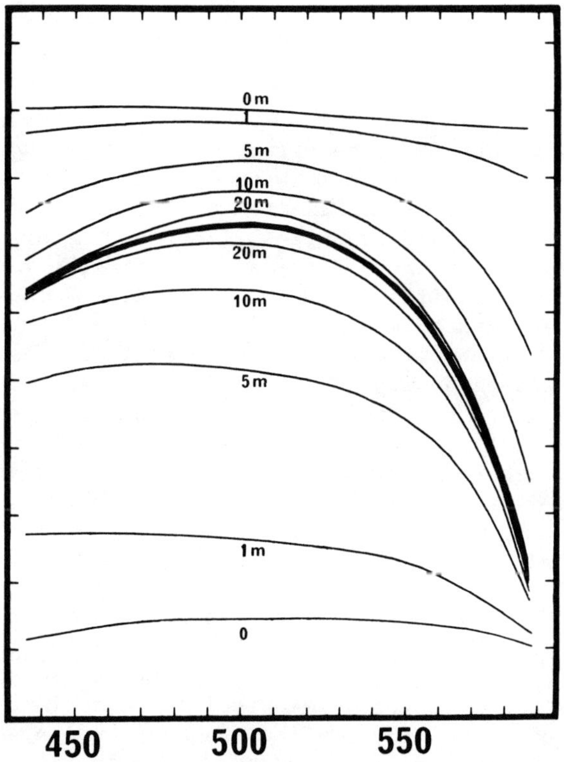

FIG. 3. The way that the spectral radiance of a dark grey and a white object become progressively more similar to that of the water background as their horizontal distance from the observer increases. (After Lythgoe, 1971).

involved, this difference can be expressed as contrast C.

$$C = \frac{t_r^N - b^N}{b^N} \qquad - 3$$

When C falls below some threshold value the object becomes invisible. The numerical value of the threshold depends, amongst other things, on the shape of the object, the sharpness of its outlines, and the adaptive state of the eye.

In colour vision there must be at least two receptor systems with differing spectral sensitivity. In man, where there are three such mechanisms, the observed psychophysical data can be well described by the Stiles-Helmholtz line element (Stiles, 1946; Wyszecki & Stiles, 1967),

$$(ds)^2 = (\frac{\Delta R}{\rho R})^2 + (\frac{\Delta G}{\gamma G})^2 + (\frac{\Delta B}{\beta B})^2 \quad - 4$$

where R, G and B represent the number of quanta absorbed by the R, G and B outer segments from the direction of the background and ΔR, ΔG, ΔB are the differences between R, G and B and the number of quanta absorbed by each mechanism from the direction of the object. ρ, γ, β are constants added by Stiles, 1946, to the original Helmholtz equation and represents the Weber fractions for the red, green and blue fundamental mechanisms. When ds fall below some liminal value the two radiances cannot be distinguished either by brightness or hue. The Stiles-Helmholtz line element postulates that the three mechanisms act independently of each other. In fishes the fundamental mechanisms are certainly not independent. However, neither are they independent in man. Nevertheless, the Stiles-Helmholtz line element successfully predicts the human hue discrimination curve (Wyszecki & Stiles, 1967).

The light reaching the eye from an object is greatly modified in its spectral distribution by the absorptive properties of the water itself. Yet the human observer is often able to assign the same colour to the object under water as he would at the surface (Kinney, Luria & Weitzman, 1967). Both behavioural evidence (Burkamp, 1923) and electrophysiological evidence (MacNichol, Wolbarsht & Wagner, 1961; Spekreijse, Wagner & Wolbarsht, 1972) suggest that some similar process takes place in fishes as well.

Inherent Visibility of Colours

On land the bandwidth of the light falling upon the retina is much greater than the absorption spectrum of the individual visual pigments. Yet even in quite shallow water (around 5m in coastal Jerlov 5 water, the ambient light is more restricted in bandwidth than the spectral absorption of the visual pigments in the retina (Lythgoe, 1966). This profoundly influences the appearance of different colours under water and, in effect, means that the spectral transmission properties of the water are more important than the absorption properties of the visual pigments in the cones. Yellows, oranges and reds are particularly suitable for analysis because they generally possess simple reflectance curves (Fig. 4) that reflect strongly at long wavelengths and the shorter wavelengths are absorbed. Of these colours yellow needs some explanation for our language is not well suited to understand it. In man the sensation of yellow is gained when the signal from the red and green absorbing cones are about equal and there is little or no signal from the blue. To our eyes the sensation of yellow is the same whether the entire spectrum minus blue is present or whether, as in the yellow sodium bands, the red and green sensitive cones are stimulated about equally.

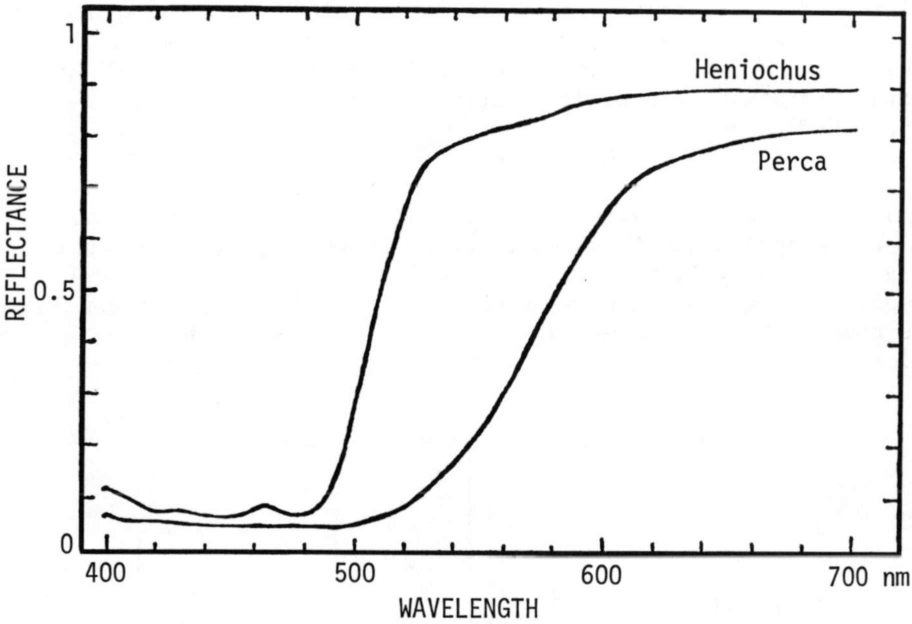

FIG. 4. The diffuse spectral reflectance of the caudal fin of the Perch, Perca fluviatilis, and the Whimple fish, Heniochus acuminatus. (Lythgoe & Northmore, 1973).

A red object in moderately deep ocean water reflects no light in the only region of the spectrum where ambient light is present. Similarly, a yellow object in an equivalent optical depth of green water has a flat spectral reflectance curve in the relevant part of the spectrum and is thus indistinguishable from grey. The spectral reflectance of the red object in green water and the yellow object in blue water change rapidly with wavelength in the relevant part of the spectrum and will thus appear different in colour from a grey or black background.

A colour that can be distinguished from some shade of grey underwater must possess a spectral reflectance curve that changes rapidly with wavelength in a region of the spectrum where there is sufficient light energy for photopic vision.

It can be seen from equation 2 and Fig. 2 that as the range between an object and the observer increases under water, so the apparent spectral radiance of the object becomes nearer and nearer to that of the background until finally the object becomes indistinguishable from the background in both colour and brightness. If the water contains many large suspended particles the spectral radiance of the spacelight is liable to be much broader than in clear water where scatter not selected for wavelength is less. Under these conditions, some underwater observations suggest that optically distant objects are discriminated by hue differences rather than brightness.

The way that the spectral radiance of an idealised yellow and a red object gradually become transformed to the horizontal background spacelight is shown in Fig. 5. The apparent spectral radiance of the yellow object in blue Gulf Stream water and the red object in green reservoir water remains different in shape (and thus difference in hue at a greater distance) than does the yellow in green water and the red in blue water.

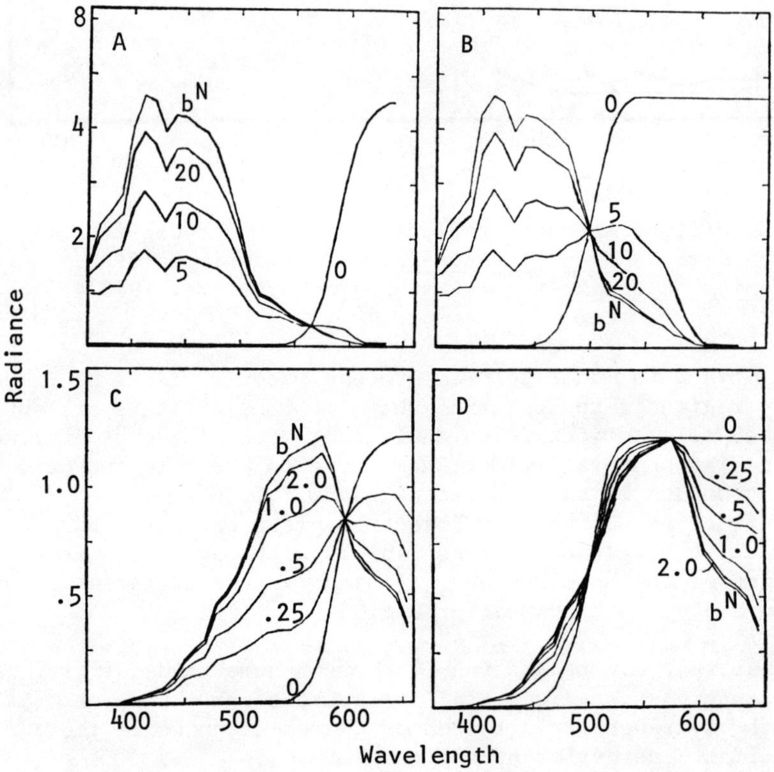

FIG. 5. The change in spectral radiance of a yellow and a red object as they recede horizontally from the eye into blue (Gulf Stream) water, and green (reservoir water). $_b N$ represents the spectral radiance of the water background. Distances are in metres. A, red in blue water; B, yellow in blue water; C, red in green water; D, yellow in green water. (Lythgoe & Northmore, 1973).

The relationship between α_λ and K_λ that have to be used in calculating the data in Fig. 5 is only an estimate and does not warrant the numerical treatment outlined for the visibility of colours at different depths. Nevertheless it is interesting to note that many inshore and freshwater fishes that live in green and yellow-green water possess brilliant red patches that have a clear behavioural significance. The red belly and flanks of the Stickleback, Gasterosteus aculeatus, the Char, Salvelinus alpinus, and the Perch, Perca fluviatilis, all intensify in the mating season.

The water that surrounds coral reefs is generally clear and blue, and many reef-dwelling fishes possess brilliant yellow markings. The yellow markings of the Butterfly fish, Chaetodon lunula, intensify in agonistic display (Hamilton & Peterman, 1971) and the yellow iris of the Damsell fish, Pomacentrus jenkinsi, is also implicated in display (Rasa, 1969).

The optical characteristics of the water that must be known for anything like a quantitative analysis of the visibility of colours at different visual ranges and depths are the narrower beam attenuation coefficient, α , the broad beam attenuation coefficient, K, and the water background radiance, $_bN$. Even now these three quantities have not been measured for any one body of water at the same time. However, if we consider a target held close to the eye consisting of a coloured central area surrounded by a grey area that is held close to the eye, neither $_bN$ nor α need enter the calculations and only K is required. A question that can be posed is down to what depth can the coloured area and the grey area be distinguished from each other by reason of hue alone.

In a computer assisted calculation (Lythgoe & Northmore, 1973) an attempt was made to give some quantitative expression to the proposition that red can be seen deeper than yellow in green coastal water and yellow deeper than red in blue oceanic water, by combining equations 1, 2 and 4.

The surface irradiance from which all these calculations are ultimately derived were in absolute quantum energy units (see Dartnall, this volume). The three fundamental response functions (equation 4) were equated with the quantum absorption of a rhodopsin of optical density 0.4 (see Lythgoe, 1966, 1972, for details), ds was taken as 0.02. The sensitivity of each fundamental mechanism could be independently varied by any factor from 0.1 to 10 as could their Weber fractions. The visual pigments used in the calculations had λ_{max} of 714, 667, 625, 588, 556, 526, 500, 476, 455, and 435. These pigments are separated by equal frequency intervals and all combinations of three separate pigments were considered.

In no case did we find exceptions to the rule that for any particular eye yellow can be seen deeper than red in blue ocean water and red deeper than yellow in green coastal water. This rule also remained true for the full range of Weber fractions used except in the trivial case when a very small Weber fraction was attributed to a response function where wavelength of maximum sensitivity was near the wavelength where the background and target radiance were equal, whilst the Weber fractions of the other two mechanisms were very large. Under these conditions, the computed maximum depth for colour vision was in any case reduced by more than 90%. The results of some of these calculations are shown in Table 1.

Table I

Oceanic			Coastal		
λ_{max}	Yellow	Red	λ_{max}	Yellow	Red
435 526 588	176m	39m	435 526 588	11m	16m
435 455 625	167m	42m	526 556 625	4m	19m

The effects of chromatic adaptations can be simulated by reducing the sensitivity of one of the three mechanisms to 10% of its previous value. This has only a small effect on the depth that red or yellow can be distinguished in hue from a grey.

At first sight it seems reasonable to ask what are the best three visual pigments needed to discriminate two radiances using colour vision. However, on further consideration it becomes clear that any discrimination between two particular hues can best be accomplished by using two pigments rather than three or some larger number.

In the calculation based on the Stiles-Helmholtz line element outlined above, it is necessary to postulate a minimum wave number interval between each of the three pigments otherwise two of the three pigments best suited to make a particular discrimination will be identical.

A glance at Table I will show that the best set of visual pigments to make a particular colour discrimination are usually not those best suited to discriminate another. In a very simple situation, such as the discrimination of two colours of photophores in deep sea fishes, two visual pigments are needed. Possible examples of this have recently been described for _Pachystomias_ by Denton, Gilpin-Brown & Wright (1970) and for _Aristostomias_ by O'Day & Fernandez (1974).

More than two pigments are presumably needed to make the more complex discriminations met on land. Perhaps three cone pigments are the best compromise between good hue discrimination in visually important parts of the spectrum and a reasonably fine hue discrimination throughout the radiation band suitable for vision.

Opponent Colour Mechanisms

In the goldfish, at least, colour opponent mechanisms seem to be of great importance in coding the response from the three cone types (Daw, 1968; MacNichol _et al_. 1961; Spekreijse, 1972). The organisation of the photopic and photopic-scotopic retinal processess in the retina are further discussed by Easter (1975), Powers (1975) and Raynauld (1975) in this volume. The possession of a strongly represented colour-opponent system does not mean that the Stiles-Helmholtz line element is inappropriate for predicting the possible discrimination between two nearly similar radiances. It does mean that trichromatic theory is unhelpful in predicting visual behaviour at suprathreshold levels.

Of considerable interest to colour vision ecologists is the fact it is the red-green opponent mechanisms that are well represented in the goldfish eye (Daw, 1968), and perhaps red-blue opponents also (Spekreijse _et al_. 1972). The predominance of the red-green opponent mechanism is also highlighted by Easter (this volume) by both electrophysiological and behavioural work and by Powers and Easter (this volume) using behavioural techniques. Important red-green opponence has also been shown by Northmore & Muntz (1973) by behavioural techniques in the Rudd (_Scardinius erythrophthalmus_).

It may be objected that the natural habitat of the goldfish is the goldfish bowl and few ecological inferences can be drawn from it. This objection may or may not be justified, but the Rudd certainly is a wild species presumably reasonably well adapted to its photic environment.

Fresh water, although immensely variable in colour, rarely looks red and it is not at once obvious why these two freshwater fish should possess red-green rather than blue-green opponent mechanisms. As Dartnall has pointed out in this volume, it is necessary actually to calculate the number of quanta captured by any particular visual pigment and that unless data are presented in the correct units "eyeball" comparison of spectral radiance and absorption data are most misleading. Calculations along these correct lines show that for Jerlov 9 yellow-green estuarine water (Jerlov, 1968) the rhodopsin that absorbs the greatest proportion of the available quanta in fact has a λ_{max} near to 600 nm. In Jerlov 5 coastal water a pigment of λ_{max} between 530 nm and 570 nm, according to depth, absorbs the most quanta (Lythgoe, 1966, 1972).

The colour of the ambient light in fresh water is very variable since an influence on its colour are the yellow products of vegetable decay (see above) that are washed out of the soil into the water when it rains. These yellow substances absorb strongly in the blue and are nearly transparent in the red. Thus changes in water transparency are much greater at short wavelengths. This is shown well in the spectral transmission curves for the different freshwater lakes shown in Fig. 1.

At any particular depth sensitivity in the red will vary less than it does at shorter wavelengths and under conditions of high turbidity associated with rainy weather the red-absorbing cones will capture the most quanta.

The blue sensitivity will show the greatest variation and there will be an intermediate variation in green sensitivity.

Colour constancy is presumably important to fishes since colours are important behavioural cues, yet the colour of the ambient light varies widely with depth, weather and season. A red-green opponent mechanism is much less sensitive to variations in the ambient light than a blue-red or a blue-green system would be. It is presumably more difficult to achieve colour constancy if the relative signal strength from the two cone types in a single opponent system are subject to large variations due to changes in the visual environment.

REFERENCES

Burkamp, W. (1923) Versuche über Farbenwiedererkennen der Fische, Z. Sinnesphysiol. 55; 133 - 170.

Clarke, G.L. & James, H.R. (1939) Laboratory analysis of the selective absorption of light by sea water. J. Opt. Soc. Amer. 29; 43 - 55.

Daw, N.W. (1968) Colour-coded ganglion cells in the goldfish retina: extension of their receptive fields by new stimuli. J. Physiol. 197; 507 - 592.

Denton, E.J., Gilpin-Brown, J.B. & Wright, P.G. (1970) On the "Filters" in the photophores of mesopelagic fish and on a fish emitting red light and especially sensitive to red light. Proc. Physiol. Soc. 199; 72 - 3.

Duntley, S.Q. (1962) Underwater visibility. In: "The Sea" I, edited by M.N. Hill, pp. 452 - 455, Interscience, New York

Duntley, S.Q. (1963) Light in the sea. J. Opt. Soc. Am. 53; 214 - 233.

Fogg, G.E. & Boalch, G.T. (1958) Extracellular products in pure cultures of a brown alga. Nature, 181; 789 - 790.

Halldal, P. (1974) Light and photosynthesis of different marine algal groups. In: Optical Aspects of Oceanography, edited by N.G. Jerlov and E. Steeman Nielsen, Academic Press, New York, pp. 345 - 359.

Hamilton, W.J. & Peterman, R.M. (1971) Countershading in the colourful reef fish Chaetodon lunula: concealment, communication or both. Anim. Behav. 19; 357 - 364.

Jerlov, N.G. (1968) Optical Oceanography, Elsevier, New York.

Kalle, K. (1938) Zum Problem der Meereswasserfarbe. Ann. Hydrol. Marine Mitt. 66; 1 - 13.

Kalle, K. (1966) The problem of Gelbstoffe in the sea. Oceanog. Marine Biol. Ann. Rev. 4; 91 - 104.

Kinney, J.S., Luria, S.M. & Weitzman, D.O. (1967) Visibility of colours under water. J. Opt. Soc. Am. 57; 802 - 809.

Lythgoe, J.N. (1966) Visual pigments and underwater vision. In Light as an ecological factor, edited by R. Bainbridge, G.C. Evans, O. Rackham, Blackwell, Oxford.

Lythgoe, J.N. (1968) Visual pigments and visual range underwater. Vis. Res. 8; 997 - 1012.

Lythgoe, J.N. (1971) Vision, in Underwater Science, edited by J.D. Woods and J.N. Lythgoe, 103 - 139. Oxford Univ. Press, London.

Lythgoe, J.N. (1972) The adaptation of visual pigments to their photic environment, in Handbook of Sensory Physiology, Vol. VII/1, pp. 566 - 603. Edited by H.J.A. Dartnall; Springer-Verlag, New York.

Lythgoe, J.N. & Northmore, D.P.M. (1973) Colours underwater, in Colour '73, pp. 77 - 98. Adam Hilger, London.

MacNichol, E.F., Wolbarsht, M.L. & Wagner, H.G. (1961) Electrophysiological evidence for a mechanism of color vision in the goldfish. In: Light & Life, edited by W.D. McElroy, and B. Glass, John Hopkins Press, Baltimore.

Northmore, D. P.M. & Muntz, W.R.A. (1973) Effects of stimulus size on spectral sensitivity in a fish, (Scardinius erythrophthalmus) measured with a classical conditioning paradigm. Vision Res. 14; 503 - 514.

O'Day, W.T. & Fernandez, H.R. (1974) Aristostomias scintillans (Malacosteidae): A deep-sea fish with visual pigments apparently adapted to its own bioluminescence. Vis. Res. 14; 545 - 550.

Rasa, O.A.E. (1969) Territoriality and the establishment of visual cues in Pomacentrus jenkinsi (Pisces, Pomacentridae). Z. Tierpsychol. 26; 825 - 845.

Shapiro, J. (1957) Chemical and biological studies of the yellow organic acids of lake water. Limnol. Oceanogr. 2; 161 - 179.

Smith, R.C. & Tyler, J.E. (1967) Optical properties of clear natural water. J. Opt. Soc. Amer. 57; 589 - 95.

Spekreijse, H., Wagner, H.G. & Wolbarsht, M.L. (1972) Spectral and spatial coding of ganglion cell responses in goldfish retina. J. Neurophysiol. 35; 73 - 86.

Stiles, W.S. (1946) A modified Helmholtz line element in brightness-colour space. Proc. Phys. Soc. (Lond.) 58; 41

Sullivan, S.A. (1963) Experimental study of the absorption in distilled water, artificial sea water and heavy water in the visible region of the spectrum. J. Opt. Soc. Am. 53; 962 - 967.

Talling, J.F. (1970) Generalized and specialized features of phytoplankton as a form of photosynthetic cover, in Prediction and Measurement of photosynthetic productivity, pp. 431 - 445. Centre for Agricultural publishing and documentation, Wagenengen.

Tyler, J.E. & Preisendorfer, R.W. (1962) In: Transmission of energy within the sea, The Sea, Vol. 1. pp. 397 - 451, edited by M.N. Hill. Interscience, New York.

Wyszecki, G. & Stiles, W.S. (1967) Colour Science Wiley, New York.

Tyler, J.E. & Smith, R.C. (1970) Measurements of spectral irradiance under water. Gordon & Breach, New York.

RHODOPSIN - PORPHYROPSIN CHANGES IN PAIRED-PIGMENT FISHES

D.D. BEATTY

Department of Zoology, University of Alberta

Edmonton 7, Alberta, T6G 2E1

O Bottom, thou art changed!
A Midsummer Night's Dream, Act. III, Sc. 1

INTRODUCTION

Visual pigments have been characterised for over 400 species of fishes (Lythgoe, 1972; Munz and McFarland, 1973), of which approximately one-sixth are paired-pigment species. The paired-pigment species are those having both rhodopsin and porphyropsin in the retina either together or sequentially. Except for the marine lamprey, *Petromyzon marinus*, all the known paired-pigment fishes are teleosts. Moreover few exclusively marine species have paired-pigments, most having only rhodopsins and a few only porphyropsins. Marine species with paired-pigments are mainly in the families Labridae and Scaridae. Among teleosts which are marine-freshwater (euryhaline) or exclusively freshwater species almost one-half of those studied have paired-pigments (Lythgoe, 1972; Schwanzara, 1967). The majority of these are in the Cyprinidae, Characidae and Cichlidae for tropical waters and in the Salmonidae and Cyprinidae for temperate waters. The cyprinids and characids are considered as primary freshwater fishes (inhabit only freshwater), the cichlids are secondary freshwater fishes (usually inhabit freshwater but may live in brackish waters) and the salmonids are peripheral freshwater fishes (live in freshwater but may live in brackish water or even full sea water) (Schwanzara, 1967).

Natural changes in the proportions of rhodopsin and porphyropsin occur seasonnally or at particular times in the life cycle of several paired-pigment species. Because the paired-pigments of a species presumably are based on the same opsin (Bridges, 1972) any change in the rhodopsin: porphyropsin ratio apparently involves an alteration in the availability and/or utilization of the two prosthetic groups, retinaldehyde (for rhodopsin) and 3-dehydroretinaldehyde (for porphyropsin). In general, a reciprocal relationship exists between the amounts of rhodopsin and porphyropsin; that is, if rhodopsin increases, porphyropsin decreases and vice versa (Bridges, 1972). This suggests that a proportion change involves a direct conversion by substitution of the alternate prosthetic group. An exception to this is the report by Allen (1971), that the proportion of porphyropsin in the redside shiner, *Richardsonius balteatus* is positively correlated to the total amount of visual pigment.

For most paired-pigment species, the identification and characterisation of the two visual pigments have resulted from sampling a population at a single time period and often required pooled retinas from several individuals. Thus, for most species, we have no information regarding possible rhodopsin: porphyropsin ratio changes in a population over time nor the individual variation in the proportions at any one time.

NATURAL CHANGES

The earliest evidence of a change in the composition of visual pigment of a fish came from studies on the marine lamprey. Lampreys migrating upstream from seawater to spawn have porphyropsin (Wald, 1957) and downstream migrants (landlocked) to the Great Lakes have rhodopsin (Crescitelli, 1956; Wald, 1957). Downstream migrants to the ocean have never been examined to determine if they too have rhodopsin nor have upstream migrants from the Great Lakes to determine whether they have porphyropsin. The suggestion is that there is a switch from rhodopsin to porphyropsin during the spawning migration in this species. Subsequently several studies have revealed that certain paired-pigment species have naturally occuring changes in their visual pigments. These natural changes have been related to the spawning migrations of euryhaline species and to the time of the year, the latter suggesting an effect of changing photic and/or temperature environments. During the spawning migration of two catadromous eels, the European eel, *Anguilla anguilla* (Carlisle and Denton, 1959; Wald, 1960) and the American eel, A. rostrata (Beatty unpublished observations) there is a shift from mainly porphyropsin to rhodopsin and then a further change to a new deep-sea rhodopsin. Among anadromous species, four species of Pacific salmon, genus *Oncorhynchus* (Beatty, 1966) and the rainbow smelt, *Osmerus eperlanus mordax* (Bridges and Delisle, 1974) change from predominantly rhodopsin to nearly pure porphyropsin during the spawning migration from seawater to freshwater. Pacific salmon are fall spawners and rainbow smelt are spring spawners.

Interestingly, Bridges and Delisle (1974) found that rainbow smelt on a second non-spawning upstream migration in the fall also have mainly porphyropsin. To complicate the picture, one landlocked population of rainbow smelt had nearly pure rhodopsin even at the time of spawning, whereas in four other lakes they had nearly pure porphyropsin and in yet two other lakes they had gross mixtures of the two pigments at the time of spawning. Beatty (1966) observed that both sea-run and land-locked sockeye salmon (kokanee), *Onco - rhynchus nerka*, (often regarded as the least anadromous of the Pacific salmon) maintained high proportions of rhodopsin during the spawning migration.

Pigment changes related to the seasons were first observed in the rudd, *Scardinius erythrophthalmus* (Dartnall et al., 1961). Rudd had maximum proportions of porphyropsin in the winter and minimum amounts in the summer. The relative proportions of the two pigments could be altered experimentally, darkness favouring more porphyropsin and light favouring less porphyropsin with the changes being reversible. Bridges and Yoshikami (1970a) confirmed this but also observed that the seasonal response was age dependent. Older fish (4 years) had much higher proportions of porphyropsin in the summer than did the younger fish. Comparable seasonal patterns have also been observed in the golden shiner, *Notemigonus crysoleucas boscii* and a poecilid, *Belonesox belizanus* (Bridges, 1965a), juvenile coho salmon, *Oncorhynchus kisutch* and king salmon, *O. tshawytscha* (Beatty, 1966), the burbot, *Lota lota* (Beatty, 1969a), the redside shiner, *Richardsonius balteatus* (Allen, 1971) another subspecies of golden shiner N.c. *aureatus* (Allen and McFarland, 1973) and cutthroat trout, *Salmo clarkii* (Allen et al., 1973). All these studies suggest that light quantity (duration and intensity) and/or light quality varies throughout the year and somehow influences the proportion of rhodopsin and porphyropsin, perhaps in an adaptive way.

ARTIFICALLY INDUCED CHANGES

Experiments designed to test whether the photic environment is a primary factor in inducing visual pigment ratio changes have produced results not always in agreement among the species tested. The rudd responds to continuous darkness by increasing porphyropsin and to light by increasing rhodopsin (Dartnall et al., 1961; Bridges and Yoshikami, 1970a). Bridges and Yoshikami (1970a) demonstrated that the eyes of rudd occluded with an opaque cap had more porphyropsin than the contralateral non-occluded eyes. This suggested that the response to light is localised within the individual eye and is not regulated by a central control mechanism. Bridges and Yoshikami (1970b) next reported that when rhodopsin replaced porphyropsin during light treatment of rudd, retinol replaced 3-dehyroretinol in the pigment epithelium and the reverse occurred in darkness. From this, they concluded that the pigment epithelium determines the proportions of rhodopsin and porphyropsin by regulating the availability of the two prosthetic groups. Light-induced changes in the proportions of rhodopsin and porphyropsin comparable to those occurring in the rudd have also exhibited by the poecilid, *B. belizanus* (Bridges, 1965b) and two subspecies of golden shiner, *N. c. boscii* and *N. c. aureatus* (Allen and McFarland, 1973). On the other hand, Allen (1971) reported that, for the redside shiner, light in a light/dark cycle favoured an increase in porphyropsin whereas darkness either had no appreciable effect or a decrease in porphyropsin, if porphyropsin were initially high. In addition, Allen (1971) concluded that continuous light also depressed the formation of porphyropsin. In continuous light, blue light was most effective in decreasing porphyropsin and in a light-dark cycle blue light was also most effective in promoting a porphyropsin increase (Allen, 1971). Allen (1971) concluded overall that photoperiod is the most important factor but intensity and quality of the light is also important, these factors interacting to determine a "steady state level" of porphyropsin "regardless of the initial percentage". In the rainbow trout, *Salmo gairdneri,* and the brook trout, *Salvelinus fontinalis,* darkness favoured rhodopsin and light favoured porphyropsin (Allen et al., 1973). The cutthroat trout also appeared to respond to continuous light by increasing porphyropsin; the effect of darkness was not tested (Allen et al., 1973). The brown trout, *Salmo trutta,* however, did not respond maintaining high porphyropsin under both photic conditions (Allen et al., 1973). Thus there are different patterns of response among species to artificially imposed photic conditions. Even though artificial conditions of continuous light or continuous darkness may induce visual pigment changes in a species, the biological significance of this is certainly open to question.

A second environmental factor which has recently been considered is temperature. Groups of juvenile golden shiner (*N.c. boscii*) held at two temperatures (7° and 21° C) had much higher proportions of porphyropsin at 7° regardless of whether the photic conditions were natural outdoor light, indoor blue light (15 L/9D) or continuous darkness (Allen and McFarland, 1973). Seasonal changes in the proportion of porphyropsin in the redside shiner (Allen, 1971) and the cutthroat trout (Allen et al., 1973) also seem to relate to seasonal changes in temperature. Thus there is a possibility that temperature may be an important factor in explaining comparable seasonal changes in visual pigment ratios among species which may respond differently to artificial photic conditions (the rudd, the golden shiner and *Belonesox* vs the cutthroat trout and the redside shiner) (Allen et al., 1973).

Unfortunately, it is extremely difficult to duplicate, in the laboratory, the variable photic and thermal environments of fishes. Therefore, at present, we cannot be definitive as to the causal environmental factors pertaining to visual pigment changes.

Two prevailing problems in studies involving changes in visual pigment ratios are the variability among individuals at any one time (Bridges 1964, 1965a; Bridges and Yoshikami, 1970a; Beatty, 1966) and possible differences in the proportions of rhodopsin and porphyropsin within various regions of the retina (Muntz and Northmore, 1971; Denton et al., 1971; Reuter et al., 1971). The former problem relates to the significance of the comparisons between control and experimentally treated groups or between different experimental groups. A possible solution to this problem is to remove one eye and measure the proportions of the two pigments prior to experimental manipulation. This method has been used with success on the rainbow trout, kokanee salmon, coho salmon and Atlantic salmon, *Salmo salar*, (Jacquest and Beatty, 1972; Beatty unpublished observations) and relies on the presence of comparable proportions of pigment between the two eyes of an individual at any one time (Bridges, 1964; Munz and Beatty, 1965; Bridges and Yoshikami, 1970a; Jacquest and Beatty, 1972). It also requires that a one-eyed fish will respond similarly as it would with both the eyes. This has not been adequately tested. The problem relating to distribution within the retina would occur if care is not taken to assure that the entire retina is always removed for analysis.

What other factors besides the spawning migrations, seasons of the year, experimental light and/or temperature conditions, individual variation and the region of the retina from which the visual pigment extract is made, are related to the proportion of rhodopsin and porphyropsin obtained from various paired-pigment species?

Little is known regarding the importance of dietary carotenoids on visual pigment composition. Rainbow trout fed walleye (*Stizostedion vitreum*) liver, containing mostly 3-dehydroretinol and little retinol, had increases in porphyropsin when compared to commercial pellet fed controls (Jacquest and Beatty, 1972). Kokanee salmon receiving injections of the saponified oil from walleye liver also had increased proportions of porphyropsin (Beatty, 1972). Rudd containing mostly porphyropsin when injected with retinyl acetate maintained the high prophyropsin if kept in darkness (Bridges and Yoshikami, 1970b). It is known that paired-pigment species may have visual pigment ratios which are quite unrelated to the proportions of retinol and 3-dehydroretinol in the liver (Beatty, 1966; Bridges and Yoshikami, 1970 b, c). Because the dietary carotenoid of most paired-pigment species is not likely to alter substantially during the year it is not a major factor to consider at this time.

POSSIBLE NEUROENDOCRINE MECHANISMS

The last possibility I wish to discuss, on the basis of evidence to date, is the possible role of the neuroendocrine system as a central control mechanism through which various environmental factors could influence visual pigment changes.

The first suggestion that a hormone can influence visual pigment composition came from studies on bullfrog tadpoles, *Rana catesbeiana* (Wilt 1959a, b). Thyroxine, known to induce anatomical changes of metamorphosis, promoted a marked increase in rhodopsin in the retina and retinol in pigment epithelium. Insertion of a pellet containing thyroxine into one eye caused a greater increase in rhodopsin in that eye compared to the contralateral eye, suggesting an ocular site for hormone action. Subsequently, Ohtsu et al., (1964) demonstrated that premetamorphic tadpole eyes in organ culture converted exogenous ^{3}H-retinol to labelled 3-dehydroretinaldehyde. Unfortunately they did not test whether thyroxine would retard this carotenoid conversion in isolated eyes. However, the conversion of exogenous ^{3}H-retinol to labelled 3-dehydroretinaldehyde decreased in intact eyes during thyroxine induced and naturally occurring metamorphosis. In isolated eyes of a species of sunfish, *Lepomis*, ^{3}H-retinol was converted to 3-dehydroretinaldehyde and the presence of triiodothyronine (T_3) in the injected ^{3}H-retinol depressed this conversion (Naito and Wilt, 1962). Unknown to Naito and Wilt, *Lepomis* is not a paired-pigment species, having only porphyropsin; thus the significance of the thyroid hormone-induced reduction of the conversion of retinol to 3-dehydroretinaldehyde is not evident. There was, however the conclusion by this group of workers that the eye is the site for carotenoid prosthetic group interconversions and that thyroid hormones depressed the conversion of retinol or retinaldehyde to 3-dehydroretinol or 3-dehydroretinaldehyde and thus favoured rhodopsin formation.

The first test of the effect of thyroid hormone on the pigment ratios of a paired-pigment fish was reported by Munz and Swanson (1965). Rainbow trout held in water containing thyroxine had increases of porphyropsin. In a more detailed report, Beatty (1969b) found that intraperitoneal injections of L-thyroxine (T4) and T_3 caused substantial increases of porphyropsin (as high as 99%) in juvenile kokanee salmon, a species normally having mostly rhodopsin (>80%). Other studies subsequently showing a similar response to thyroid hormones are reported for coho salmon (Beatty, 1972), rainbow trout (Jacquest and Beatty, 1972; Cristy, 1974), redside shiner (Allen, 1971) and Atlantic salmon parr and smolt (Beatty, unpublished observations). Thyroxine treatment of rudd (Bridges and Yoshikami, 1970a) and rainbow smelt (Bridges and Delisle, 1974) did not cause any significant increase in porphyropsin-rich fish. So far no study has shown that thyroid hormone favours rhodopsin as it does in bullfrogs.

The levels of thyroid hormone used to induce the rhodopsin to porphyropsin conversion are probably unnaturally high (we have no information on the circulating blood levels). Thus, the pigment change may merely represent a response to pharmacological doses with no biological significance to the natural situation. On the other hand, Beatty (1972) reported that kokanee salmon receiving bovine thyroid stimulating hormone (TSH) had increases in porphyropsin comparable to those receiving low doses of thyroid hormone. This suggests that the fish can produce and release sufficient endogenous thyroxine under TSH stimulation to induce the visual pigment change.

Another hormone recently implicated in visual pigment change in the rainbow trout is prolactin (Cristy, 1974). Ovine prolactin also causes an increase in porphyropsin. This result however may not be a direct effect of prolactin because mammalian prolactin may cause TSH secretion in certain fish and thereby stimulate the thyroid gland.

Nonetheless three hormones (thyroid hormone, thyrotropic hormone and prolactin) are effective in promoting visual pigment changes in certain fishes. One of these, thyroid hormone, has been implicated in a wide range of functions in teleosts including metamorphosis, reproduction, certain aspects of metabolism, the central nervous system, osmoregulation and others (Sage, 1973). Likewise, prolaction is involved in ionic and osmotic regulation of certain teleosts especially in freshwater fish or euryhaline fish adapting to freshwater (Johnson, 1973; Schreibman et al., 1973). There is some evidence for a hypothalamus-pituitary-thyroid axis operating in fish to control thyroid function in response to intrinsic and extrinsic factors (reviewed by Gorbman, 1969; also see Schreibman, et al., 1973; Peter, 1973). Nonetheless as Peter (1973) and Gorbman (1973) have commented, there is not sufficient basic information on the control mechanism for the synthesis and release of any pituitary hormone and its feedback regulation to make definitive conclusions.

We are now left with the challenge of determining whether the neuroendocrine system is intimately involved in visual pigment changes in species such as the salmonids and cyprinids. Techniques for measuring thyroid activity and circulating levels of thyroid hormone are available (Higgs and Eales, 1973; Eales, 1974). Likewise it is now possible to obtain information regarding the hypothalamic and pituitary centers related to TSH synthesis and release (Schreibman et al., 1973; Peter, 1973). Perhaps in the not too distant future we shall have some answers to this question.

In conclusion, we cannot rule out a direct effect of light, or some other environmental factor, in the eye or an indirect effect via a purely neuronal mechanism as being an essential feature of visual pigment changes in certain species. However the mechanism is controlled, it relates to a genetic capability of the fish to have both rhodopsin and porphyropsin and then the ability to alter the availability and/or utilisation of the two prothetic groups in forming rhodopsin and porphyropsin. For a more comprehensive treatment of this subject, refer to the excellent review by Bridges (1972).

REFERENCES

Allen, D.M. (1971) Photic control of the proportions of two visual pigments in a fish. Vision Res. 11: 1077-1112.

Allen, D.M. and McFarland, W.N. (1973) The effect of temperature on rhodopsin-porphyropsin ratios in fish. Vision Res. 13: 1303-1309.

Allen, D.M., McFarland, W.M., Munz, F.W., and Poston, H.A. (1973) Changes in the visual pigments of trout, Can. J. Zool. 51: 901-914.

Beatty, D.D. (1966) A study of the succession of visual pigments in Pacific salmon (*Oncorhynchus*). Can. J. Zool. 44: 429-455.

Beatty, D.D. (1969a) Visual pigments of the burbot, Lota lota, and seasonal changes in their relative proportions. Vision Res. 9: 1173-1183.

Beatty, D.D. (1969b) Visual pigment changes in juvenile kokanee salmon in response to thyroid hormones. Vision Res. 9: 855-864.

Beatty, D.D. (1972) Visual pigment changes in salmonid fishes in response to exogenous L-thyroxine, bovine TSH, and 3-dehydroretinol. Vision Res. 12: 1947-1960.

Bridges, C.D.B. (1964) Variation of visual pigment amongst individuals of an American minnow, *Notemigonus crysoleucas boscii*. Vision Res. 4: 233-239.

Bridges, C.D.B. (1965a) Variability and relationships of fish visual pigments. Vision Res. 5: 239-251.

Bridges, C.D.B. (1965b) Visual pigments of a fish exposed to different light-environments. Nature (Lond) 206: 1161-1162.

Bridges, C.D.B. (1972) The rhodopsin-porphyropsin visual system. In Handbook of Sensory Physiology VII-1A.Photochemistry of Vision, edited by H.J.A. Dartnall, pp. 418-480, Springer-Verlag, N.York.

Bridges, C.D.B. and Delisle, C.E. (1974) Postglacial evolution of the visual pigment of the smelt, *Osmerus eperlanus mordax*. Vision Res. 14: 345-356.

Bridges, C.D.B. and Yoshikami, S. (1970a) The rhodopsin-porphyropsin system in freshwater fishes-1. Effects of age and photic environment. Vision Res. 10: 1315-1332.

Bridges, C.D.B. and Yoshikami, S. (1970b) The rhodopsin-porphyropsin system in freshwater fishes-2. Turnover and interconversion of visual pigment prosthetic groups in light and darkness: role of the pigment epithelium. Vision Res. 10: 1333-1345.

Bridges, C.D.B. and Yoshikami, S. (1970c) Distribution and evolution of visual pigments in salmonid fishes. Vision Res. 10: 609-626.

Carlisle, D.B. and Denton, E.J. (1959). On the metamorphosis of the visual pigments of *Anguilla anguilla* (L) J. Mar. Biol. Ass. U.K. 38: 97-102.

Crescitelli, F. (1956). The nature of the lamprey visual pigment. J. Gen. Physiol. 39: 423-435.

Cristy, M. (1974) Effects of prolactin and thyroxine on the visual pigments of trout, *Salmo gairdneri*. Gen. Comp. Endocrinol 23: 58-62.

Dartnall, H.J.A., Lander, M.E., and Munz, F.W. (1961) Periodic changes in the visual pigment of a fish. In Progress in Photobiology, edited by B. Chr. Christensen and B. Buchmann, pp. 203-213, Elsevier, Amsterdam, 1961.

Denton, E.J., Muntz, W.R.A. and Northmore, D.P.M. (1971) The distribution of visual pigment within the retina in two teleosts. J. Mar. Biol. Ass. U.K. 51: 905-915.

Eales, J.G. (1974) Creation of chronic physiological elevations of plasma thyroxine in brook trout *Salvelinus fontinalis* (Mitchell) and other teleosts. Gen. Comp. Endocrinol. 22: 209-217.

Gorbman, A. (1969) Thyroid function and its control in fishes. In Fish Physiology, edited by W.S. Hoar and D.J. Randall, pp. 241-274 Academic Press, New York.

Gorbman, A. (1973) Comments on "Neuroendocrinology of Teleosts" by R.E. Peter. Am. Zool. 13: 757-758.

Higgs, D.A. and Eales, J.G. (1973) Measurement of circulating thyroxine in several freshwater teleosts by competetive binding analysis. Can. J. Zool. 51: 49-53.

Jacquest, W.L. and Beatty, D.D. (1972) Visual pigment changes in the rainbow trout, *Salmo gairdneri*. Can. J. Zool. 50: 1117-1126.

Johnson, D.W. (1973) Endocrine control of hydromineral balance in teleosts. Amer. Zool. 13: 799-818.

Lythgoe, J.N. (1972) List of vertebrate visual pigments. In Handbook of Sensory Physiology VII-1A. In Photochemistry of Vision, edited by J.H.A. Dartnall, pp. 604-624, Springer-Verlag, N.York.

Muntz, W.R.A. and Northmore, D.P.M. (1971) Visual Pigments from different parts of the retina in rudd and trout. Vision Res. 11: 551-561.

Munz, F.W. and Beatty, D.D. (1965) A critical analysis of the visual pigments of salmon and trout, Vision Res. 5: 1-17.

Munz, F.W. and Swanson, R.T. (1965) Thyroxine-induced changes in the proportions of visual pigments. Am. Zool. 5: 683.

Munz, F.W. and McFarland, W.M. (1973) The significance of spectral position in the rhodopsins of tropical marine fishes. Vision Res. 13: 1829-1874.

Naito, K. and Wilt, F.H. (1962) The conversion of vitamin A to retinene$_2$ in a freshwater fish. J. Biol. Chem. 237: 3060-3064.

Ohtsu, K., Naito, K. and Wilt, F.H. (1964) Metabolic basis of visual pigment conversion in metamorphosing *Rana catesbeiana*. Dev. Biol. 10: 216-232.

Peter, R.E. (1973) Neuroendocrinology of teleosts. Am. Zool. 13: 743-755.

Reuter, T.E., White, R.H. and Wald, G. (1971) Rhodopsin and porphyropsin fields in the adult bullfrog retina. J. Gen. Physiol. 58: 351-371.

Sage, M. (1973) The evolution of thyroid function in fishes. Am. Zool. 13: 899-905.

Schreibman, M.P., Leatherland, J.F. and McKeown, B.A. (1973) Functional morphology of the teleost pituitary gland. Am. Zool. 13: 719-742.

Schwanzara, S.A. (1967) The visual pigments of freshwater fishes. Vision Res. 7: 121-148.

Wald, G. (1957) The metamorphosis of visual systems in the sea lamprey. J. Gen. Physiol. 40: 901-914.

Wald, G. (1960) The distribution and evolution of visual systems. In Comparative Biochemistry Vol. 1, edited by M. Florkin and H. Mason, pp. 311-345, Academic Press, New York.

Wilt, F.H. (1959a) The differentiation of visual pigments in metamorphosing larvae of *Rana catesbeiana*. Dev Biol. 1: 199-233.

Wilt, F.H. (1959b) The organ specific action of thyroxin in visual pigment differentiation. J. Embryol. Exp. Morphol. 7: 556-563.

SOME PROBLEMS OF DEEP-SEA FISH EYES

N.A. LOCKET

Institute of Ophthalmology, University of London

Judd Street, London WC1H9QS, U.K.

I can call spirits from the vasty deep
Henry IV, Part 1, Act III, Sc. 1

The eyes of deep-sea teleost fishes show a number of peculiar features, some of which appear restricted to few or even single species, while others occur more widely among fishes of diverse kinds. In some cases the adaptations may be tentatively correlated with depth, but in others there is no clear association with the habitat.

Some adaptations appear bizarre, and in others it is hard to understand how they may affect the vision of the fish which bears them. In the following examples two sorts of problem are considered; those in which the fishes' solution to a problem imposed by the environment can be described, and those in which the adaptation poses some questions to the investigator.

Workers familiar with the retinal structure of shallow water teleosts know well how widely they vary. Teleost retinae are commonly highly organised and contain, as a rule, several types of visual cells. The rods are usually smaller than the cones, and more numerous, but in some examples, e.g. Ictalurus and the other catfishes, the rods are larger than the cones, as in amphibians. The cones themselves are commonly of more than one type. There are single cones and multiple cones, equal twins and unequal doubles. Triple and quadruple cones also are known. In most cases the predominant types are the single and twin or double, which are often arranged in geometrical patterns. The outer segments of the cones are usually about 20-30μm in length, accounting for perhaps a quarter of the retinal thickness. The neural layers are prominently developed with the ganglion cells abundant and the layering well defined. In deep-sea fishes cones are exceptional. Most of them have only rods. These are closely packed and contribute largely to the bulk of the retina.

In some cases the rods are present in a single bank and their outer segments are long. Examples are Sternoptyx, a silvery-sided hatchet fish, and the searsid family, drab grey or black fishes, mostly living below 1000m.

In a Sternoptyx eye the outer segments contributed 41% of the retinal thickness, compared to 15% in a goldfish eye of the same dimensions. This difference might be due to a greater development of the other elements in the goldfish, but the Sternoptyx rods are of greater absolute length, and are proportionately longer.

Sternoptyx and the searsids certainly have long outer segments, but they are dwarfed by those of Diretmus. The most unusual retina of this species includes a patch where rod outer segments 600μm long are present in a retina 900μm thick. (Munk, 1966a)

The presence of long outer segments raises some questions about how they may function. The visual pigment is located in the outer segment, and in deep-sea fishes it is present in high density. (Denton and Warren, 1957). In certain animals the visual pigment is regenerated, the process involving the pigment epithelium. (Wald, 1936, Dowling, 1960). There is also evidence from radioautographic studies in frogs and rats that the outer segment is formed at its vitread end, and eroded at its sclerad end. (Young and Bok, 1969 for references).

In many deep-sea fishes the pigment epithelium is weakly developed, without the extensive processes which protrude between the outer segments in shallow water fishes. The pigment epithelium surface is more or less flat, so that the area of contact with the outer segment is practically confined to the tip. What little light there is at depth, falls on the vitread end of the outer segment and may be largely absorbed by the pigment at that end. Thus the pigment broken down, and so needing regeneration, is some distance from the pigment epithelium, and it is not clear how regeneration might take place.

The work of Young and his colleagues showed that the labelled outer segment material was eventually shed from the sclerad end, to appear in the pigment epithelium cells as phagosomes, which eventually disappear after lysis within the cell. The presence of phagosomes in the epithelium of deep-sea fishes suggests that the process of outer segment phagocytosis may take place here too. If so, the outer segment is presumably formed at the vitread end during life, as in the species studied by Young. His radio-autography showed the labelled amino acids first in the inner segment, then appearing to pass through the cilium and finally forming a band in the outer segment. It was suggested that the label does indeed pass through the cilium, sometimes regarded as the sole connexion between the inner and outer segments. (Young, 1968).

The junction of inner and outer segments in several species have features to suggest that disc formation does take place, and that it is a complicated process involving more than transport of substances through the cilium. The visual pigment is located in the discs, and thus new pigment may be formed throughout life. One may speculate whether such pigment formation might be sufficient for all requirements if breakdown is slow, which it may be in conditions of low illumination and high pressure. If this is so the involvement of the pigment epithelium in visual pigment regeneration need not be invoked, though it could still remove the outer segment tips, and return their contained metabolites to the circulation.

The length of an outer segment will be determined by the rates at which it is formed and destroyed. Some fishes show very different outer segment lengths in different areas of the retina. Examples are Diretmus (Munk, 1966a) which has an area containing the longest rods recorded, and Scopelarchus sagax (but not the closely related S. guntheri) (Locket 1971). In S. sagax there is an area in the main retina where the rod length changes from 40 to 200μm over the width of a few receptors. The other retinal elements do not appear in any way different where they are related to the long rods, and the reason for and implications of such a localised area remain obscure.

The properties of the fish lens have been demonstrated by Pumphrey (1961). The lens is spherical, and the ratio of the lens radius to its focal length is 1;2.55, known as Matthiessen's ratio. A lens of these proportions produces an exit cone of light which subtends 46.5° at its focal point. If this focal point is assumed to be at the vitread end of the outer segment, then if the whole retina were a homogeneous medium the light would have diverged to cover a circle 130μm in diameter in the length of the rods in an example studied of the searsid fish Platytroctes apus. Such a circle would contain some 770 outer segments. But the outer segments and the fluid between them are not a homogeneous medium: Nicol (1969) quotes Sidman (1957) who found the refractive index of certain outer segments to be 1.41, and that of the surrounding medium 1.336. These figures give a critical angle of 71° for an outer segment, and a 42° cone of light focussed on the vitread face would be confined to that outer segment by total internal reflection. This figure agrees well with the 46° cone of light subtended at the focus of the lens, so that light will tend to be so confined; however long the outer segment is, it will be optically isolated from its neighbours.

In deep-sea retinae it is probably advantageous to receive information from as much of the incoming light as possible. Ideally the whole area of the retina would be occupied at an appropriate level by outer segments, but in reality these are cylindrical, not hexagonal as they would have to be to occupy all the space. Even in an array of packed circles in mutual contact the circles occupy 88.6% of the whole area. Figure 1. If the receptors were more tightly packed there would be larger areas of contact, and the light would not be well confined by total internal reflexion.

A common feature in deep-sea retinae, surprisingly overlooked by Brauer (1908), the first to study deep-sea fish eyes

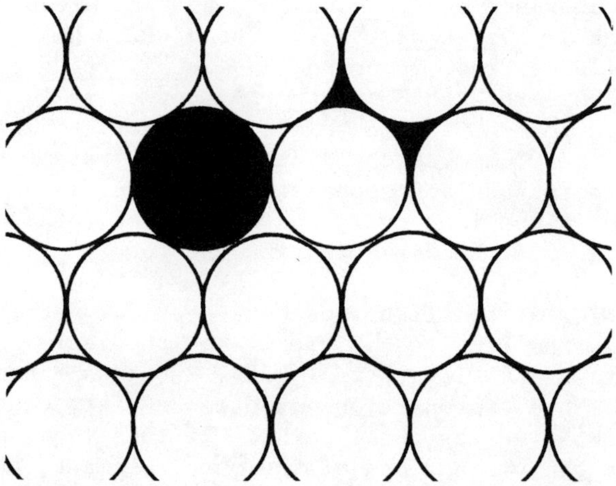

FIG. 1. Uniform cylindrical rods when packed as tightly as possible take up this formation. The repeating unit in such a pattern is a circle with two triangular spaces. The circle accounts for 88.6% of that area. The points of contact between adjacent rods are minimal, and light will be confined to each by total internal reflexion.

in detail, is the arrangement of receptors in multiple banks. The receptors in shallow water fishes are often at different levels, and the arrangement may change with illumination, giving the familiar retinomotor responses. In these cases the receptors are of different families, i.e. rods and cones, but those of deep-sea fishes are all rods. Though Rochon-Duvigneaud (1943) shows the arrangement in a figure he does not comment upon it, and it was not until 1953 that Vilter drew attention to it. The feature has since been found in many species; Munk (1966b) lists 14, and points out that the feature may occur apparently without regard for taxonomic relationships. This led him to suggest that it is an adaptive feature, a view supported by further examination of unrelated species.

The arrangement in multiple banks occurs only sclerad to the outer limiting membrane; (Figure 2) the myoids project sclerad to this level, and in the most vitread bank of rods they are short, leading directly into the ellipsoid which in turn bears the outer segment. These vitread rods commonly form a regular and apparently uniform bank. Other rods, having the same diameter as the vitread at the outer limiting membrane, narrow abruptly sclerad to it and form elongated threadlike myoids. These thin myoids pass between the outer segments of the vitread bank, and widen to join the ellipsoid just beyond the outer segment tips. The ellipsoids in turn bear the outer segments of the sclerad bank (Figure 3).

Though two banks are considered here, there are cases in which more than two are present. The example described by Vilter had three banks over most of the retina, and six banks at its shallow fovea. The number of banks may vary not only from species to species, and with location in the retina, but also between specimens within a species. Munk (1966b) noted three banks in the main retina of a specimen of Stylophorus 140mm long, but I found five banks in a specimen 215mm in length. In some fishes showing the feature, e.g. Stylophorus, Diretmus, the rods in the different layers are of increasing diameter from vitread to sclerad. In others, including Stomias boa ferox (Munk 1963) and the melamphaeid Poromitra nigrofulvus (Locket 1969) the rods are of the same size.

The possession of multiple banks is presumably advantageous to the fishes which have them, but their exact function has remained uncertain despite ingenious attempts to elucidate it. Vilter believed they represented a static accommodatory device, the sclerad bank receiving a clear image from near objects and the vitread a clear image from far ones. Munk (1966b) discussed this theory: The retinal thickness is not sufficient to give the depth of field which would be required, and furthermore the fishes

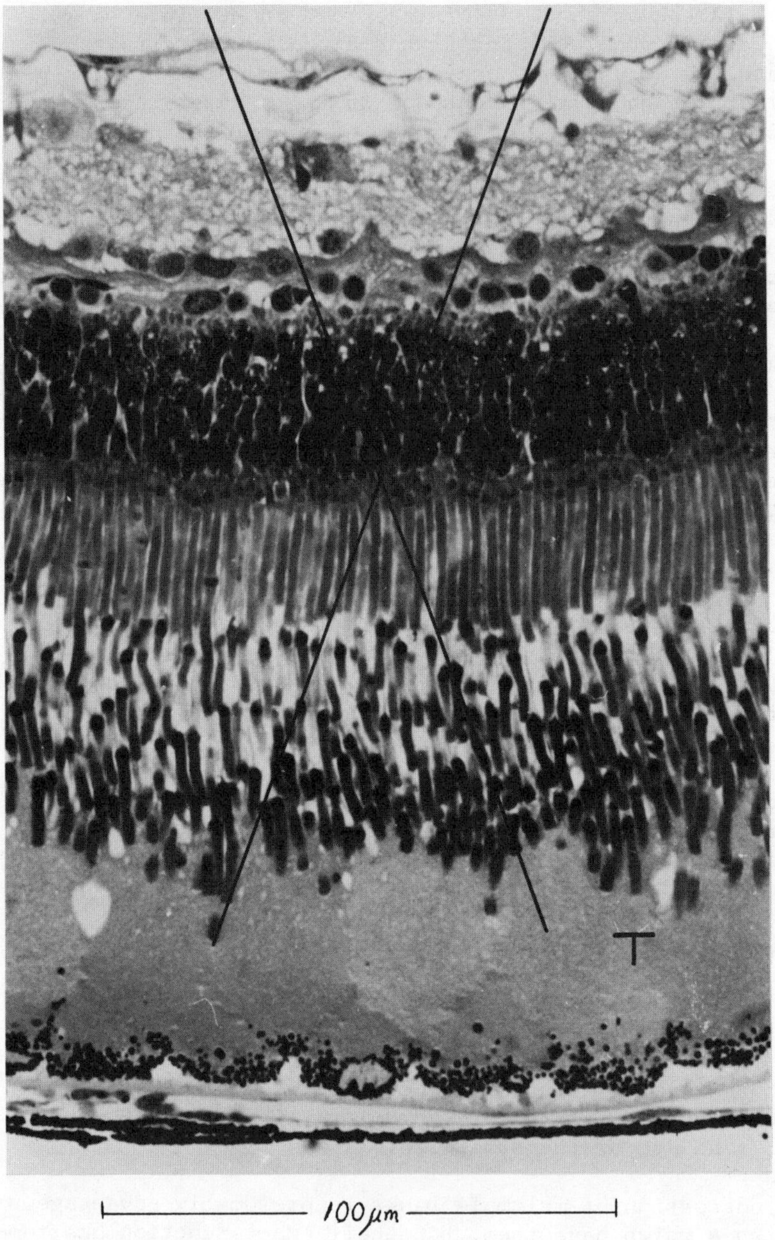

FIG. 2. Retina of Malacosteus niger. The receptors are all rods, and are arranged in a vitread compact bank and a more diffuse sclerad bank. Some of the threadlike myoids of the sclerad bank can be seen. The crossed lines represent the cone of light from the lens focussed at the

with multiple banks which he studied have a normally developed lens muscle. Vilter's proposition would require a retina which could distinguish between focussed and non-focussed images in up to six superposed banks, and he made no suggestion as to how this might occur.

Duke-Elder (1958) proposed that the multiple arrangement might increase retinal sensitivity, in the same way that long rods might, by increasing the light path length through visual pigment. Munk suggested that the arrangement might increase sensitivity by increasing the convergence ratio from receptors to ganglion cells, but cell counts did not support this view.

For vision to occur in very low illumination certain conditions must be satisfied. Vision implies that the stimulus is perceived, and this requires that ganglion cell impulses reach the brain. The ganglion cell discharge is evoked by the summated action of the receptors, transmitted and modified by the other retinal neurons. Considering the receptor itself, an input of some quanta of light is trapped by the visual pigment, which undergoes a photochemical reaction. The disturbance in the outer segment due to this reaction in some way affects the vitread portion of the receptor, altering conditions in the synapse so as to affect the cells making contact with that receptor.

Consider some implications of this in the case of multiple bank retinae: Light is focussed by the lens, probably at the level of the outer limiting membrane; at this level all the receptors are of the same diameter. Light to the vitread bank of rods passes through the myoid and ellipsoid to the base of the outer segment, all of nearly uniform diameter.

Miller and Snyder (1972) have considered the optical properties of the myoids of frog receptors, and state that the wavelength preferentially transmitted will depend on the diameter of the myoid, due to light guiding effects. If an optical fibre's diameter is of the order of 12 times the wavelength, light will be guided by total internal reflexion; for the blue light of deep ocean waters the corresponding diameter would be about 5.6μm. If the fibre is smaller, then geometrical optics cease to apply, and waveguide action must be considered.

(FIG. 2. cont.) outer limiting membrane. If the light continued undeviated it would spread out to cover many rods in the thickness of the outer segment zone. This species has a retinal tapetum, T, which contains droplets of a red lipid.

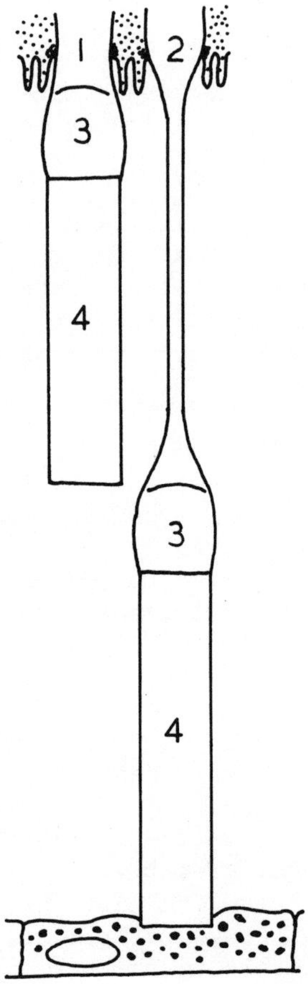

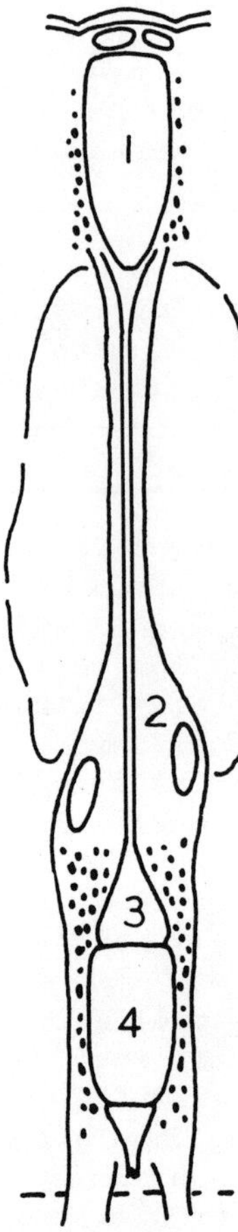

FIG. 3. (left) Diagram of the multiple bank arrangement of rods. The vitread bank have short myoids and light can reach the outer segment with little disturbance. The myoid of the sclerad rod is the same diameter as the vitread at the outer limiting membrane. It narrows to a thread which passes between the outer segments of the vitread bank, and widens to the ellipsoid sclerad to those outer segments. The image is believed to be focussed at the outer limiting membrane, and light for the sclerad bank conducted along the narrow myoid. 1, Myoid of vitread rod. 2, Myoid of sclerad rod. 3, Ellipsoid. 4, Outer segment.

FIG. 4. (right) Diagram of ommatidium in euphausiid crustacean. Light is focussed by the corneal facet and crystalline cone into a hyaline filament, which conducts it to the rhabdome. 1, Crystalline cone. 2, Retinular cells. 3, Hyaline filament. 4, Rhabdome.

The myoids of the vitread rods are about 2.5μm in diameter, but are short and will probably offer little obstacle to light traversing them. The myoids of the sclerad banks are both longer and thinner. They are about 0.3um in diameter and may be hundreds of microns long. Though they will not behave as perfect light guides at these dimensions they are still likely to have significant light guiding properties. If this is so, the more sclerad banks of rods may be regarded as optically 'wired' to the image plane at the outer limiting membrane, and they will receive a share of light equal to that of the short rods less any lost due to light guiding imperfections.

In such a case the retinal grain will be determined by the packing at the outer limiting membrane, and not by the outer segment diameter. A given area of multibank retina will contain more rods of a given diameter than a single bank one; the opportunity for high convergence within a small area could lead to improved resolving power and sensitivity.

Indirect evidence that the long myoids could function in the way proposed is provided by the closely parallel situation in the compound eyes of euphausiids (Kampa, 1965). In each ommatidium of these crustaceans the light is focussed by the crystalline cone into a hyaline filament 350μm long and about 0.5μm diameter, which conducts it to the rhabdome. The filaments are expanded at each end, but the narrow portion is of comparable diameter to the myoids now considered. Figure 4.

Besides conducting light, the long myoids must convey the disturbance due to the photochemical reaction to the perikaryon and ultimately the synapse. The myoid is part of a neuron, and at 0.3μm comparable in diameter to a dendrite or an axon. Those elements are usually surrounded by glial or Schwann cell cytoplasm; the myoids lie in the ventricular fluid between the outer limiting membrane and the pigment epithelium.

The relationship between the rods and pigment cells has been mentioned above in connexion with long outer segments. In the multiple bank arrangement the outer segments of the outermost bank are in contact with the epithelium, and so might be capable of pigment regeneration if this is epithelium-mediated, but those in other banks are separated from the epithelium by distances of tens of microns. It seems unlikely that regenerative processes could continue in that case. The epithelial cells of multiple bank retinae may contain phagosomes, but how fragments of outer segment could become detached and traverse the ventricular space is not clear. It might be that only the outer bank have their outer segments eroded; in that case one would expect the inner banks to be longer than the outer, but this is not necessarily the case.

REFERENCES

Brauer. A. (1908). Die Tiefseefische. Wiss. Ergebn. dt. Tiefsee-Exped. 'Valdivia', 15: 2. Anat. Teil.

Denton, E.J. and Warren, F.J. (1957). The photosensitive pigments in the retinae of deep-sea fish. J. Mar. Biol. Ass. U. K., 36; 651-662

Dowling, J. (1960). Chemistry of visual adaptation in the rat. Nature, (Lond) 188: 114-118.

Duke-Elder, S. (1958). System of Ophthalmology. I. The eye in evolution. Kimpton, London.

Kampa, E. M. (1965). The euphausiid eye; a reevaluation. Vision Res. 5: 475-481.

Locket, N.A. (1969). The retina of Poromitra nigrofulvus Garman, an optical and electron microscope study. Expl. Eye. Res. 8: 265-275.

Locket, N.A. (1971). Retinal anatomy in some scopelarchid deep-sea fishes. Proc. R. Soc. Lond. B., 178: 161-184.

Miller, W.H. and Snyder, A.W. (1972). Optical function of myoids. Vision Res. 12: 1841-1848.

Munk, O. (1963). The eye of Stomias boa ferox. Vidensk. Medd. Dan. Naturhist. Foren KBH. 125: 353.

Munk, O. (1966a). On the retina of Diretmus argenteus Johnson, 1863 (Diretmidae, Pisces). Vidensk. Medd. Dan. Naturist. Foren. KBH. 129: 73-80.

Munk, O. (1966b). Ocular anatomy of some deep-sea teleosts. Dana. Rep., 70; 71pp.

Nicol, J.A.C. (1969). The tapetum lucidum of the sturgeon. Contrib. Mar. Sci., 14: 7-18.

Pumphrey, R.J. (1961). Concerning vision. In: The cell and the organism. Essays presented to Sir James Gray. Edited by J.A. Ramsay and V.B. Wigglesworth, p. 193-208. Cambridge University Press.

Rochon-Duvigneaud, A. (1943). Les yeux et la vision des vertébrés. Masson. Paris.

Sidman, R.L. (1957). The structure and concentration of solids in photoreceptor cells studied by refractometry and interference microscopy. J. Biophys. Biochem. Cytol., 3: 15-30.

Vilter, V. (1953). Existence d'une rétine à plusieurs mosaïques photoreceptrices chez un poisson abyssal bathypelagique, Bathylagus benedicti. C.R. Seances Soc. Biol., 147; 1937-1939.

Wald, G. (1935-36). Carotenoids and the visual cycle. J. Gen. Physiol. 19: 351.

Young, R.W. (1968). Passage of newly formed protein through the connecting cilium of retinal rods in the frog. J. Ultrastruct. Res. 23: 462-473.

Young, R.W. and Bok, D. (1969). Participation of the retinal pigment epithelium in the rod outer segment renewal process. J. Cell Biol. 42: 392-403.

PROSPECTS IN THE STUDY OF INTERRELATIONSHIPS BETWEEN VISION AND BIOLUMINESCENCE

Michel ANCTIL*

Département de Biologie, Université de Montréal

C.P. 6128, Montréal 101, Canada

Having call'd them from the deep! O, still
Thy deafening, dreadful thunders; gently quench
Thy nimble sulphurous flashes!
Pericles, Act. II. Sc. I

INTRODUCTION

Bioluminescence is not only potentially detected by the visual apparatus of an animal, but it is also possible that the photoreceptive process of this animal serves in specific cases to exert control on its own luminescent output. In recent years a large body of information has been collected on the visual and, to a lesser extent, the photogenic processes in fish. But in what ways these processes affect each other in a given set of environmental conditions remains highly conjectural. We are faced with the unconfortable prospect of proposing integrative mechanisms for which evidence is tenuous, or absent altogether. In the present review I shall pay attention to a few selected aspects of vision and bioluminescence in fish, emphasising their relationships and exploring some approaches that might help clarify them. Some of the topics discussed here have been previously reviewed under various perspectives (Clarke and Denton, 1962; Nicol, 1967, 1969; Tett and Kelly, 1973).

* Present address: Department of Biological Sciences, University of California, Santa Barbara, Calif. 93106, U.S.A.

HISTORICAL SURVEY AND THE DIOPTRIC APPARATUS

It is interesting to note that early students of luminescent fish confusedly associated visual and light organs. The serial photophores of many deep-sea fishes were wrongly named "accessory eyes" because of their serial distribution along the body (Leuckart, 1865; Ussow, 1879; Bell, 1881; see Anctil, 1971, for more references). The structure of these organs was even compared with that of the photophores of euphausiid shrimps, which in turn were believed to be ocellar eyes. Conversely, retinal and ocular anatomical features were mistaken for light organs in two species of bathypelagic fish (Moseley, 1887; see Munk, 1968). Some fish eyes were said to emit light. This was an error and the light was later ascribed to the presence of a tapetal reflector backing the retina (Dubois, 1924).

Nevertheless, the fact that early histologists were deceived by the appearance of the photophores, or that they labelled them "eye-like" (Leydig, 1881), points to the superficial eye-photophore similarities in many self-luminous fishes. Such striking structural similarities were reviewed by Buck (1961) for crustaceans and cephalopods; he speculated on the origin of invertebrate photophores and was inclined to believe that they had developed secondarily from eyes. The case for close similarity is less impressive in fish; yet the dioptric organisation of some dermal photophores in stomiatoid, myctophid and midshipman fishes is reminiscent of that found in the vertebrate eye: pigment and reflector backings, lens, corneal window. This convergence obviously stems from similar optical requirements. However, this applies only to a few types of light organs among those present in the various luminescent fish groups. Detailed, updated studies on the ontogeny of the dioptric elements of fish photophore are needed for their reliable comparison with the presumably analogous ocular structures.

Providing that high selective pressures might have worked similarly in ocular and light organs so as to make them take advantage of a common environmental parameter, light, having well-defined physical characteristics, it has to be demonstrated that such dioptric structures are functional in photogeny as they are in photoreception. The indirect evidence for this is derived from close examination of the pigment screens, reflector layers and lenticular modifications, and the way these components are set up in the predominantly ventral photophores to collimate the photoemission and distribute it within a certain angle of radiation (Bassot, 1966; Denton et al., 1969; Lawry, 1973). Direct observations and photography of lanternfish luminescence intercepted from various angles also support the view that the photophore dioptric accessories may contribute to the direction of photoemission (Anctil and Gruchy, 1970; Lawry, 1973). Only recently have controlled experiments permitted precise measurements of the angular distribution of photoemission in hatchet-fish, *Argyropelecus*, and viperfish, *Chauliodus* (Denton et al., 1972). It was determined that displacing the vertical axis of the fish by, say, 30°, would bring about an approximate 60% reduction of the luminescence recorded. The significance of these results will be discussed in a wider perspective below it is sufficient for now to point out that such data characterise reasonably well an efficient dioptric system. This approach must be extended to other luminescent fish group prior to any evaluation of the optical efficacy and overall ecological significance of these arrangements.

THE ECOLOGICAL PERSPECTIVE

All but one of the theories explaining the utility of fish luminescence (see Nicol, 1967, 1969 for a review) require the intervention of visual processes to be effective. It could hardly have been otherwise unless one considers the phenomenon as merely a biochemical oddity. Some functions (as luminescent lures, predator repellents), which are plausible but have no direct evidence, require that the eye of the interacting fish or invertebrate acts as a simple photosensitive organ, thus leading to a characterised response. Others (as intraspecific signals, sources of illumination for sight) require that the eye of the fish be involved in image formation. Unfortunately nothing is known about the visual acuity of the eyes of luminescent fish. Most luminescent fishes are mesopelagic forms which are so difficult to keep alive under experimental conditions that even simple photobehavioural studies present a challenge. Once this challenge is met, revealing studies on their optokinetic reactions and visual discrimination will help us understand the overall significance of bioluminescence in midwater communities.

So far as is known only rods are present in the retinal photoreceptor layer of deep-sea luminescent fish (Brauer, 1908; Munk, 1966; Locket, 1969). The rod outer segments are usually very long and the rod axonic terminals summate to a great extent with the relatively few bipolar cells. These are two indications of a very sensitive retina with poor resolution. Some of these retinas possess a restricted foveal area which may improve image resolution. All this is in keeping with the expected light reactions of the fish in a pelagic community provided with multi-purpose bioluminescence. Observations from submersibles indicated that lanternfishes (Myctophidae) are capable of phototropic responses, negative or positive (Backus et al., 1968; Barham, 1966, 1970; Church, 1970). However, these responses are not necessarily related to expected stimulus levels under natural conditions. Carefully planned submersible dives should reveal more about the photic responses or visual tactics of these animals.

One theory which has been much publicised in the last decade capitalises on the fact that the serial light organs of fish occur mostly on the ventrolateral and/or ventral surface of the body, thus possibly allowing the photoemission to match the downwelling ambient light and obliterate the ventral silhouette of the fish (Fraser, 1962; Clarke, 1963; McAllister, 1967; Nicol, 1969). There is no direct evidence for this but the data on the intensities of lanternfish luminescence and of daylight penetrating midwater depths (Nicol, 1958; Clarke et al., 1962; Clarke, 1966). However, Denton et al. (1972) provide data on the angular distribution of ventrally directed luminescence in the stomiatoids, *Argyropelecus* and *Chauliodus*, which lend credibility to the theory. The larger mesopelagic stomiatoid fishes (themselves luminescent) appear to be the dominant consumers of such luminescent fishes as lanternfishes, gonostomatids, deep-sea smelts, melamphaeids, etc. (Borodulina, 1972). It is striking that the retina of the potential prey is in general, more highly organised than that of its predator. The smaller mesopelagic fishes possess retinas with unusually long, slender, densely packed rods, or multiple layers of rod outer segments (Vilter, 1951, 1953, 1954a, 1954b, 1954c; Munk, 1966; Locket, 1969), and, as mentioned above, in many cases there are remarkable foveal specialisations and rich neuronal integrative networks. It has been suggested that the shallow pure-rod foveas of deep-sea fishes serve for movement perception and fixation at the expense of the retinal resolving power (Pumphrey, in Munk, 1966). In this respect, the smaller mesopelagic fish may be adequately equipped visually to pursue moving planktonic prey, luminescent or not, that constitute their diet (Paxton, 1967; Cailliet, 1972). The larger mesopelagic fishes possess retinas with one, two or three superimposed layers of short and loosely arranged rods, and their neuronal layers appear to be poorly developed (Brauer, 1908; Munk, 1966). No foveal structure is present in any of them. Although apparently both kinds of mesopelagic fishes are apt to detect minimal amounts of light, either daylight at midwater depths or point-sources of bioluminescence (Clarke and Denton, 1962), it seems that the smaller mesopelagic teleosts have better visual capabilities than the large stomiatoids with which they interact. The protective advantage conferred by ventral luminescence camouflage thus appears to be greater for planktonic crustacean forms than for small midwater teleosts in regard to their respective predators. Borodulina's data (1972) show that by far the main diet component of the large stomiatoids consists of luminescent fishes (lanternfishes, gonostomatids). Moreover, all members of one of the largest stomiatoid family, Melanostomiatidae, appear to prey exclusively on crustaceans, not fish. Although luminescent fishes such as lanternfishes, some gonostomatids and hatchetfishes form an overwhelming majority of the oceanic piscine fauna upon which they feed, there is no clear indication that stomiatoids may discriminate, visually or otherwise, between presumably camouflaging and non-luminescent preys. Other sensory cues than visual may be involved, and protective behaviour such as schooling by lantern fishes at night (Backus et al., 1968), need to be considered.

RETINAL MONITORING OF PHOTOEMISSION INTENSITY - A LIKELY MECHANISM?

Hypothetically, the predominantly ventral luminescence of squids, crustaceans and fish may serve to obliterate the silhouette by matching the intensity of light emission with that of downwelling light (Clarke, 1963; Nicol, 1969; Hastings, 1971; Young, 1973). This ingenious theory may be applied to both the prey and the predator, for different advantages.

Recently a mechanism has been proposed to explain how information on ambient light intensity could be processed through the midwater fish retina, which in turn would allow adjustment of the photophore light intensity to proper countershading effect (Lawry, 1974). *Tarletonbeania crenularis* is a vertically migrating mesopelagic fish belonging to the large lanternfish family whose members usually are endowed with serial discrete photophores, mostly located on the ventral aspect of the body. Lawry (1974) focussed attention on a small supraorbital photophore that may shine light onto the ventral retina after the light rays from the photophore are bent by the lens and channelled through a special "aphakic aperture" (Fig. 1). He proposes a mechanism by which downwelling light rays and orbital photophore light rays enter through the aphakic aperture. If light from these two sources were of different intensity, e.g., if orbital luminescence is stronger, the fish retina might show a spot in the aphakic visual field and the fish might adjust its luminescent activity to the extent necessary to obliterate the spot (see Fig. 1). This provides a very interesting and relatively simple model of photometric standard for monitoring ambient light levels and luminescent activity. However, the hypothesis is based on suggestive morphological and dioptric features and lacks experimental evidence. Young (1973) proposed a similar feedback mechanism for ventral countershading between extraocular photoreceptors and photophores in mesopelagic squids.

Nicol (1969) previously suggested that small orbital photophores may shine directly into the eye, thus providing a monitoring device to adjust luminescent output. His hypothesis stems from the morphological features described by Brauer (1908) in the bristlemouth *Cyclothone braueri* and the hatchet-fish *Argyropelecus affinis*. However, it may not apply to these species when the topographical relations between the orbital photophore and the eye are considered in *Cyclothone signata* (Nusbaum-Hilarowicz, 1920). In this case,

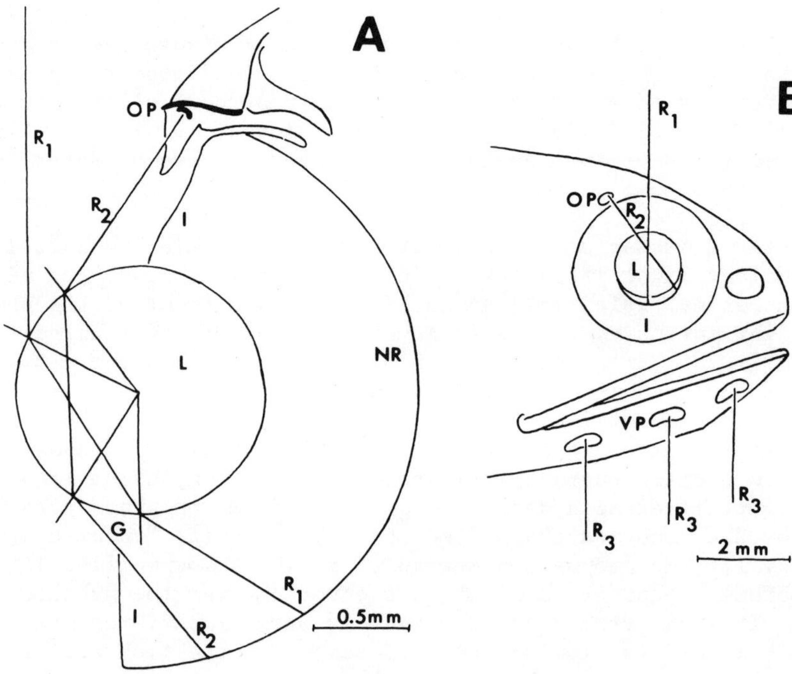

FIG. 1. Hypothetical model of the optical pathways followed by light (R_1, R_2) from the dorsal visual field (VF) and from the supraorbital photophore (OP) of *Tarletonbeania crenularis* through the dioptric media of the eye. A, cross-section of the eye and photophore; B, lateral view of the fish head. G, aphakic aperture; I, iris; L, lens; NR, retina; R_3, light output from ventral photophores, VP. Reprinted with permission from Lawry (1974), and Nature (Mac-Millan Journals, Ltd).

the pathway for the light theoretically shining from the infraorbital photophore is not oriented towards the retina but rather seems to follow an oblique course in the dorso-rostral direction and is likely to be deflected externally by the orbital shelf. It would be hazardous to speculate on the function of this arrangement as providing illumination or light-adaptation to the immediate visual field of the fish, especially since the ability of *Cyclotone* to luminesce at all is seriously questioned (Bassot, 1966).

VISUAL PIGMENTS AND SPECTRAL DISTRIBUTION OF PHOTOEMISSION

There is a good correlation between the spectral characteristics of visual pigments of marine fish and of their environmental light regimes (see Lythgoe, 1972). In a large mumber of mesopelagic forms, the scotopic visual pigments absorb maximally in the blue at 480-490 nm (Denton and Warren, 1957; Munz, 1958), whereas the rhodopsin maxima shift to 505-510 nm in epipelagic or coastal teleosts (Munz,1964). It is necessary to determine if such a correlation holds between the spectral characteristics of deep-sea fish visual pigments and bioluminescence. This matching is a requisite if fish eyes are to detect bioluminescence effectively.

The evidence that this is the case has been established from a few data on colour and spectral curves of photoemission (see Nicol, 1969 for a review). It brings out that the luminescence of deep-sea teleosts is generally blue or blue-green. The emission spectrum curve of a lanternfish, *Myctophum punctatum*, shows a peak at 470 nm (Nicol, 1960a), which is only a slight shift from the visual pigment maximal absorption value (Denton and Warren, 1957) and allows for an efficiency of about 75% in the photoreception-photoemission interaction (Nicol, 1969). As pointed out earlier, such an efficiency is critical when discussing any reciprocal effect of vision and bioluminescence from an ecological view-point.

The light produced by the discrete photogenic tissue of fish photophores often has physical characteristics that do not match either the ambient natural light or the visual system. In order to harmonise the luminescent output with these parameters, dioptric devices such as reflectors, lenses, pigment sheaths may act to intensify and distribute the photoemission so that the intensity and direction of the light fit the ambient light regime or the functional limitations of the visual system. So is the case with the colour of the photoemission. The colour of the light produced may be at odds with that of ambient penetrating daylight, or such that it cannot generate a response from the retina. In these cases a colour filter has been apparently provided. In the hatchet-fishes *Argyropelecus* and *Sternoptyx*, the specific tissue which is said to act as a filter contains a pigment tentatively identified as reduced cytochrome c, and might be responsible for the blue-green colour of the light emitted by the photophores (Denton et al., 1970). Some larger stomiatoid fishes such as *Pachystomias* and *Malacosteus* possess large subocular light organs (see Nicol, 1960b for their histology) which are covered with red tissues apparently transmitting a red-coloured luminescence. The eyes of these fishes have photosensitive retinal pigments absorbing maximally at 575 nm, or have a red tapetum lucidum (Denton et al., 1970). From *Aristostomias scintillens* two visual pigments were extracted, one absorbing maximally at 551 nm (in the yellow-green) and the other at 526 nm (in the green) (Fig. 2; see O'Day and Fernandez, 1974). These pigments appear to be especially qualified to detect the red luminescence of the large suborbital light organ and the green luminescence of the postorbital light organ in this fish. It is noteworthy that the green-sensitive photopigment ($P526_1$) has an absorption maximum about 25 to 50 nm higher than those of photopigments extracted from other midwater fishes (O'Day and Fernandez, 1974). As suggested by these authors, it might imply that this photopigment is used more specifically to detect the postorbital's light emission rather than ambient transmitted blue light, or the blue-green luminescence from other mesopelagic forms. The possible implications of these matchings, except for intraspecific recognition and predation upon red and brown crustaceans or fishes at midwater depths, are still not clear.

PHOTIC INHIBITION AND STIMULATION OF LUMINESCENT DISPLAY

The effect of light on the luminescent response of fish needs yet to be investigated; such effects could be mediated through the eyes. Beebe and Vander Pyl (1944) observed that the weak phosphorescent light of a wrist-watch induced strong luminescent flashes from the caudal light organs of a lanternfish, but stronger light stimuli were ineffective. The available data on the subject collected from other luminescent forms may serve as guidelines towards an understanding of the possible mechanisms involved.

Luminescent flashes occurring spontaneously or induced by electrical stimulation of the brain can be suppressed in fire-flies upon electrical stimulation of the eyes (Case and Buck, 1963). Furthermore, upon illumination of the eyes, luminescent flashing of the fire-fly is inhibited to an extent that is light-intensity dependent (Case and Trinkle, 1968). However, there is no inhibition of the flashing triggered by direct excitation of the light organ, suggesting that a central neural mechanism is involved. In the euphausiid shrimps, *Meganyctiphanes* and *Thysanoessa*, both stimulatory and inhibitory influences on luminescence have been reported upon previous exposure to brief flashes of light (Kay, 1965, 1966); the initial photoflash induces a luminescent response, and a second

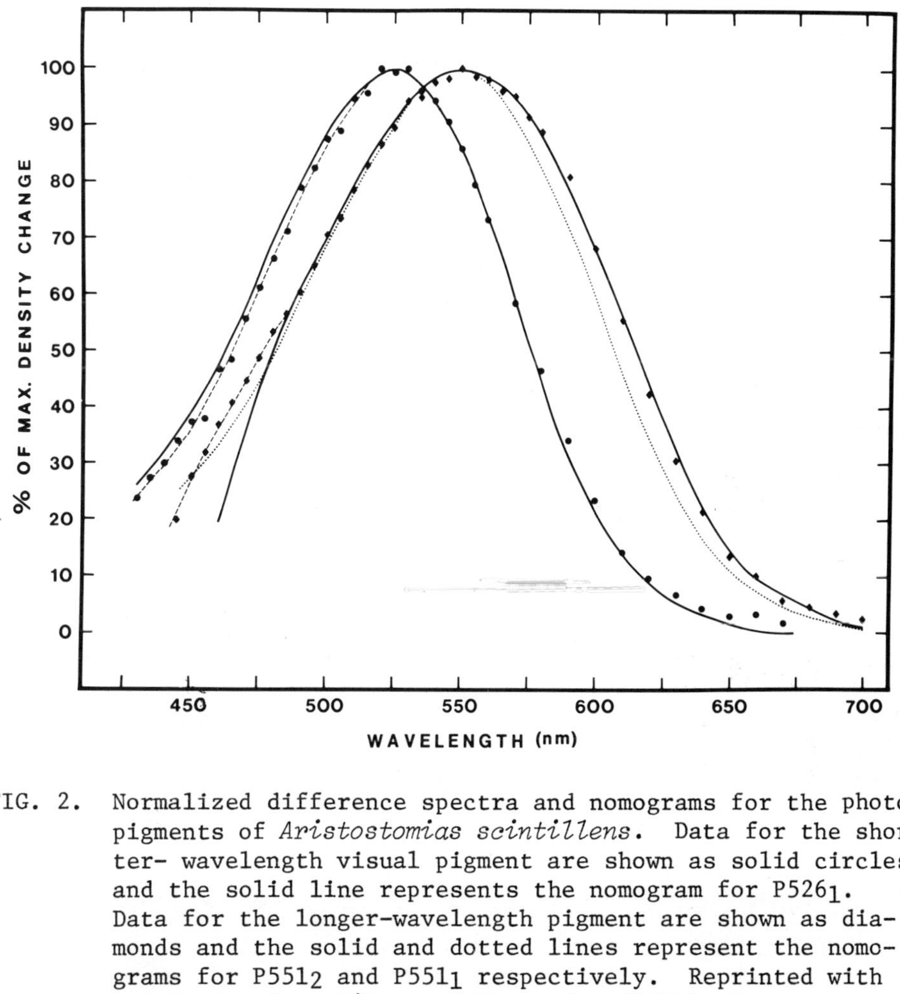

FIG. 2. Normalized difference spectra and nomograms for the photopigments of *Aristostomias scintillens*. Data for the shorter- wavelength visual pigment are shown as solid circles and the solid line represents the nomogram for $P526_1$. Data for the longer-wavelength pigment are shown as diamonds and the solid and dotted lines represent the nomograms for $P551_2$ and $P551_1$ respectively. Reprinted with permission from O'Day and Fernandez (1974), Pergamon Press Ltd.

flash delivered during that response temporarily inhibits the luminescent output. This flash-stimulated luminescence is dependent on temperature and seasonal cycles (Tett, 1969, 1972).

It is apparent that interactions between photoreception and photoemission primarily involve the flash type of response in these luminescent animals. Since luminescent flashes displayed in a regular rhythm are believed to play a significant role in social communication, particularly in fire-flies, it has been proposed that photic inhibition, even in the form of flashing from another male fire-fly nearby, might serve to bring the signalling patterns of various individuals into synchrony and so prevent confusion (Case and Trinkle, 1968). A mechanism to account for such a synchronisation has been suggested for fire-flies responding to rhythmic flashes of pacer light (Hanson et al., 1971); it implies that the pacer signal resets the presumed flash-timing oscillator in the brain which mediates the flash sequence entrainment. This hypothetical model may be usefully kept in mind whenever such approaches are exploited to resolve the precise, rhythmic flashing of the lantern-fish's caudal light organs (see Barnes and Case, 1974).

In addition to studies on the flash-stimulated and flash-inhibited responses in euphausiid luminescence mentioned above, the response of marine bioluminescence in situ to pulsed light has been investigated (Neshyba, 1967), although the organisms presumably involved could not be identified. An overall increase in bioluminescent activity was reported which was dependent on colour (maximal for white light stimuli) and intensity ($1.4 \times 10^{-5} \mu W/cm^2$) of stimulation, and the pulse rate threshold for the response ranged from 1/50 sec. to 1/10 sec. at a depth of 700 metres. However, the usual flash-type response was not primarily involved in these increments of luminescent activity, and there has been no mention of possible effects of pulsed artificial light on overall flash rhythmicity in the ocean. Tett and Kelly (1973) expressed doubts that these in situ responses may be due to "re-fluorescence by living or dissolved organic materials in the water."

CONCLUDING COMMENTS

There is very little information on the influence of luminescent activity upon the visual systems of animals, luminescent or not. There is the remarkable instance of the apparently converging features of the fire-fly and deep-sea fish eyes, in which the photosensitive structures (rhabdomes in fire-fly, rod outer segments in fish) form superimposed layers (Munk, 1966; Horridge, 1969). This appears to be devised to favour maximal photic sensitivity under dim light conditions, but it would be hazardous to point at luminescent sources as being responsible for the development of such systems. It is also not easy to determine whether the photopigment constituents of these rhabdomes or rod outer segments really adapt their spectral sensitivities in order to match the spectral properties of the animal's own luminescence. What is needed is the right experimental approach to show how these interactions might work. Since most luminescent fishes are fragile mesopelagic dwellers (200 - 1000m deep), it would be impractical to design laboratory behavioural studies dies that cannot, even remotely, reproduce their three-dimensional environment. It is wise, therefore, to rely on biochemical and electrophysiological experiments to determine the retinal sensitivity and visual performance of the fish eye such as has been done in luminescent crustaceans (Boden et al., 1961).

Visually mediated luminescent responses have yet neither been found nor investigated experimentally in fish. From the various speculations and indirect evidence mentioned in previous sections, three optical stimulus-bioluminescent response relationships need experimental confirmation: (1) a retinal monitoring system using an ocular photophore as a photometric standard to adjust luminescent output to ambient downwelling light; (2) an optically stimulated and/or inhibited bioluminescent system via a photoflash or continuous light stimulus; and (3) a visually mediated synchrony of inter-individual bioluminescent flash responses. The type of luminescence displayed by the fish determines which of these categories is fitted for experimental analysis: lanternfish and other mesopelagic fish photophores in the first category, midshipman (*Porichthys*) and various mesopelagic fish photophores in the second, and lanternfish caudal luminous organs in the third. Since most of these photogenic structures are neuroeffectors, and provided these three response types are eventually unveiled in luminescent fish, it would be of certain interest to investigate the neural integrative mechanisms involved. Which relay centers are involved in the central nervous system? Are we dealing with complex neuro-humoral pathways, involving neurosecretion, or are there rather rapid and closely connected neurotransmissive mechanisms? In this regard, we have for comparison the previous studies of Kay (1966) on the photoflash stimulated and inhibited luminescence of the euphausiid shrimp *Meganyctiphanes*, in which it is postulated that the flash inhibition is a low-threshold on-response to increasing light, mediated by closely connected neural circuits to the photophores, whereas flash excitation is said to be a higher threshold off-response to decreasing light, mediated by slow hormonal or neurosecretory pathways.

ACKNOWLEDGEMENTS

The author is grateful to Drs. W.T. O'Day and J.V. Lawry, Jr., for allowing him to reproduce their figures. Thanks are due to

Dr. O'Day for allowing me to read his manuscript in press. Drs. J.F. Case and W.T. O'Day kindly made suggestions to improve this paper.

REFERENCES

Anctil, M. (1971). A bibliography on bioluminescence in fishes. Sci. Rep. Yokosuka City Mus. 18: 1-14.

Anctil, M. and Gruchy, C.G. (1970). Stimulation and photography of bioluminescence in lanternfishes (Myctophidae). J. Fish. Res. Board Can. 27: 826-829.

Backus, R.H., Craddock, J.E., Haedrich, R.L., Shores, D.L., Teal, J.M., Wing, A.S., Mead, G.M. and Clarke, W.D. (1968). *Ceratoscopelus maderensis:* peculiar sound-scattering layer identified with this myctophid fish. Science 160: 991-993.

Barham, E.G. (1966). Deep scattering layer migration and composition: observations from a diving saucer. Science 151: 1399-1403.

Barham, E.G. (1970). Deep-sea fishes: lethargy and vertical orientation. In Proc. Intern. Symp. Biological Sound Scattering in the Ocean (G.B. Farquhar, ed.). Maury Center for Ocean Science, Washington. pp. 100-118.

Barnes, A.T. and Case, J.F. (1974). The luminescence of lanternfishes (Myctophidae): spontaneous activity and responses to mechanical, electrical and chemical stimulation. J. Exp. Mar. Biol. Ecol. 15: 203-221.

Bassot, J.M. (1966). On the comparative morphology of some luminous organs. In: Bioluminescence in Progress, edited by F.H. Johnson and Y. Haneda. Princeton University Press, Princeton, N.J. pp. 557-610.

Beebe, W. and Vander Pyl, M. (1944). Eastern Pacific expeditions of the New York Zoological Society. XXXIII. Pacific Myctophidae (fishes). Zoologica (N.Y.) 29: 59-95.

Bell, F.J. (1881). The eye-like spots of fishes. Pop. Sci. Rev. 20, N.S. 5, 221-234.

Boden, B.P., Kampa, E.M. and Abbott, B.C. (1961). Photoreception of a planktonic crustacean in relation to light penetration in the sea. In: Progress in Photobiology, edited by B.C. Christensen, and B. Buchmann, Elsevier, Amsterdam. pp. 189-196.

Borodulina, O.D.(1972). The feeding of mesopelagic predatory fish in the open ocean. J. Ichthyol. 12, 692-702.

Brauer, A. (1908). Die Tiefseefische. II. Anatomischer Teil. Wiss. Ergebn. deut. Tiefsee-Exped. "Valdivia", Bd. 15. Gustav Fischer verlag, Jena.

Buck, J. (1961). Speculations on the interrelations and evolution of photic organs. In: Light and life, edited by W.D. McElroy and B. Glass. Johns Hopkins Press, Baltimore, Md. pp. 754-770.

Cailliet, G.M. (1972). Feeding habits and distribution of two deep-sea fishes off Santa Barbara, California. Ph.D. Thesis, Univ. of California, Santa Barbara.
Case, J.F. and Buck, J.B. (1963). Control of flashing in fireflies. II. Role of central nervous system. Biol. Bull. 125: 234-250.
Case, J. and Trinkle, M.S. (1968). Light-inhibition of flashing in the firefly *Photuris missouriensis*. Biol. Bull 135: 476-485.
Church, R. (1970). Bioluminescence: the sea's living light. Oceans Mag. 3: 20-29.
Clarke, G.L. and Denton, E.J. (1962). Light and animal life. In: The Sea, edited by M.N. Hill, Vol. 1. Interscience Publ., New York pp. 456-468.
Clarke, G.L., Conover, R.J., David, C.N. and Nicol, J.A.C. (1962). Comparative studies of luminescence in copepods and other pelagic marine animals. J. Mar. Biol. Ass. U.K. 42: 541-564.
Clarke, W.D. (1963). Function of bioluminescence in mesopelagic organisms. Nature 198: 1244-1246.
Clarke, W.D. (1966). Bathyphotometric studies on the light regime of organisms of the deep scattering layers. Final Rept., G.M. Defense Lab., Sea Operations Dept., Santa Barbara, California.
Denton, E.J. (1971). Reflectors in fishes. Sci. Am. 224: 65-72.
Denton, E.J. and Warren, F.J. (1957). The photosensitive pigments in the retinae of deep-sea fish. J. Mar. Biol. Ass. U.K. 36: 651-662.
Denton, E.J., Gilpin-Brown, J.B. and Roberts, B.L. (1969). On the organization and function of the photophores of *Argyropelecus*. J. Physiol. 204: 38P-39P.
Denton, E.J., Gilpin-Brown, J.B. and Wright, P.G. (1970). On the 'filters' in the photophores of mesopelagic fish and on a fish emitting red light and especially sensitive to red light. J. Physiol. 208: 72P-73P.
Denton, E.J., Gilpin-Brown, J.B. and Wright, P.G. (1972). The angular distribution of the light produced by some mesopelagic fish in relation to their camouflage. Proc. Soc. Lond. Ser B 182: 145-158.
Dubois, R. (1924). La pseudoluminescence et le rôle du tapis chez certains poissons. C.R. Acad. Sci. Paris 178: 1030-1032.
Fraser, J. (1962). Nature adrift. The story of marine plankton. C.T. Foulis, London.
Hanson, F.E., Case, J.F., Buck, E. and Buck, J. (1971). Synchrony and flash entrainment in a New Guinea firefly. Science 174: 161-164.
Hastings, J.W. (1971). Light to hide by: ventral luminescence to camouflage the silhouette. Science 173: 1016-1017.
Horridge, G.A. (1969). The eye of the firefly *Photuris*. Proc. R. Soc. Lond. Ser B 171: 445-463.
Kay, R.H. (1965). Light-stimulated and light-inhibited bioluminescence of the euphausiid *Meganyctiphanes norvegica* (G.O. Sars). Proc. R. Soc. Lond. Ser B 162: 365-386.

Kay, R.H. (1966). The inhibition of optically stimulated bioluminescence in *Meganyctiphanes norvegica*. In Some Contemporary Studies in Marine Science, edited by H. Barnes. George Allen & Unwin, London. pp. 421-427.

Lawry, J.V., Jr. (1973). Dioptric modifications of the scales overlying the photophores of the lantern fish, *Tarletonbeania crenularis* (Myctophidae). J. Anat. 114: 55-63.

Lawry, J.V., Jr. (1974). Lantern fish compare downwelling light and bioluminescence. Nature 247: 155-157.

Leuckart, R. (1865). Ueber muthmassliche Nebenaugen bei einem Fische, *Chauliodus sloani*. Ber. Versamml. deut. Naturf. Aerz. (1864) 39, 153-155.

Leydig, F. (1881). Die augenahnlichen Organe der Fische. E. Strauss, Bonn.

Locket, N.A. (1969). The retina of *Poromitra nigrofulvus* (Garman)- An optical and electron microscope study. Exp. Eye Res. 8, 265-275.

Lythgoe, J.N. (1972). The adaptation of visual pigments to the photic environment. In Handbook of Sensory Physiology V11/1. Photochemistry of Vision, edited by H.J.A. Dartnall. Springer-Verlag, New York, pp. 566-603.

McAllister, D.E. (1967). The significance of ventral bioluminescence in fishes. J. Fish. Res. Board Can. 24: 537-554.

Mosely, H.N. (1887). Report on the structure of the peculiar organs on the head of *Ipnops*. Challenger Rep., Zool. 22: 269-276.

Munk, O. (1966). Ocular anatomy of some deep-sea teleosts. Dana-Rep. No. 70: 1-62.

Munk, O. (1968). On the eye and the so-called preorbital light organ of the isospondylous deep-sea fish *Bathylaco nigricans* Goode and Bean, 1896. Galathea Rep. 9: 211-218.

Munz, F.W. (1958). Photosensitive pigments from retinas of certain deep-sea fishes. J. Physiol. 140: 220-235.

Munz, F.W. (1964). The visual pigments of epipelagic and rocky-shore fishes. Vision Res. 4: 441-454.

Neshyba, S. (1967). Pulsed light stimulation of marine bioluminescence in situ. Limnol. Oceanogr. 12: 222-235.

Nicol, J.A.C. (1958). Observations on luminescence in pelagic animals. J. Mar. Biol. Ass. U.K. 37: 705-752.

Nicol, J.A.C. (1960a). Spectral composition of the light of the lanternfish, *Myctophum punctatum*. J. Mar. Biol. Ass. U.K. 39: 27-32.

Nicol, J.A.C. (1960b). Studies on luminescence. On the subocular light-organs of stomiatoid fishes. J. Mar. Biol. Ass. U.K. 39: 529-548.

Nicol, J.A.C. (1967). The luminescence of fishes. Symp. Zool. Soc. Lond. 19: 27-55.

Nicol, J.A.C. (1969). Bioluminescence. In Fish Physiology, edited by W.S. Hoar and D.J. Randall, Vol. 3. Academic Press, New York. pp. 355-400.

Nusbaum-Hilarowicz, J. (1920). Etudes d'anatomie comparée sur les poissons provenant des campagnes scientifiques de S.A.S. le Prince de Monaco. Résult. Camp. scient. Prince Albert I, Deuxième Partie, Fasc. 58, 1-115.

O'Day, W.T. and Fernandez, H.R. (1974). *Aristostomias scintillens* (Malacosteidae): a deep-sea fish with visual pigments apparently adapted to its own bioluminescence. Vision Res. 14: 545-550.

Paxton, J.R. (1967). Biological notes on Southern California lanternfishes (Family Myctophidae). Calif. Fish & Game 53: 214-217.

Tett, P.B. (1969). The effects of temperature upon the flash-stimulated luminescence of the euphausiid *Thysanoessa raschi*. J. Mar. Biol. Ass. U.K. 49: 245-258.

Tett, P.B. (1972). An annual cycle of flash induced luminescence in the euphausiid *Thysanoessa raschi*. Mar. Biol. 12: 207-218.

Tett, P.B. and Kelly, M.G. (1973). Marine bioluminescence. Oceanogr. Mar. Biol. Ann. Rev. 11: 89-173.

Ussow, M. (1879). Ueber den Bau der so genannten augenaehnlichen Flecken einiger Knochenfische. Bull. Soc. Imp. Nat. Moscou 54: 79-115.

Vilter, V. (1951). Bases cyto-architectoniques de l'acuité visuelle chez un poisson abyssal, le *Lampanyctus crocodilus*. C.R. Soc. Biol. Paris 145: 52-54.

Vilter, V. (1953). Existence d'une rétine à plusieurs mosaïques photoréceptrices chez un poisson abyssal bathypélagique, *Bathylagus benedicti*. C.R. Soc. Biol. Paris 147: 1937-1939.

Vilter, V. (1954a). Différentiation fovéale dans l'appareil visuel d'un poisson abyssal, le *Bathylagus benedicti*. C.R. Soc. Biol. Paris 148: 59-63.

Vilter, V. (1954b). Interprétation biologique des trames photoréceptrices superposées de la rétine du *Bathylagus benedicti*. C.R. Soc. Biol. Paris 148: 327-330.

Vilter, V. (1954c). Relations neuronales dans la fovea à bâtonnets du *Bathylagus benedicti*. C.R. Soc. Biol. Paris 148: 466-469.

Young, R.E. (1973). Information feedback from photophores and ventral countershading in mid-water squid. Pac. Sci. 27: 1-7.

THE ROLE OF THE PSEUDOBRANCH AND CHOROID RETE MIRABILE IN FISH VISION

D. D. BEATTY

Department of Zoology, University of Alberta

Edmonton 7, Alberta, T6G 2E1

Where the heart lies, let the brain lie also.

Browning

INTRODUCTION

The choroid rete mirabile is a large horseshoe-shaped, gland-like structure located around the optic nerve in the choroid layer of the eye of many species of fishes. This rete consists of thousands of closely packed and parallel arterial and venous capillaries with a countercurrent flow of blood. Arterial blood from the rete is supplied to the choriocapillaris underlying the retina, the venous blood draining back through the venous capillaries (Barnett, 1951). The ophthalmic artery (the efferent artery of the pseudobranch) supplies blood to the rete and thus is the major source of blood for the choriocapillaris. In addition, the retinal artery (from the internal carotid) supplies some blood to the choriocapillaris via the lentiform body which also has a rete type organization (Barnett, 1951). The choroid rete, together with the pigment cell epithelium and choriocapillaris, maintains high oxygen tensions in the mainly avascular teleost retina (Wittenberg and Wittenberg, 1962, 1974).

The pseudobranch in many teleosts has the general appearance of a hemibranch, being the remnant of the first gill arch. In these species it hangs freely on the inner surface of the opercle and is bathed by respiratory water. In others, the filaments of the pseudobranch are partly fused and often covered by epithelium. In some the filaments are entirely fused and the entire pseudobranch lies embedded in connective tissue (Wittenberg and Haedrich, 1974). The freely hanging type of pseudobranch is often referred to as the "free" pseudobranch and the latter two as the "glandular" pseudobranchs because of the fusion of the filaments and the minimum chance for exchanges with the respiratory water. Both types have characteristic pseudobranch-type cells, which are probably absent in gills, and the "free" type, at least, has chloride-type cells comparable to those of the gills (Harb and Copeland, 1969). Pseudobranch cells contain large amounts of the enzyme, carbonic anhydrase, as does the choroid-retina (Maetz, 1956; Hoffert, 1966). The pseudobranch receives oxygenated blood from the first efferent gill artery, the blood entering the pseudobranch capillaries before going to the ophthalmic artery.

Wittenberg and Haedrich (1974) have presented an excellent review, including previously unpublished information, on the distribution and relationship of the choroid rete, the swim bladder rete and the pseudobranch. I shall summarize some of their most interesting conclusions before considering the function of the choroid rete and the pseudobranch.

The choroid rete, with the exception of the bowfin, *Amia*, is found only in teleosts, being absent in all other fish so far. Among teleosts, it is especially common in the Acanthopterygii. It is absent in all Anguilliformes (eels), known to have vascular retinas (Walls, 1942).

Although there is some suggestion that the presence of the choroid rete is associated with the depth of the habitat, this is not always the case. Some surface dwelling species lack rete and certain bottom species have it, though it may be reduced in size. Among major groups of fish such as the Gadiformes and Salmoniformes, nearly all species of certain suborders (Gadoidei, Salmonoidei, and Argentinoidei) have a rete, whereas in other suborders (Ophidioidei, and Stomiatoidei) it is generally absent.

There is no correlation between the presence of a choroid rete and a swim bladder rete. As an example, many eels have well developed swim bladder retia. Likewise, many species with large choroid rete have poorly developed or no swim bladder rete. This is not surprising, for although both are involved in gas secretion they probably function under totally different requirements.

Pseudobranchs are present in most teleosts, and their homologue occurs in elasmobranchs, chondrosteans and holosteans but is absent in sarcopterygians. It may be highly developed in teleosts lacking choroid rete, however, this has been observed only in marine species so far. Marine teleosts not having pseudobranchs either lack or have poorly developed choroid rete. Teleosts lacking pseudobranchs but having highly developed choroid rete are always freshwater species. This suggests that certain freshwater species do not require a pseudobranch for the functioning of the choroid rete (Wittenberg and Haedrich, 1974).

FUNCTION OF THE PSEUDOBRANCH

A number of functions have previously been attributed to the pseudobranch. These include ion regulation, supplementary gas exchange, regulation of blood pressure or fluids in the eye, an endocrine function (Parry and Holliday, 1960) and sensory receptors for regulation of ventilation (Laurent and Rouzeau, 1972). Parry and Holliday suggested an endocrine function related to chromatophores because removal of the pseudobranch or stoppage of its blood supply causes body darkening. This has not been substantiated and can now be explained equally well by the blinding which occurs when both pseudobranchs are removed. Its role in regulation of ventilation during hypoxia or hyperoxia has not been shown (Randall and Jones, 1973).

A more likely explanation set forth by Wittenberg and Haedrich (1974) is that arterial blood is altered, perhaps by decreasing carbon dioxide, in pseudobranch capillaries prior to going to the choroid rete. Thereby, any concentrating mechanism associated with the counter-current multiplication system of the rete for oxygen secretion in the eye will not also create excessively high concentrations of carbon dioxide which when hydrated may be harmful. Maetz (1956) reported that administration of acetazolamide (a potent inhibitor of carbonic anhydrase) to perch caused a drop in pH of the vitreous humor from the normal 7.6 to 6.6, implying a marked increase in carbon dioxide tension in the eye. It is unlikely that the pseudobranch provides carbonic anhydrase or some other factor for the choroid rete. Instead, the high levels of this enzyme in the pseudobranch suggests a role in acid-base balance and perhaps the ion regulation involved in bicarbonate-chloride exchange between blood and water.

Although Wittenberg and Haedrich conclude that the pseudobranch is not required for developing high oxygen tensions in the swim bladder or the eye, the latter is still open to question. Oxygen tensions have not been systematically measured in teleosts that have choroid rete in the eye, but lack pseudobranchs.

Barnett (1951) concluded that the lentiform body rete may supply blood from the retinal artery to the choriocapillaris. However, the conclusion of Fairbanks et al. (1969) that the relatively high oxygen tension in the ipsilateral eye of rainbow trout with unilateral pseudobranchectomy was due to the lentiform body rete is not entirely justified. In a study on Atlantic salmon, *Salmo salar* in which I have collaborated with Dr. R.L. Saunders and Dr. C.M. Ballintijn, we have observed that an arterial crossover exists such that blood from a single intact pseudobranch can be shunted across to the ophthalmic artery of the contralateral eye whose ipsilateral pseudobranch has been removed.

FUNCTION OF THE CHOROID RETE MIRABILE

All the existing evidence suggests that the choroid rete is essential for maintaining high oxygen tensions in the eye of fishes (Wittenberg and Wittenberg, 1974). So far, measurements have shown that the ocular oxygen pressures in fishes lacking a choroid rete, and in other groups of vertebrates, are below that of arterial blood. Teleosts with small rete have lower oxygen tensions than those with large well-developed rete. The teleosts with large rete have oxygen tensions from 250 to 800 mm Hg (Wittenberg and Wittenberg, 1974) comparged to 85-100mm Hg for arterial blood.

Presumably the high ocular oxygen pressures are required to overcome the large diffusion distances in the avascular retina (Wittenberg and Wittenberg 1974) which has a very high oxygen consumption (Hoffert and Fromm, 1972). Wittenberg and Wittenberg (1974), in an elaboration of an earlier proposal for the swim bladder rete (Kuhn et al., 1963), have described a mechanism for the functioning of the choroid rete counter-current multiplier which involves movement of water from the arterial to the venous capillaries.

In our own studies (Beatty, Saunders and Ballantijn, Unpublished observations), we have found high oxygen tensions in the vitreous humor adjacent to the retina in the Atlantic salmon. Measurements of extractable visual pigment have shown that bilateral pseudobranchectomy causes marked decreases in pigment concentration within a few hours and almost total absence within 9-12 hours. Unilateral pseudobranchectomy does not have this effect in either eye. Within three hours, bilateral pseudobranchectomy causes blindness and complete body darkening, unilateral pseudobranchectomy having little or no effect. If one eye and the contralateral pseudobranch are removed there is no decrease in visual pigment in the intact eye over time nor does the fish darken or show blindness. This suggests that blood to the choroid rete of the intact eye is supplied by the intact contralateral pseudobranch via the crossover previously mentioned. Bilateral pseudobranchectomy as expected abolishes the high ocular oxygen tension whereas unilateral pseudobranchectomy has no apparent effect on the oxygen tension in either eye.

Acetazolamide injected into fish has effects comparable to bilateral pseudobranchectomy in agreement with the results obtained for rainbow trout by Fairbanks et al. (1969).

We now have the problem of demonstrating whether the pseudobranch is essential for functioning of the gas secreting mechanism of the eye and for the general physiological function of vision. This cannot be concluded from its presence or absence in a certain species nor from its removal because of the interruption of the major blood supply to the choriocapillaris. Only by shunting arterial blood around the pseudobranch or by artificial perfusion of the choroid rete can we get at the answer. Both of these approaches are possible and warrant being done.

From a practical standpoint, high gas tension in the teleost eye can cause gas bubbles and exopthalmia in fish held for experimental study (Dehadrai, 1966). Likewise any damage to the pseudobranchs from gill parasites may severely affect the visual system.

REFERENCES

Barnett, C.H. (1951). The structure and function of the choroidal gland of teleostean fish. J. Anat. 85: 113-119.

Dehadrai, P.V. (1966). Mechanism of gaseous exophthalmia in the Atlantic cod, *Gadus morhua* L. J. Fish. Res. Board Can. 23: 909-914.

Fairbanks, M.B., Hoffert, J.R. and Fromm, P.O. (1969). The dependence of the oxygen-concentrating mechanism of the teleost eye (*Salmo gairdneri)* on the enzyme carbonic anhydrase. J. Gen. Physiol. 54: 203-211.

Harb, J.M. and Copeland, D.E. (1969). Fine structure of the pseudobranch of the flounder *Paralichthys lethostigma*. Z. Zellforsch. 101: 167-174.

Hoffert, J.R. (1966). Observations on ocular fluid dynamics and carbonic anhydrase in tissues of lake trout (*Salvelinus namaycush*) Comp. Biochem. Physiol. 17:107-114.

Hoffert, J.R. and Fromm, P.O. (1972). Teleost retinal metabolism as affected by acetazolamide. Proc. Soc. Exp. Biol. Med. 139: 1060-1064.

Kuhn, W., Ramel, A., Kuhn, H.J. and Marti, E. (1963). The filling mechanism of the swimbladder. Generation of high gas pressures through hairpin counter-current multiplication. Experentia 19: 497-511.

Laurent, P. and Rouzeau, J.D. (1972). Afferent neural activity from pseudobranch of teleosts. Effects of PO_2, pH, osmotic pressure and Na^+ ions. Respir. Physiol. 14: 307-331.

Maetz, J. (1956). Le rôle biologique de l'anhydrase carbonique chez quelques téléostéens. Bull. Biol. Fr. Belg. Supp. 40: 1-129.

Parry, G. and Holliday, F.G.T. (1960). An experimental analysis of the function of the pseudobranch in teleosts. J. Exp. Biol. 37: 344-354.

Randall, D.J. and Jones, D.R. (1973). The effect of deafferentiation of the pseudobranch on the respiratory response to hypoxia and hyperoxia in the trout *Salmo gairdneri*. Respir. Physiol. 15: 291-301.

Walls, G.L. (1942). The vertebrate eye and its adaptive radiation. Cranbook Inst. Sci. Bull. No. 19 Bloomfield Hills, Michigan (reprinted by Hafner Publishing Co., New York, 1963) 785 pp.
Wittenberg, J.B. and Haedrich, R.L. (1974). The choroid rete mirabile of the fish eye. II. Distribution and relation to the pseudobranch and to the swimbladder rete mirable. Biol. Bull. 146: 137-156.
Wittenberg, J.B. and Wittenberg, B.A. (1962). Active secretion of oxygen into the eye of fish. Nature (Lond). 194: 106-107.
Wittenberg, J.B. and Wittenberg, B.A. (1974). The choroid rete mirabile of the fish eye. I. Oxygen secretion and structure: Comparison with the swimbladder rete mirabile. Biol. Bull. 146: 116-136.

QUANTITATIVE CHANGES OF SYNAPTIC RIBBONS IN THE CONE PEDICLES OF *NANNACARA*: LIGHT DEPENDENT OR GOVERNED BY A CIRCADIAN RHYTHM?

H.-J. WAGNER

Lehrstuhl für Morphologie und Anatomie der
Universität Regensburg, Universitätstr. 31
D-84 Regensburg, W. Germany

My noble uncle, do you know the cause?
Romeo and Juliet, Act I, Sc. I

INTRODUCTION

Receptor cells may exhibit a special kind of synapse, which on the presynaptic side, is characterized by numerous clear synaptic vesicles and a multilayered structure, the synaptic ribbon. Ribbon synapses of the retina occur in terminals of visual and bipolar cells. Quantitative assessment of these synapses showed no alterations of the number of synaptic vesicles in response to light or dark adaptation (Mountford, 1963; Cragg, 1972). In the present communication, attention is focused on the synaptic ribbons in the receptor cells of the retina. It will be shown that their number changes under different light conditions. An attempt is made to determine the factors leading to morphological alterations of the synaptic ribbons.

MATERIAL AND METHODS

Adult cichlid fish (*Nannacara anomala*, Regan, 1905) with a minimum body length of 4 cm. and an eye diameter of 3.5 mm. were kept under a constant day-night-rhythm of 14hrs. light and 10hrs. darkness for at least 4 weeks prior to the experiments. Preparation and fixation of the eyes were carried out in the light with light adapted animals and under very dim red light (Ilford S 902 filter) with dark adapted fish. The eye cups were bisected and immersed in

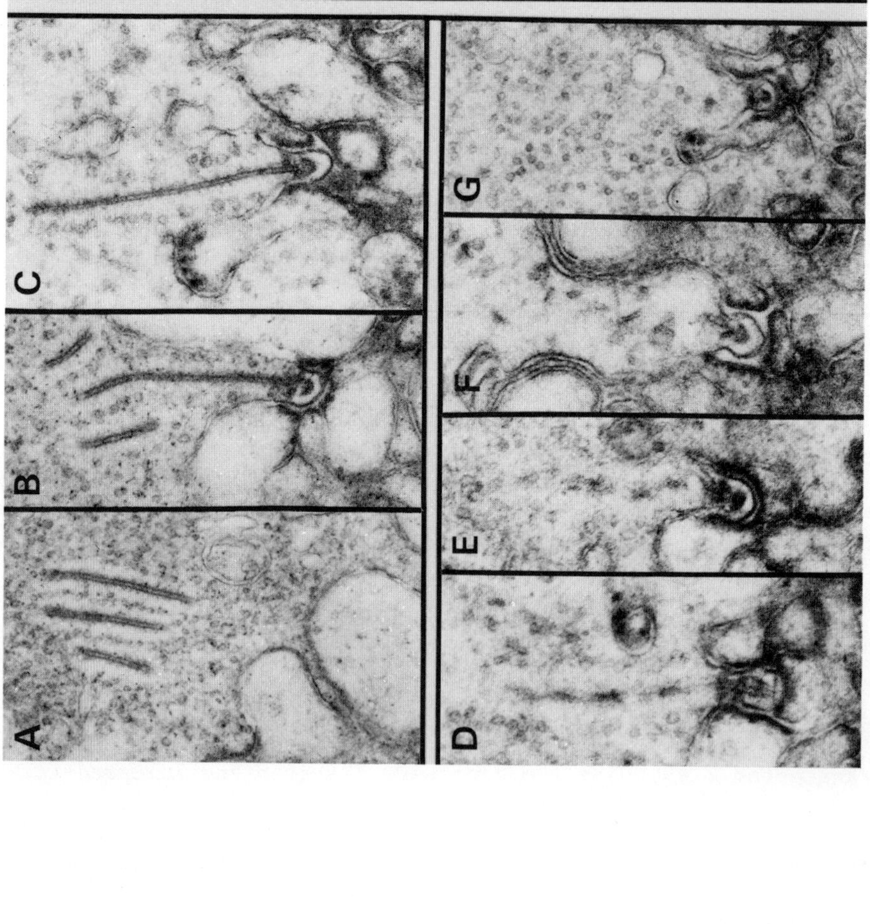

Figure 1

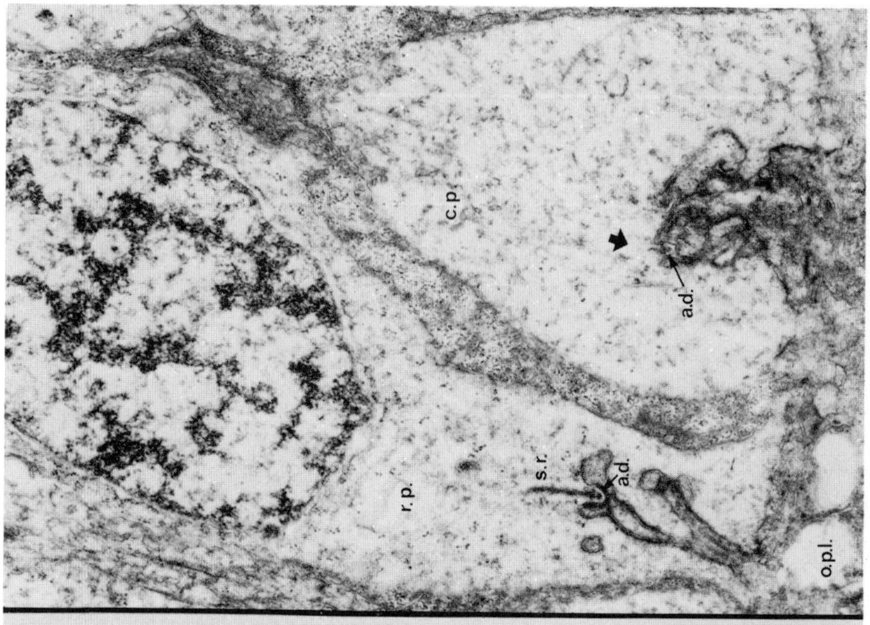

Figure 2

cold Veronal-acetate-buffered glutaraldehyde (4%, 2 hrs) and post-fixed in 1% OsO_4 (2 hrs). Subsequently, the retinal tissue was isolated , trimmed to convenient size, and embedded in Araldite. Ultra-thin sections were cut on a Reichert OmU 3 microtome and double stained with uranyl acetate and lead citrate. Electron microscope: Siemens 101.

RESULTS

In vertical sections of receptor pedicles, the ribbon synapses are characterised by three features: first, the synaptic ribbon, a trilaminate structure, 200 to 300 Å thick and of variable length. It is situated in the middle of a wedge-shaped projection of the receptor cytoplasm, the "synaptic ridge", the vitread edge of which shows a ring-like protrusion. In the middle of this ring, the second characteristic of the ribbon synapse is found: a round, electron dense element, called "arciform density" (Ladman, 1958). The last attribute is the triad-configuration of the postsynaptic elements consisting of one bipolar cell and two horizontal cell processes (Dowling and Boycott, 1966). Vertically sectioned cone pedicles contain an average of 3-4 ribbon synapses, whereas in rods rarely more than one ribbon synapse is encountered. In the first experiment, the retinas of three-light adapted fish were compared to those of three dark adapted animals. Assessment of cone pedicles revealed normal ribbon synapses in the light (Fig. 1C), but a marked reduction of the number of synaptic ribbons in the dark adapted eyes. The place of the ribbon in the cone cytoplasm is then taken by numerous synaptic vesicles which are no longer found in the orderly array of two parallel rows (Fig. 1G). For the quantitative evaluation, the number of synaptic ridges containing a synaptic ribbon was expressed as percentage of the total of ridges counted in cone pedicles. Thus, in light adapted retinas, 100% of the rid-

---

FIG. 1. Stages of synaptic ribbons during a normal 24 hours-cycle. x 40,000.

A, B: ribbon-field-like stages during proliferation phase.

C: normal ribbon

D, F: ribbons in decomposition during degradation phase

G: dark adapted ribbon synapse without ribbon.

FIG. 2. Dark adapted receptor terminals of the *Nannacara* retina. In the cone pedicle (c.p.) there is no ribbon in the synaptic ridge of the ribbon synapse (arrow) whereas the rod (r.p.) ribbon synapse is complete. o.p. l. outer plexiform layer; a. d. arciform density; s.r. synaptic ribbon; x x 16,000.

ges included a ribbon, whereas, after dark adaptation, only 19.4% of the ridges showed a ribbon. These findings would indicate that, in cone pedicles, synaptic ribbons are formed and degraded according to the functional state of the visual cell (Wagner, 1973).

In contrast to the fluctuations of the synaptic ribbons in cones, there is no reduction of the ribbons in the ribbon synapses of rod pedicles during the night (Fig. 2). From this difference in the behavior of the ribbons between these two classes of receptor cells one can safely assume that the above-mentioned results are not due to methodical faults.

In the second experiment, the amount of synaptic ribbons in cone pedicles was studied during a normal 24 hours-period of 14hrs. light and 10hrs.darkness, in order to elucidate the factors that govern the fluctuations within the ribbon synapses. Beginning at 6 a.m., at least every 2 hours 2 fishes were sacrified and the number of ribbons assessed as above. In 34 animals about 12,000 ridges were counted. Fig.3 shows that the curve resulting from this test may be divided into four stages: During phase 1 (5.30 to 6.30 h) the number of ribbon-containing ridges increases rapidly from about 10 to 80%; phase 2 includes a plateau between 90 and 100% lasting until 4.00 h; phase 3 shows a slow reduction in two steps resulting in phase 4, a second plateau between 0 and 10% from 0.00 to 5.30 h. The two plateau phases are approximately in accordance with the periods of light and darkness suggesting that synaptic ribbons are present during daytime and absent at night. The onset of both increase and reduction occurs, however, at least 30 min. prior to the beginning of light or darkness. This is the first indication that the absence or the presence of light cannot be regarded as the only factors governing the metabolism of the synaptic ribbons.

For a more precise definition of the principles of this regulation, another group of animals was kept under constant darkness for 48 hrs, beginning at 6.00 h. in the morning. During this time, 54 fishes were examined and a total of ca. 27,000 ridges counted. The percentage of ribbon-containing ridges was plotted on the graph Fig. 4. The diagram exhibits two 24 hour-cycles which are very similar to each other. Again the four phases of increase, plateau, decrease, and a second plateau of the number of ribbon-containing synaptic ridges can be distinguished. In this experiment, proliferation and reduction of the ribbons take place independently of the external light conditions. Therefore, one must conclude that they are governed by an endogenous circadian rhythm. It should be noted that, under constant darkness, the level of the plateau phase after the increase does not exceed the 50% mark, whereas, under normal light conditions it reaches about 100% (Fig. 3). That means that the influence of light or darkness on the metabolism of the synaptic ribbons affects only the definitive amount of ribbons which are formed or degraded, but not the initiation of the processes themselves.

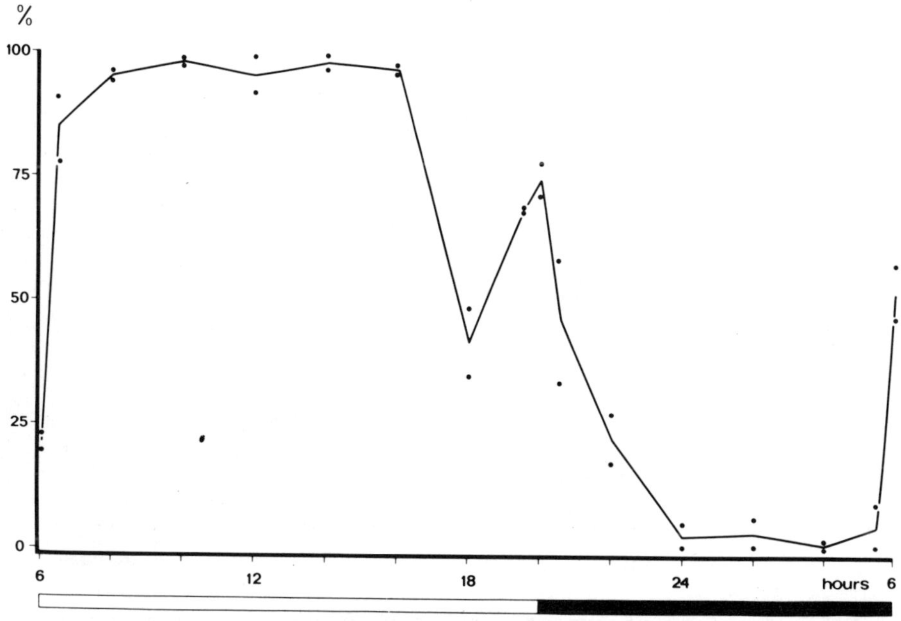

FIG. 3. Fluctuation of synaptic ribbons in cone pedicles during a normal 24 hours-cycle of 14 hours light and 10 hours darkness. The number of ribbon-containing synaptic ridges is expressed as percentage of the total of ridges counted (about 12,000).

Closer examination of the experimental material shows that presence and absence of synaptic ribbons during a 24 hour-cyle is not an all-or-none phenomenon. Several transitional stages can be observed. During the proliferation period (phase 1), one finds ribbons in a normal position being accompanied by one or two other shorter ribbons, which are arranged parallel to the first but do not reach down into the ridge (Fig. 1A, B). This finding bears close similarity to ribbon fields of pinealocytes (Vollrath, 1973). When the ribbons are disappearing in the evening (phase 3), there are numerous stages showing ribbons with fading outlines or those that seem to fall into pieces (Fig. 1D, G).

DISCUSSION

The ribbon is a presynaptic part of a structural complex where synaptic transmission is known to take place. It has been shown

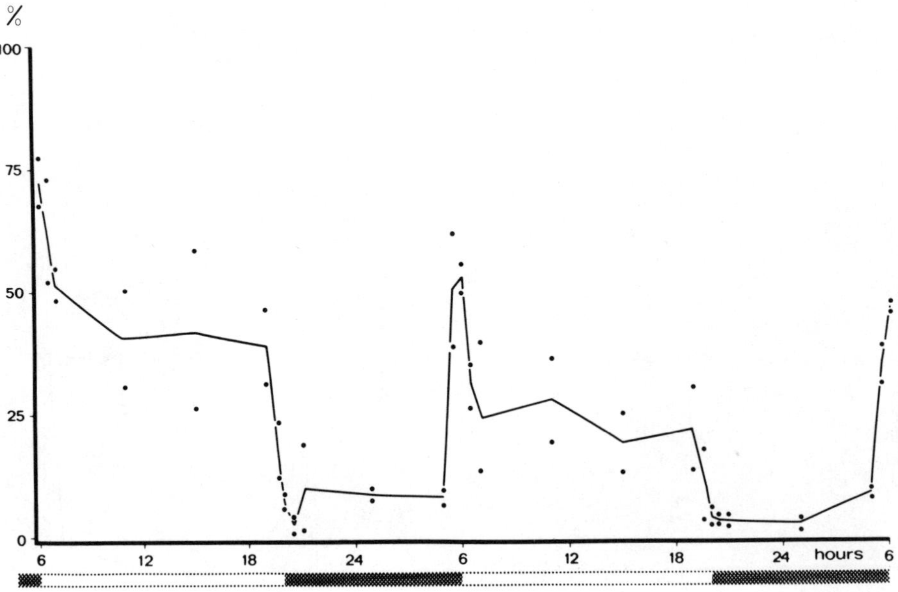

FIG. 4. Fluctuations of synaptic ribbons in cone pedicles during a period of 48 hours of constant darkness. The number of ribbon containing synaptic ridges is expressed as percentage of the total ridges counted (about 27,000).

that, in the course of ontogeny, it is a requisite for the generation of the b-wave of the ERG (Nilsson and Crescitelli, 1969, 1970); its exact role is, however, open to speculation. Our observations are the first indication that, under physiological conditions, synaptic ribbons can be degraded and restituted, suggesting a high degree of morphological plasticity of the receptor-bipolar synapse. Pharmacological experiments with monoamine depleting drugs on the synaptic bars of the inner ear of the bullfrog (Osborne and Thornhill, 1972) led to the hypothesis that these structures might function as storage sites for transmitter substances. As for the visual cells, electrophysiological reactions to light and darkness revealed a continuous transmitter release in the dark which is turned off in the light (Werblin and Dowling, 1969). In this context, the present findings may be interpreted as follows: Dark adaptation leads to the depletion of the transmitter stores, which, in our material can be related to the disappearance of the synaptic ribbons; in the light, however, release of transmitter is shut off, the storage sites are replenished and the ribbons become visible.

Endogenous circadian influences on photoreceptor organs have been observed in vertebrates as well as in invertebrates. In teleosts, this phenomenon concerns the retinomotor responses of pigment epithelium and visual cells and was reported for rhythmically living species by Studnitz (1952) and Ali (1961), whereas in species with non-rhythmical behaviour, no such changes could be observed (Ali, 1959). As for the species studied in the present experiments, it exhibits, like most cichlid fishes, marked differences of activity under light and dark conditions. Furthermore, investigations of the mammalian pineal showed that, in pinealocytes, specifically the synaptic ribbons, were subjected to circadian changes in number and localisation (Vollrath, 1973).

The present results indicate that endogenous circadian changes of the metabolism of visual receptor cells must be taken into account in experiments with light and dark adapted retinas. In addition, the demonstration on an endogenous rhythm is of particular importance for experiments on biological rhythmicity because the eye is often used as a timer for a number of periodic functions of the organism.

REFERENCES

Ali, M.-A. (1959) The ocular structure, retinomotor and photobehavioural responses of juvenile Pacific salmon. Can. J. Zool. 37: 965-996.

Ali, M.-A. (1961) Histophysiological studies on the juvenile Atlantic salmon (*Salmo salar*) retina. Can. J. Zool. 39: 511-526.

Cragg, B.G. (1972) Plasticity of Synapses. In: The structure and function of nervous tissue. IV, pp. 1-60, edited by Bourne, G. H., Academic Press, New York.

Dowling, J.E. and Boycott, B.B. (1966) Organisation of the primate retina: electron microscopy. Proc.R.Soc. B. 166: 80-11.

Ladman, A.J. (1958) The fine structure of the rod bipolar cell synapse in the retina of the albino rat. J. Biophys, Biochem. Cytol. 4: 459-466.

Mountford, S. (1963) Effects of light and dark adaptation of the vesicle population of receptor-bipolar synapses. J. Ultrastruct. Res. 9: 403-418.

Nilson, S.E.G. and Crescitelli, F. (1969) Changes in ultrastructure and electroretinogram of bullfrog retina during development. J. Ultrastruct. Res. 27: 45-62.

Nilson, S.E.G. and Crescitelli, F. (1970) A correlation of ultrastructure and function in the developing retina of the frog tadpole. J. Ultrastruct. Res. 30: 87-102.

Osborne, M.P. and Thornhill, R.A. (1972) The effect of monoamine depleting drugs upon the synaptic bars in the inner ear of the bullfrog. Z. Zellforsch. 127: 347-355.

Studnitz, G. von (1952) Physiologie des Sehens. (Retinale Primärprozesse). Akademische Verlagsgesellschaft, Leipzig, 2e ed.

Vollrath, L. (1973) Synaptic ribbons of a mammalian pineal gland:

circadian changes. Z. Zellforsch. 145: 171-183.
Wagner, H.-J. (1973) Darkness-induced reduction of the number of synaptic ribbons in fish retina. Nature, N.B. 246: 53-55.
Werblin, F.S. and Dowling, J.E. (1969) Organisation of the retina of the mudpuppy. *Necturus maculosus*. II. Intracellular recording. J. Neurophysiol. 32: 339-355.

ACKNOWLEDGMENT

This work was supported by Grant No. Wa 348/1 of the Deutsche Forschungsgemeinschaft.

7. BEHAVIOURAL ASPECTS

PSYCHOPHYSICAL METHODS FOR INVESTIGATIONS OF VISION IN FISHES

D.P.M. NORTHMORE and Dean YAGER

Institute for Neurosciences and Behavior
University of Delaware
and State College of Optometry
State University of New York

Ay, while you live, draw your neck out o' th' collar.
Romeo and Juliet, Act I, Sc. 1

INTRODUCTION

A great deal of information about fish visual systems has been obtained through physiological, anatomical, and biochemical studies; however, very few quantitative data have been obtained about the visual discriminating capacities of the same species on which these studies have been conducted. Historically, psychophysical investigations have pointed the way for a variety of physiological investigations. The input - output relations revealed by psychophysics have led to theoretical models which attempt quantitatively to account for these relations; these models in turn suggest the expected form of physiological functions. A notable example of this logic is the prediction by Hering of opponent-colour-coding many decades before it was demonstrated at the electrophysiological level. The ultimate task for investigators of any sensory system is to develop linking hypotheses among psychophysics, electrophysiology, photochemistry, and anatomy.

In this paper we shall outline several broad categories of psychophysical methods used to investigate visual functions, as well as suggest some methods that have yet to be used on fish. At the same time we shall attempt to acquaint the reader with many of the terms and concepts common to animal behavioural studies. The methods are all concerned with devising ways of inducing an animal to report on the presence-absence, sameness-difference, magnitude, or degree of similarity, of one or more stimuli; we shall illustrate these methods with some specific examples. Reviews of animal psychophysical methods have been published earlier by Herter (1953), and Blough and Yager (1972).

METHODS

Reflex Methods

It is undeniably attractive to use a fish's innate stereotyped responses to visual stimuli as a means of measuring its powers of detection and discrimination. Of the many responses that might be employed, the most useful are those that do not wane with repeated elicitation.

One such visual response is the dorsal light reflex which is shown by the tilting of a fish about its longitudinal axis in the direction of a lateral source of light, the angle of tilt varying in a monotonic fashion with the difference in stimulation of the two eyes (Thibault, 1949) (see fig. 1b). Making use of this fact, Silver (1974) used her subjects, neon tetras, as null detectors to perform photometric matching between a constant white light and different monochromatic lights. The white and coloured lights were placed on opposite sides of the fish (fig. 1a) and the relative radiances of the coloured lights required to keep the fish upright were measured to give light-adapted spectral sensitivity curves.

Optomotor responses are usually studied in the laboratory by rotating an array of stripes around a fish contained in a stationary glass cylinder. If the stripes rotate at more than some threshold speed, a susceptible fish subject will either swim along with the stripes or orient to them like a compass needle orients to a magnet (Harden Jones, 1963). Using monochromatic light to illuminate or project the stripes, spectral sensitivity curves have been derived for sunfish (Grundfest, 1932), goldfish (Cronly-Dillon and Muntz, 1965) and sticklebacks (Cronly-Dillon and Sharma, 1968). Flicker fusion frequencies as a function of radiance have been measured for sunfish by determining the maximum velocity of rotation to give an optomotor response (Wolf and Zerrahn-Wolf, 1936). Whereas midwater fishes, especially pike and sunfish, are good responders, the optomotor responses of bottom-dwelling fishes are difficult to observe. However, even sedentary fishes make eye movements in response to stripe rotations and Easter's (1971) technique (see Easter this volume) for the recording of these movements might be used to good effect.

Other untrained behaviours dependent upon light have suggested themselves: Blaxter (1964) measured the least amount of light required by herring for feeding, avoiding a barrier, and phototaxis,

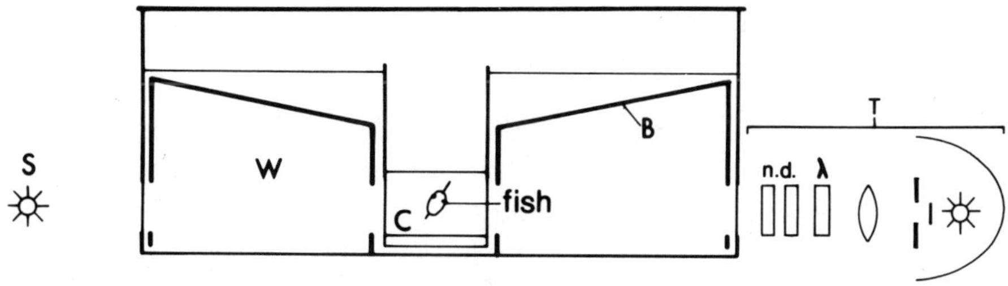

a

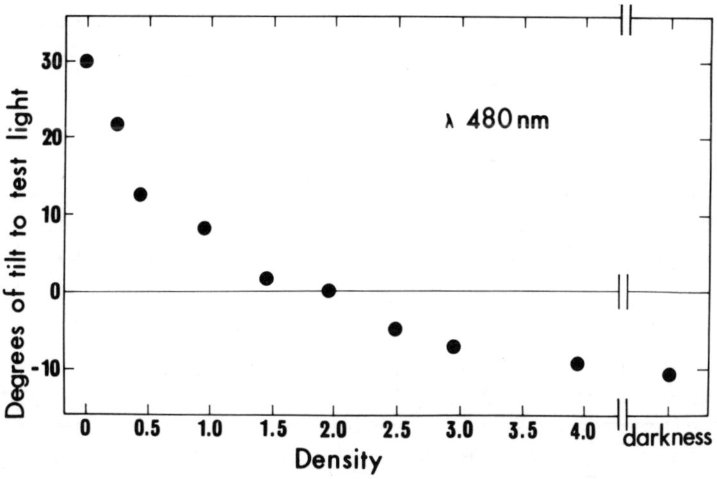

b

Fig. 1. a) Apparatus for photometric matching using the dorsal light reflex. The fish is contained in the test cell, C, submerged in a water bath, W, fitted with light baffles, B. S is a constant white standard light, and T provides the monochromatic test light controlled by neutral density filters, n.d., and by an interference filter, λ. The experimenter measures the radiance of the test light required to keep the fish upright. b) Angle of tilt of a neon tetra towards a test light of 480 nm as a function of the density of neutral filters in the beam. (From Silver, 1974).

and obtained different spectral and absolute sensitivities for each of the three activities.

The appeal of reflex methods is less apparent when it comes to recording responses; their appearance may be capricious or hard to observe, and the experimenter is often forced to rely on highly subjective judgements. For the most part, the responses used in the following training methods have been selected for the convenience of the experimenter.

Training Methods: Respondent

These methods, which are also called Pavlovian or classical conditioning methods, depend upon establishing a conditioned response (CR) by correlating the occurrence of a visual stimulus (CS) with another stimulus (UCS) that reliably elicits some conveniently measurable response (UCR). The CR is then used to indicate whether the animal can discriminate the CS from other stimuli, be they stimuli which have not been correlated with the UCS, or simply the absence of CS.

The responses most commonly conditioned for the purposes of fish psychophysics are the shock-induced increases in swimming activity, and the inhibition of respiration and heart rate, although other responses may be used such as the closure of the nictitating membrane in sharks (see Gruber, this volume).

The conditioning of gross activity is most simply achieved by confining the fish in a small tank equipped with carbon or stainless steel electrodes placed at either end for delivering brief AC shocks for the UCS, and a detector of water currents such as a paddle or self-heating thermistor for registering the CR (Bull, 1928; Horner, Longo & Bitterman, 1960; Northmore & Muntz, 1974). In a typical experiment, the CS, say a light, would be turned on for 10 seconds terminating in a 0.5 sec shock. Once established, the activity CR normally occurs before the shock so that an indication of responding can be obtained on every trial.

Activity cannot be readily conditioned in some fishes when closely confined, presumably because their natural response under these conditions is to freeze. Nevertheless, experiments in vision often require that the fish be restrained so that the angular subtense and position of stimuli in the visual field can be more closely controlled. For this reason, respondent conditioning of heart rate and respiration have been used for studies of spatial summation (Hester, 1968; Northmore, 1974) and visual field defects caused by lesions (Schwassmann and Krag, 1970).

In order to record the electrocardiogram uncontaminated with muscle potentials and movement artifacts, it is necessary to implant recording electrodes close to the heart (Roberts, Wright and Savage, 1973; Powers and Easter, this volume). In addition to the risk of injury and infection, the drawback to the procedure is that the electrodes work loose and fall out. However, the use of respiration as the response indicator in the technique employed by Northmore (1974), avoids the need for surgery and enables individual fish to be studied for prolonged periods. Figure 2a shows the subject, in this case a goldfish, restrained between two thin sheets of transparent polyethylene clamped by Plexiglass formers. The sheet to the rear of the gill covers is perforated to allow both the expired water to escape, and the shock current to pass between two carbon electrodes placed either side of the body. To reduce movements still further, it may be necessary to staple the sheets together to conform to the outline of the fish's body. The rhythmic water movements set up by the fish's breathing are detected by a small glass-encapsulated thermistor flexibly mounted in front of the mouth. The thermistor is heated by an electric current passed through it, and the cooling due to water movements causes a change in its resistance. The resulting voltage variations across the thermistor are recorded on a polygraph.

Because heart rate and breathing rate are highly correlated, at least in goldfish (Scobie, 1973), psychophysical measures obtained from their conditioning should be closely equivalent. The frequency of both activities are suppressed not only by noxious stimuli but also by novel stimuli, even before conditioning training is begun. This readiness to respond probably accounts for the rapid acquisition of the CR with these methods, often less than 10 trials (Woodward, 1971; Scobie, 1973), compared to the slow acquisition of conditioned activity (Horner et al., 1960).

The strength of the conditioned response can be assessed from a suppression ratio derived by comparing the heart or respiration rate during the CS with the rate immediately preceeding it, e.g. (pre CS rate - CS rate)/(pre CS rate). Alternatively, interbeat or interbreath intervals may be measured between trials in order to compute a criterion interval such that the probability of a longer interval is, say, only 5%. A response is counted if there is a pause in breathing or heart rate during the CS longer than the criterion interval.

Training Methods: Operant

In operant conditioning, the animal's behaviour is maintained by administering reinforcement of rewards immediately following the desired response.

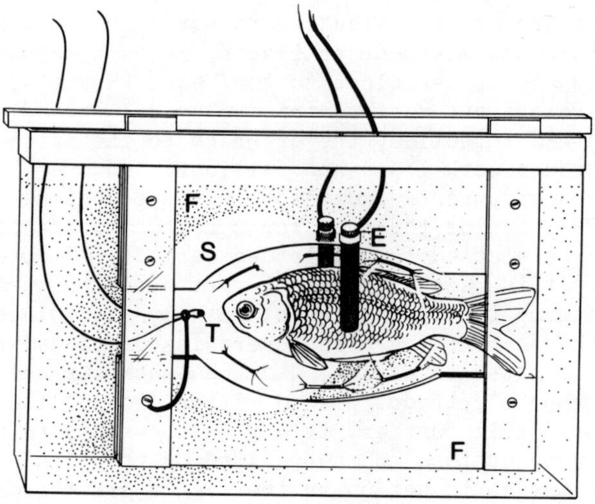

a

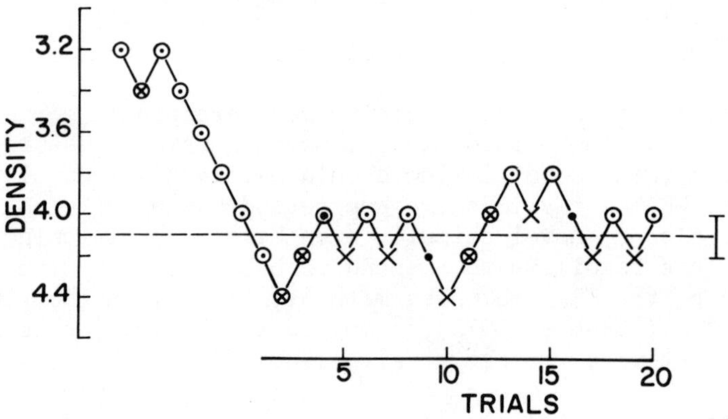

b

Fig. 2. a) Apparatus for determining thresholds using respondent conditioning of respiration. The goldfish is restrained between two sheets of transparent plastic clamped between Plexiglass formers, FF. Staples around the fish help to reduce movements. Respiration is detected by a thermistor, T, and shock is administered through carbon electrodes, E. The CS consists of a circular stimulus patch, S, projected for 5 sec onto a tracing paper screen 15 cm from the right eye. b) Tracking record from a respiratory conditioning experiment. Points represent CRs, crosses failures to give CRs or 'misses ', and circles trials in which shock is paired with the CS. Initially the CS is presented about 1 long unit above

Since the early investigations of fish colour vision by Reeves (1919), who used an aquatic version of the Yerkes-Watson discrimination box, the two-choice procedure has been applied to the study of fishes' powers of discriminating colours and shapes (Muntz and Cronly-Dillon, 1966; Herter, 1953; Sutherland, 1968), and for measuring acuity (Weiler, 1966). The simple hand-operated apparatus has since given way to automated machines more suited to the task of presenting large numbers of trials for the determination of psychophysical thresholds. An apparatus which was used by Yager (1967) with goldfish to obtain spectral sensitivity and saturation functions is representative of the automated method for determining thresholds and is very similar to that used by Volkmann, Zametkin and Stoykovich (1974; see Volkmann, this volume) for studying the discrimination of bar orientation. At one end of a fish tank two clear plastic panels are suspended from rods connected to pressure-sensitive transducers. On each trial the positive stimulus, the stimulus to which responding is rewarded, appears upon a screen behind one of the panels. If the fish responds correctly by pushing this panel, the stimulus is extinguished and a 'cue light' is immediately turned on, signalling the arrival of a food pellet dropped into the tank. A push on the wrong panel merely turns the stimulus off. Good discrimination depends upon the fish being oriented towards the response panels as it makes its choice. For this purpose, Yager trained his fish to make an 'observing response' by pushing a third panel at the opposite end of the tank in order to turn on the stimulus.

Fishes of different temperaments and habits may require a somewhat modified testing situation. Because the rudd (Scardinius erythrophthalmus) is more nervous than the goldfish, Muntz and Northmore (1970) moved the testing apparatus to the fish's home tank to avoid the trauma of netting. It was also found that the rudd's panel pushing lacked the force of the goldfish so that photoelectric cells had to be used to monitor the fish's entrance into a chamber to which the stimuli were projected (see Muntz, this volume). It is a sound principle in operant methods to reinforce correct responses close to the positive stimulus because it makes training easier. Therefore the rudd were rewarded with tubifex worms expelled directly into the response chambers. Also for rapidity of learning, the observing response consisted of passing

Fig. 2. (contd) threshold and its radiance in successive trials is reduced by 0.2 log unit following a CR, and increased by the same amount following a miss. Except for the initial trials, shock is paired with the CS only following a miss. The last 20 trials are used to estimate the threshold (dashed line) and 95% confidence interval (vertical bar).

through a rectangular doorway situated in the centre of the tank. This response did not have to be explicitly trained because rudd normally swim to and fro in a tank.

In a two-choice discrimination situation for the shanny (Blennius pholis), a littoral marine fish, Wainwright (1973) made use of the fact that these fish are negatively buoyant and spend much of their time resting on the bottom of a tank. The two stimulus patches lay in a horizontal plane just below the water surface so that the fish swam vertically to make their discrimination and then fell back to the bottom of the tank in order to set up the next trial.

When confronted with a difficult discrimination, fish often develop a strong preference for the stimulus panel on one side. It is then necessary to institute a correction procedure such that the positive stimulus is repeatedly presented on the non-preferred side until the fish switches sides and obtains reinforcement.

Single Response Methods

Although discrimination is best studied in the two-choice situation where the positive and negative stimuli are presented simultaneously, there are applications for methods in which the stimuli are presented successively on a single panel, particularly in the study of stimulus generalization (see Suprathreshold Response Measures) (Yarczower and Bitterman, 1965; Yager, Buck and Duncan, 1971).

Conditioned suppression, a relatively new technique combining operant and respondent conditioning has proved valuable for obtaining psychophysical thresholds in various animals (Smith, 1970). The paradigm has been shown to work with goldfish (Geller, 1964) and might be implemented for psychophysics as follows. The fish is first trained to push a panel for intermittent food rewards. When a stable rate of pushing has been established, a visual stimulus of say 30 sec duration is projected at random intervals upon the panel and is paired with unavoidable shock in the manner of classical conditioning. After a number of pairings, the rate of panel pushing is depressed or completely inhibited during the stimulus. The suppression effect can be quantified by the same suppression ratio suggested for cardiac or respiratory conditioning. The conditioned suppression technique has much in common with the methods based on heart rate and respiratory conditioning and appears to obey the laws of respondent conditioning (Kamin, 1965). One advantage of the method is that the orientation of the fish to the stimulus is assured without having to restrain it.

AVOIDANCE METHODS

Goldfish can be trained to avoid electric shock by swimming out of a compartment in which a warning stimulus is presented. An automated two-compartment apparatus (shuttle box) was described by Horner et al., (1961) which has been used for some visual psychophysical experiments. The two compartments are divided by a hurdle which comes within a few centimetres of the water surface and is intended to discourage the fish from crossing from one side to the other. However, when shocked, the fish can escape over it and after training this response comes to be elicited by the visual stimulus. The apparatus has been used to obtain a scotopic spectral sensitivity curve for a cichlid fish (Tavolga and Jacobs, 1971), and in a generalization experiment to show that an appropriate mixture of yellowish-red and yellowish-green light resembles a spectral yellow for a carp (Oyama and Jitsumori, 1973). The difficulty with the method seems to be that fish cross the hurdle spontaneously giving an unstable baseline of behaviour. Furthermore, the maximum probability of responding is not much better than 0.8, even with discriminable stimuli.

DATA ANALYSIS

Psychometric Functions and Signal Detection Theory

Measuring a visual threshold usually involves recording the percentage of 'appropriate', or as we shall call them, correct responses to each of a set of different stimulus values distributed along some dimension (e.g. radiance, wavelength differences etc), each stimulus value being presented several times. A psychometric function is a plot of these percentages against the corresponding stimulus values. Psychometric functions are typically sigmoid curves rising from some baseline of percent correct response at low stimulus values, to 100% at high values. From a practical point of view, these functions are important in showing the degree to which an animal's response is under stimulus control; shallow and irregular functions probably mean that the animal is not attending to the stimuli, or for some other reason is performing erratically.

Conventionally, thresholds are estimated from psychometric functions by finding the stimulus value which gives a percent correct response lying midway between the baseline and 100%. In the two-choice situations, the baseline must be at the 50% level, but in single response and respondent conditioning experiments when the fish in effect answers 'yes' or 'no' to the question 'did you see anything?', the baseline could be at almost any percentage level, depending on the animal's bias to respond 'yes' in the absence of a stimulus.

If, in the yes-no methods, the baseline represents the probability of correct responding purely by chance (i.e. guessing), then the estimate of the threshold should be unaffected by variations in the baseline rate. However, this is generally found not to be the case, and the implications of the baseline rate which animal psychophysicists have tended to ignore have to be re-examined (Wright, 1974). A successful interpretation is offered by signal detection theory which postulates that in the absence of a stimulus, random events in the nervous system are occasionally of sufficient magnitude to elicit a response by exceeding some internal criterion level. Thus the baseline rate is thought of not just as the result of guessing but rather neural noise mimicking the effect of a real stimulus. By manipulating such factors as the a priori probability of a stimulus, the reinforcement contingencies, or the motivation, it is possible to alter the animal's response bias and with it the baseline rate. Signal detection theory explains this effect by supposing that the criterion is shifted by these manipulations. The theory also suggests measures of the discriminability of a stimulus that are independent of response bias (e.g., d'. See Wright, 1974). Some animal studies have tended to confirm the invariance of these measures under various conditions of stimulus probability and reinforcement contingency (Hack, 1963; Nevin, 1970; Yager and Duncan, 1971).

The great advantage of the two-choice methods is that a moderate bias in favour of one or other response panel need not be considered in determining thresholds because the baseline is fixed at 50%.

Stimulus Ordering and Data Acquisition

The experimenter is faced with several choices as to how stimuli should be presented in order to generate psychometric functions. A traditional method, and one that is simple to implement with automated apparatus is the method of constant stimuli; a set of stimulus values is chosen so as to span the threshold and presented in blocks of equal value or in some other predetermined order (Yager and Thorpe, 1970).

Following Blough's (1958) highly successful psychophysical experiments on pigeons, methods involving the adjustment of stimulus values according to the animal's discrimination performance have become increasingly popular. These so-called stimulus tracking methods are efficient because they can be designed to concentrate trials around a particular percentage level on the psychometric function. Other advantages are that temporal variations in threshold can be followed, as for example during dark-adaptation, and that stimulus control may be monitored continuously. However, tracking has been successfully accomplished in a two-choice situation with fish by adopting a procedure whereby the stimulus intensity decreases by 0.2 log unit following two consecutive correct responses, but increases by the same amount following every incorrect response (Muntz and Northmore, 1970, 1973). It can be shown that with this procedure, stimulus intensity fluctuates around the 71% level of correct responding. Many variants on the theme may be devised and it is reassuring to find that one used by Volkmann et al. (1974) yielded essentially identical psychometric functions to those obtained with the method of constant stimuli.

A recurring problem in animal psychophysics is what to do about the reinforcement of responses near threshold without losing stimulus control. As we have seen, a 50% reinforcement rate for subthreshold stimuli in the two-choice situations can be sufficient to maintain a side preference. Similarly, shock reinforcement delivered with subthreshold stimuli in respondent conditioning, conditioned suppression, and avoidance paradigms, is liable to lead to loss of stimulus control by disrupting baseline behaviour. A method that has yielded reliable thresholds with respiratory conditioning (see fig. 2b) involves a simple tracking procedure in which respiratory pauses during the CS lead to a less intense CS on the next trial, and the absence of a pause, a miss, to a more intense CS. Shock is also made contingent upon responses by delivering it only on trials following a miss. As a result, shock-paired stimuli tend on the whole to be suprathreshold. Each session starts with stimuli well above threshold and all are paired with shock.

Suprathreshold Response Measures

So far, the experimental methods have been discussed with the aim of measuring visual thresholds. In practice, an animal's responses in the experimental situation are usually graded in some way that allows us to make inferences about the quality of the stimulus, other than whether or not it is detectable. The strength of an innate response, the dorsal light reflex, is related monotonically to stimulus intensity (fig. 1b), and an orderly relation was found between log stimulus radiance and the vigour of the CR during respondent conditioning of a fish's activity (Northmore, unpublished observations). Operant conditioning too, can be made to show a graded aspect by recording the latency of responses. Without any specific training, latency generally decreases with increasing stimulus radiance (Moody, 1970). It is not possible, of course, to say that a particular function relating response magnitude or latency with stimulus radiance describes the psychophysical function' which is supposed to relate brightness with radiance. It is a fair inference, however, that two stimuli of different colour are equal in some attribute, which we might call brightness, if they both give the same response latencies or magnitudes.

In the preceeding instances, the graded, intensity-dependent nature of the responses were untrained 'stimulus intensity dynamisms' (Kamin, 1965). However, response magnitude can be influenced by training subjects in operant or respondent conditioning to respond to only one value of a stimulus dimension. Once stable responding to this stimulus has been established, the strength or latency of response is tested with a range of other stimuli while reinforcement is witheld. The change in response strength or latency as a function of the stimulus value is known as a generalization gradient, and is interpreted as a measure of similarity of the test stimuli to the training stimulus.

Generalization testing using respondent conditioning of heart rate was done by McCleary and Bernstein (1959) who demonstrated that goldfish could be trained to respond differentially either to different wavelengths but not different intensities, or vice versa. Respiratory conditioning also promises to be useful since Fay (1970) obtained steep auditory generalization gradients about different frequencies of training tone with this method.

Single response operant techniques for goldfish, similar to those widely used with pigeons and rats have enabled generalization gradients to be obtained for wavelength (Yarczower and Bitterman, 1965) and radiance (Yager and Duncan, 1971). Yager et al. (1971) used the method to show that the brightness of a given radiance of light for a goldfish is unaffected by the fish's body temperature.

When using suprathreshold response measures, it should be borne in mind that the results will be distorted by any innate preference or aversion which operates over part of the stimulus continuum studied.

CONCLUSIONS

One of the criteria for judging the success of a method for doing psychophysics is the maximum sensitivity that can be achieved with it. The results for the cyprinids, goldfish and rudd, show that respectably low absolute thresholds, at least comparable to human scotopic thresholds, are attainable both with the two-choice operant method (Yager, 1966; Muntz and Northmore, 1973) and with respondent conditioning (Powers and Easter, this volume). These same methods also yield photopic, or increment thresholds which are again similar to those of humans, if not lower at long wavelengths (Yager, 1967; Northmore, 1974). Furthermore, the 'envelope' type of photopic spectral sensitivity found for the rudd in the two-choice situation (Muntz, this volume) is the result expected of a three receptor system operating at its highest sensitivity. Phototaxes, optomotor response and other reflexes should not be expected to yield similar sensitivities in psychophysical methods based upon them (Blaxter, 1964), although the optomotor response does appear to be a sensitive indicator; sunfish can detect rotating stripes at lower illuminations than humans (Wolf and Zerrahn-Wolf, 1936).

Judging from psychometric functions and tracking records (e.g. fig. 2b), the precision of threshold data obtainable with training methods, particularly in the respondent experiments (Northmore and Muntz, 1974) can be remarkably good, the latter giving steeper psychometric functions than two-choice operant methods (Yager, 1967; Muntz and Northmore, 1970). Where reliability is concerned, avoidance methods have a poor reputation, both because of variable baseline behaviour and because the maximum response rate rarely reaches 100%. It is difficult to generalise about the accuracy of reflex methods because this will usually be limited by the accuracy with which the experimenter can record responses.

Ideally, the validity of a result obtained with a given psychophysical method should be checked against the results of different kinds of experiment designed to measure the same thing. However, in the absence of systematic comparisons of this sort, it should be possible to show that the results of the method conform to well established laws such as Weber's and Ricco's, and this has been done with both the two choice operant (Muntz and Northmore, 1971) and respondent methods (Hester, 1968; Northmore, 1974). In some instances psychophysical and physiological data can be related, as for example between the rudd's scotopic spectral sensitivity and its pigment absorption spectrum derived from retinal densitometry (Denton, Muntz and Northmore, 1971). A further example is the prediction of the outcome of a wavelength generalization experiment from the properties of goldfish retinal ganglion cells (Powers and Easter, this volume). Reflexes tend to be triggered by a more or less limited set of visual stimuli, as Schlieper (1927) found when he failed to elicit optomotor responses using moving stripe patterns exhibiting only contrasts of colour, although his animals were shown to be capable of colour discrimination by other tests. Thus the results of a psychophysical method based on a visual reflex cannot be assumed to predict the behaviour in some other psychophysical experiment. Nevertheless, these results may be the appropriate ones to apply to the visual behaviour of fish in their natural habitat.

ACKNOWLEDGEMENT

Financial support for D.P.M. Northmore obtained from the Delaware Institute for Medical and Educational Research, is gratefully acknowledged.

REFERENCES

Blaxter, J.H.S. (1964) Spectral sensitivity of the herring, Clupea harengus L. J. Exp. Biol. 41: 155-162.
Blough, D.S. (1958) A method for obtaining psychophysical thresholds from the pigeon. J. Exp. Anal. Behav. 1: 31-43.
Blough, D.S. and Yager, D. (1972) Visual psychophysics in animals. In: Handbook of Sensory Physiology, Vol. VII/4. Edited by D. Jameson and L. Hurvich. Springer-Verlag, Berlin. 732-763.
Bull, H.O. (1928) Studies on conditioned responses in fishes. Part I. J. Mar. Biol. Assoc. U.K. 15: 485-533.
Cronly-Dillon, J.R. and Muntz, W.R.A. (1965) The spectral sensitivity of the goldfish and the clawed tadpole under photopic conditions. J. Exp. Biol. 42: 481-493.
Cronly-Dillon, J.R. and Sharma, S.C. (1968) Effect of season and sex on the photopic spectral sensitivity of the three-spined stickleback. J. Exp. Biol. 49: 679-687.
Denton, E.J., Muntz, W.R.A. and Northmore, D.P.M. (1971) The distribution of visual pigments within the retina in two teleosts. J. Mar. Biol. Ass. U.K. 51: 905-915.
Easter, S.S. (1971) Spontaneous eye movements in restrained goldfish. Vision Res. 11: 33-343.
Fay, R.F. (1970) Auditory frequency generalization in the goldfish (Carassius auratus). J. Exp. Anal. Behav. 14: 353-360.
Geller, I. (1964) Conditioned suppression in goldfish as a function of shock-reinforcement schedule. J. Exp. Anal. Behav. 7: 345-349.
Grundfest, H. (1932) The sensibility of the sun-fish, Lepomis, to monochromatic radiation of low intensities. J. Gen. Physiol. 15: 307-328.
Hack, M.H. (1963) Signal detection in the rat. Science. 193: 758-759.
Harden Jones, F.R. (1963) The reaction of fish to moving backgrounds. J. Exp. Biol. 40: 437-446.
Hester, F.J. (1968) Visual contrast thresholds of the goldfish (Carassius auratus). Vision Res. 8: 1315-1336.
Herter, K. (1953) Die Fischdressuren und ihre sinnesphysiologischen Grundlagen. Akademie-Verlag, Berlin.
Horner, J.L., Longo, N. and Bitterman, M.E. (1960) A classical conditioning technique for small aquatic animals. Am. J. Psychol. 73: 623-626.
Horner, J.L., Longo, N. and Bitterman, M.E. (1961) A shuttle box for fish and a control circuit of general applicability. Am. J. Psychol. 74, 114-120.
Kamin, L.J. (1965) Temporal and intensity characteristics of the conditioned stimulus. In Classical Conditioning: A Symposium. Edited by W.F. Prokasy. p. 118-147. Appleton-Century-Crofts, New York.
McLeary, R.A. and Bernstein, J.J. (1959) A unique method for control of brightness cues in the study of colour vision in fish. Physiol. Zool. 32: 284-292.

Moody, D.B. (1970) Reaction time as a sensory function. In: Animal Psychophysics. Edited by W.C. Stebbins, p. 277-302, Appleton-Century-Crofts, New York.
Muntz, W.R.A. and Northmore, D.P.M. (1970) Vision and visual pigments in a fish, Scardinius erythrophthalmus (the rudd). Vision Res. 10: 281-298.
Muntz, W.R.A. and Northmore, D.P.M. (1971) The independence of the photopic receptor systems underlying visual thresholds in a teleost. Vision Res. 11: 861-876.
Muntz, W.R.A. and Northmore, D.P.M. (1973) Scotopic spectral sensitivity in a teleost fish (Scardinius erythrophthalmus) adapted to different daylengths. Vision Res. 13: 245-252.
Nevin, J.A. (1970) On differential stimulation and differential reinforcement. In: Animal Psychophysics. Edited by W.C. Stebbins, p. 401-423. Appleton-Century-Crofts, New York.
Northmore, D.P.M. (1974) Area-threshold functions in goldfish. Paper presented to the Association for Research in Vision and Ophthalmology, Sarasota.
Northmore, D.P.M. and Muntz, W.R.A. (1974) Effects of stimulus size on spectral sensitivity in a fish (Scardinius erythrophthalmus), measured with a classical conditioning paradigm. Vision Res. 14: 503-514.
Oyama, T. and Jitsumori, M. (1973) A behavioural study of colour mixture in the carp. Vision Res. 13: 2299-2308.
Reeves, C.D. (1919) Discrimination of light of different wavelengths by fish. Behav. Monog. vol. 4: No. 3, p. 106.
Roberts, M.G., Wright, D.E. and Savage, G.E. (1973) A technique for obtaining the electrocardiogram of fish. Comp. Biochem. Physiol. 44: 665.
Schlieper, C. (1927) Farbensinn der Tiere und optomotorische Reaktionen. Z. Vgl. Physiol. 12: 1-77.
Schwassmann, H.O. and Krag, M.H. (1970) Relation of visual field defects to retinotectal topography in teleost fish. Vision Res. 10: 29-42.
Scobie, S.R. (1973) Unconditional stimulus intensity and cardiac conditioning in the goldfish (Carassius auratus). Physiol. Behav. 11: 31-34.
Silver, P.H. (1974) Photopic spectral sensitivity of the neon tetra (Paracheirodon innesi Myers) found by the use of a dorsal light reaction. Vision Res. 14, 329-334.
Smith, J.C. (1970) Conditioned suppression as an animal psychophysical technique. In Animal Psychophysics. Edited by W.C. Stebbins, p. 125-159. Appleton-Century-Crofts, New York.
Sutherland, N.S. (1968) Shape discrimination in the goldfish. In: The Central Nervous System and Fish Behaviour. Edited by D. Ingle, p. 35-50. The University of Chicago Press, Chicago.
Tavolga, W.N. and Jacobs, D.W. (1971) Scotopic thresholds for monochromatic light in the cichlid fish, Tilapia heudolotii macrocephala. Vision Res. 11: 713-717.

Thibault, C. (1949) Action de la lumière blanche et monochromatique sur la posture des poissons téléostéens. Arch. Sci. Physiol. 3: 101-124.

Volkmann, F.C., Zametkin, A.J. and Stoykovich, C.A. (1974) Visual discrimination of orientation by the goldfish (Carassius auratus). J. Comp. Physiol. Psych. 86: 875-882.

Wainwright, A. (1973) Spectral sensitivity of the shanny (Blennius pholis). Unpublished doctoral thesis. University of Sussex, England.

Weiler, I.J. (1966) Restoration of visual acuity after optic nerve section and regeneration in Astronotus ocellatus. Exp. Neurol. 15: 377-386.

Wolf, E. and Zerrahn-Wolf, G. (1936) Threshold intensity of illumination and flicker frequency for the eye of the sun-fish. J. Gen. Physiol. 19: 495-502.

Woodard, W.T. (1971) Classical respiratory conditioning in the fish: CS intensity. Am. J. Psychol. 84: 549-554.

Wright, A.A. (1974) Psychometric and psychophysical theory within a framework of response bias. Psychol. Rev. 81: 322-347.

Yager, D. (1966) Behavioural measures and theoretical analysis of spectral sensitivity and spectral saturation in the goldfish Carrassius auratus. Unpublished Doctoral dissertation, University of Pennsylvania.

Yager, D. (1967) Behavioural measures and theoretical analysis of spectral sensitivity and spectral saturation in the goldfish, Carassius auratus. Vision Res. 7: 707-727.

Yager, D. (1968) Behavioural measures of the spectral sensitivity of the dark-adapted goldfish. Nature 220: 1052-1053.

Yager, D., Buck, S. and Duncan, I. (1971) Effects of temperature on the visually evoked tectal potential and brightness perception in goldfish. Vision Res. 11: 849-860.

Yager, D. and Duncan, I. (1971) Signal detection analysis of luminance generalization in goldfish using latency as a graded response measure. Percept. and Psychophys. 9: 353-355.

Yager, D. and Thorpe, S. (1970) Investigation of goldfish colour vision. In: Animal Psychophysics. Edited by W.C. Stebbins, p. 259-275. Appleton-Century-Crofts, New York.

Yarczower, M. and Bitterman, M.E. (1965) Stimulus generalization in the goldfish. In: Stimulus Generalization. Edited by D. J. Mostofsky, p. 179-192. Stanford University Press.

BEHAVIOURAL STUDIES OF VISION IN A FISH AND POSSIBLE RELATIONSHIPS TO THE ENVIRONMENT

W.R.A. MUNTZ

Lab. of Exptl. Psychology, University of Sussex

Brighton, BN1 9QG, England

> I cannot be offended with my trade. Please you to name it.
>
> Pericles, Act IV, Sc. 5

INTRODUCTION

Over the past years a considerable body of behavioural data has been accumulated on the vision of the cyprinid fish *Scardinius erythophthalmus* (the rudd), and we now probably know as much about the visual capabilities of this species as of any other fish (Muntz and Northmore, 1970, 1971, 1973; Northmore and Muntz, 1974). The present chapter summarises these data, and discusses some of the problems they raise, including possible relationships to physiology and the visual tasks faced by the animals in their natural environment. A variety of behavioural methods were used. Since these yielded very different results they are briefly described here; a more detailed review of the problems involved in obtaining sensory data by behavioural means is given in the previous paper (Yager and Northmore, this volume).

SPECTRAL SENSITIVITY UNDER PHOTOPIC CONDITIONS

The sensitivity of the rudd to different monochromatic lights superimposed on tungsten light backgrounds has now been determined under four different sets of conditions, and in each case a different spectral sensitivity curve was obtained. The different experimental conditions and results were as follows:

(i) Two-choice discrimination situation. In this experiment (Muntz and Northmore, 1970) an opaque black response panel with two apertures was placed at one end of the fish's home tank. At the beginning of each trial a tungsten light background of fixed intensity was projected onto a screen behind both apertures, and a monochromatic stimulus light of variable intensity was superimposed on the background light of one side. The fish's task was to enter the aperture on the same side as the monochromatic stimulus light, and correct responses were reinforced by delivering two or three Tubifex worms into a small chamber immediately behing the aperture. Incorrect responses resulted in the trial being terminated without reward.

A second panel with a 9.5 x 5.5 cm rectangular hole in it was placed in the centre of the tank. A trial was initiated by the fish swimming through this hole into the far end of the tank (the observing response). Movements of the fish into the response apertures and through the hole of the middle panel were monitored with photocells. Thresholds were determined with a tracking procedure.

The fish learnt very rapidly in this situation and gave highly reliable results, with the 95% fiducial limits for the individual thresholds typically of the order of 0.25 log units. The resulting spectral sensitivity curves for fish adapted to 12 and 20 hr daylengths are shown in figure 1.

The first striking feature about these curves is the very high short wavelength sensitivity that was found. Very similar results had previously been obtained by Yager (1967), using goldfish and a two-choice situation resembling the one used with rudd, and by Thibault (1949), using the dorsal light reaction with carp. The maximum of the short wavelength sensitivity is at shorter wavelengths (430 nm or less) than the λ max of the blue receptors of the goldfish (Marks, 1965) or carp (Tomita et al., 1967). Various possible reasons for this, such as fluorescence of the lens and optic media, have been considered by both Yager and by Muntz and Northmore, without however reaching any definite conclusion. As we shall see below, high short wavelength sensitivity does not occur in all situations.

In order to understand the form of the spectral sensitivity curve we need to know the spectral characteristics of the receptors that contribute to it. Unfortunately, this information is not available for the rudd. Nevertheless it seems likely that the results can be best explained by assuming that the threshold at any wavelength depends on the single receptor mechanism that is most sensitive at that wavelength (the envelope model). Details of the evidence for this statement may be found in Muntz and Northmore (1970, 1971) and will not be reviewed here at an length. Briefly, they are as follows.

(a) The data can be adequately fitted with the envelope model by assuming three receptors conforming to known properties of visual pigments and having their λ max at 430, 507, and 620 nm.

(b) Increment thresholds to tungsten light stimuli superimposed on tungsten light backgrounds were determined in the same apparatus. The mean value of the Weber fraction (four fish) was 7.5%. Using the envelope model it is possible to predict the Weber fraction to tungsten light stimuli from the results obtained with monochromatic stimuli: the predicted value differed from the experimental value by only 0.15 log units.

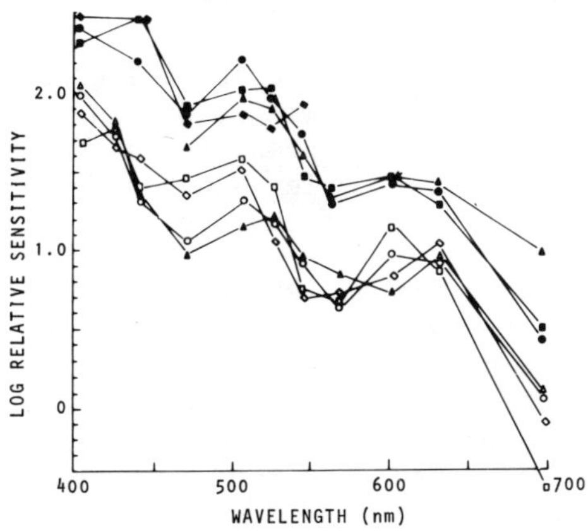

FIG. 1. Two-choice discrimination data for individual fish adapted to 12 hr (filled symbols) and 20 hr (empty symbols) daylengths. The curves have been shifted on the vertical axis to minimise individual differences in absolute sensitivity, and the two sets of curves have been displaced vertically by an arbitrary amount for clarity.

(c) By obtaining thresholds to monochromatic stimuli superimposed on different monochromatic backgrounds it was possible to isolate the different spectral classes of receptor from each other almost completely (Muntz and Northmore, 1971), in a manner analogous to the increment threshold experiments performed by Stiles (1959) on human subjects. The results show that at threshold the receptors are acting to a large degree independently of each other, as required by the envelope model.

A final point about the results of this experiment is that the shape of the spectral sensitivity curve was unaffected by daylength. This is surprising, for the changes in daylength used had a marked effect on the extractable visual pigments, with, as was originally shown by Dartnall et al. (1961), longer daylengths resulting in higher proportions of the A_1-based pigment (see papers by Beatty and Muntz in this volume for further information on how daylength and other environmental factors often affect the visual pigments of fishes). If such an effect were also occurring for the cone pigments it should show up in the spectral sensitivity curves. The simplest explanation of the lack of such an effect is that daylength affects the visual pigments of the rods only, a suggestion that has also been made by Bridges and Yoshikami (1970) on the basis of the different ways in which the outer segment discs of the rods and cones are formed (e.g. Young, 1971). It should be remembered, however, that the behavioural experiment measured visual thresholds, and that it is possible that when suprathreshold stimuli are used less sensitive photopic receptors come into play, which do show an effect of daylength. Measuring sensitivity using the electroretinogram, which requires much more intense stimuli, did in fact reveal an effect of daylength under photopic conditions (Northmore and Muntz, 1970), though the results were complex and could not be interpreted directly in terms of visual pigment changes.

(ii) Classical conditioning. Photopic spectral sensitivity curves have also been obtained using a classical conditioning procedure (Northmore and Muntz, 1974). The fish were placed in a small tank and the stimuli, which were stationary and moving bars of light, were projected on to a horizontal screen 250 mm above the water surface. The unconditioned stimulus was electric shock, and both the conditioned and unconditioned responses were increases in swimming activity. The experimental tank contained a self-heated thermistor; changes in swimming activity generated water currents which altered the temperature of the thermistor, and could therefore be monitored as changes in the thermistor's resistance.

In this situation the results were strongly dependent on the size of the stimulus, as is shown in fig. 2. The upper curve in this figure was obtained with moving bars of light, the width of which subtended approximately 4.75°. The lower curve was obtained with stimuli that illuminated the whole of the screen, and subtended approximately 40° at the level of the fish. Subsequent tests showed that the difference depended on the size of the stimulus, and not on whether they where moving or stationary. The two curves are in the correct position relative to each other on the vertical axis. Not only is the shape of the two curves different from each other, the fish were also considerably less sensitive when the larger stimuli were used.

The upper set of data in the figure, obtained with the 4.75° bar of light, can be fairly well fitted by an envelope model. This is indicated by the continuous lines on this part of the figure, which give the spectral sensitivities of three assumed receptor

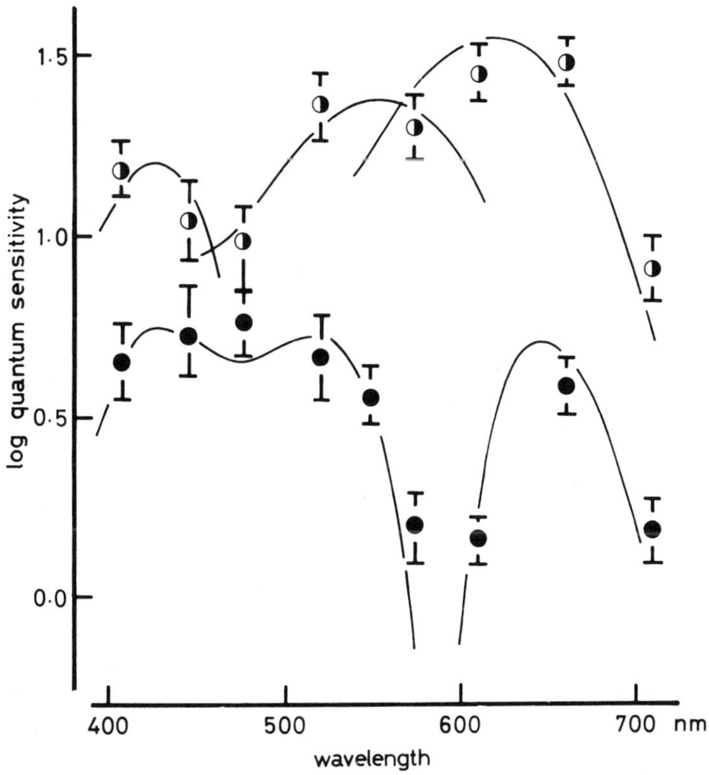

FIG. 2. Results obtained with the classical conditioning procedure. The half filled circles show the thresholds for a 4.75° moving bar stimulus with 95% limits, the filled circles the thresholds for diffuse light stimuli. The continuous lines have been fitted as described in the text.

systems. The fit is not perfect: in particular the red peak of the curve appears to be somewhat displaced towards long wavelengths in comparison to the hypothesised 620 nm receptor. It is likely that this reflects some degree of interaction between the receptor systems (see below). Nevertheless the mismatch is not large, and to a first approximation an envelope model can account for the data.

It is very clear, however, that an envelope model cannot account for the lower set of data on the figure. In this case the red peak is very narrow and considerably displaced towards long wavelengths, and there is a marked trough in the sensitivity curves at about 580-600 nm. These characteristics suggest that some form of inhibition is occurring between the red and green receptor mechanisms. The curve drawn through the lower set of points has been computed for each wavelength λ from the equation:

$$r = k_1(A\lambda.I_\lambda)^P - k_2(B_\lambda.I_\lambda)^P + k_3(C_\lambda.I_\lambda)^P$$

by finding the intensity of light necessary to produce a criterion value of the response r, of arbitrary magnitude. In this equation I_λ is the intensity of light at wavelength λ from the equation:

$$r = k_1(A_\lambda.I_\lambda)^P - k_2(B_\lambda.I_\lambda)^P + k_3(C_\lambda.I_\lambda)^P$$

value of the response r, of arbitrary magnitude. In this equation I_λ is the intensity of light at wavelength λ; A_λ, B_λ, and C_λ are the absorbances of the assumed pigments of the three receptors, which are $VP620_2$, $VP535_2$, and $VP430_1$ respectively; k_1; k_2, and k_3 are three constants with values of 1, 1.2, and 0.6; and p is a compressive power function of the type described by Barlow (1953) and Easter (1968), and has a value of 0.6.

This explanation of the results assumes that with the smaller stimuli the receptors respond independently of each other according to an envelope model, while with larger stimuli the red and green receptors interact with each other in an inhibitory fashion. Small stimuli are detected when they deliver a sufficient number of extra quanta to any receptor system, which is analogous to a change in brightness; whereas large stimuli are detected if they cause a change in the relative output of different receptor systems, which is analogous to a change in colour. The results can be explained in neurophysiological terms if we assume that the rudd retina has ganglion cells with inhibitory surrounds of the type described, for example, by Daw (1968) in the goldfish. The larger stimuli invade these surrounds, both reducing the absolute sensitivity and resulting in inhibitory interactions between the different spectral classes of receptor. This neurophysiological explanation is given in more detail in Northmore (1973) and in Northmore and Muntz (1974).

(iii) Innate escape response. The rudd shows an innate escape response to overhead moving stimuli, which has also been used by Northmore (1973) to measure spectral sensitivity under photopic conditions. The stimuli were moving overhead bars of light superimposed on a tungsten light background, very similar to those used in the conditioning experiments. The responses of the fish, which were recorded photographically, consisted in diving to the bottom of the tank and turning against the direction of movement of the stimulus. The results are shown in fig. 3. Although the stimuli were very similar to those used in part of the conditioning experiment, the results are very different: maximal sensitivity occurred at about 640 nm and no response at all could be elicited below 540 nm with the stimulus intensities that were available. Northmore (1973) has shown that these results also can be explained in physiological terms if it is assumed that the behaviour is dependent

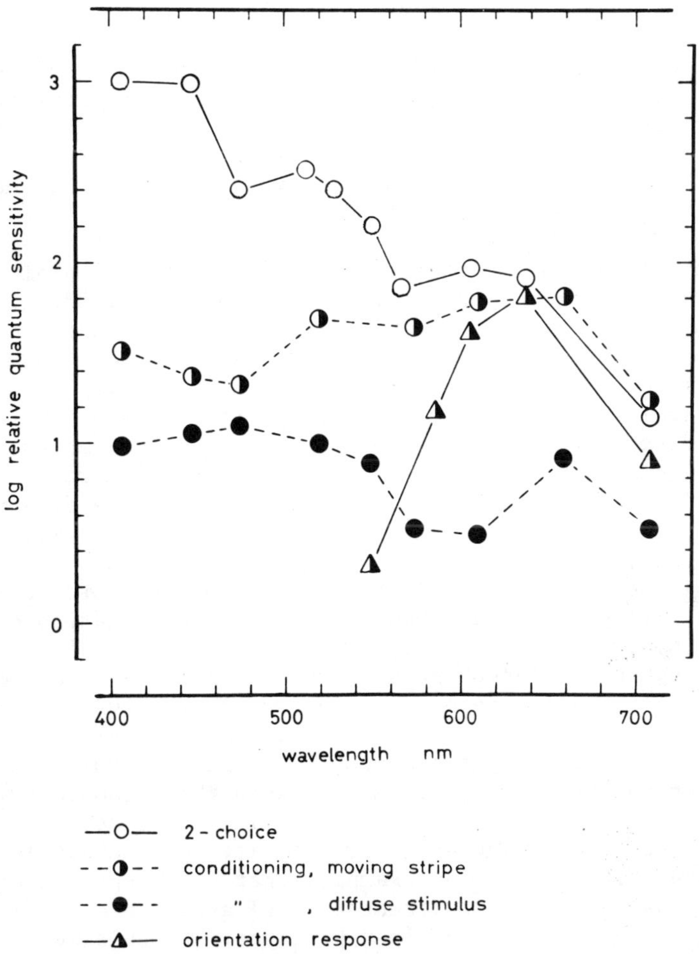

FIG. 3. Photopic spectral sensitivity curves for the rudd. The empty circles show the two-choice discrimination data, the half-filled triangles the escape response data, the half-filled circles the conditioning data for 4.75° bars, and the filled circles the conditioning data for diffuse stimuli.

on the long wavelength part of the output of red-green opponent units.

The four separate photopic sensitivity curves that have been obtained are shown together in fig. 3. It is clear that in fishes, as in many other animals (Muntz, 1974), a wide variety of spectral sensitivity curves can be obtained depending on the experimental conditions. At a practical level this means that data obtained in a given situation cannot be used with any confidence when the conditions are changed. It has been common practice, for example, to test for colour vision in animals by measuring the spectral sensitivity, and then to use these results to equate different spectral stimuli for brightness. Using such a procedure with the rudd would however raise the problem of which (if any) of the curves shown in fig. 3 is the appropriate one to use.

As summarised above, attempts have been made to explain all the four spectral sensitivity curves that have been obtained, either by assuming that the different receptor mechanisms act independently of each other (the envelope model), or by assuming some form of inhibitory interaction between them. Such attempts imply that the incoming visual information is analysed by the animals in different ways, depending on the experimental conditions. In some cases the type of analysis that occurs clearly depends on the form of the stimulus. The difference between the two sets of data obtained with the conditioning procedure, which depends on stimulus size, is an example of this, and can be explained if we assume that the large stimuli invade the inhibitory surrounds of concentric receptor fields (see above). In other cases, however, such an explanation does not appear adequate. In particular, it is difficult to explain in this way why the short wavelength sensitivity is higher in the two-choice discrimination situation than it is in the other situations. This finding is not restricted to the rudd: in goldfish also high short wavelength sensitivity appears strongly in the two-choice discrimination situation (Yager, 1967), but not when the optomotor response is used (Cronly-Dillon and Muntz, 1965). There are several possible reasons why high short wavelength sensitivity only occurs in certain circumstances. It may be, for example, that different receptor mechanisms, possibly even different retinal ganglion cells, are involved in the different behaviour patterns. If this were the case there is no reason why some of the mechanisms should not receive a greater input from blue receptors than others. Another possibility is that some behavioural methods are more sensitive than others yielding reliable data at lower light levels, and that the different receptor mechanisms have different response-intensity relationships. It may be, for example, that the blue-receptor mechanism is more sensitive than the others, but that the rate at which its response increases with intensity is slower than it is for the other mechanisms. Responses to low intensity lights would then be dominated by the blue mechanism, and those to higher intensities by the red and green mechanisms. The fact that blue sensitivity only appears in the two-choice discrimination situation can then be accounted for by assuming that this method is more sensitive than the others, so that the animals are responding to lower light levels. In this context it may be relevant that some of the electroretinogram results of Northmore and Muntz (1970), obtained with rudd, could best be explained by assuming exactly this relationship between the sensitivities and response-intensity characteristics of the blue and other mechanisms. It is also of interest that high blue sensitivity was also reported for a freshwater turtle in a two-choice discrimination situation (Muntz and Sokol, 1967), but did not appear in an experiment in which a conditioning paradigm was used (Granda et al., 1972). An experiment by Sokol and Muntz (1966) has shown that under appropriate experimental conditions this animal is most responsive to short wavelengths when low intensity stimuli are used, but comes to be most responsive to long wavelengths when the intensities are higher; so here again the appearance of high short wavelength sensitivity may depend on the sensitivity of the method. Further work is, however, clearly needed on this point.

SPECTRAL SENSITIVITY UNDER SCOTOPIC CONDITIONS

The main problem of scotopic vision is one of sensitivity. Since many of the more complex visual functions, such as colour vision, movement detection, &c., necessarily involve neural inhibition and therefore a loss of sensitivity, we might expect them to be less developed or absent under dark adapted conditions, and for scotopic vision to be relatively simple compared to photopic vision. This does indeed appear to be the case. The dark adapted spectral sensitivity curve of the rudd, measured in a two-choice discrimination situation, can be directly related to the spectral absorbance of the visual pigments at their known retinal concentration (Muntz and Northmore, 1973; Muntz, this volume), and to change with daylength as we should expect from our knowledge of how daylength affects the visual pigments of this species. When dark-adapted rudd are at least as sensitive as humans at their wavelength of maximal sensitivity, and more sensitive at the spectral extremes.

RELATIONSHIP TO THE ENVIRONMENT

The apparent contrast between an object and its background is smaller underwater than it is in air. This occurs because the image-forming light from the object is attenuated in its passage through the water, by being absorbed and scattered out of the light beam, while at the same time the water between the object and the observer scatters diffuse veiling light into the eye. As the distance between the object and the observer increases these factors cause its contrast with the background to decrease exponentially, until it finally disappears at the point where its contrast falls below the threshold of the observer. For fish contrast thresholds are therefore very important, since they determine the greatest range at which objects can be detected. All the photopic experiments summarised in this paper measured visual thresholds to spectral stimuli super-imposed on tungsten light backgrounds, and are therefore measures of the smallest detectable contrasts between the stimulus and the background under the various conditions that were used. They should therefore be fairly directly related to the distance at which objects can be seen: high sensitivity should mean that these conditions would result in a greater visual range.

In nature the background does not, of course, resemble tungsten light, but depends on the spectral transmission properties of the water. Rudd characteristically live in slow moving and relatively productive bodies of water. Measurements in a lake inhabited by this species show that the spectral characteristics are very variable, but that the water is always quite highly coloured, with the transmission maximal at around 570-580 nm and falling off at both longer and shorter wavelengths (Muntz and Wainwright, unpublished). Inspection of fig. 3 shows that none of the sensitivity curves match such a spectral distribution of light: there is in fact a tendency for the fish to be less sensitive at 570-580 nm than at other wavelengths. This is perhaps not surprising since the function of photopic vision is not to detect the water background itself, but to detect various objects seen against it. As has been emphasised by Lythgoe (1968), in such cases different visual tasks are best served by different spectral sensitivities, and having a sensitivity that matches the background is not always optimal.

The most visible objects underwater will presumably be large objects that have a high intrinsic contrast with the background. Such objects will be visible at fairly large distances, so that light coming from them will have travelled through a long path of water. Considerable veiling light will also have been scattered into the beam, and the result of these factors will be that at the point where such objects disappear the light coming from them will have almost the same spectral characteristics as the water background. Under photopic conditions almost any spectral sensitivity will be equally effective in this case, provided Weber's law holds, since changing the spectral sensitivity will change the response to both object and background equally, thus leaving the contrast unchanged. Weber's law has been shown to hold for the rudd under photopic conditions (Muntz and Northmore, 1971).

Smaller objects, and objects that have low intrinsic contrasts, will disappear at much smaller distances, and while they remain visible the light coming from them will remain spectrally distinct from that of the background. If the objects are brighter than the background contrast will be maximised by having the receptors as insensitive as possible to the water background, so that this will appear as dark as possible in comparison to the brighter stimulus. The spectral positions and high sensitivities of the blue and red receptors of the rudd would appear well adapted for this. By an analogous argument intrinsically less visible objects that are darker than the background will be best detected by receptors with sensitivities matching the background. The results that have been obtained might suggest that the rudd is not well adapted for such cases. However, all the results were obtained with bright stimuli, and it is quite possible that with dark stimuli a different sensitivity function would emerge.

These arguments are an application of Lythgoe's (1968) contrast sensitivity hypothesis to vision under photopic conditions. The benefits of offset sensitivities in fact appear greater for photopic than scotopic vision, since sensitivity is not crucial in the former case. Thus the argument suggests that under photopic conditions sensitivities that do not match the background are an advantage for nearby bright objects, a disadvantage for nearby dark objects, and neither an advantage or a disadvantage in all other cases. Under scotopic conditions, where catching as many quanta as possible is a prime necessity, offset sensitivities are an advantage for nearby bright objects, but a disadvantage in all other circumstances.

It should be emphasised that any attempt to relate the visual capabilities of the rudd to vision under natural conditions must at present be very tentative. The experimental results summarised in this chapter were all obtained under laboratory conditions, and may be irrelevant to the animals' behaviour in their normal environment. Some of the data also imply that there are quite complex interactions between the nature of the task and the type of coding that occurs. Is it the case, for example, that under natural conditions rudd detect large objects by their colour, and small objects because they appear brighter or darker than the background, as the results of the conditioning experiments suggest? Similarly, if the appearance of short wavelength sensitivity depends on the sensitivity of the method, as suggested above, does this mean that red objects are more conspicuous for the fish when nearby, but that blue objects are visible further away? Questions such as these must ultimately be answered through behavioural experiments carried out under more natural conditions. Until such data are obtained the possible significance of laboratory experiments should be viewed with great circumspection.

REFERENCES

Barlow, H.B . (1953) Summation and inhibition in the frog's retina. J. Physiol. 119: 69-88.

Bridges, C.D.B., and Yoshikami, S. (1970). The rhodopsin-porphyropsin system in freshwater fishes - 2. Turnover and interconversion of visual pigment prosthetic groups in light and darkness: Role of the pigment epithelium. Vision Res. 10: 1333-1346.

Cronly-Dillon, J.R., and Muntz, W.R.A. (1965) The spectral sensitivity of the goldfish and the clawed toad tadpole under photopic conditions. J. Exp. Biol. 42: 481-493.

Dartnall, H.J.A., Lander, M.R., and Munz, F.W. (1961) Periodic changes in the visual pigment of a fish. In: Progress in Photobiology, edited by B.C. Christensen & B. Buchmann, Elsevier, Amsterdam.

Daw, N.W. (1968) Colour-coded ganglion cells in the goldfish retina: extension of their receptive fields by means of new stimuli. J. Physiol. Lond. 197: 567-592.

Easter, S.S. (1968) Excitation in the goldfish retina: evidence for a non-linear intensity code. J. Physiol. Lond. 195: 253-271.

Granda, A.M., Maxwell, J.H., and Zwick, H. (1972) The temporal course of dark adaptation in the turtle, Pseudemys, using a behavioural avoidance paradigm. Vision Res. 12: 653-672.

Lythgoe, J.N. (1968) Visual pigments and visual range underwater. Vision Res. 8: 997-1012.

Marks, W.B. (1965) Visual pigments of single goldfish cones. J. Physiol. Lond. 178: 14-32.

Muntz, W.R.A. (1971) Sensory processes and behaviour. In: Psychobiology, edited by J.L. McGaugh. Academic Press, New York.

Muntz, W.R.A. (1974). Comparative aspects in behavioural studies of vertebrate vision. In: The Eye, vol. 6, edited by H. Davson & L.T. Graham. Academic Press, New York.

Muntz, W.R.A. and Northmore, D.P.M. (1970) Vision and visual pigments in a fish, *Scardinius erythrophthalmus* (the rudd). Vision Res. 10: 281-295.

Muntz, W.R.A., and Northmore, D.P.M. (1971) The independence of the photopic receptor systems underlying visual thresholds in a teleost. Vision Res. 11: 861-876.

Muntz, W.R.A. and Northmore, D.P.M. (1973) Scotopic spectral sensitivity in a teleost fish (*Scardinius erythrophthalmus*) adapted to different daylengths. Vision Res. 13: 245-252.

Northmore, D.P.M. (1973) Spectral sensitivity of the rudd (*Scardinius erythrophthalmus*). Unpublished D. Phil. Thesis. University of Sussex.

Northmore, D.P.M., and Muntz, W.R.A. (1970) Electroretinogram determinations of spectral sensitivity in a teleost fish adapted to different daylengths. Vision Res. 10: 799-816.

Northmore, D.P.M., and Muntz, W.R.A. (1974) Effects of stimulus size on spectral sensitivity in a fish (*Scardinius erythrophthalmus*) measured with a classical conditioning paradigm. Vision Res. 14: 503-514.

Sokol, S., and Muntz, W.R.A. (1966) The spectral sensitivity of the turtle, *Chrysemys picta picta*. Vision Res. 6: 285-292.

Stiles, W.S. (1959) Color vision: the approach through increment threshold sensitivity. Proc. Nat. Acad. Sci. 45: 100-114.

Thibault, C. (1949) Action de la lumière blanche et monochromatique sur la posture des poissons téléostéens. Arch. des sciences

Physiol. 3: 101-124.
Tomita, T., Kaneko, A., Murakami, M., and Paulter, E.L. (1967). Spectral response curves of single cones in the carp. Vision Res. 7: 519-531.
Yager, D. (1967) Behavioural measures and theoretical analysis of spectral sensitivity and spectral saturation in the goldfish, *Carassius auratus*. Vision Res. 7: 707-727.
Young, R.W. (1971) An hypothesis to account for a basic distinction between rods and cones. Vision Res. 11: 1-6.

DO FISH HAVE AN ABSOLUTE SENSE OF LIGHT INTENSITY?

J.H.S. BLAXTER

Dunstaffnage Marine Research Laboratory, P.O. Box 3

Oban, Argyll, PA34 4AD, Scotland

To answer that, were to confess to you.
Romeo and Juliet, Act IV, Sc. I

INTRODUCTION

The appreciation of intensity is based on a number of physiological characteristics. One of the most striking attributes of the sense organs, especially of sight and hearing, is their versatility in terms of the range of intensity which they can appreciate, of the order of 12 logarithmic units from the minimum threshold to the maximum level "bearable". Fechner's Law states that the "sensation" is proportional to the logarithm of the stimulus intensity (Fig. 1A), which means that visual or auditory perception goes through a process opposite to amplification. Putting it very simply, the sensory system in some way reduces the range of "sensation" perceivable from about one million million to only 12 times. In the case of vision the versatility of the eye is aided by the presence in many animals of rods and cones with a mechanism of dark-and light-adaptation.

Within the range of intensity perception, according to Weber's Law (see Fig. 1B) the increment of intensity discriminable (ΔI) is a constant proportion of the reference intensity (I) i.e. $\Delta I/I$ (the Weber Fraction) = k. In fact, tests show that Fechner's and Weber's Law only hold over limited ranges of intensity, but for our purposes they are useful concepts when considering whether fish have an absolute intensity sense.

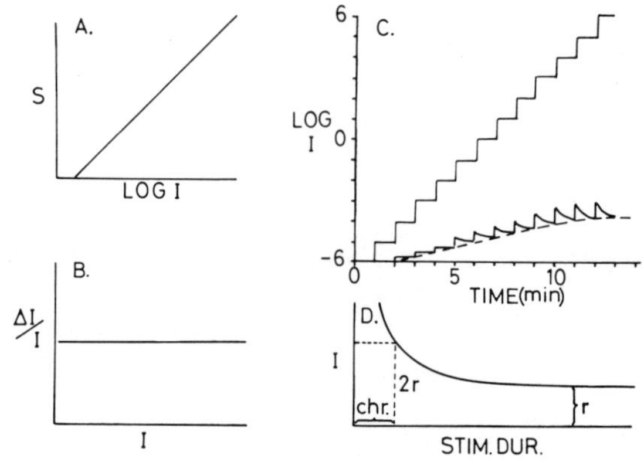

FIG. 1. A.The "sensation" (S) of a stimulus plotted against the logarithm of the intensity (I) - Fechner's Law. B. The increment (ΔI) of the intensity (I)discriminated plotted against I - Weber's Law. C. Adaptation of a sense organ; stepwise increase in intensity I shown on upper plot result in increasing adaptation shown on lower plot. D. The importance of stimulus duration-chronaxie (Chr.) is the time required for the stimulus to be effective at twice the rheobase (r) value.

At the low intensity end of the scale, perception at the threshold depends greatly on the intensity characteristics of the stimulus, that is not only on the intensity itself but on the duration of the stimulus and, in the case of light, the size of the field. The time factor is expressed in the concept of chronaxie i.e. the stimulation duration required for excitation at twice the rheobase (minimum threshold) value (see Fig. 1D).

Another feature of the sense organs is their tendency, apart from the proprioceptors, to adapt (Fig. 1C). This seems to be in part a "central" phenomenon and is not a sign of fatigue.

The application of Fechner's Law and implicit in it, that of Weber's Law, is that the eye is not very accurate at perceiving the level of intensity and has limited ability to discriminate light stimuli of closely-similar intensity, especially where the lights are bright. This in itself will make absolute intensity perception inaccurate. Couple to it the phenomenon of adaptation, and the presence of such a capability seems to be more difficult still.

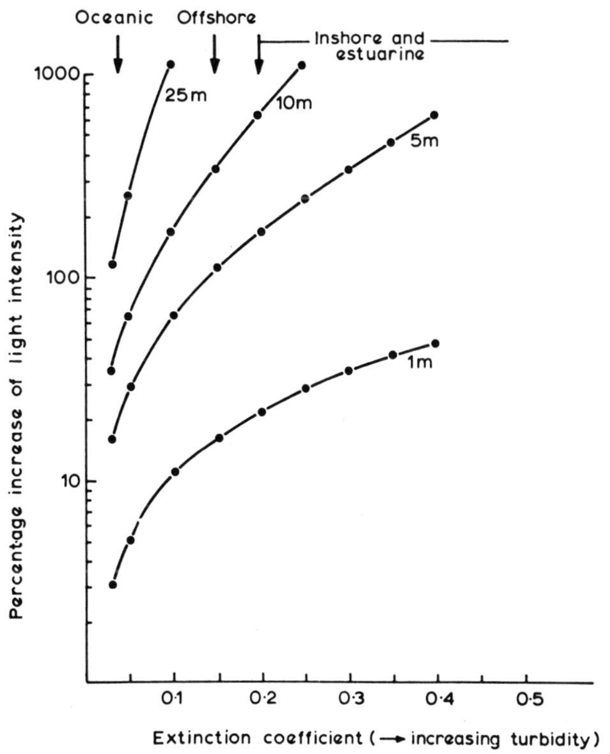

FIG. 2. The percentage increase in light intensity following upward movements of 1m, 5m, 10m and 25m in water of different turbidity (Blaxter, 1970).

Let us consider how light intensity levels are transmitted to the brain. The initial depolarisation of the sensory cell will be caused by the breakdown of visual pigment by quanta. At low light intensities summation will occur in the rods to assist in this process. Above threshold, as the intensity increases, there will presumably be an increase in firing rate in the optic nerve, both within single nerves from individual groups of rods and as a result of recruitment of higher threshold rod systems. At higher light intensities the rods will be phased out, at least in part due to retinomotor changes which cause the rods to be invested by masking pigment. The low-threshold cones will then start to be stimulated giving the first chromatic sense. Gradually other cones will be recruited until the general cone system is functioning.

While recruitment of visual cells, together with a duplex retinal structure and retinomotor responses, give the fish eye its versatility, the built-in process of adaptation is also underway as the eye perceives light. This process of adaptation is imperfectly known. It is presumably in part due to the state of decomposition or regeneration of the visual pigment, to the retinomotor changes which may sometimes be delayed if the changes of light intensity are rapid, and to other more central phenomena, both in the retina and possibly in the brain.

ABSOLUTE LIGHT INTENSITY PERCEPTION AND FISH BEHAVIOUR

One might ask-what use is an absolute perception of light intensity - if a light is "unpleasantly" bright or "unacceptably" dim, the eye will adapt in an appropriate way to make use of the light-or at least that is how the human eye seems to function? The human eye is almost useless as a photometer - in photometry its only value is in the simultaneous comparison of light sources.

In fish, illumination seems to be used to monitor depth, in other words it is being used proprioceptively. Although the depth range of any particular species is quite wide at any one time of day, there is echo-sounding evidence that populations of a species are holding depth quite well at any one time, and not making short-term vertical movements in response to short-term changes in cloud cover. At dusk many species rise to the surface, remain there during the night and move down to the same day time depth at dawn. This diel vertical migration in fish has been the subject of scientific reviews (e.g. Woodhead, 1966; Blaxter, 1970, 1975).

Although depth monitoring may also be brought about by sensory information from the swim bladder, at least in physoclists (Tytler and Blaxter, 1973), the swim bladder is an especially dynamic structure and the processes of gas secretion and absorption are often likely to be lagging behind the "resting" position for the depth in question.

The aquatic environment is particularly suited to depth-holding by the evaluation of light intensity since absorption and scattering result in light decreasing exponentially with depth (Fig. 2), thus short vertical excursions by fish will result in large changes of light intensity. If the fish are to move to the surface at dusk and away from it at dawn, although the rate of change of light intensity at any one depth will be the same as at the surface, only limited testing excursions, especially in turbid water, may enable a fish to keep within a "preferendum".

THE PREFERENDUM

Echo-sounding records show that vertically migrating fish rarely respond to changes in cloud cover. Only if drastic changes of light occur as in an eclipse (see Schüler 1954, and review by Blaxter, 1975) are responses seen (Fig. 3). It is likely that these responses only occur if the light falls to a level which might itself trigger dusk vertical migration.

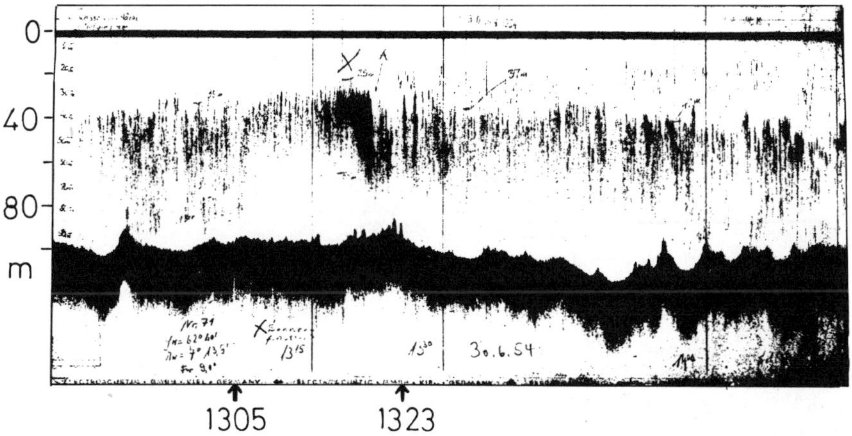

FIG. 3. Echo-trace showing change in depth of herring near the Faroe Islands during an eclipse from 1305-1323h (Schüler, 1954).

A number of workers have measured the light intensity at which diel vertical migration commences at dusk and dawn. The ascent and descent of North Sea herring was found by Postuma (1957) to occur near 10^{-1}mc, and at 10^{-1} - 10^{1} by Blaxter and Parrish (1965). Nomura (1960) reported that the dawn descent of Japanese sardines occurred at 10^{-2} - 10^{-1}mc. The movement seems to take place within the range of light intensity at which dark and light adaptations occur i.e. 10^{-3} - 10^{0}mc (see Fig. 4).

During such vertical movement there is evidence (e.g. Postuma, (1957) that fish follow a preferendum rather well (Fig.5). A closer inspection of the movements of deep scattering layers in relation to isolumes (Figs. 6 and 7) shows the appearance of over- and under-compensation as if brightness discrimination is inadequate for a precise following of a particular intensity. Over-and under compensation suggest that the organisms are "hunting" round a preferendum. In other words they are returning to near a certain level of illumination, suggesting an absolute appreciation of intensity. Some other scattering layers will "pass across" isolumes into intensities quite different from those at the beginning of the

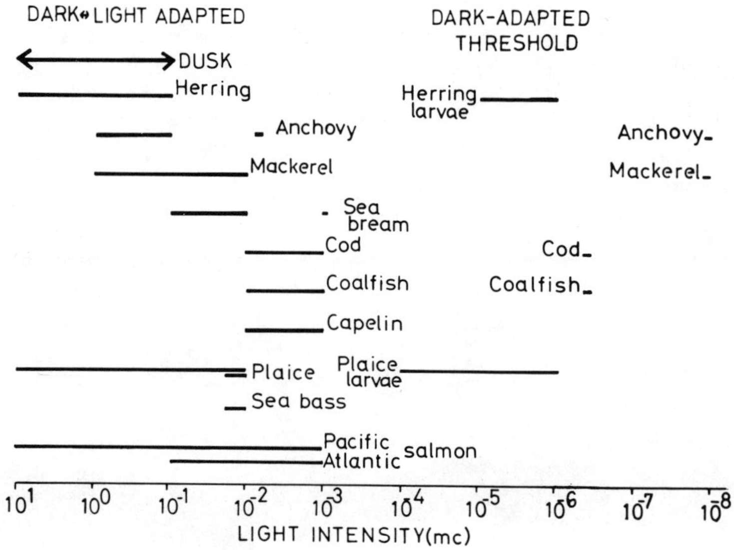

FIG. 4. Range of light intensity over which dark-and light-adaptation take place and the absolute dark-adapted threshold (Blaxter, 1975). All species shown might be involved in vertical migration.

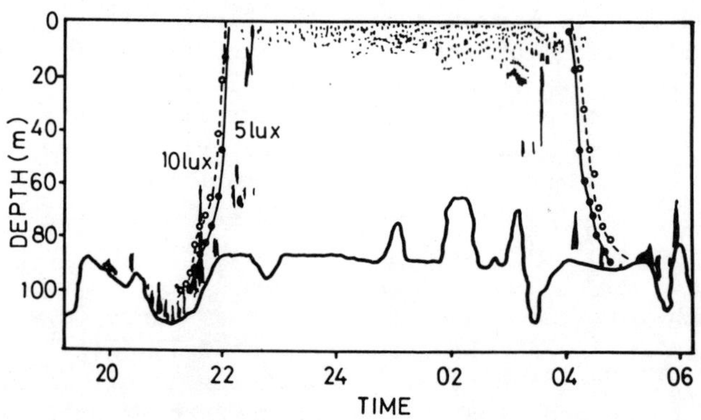

FIG. 5. Vertical movements of herring in the North Sea in relation to isolumes (Postuma, 1957).

migration. It remains to identify many of these layers (although some are likely to comprise myctophid fishes) and so the reasoning must be speculative; but it seems likely that some of the organisms follow isolumes of 10^{-2} - 10^{-1} μW/cm^2 which are within the range of intensities for light-to dark-adaptation.

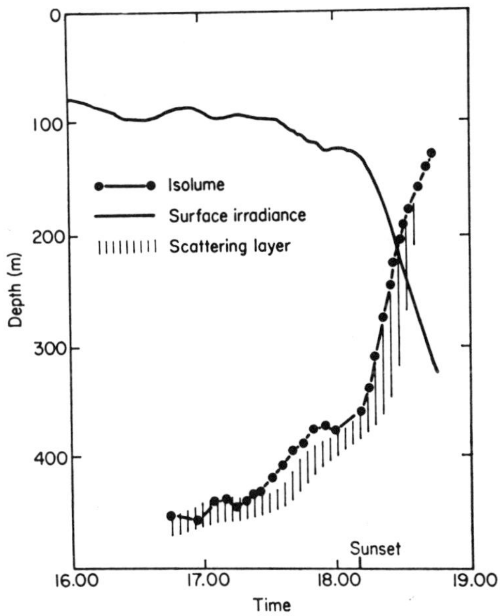

FIG. 6. Movement of a scattering layer in relation to a 10^{-1} μW/cm^2 isolume in the N.E. Atlantic (Boden and Kampa, 1967).

ABSOLUTE INTENSITY PERCEPTION

We may postulate that at least a broad reference point in the range of intensity perception occurs at dark/light adaptation. Not only are there changes in colour vision and acuity of perception as rods or cones dominate, but the concomitant retinomotor responses represent a marked anatomical change in retinal structure (Fig. 8). These might provide the most useful and easily recognised reference point in the visual range.

The fish itself must also "recognise" the light intensity at which the eye is to dark-or light-adapt. This process seems to be little known. The evidence that exists suggests that the control of retinomotor responses is at least partly intrinsic to the eye (Ali, 1964), so that the eye itself must have some inbuilt device for measuring absolute intensity, which influences the retinomotor responses. More work is needed on the flexibility of light-and dark-adaptations to the experience of light over the previous few

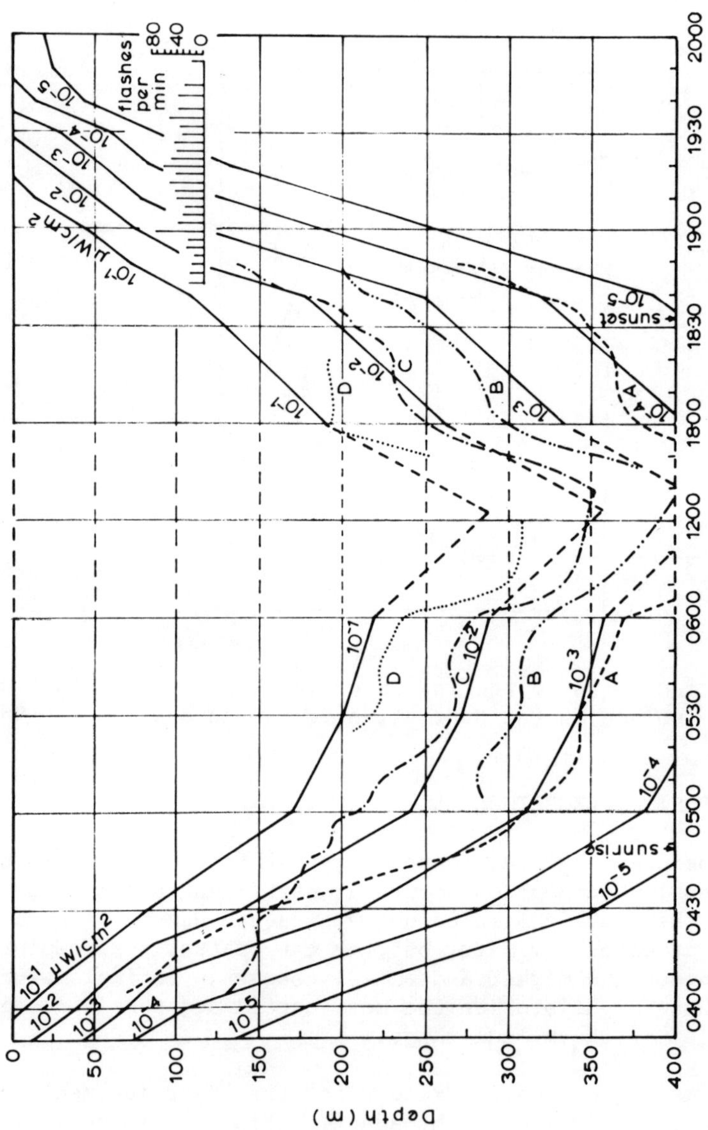

Fig. 7. Movement of scattering layer in relation to isolumes in the N.W. Atlantic (Clarke and Backus, 1964).

Dark-adapted	←——————→	Light-adapted
Low intensity scotopic vision by rods High sensitivity More sensitive in the blue (though not seen as a colour) No colour vision Low acuity (much summation) Low frequency for flicker-fusion	←—Purkinje—→ Shift	High intensity photopic vision by cones Low sensitivity More sensitive in green-yellow Colour vision High acuity (less summation) High frequency for flicker-fusion
	←—Change in many behaviour patterns—→	

FIG. 8. Diagram showing changes in visual physiology in the dark- and light adapted eye.

hours before it can be said how precise the absolute response is. It is possible that the light intensity at which light-dark adaptation takes place may be modifiable by a particular light environment beforehand. There is insufficient evidence of marked diel rhythms in the retinomotor responses (see Blaxter, 1970) which could bring about light/dark adaptation irrespective of ambient light. Ali (1959), however, suggests that fish may prolong the time for dark-and light-adaptation at dusk and dawn by vertical migration and that over-rapid changes of light in shallow water could cause a period of night-blindness.

THRESHOLD FOR VISION

Another convenient point in the range of light intensity perceived is at the ultimate lowest threshold. Given full time to dark-adapt and the rather broad-field type of light stimulus at considerable depths, it seems likely that the fish eye can reach its full potential sensitivity. Clarke and Denton (1962) suggest that the deep sea fish have a threshold 10^{-9}-10^{-10} $\mu w/cm^2$, possibly just reacting to light at 900-1000 m in clear water. This would provide an ideal boundary for deep sea fish to follow by day and night. Some thresholds in other vertically migrating fish are shown in Fig. 4. Even with eyes of lower sensitivity the same hypothesis could apply in shallower water i.e. fish have a preferendum very near their absolute threshold. Such a preferendum would be less useful than one based on light/dark adaptation thresholds, since much of the light-determined behaviour exhibited by still shallower species would not be possible.

CONCLUSIONS

While it seems unlikely that fish, or other vertebrates, have a good sense of absolute intensity perception, at least some sensory information is given by the anatomical state of the iris of higher vertebrates, by the retinal masking pigment or rod and cone myoids in teleosts, and by the tapetal masking pigment of the elasmobranchs (Denton and Nicol, 1964; Nicol, 1965). Dowling (1967) also gives evidence that visual adaptation may be in the bipolar layer. Thus there may be subjective adaptation of the eye although there is sufficient information within the eye to give a measure of absolute intensity. Presumably the central nervous system can also obtain information from the firing rate and recruitment of fibres in the optic nerve, especially if part of the adaptation process is "central" i.e. applied subsequently to the receipt of afferent impulses from the eye. Whether further information can be obtained from the chemical state of the visual pigment does not seem to be known. It is certainly clear from Dowling (1967) that the threshold for vision is not dependent entirely on the degree of photochemical breakdown of rhodopsin.

Aquatic animals seem to have a special requirement for depth-holding and with water having the useful quality of exponential reduction of light with depth, one might expect at least a crude sense of absolute light intensity to be functioning in pelagic animals, especially in the many species which make diel vertical migrations.

REFERENCES

Ali, M.A. (1959) The ocular structure, retinomotor and photobehavioural responses of the juvenile Pacific salmon. Can. J. Zool. 37: 965-996.

Ali, M.A. (1964) Retinomotor responses in the enucleated eyes of the brown bullhead *Ictalurus nebulosus* and the goldfish *Carassius auratus*. Revue Can. Biol. 23: 55-66.

Blaxter, J.H.S. (1970) 2.Light.2.3Animals.2.32.Fishes. In Marine Ecology Vol. Pt. 1 pp. 213-320, edited by O. Kinne, Wiley Interscience, London.

Blaxter, J.H.S. (1975) The role of light in the vertical migration of fish - a review. Symp. Br. Ecol. Soc. (in press).

Blaxter, J.H.S. and Parrish, B.B. (1965) The importance of light in shoaling, avoidance of nets and vertical migration by herring. J. Cons. Perm. Int. Explor. Mer. 30: 40-57.

Boden, B.P. and Kampa, E.M. (1967) The influence of natural light on the vertical migration of an animal community in the sea. Symp. Zool. Soc. Lond. 19: 15-26.

Clarke, G.L. and Backus, R.H. (1964) Interrelations between vertical migration and deep scattering layers, bioluminescence and changes in daylight in the sea. Bull. Inst. Oceanogr. Monaco 64: (1318), 36pp.

Clarke, G.L. and Denton, E.J. (1962) Light and animal life. In the Sea Vol. 1, pp. 456-468, edited by M.N. Hill, Wiley, New York.

Denton, E.J. and Nicol, J.A.C. (1964) The chorioidal tapeta of some cartilaginous fishes (Chondrichthyes). J. Mar. Biol. Ass. U.K. 44: 219-258.

Dowling, J.E. (1967) The site of visual adaptation. Science 155: 273-279.

Nicol, J.A.C. (1965) Migration of chorioidal tapetal pigment in the spur dog *Squalus acanthias*. J. Mar. Biol. Ass. U.K. 45: 405-427.

Nomura, M. (1960) Some knowledge on behaviour of fish schools. Proc. Indo-Pacific Fish. Coun. 8th session (111), 95-96.

Postuma, K.H. (1957) The vertical migration of feeding herring in relation to light and vertical temperature gradient. Cons. Perm. Int. Explor. Mer, Herring Cttee. (mimeo).

Schüler, F. (1954) Über die echographische Aufzeichnung des Verhaltens von Meeresfischen während der Sonnenfinsternis vom 30 Juni 1954. Dt. Hydrogr. Z. 7: 141-143.

Tytler, P. and Blaxter, J.H.S. (1973) Adaptation by cod and saithe to pressure changes. Neth. J. Sea. Res. 7: 31-45.

Woodhead, P.M.J. (1966) The behaviour of fish in relation to light in the sea. Oceanogr. Mar. Biol. Ann. Rev. 4: 337-403.

BEHAVIOURAL STUDIES OF THE DISCRIMINATION OF VISUAL ORIENTATION AND MOTION BY THE GOLDFISH

Frances C. VOLKMANN

Clark Science Center, Smith College

Northampton, Massachusetts, U.S.A.

Though this be madness yet there is method in it.
Hamlet, Act II, Sc. 2

INTRODUCTION

The discovery and investigation of stimulus analysing mechanisms at the cellular level in a variety of species has led quite naturally to a search for systematic relations between the responses of these analysers and the behaviour of the animals in whose brains they reside. In an early approach, Sutherland hoped to correlate variations in behaviour with variations in the proportion of different types of receptive fields present in the brains of different species (1963, p. 122). For example, animals with retinal, tectal, or cortical receptive fields which respond predominantly to stimuli oriented in the horizontal or vertical, as opposed to oblique, directions, should show corresponding asymmetries in behavioural discriminations of visual tilt, termed the *oblique effect* (Appelle, 1972). Or, animals in which cells predominate which respond maximally to stimuli which move across their receptive fields in a particular direction (Jacobson and Gaze, 1964; Cronly-Dillon, 1964) should show behaviourally the effects of this organisation (Ingle, 1968).

Although it now seems unlikely that discrimination can be described on the basis of simple proportions of receptive field types, it is still useful to look for (perhaps more complex) behavioural-physiological correlations; and in doing so one legitimate guide for behavioural research is an analysis of unit responses to highly specific stimulus configurations (Lettvin, et al., 1961).

My colleagues and I have been studying behavioural responses of goldfish to two classes of stimulus variables which are of primary importance neurophysiologically: the orientation in space and the motion of visual stimuli.

METHOD

The experiments employ a discrete-trial simultaneous discrimination procedure developed from that used by Yager and by Yager and Thorpe in their studies of goldfish colour vision (Yager and Thorpe, 1970), and by Sutherland (1968) in his work on shape discrimination in goldfish. The details of the method have been described elsewhere (Volkmann, 1972; Volkmann, et al., 1974). Figure 1 shows the general experimental arrangements. The eight large (20-25 cm) comets that participated in these experiments were housed individually in 5 1/2 gal. tanks and maintained for a period of about two years on approximately 50 Noyes 20-mg. fish pellets per day. The experimental apparatus fit over the fish's living tank and placed the animal in a white environment which excluded all extraneous visual stimuli. At the front of the tank were two 5-cm diameter response keys which the fish could hit with its nose. They were located directly behind two circular holes cut in a vertically oriented opaque Plexiglas baffle which prevented the fish from striking the keys with its side or tail in swimming by them. A third (observing) key was similarly mounted at the back of the tank. Discriminative stimuli were projected onto the choice keys; the observing key was illuminated at appropriate times by a light located just behind it outside of the tank.

A Gerbrands automatic feeder mounted on the top of the apparatus delivered reinforcement pellets, which dropped into the water just above the response keys and were held in a restricted area of the tank by an inverted Pyrex funnel. A mesh false bottom in the tank prevented the fish from obtaining noncontingent reinforcement if an uneaten pellet happened to sink.

At the beginning of each trial, the observing key was illuminated, and the fish could initiate the trial by hitting the key with its nose. This observing response turned off the light behind the observing key and caused the two discriminative stimuli to be projected onto the choice keys at the front of the tank. The fish then had to turn around, swim forward toward the discriminative stimuli, and respond by hitting one of the choice keys. A correct response turned off the stimuli and was reinforced by a red magazine light and a Noyes pellet. An incorrect response simply turned off the stimuli. After a 4-sec. inter-trial interval the observing key light came on again and the fish could initiate the next trial by making another observing response. On every trial the stimulus serving as the S^D

FIG. 1. The experimental environment used in the present research. The near side of the white surround has been removed for photography.

or positive stimulus appeared on one choice key, and some other stimulus, serving as the S^{Δ}, or negative stimulus, appeared on the other key. The fish were trained to respond to the S^D, and were reinforced with a food pellet for such responses. The right-left position of the S^D on the keys was varied in an irregular order. A fish typically ran for about 60 trials per day, and received most of its daily ration of food in the apparatus. Running times varied from about 3/4 - 2 hours.

VISUAL ORIENTATION

The present studies investigated differential sensitivity with respect to visual tilt in the goldfish by obtaining psychophysical thresholds. Previous experiments had already provided data on relative acquisition rates and performances on transfer tests (Mackintosh and Sutherland, 1963; Sutherland, 1968; 1969). In addition, our studies were designed to compare, by means of these thresholds, (1) the effects of using two standard psychophysical methods, those of constant stimuli and stimulus tracking, and (2) the relative size of the threshold when the standard stimulus was vertical versus when it was oblique, i.e. the *oblique effect* under conditions which were free from the possibility of mirror-image confusions.

The grating-like stimuli consisted of black bars 5 mm wide and 10 mm apart, which filled the entire circular choice keys in length. Their orientation varied between vertical and 45 deg in the upper right quadrant, that is, between 12:00 and 1:30 on a clock face. Eight to 12 intervening values of stimulus orientation were used. In the method of constant stimuli each value was paired with the S^D in an irregular order; in the stimulus tracking method correct responses were followed on the next trial by a stimulus value slightly closer to the S^D, and incorrect responses by a stimulus value farther removed from the S^D (von Békésy, 1947; Koh and Teitlebaum, 1961). One group of animals worked with a vertical S^D, and thresholds were determined for them using both psychophysical methods and two reinforcement schedules, continuous reinforcement, (CRF) and reinforcement following every second correct response (FR2). A second group of animals worked with a 45 deg S^D; their thresholds were determined for constant stimulus and CRF procedures only.

The psychometric functions for all animals appear in Fig. 2. The number of trials per plotted point range from about 25 to about 80. The functions show orientation thresholds (75% correct discrimination) ranging between about 15 and 25 deg of visual tilt. The goldfish clearly can discriminate visual orientation, but according to the present data, its thresholds are rather large. It is impossible to say, of course, whether these thresholds represent discriminatory limits for the fish, but, as Fig. 2 also shows, they were independent of the psychophysical method used to obtain them. There are no differences in the results obtained with the constant stimulus method and with the tracking method. Perhaps the most interesting finding shown in Fig. 2 is that the goldfish performs equally well with either the vertical or the 45 deg S^D; in short, contrary to what one might expect (Appelle, 1972), the goldfish does not show the *oblique effect*. On the other hand, one can make a case for the lack of adaptive value for the oblique effect in the goldfish, and for the lack of possible tuning of visual units to the horizontal and vertical by early visual experience (Volkmann, 1974).

Single units in the optic nerve and tectum of the goldfish show orientation specificity (Jacobson and Gaze, 1964; Daw and Beauchamp, 1972), but sufficient data are not yet available to yield a systematic distribution of the proportions of these units sensitive to selected stimulus orientations. The present data would suggest that either the distribution ought to be random or the fish must use a mechanism other than simple proportion of selectively tuned cells to discriminate visual tilt.

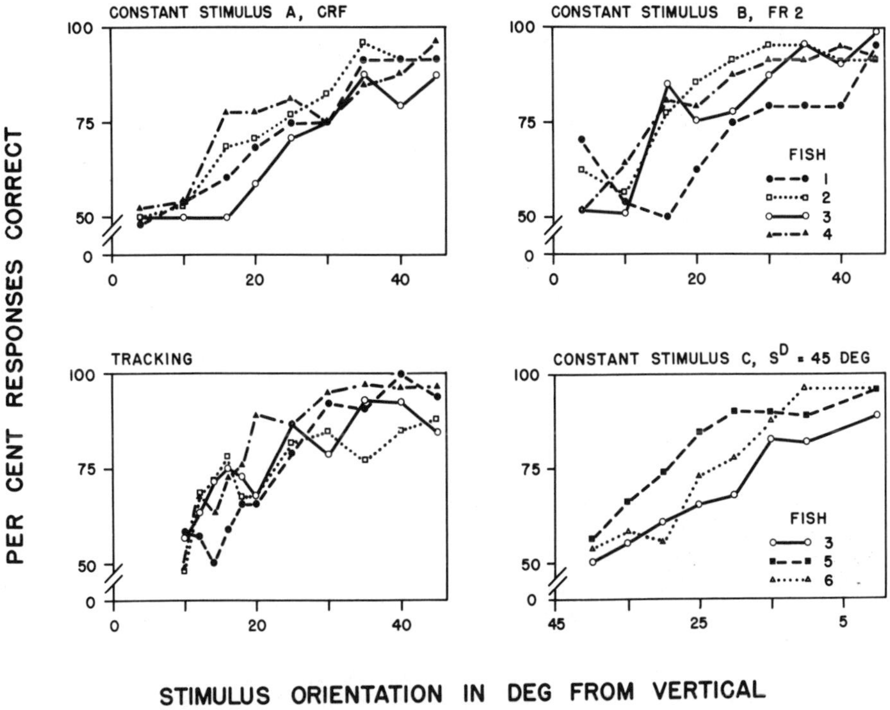

FIG. 2. Individual data from six goldfish showing percentages of responses correct as a function of stimulus orientation in degrees from the vertical. Vertical = 0 deg.

VISUAL MOVEMENT

Our work on behavioural responses of the goldfish to moving stimuli is still in its early stages. This paper represents, in a sense, a progress report, pointing out some aspects of the behaviour that are interesting and some that are puzzling, and seeking neurophysiological and ethological correlations for the behaviour observed.

Electrophysiological studies have shown directional movement detectors to be present in the retina and in the tectum (Jacobson and Gaze, 1964; Cronly-Dillon, 1964; Zenkin and Pigarev, 1969; and Drs. Galand and Schwassmann at this conference). Behaviourally, any fisherman can attest to the importance of stimulus motion, as he seeks to present the "big one" with a lure of just the right colour, size, brightness, and motion. Observers of fish behaviour have gathered a wealth of information about stimulus configurations that signal prey or predator to a species; movement is of primary importance in a large proportion of them.

Nevertheless there are few systematic studies of motion perception in the fish, and they point to a complex set of stimulus-response relations. In one such study, Ingle (1967, 1968) investigated relative sensitivity to stimuli moving in the nasal versus the temporal direction in the monocular lateral visual field of immobilized goldfish. Using cardiac deceleration as the dependent variable, he found an interaction between the number of dots in the stimulus field, the velocity, and the direction of motion: the fish responded most strongly to a single dot moving at 12 deg/sec nasally, to a field of dots moving temporally at the same velocity, and to a single dot moving temporally at 3 deg/sec. Ingle has sought to account for these differences in terms of a two-process model of feed forward loops. The fish presumably uses a "change of position" analysis to orient toward moving prey (single dot moving nasally) and a continuous monitoring based on velocity analysis to track the "flow" of surfaces or to aim his final lunge and snap (dot field or slowly moving single dot) (see also Held, 1968; Trevarthan, 1968).

In the present research we wished to look at two different types of stimulus motion that, on neurophysiological grounds, would be expected to excite the receptive fields of many units. The first type consisted of a small rotatory motion in which stimuli moved at various rates in elipses with radii of 2-4 mm. Such motions incorporated successively every direction in the vertical frontal plane. The second type consisted of a 4-cm horizontal bi-directional sweep, or oscillation, in which stimuli moved repetitively back and forth across one or both response keys at predetermined velocities.

Using these two types of motion, we have conducted pilot studies concerned with a) the effects of stimulus motion on other visual discriminations, b) the discrimination of stimulus motion versus non-motion and the detection of moving stimuli, and c) preferences for motion versus non-motion of a variety of visual stimuli. In all of these studies we used the basic experimental arrangements described above, with the freely swimming goldfish.

Table I summarises the results of this work. The five discriminations considered in Part A may be reviewed as in Table I.

TABLE I

Summary of Behavioural Responses to Stimulus Motion

EXPERIMENT	STIMULI	TYPE OF MOTION	SPEED	TOTAL TRIALS	FISH #	RESPONSES MOVT	RESPONSES STILL
		A. DISCRIMINATION EXPERIMENTS					
1	H vs V bar	Rotatory	36 RPM	560	4	(% correct)	
	6x35 mm					100	96
	2x10					86	71
	6x35	Rotatory	115 RPM	224	4	96	96
	2x10					75	82
2	Tilt, stripes both stimuli moving or still	Rotatory	36 RPM	384	3	No threshold differences; general depression of performance	
3	Vertical stripes moving or still	Horizontal Bi-directional sweep	Variable, 0.5-10 cm/sec	650	3,4	No discrimination	
4	Single dot 1,2,3,5,10 mm	Rotatory	36 RPM	530	4,6	No discrimination	
5	Multiple dots vs blank field	Rotatory	36 RPM	256	4	(% correct)	
	1 mm					67	62
	2 mm					82	75
	3 mm					90	90
		B. PREFERENCE EXPERIMENTS					(% still)
	Large redundant dots	Horizontal Bi-directional sweep	7.5 cm/sec	78	3,4,6		51
			1.0 cm/sec	78	3,4,6		46
	Small redundant dots	Horizontal Bi-directional sweep	7.5	77			58
			1.0	78			50
	Large single dot		7.5	73			44
			1.0	78			38
	Small single dot		7.5	72			46
			1.0	72			51
	Vertical stripes		7.5	75			72
			1.0	75			49
	Small white dot on dark field		7.5	69			57
			1.0	70			53
	Red/Green Control	None		718			98 correct

Previous work (Volkmann, 1972) had shown that accuracy of a horizontal-vertical discrimination varied with the size of the single dark-bar stimuli. In experiment 1, we asked whether moving the stimuli would serve an orienting or arousing function and thus improve the discrimination. The choice was always between a horizontal and a vertical bar of the same size, and both stimuli were either stationary or moved at the same rate, 36 or 115 RPM. Results showed a slight but non-significant increase in correct responding to stimuli moving at 36 RPM over either stationary or more rapidly moving stimuli.

In experiment 2, orientation thresholds similar to those reported earlier in this paper were obtained while both the S^D and the S^Δ underwent rotatory motion. Performance was generally depressed for all stimuli in comparison to sessions in which no stimulus motion occurred, and no reliable differences were found between thresholds for moving stripes versus thresholds for stationary stripes when both stimulus conditions occurred concurrently in a single session.

In experiments 3 and 4, we turned to the discrimination of these two types of motion per se. The goldfish were asked to discriminate between two identical stimuli, one of which was moving. The fish were not able to reach the 75% threshold level of correct responding. In experiment 5, the fish was required to choose a field of multiple dots over a blank field, when the dots moved together or remained stationary. The results given in Table I indicate that correct performance in general was not high, and that stimulus motion was not a salient variable.

The preference experiments summarised in Part B of Table 1 were all conducted together. They sought to determine whether there were classes of stimulus form and movement combinations which the goldfish would tend to approach or avoid. Since we did not wish to alter response probabilities by reinforcing selected responses, no reinforcement was given on any preference trial. Rather, these trials were interspersed with trials requiring a simple red-green discrimination, on which the animals were reinforced for hitting the red key. This arrangement kept the fish working and presumably observing both choice keys. On preference trials, any one of the stimuli listed on the Table appeared on both keys, with the stimulus on one key moving.

The results were originally startling to us. The fish performed the red-green discrimination with high accuracy and almost no hesitation, hitting either choice key as the red stimulus appeared on it. Concurrently,they appeared to pay no attention whatsoever to the stimuli on the preference trials, usually showing a strong position habit whenever these stimuli appeared. The only significant preference effect in the experiment was the avoidance of rapidly oscillating vertical stripes.

The interpretation of data such as these may be puzzling indeed: we are faced with a situation in which healthy and well-trained animals perform some discriminations almost flawlessly in a given experimental situation (red-green), other discriminations at least consistently (visual tilt), and yet fail to perform still other discriminations which are equally justifiable on feature detector grounds and equally obvious to human experimenters.

One must be careful, of course, in attributing poor performance to limits in discriminatory capacity. Nevertheless, given the context of the present experiments we are inclined to regard oscillatory and rotatory motion of the types used here as very poor stimuli for the goldfish, whereas uni-directional motion is clearly a powerful stimulus. One might speculate that such stimulus selectivity may be adaptive: prey or predator moving through the water are noticed immediately, while the repetitive swaying back and forth of weeds, grasses, and shadows are "tuned out" by the visual system.

In the toad, Ewert (see Ingle, 1973) has demonstrated that animals with appropriate pretectal lesions do not habituate to repeated stimulation such as normal toads do (Lettvin, et al., 1961), and Ingle (1973) has shown that tectal units in the frog are "disinhibited" by prectectal lesions, so that they can no longer "ignore" their retinal input. Whether a similar inhibitory mechanism might exist for the goldfish is an interesting question.

At a more theoretical level, Sekular and Levinson (1974) have presented a simple model of the initial stages of human motion perception. At the first stage in their model is an analysis of the spatial frequency and orientation properties of the stimulus. Next come motion analysers, tuned to specific directions of motion. The output of each analyser goes to a separate detector. A feature of the model which is interesting here is that the detectors exert mutual inhibitory effects upon one another, similar to those observed in the responses of *limulus* ommatidium to light (Ratliff, 1961). Thus oscillatory or rotatory motion might be expected to stimulate many detectors, thus producing a larger net inhibitory effect than would unidirectional motion.

These speculations on adaptive stimulus selectivity, pretectal inhibition, and mutually inhibitory detectors are not inconsistent with one another. They represent an effort to retain a neurophysiological account of the basic classes of discriminatory behaviour relating to orientation and motion, but an account which goes beyond simple proportions of receptive field types to a complex interplay among them.

ACKNOWLEDGEMENT

This research was supported by Grant No. GB-30884 from the National Science Foundation. The following persons collaborated with the author on various aspects of this work: N.E. Judge, on movement; A.J. Zametkin, C.A. Stoykovich, and Ann Pufall on orientation; John Volkmann on apparatus design and construction.

REFERENCES

Appelle, S. (1972). Perception and discrimination as a function of stimulus orientation: The "oblique effect" in man and animals. Psychol. Bull. 78: 266-278.

Cronly-Dillon, J.R. (1964). Units sensitive to direction of movement in goldfish optic tectum. Nature, 203: 214-215.

Daw, N.W. and Beauchamp, R.D. (1972). Unusual units in the goldfish optic nerve. Vision Res. 12: 1849-1856.

Held, R. (1968). Dissociation of visual functions by deprivation and rearrangement. Psychol. Forsch. 31: 338-348.

Ingle, D. (1967) Two visual mechanisms underlying the behavior of fish. Psychol. Forsch. 31: 44-51.

Ingle, D. (1968). "Spatial dimensions of vision in fish". In: The Central nervous system and fish behavior, edited by D. Ingle, University of Chicago Press, Chicago.

Ingle, D. (1973). Disinhibition of tectal neurons by pretectal lesions in the frog. Science, 180: 422-424.

Jacobson, M. and Gaze, R.M. (1964). Types of visual response from single units in the optic tectum and optic nerve of the goldfish. Q.J.Exp. Physiol. 49: 199-209.

Koh, S. and Teitlebaum, P. (1961). Absolute behavioral taste thresholds in the rat. J. Comp. Physiol. Psychol. 54: 223-229.

Lettvin, J.Y., Maturana, H.R., Pitts, W.H., and McCulloch, W.S. (1961). "Two remarks on the visual system of the frog". In: Sensory Communication, edited by W.A. Rosenblith, John Wiley, New York.

Mackintosh, J., and Sutherland, N.S. (1963). Visual discrimination by the goldfish: The orientation of rectangles. Anim. Behav. XI: 135-141.

Ratliff, F. (1961). "Inhibitory interaction and the detection and enhancement of contours". In: Sensory Communication, edited by W.A. Rosenblith, John Wiley, New York.

Sekuler, R. and Levinson, E. (1974). Mechanisms of motion perception. Psychologia-An International Journal of Psychology in the Orient, XVII: 38-49.

Sutherland, N.S. (1963). "Shape discrimination and receptive fields." Nature, 197: 118-122.

Sutherland, N.S. (1968)."Shape discrimination in the goldfish." In: The Central Nervous System and Fish Behavior, edited by D. Ingle, University of Chicago Press, Chicago.

Sutherland, N.S. (1969). Shape discrimination in rat, octopus, and goldfish: A comparative study. J. Comp. Physiol. Psychol. 67: 160-176.

Trevarthan, C. (1968). "Vision in fish: The origins of the visual frame for action in vertebrates". In: The Central Nervous System and Fish Behavior, edited by D. Ingle, University of Chicago Press, Chicago.

Volkmann, F.C. (1972). An improved experimental situation for studying visual psychophysics in the goldfish, with data on the discrimination of the vertical. Percept. Mot. Skills, 35: 271-280.

Volkmann, F.C., Zametkin, A.J., and Stoykovich, C.A. (1974). Visual discrimination of orientation by the goldfish (*Carassius auratus*). J. Comp. Physiol. Psychol. 86: 875-882.

von Békésy, G. (1947). A new audiometer. Acta Oto-Laryngol., 35: 411-422.

Yager, D. and Thorpe, S. (1970). "Investigations of goldfish color vision". In: Animal Psychophysics: The design and conduct of sensory experiments, edited by W.C. Stebbins, Appleton-Century-Crofts, New York.

Zenkin, G.M. and Pigarev, I.N. (1969). Detektornye svoisti ganglioznykh kletok setchatki shchuki. Biofizika, 14: 722-730.

A BEHAVIOURAL TEST OF ROD-RED CONE CONVERGENCE IN THE GOLDFISH RETINA

M.K. POWERS and S.S. EASTER, Jr.

Depts. of Psychology & Zoology, University of Michigan

Ann Arbor, Michigan 48104, U.S.A.

A B C D   Gol'fish?
M N O   Gol'fish.
S D R   Gol'fish!
R D R   Gol'fish.

Dartnall

The experiments described here represent a behavioural test of the physiological model of the goldfish retina, proposed by Raynauld (1972), in which rods and long wavelength sensitive cones are connected by similar, perhaps identical, neural pathways to the ganglion cells. The model was suggested by Raynauld's observation that the ganglion cells responded in the same way to long wavelength light at photopic levels of illumination as they did to any wavelength at scotopic levels. This was confirmed by Beauchamp and Daw (1972), and supported by the anatomical report of Scholes and Morris (1973). If we can accurately predict behaviour from the cellular electrophysiology, then we can be more confident that the small number of ganglion cells that have been sampled by microelectrodes are representative of the tens of thousands whose axons carry the output of the entire retina. Moreover, a test of the model bears on current theories of underwater vision (Lythgoe, 1968; Easter, 1974), in which the long wavelength end of the spectrum is assigned special importance because of the higher contrast available there (Duntley, 1962). Perhaps the convergence of rods and red cones represents a physiological exploitation of this physical fact. This possibility is supported by the observation that the rod-red cone convergence does *not* exist in either the monkey (Wiesel and Hubel, 1966) or the cat (Pearlman and Daw, 1970), the only other animals whose visual systems have been examined in a way to reveal it.

Fig. 1 illustrates the impact of a rod-red cone link on the whole set of signals ascending the goldfish optic nerve. It gives the signs of the centre-mediated responses of the four most common ganglion cell types when light- or dark-adapted. The set of messages from these four classes of cells provides most if not all of the contrast and wavelength information available to the fish's brain. Therefore, the fish's decisions about the nature of a particular visual stimulus must be made on the basis of the set of signals it evokes from these four classes of cells. Because red photopic increments and scotopic increments of any wavelength produce the same set (columns 1 and 5 in Fig. 1), they should appear the same to the fish. Operationally, this argument suggests that animals trained under scotopic conditions to respond to increments and ignore decements will respond to red increments more often than to green ones under photopic conditions. We find that they do.

STIMULUS

	Light Adapted				Dark Adapted	
	L↑	S↑	L↓	S↓	↑	↓
Color Opponent L+/S−	+	−	−	+	+	−
Contrast +	+	+	−	−	+	−
Color Opponent L−/S+	−	+	+	−	−	+
Contrast −	−	−	+	+	−	+

Fig. 1. Sets of receptive field centre responses for four common types of goldfish retinal ganglion cells under different stimulus conditions. Upward arrow: increase in stimulus intensity above background level. Downward arrow: decrease in intensity. L: long wavelength (red) light. S: shorter wavelengths (green and blue). Plus sign: excitation. Minus sign: inhibition.

First we describe the determination of absolute visual threshold, which allowed us to give scotopic stimuli. Then we describe the scotopic training procedure and the photopic tests. Our technique is similar to others that have been effective in producing conditioned heart-rate deceleration to visual stimuli (Otis, Cerf and Thomas, 1957; McCleary and Bernstein, 1959; Hester, 1968).

Goldfish with electrodes chronically implanted in the thoracic cavity were restrained in a darkened, aerated aquarium. The electrocardiogram (EKG) was amplified and displayed by conventional means. Photic stimuli ($\lambda = 538$ nm) were projected from outside the aquarium onto a diffusing screen close to the fish's right eye. Stimulus duration was 10 sec. A brief (100 msec) tail shock was delivered through grid electrodes at the end of the stimulus period. Shock voltage was adjusted so that the fish clearly felt it (2 to 10 V). Interstimulus intervals were 30 sec to 1 min.

The time between successive QRS waves of the EKG was monitored (see Fig. 2) while the fish was not being stimulated. The mean and standard deviation of this "resting" interbeat interval (IBI) were computed. The delivery of the light flash was synchronous with one heart beat, and the first IBI following light onset was measured. If it was more than two standard deviations (SD) greater than the mean resting IBI, detection was said to have occurred. Resting IBI's were typically 1 sec; decelerated IBI's ranged as high as 10 sec. with average SD 0.32 sec.

Determination of absolute visual threshold began by training a fish to respond to a clearly visible stimulus ($\lambda = 538$ nm) on a completely dark background. Training usually required 10 to 20 trials. When training was complete, a modified staircase psychophysical method was used to establish a ychometric curve of the percent of trials in which a stimulus was detected as a function of its intensity. Threshold was defined as the intensity that was detected 50% of the time.

Fig. 2 shows that at absolute visual threshold there is only one quantum incident for every 90 rods. This determination of absolute threshold is an upper limit; because quantum efficiency is less than one and the pigment density is finite, the ratio of effective quanta to rods is certainly smaller. Therefore any given rod received either one quantum or, more likely, none. This result strongly implies that the threshold we have measured is the absolute threshold for vision in the goldfish (Hecht, Schlaer and Pirenne, 1942; Raynaud, 1972).

Knowing the absolute visual threshold for goldfish allowed us to select stimuli that were below the range of sensitivity of the cones. Thus we could be sure that we were stimulating only rods during the scotopic training phase of the next experiment.

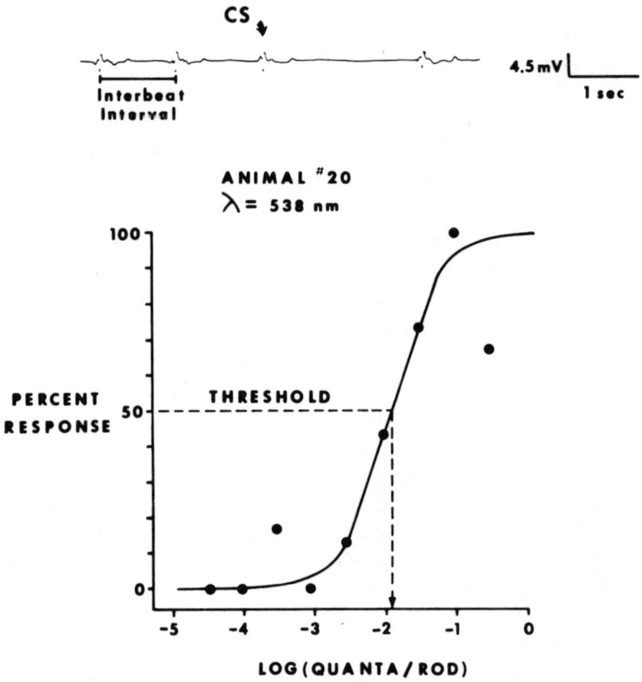

Fig. 2. Behaviourally determined psychometric function defining absolute visual threshold in the goldfish. Inset: EKG record from a trained fish, illustrating cardiac deceleration following conditional stimulus (CS) presentation.

Fish with EKG electrodes were restrained and placed in a light-tight enclosure as described before. For training sessions, each fish was adapted for 30 minutes to a diffuse field (λ = 538 nm) about 1 log unit above absolute threshold. The use of a discrimination task for this experiment, instead of the detection task used to determine absolute threshold, decreased the likelihood that an animal would respond to brightness cues during photopic testing. The discriminative stimulus (S+) was a sudden increase in intensity of the field. The nondiscriminative stimulus (S-) was a sudden decrease in intensity. Both increments and decrements were .5 log units for most fish; for some later animals they were increased to 1 log unit. Interstimulus intervals, as before, were 30 sec to 1 min. In subsequent photopic testing both S+ and S- were increments, but of different wavelengths. In training as well as testing, shock was paired with the S+ but never with the S-. Each session was 20 trials long with 15 S- and 5 S+ presentations in random sequence. Fish were frequently trained more than once per day. The response criterion for all trials was the same as before. Fish were trained until they responded on at least 80% (4 out of 5) of the S+ trials for two consecutive training sessions. (One fish out of the four trained did not reach this criterion in 9 sessions, but was tested nonetheless.)

Photopic tests were given at the completion of scotopic training. The procedure was identical to scotopic training except that the fish was adapted to a white photopic field (at least 7.4 log units above absolute threshold) and the stimuli were red (λ =637 nm) and green (λ = 538 nm) increments about 1 log unit more intense than the background. For two of the animals the S+ was red during testing and the S- green; the reverse was true of the other two. Response criteria were the same as for scotopic training.

Fig. 3 shows the results of this experiment. The mean performance of all four fish on the last three days of scotopic training shows a clear discrimination of the colourless increment from the decrement. Regardless of which wavelength was reinforced during photopic testing, however, all four fish responded more often and more strongly to red than to green, as the model predicts. Control experiments showed that there was no innate tendency to respond more to the red than to the green.

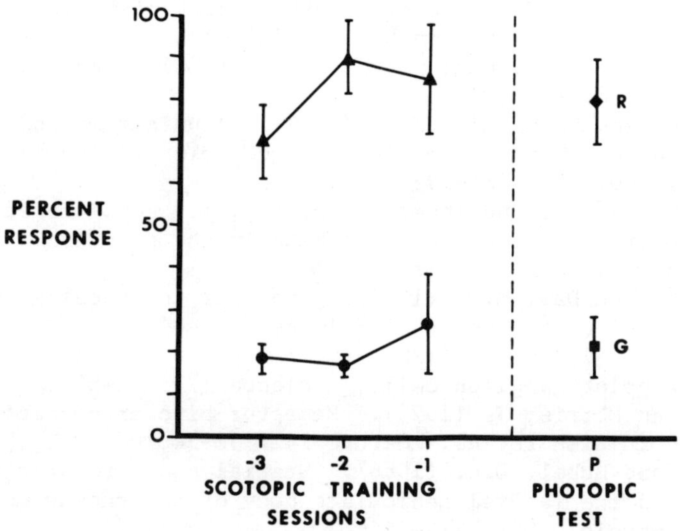

Fig. 3. Results of the behavioural test of rod-red cone convergence. Triangles: response to increments of light in scotopic range. Circles: response to scotopic decrements. Diamond: response to red increment at photopic levels. Square: response to photopic green increment. Each point is the mean performance of four animals on at least five trials each. Error bars are ± one standard error.

In summary, we have behaviourally determined the absolute scotopic threshold of unanesthetised goldfish, and supported Raynauld's (1972) model of rod-red cone convergence in the teleost retina. The work is still in progress.

ACKNOWLEDGMENTS

This research was supported by Grant EY-00168 to S. S. E. from the United States Public Health Service. We wish to thank W. M. Schoel for valuable help with EKG electrode implantation and recording in the early stages of this work.

REFERENCES

Beauchamp, R.D. and Daw, N.W. (1972). Rod and cone input to single goldfish optic nerve fibres. Vision Res. 12: 1201-1212.

Duntley, S.Q. (1962). Underwater visibility. In: The Sea, Vol. 1, edited by M.N. Hill. Interscience: New York.

Easter, S.S. (1974). Retinal specializations for aquatic vision: theory and facts. (This volume).

Hecht, S., Schlaer, S. and Pirenne, M.H. (1942). Energy, quanta and vision. J. Gen. Physiol. 25: 819-840.

Hester, F.J. (1968). Visual contrast thresholds of the goldfish (*Carassius auratus*). Vision Res. 8: 1315-1335.

Lythgoe, J.N. (1968). Visual pigments and visual range underwater. Vision Res. 8: 997-1012.

McCleary, R.A. and Bernstein, J.J. (1959). A unique method for control of brightness cues in study of color vision in fish. Physiol. Zool. 32: 284-292.

Otis, L.S., Cerf, J.A. and Thomas, G.J. (1957). Conditioned inhibition of respiration and heart rate in the goldfish. Science 126: 263-264.

Pearlman, A.L. and Daw, N.W. (1970). Opponent color cells in cat lateral geniculate nucleus. Science 167: 84-86.

Raynauld, J.P. (1972). Goldfish retina: Sign of the rod input in opponent color ganglion cells. Science 177: 84-85.

Scholes, J. and Morris, J. (1973). Receptor-bipolar connectivity patterns in fish retina. Nature 241: 52-54.

Wiesel, T.N. and Hubel, D.H. (1966). Spatial and chromatic interactions in the lateral geniculate body of the rhesus monkey. J. Neurophysiol. 29: 1115-1156.

THE AREA OF THE STIMULUS ELICITING RESPONSE AS A FACTOR IN MIRROR IMAGE REVERSAL

ALEX. CAMPBELL

Neuropsychology Laboratory, McLean Hospital

Belmont, MA 02178 U.S.A.

> Methinks I see these things with parted eye,
> When everything seems double.
> A Midsummer Night's Dream, Act IV, Sc. I

INTRODUCTION

When a goldfish is trained monocularly to make an avoidance response to one of a pair of left-right mirror images, testing via the untrained eye shows a reversal of training. For example, after training to avoid ⊣ but not ⊢, testing the opposite eye shows an avoidance of ⊢, but not ⊣, (Ingle, 1967). This phenomenon, interocular/interhemispheric reversal of mirror images, does not occur when up-down mirror images are used or when the stimuli subtend less than 10° of visual angle, (Campbell, 1971).

Noble, (1966 and 1968), demonstrated a similar interocular reversal of mirror images in chiasm sectioned monkeys and proposed that homotopic commissural connexions caused a reversal resulting in two unmatched engrams. Based on work with pigeons, Beale and Corballis (1967 & 1968), presented a behavioural hypothesis referred to as "beak shift", which proposed that in pecking at the response plate, the birds masked part of the stimulus with their head and beak, thus turning a left-right mirror image discrimination into a simple up-down problem. In later experiments, (Corballis and Beale, 1970 & 1971), where they found that split brain birds trained binocularly showed no mirror image reversal when tested for stimulus generalisation, they modified their hypothesis, combining it with Noble's.

Hamilton et al., (1973 a & b), extend the masking theory into one of restricted attention, which explains mirror image reversal entirely within a behavioural framework. They present evidence showing that their chiasm sectioned monkeys were indeed responding to only part of the stimulus and that their results are incompatible with the theory of homotopic connexions.

EXPERIMENTS

These experiments present a parallel in the fish, where the lateral position of the eyes eliminates the possibility of masking (by the pigeon's beak), while the complete crossing of the optic nerves eliminates the need for surgery, which in the monkey causes a bilateral hemianopia. The experimental conditions and methods have been described elsewhere (Campbell, 1971). Using shock avoidance technique, the fish were trained monocularly on two mirror image discriminations, sideways T's and sideways V's, each stimulus subtending 15° of visual angle. When the positive stimulus was presented in the lateral field, the fish was required to swim from the start box into an identical goal box, and to withhold this response in the presence of the negative shape, (i.e. the mirror image of the cue stimulus). Criterion was 80% or better in the last twenty trials with no errors in the last ten. After the fish reached criterion on the untrained eye was tested.

In the first group the fish were tested for interocular transfer in the normal manner, ten trials with the same pair used in training, but with one of the stimuli either showing only the temporal half or only the nasal half of the pattern. Each fish was given two problems and sixteen Oscars and sixteen goldfish were used.

In the second group part stimuli only were used for the transfer tests (Fig. 1). Five test trials were given in a pseudo-random order, then reverting to the trained eye, the fish was again brought to criterion. The untrained eye was then given a further five test trials and this was repeated until each fish has completed five test trials to each of the test stimuli. Eight Oscars and eight goldfish were used.

RESULTS

When the naive eye was tested with the pairs of stimuli, i.e., one of them showing only half of the original pattern, the results were around chance level when the nasal half of one of the stimuli was missing. When the temporal half of one of the stimuli was missing, the subjects performed as well, or nearly as well, as to a pair of the original stimuli. This occurred even when half of the stimulus was moved from its original position to the opposite half of the disc, i.e., when what was the nasal half of the training stimulus was shown on the temporal half of the test stimulus. This occurred with both problems in both orientations.

In the second group where they were tested on part stimuli only, the fish made avoidance responses to the nasal half of the pattern they were trained to avoid, regardless of whether it was placed on the nasal or temporal half of the test stimulus. They also avoided the temporal half of the negative stimulus when it was a mirror image of the nasal half of the positive or when it was the same as the nasal half of the positive. In both problems they did not avoid the nasal half of the negative stimulus or the temporal half of the positive one.

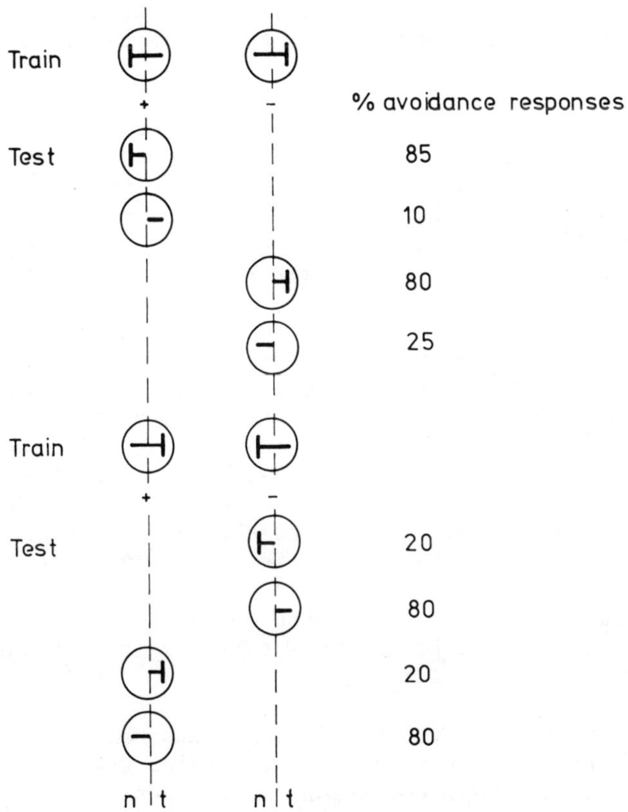

FIG. 1. In both Groups One and Two sideways T's and sideways V's were used. This figure shows results from the subjects in Group Two and it can be seen that the nasal half of the stimulus, or its mirror image, elicits avoidance as trained, even when it is on the temporal half of the test stimulus. n = nasal, t = temporal.

DISCUSSION

It is obvious that in the first group, removing the temporal half of one of the stimuli makes no differences to the scores recorded when testing the naive eye. In the second group where the fish were tested on a wide range of part stimuli only, they avoided those stimuli which were most like the nasal half of the original positive stimuli.

It cannot be argued that the fish were responding to the more complex part of the figure as half of the time this was the positive one and the other half of the time the negative one. With one of the problems, (the sideways T's), it may be argued that if the fish was responding to only part of the stimulus it turned the problem into a horizontal vs vertical discrimination, but it can be seen from what the fish avoided during testing that if it was only part of the stimulus it avoided it was the nasal half, or the part most resembling it, (its mirror image).

As both left/right orientations of both figures were tested and in each case similar results obtained, it seems that it is the nasal part of the stimulus which elicits the response regardless of whether this contains the most simple or the most complex part of the figure. With the small shapes the fish must be able to attend to the whole pattern and therefore normal transfer occurs, (Campbell, 1971).

It has been discussed whether such limitation is "restricted attention" or "selective attention", where selective attention means that with a completely unrestricted view of the stimulus, the subject responds to only part of it, (Hamilton et al., 1973b). In pigeons and in chiasm sectioned monkeys, restricted attention would be sufficient explanation for mirror image reversal, as both have restrictions on their field of view resulting from their response or to prior surgery. In these experiments, however, the stimuli, alghough presented slightly caudal to the fish's eye, were well within the visual field and there were no observable restrictions of the fish's view. This points to a selective rather than a restrictive process.

It is important to note at this point that alghough there seems to be an explanation here for interocular reversal, the fish still display mirror image confusion, in the absence of the appropriate stimulus the fish responds to a mirror image of that stimulus.

This data provides further evidence that interhemispheric reversal of mirror images can be adequately explained within a behavioural framework without reference to anatomical connectivity.

There seems to be a selective attention process which, in the fish at least, results in the nasal half of visual stimuli eliciting the strongest response.

*Part of this study is contained in a thesis presented for the degree of M. Phil. at University College London.

REFERENCES

Beale, I.L., and Corballis, M.C., (1968) Beak shift: An explanation for interocular mirror image reversal in pigeons. Nature 220: 82-83.

Campbell, A. (1971) Interocular transfer of mirror images by goldfish. Brain Res. 33: 486-490.

Corballis, M.C., and Beale, I.L. (1970) Bilateral symmetry and behaviour. Psychol. Rev. 17: 451-464.

Corballis, M.C., and Beale, I.L., (1971) On telling left from right. Sci. Am. 224: 96-104.

Hamilton, C.R., Tieman, S.B. and Winter, H.L. (1973a) Optic chiasm section affects discriminability of asymetric patterns by monkeys. Brain Res. 49: 427-431.

Hamilton, C.R., Tieman, S.B., and Brody, B.A. (1973b) Interhemispheric comparison of mirror image stimuli by chiasm sectioned monkeys. Brain Res. 58: 415-425.

Ingle, D. (1967) Two visual mechanisms underlying the behaviour of fish. Psychol. Forsch. 31: 44-51.

Noble, J. (1966) Mirror images and the forebrain commissures of the monkey. Nature 211: 1263.

Noble, J. (1968) Paradoxical interocular transfer of mirror image discriminations in the optic chiasm sectioned monkey. Brain Res. 10: 127-151.

8. APPLIED ASPECTS

FISH VISION AND APPLIED RESEARCH

J.H.S. BLAXTER

Dunstaffnage Marine Research Laboratory, P.O. Box 3

Oban, Argyll, PA34 4AD, Scotland

O, teach me how I should forget to think.
Romeo and Juliet, Act. I, Sc. 1

INTRODUCTION

Light may be perceived by fish via the eye, the pineal and sometimes the skin. With the characteristics of direction, intensity, wavelength and period, light may be involved in many functions of life in fish (Fig. 1). Direction is probably the most useful attribute since it enables animals to perceive very precisely the direction of food, predators or other members of the species.

A study of fish vision as part of an applied research programme centres mainly on the role of vision in three types of behaviour related to capture of fish.

1) Visual attraction of fish to food, lures and light.

2) Visual reaction of fish to nets.

3) The effect of light-controlled behaviour on fish distribution and therefore on fishing strategy.

    a. Schooling.

    b. Activity.

    c. Vertical migration.

    d. Horizontal migration.

Visual behaviour may also be involved in guiding fish for conservation purposes.

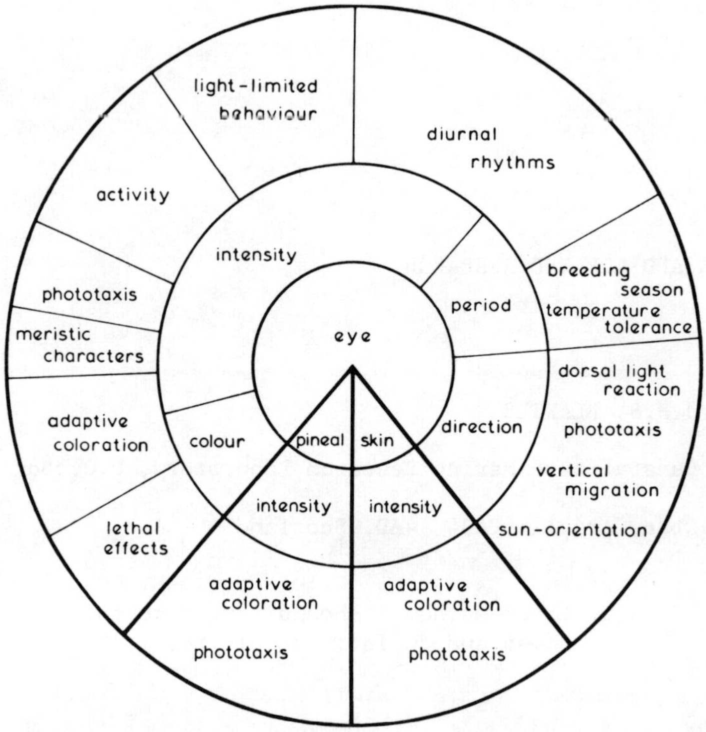

FIG. 1. Diagram showing functioning of the eye in fish (from Blaxter, 1970).

VISUAL ATTRACTION TO FOOD, LURES AND LIGHT

Many species of fish feed by sight. This is most easily demonstrated by experiments where feeding intensity is measured at different light intensities (see review by Blaxter, 1970) and a threshold for feeding established, as shown in Fig. 2.

Feeding may also be examined by sampling a natural fish population over a 24-hour period and measuring the amount of recently caught food in the stomach. An example of the results of such a study is given in Fig. 3.

Feeding, where it is visual, becomes reduced in the twilight period at dusk and dawn (see Fig. 4) with the threshold tending to centre round 10^{-1} m.c. This suggests that the fish are feeding by cone vision and that the acuity of rod vision is inadequate for feeding at lower light intensities. The use of lures such as live bait in tuna fishing or for angling, in so far as they stimulate the eye, are more likely to be useful by day than by night and in clear water. Murphy (1959), for instance, found a tendency for albacore

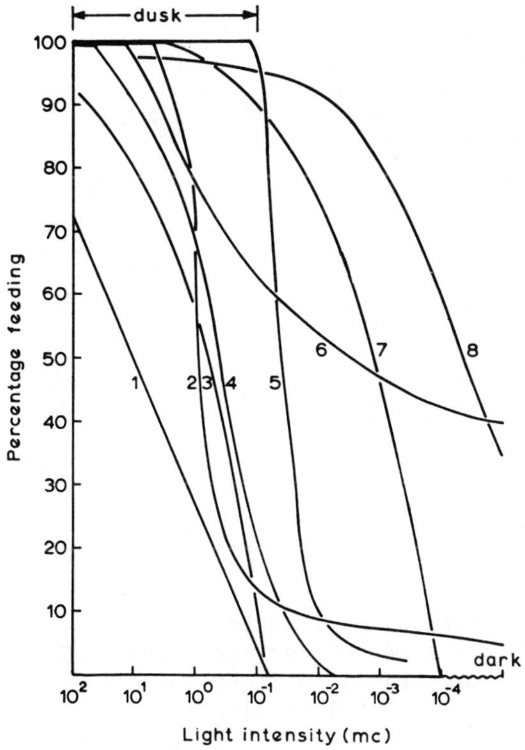

FIG. 2. Percentage feeding of fish at different light intensities (from Blaxter, 1970). 1. herring larvae feeding on *Balanus nauplii*, 2. bleak, 3. herring larvae feeding on *Artemia nauplii*, 4. cod, 5. pike, 6. carp, 7. coho salmon, 8. minnow.

catches to worsen in turbid water when trolling (Fig. 5).

The situation is, however, more complex than this (see Table 3) because feeding is also combined with vertical migration, itself a light-dependent behaviour pattern, with periods of feeding concentrated at dusk and dawn. This will influence the strategy of fishing, especially in terms of time and depth for setting the nets.

Teleosts, where they have been adequately investigated (see Walls, 1963) also have colour vision and this will again be of importance in deciding on appropriate lures. Coloured lures will only be of value at light intensities above the cone threshold of about 0.1 m.c. (see Blaxter 1970 and Blaxter in this volume - paper on absolute thresholds).

The response of fish to artificial lights in fishing has been

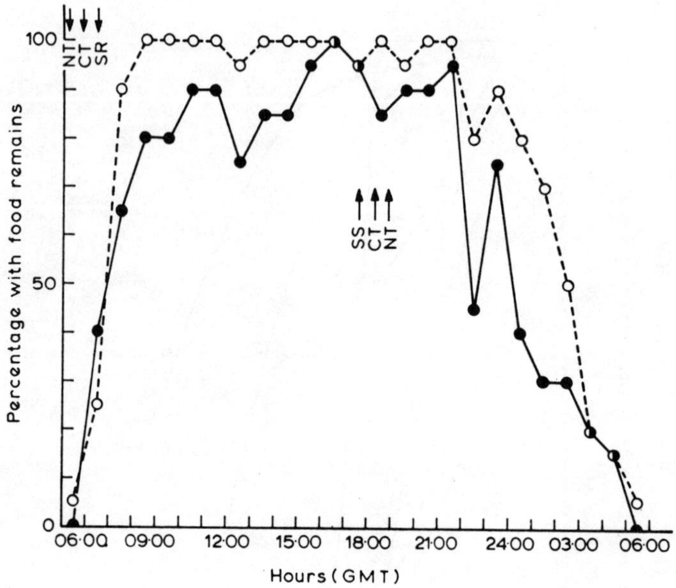

FIG. 3. Feeding periodicity of plaice 0 -- 0 and sand eel ● - ● at different times of day and night. NT - nautical twilight, CT - civil twilight, SR - sunrise, SS - sunset (from Ryland, 1964).

reviewed by Schärfe (1953) and again by Blaxter (1970). Fish may be attracted to lights above and below the surface and subsequently speared, hooked, stunned by explosives, or netted. They may be further attracted electrically to the intakes of pumps and sucked on deck. Verheijen (1958) was of the opinion that the response to light was a mass disorientation caused by an unnatural point source of light underwater. Other workers attributed it to conditioning, curiosity or phototaxis. It is, of course, essentially an unnatural reaction but some species may collect at a distance from the light, in what seems to be an optimum intensity, and so it is difficult to consider it always as a mass disorientation.

Light-fishing may be especially effective in schooling species where individuals can "lock-on"to each other at light intensities above the threshold required for schooling. The requirement for a light fishery is to accumulate fish, hence acclimation is undesirable.

Fish will also be attracted to floating objects (Hunter and Mitchell, 1968) and accumulate under them. This is almost certainly a visual response capable of development in fishing practice.

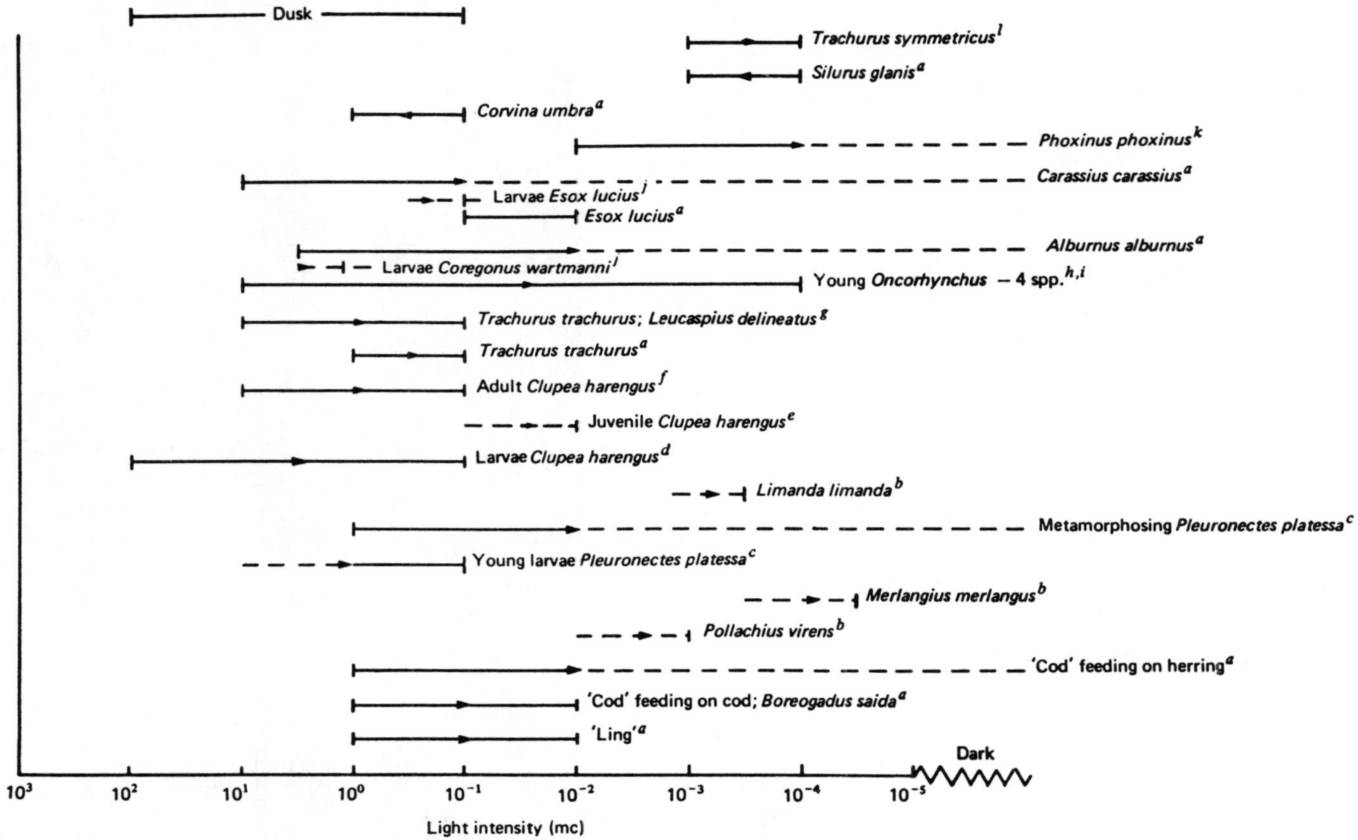

FIG. 4. Range of light intensity over which feeding activity declines (from Blaxter, 1970).

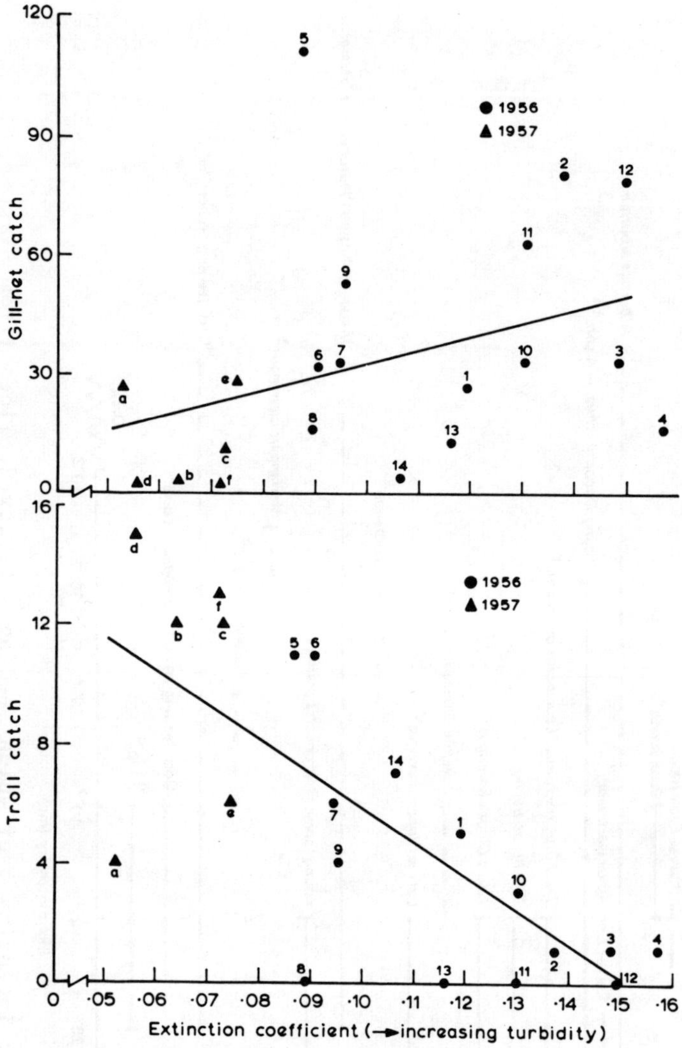

FIG. 5. Catches of albacore by gill net and troll at different turbidity of water (after Murphy, 1959).

VISUAL REACTION TO NETS

Responses to nets may be of importance in two ways. First, the fish should be herded by the net, for example by the wings of a trawl or seine, and collected at a point where they are finally caught; in other words certain parts of the net should be conspicuous. Second, the fish should not respond to the net, for example to gill or entangling nets, where the process of capture is essentially due to the net being poorly visible.

Blaxter and Parrish (1966) measured the light intensity threshold, in aquaria,for herding of fish by netting of different twine and mesh size, other parts of nets such as ground ropes and by underwater lights and airbubble screens. They found a great variation in the threshold (see Table 1 and Fig. 6) which could be related to the visibility of the device concerned. According to Hemmings (1966), brightness rather than colour seems to be important in underwater visibility. Parrish et al (1967) and Beamish (1967) photographed fish on the sea bed in front of trawls in "visual" and "non-visual" conditions and found good orientation in both circumstances, suggesting that in normal fishing other stimuli such as the sound of the net or its pressure wave may play a role in orientation.

Model	*Clupea harengus*	*Gadus morhua*	*Melanogrammus aeglefinus*
Ground rope and attachments	0·1–0·55	0·07–0·6	0·01–0·5
Panels of netting	0·08	?0·3–0·5	0·3
Underwater lights	0·005–0·66	0·005–0·7?	0·7
Air bubble screen	0·09–0·25	—	0·45

TABLE 1. Light intensity threshold in m.c. for herding of herring cod and haddock by various "devices" (Blaxter and Parrish 1966).

With gill nets the situation is a little easier to detetermine The data assembled in Table 2 show how important net visibility is for entangling fish. It seems unlikely that the lateral line system is effective in warning fish of the presence of a net and warning is, in fact, most likely to be given by other fish struggling in the net. Transparent nets are very effective for catching fish and may only become less effective if saturated by caught fish or shown up by luminescence in the water. Blaxter and Parrish (1965) measured the light intensity threshold for herring swimming into gill nets of different colour and found a relationship between capture and visibility of the net in terms of colour and background contrast (Table 3). Murphy (1959) applied such thinking to catches of albacore tuna and found a tendency for gill nets to be more effective in turbid water (Fig. 5).

LIGHT-CONTROLLED BEHAVIOUR AND DISTRIBUTION

Light is involved in fish behaviour and therefore locomotion and distribution. Hence the availability of fish for capture is controlled to some extent by light.

Schooling. The school is here defined (Breder, 1959) as a group of fish polarised or orientated to one another with the members evenly spaced and swimming at similar speeds. Shaw (1962)

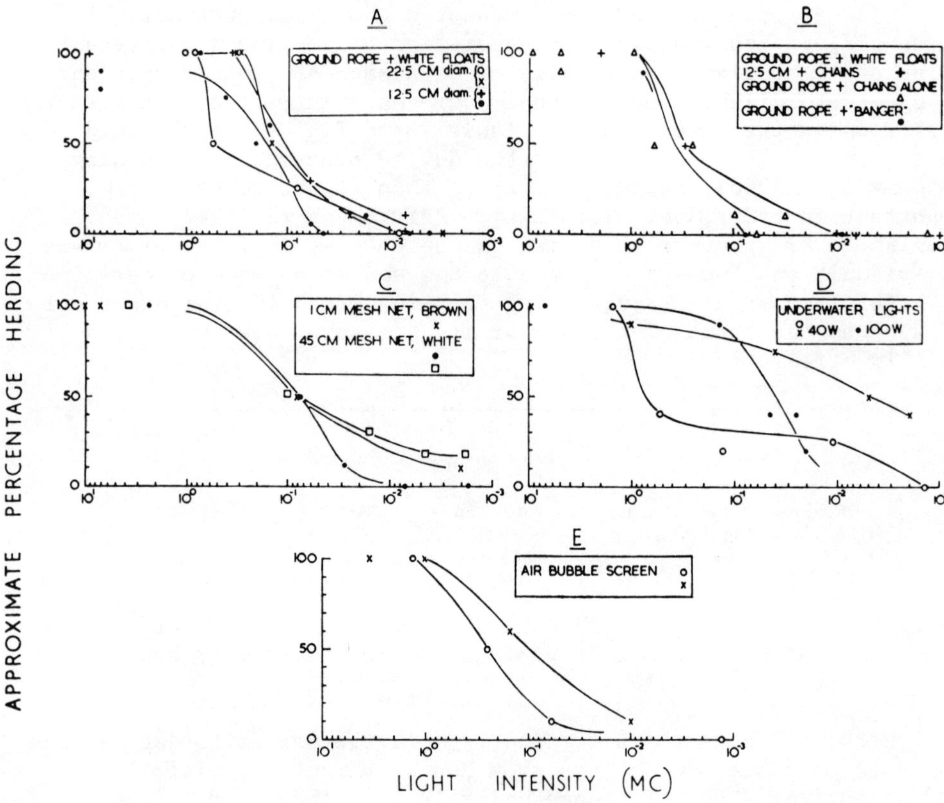

FIG. 6. Herding of herring by various "devices" at different light intensities (Blaxter and Parrish, 1966).

estimated that about 2000 or more species of marine fish schooled, especially the planktonophage pelagic species like the clupeids, engraulids and also the mugilids, scombrids and trachurids.

Most species investigated school optically (see Fig. 7) with a threshold near 10^{-1} m.c. There is an interesting paradox here in that silvery-sided schooling fish are very difficult to see when viewed laterally, as by other members of the school (Denton and Nicol, 1966). This invisibility depends on a homogeneous distri-

Species	Result	Author
Clupea harengus (herring)	Taken by day with monofilament nylon nets.	von Brandt (1954)
Clupea harengus	Ability to avoid nets in tanks is least when nets are of low visibility or twine is very thin; threshold twine diameter 0·2 mm.	Mohr (1961a)
Clupea harengus	Monofilament nylon nets very effective in catching fish in tanks. Cannot avoid nets with opaque plastic over eyes (see also Table 2-21).	Blaxter and Parrish (1965)
Anchovy	Gill nets less effective in clear water.	Aslanova (1958)
Gadus morhua (cod)	Avoid gill nets by day.	Sundnes (1964)
Gadus morhua	Greater catches by perlon nets.	Schwarz (1959)
Thunnus alalunga (albacore)	Gill nets seem more effective in turbid water, visual lures (trolling) better in clear water (see also Fig. 2-42).	Murphy (1959)
Lateolabrax japonicus (sea bass), *Konosirus punctatus* (gizzard shad)	Better fishing when illumination poor, nets dull in colour, water turbid.	Nomura (1959)
Coregonids (white fish)	Monofilament nylon nets take seven times, and multifilament nylon four times, the catch with cotton nets.	Molin (1959)
Oncorhynchus spp. (Pacific salmon)	Monofilament nets outfish multifilament by at least 2:1.	Larkins (1963–1964)
Perca fluviatilis (perch), *Rutilus rutilus* (roach)	Better catches with inconspicuous nets (also tested by optomotor response to netting).	Steinberg (1961)
Salmon	Entangling properties best in the order grey = blue-green > brown.	Koike (1958)
Trachurus trachurus (horse mackerel), *Atherina elymus* (sand smelt), *Cyprinus carpio* (carp)	Experimental observation that *barrier* effect is best in the order red > yellow > green.	Kanda and co-authors (1958)

TABLE 2. Visibility effects in fish capture (see Blaxter, 1970 for references).

bution of light from all azimuths, hence the school will find it easier to keep together visually at dusk and dawn when the sun is nearest the horizon. There is obviously an adaptive "conflict" between the needs of camouflage and visual contact in the school.

The strategy of fishing may be determined by schooling in terms of net size, design and speed of tow. Generally speaking the really large catches of fish, such as the Peruvian anchoveta, come from pelagic schooling species. If schools break up at night the

Experimental condition	Colour of twine	Background		
		Grey	White	Black
Small tank	Black	1×10^{-3}	—	—
	White	$< 10^{-3}$	—	—
Large tank	Black	—	8×10^{-3}	3×10^{-2}
	White	—	3×10^{-3}	2×10^{-3}

TABLE 3. Light intensity thresholds in m.c. for herring swimming into gill nets (Blaxter and Parrish, 1965).

density of fish will become reduced and their response to nets may well change.

Activity. Fish may become less active in the dark and cycles of feeding and locomotor activity and of vertical migration (see later) are a common feature of fish behaviour (see Blaxter, 1970). The reported instances of fish sleeping have never been supported by a definition of what is meant by sleep. It is clear, however, that inactivity at night may lead to greater vulnerability to capture.

Vertical migration. The changes in vertical distribution of fish y day and night have been reviewed by Woodhead (1966) and Blaxter (1970). The upward movement to the surface at dusk and downard movement at dawn seems, for example in herring, to be related to light intensity (Fig. 8) and, as shown by Postuma (1957), to the "following" of isolumes. These vertical movements, which are widespread in both marine and freshwater fish, clearly influence the availability of the species to capture by different gear. The "following" of isolumes is a function of brightness discrimination (see chapter by Blaxter in this volume on absolute thresholds). The implications of activity changes and diel vertical migration are demonstrated in Table 4 which shows how the catches of fish by trawl on the sea bed vary greatly by day and night. The variation in catch may be due partly to presence or absence (availability) but also to better net avoidance (vulnerability) or to better herding by certain parts of the net by day.

Horizontal migration. Jones (1968) has reviewed the role of light and vision in fish navigation. The azimuth of the sun and a sense of time in the fish may well play a part in fairly crude orientation across long distances. This will also affect fish distribution, especially of species like salmonids, and influence fishing strategy. Salmonids and other species may well identify their home rivers and streams by memory of visual reference points. If these could be identified they might give indications of the best to fish.

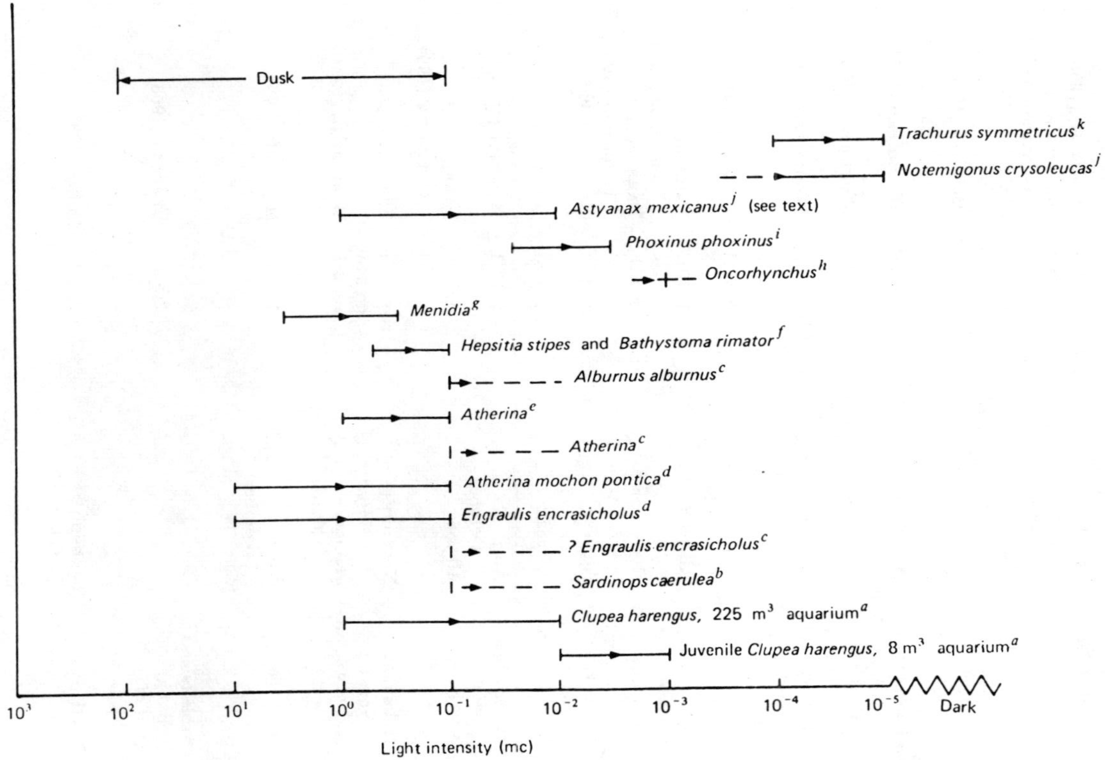

FIG. 7. Range of light intensity over which schooling activity declines (from Blaxter, 1970).

Species	Area	Season	Catch higher Day	Catch higher Night	Ratio	Author
Gadus morhua (cod)	N. Norwegian coast	January–March	√		av. 2:1	WOODHEAD (1964)
	Faroe Islands	April		√	4–5:1	JONES, R. (1956)
	E. Scottish coast	November, January	√		2–3:1	PARRISH and co-authors (1964)
	W. Scottish coast	April–October	√		2–3:1	BAGENAL (1958)
	St. Lawrence, Cape Breton	All year	√		av. 1·9:1	BEAMISH (1966)
Melanogrammus aeglefinus (haddock)	Barents Sea (Skolpen Bank)	November	√		av. 4:1	WOODHEAD (1964)
	Barents Sea (N. Deeps)	November		√	av. 2:1	WOODHEAD (1964)
	N. Norwegian coast	January–March	√		av. 2·8:1	WOODHEAD (1964)
	Faroe Islands	April		√	4–5:1	JONES, R. (1956)
	W. Scottish coast	April–October	√		5–33:1	BAGENAL (1958)
	E. Scottish coast	All year		√	1–10:1	PARRISH and co-authors (1964)
	North Sea area	December–February	√		2–5:1	PARRISH and co-authors (1964)
	Sable–Emerald Banks	All year	√		av. 1·2:1	BEAMISH (1966)
	Cape Breton	All year	√		av. 1·5:1	BEAMISH (1966)
Merlangius merlangus (whiting)	W. Scottish coast	April–October	√		2–17:1	BAGENAL (1958)
	North Sea area	All year		√	1–3:1	PARRISH and co-authors (1964)
Pollachius virens (coalfish saithe)	Iceland and Norway	November–March	√	√	depending on age	SCHMIDT (1955)
	N. Norwegian coast	January–March	√		av. 2:1	WOODHEAD (1964)
Merluccius merluccius (hake)	Irish Sea	June–December	√		2–6:1	HICKLING (1933)
Sebastes marinus (redfish)	Sable Island Bank Gulf of St. Lawrence	?	√		2·8:1	BEAMISH (1966)
Clupea harengus (herring)	North Sea	September–October	√		av. 7:1	LUCAS (1936)

TABLE 4. Diurnal variations in catch by bottom trawl (see Blaxter, 1970 for references).

Species	Area	Season	Catch higher Day	Catch higher Night	Ratio	Author
Pleuronectes platessa (plaice)	W. Scottish coast	April–October	√		2–4:1 in 4 out of 5 expts.	BAGENAL (1958)
	E. Scottish coast	All year		√	av. 2:1	PARRISH and co-authors (1964)
	North Sea (Haddock Bank)	April–May	√		av. 1·4:1	WOODHEAD (1964)
	North Sea	All year except January, February	√		1·2–1·3:1	DE GROOT (1964)
	North Sea	All year		√	av. 1·3:1	BOEREMA (1964)
	North Sea (off Sylt)	All year	√		av. 1·4:1	HEMPEL (1964)
Hippoglossoides platessoides (American 'plaice')	Cape Breton and Sable Island	?	√		av. 1·8:1	BEAMISH (1966)
Hippoglossoides platessoides (long rough dab)	W. Scottish coast	April–October	√		2–5:1	BAGENAL (1958)
Hippoglossoides platessoides and *Limanda limanda* (common dab)	E. Scottish coast	All year		√	av. 4:1	PARRISH and co-authors (1964)
Limanda ferruginea (yellowtail)	Sable Island and Emerald Bank	?		√	av. 2·5:1	BEAMISH (1966)
Glyptocephalus cynoglossus (witch)	W. Scottish coast	April–October	√		up to 3:1 in 4 out of 5 expts	BAGENAL (1958)
Glyptocephalus cynoglossus (American grey sole)	Scatari Bank	February		√	av. 2·6:1	BEAMISH (1966)
Solea solea (sole)	North Sea	?		√	av. 2·6:1	WOODHEAD (1964)
	North Sea	All year		√	about 10:1	BOEREMA (1964)
Microstomus kitt (lemon sole)	E. Scottish coast	All year		√	av. 3:1	PARRISH and co-authors (1964)
	Faroe Islands	April		√	20:1	JONES, R. (1956)
Pseudopleuronectes americanus (winter flounder)	Bay of Fundy	November		√	av. 1·4:1	BEAMISH (1966)

TABLE 4. Diurnal variations in catch by bottom trawl (see Blaxter, 1970 for references).

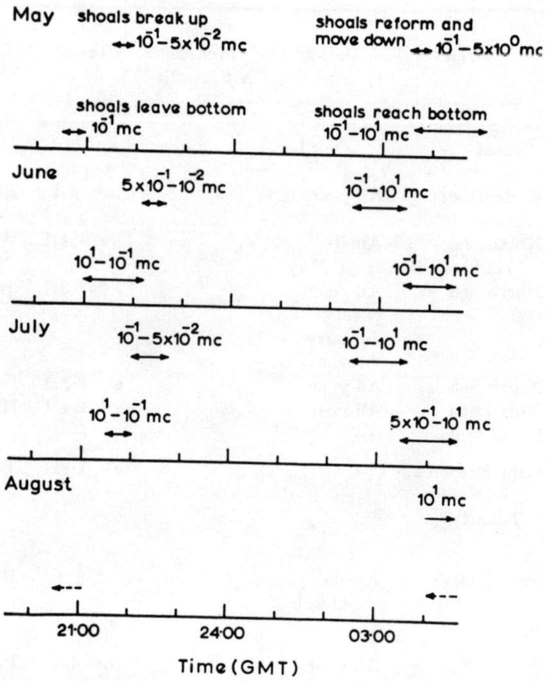

FIG. 8. Light intensities for events in the vertical migration of herring in different months in the North Sea (from Blaxter and Parrish, 1965).

FISH GUIDING

The deflection of salmonids from the spillways of power stations by means of visual barriers such as lights, or barriers involving light and sound, such as chains or air bubbles, have been reviewed by Brett and Alderdice (1958). Although the results of deflection experiments were inconsistent they showed a general repelling effect of visual stimuli.

DISCUSSION

Light-controlled behaviour such as feeding and schooling, net avoidance and vertical migration seem to become extinct at light intensities near 10^{-1} m.c. equivalent to late dusk or early dawn and to the light when the main light/dark adaptation processes occur in the retina. The rods appear to play a much smaller part in behaviour than do the cones. The importance of this in terms of applied research can be looked at in a geographical sense in two ways. First, the time available for light-controlled behaviour will

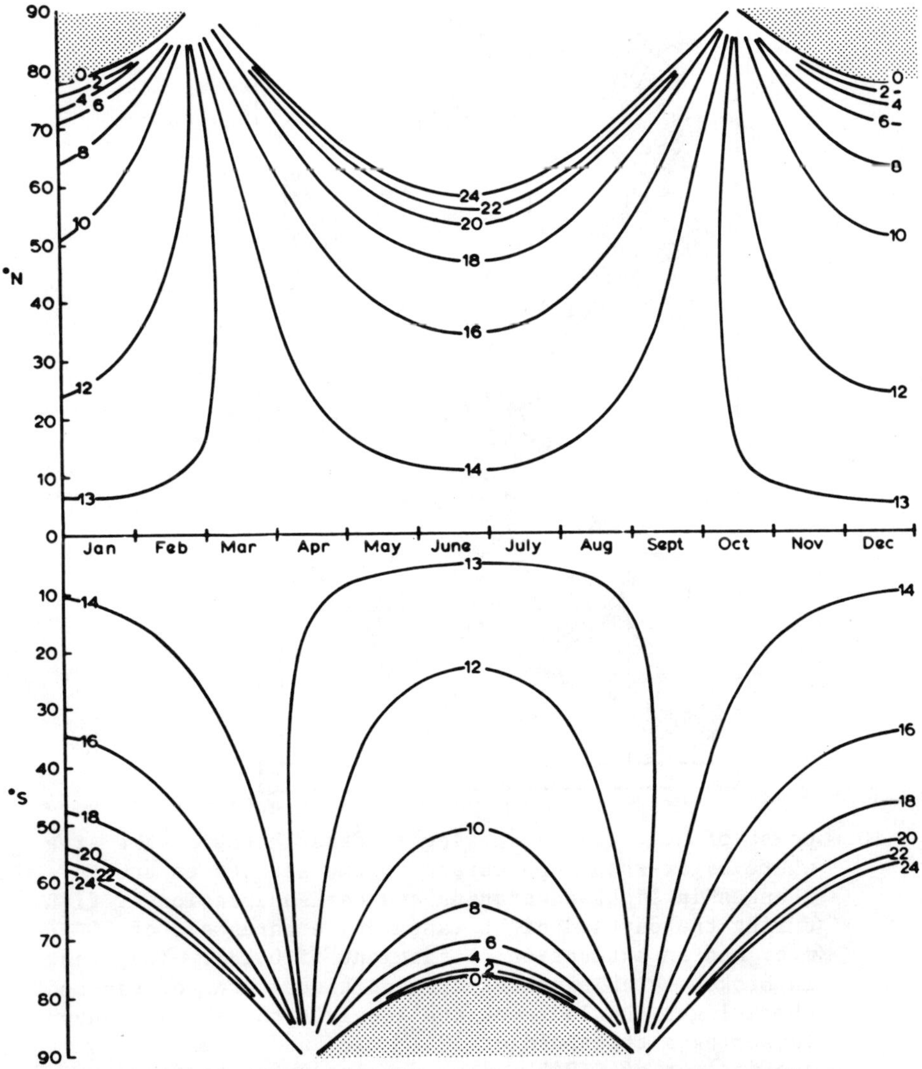

FIG. 9. Number of hours of daylight when the light intensity is above 0.1 m.c. at the surface at different latitudes and seasons.

vary greatly with season and latitude as shown in Fig.9, where the hours (at the surface) when the light intensity is above 10^{-1} m.c. are given. Clearly in higher latitudes there will be much greater seasonal variations in the time available for feeding, schooling and other light-limited behaviour.

In terms of depth, the extinction of light will vary greatly with water transparency. As an exercise to demonstrate this the

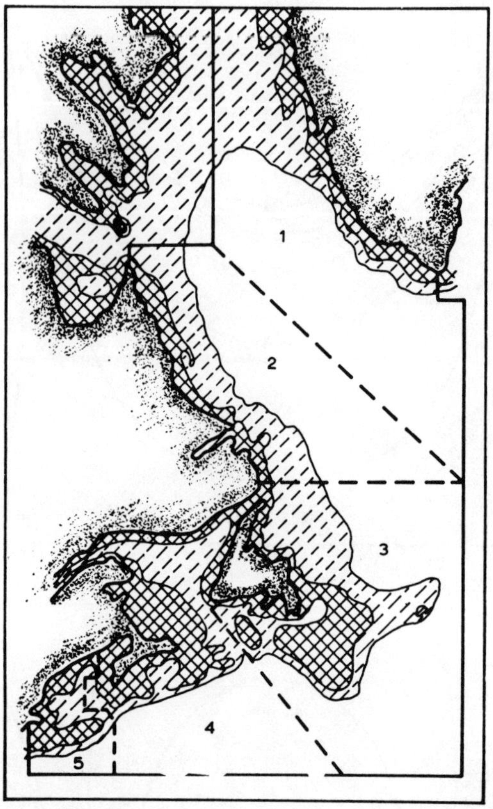

FIG. 10. Coast of N.E. Canada showing in cross hatching the area where light should penetrate to the sea bed to cause changes in light-adaptation and dark-adaptation of fish during the day and night (assuming a threshold of 10^{-1} m.c. and an extinction coefficient of 0.13). The area in broken hatching shows the additional area of sea bed where light should reach the absolute threshold at noon (assuming a threshold of 10^{-10} m.c. and an extinction coefficient of 0.045). The numbers refer to fishery areas (from Blaxter, 1970).

area of the sea bed off N.E. Canada where light should penetrate sufficiently at least during a part of the day for light-adaptation of the retina is mapped in Fig. 10. A further area of deeper water is also shown where enough light should penetrate at least for a short time, to reach the lowest rod threshold. The vast area of the sea bed where no effective light penetrates at all becomes obvious on inspection of the figure.

REFERENCES

Beamish, F.W.H. (1967) Photographic observations on reactions of fish ahead of otter trawls. FAO Fisheries Rep. No 62, Vol. 3, 511-521.
Blaxter, J.H.S. (1970) 2.Light 2.3. Animals. 2.32. Fishes. In Marine Ecology Vol. 1, Pt. 1. pp. 213-320 (ed. O. Kinne). Wiley Interscience, London.
Blaxter, J.H.S. and Parrish, B.B. (1965) The importance of light in shoaling, avoidance of nets and vertical migration by herring. J. Cons. Int. Explor. Mer, 30: 40-57.
Blaxter, J.H.S. and Parrish, B.B. (1966) The reaction of marine fish to moving netting and other devices in tanks. Mar. Res. No 1, 15 pp.
Breder, C.M. Jr. (1959) Studies on social groupings in fishes. Bull. Am. Mus. Nat. Hist. 117: 397-481.
Brett, J.R. and Alderdice, D.F. (1958) Research on guiding young salmon at two British Columbia field stations. Fish. Res. Board Bull. 117: 75 pp.
Denton, E.J. and Nicol, J.A.C. (1966) A survey of reflectivity in silvery teleosts. J. Mar. Biol. Ass. U.K., 46: 685-722.
Hemmings, C.C. (1966) Factors influencing the visibility of objects underwater. Symp. Br. Ecol. Soc. 6: 359-374.
Hunter, J.R. and Mitchall, C.T. (1968) Field experiments on the attraction of pelagic fish to floating objects. J. Cons. Int. Explor. Mer. 31: 427-434.
Jones, F.R.H. (1968) Fish migration. Arnold, London.
Murphy, G.I. (1959) Effect of water clarity on albacore catches. Limnol. Oceanogr. 4: 86-93.
Parrish, B.B., Blaxter, J.H.S., Pope, J.A. and Osborn, R. (1967). Underwater photography of fish behaviour in response to trawls. F.A.O. Fisheries Rep. No 62, Vol. 3: 873-884.
Postuma, K.H. (1957) The vertical migration of feeding herring in relation to light and vertical temperature gradient. Cons. Perm. Int. Explor. Mer, 1957 Meeting, Herring Cttee (Mimeo).
Ryland, J.S. (1964) The feeding of plaice and sand eel larvae in the southern North Sea. J. Mar. Biol. Ass. U.K. 44: 343-364.
Schärfe, J. (1953) Über die Verwendung künstlichem Lichtes in der Fischerei. Prot. Fisch. Tech. 8: 1-29.
Shaw, E. (1962) The schooling of fishes. Scient. Am. 206: 126-138.
Verheijen, F.J. (1958) The mechanisms of the trapping effect of artificial light sources upon animals. Archs. néerl. Zool. 13: 1-107.
Walls, (1963) The vertebrate eye. New. Ed. Hafner, New York.
Woodhead, P.M.J. (1966) The behaviour of fish in relation to light in the sea. Oceanogr. Mar. Biol. Ann. Rev. 4: 337-403.

OPHTHALMIC DISEASES OF FISHES

THOMAS W. DUKES

P.O. Box 9

Kemptville, Ont. KOG 1JO, Canada

And now remains
That we find out the cause of this effect,
Or rather say, the cause of this defect,
For this effect defective comes by cause.
Hamlet, Act II, Sc. 2

Fishes have more ocular structural variation than all other vertebrates together and some have developed patterns of camouflage such as skin spots on the dorsal part of the cornea as in *Epinephelus guttatus*. Others have developed shields to protect the eye as in *Selar crumenophthalmus*. People are generally familiar with the eye migration of flat fishes. The eyeball of Anableps of South America is adapted to seeing, both below and above the water simultaneously and for this purpose the fish has two corneas and two retinas in each globe. Many deep-sea fishes such as *Dolicopteryx* have varied ocular structure. Due to limitation of time and space only a few examples of eye diseases are presented.

TECHNIQUES

A clinical examination of the eye can be very revealing. The eye is normally kept level when a fish is rotated on the body axis. It should also be noted if the eyes are sunken, or protruded, and observations should be described with accuracy since, for example, corneal opacity and lens opacity can be confused by the unwary observer. An examination can be performed under anaesthesia, or immediately after death of the fish, using either standard techniques for an ophthalmoscope or just a hand lens and flashlight. A dissecting microscope is very useful for examination, or when trimming the eye for processing after fixation, in order to make accurate gross descriptions.

The general pathologic changes occurring in other organs apply to the eye but certain factors are peculiar to it. Most disease processes are disastrous to transparency and hence to the function of the eye. The avascularity of some ocular tissues (cornea and lens) makes them prone to degeneration and also unresponsive in the inflammatory response to infection, whereas the rich vascular network of the uvea makes the eye susceptible to involvement in systemic diseases. The mixture of a dense fibrous sclera (through which fixatives permeate slowly), and a delicate retinal tissue with a high metabolic rate (which must be fixed rapidly for the best results), make processing difficult. To obtain the best results the eye must be removed rapidly and carefully, and fixed promptly. The choice of fixatives is important since rapid penetration is necessary. Bouin's or Zenker's fixatives are the best for routine use. It is unnecessary and even harmful to make holes in the globe or inject fixatives into it unless only parts of the eye are to be examined as in electron microscopy.

EXOPHTHALMUS (POP-EYE)

This is the most frequently reported change in the fish eye. This condition can be due to many causes and cannot be judged as a very specific condition. This disease has been seen in fish affected with various infectious diseases. Hormonal problems have also been suggested (testosterone and thyroid hormone) (Matty et al 1958). Mechanical injury and gas bubbles in the eye due to blows to the head could also produce exophthalmus.

Fluke metacercaria can produce exophthalmus. (Sinderman & Rosenfield, 1954; Millemann & Knapp, 1970). Frequently these diseases are found subretinally or retrobulbarly and have been associated with extensive fluid or hemorrhages behind the globe. The exophthalmus may be due to edema being the result of kidney damage by the parasite, as it occurs in other diseases affecting the kidney such as bacterial kidney disease, viral hemorrhagic septicaemia and Sacramento River Chinook disease.

Exophthalmus has also been seen in infectious pancreatic necrosis and infectious hematopoietic necrosis. Numerous punctiform extravasations were seen in the periocular connective tissues.

Aeromonas, pseudomonas, flavobacteria and vibrio infections are frequently associated with exophthalmus. These cases are usually a type of bacterial panophthalmitis or hemorrhage from a septicaemia. Kluge (1965) described an exophthalmus produced in mollies by a flavobacterium inducing granulomas in the optic nerve, orbit and retina. About 80% of the fish had unilateral or bilateral exophthalmus. Lesions were characterised by a central mass of macrophages surrounded by a zone of lymphocytes one or two cells thick which in turn was surrounded by connective tissue. Lesions were found in various stages of development with the centre frequently having undergone coagulation necrosis extending to the fibrous connective tissue.

Acid-fast bacteria can be demonstrated in smears from the eyes of fish affected with mycobacteriosis. Nigrelli and Vogel (1963) cited the eye as the organ involved. The typical granulomatous reaction found in the rest of the body has been seen histologically. The similarities of tuberculosis and nocardiosis must not be overlooked when confronted with this type of lesion.

Rock and Nelson (1965) described blindness in varying degrees of exophthalmus in Catfish affected by Aeromonas liquefaciens. In an experiment involving an inoculation of tumour tissue into fishes studied by the author, Aeromonas liquefaciens was isolated from protruding eyes seen in the experimental sunfish. The unilateral eye lesion was the only change until the fish was "in-extremis". The eye was disoriented internally and had numerous inflammatory cells and some exudate. The retina was completely detached and chiefly destroyed. The choroidal gland was very engorged with blood and inflammatory cells were numerous. The inflammation also involved the cornea and extraglobal tissues and other fish became contaminated. In these specimens initially there was marked vascular engorgement with hemorrhage in the uvea and then infiltration of inflammatory cells and necrosis. The cornea became oedematous, then ulcerated and there was proteinaceous exudate in the anterior, posterior and vitreous chambers. The choroid was thickened by fluid and the retina became detached locally. Globoid cells were evident in the lens.

An exophthalmus due to nitrogen gas bubbles associated with overaeration was described by Marsh and Gorham (1905). This possible that changes in hydrostatic pressure, as a fish is moved from deep water to shallow water, could also be involved? Atlantic cod, Gadus morhua, held in captivity often develop unilateral or bilateral exophthalmus characterised by the accumulation of gas bubbles in the eye. Some preliminary work was performed on pop-eyed cod by Dehadrai in 1966. He suggested that the cause was a malfunctioning of the choroid gland-pseudobranch complex that is rich in carbonic anhydrase and which is involved in gas secretion in the choroid gland of cod. The choroid gland of the exophthalmic eye of the cod was hypertrophic and composed a spongy, dark mass of red tissue around the retina. Histologically vessels were seen to be dilated. Grossly, in specimens seen by this author, the cornea was greatly enlarged and thin, with the anterior chamber greatly increased in size and there were large cystic spaces in the choroid. Histologically in these eyes there were accumulations of reticulo-endothelial cells and some macrophages containing brownish pigment, vascular walls of the choroid gland were thickened. The retinas were detached and there was some inflammatory cells in the vitreous. Hemorrhage was evident in anterior parts of the eye.

CATARACT (WHITE EYE, OPAQUE EYE)

This is the second most frequently described ophthalmic condition but again specific pathological descriptions were lacking in most cases. Care must be taken to avoid confusing disease of the cornea with that of the lens. Opaque eyes were seen associated with riboflavin-deficient diets. Corneal vascularisation, cloudy lens, ulceration and photophobia were evident in riboflavin deficient fish (Halver 1953). Feeding Thioacetamide for 12 months to induce hepatoma resulted in cataracts that were irreversible after withdrawal of the carcinogen (Sallmann et al 1966). Allison (1962, 1963) was able to induce this disease by using ultraviolet light resembling that work performed in other animals. There has been an increase in cataracts in lake trout due to the increased demands for hatchery production. The reason being probably increased light (ultraviolet) exposure of fish raised in shallow water of raceways which is the contrary to their natural deep water environment of low light intensity. The lens capsule was thickened with minimal epithelial proliferation in sections of eyes examined by this author. Inflammatory cells were occasionally present in the various eye chambers.

Parasitic cataracts are commonly seen. The trematode parasites of Strigidae are frequently found in the lens or other areas of the eye. The metacercaria of one species, Diplostromum flexicaudum, has a wide distribution. It has been found in Lake of the Woods, Ontario, in 16 species of fishes, in Lake Huron in 20 species, in Lake Ontario in 21 species and in Lake Erie in 20 species (Dechtiarencho, 1972). The metacercaria were found in the lens. These parasites, when they occur in small number, can be seen by careful examination as small white lines with the naked eye, hand lens or ophthalmoscope. More severely affected eyes were diffusely opaque. Histologically there was liquefaction of the lens at the periphery initially with no apparent host reaction to the parasites unless secondary ocular injury occurred. Globoid cells were seen on some occasions in the lens. A detailed description of the lesions was published by Ashton et al (1969). Ali and Hanyu (1964) described lens flukes in the eyes of brown bullheads. Multiple lenses were also described. The parasitised multiple lenses were rougher than unaffected multiple lenses. Histologically various stages of degenerate were seen. It was not known whether the multiple lenses were congenital or acquired lesion.

OTHER PARASITIC DISEASES OF THE EYE

In addition to the aforementioned fluke-induced cataract, several other parasites are involved with the eye. Hall and Iversen (1967) described a myxosporidian *Henneguya lagodon* in *Lagodon rhomboides*, the Florida pinfish. Cysts occurred in tissues surrounding the eyes, often forming conspicuous external bulges. Pflugfelder (1952) described *Glugea pseudo-tumifaciens* in cells of connective tissue. Nodules were formed in ovary, liver, spleen, kidney, central nervous system and the eye. Elkan (1962) described a *Dermocystidium* (microsporidian) in 2 species of sticklebacks. Cylindrical cysts covered the whole body of the fish including the cornea of the eye and extended into the orbital tissues. Spores had a large inclusion body but no lid. Perch or other fish living in the same area were found by Sheer (1957). This parasite may be identical to *Dermocystidium cuticulare*. Dechtiarenko (1972) has seen a *Myxosoma scleroperca*, in perch that appears similar to the above-mentioned cases. The large myxosporidial cysts were present in the scleral cartilage and adjacent periocular tissues. Four of five cysts were frequently found in one eye. Although there was no reaction to the cysts, degenerative changes were evident in the lens and the retina was detached secondarily.

Baudouin (1904) described a copepod, *Lernaeenicus sprattae* that occurs embedded in the eye, or occasionally in dorsal body muscle of sprat and sardines in Europe. Bennett (1967) described another ocular copepod belonging to genus *Bomolochus*. A crustacean, *Artystone trysibia*, was described recently from the orbit of a South American fish (Huizinga, 1972) Ali (1974) has observed a copepod, *Phrixcephalus cincinnatus* in the eye of the flounder, *Atheresthes stomias*. *Ichthyophthirius multifiliis* has been found under the corneal epithelium at the outer edge of the corneal lamina propria.

ENDOPHTHALMUS

This can be seen as an artifact after death but has also been associated with emaciation and with certain hemoprotozoans.

MISCELLANEOUS CONDITIONS

A. Trauma. Injuries can lead to hemorrhage and dislocation of the lens.

B. Neoplasia. *Melanosarcoma* with loss of the eyes have been described by Levine and Gordon (1946). A *fibrosarcoma* was seen below the eye of one fish caught commercially in Ontario. *Lymphosarcoma* in the eye of Atlantic Cod was described by Wolke and Wyand (1969).

C. Fungus. Saprolegnia fungi frequently attack the eyes but are opportunists in already damaged eyes. Chemicals and trauma are frequent causes of initial damage. Fungal hyphae are frequently found during histologic examination of corneal lesions and may extend to the lens and other tissues and also to the brain. Many inflammatory cells and oedema together with hyphae were evident in the cornea.

D. Lymphocystis. Lymphocystis, a viral induced hypertrophy, studied for many years in other tissues (Lawler et al 1974) has recently been described in the eye. Lymphocystis cells were found as a retrobulbar mass causing exophthalmus and as variably sized modules in the cornea, optic nerve, choroid and iris of other eyes.

E. Malformations. Pigmentation of the underside of some flat fishes is associated with improper migration of the eye and some individuals have the wrong eye for that species migrated, making taxonomy more challenging. The snub nose malformation seen in some fishes can be severe enough to alter the normal ocular structures. Cyclopia and microphthalmia have been associated with chemicals and altered temperatures. There are numerous malformed eyes that are "normal" for certain species or strains (eg. telescope eye goldfish). Multiple lenses that may have been hereditary or secondary to parasites were described by Ali and Hanyu (1964).

REFERENCES

Ali, M.A. (1974). Personal Communication.

Ali, M.A. and I. Hanyu (1964). Occurrence of Multiple Lenses in the eye of brown bullheads (Ictalurus nebulosus) Copeia: 4: 704-705.

Allison, L.N. (1962). Cataract among hatchery reared lake trout. Prog. Fish-Cult. 24:155.

Allison, L.N. (1963). Cataract in hatchery lake trout. Trans. Am. Fish Soc. 92: 34-38.

Ashton, N., N. Brown and D. Easty (1969). Trematode cataract in freshwater fish. J. Small Anim. Pract. 10: 471-478.

Baudouin, M. (1904). Le Lernaeenicus sprattae, parasite de la Sardine en Vendee. Comptes Rendus, 998-1000.

Bennet, P.S. (1967). On Bomolochus jonesi parasitic on the eye of the Indian mackerel Rastrelliger kangurta. J. Mar. Biol. Ass. India 9: 132-136.

Dechtiarenko, A. (1972) personal communication.

Dehadrai, P.V. (1966). Mechanism of gaseous exophthalmia in the Atlantic Cod Gadus morhua L. J. Fish Res. Board Can. 23: 909-914.

Elkan, E. (1962). Dermocystidium gasterostei n. sp. a parasite of Gasterosteus aculeatus L. and Gasterosteus pungitius L. Nature, (Lond). 196: 958-960.

Hall, D.L. and E.S. Iversen (1967). Henneguya lagodon, a new species of myxosporidian parasitizing the pinfish Lagodon rhomboides. Bull. Mar. Sci. 17: 274-279.

Halver, J.E. (1953). Ph.D. Thesis, U. of Washington.

Huizinga, H.W. (1972). Pathobiology of Artystone trysibia Schioedte (Isopoda: Cymothoidae), and endoparasitic isopid of South American freshwater fishes. J. Wildl. Dis. 8: 225-232.

Kluge, J.P. (1965). A granulomatous disease of fish produced by flavobacteria. Path. Vet. 2: 545-552.

Lawler, A.R., H.D. Howse and D.W. Cook (1974). Silver Perch, Bairdiella chrysura. New host for lymphocystis, Copeia. 3: 266-269.

Levine, M. and M. Gordon (1946). Ocular tumors with exophthalmia in Xiphophorin Fishes. Cancer Res. 6: 197-204.

Marsh, M.C. and F.P. Gorham (1905). The gas disease in fishes. Rep. U.S. Bur. Fish, for 1904. pp. 343-276.

Matty, A.J., D. Menzel and J.E. Bordach, (1958). The production of exophthalmus by androgens in two species of teleost fish. J. Endocrinol. 17: 314-318.

Millemann, R.E. and S.E. Knapp (1970). Pathogenicity of the "Salmon poisoning" trematode Nanophyetus salmincola to Fish. In: Symposium on Diseases of Fishes and Shellfishes, Am. Fish Soc. 209-217.

Nigrelli, R.F. & H. Vogel (1963). Spontaneous tuberculosis in fishes and other cold-blooded vertebrates with special reference to Mycobacterium fortuitum from fish and human lesions. Zoologica 48: 131-143.

Pflugfelder, O. (1952). Ein neuer Fischparasit aus der Gruppe der Mickrosporidien Die Aquar und Terrarien, 5.

Rock, L.F., and H.M. Nelson (1965). Channel catfish and gizzard shad mortality caused by Aeromonas Liquefaciens. Prog. Fish-Cult. 27: 138-141.

Sallmann, L. et al (1966). Thioacetamide-induced cataract with invasive proliferation of the lens epithelium in rainbow trout. Cancer Res 26: 1819-1825.

Sheer, D. (1957). Die Fischparasiten der Haplosposidiengattung Dermocystidium. Z Fischerei 6: 127-134.

Sindermann, C.J. and H. Rosenfield (1954). Diseases of Fishes of the Western North Atlantic, III, Nortalities of sea herring (Clupea harengus) caused by larval trematode invasion. Maine Dept. Sea Shore Fish. Res. Bull. No. 21, 16 pp.

Wolke, R.E. and D.S. Wyand (1969). Ocular lymphosarcoma in an Atlantic Co. Bull. Wildl. Dis. Assoc. 5: 401-403.

CONTRIBUTORS AND PARTICIPANTS

AGHION, J., Département de Botanique, Université de Liège, Sart Tilman, B-4000, Liège, Belgique.

ALI, M.A., Département de Biologie, Université de Montréal, C.P. 6128, Montréal, Canada.

*ANCTIL, M., Département de Biologie, Université de Montréal, C.P. 6128, Montréal, Canada.

BEATTY, D.D., Department of Zoology, University of Alberta, Edmonton 7, Alberta, Canada.

BLAXTER, J.H.S., Scottish Marine Biological Association, Dunstaffnage Marine Research Laboratory, P.O. Box 3, Oban PA34 4AD, Argyll, Scotland.

BOWMAKER, J.K., MRC Vision Unit, Centre for Research on Perception and Cognition, University of Sussex, Falmer, Brighton, BN1 9QG, Sussex, U.K.

CAMPBELL, A., Neuropsychology Laboratory, McLean Hospital, Belmont, MA. 02178, USA.

CHAMPLAIN, J. de, Centre de Recherche en Sciences Neurologiques, Université de Montréal, C.P. 6128, Montréal, Canada.

COHEN, J., Division of Biology and Living Resources, Rosenstiel School of Marine and Atmospheric Sciences, University of Miami, Virginia Key, Fla, 33149, USA.

COUILLARD, P., Département de Biologie, Université de Montréal, C.P. 6128, Montréal, Qué.

DALTON, S.A., Department of Zoology, University of Florida, Gainesville, Florida 32611, USA.

DARTNALL, H.J.A., MRC Vision Unit, Centre for Research on Perception and Cognition, University of Sussex, Falmer, Brighton, BN1 9QG, Sussex, U.K.

DELISLE, C., Centre de Recherche Noranda, 240 Boul. Hymus, Pointe-Claire 730, Qué. Canada.

DELMELLE, M.C., Département de Physique Atomique et Moléculaire, Université de Liège, Sart Tilman, 4000 Liège, Belgique.

DESCARRIES, L., Centre de Recherche en Sciences Neurologiques, Université de Montréal, C.P. 6128, Montréal, Canada.

DESMARAIS, M., Département de Physiologie, Université de Montréal, C.P. 6128, Montréal, Canada.

*DETWILLER, P.B., Laboratory of Neurophysiology NINDS, National Institute of Health, Bethesda, Md. 20014, USA.

DRUJAN, B.D., Laboratory of Neurochemistry, Instituto Venezolano de Investigaciones Cientificas, Apartado 1827, Caracas 101, Venezuela.

DUKES, T.W., Box 9, Kemptville, Ontario, Canada.

EASTER, S.S., Department of Zoology, University of Michigan, Ann Arbor, Mich. 48104, USA.
GAILLARD, F.P., Laboratoire de Neurophysiologie, Faculté des Sciences, Avenue du Recteur Pineau, 86022 Poitiers, France.
GALAND, G., Laboratoire de Neurophysiologie et Psychophysiologie, Faculté des Sciences Fondamentales et Appliquées, 40, Ave. du Recteur Pineau, 86022, Poitiers, France.
GALEANO, C., Département de Physiologie et Pharmacologie, Université de Sherbrooke C.H.U., Sherbrooke, Qué. Canada.
GRUBER, S.H., Division of Biology and Living Resources, Rosenstiel School of Marine and Atmospheric Sciences, University of Miami, Virginia Key, Fla, 33149, USA.
HÁROSI, F.H., Room 2C02, Building 36, National Institute of Health, Bethesda, Md. 20014, USA.
HOCHANDANI, S., Groupe de Recherche en Biophysique, Département de Chimie-Biologie, Université du Québec, Trois-Rivières, P.Q. Canada.
JASPER, H.H., Groupe de Recherche en Sciences Neurologiques, Université de Montréal, C.P. 6128, Montréal, Canada.
*JODLOWSKY, M., Laboratory of Neurochemistry, Instituto Venezolano de Investigationes Cientificas, Apartado 1827, Caracas 101, Venezuela.
JOHNS, P.R., Department of Zoology, University of Michigan, Ann Arbor, Mich. 48104, USA.
KOSSI, C., Groupe de Recherche en Biophysique, Département de Chimie-Biologie, Université du Québec, Trois-Rivières, P.Q. Canada.
LARSON, W.L., Ecole d'Optométrie, Université de Montréal, C.P. 6128, Montréal, Canada.
LEBLANC, R.M., Groupe de Recherche en Biophysique, Département de Chimie-Biologie, Université du Québec, Trois-Rivières, P.Q. Canada.
LEPORE, F., Département de Psychologie, Université de Montréal, C.P. 6128, Montréal, Canada.
LEVIS, R.A., Jules Stein Eye Institute, UCLA Medical Center, Los Angeles, CA. 90024,University of California, USA.
*LIEGE, B., Laboratoire de Neurophysiologie et Psychophysiologie, Faculté des Sciences Fondamentales et Appliquées, 40, Ave. du Recteur Pineau, 86022, Poitiers, France.
LOCKET, N.A., Institute of Ophthalmology, University of London, Judd Street, London WC1H 9QS, U.K.
LOEW, E.R., MRC Vision Unit, Centre for Research on Perception and Cognition, University of Sussex, Falmer, Brighton, BN1 9QG, Sussex, U.K.
LYTHGOE, J.N., MRC Vision Unit, Centre for Research on Perception and Cognition, University of Sussex, Falmer, Brighton, BN1 9QG, Sussex, U.K.
MAILLOUX, G., Département de Physiologie, Université de Montréal, C.P. 6128, Montréal, Canada.

MUNTZ, W.R.A., Laboratory of Experimental Psychology, University of Sussex, Falmer, Brighton, BN1 9QG, Sussex, U.K.
NIAZI, I.A., Department of Zoology, University of Rajasthan, Jaipur-4 (Raj.), India.
NICOL, J.A.C., Marine Science Institute, University of Texas, Port Aransas, Texas 78373, USA.
NORTHMORE, D.P.M., Institute for Neurosciences and Behavior and Department of Psychology, University of Delaware, Newark, Delaware 19711, USA.
*PONTUS, M., Département de Physique Atomique et Moléculaire, Université de Liège, Sart Tilman, 4000 Liège, Belgique.
POWERS, M.K., Departments of Psychology and Zoology, University of Michigan, Ann Arbor, Mich. 48104, USA.
PUIL, E., Groupe de Recherche en Sciences Neurologiques, Université de Montréal, C.P. 6128, Montréal, Canada.
RAYNAULD, J.P., Département de Physiologie, Université de Montréal, C.P. 6128, Montréal, Canada.
SADOGLU, P., Division of Biological and Medical Sciences, Box G, Brown University, Providence, R.I. 02912, USA.
SCHOEBITZ, K., Anatomisches Institut, Abt. f. Neuroanatomie, Universitaets-Krankenhaus Eppendorf, Martinstrasse 52, 2 Hamburg 20, West Germany.
SCHWASSMANN, H.O., Department of Zoology, University of Florida, Gainesville, Fla. 32601, USA.
SEIM, T., Institute of Physics and Biophisics, University of Oslo, P.O. Box 1048, Blindern, Oslo 3, Norway.
SHARMA, S.C., Department of Ophthalmology, New York Medical College, New York, N.Y. 10029, USA.
SIVAK, J.G., School of Optometry, University of Waterloo, Waterloo, Ontario, Canada.
SOLIOZ, J.N., Département de Physiologie, Université de Montréal, C.P. 6128, Montréal, Canada.
STELL, W.K., Jules Stein Eye Institute, UCLA Medical Center, Los Angeles, CA. 90024, USA.
TAYLOR, D., Section of Ecology and Systematics, Bldg. 6, Langmuir Lab., Cornell University, Ithaca, N.Y. 14850, USA.
TOMITA, T., Department of Ophthalmology and Visual Science, Yale University School of Medicine, New Haven, Conn. 06510, USA.
VALBERG, A., Institute of Physics and Biophysics, University of Oslo, P.O. Box 1048, Blindern, Oslo 3, Norway.
VANEGAS, H., Departamento de Fisiología, Centro de Biofisica y Bioquímica, Instituto Venezolano de Investigaciones Científicas, Apartado 1827, Caracas 101, Venezuela.
VAN GELDER, N.M., Groupe de Recherche en Sciences Neurologiques, Université de Montréal, C.P. 6128, Montréal, Canada.
VOLKMANN, F.C., Clark Science Center, Smith College, Northampton, Mass. 01060, USA.
*WAGNER, H.G., Laboratory of Neurophysiology, NINDS, National Institute of Health, Bethesda, Md. 20014, USA.

WAGNER, H.J., Lehrstuhl für Morphologie und Anatomie der Universität Regensburg, Fachbereich Biologie, Universitätsstr. 31, D-8400 Regensburg, West Germany.

WALES, W., University of Stirling and Scottish Marine Biological Association, Dunstaffnage Marine Laboratory, P.O. Box 3, Oban PA34 4AD, Argyll, Scotland.

WHITT, G.S., Department of Zoology, University of Illinois, Urbana, Ill. 61801, USA.

*WOLBARSHT, M.L., Department of Ophthalmology, Duke University Medical Center, Durham, N.C., USA.

YAGER, D., State College of Optometry, State University of New York, 122 East, 25th Street, New York, N.Y., USA.

ZYZNAR, E., Département de Biologie, Université de Montréal, C.P. 6128, Montréal, Canada.

*Did not attend the Advanced Study Institute.

AUTHOR INDEX

Page numbers in italics refer to the bibliography

SUBJECT INDEX